ROBOTICS AND AUTOMATION HANDBOOK

ROBOTICS AND AUTOMATION HANDBOOK

Edited by
Thomas R. Kurfess Ph.D., P.E.

CRC PRESS

Boca Raton   London   New York   Washington, D.C.

Library of Congress Cataloging-in-Publication Data

Robotics and automation handbook / edited by Thomas R. Kurfess.
   p. cm.
   Includes bibliographical references and index.
   ISBN 0-8493-1804-1 (alk. paper)
   1. Robotics--Handbooks, manuals, etc. I. Kurfess, Thomas R.

TJ211.R5573 2000
629.8'92—dc21   2004049656

This book contains information obtained from authentic and highly regarded sources. Reprinted material is quoted with permission, and sources are indicated. A wide variety of references are listed. Reasonable efforts have been made to publish reliable data and information, but the author and the publisher cannot assume responsibility for the validity of all materials or for the consequences of their use.

Neither this book nor any part may be reproduced or transmitted in any form or by any means, electronic or mechanical, including photocopying, microfilming, and recording, or by any information storage or retrieval system, without prior permission in writing from the publisher.

All rights reserved. Authorization to photocopy items for internal or personal use, or the personal or internal use of specific clients, may be granted by CRC Press, provided that $1.50 per page photocopied is paid directly to Copyright Clearance Center, 222 Rosewood Drive, Danvers, MA 01923 USA. The fee code for users of the Transactional Reporting Service is ISBN 0-8493-1804-1/05/$0.00+$1.50. The fee is subject to change without notice. For organizations that have been granted a photocopy license by the CCC, a separate system of payment has been arranged.

The consent of CRC Press does not extend to copying for general distribution, for promotion, for creating new works, or for resale. Specific permission must be obtained in writing from CRC Press for such copying.

Direct all inquiries to CRC Press, 2000 N.W. Corporate Blvd., Boca Raton, Florida 33431.

Trademark Notice: Product or corporate names may be trademarks or registered trademarks, and are used only for identification and explanation, without intent to infringe.

Visit the CRC Press Web site at www.crcpress.com

© 2005 by CRC Press

No claim to original U.S. Government works
International Standard Book Number 0-8493-1804-1
Library of Congress Card Number 2004049656
Printed in the United States of America 1 2 3 4 5 6 7 8 9 0
Printed on acid-free paper

Preface

Robots are machines that have interested the general population throughout history. In general, they are machines or devices that operate automatically or by remote control. Clearly people have wanted to use such equipment since simple devices were developed. The word *robot* itself comes from Czech *robota*, "servitude, forced labor," and was coined in 1923 (from dictionary.com). Since then robots have been characterized by the media as machines that look similar to humans. Robots such as "Robby the Robot" or *Robot* from the Lost in Space television series defined the appearance of robots to several generations. However, robots are more than machines that walk around yelling "Danger!" They are used in a variety of tasks from the very exciting, such as space exploration (e.g., the Mars Rover), to the very mundane (e.g., vacuuming your home, which is not a simple task). They are complex and useful systems that have been employed in industry for several decades. As technology advances, the capability and utility of robots have increased dramatically. Today, we have robots that assemble cars, weld, fly through hostile environments, and explore the harshest environments from the depths of the ocean, to the cold and dark environment of the Antarctic, to the hazardous depths of active volcanoes, to the farthest reaches of outer space. Robots take on tasks that people do not want to perform. Perhaps these tasks are too boring, perhaps they are too dangerous, or perhaps the robot can outperform its human counterpart.

This text is targeted at the fundamentals of robot design, implementation, and application. As robots are used in a substantial number of functions, this book only scratches the surface of their applications. However, it does provide a firm basis for engineers and scientists interested in either fabrication or utilizing robotic systems. The first part of this handbook presents a number of design issues that must be considered in building and utilizing a robotic system. Both issues related to the entire robot, such as control and trajectory planning and dynamics are discussed. Critical concepts such as precision control of rotary and linear axes are also presented at they are necessary to yield optimal performance out of a robotic system. The book then continues with a number of specialized applications of robotic systems. In these applications, such as the medical arena, particular design and systems considerations are presented that are highlighted by these applications but are critical in a significant cross-section of areas. It was a pleasure to work with the authors of the various sections. They are experts in their areas, and in reviewing their material, I have improved my understanding of robotic systems. I hope that the readers will enjoy reading the text as much as I have enjoyed reading and assembling it. I anticipate that future versions of this book will incorporate more applications as well as advanced concepts in robot design and implementation.

The Editor

Thomas R. Kurfess received his S.B., S.M., and Ph.D. degrees in mechanical engineering from M.I.T. in 1986, 1987, and 1989, respectively. He also received an S.M. degree from M.I.T. in electrical engineering and computer science in 1988. Following graduation, he joined Carnegie Mellon University where he rose to the rank of Associate Professor. In 1994 he moved to the Georgia Institute of Technology where he is currently a Professor in the George W. Woodruff School of Mechanical Engineering. He presently serves as a participating guest at the Lawrence Livermore National Laboratory in their Precision Engineering Program. He is also a special consultant of the United Nations to the Government of Malaysia in the area of applied mechatronics and manufacturing. His research work focuses on the design and development of high precision manufacturing and metrology systems. He has chaired workshops for the National Science Foundation on the future of engineering education and served on the Committee of Visitors for NSF's Engineering Education and Centers Division. He has had similar roles in education and technology assessment for a variety of countries as well as the U.N.

His primary area of research is precision engineering. To this end he has applied advanced control theory to both measurement machines and machine tools, substantially improving their performance. During the past twelve years, Dr. Kurfess has concentrated in precision grinding, high-speed scanning coordinate measurement machines, and statistical analysis of CMM data. He is actively involved in using advanced mechatronics units in large scale applications to generate next generation high performance systems. Dr. Kurfess has a number of research projects sponsored by both industry and governmental agencies in this area. He has also given a number of workshops, sponsored by the National Science Foundation, in the areas of teaching controls and mechatronics to a variety of professors throughout the country.

In 1992 he was awarded a National Science Foundation Young Investigator Award, and in 1993 he received the National Science Foundation Presidential Faculty Fellowship Award. He is also the recipient of the ASME Pi Tau Sigma Award, the SME Young Manufacturing Engineer of the Year Award, the ASME Gustus L. Larson Memorial Award and the ASME Blackall Machine Tool and Gage Award. He has received the Class of 1940 W. Howard Ector's Outstanding Teacher Award and the Outstanding Faculty Leadership for the Development of Graduate Research Assistants Award while at Georgia Tech. He is a registered Professional Engineer, and is active in several engineering societies, including ASEE, ASME, ASPE, IEEE and SME. He is currently serving as a Technical Associate Editor of the *SME Journal of Manufacturing Systems*, and Associate Editor of the *ASME Journal of Manufacturing Science and Engineering*. He has served as an Associate Editor of the *ASME Journal of Dynamic Systems, Measurement and Control*. He is on the Editorial Advisory Board of the *International Journal of Engineering Education*, and serves on the board of North American Manufacturing Research Institute of SME.

Contributors

Mohan Bodduluri
Restoration Robotics
Sunnyvale, California

Wayne J. Book
Georgia Institute of Technology
Woodruff School of
 Mechanical Engineering
Atlanta, Georgia

Stephen P. Buerger
Massachusetts Institute of
 Technology
Mechanical Engineering
 Department
North Cambridge,
 Massachusetts

Keith W. Buffinton
Bucknell University
Department of Mechanical
 Engineering
Lewisburg, Pennsylvania

Francesco Bullo
University of Illinois at
 Urbana-Champaign
Coordinated Science
 Laboratory
Urbana, Illinois

Gregory S. Chirikjian
Johns Hopkins University
Department of Mechanical
 Engineering
Baltimore, Maryland

Darren M. Dawson
Clemson University
Electrical and Computer
 Engineering
Clemson, South Carolina

Bram de Jager
Technical University of
 Eindhoven
Eindhoven, Netherlands

Jaydev P. Desai
Drexel University
MEM Department
Philadelphia, Pennsylvania

Jeanne Sullivan Falcon
National Instruments
Austin, Texas

Daniel D. Frey
Massachusetts Institute of
 Technology
Mechanical Engineering
 Department
North Cambridge,
 Massachusetts

Robert B. Gillespie
University of Michigan
Ann Arbor, Michigan

J. William Goodwine
Notre Dame University
Aerospace and Mechanical
 Engineering Department
Notre Dame, Indiana

Hector M. Gutierrez
Florida Institute of Technology
Department of Mechanical and
 Aerospace Engineering
Melbourne, Florida

Yasuhisa Hirata
Tohoku University
Department of Bioengineering
 and Robotics
Sendai, Japan

Neville Hogan
Massachusetts Institute of
 Technology
Mechanical Engineering
 Department
North Cambridge,
 Massachusetts

Kun Huang
University of Illinois at
 Urbana-Champagne
Coordinated Sciences
 Laboratory
Urbana, Illinois

Hodge E. Jenkins,
Mercer University
Mechanical and Industrial
 Engineering Department
Macon, Georgia

Dragan Kostić
Technical University of
 Eindhoven
Eindhoven, Netherlands

ix

Kazuhiro Kosuge
Tohoku University
Department of Bioengineering and Robotics
Sendai, Japan

Kenneth A. Loparo
Case Western Reserve University
Department of Electrical Engineering and Computer Science
Cleveland, Ohio

Lonnie J. Love
Oak Ridge National Laboratory
Oak Ridge, Tennessee

Stephen J. Ludwick
Aerotech, Inc.
Pittsburgh, Pennsylvania

Yi Ma
University of Illinois at Urbana-Champagne
Coordinated Sciences Laboratory
Urbana, Illinois

Siddharth P. Nagarkatti
MKS Instruments, Inc.
Methuen, Massachusetts

Mark L. Nagurka
Marquette University
Department of Mechanical and Industrial Engineering
Milwaukee, Wisconsin

Chris A. Raanes
Accuray Incorporated
Sunnyvale, California

William Singhose
Georgia Institute of Technology
Woodruff School of Mechanical Engineering
Atlanta, Georgia

Mark W. Spong
University of Illinois at Urbana-Champagne
Coordinated Sciences Laboratory
Urbana, Illinois

Maarten Steinbuch
Technical University of Eindhoven
Eindhoven, Netherlands

Wesley L. Stone
Valparaiso University
Department of Mechanical Engineering
Wanatah, Indiana

Ioannis S. Vakalis
Institute for the Protection and Security of the Citizen (IPSC) European Commission
Joint Research Centre I
Ispra (VA), Italy

Miloš Žefran
University of Illinois
ECE Department
Chicago, Illinois

Contents

1. The History of Robotics
   Wesley L. Stone 1-1

2. Rigid-Body Kinematics
   Gregorg S. Chirikjian 2-1

3. Inverse Kinematics
   Bill Goodwine 3-1

4. Newton-Euler Dynamics of Robots
   Mark L. Nagurka 4-1

5. Lagrangian Dynamics
   Miloš Žefran and Francesco Bullo 5-1

6. Kane's Method in Robotics
   Keith W. Buffinton 6-1

7. The Dynamics of Systems of Interacting Rigid Bodies
   Kenneth A. Loparo and Ioannis S. Vakalis 7-1

8. D-H Convention
   Jaydev P. Desai 8-1

9. Trajectory Planning for Flexible Robots
   William E. Singhose 9-1

10. Error Budgeting
    Daniel D. Frey 10-1

11. Design of Robotic End Effectors
    Hodge Jenkins 11-1

12. Sensors
    Jeanne Sullivan Falcon 12-1

13	Precision Positioning of Rotary and Linear Systems *Stephen Ludwick* 13-1
14	Modeling and Identification for Robot Motion Control *Dragan Kostić, Bram de Jager, and Maarten Steinbuch* 14-1
15	Motion Control by Linear Feedback Methods *Dragan Kostić, Bram de Jager, and Maarten Steinbuch* 15-1
16	Force/Impedance Control for Robotic Manipulators *Siddharth P. Nagarkatti and Darren M. Dawson* 16-1
17	Robust and Adaptive Motion Control of Manipulators *Mark W. Spong* 17-1
18	Sliding Mode Control of Robotic Manipulators *Hector M. Gutierrez* 18-1
19	Impedance and Interaction Control *Neville Hogan and Stephen P. Buerger* 19-1
20	Coordinated Motion Control of Multiple Manipulators *Kazuhiro Kosuge and Yasuhisa Hirata* 20-1
21	Robot Simulation *Lonnie J. Love* 21-1
22	A Survey of Geometric Vision *Kun Huang and Yi Ma* 22-1
23	Haptic Interface to Virtual Environments *R. Brent Gillespie* 23-1
24	Flexible Robot Arms *Wayne J. Book* 24-1
25	Robotics in Medical Applications *Chris A. Raanes and Mohan Bodduluri* 25-1
26	Manufacturing Automation *Hodge Jenkins* 26-1

Index I-1

1
The History of Robotics

1.1 The History of Robotics **1**-1
    The Influence of Mythology • The Influence of Motion Pictures
    • Inventions Leading to Robotics • First Use of the Word *Robot*
    • First Use of the Word *Robotics* • The Birth of the Industrial
    Robot • Robotics in Research Laboratories
    • Robotics in Industry • Space Exploration • Military and Law
    Enforcement Applications • Medical Applications
    • Other Applications and Frontiers of Robotics

Wesley L. Stone
Western Carolina University

1.1 The History of Robotics

The history of robotics is one that is highlighted by a fantasy world that has provided the inspiration to convert fantasy into reality. It is a history rich with cinematic creativity, scientific ingenuity, and entrepreneurial vision. Quite surprisingly, the definition of a robot is controversial, even among roboticists. At one end of the spectrum is the science fiction version of a robot, typically one of a human form — an android or humanoid — with anthropomorphic features. At the other end of the spectrum is the repetitive, efficient robot of industrial automation. In ISO 8373, the International Organization for Standardization defines a robot as "an automatically controlled, reprogrammable, multipurpose manipulator with three or more axes." The Robot Institute of America designates a robot as "a reprogrammable, multifunctional manipulator designed to move material, parts, tools, or specialized devices through various programmed motions for the performance of a variety of tasks." A more inspiring definition is offered by Merriam-Webster, stating that a robot is "a machine that looks like a human being and performs various complex acts (as walking or talking) of a human being."

1.1.1 The Influence of Mythology

Mythology is filled with artificial beings across all cultures. According to Greek legend, after Cadmus founded the city of Thebes, he destroyed the dragon that had slain several of his companions; Cadmus then sowed the dragon teeth in the ground, from which a fierce army of armed men arose. Greek mythology also brings the story of Pygmalion, a lovesick sculptor, who carves a woman named Galatea out of ivory; after praying to Aphrodite, Pygmalion has his wish granted and his sculpture comes to life and becomes his bride. Hebrew mythology introduces the golem, a clay or stone statue, which is said to contain a scroll with religious or magic powers that animate it; the golem performs simple, repetitive tasks, but is difficult to stop. Inuit legend in Greenland tells of the Tupilaq, or Tupilak, which is a creature created from natural materials by the hands of those who practiced witchcraft; the Tupilaq is then sent to sea to destroy the enemies of the creator, but an adverse possibility existed — the Tupilaq can be turned on its creator if the enemy knows witchcraft. The homunculus, first introduced by 15th Century alchemist Paracelsus, refers to a small human form, no taller than 12 inches; originally ascribed to work associated with a golem, the homunculus became synonymous with an inner being, or the "little man" that controls the thoughts of a human. In 1818, Mary Wollstonecraft Shelley wrote *Frankenstein*, introducing the creature created by scientist Victor Frankenstein from various materials, including cadavers; Frankenstein's creation is grossly misunderstood, which leads to the tragic deaths of the scientist and many of the loved ones in his life. These mythological tales, and many like them, often have a common thread: the creators of the supernatural beings often see their creations turn on them, typically with tragic results.

1.1.2 The Influence of Motion Pictures

The advent of motion pictures brought to life many of these mythical creatures, as well as a seemingly endless supply of new artificial creatures. In 1926, Fritz's Lang's movie "Metropolis" introduced the first robot in a feature film. The 1951 film "The Day the Earth Stood Still" introduced the robot Gort and the humanoid alien Klaatu, who arrived in Washington, D.C., in their flying saucer. Robby, the Robot, first made his appearance in "Forbidden Planet" (1956), becoming one of the most influential robots in cinematic history. In 1966, the television show "Lost in Space" delivered the lovable robot B-9, who consistently saved the day, warning Will Robinson of aliens approaching. The 1968 movie "2001: A Space Odyssey" depicted a space mission gone awry, where Hal employed his artificial intelligence (AI) to wrest control of the space ship from the humans he was supposed to serve. In 1977, "Star Wars" brought to life two of the most endearing robots ever to visit the big screen — R2-D2 and C3PO. Movies and television have brought to life these robots, which have served in roles both evil and noble. Although just a small sampling, they illustrate mankind's fascination with mechanical creatures that exhibit intelligence that rivals, and often surpasses, that of their creators.

1.1.3 Inventions Leading to Robotics

The field of robotics has evolved over several millennia, without reference to the word *robot* until the early 20th Century. In 270 B.C., ancient Greek physicist and inventor Ctesibus of Alexandria created a water clock, called the clepsydra, or "water-thief," as it translates. Powered by rising water, the clepsydra employed a cord attached to a float and stretched across a pulley to track time. Apparently, the contraption entertained many who watched it passing away the time, or stealing their time, thus earning its namesake. Born in Lyon, France, Joseph Jacquard (1752–1834) inherited his father's small weaving business but eventually went bankrupt. Following this failure, he worked to restore a loom and in the process developed a strong interest in mechanizing the manufacture of silk. After a hiatus in which he served for the Republicans in the French Revolution, Jacquard returned to his experimentation and in 1801 invented a loom that used a series of punched cards to control the repetition of patterns used to weave cloths and carpets. Jacquard's card system was later adapted by Charles Babbage in early 19th Century Britain to create an automatic calculator, the principles of which later led to the development of computers and computer programming. The inventor of the automatic rifle, Christopher Miner Spencer (1833–1922) of Manchester, Connecticut, is also credited with giving birth to the screw machine industry. In 1873, Spencer was granted a patent for the lathe that he developed, which included a camshaft and a self-advancing turret. Spencer's turret lathe took the manufacture of screws to a higher level of sophistication by automating the process. In 1892, Seward Babbitt introduced a motorized crane that used a mechanical gripper to remove ingots from a furnace, 70 years prior to General Motors' first industrial robot used for a similar purpose. In the 1890s Nikola Tesla — known for his discoveries in AC electric power, the radio, induction motors, and more — invented the first remote-controlled vehicle, a radio-controlled boat. Tesla was issued Patent #613.809 on November 8, 1898, for this discovery.

1.1.4 First Use of the Word *Robot*

The word *robot* was not even in the vocabulary of industrialists, let alone science fiction writers, until the 1920s. In 1920, Karel Čapek (1890–1938) wrote the play, *Rossum's Universal Robots*, commonly known as *R.U.R.*, which premiered in Prague in 1921, played in London in 1921, in New York in 1922, and was published in English in 1923. Čapek was born in 1890 in Malé Svatonovice, Bohemia, Austria-Hungary, now part of the Czech Republic. Following the First World War, his writings began to take on a strong political tone, with essays on Nazism, racism, and democracy under crisis in Europe.

In *R.U.R.*, Čapek's theme is one of futuristic man-made workers, created to automate the work of humans, thus alleviating their burden. As Čapek wrote his play, he turned to his older brother, Josef, for a name to call these beings. Josef replied with a word he coined — *robot*. The Czech word *robotnik* refers to a peasant or serf, while *robota* means drudgery or servitude. The Robots (always capitalized by Čapek) are produced on a remote island by a company founded by the father-son team of Old Rossum and Young Rossum, who do not actually appear in the play. The mad inventor, Old Rossum, had devised the plan to create the perfect being to assume the role of the Creator, while Young Rossum viewed the Robots as business assets in an increasingly industrialized world. Made of organic matter, the Robots are created to be efficient, inexpensive beings that remember everything and think of nothing original. Domin, one of the protagonists, points out that because of these Robot qualities, "They'd make fine university professors." Wars break out between humans and Robots, with the latter emerging victorious, but the formula that the Robots need to create more Robots is burned. Instead, the Robots discover love and eliminate the need for the formula.

The world of robotics has Karel and Josef Čapek to thank for the word *robot*, which replaced the previously used *automaton*. Karel Čapek's achievements extend well beyond *R.U.R.*, including "War With The Newts," an entertaining satire that takes jabs at many movements, such as Nazism, communism, and capitalism; a biography of the first Czechoslovak Republic president, Tomáš Masaryk; numerous short stories, poems, plays, and political essays; and his famous suppressed text "Why I Am Not a Communist." Karel Čapek died of pneumonia in Prague on Christmas Day 1938. Josef Čapek was seized by the Nazis in 1939 and died at the Bergen-Belsen concentration camp in April 1945.

1.1.5 First Use of the Word *Robotics*

Isaac Asimov (1920–1992) proved to be another science fiction writer who had a profound impact on the history of robotics. Asimov's fascinating life began on January 2, 1920 in Petrovichi, Russia, where he was born to Jewish parents, who immigrated to America when he was three years old. Asimov grew up in Brooklyn, New York, where he developed a love of science fiction, reading comic books in his parents' candy store. He graduated from Columbia University in 1939 and earned a Ph.D. in 1948, also from Columbia. Asimov served on the faculty at Boston University, but is best known for his writings, which spanned a very broad spectrum, including science fiction, science for the layperson, and mysteries. His publications include entries in every major category of the Dewey Decimal System, except for Philosophy. Asimov's last nonfiction book, *Our Angry Earth*, published in 1991 and co-written with science fiction writer Frederik Pohl, tackles environmental issues that deeply affect society today — ozone depletion and global warming, among others. His most famous science fiction work, the *Foundation Trilogy*, begun in 1942, paints a picture of a future universe with a vast interstellar empire that experiences collapse and regeneration. Asimov's writing career divides roughly into three periods: science fiction from approximately 1940–1958, nonfiction the next quarter century, and science fiction again 1982–1992.

During Asimov's first period of science fiction writing, he contributed greatly to the creative thinking in the realm that would become robotics. Asimov wrote a series of short stories that involved robot themes. *I, Robot*, published in 1950, incorporated nine of these related short stories in one collection — "Robbie," "Runaround," "Reason," "Catch That Rabbit," "Liar!," "Little Lost Robot," "Escape!," "Evidence," and "The Evitable Conflict." It was in his short stories that Asimov introduced what would become the "Three Laws of Robotics." Although these three laws appeared throughout several writings, it was not until "Runaround," published in 1942, that they appeared together and in concise form. "Runaround" is also the first time that the word *robotics* is used, and it is taken to mean the technology dealing with the design, construction, and operation of robots. In 1985 he modified his list to include the so-called "Zeroth Law" to arrive at his famous "Three Laws of Robotics":

Zeroth Law: A robot may not injure humanity, or, through inaction, allow humanity to come to harm.
First Law: A robot may not injure a human being, or, through inaction, allow a human being to come to harm, unless this would violate a higher order law.
Second Law: A robot must obey the orders given to it by human beings, except where such orders would conflict with a higher order law.
Third Law: A robot must protect its own existence, as long as such protection does not conflict with a higher order law.

In "Runaround," a robot charged with the mission of mining selenium on the planet Mercury is found to have gone missing. When the humans investigate, they find that the robot has gone into a state of disobedience with two of the laws, which puts it into a state of equilibrium that sends it into an endless cycle of running around in circles, thus the name, "Runaround." Asimov originally credited John W. Campbell, long-time editor of the science fiction magazine *Astounding Science Fiction* (later renamed *Analog Science Fiction*), with the famous three laws, based on a conversation they had on December 23, 1940. Campbell declined the credit, claiming that Asimov already had these laws in his head, and he merely facilitated the explicit statement of them in writing.

A truly amazing figure of the 20th Century, Isaac Asimov wrote science fiction that profoundly influenced the world of science and engineering. In Asimov's posthumous autobiography, *It's Been a Good Life* (March 2002), his second wife, Janet Jeppson Asimov, reveals in the epilogue that his death on April 6, 1992, was a result of HIV contracted through a transfusion of tainted blood nine years prior during a triple-bypass operation. Isaac Asimov received over 100,000 letters throughout his life and personally answered over 90,000 of them. In *Yours, Isaac Asimov* (1995), Stanley Asimov, Isaac's younger brother, compiles 1,000 of these letters to provide a glimpse of the person behind the writings. A quote from one of those letters, dated September 20, 1973, perhaps best summarizes Isaac Asimov's career: "What I *will* be remembered for are the Foundation Trilogy and the Three Laws of Robotics. What I *want* to be remembered for is no one book, or no dozen books. Any single thing I have written can be paralleled or even surpassed by something someone else has done. However, my total corpus for quantity, quality, and *variety* can be duplicated by no one else. That is what I want to be remembered for."

1.1.6 The Birth of the Industrial Robot

Following World War II, America experienced a strong industrial push, reinvigorating the economy. Rapid advancement in technology drove this industrial wave — servos, digital logic, solid state electronics, etc. The merger of this technology and the world of science fiction came in the form of the vision of Joseph Engelberger, the ingenuity of George Devol, and their chance meeting in 1956. Joseph F. Engelberger was born on July 26, 1925, in New York City. Growing up, Engelberger developed a fascination for science fiction, especially that written by Isaac Asimov. Of particular interest in the science fiction world was the robot, which led him to pursue physics at Columbia University, where he earned both his bachelor's and master's degrees. Engelberger served in the U.S. Navy and later worked as a nuclear physicist in the aerospace industry.

In 1946, a creative inventor by the name of George C. Devol, Jr., patented a playback device used for controlling machines. The device used a magnetic process recorder to accomplish the control. Devol's drive toward automation led him to another invention in 1954, for which he applied for a patent, writing, "The present invention makes available for the first time a more or less general purpose machine that has universal application to a vast diversity of applications where cyclic control is desired." Devol had dubbed his invention *universal automation,* or *unimation* for short. Whether it was fate, chance, or just good luck, Devol and Engelberger met at a cocktail party in 1956. Their conversation revolved around robotics, automation, Asimov, and Devol's patent application, "A Programmed Article Transfer," which Engelberger's imagination translated into "robot." Following this chance meeting, Engelberger and Devol formed a partnership that lead to the birth of the industrial robot.

Engelberger took out a license under Devol's patent and bought out his employer, renaming the new company Consolidated Controls Corporation, based out of his garage. His team of engineers that had been working on aerospace and nuclear applications refocused their efforts on the development of the first industrial robot, named the Unimate, after Devol's "unimation." The first Unimate was born in 1961 and was delivered to General Motors in Trenton, New Jersey, where it unloaded high temperature parts from a die casting machine — a very unpopular job for manual labor. Also in 1961, patent number 2,998,237 was granted to Devol — the first U.S. robot patent. In 1962 with the backing of Consolidated Diesel Electric Company (Condec) and Pullman Corporation, Engelberger formed Unimation, Inc., which eventually blossomed into a prosperous business — GM alone had ordered 66 Unimates. Although it took until 1975 to turn a profit, Unimation became the world leader in robotics, with 1983 annual sales of $70 million and 25 percent of the world market share. For his visionary pursuit and entrepreneurship, Joseph Engelberger is widely considered the "Father of Robotics." Since 1977, the Robotic Industries Association has presented the annual Engelberger Robotics Awards to world leaders in both application and leadership in the field of robotics.

1.1.7 Robotics in Research Laboratories

The post-World War II technology boom brought a host of developments. In 1946 the world's first electronic digital computer emerged at the University of Pennsylvania at the hands of American scientists J. Presper Eckert and John Mauchly. Their computer, called ENIAC (electronic numerical integrator and computer), weighed over 30 tons. Just on the heels of ENIAC, Whirlwind was introduced by Jay Forrester and his research team at the Massachusetts Institute of Technology (MIT) as the first general purpose digital computer, originally commissioned by the U. S. Navy to develop a flight simulator to train its pilots. Although the simulator did not develop, a computer that shaped the path of business computers was born. Whirlwind was the first computer to perform real-time computations and to use a video display as an output device. At the same time as ENIAC and Whirlwind were making their appearance on the East Coast of the United States, a critical research center was formed on the West Coast.

In 1946, the Stanford Research Institute (SRI) was founded by a small group of business executives in conjunction with Stanford University. Located in Menlo Park, California, SRI's purpose was to serve as a center for technological innovation to support regional economic development. In 1966 the Artificial Intelligence Center (AIC) was founded at SRI, pioneering the field of artificial intelligence (AI), which gives computers the heuristics and algorithms to make decisions in complex situations.

From 1966 to 1972 Shakey, the Robot, was developed at the AIC by Dr. Charles Rosen (1917–2002) and his team. Shakey was the first mobile robot to reason its way about its surroundings and had a far-reaching influence on AI and robotics. Shakey was equipped with a television camera, a triangulating range finder, and bump sensors. It was connected by radio and video links to DEC PDP-10 and PDP-15 computers. Shakey was equipped with three levels of programming for perceiving, modeling, and interacting with its environment. The lowest level routines were designed for basic locomotion — movement, turning, and route planning. The intermediate level combined the low-level routines together to accomplish more difficult tasks. The highest-level routines were designed to generate and execute plans to accomplish tasks presented by a user. Although Shakey had been likened to a small unstable box on wheels — thus the name — it represented a significant milestone in AI and in developing a robot's ability to interact with its environment.

Beyond Shakey, SRI has advanced the field of robotics through contributions in machine vision, computer graphics, AI engineering tools, computer languages, autonomous robots, and more. A nonprofit organization, SRI disassociated itself from Stanford University in 1977, becoming SRI International. SRI's current efforts in robotics include advanced factory applications, field robotics, tactical mobile robots, and pipeline robots. Factory applications encompass robotic advances in assembly, parts feeding, parcel handling, and machine vision. In contrast to the ordered environment of manufacturing, field robotics involves robotic applications in highly unstructured settings, such as reconnaissance, surveillance, and explosive ordnance disposal. Similar to field robotics, tactical mobile robots are being developed for unstructured surroundings in both military and commercial applications, supplementing human capabilities, such as searching through debris following disasters (earthquakes, bombed buildings, etc.). SRI's pipeline robot, Magnetically Attached General Purpose Inspection Engine (MAGPIE), is designed to inspect natural gas pipelines, as small as 15 cm in diameter, for corrosion and leakage, navigating through pipe elbows and T-joints on its magnetic wheels.

In 1969 at Stanford University, a mechanical engineering student by the name of Victor Scheinman developed the Stanford Arm, a robot created exclusively for computer control. Working in the Stanford Artificial Intelligence Lab (SAIL), Scheinman built the entire robotic arm on campus, primarily using the shop facilities in the Chemistry Department. The kinematic configuration of the arm included six degrees of freedom with one prismatic and five revolute joints, with brakes on all joints to hold position while the computer computed the next position or performed other time-shared duties. The arm was loaded with DC electric motors, a harmonic drive, spur gear reducers, potentiometers, analog tachometers, electromechanical brakes, and a servo-controlled proportional electric gripper — a gripper with a 6-axis force/torque sensor in the wrist and tactile sense contacts on the fingers. The highly integrated Stanford Arm served for over 20 years in the robotics laboratories at Stanford University for both students and researchers.

The Stanford Cart, another project developed at SAIL, was a mobile robot that used an onboard television camera to navigate its way through its surroundings. The Cart was supported between 1973 and 1980 by the Defense Advanced Research Projects Agency (DARPA), the National Science Foundation (NSF), and the National Aeronautics and Space Administration (NASA). The cart used its TV camera and stereo vision routines to perceive the objects surrounding it. A computer program processed the images, mapping the obstacles around the cart. This map provided the means by which the cart planned its path. As it moved, the cart adjusted its plan according to the new images gathered by the camera. The system worked very reliably but was very slow; the cart moved at a rate of approximately one meter every 10 or 15 minutes. Triumphant in navigating itself through several 20-meter courses, the Stanford Cart provided the field of robotics with a reliable, mobile robot that successfully used vision to interact with its surroundings.

Research in robotics also found itself thriving on the U. S. East Coast at MIT. At the same time Asimov was writing his short stories on robots, MIT's Norbert Wiener published *Cybernetics, or the Control and Communication in the Animal and the Machine* (1948). In *Cybernetics* Wiener effectively communicates to both the trained scientist and the layman how feedback is used in technical applications, as well as everyday life. He skillfully brought to the forefront the sociological impact of technology and popularized the concept of control feedback.

Although artificial intelligence experienced its growth and major innovations in the laboratories of prestigious universities, its birth can be traced to Claude E. Shannon, a Bell Laboratories mathematician, who wrote two landmark papers in 1950 on the topic of chess playing by a machine. His works inspired John McCarthy, a young mathematician at Princeton University, who joined Shannon to organize a 1952 conference on automata. One of the participants at that conference was an aspiring Princeton graduate student in mathematics by the name of Marvin Minsky. In 1953 Shannon was joined by McCarthy and Minsky at Bell Labs. Creating an opportunity to rapidly advance the field of machine intelligence, McCarthy approached the Rockefeller Foundation with the support of Shannon. Warren Weaver and Robert S. Morison at the foundation provided additional guidance and in 1956 The Dartmouth Summer Research Project on Artificial Intelligence was organized at Dartmouth University, where McCarthy was an assistant professor of mathematics. Shannon, McCarthy, Minsky, and IBM's Nat Rochester joined forces to coordinate the conference, which gave birth to the term *artificial intelligence*.

In 1959 Minsky and McCarthy founded the MIT Artificial Intelligence Laboratory, which was the initiation of robotics at MIT (McCarthy later left MIT in 1963 to found the Stanford Artificial Intelligence Laboratory). Heinrich A. Ernst developed the Mechanical Hand-1 (MH-1), which was the first computer-controlled manipulator and hand. The MH-1 hand-arm combination had 35 degrees of freedom and was later simplified to improve its functionality. In 1968 Minsky developed a 12-joint robotic arm called the Tentacle Arm, named after its octopus-like motion. This arm was controlled by a PDP-6 computer, powered by hydraulics, and capable of lifting the weight of a person. Robot computer language development thrived at MIT as well: THI was developed by Ernst, LISP by McCarthy, and there were many other robot developments as well. In addition to these advancements, MIT significantly contributed to the field of robotics through research in compliant motion control, sensor development, robot motion planning, and task planning.

At Carnegie Mellon University, the Robotics Institute was founded in 1979. In that same year Hans P. Moravec took the principles behind Shakey at SRI to develop the CMU Rover, which employed three pairs of omni-directional wheels. An interesting feature of the Rover's kinematic motion was that it could open a door with its arm, travel a straight line through the doorway, rotating about its vertical axis to maintain the arm contact holding the door open. In 1993 CMU deployed Dante, an eight-legged rappelling robot, to descend into Mount Erebus, an active volcano in Antarctica. The intent of the mission was to collect gas samples and to explore harsh environments, such as those expected on other planets. After descending 20 feet into the crater, Dante's tether broke and Dante was lost. Not discouraged by the setback, in 1994 the Robotics Institute, led by John Bares and William "Red" Whittaker, sent a more robust Dante II into Mount Spurr, another active volcano 80 miles west of Anchorage, Alaska. Dante II's successful mission highlighted several major accomplishments: transmitting video, traversing rough terrain (for more than five days), sampling gases, operating remotely, and returning safely. Research at CMU's Robotics Institute continues to advance the field in speech understanding, industrial parts feeding, medical applications, grippers, sensors, controllers, and a host of other topics.

Beyond Stanford, MIT, and CMU, there are many more universities that have successfully undertaken research in the field of robotics. Now virtually every research institution has an active robotics research group, advancing robot technology in fundamentals, as well as applications that feed into industry, medicine, aerospace, military, and many more sectors.

1.1.8 Robotics in Industry

Running in parallel with the developments in research laboratories, the use of robotics in industry blossomed beyond the time of Engelberger and Devol's historic meeting. In 1959, Planet Corporation developed the first commercially available robot, which was controlled by limit switches and cams. The next year, Harry Johnson and Veljko Milenkovic of American Machine and Foundry, later known as AMF Corporation, developed a robot called Versatran, from the words *versatile transfer*; the Versatran became commercially available in 1963.

In Norway, a 1964 labor shortage led a wheelbarrow manufacturer to install the first Trallfa robot, which was used to paint the wheelbarrows. Trallfa robots, produced by Trallfa Nils Underhaug of Norway, were hydraulic robots with five or six degrees of freedom and were the first industrial robots to use the revolute coordinate system and continuous-path motion. In 1966, Trallfa introduced a spray-painting robot into factories in Byrne, Norway. This spray-painting robot was modified in 1976 by Ransome, Sims, and Jefferies, a British producer of agricultural machinery, for use in arc welding applications. Painting and welding developed into the most common applications of robots in industry.

Seeing success with their Unimates in New Jersey, General Motors used 26 Unimate robots to assemble the Chevrolet Vega automobile bodies in Lordstown, Ohio, beginning in 1969. GM became the first company to use machine vision in an industrial setting, installing the Consight system at their foundry in St. Catherines, Ontario, Canada, in 1970.

At the same time, Japanese manufacturers were making quantum leaps in manufacturing: cutting costs, reducing variation, and improving efficiency. One of the major factors contributing to this transformation was the incorporation of robots in the manufacturing process. Japan imported its first industrial robot in 1967, a Versatran from AMF. In 1971 the Japanese Industrial Robot Association (JIRA) was formed, providing encouragement from the government to incorporate robotics. This move helped to move the Japanese to the forefront in total number of robots used in the world. In 1972 Kawasaki installed a robot assembly line, composed of Unimation robots at their plant in Nissan, Japan. After purchasing the Unimate design from Unimation, Kawasaki improved the robot to create an arc-welding robot in 1974, used to fabricate their motorcycle frames. Also in 1974, Hitachi developed touch and force-sensing capabilities in their Hi-T-Hand robot, which enabled the robot to guide pins into holes at a rate of one second per pin.

At Cincinnati Milacron Corporation, Richard Hohn developed the robot called The Tomorrow Tool, or T^3. Released in 1973, the T^3 was the first commercially available industrial robot controlled by a microcomputer, as well as the first U.S. robot to use the revolute configuration. Hydraulically actuated, the T^3 was used in applications such as welding automobile bodies, transferring automobile bumpers, and loading machine tools. In 1975, the T^3 was introduced for drilling applications, and in the same year, the T^3 became the first robot to be used in the aerospace industry.

In 1970, Victor Scheinman, of Stanford Arm fame, left his position as professor at Stanford University to take his robot arm to industry. Four years later, Scheinman had developed a minicomputer-controlled robotic arm, known as the Vicarm, thus founding Vicarm, Inc. This arm design later came to be known as the "standard arm." Unimation purchased Vicarm in 1977, and later, relying on support from GM, used the technology from Vicarm to develop the PUMA (Programmable Universal Machine for Assembly), a relatively small electronic robot that ran on an LSI II computer.

The ASEA Group of Västerås, Sweeden, made significant advances in electric robots in the 1970's. To handle automated grinding operations, ASEA introduced its IRb 6 and IRb 60 all-electric robots in 1973. Two years later, ASEA became the first to install a robot in an iron foundry, tackling yet more industrial jobs that are not favored by manual labor. In 1977 ASEA introduced two more electric-powered industrial robots, both of which used microcomputers for programming and operation. Later, in 1988, ASEA merged with BBC Brown Boveri Ltd of Baden, Switzerland, to form ABB (ASEA, Brown, and Boveri), one of the world leaders in power and automation technology.

At Yamanashi University in Japan, IBM and Sankyo joined forces to develop the Selective Compliance Assembly Robot Arm (SCARA) in 1979. The SCARA was designed with revolute joints that had vertical axes, thus providing stiffness in the vertical direction. The gripper was controlled in compliant mode, or using force control, while the other joints were operated in position control mode. These robots were used and continue to be used in many applications where the robot is acting vertically on a workpiece oriented horizontally, such as polishing and insertion operations. Based on the SCARA geometry, Adept Technology was founded in 1983. Adept continues to supply direct drive robots that service industries, such as telecommunications, electronics, automotive, and pharmaceuticals. These industrial developments in robotics, coupled with the advancements in the research laboratories, have profoundly affected robotics in different sectors of the technical world.

1.1.9 Space Exploration

Space exploration has been revolutionized by the introduction of robotics, taking shape in many different forms, such as flyby probes, landers, rovers, atmospheric probes, and robot arms. All can be remotely operated and have had a common theme of removing mankind from difficult or impossible settings. It would not be possible to send astronauts to remote planets and return them safely. Instead, robots are sent on these journeys, transmitting information back to Earth, with no intent of returning home.

Venus was the first planet to be reached by a space probe when Mariner 2 passed within 34,400 kilometers in 1962. Mariner 2 transmitted information back to earth about the Venus atmosphere, surface temperature, and rotational period. In December 1970, Venera 7, a Soviet lander, became the first man-made object to transmit data back to Earth after landing on another planet. Extreme temperatures limited transmissions from Venera 7 to less than an hour, but a new milestone had been achieved. The Soviets' Venera 13 became the first lander to transmit color pictures from the surface of Venus when it landed in March 1982. Venera 13 also took surface samples by means of mechanical drilling and transmitted analysis data via the orbiting bus that had dropped the lander. Venera 13 survived for 127 minutes at 236°C (457°F) and 84 Earth atmospheres, well beyond its design life of 32 minutes. In December 1978, NASA's Pioneer Venus sent an Orbiter into an orbit of Venus, collecting information on Venusian solar winds, radar images of the surface, and details about the upper atmosphere and ionosphere. In August 1990, NASA's Magellan entered the Venus atmosphere, where it spent four years in orbit, radar-mapping 98% of the planet's surface before plunging into the dense atmosphere on October 11, 1994.

NASA's Mariner 10 was the first space probe to visit Mercury and was also the first to visit two planets — Venus and Mercury. Mariner 10 actually used the gravitational pull of Venus to throw it into a different orbit, where it was able to pass Mercury three times between 1974 and 1975. Passing within 203 kilometers of Mercury, the probe took over 2800 photographs to detail a surface that had previously been a mystery due to the Sun's solar glare that usually obscured astronomers' views.

The red planet, Mars, has seen much activity from NASA spacecraft. After several probe and orbiter missions to Mars, NASA launched Viking 1 and Viking 2 in August and September of 1975, respectively. The twin spacecraft, equipped with robotic arms, began orbiting Mars less than a year later, Viking 1 in June 1976 and Viking 2 in August 1976. In July and September of the same year, the two landers were successfully sent to the surface of Mars, while the orbiters remained in orbit. The Viking orbiters 1 and 2 continued transmission to Earth until 1980 and 1978, respectively, while their respective landers transmitted data until 1982 and 1980. The successes of this mission were great: Viking 1 and 2 were the first two spacecraft to land on a planet and transmit data back to Earth for an extended period; they took extensive photographs; and they conducted biological experiments to test for signs of organic matter on the red planet. In December 1996, the Mars Pathfinder was launched, including both a lander and a rover, which arrived on Mars in July of 1997. The lander was named the Carl Sagan Memorial Station, and the rover was named Sojourner after civil rights crusader Sojourner Truth. Both the lander and rover outlived their design lives, by three and 12 times, with final transmissions coming in late September 1997. In mid-2003, NASA launched its Mars Exploration Rovers mission with twin rovers, Spirit and Opportunity, which touched down in early and late January 2004, respectively. With greater sophistication and better mobility than Sojourner, these rovers landed at different locations on Mars, each looking for signs of liquid water that might have existed in Mars' past. The rovers are equipped with equipment — a panoramic camera, spectrometers, and a microscopic imager — to capture photographic images and analyze rock and soil samples.

Additional missions have explored the outer planets — Jupiter, Saturn, Uranus, and Neptune. Pioneer 10 was able to penetrate the asteroid belt between Mars and Jupiter to transmit close-up pictures of Jupiter, measure the temperature of its atmosphere, and map its magnetic field. Similarly, Pioneer 11 transmitted the first close-up images of Saturn and its moon Titan in 1979. The Voyager missions followed closely after Pioneer, with Voyager 2 providing close-up analysis of Uranus, revealing 11 rings around the planet, rather than the previously thought nine rings. Following its visit to Uranus, Voyager 2 continued to Neptune, completing its 12-year journey through the solar system. Galileo was launched in 1989 to examine Jupiter and its four largest moons, revealing information about Jupiter's big red spot and about the moons Europa and Io. In 1997, NASA launched the Cassini probe on a seven-year journey to Saturn, expecting to gather information about Saturn's rings and its moons.

The International Space Station (ISS), coordinated by Boeing and involving nations from around the globe, is the largest and most expensive space mission ever undertaken. The mission began in 1995 with U.S. astronauts, delivered by NASA's Space Shuttle, spending time aboard the Russian Mir space station. In 2001, the Space Station Remote Manipulator System (SSRMS), built by MD Robotics of Canada, was successfully launched to complete the assembly operations of the ISS. Once completed, the ISS research laboratories will explore microgravity, life science, space science, Earth science, engineering research and technology, and space product development.

1.1.10 Military and Law Enforcement Applications

Just as space programs have used robots to accomplish tasks that would not even be considered as a manned mission, military and law enforcement agencies have employed the use of robots to remove humans from harm's way. Police are able to send a microphone or camera into a dangerous area that is not accessible to law enforcement personnel, or is too perilous to enter. Military applications have grown and continue to do so.

Rather than send a soldier into the field to sweep for landmines, it is possible to send a robot to do the same. Research is presently underway to mimic the method used by humans to identify landmines. Another approach uses swarm intelligence, which is research being developed at a company named Icosystems, under funding from DARPA. The general approach is similar to that of a colony of ants finding the most efficient path through trial and error, finding success based on shear numbers. Icosystems is using 120 robots built by I-Robot, a company co-founded by robotics pioneer Rodney Brooks, who is also the director of the Computer Science and Artificial Intelligence Laboratory at MIT. One of Brooks' research interests is developing intelligent robots that can operate in unstructured environments, an application quite different from that in a highly structured manufacturing environment.

Following the tragic events of September 11, 2001, the United States retaliated against al Qaeda and Taliban forces in Afghanistan. One of the weapons that received a great deal of media attention was the Predator UAV (unmanned aerial vehicle), or drone. The drone is a plane that is operated remotely with no human pilot on-board, flying high above an area to collect military intelligence. Drones had been used by the U.S. in the Balkans in 1999 in this reconnaissance capacity, but it was during the engagement in Afghanistan that the drones were armed with anti-tank missiles. In November 2002, a Predator UAV fired a Hellfire missile to destroy a car carrying six suspected al Qaeda operatives. This strike marked a milestone in the use of robotics in military settings.

Current research at Sandia National Laboratory's Intelligent Systems and Robotics Center is aimed at developing robotic sentries, under funding from DARPA. These robots would monitor the perimeter of a secured area, signaling home base in the event of a breach in security. The technology is also being extended to develop ground reconnaissance vehicles, a land version of the UAV drones.

1.1.11 Medical Applications

The past two decades have seen the incorporation of robotics into medicine. From a manufacturing perspective, robots have been used in pharmaceuticals, preparing medications. But on more novel levels, robots have been used in service roles, surgery, and prosthetics.

In 1984 Joseph Engelberger formed Transition Research Corporation, later renamed HelpMate Robotics, Inc., based in Danbury, Connecticut. This move by Engelberger marked a determined effort on his part to move robotics actively into the service sector of society. The first HelpMate robot went to work in a Danbury hospital in 1988, navigating the hospital wards, delivering supplies and medications, as needed by hospital staff.

The capability of high-precision operation in manufacturing settings gave the medical industry high hopes that robots could be used to assist in surgery. Not only are robots capable of much higher precision than a human, they are not susceptible to human factors, such as trembling and sneezing, that are undesirable in the surgery room. In 1990, Robodoc was developed by Dr. William Bargar, an orthopedist, and the late Howard Paul, a veterinarian, of Integrated Surgical Systems, Inc., in conjunction with the University of California at Davis. The device was used to perform a hip replacement on a dog in 1990 and on the first human in 1992, receiving U.S. Food and Drug Administration (FDA) approval soon thereafter. The essence of the procedure is that traditional hip replacements required a surgeon to dig a channel down the patient's femur to allow the replacement hip to be attached, where it is cemented in place. The cement often breaks down over time, requiring a new hip replacement in 10 or 15 years for many patients. Robodoc allows the surgeon to machine a precise channel down the femur, allowing for a tight-fit between replacement hip and femur. No cement is required, allowing the bone to graft itself onto the bone, creating a much stronger and more permanent joint.

Another advantage to robots in medicine is the ability to perform surgery with very small incisions, which results in minimal scar tissue, and dramatically reduced recovery times. The popularity of these minimally invasive surgical (MIS) procedures has enabled the incorporation of robots in endoscopic surgeries. Endoscopy involves the feeding of a tiny fiber optic camera through a small incision in the patient. The camera allows the surgeon to operate with surgical instruments, also inserted through small incisions, avoiding the trauma of large, open cuts. Endoscopic surgery in the abdominal area is referred to as laparoscopy, which has been used since the late 1980's for surgery on the gall bladder and female organs, among others. Thorascopic surgery is endoscopic surgery inside the chest cavity — lungs, esophagus, and thoracic artery. Robotic surgical systems allow doctors to sit at a console, maneuvering the camera and surgical instruments by moving joysticks, similar to those used in video games. This same remote robotic surgery has been extended to heart surgery as well. In addition to the precision and minimized incisions, the robotic systems have an advantage over the traditional endoscopic procedure in that the robotic surgery is very intuitive. Doctors trained in endoscopic surgery must learn to move in the opposite direction of the image transmitted by the camera, while the robotic systems directly mimic the doctor's movements. As of 2001, the FDA had cleared two robotic endoscopic systems to perform both laparoscopic and thoracoscopic surgeries — the da Vinci Surgical System and the ZEUS Robotic Surgical System.

Another medical arena that has shown recent success is prosthetics. Robotic limbs have been developed to replicate natural human movement and return functionality to amputees. One such example is a bionic arm that was developed at the Princess Margaret Rose Hospital in Edinburgh, Scotland, by a team of bioengineers, headed by managing director David Gow. Conjuring up images of the popular 1970's television show "The Six Million Dollar Man," this robotic prosthesis, known as the Edinburgh Modular Arm System (EMAS), was created to replace the right arm from the shoulder down for Campbell Aird, a man whose arm was amputated after finding out he had cancer. The bionic arm was equipped with a motorized shoulder, rotating wrist, movable fingers, and artificial skin. With only several isolated glitches, the EMAS was considered a success, so much so that Aird had taken up a hobby of flying.

Another medical frontier in robotics is that of robo-therapy. Research at NASA's Jet Propulsion Laboratory (JPL) and the University of California at Los Angeles (UCLA) has focused on using robots to assist in retraining the central nervous system in paralyzed patients. The therapy originated in Germany, where researchers retrained patients through a very manually intensive process, requiring four or more therapists. The new device would take the place of the manual effort of the therapists with one therapist controlling the robot via hand movements inside a set of gloves equipped with sensors.

1.1.12 Other Applications and Frontiers of Robotics

In addition to their extensive application in manufacturing, space exploration, the military, and medicine, robotics can be found in a host of other fields, such as the ever-present entertainment market — toys, movies, etc. In 1998 two popular robotic toys came to market. Tiger Electronics introduced "Furby" which rapidly became the toy of choice in the 1998 Christmas toy market. Furby used a variety of different sensors to react with its environment, including speech that included over 800 English phrases, as well as many in its own language "Furbish." In the same year Lego released its Lego MINDSTORMS robotic toys. These reconfigurable toys rapidly found their way into educational programs for their value in engaging students, while teaching them about the use of multiple sensors and actuators to respond to the robot's surroundings. Sony released a robotic pet named AIBO in 1999, followed by the third generation AIBO ERS-7 in 2003. Honda began a research effort in 1986 to build a robot that would interact peacefully with humans, yielding their humanoid robots P3 in 1996 and ASIMO in 2000 (ASIMO even rang the opening bell to the New York Stock Exchange in 2002 to celebrate Honda's 25 years on the NYSE). Hollywood has maintained a steady supply of robots over the years, and there appears to be no shortage of robots on the big screen in the near future.

Just as Dante II proved volcanic exploration possible, and repeated NASA missions have proven space exploration achievable, deep sea explorers have become very interested in robotic applications. MIT researchers developed the Odyssey IIb submersible robot for just such exploration. Similar to military and law enforcement robotic applications of bomb defusing and disposal, nuclear waste disposal is an excellent role for robots to fill, again, removing their human counterparts from a hazardous environment. An increasing area of robotic application is in natural disaster recovery, such as fallen buildings and collapsed mines. Robots can be used to perform reconnaissance, as well as deliver life-supporting supplies to trapped personnel.

Looking forward there are many frontiers in robotics. Many of the applications presented here are in their infancy and will see considerable growth. Other mature areas will see sustained development, as has been the case since the technological boom following the Second World War. Many theoretical areas hold endless possibilities for expansion — nonlinear control, computational algebra, computational geometry, intelligence in unstructured environments, and many more. The possibilities seem even more expansive when one considers the creativity generated by the cross-pollination of playwrights, science fiction writers, inventors, entrepreneurs, and engineers.

2
Rigid-Body Kinematics

Gregory S. Chirikjian
Johns Hopkins University

2.1	Rotations in Three Dimensions	2-1
	Rules for Composing Rotations • Euler Angles • The Matrix Exponential	
2.2	Full Rigid-Body Motion	2-5
	Composition of Motions • Screw Motions	
2.3	Homogeneous Transforms and the Denavit-Hartenberg Parameters	2-6
	Homogeneous Transformation Matrices • The Denavit-Hartenberg Parameters in Robotics	
2.4	Infinitesimal Motions and Associated Jacobian Matrices	2-8
	Angular Velocity and Jacobians Associated with Parametrized Rotations • The Jacobians for ZXZ Euler Angles • Infinitesimal Rigid-Body Motions	

2.1 Rotations in Three Dimensions

Spatial rigid-body rotations are defined as motions that preserve the distance between points in a body before and after the motion and leave one point fixed under the motion. By definition a motion must be physically realizable, and so reflections are not allowed. If $\mathbf{X}_1$ and $\mathbf{X}_2$ are any two points in a body before a rigid motion, then $\mathbf{x}_1$, and $\mathbf{x}_2$ are the corresponding points after rotation, and

$$d(\mathbf{x}_1, \mathbf{x}_2) = d(\mathbf{X}_1, \mathbf{X}_2)$$

where

$$d(\mathbf{x}, \mathbf{y}) = ||\mathbf{x} - \mathbf{y}|| = \sqrt{(x_1 - y_1)^2 + (x_2 - y_2)^2 + (x_3 - y_3)^2}$$

is the *Euclidean* distance. We view the transformation from $\mathbf{X}_i$ to $\mathbf{x}_i$ as a function $\mathbf{x}(\mathbf{X}, t)$.

By appropriately choosing our frame of reference in space, it is possible to make the pivot point (the point which does not move under rotation) the origin. Therefore, $\mathbf{x}(\mathbf{0}, t) = \mathbf{0}$. With this choice, it can be shown that a necessary condition for a motion to be a rotation is

$$\mathbf{x}(\mathbf{X}, t) = A(t)\mathbf{X}$$

where $A(t) \in \mathbb{R}^{3 \times 3}$ is a time-dependent matrix.

Constraints on the form of $A(t)$ arise from the distance-preserving properties of rotations. If $\mathbf{X}_1$ and $\mathbf{X}_2$ are vectors defined in the frame of reference attached to the pivot, then the triangle with sides of length

$||X_1||$, $||X_2||$, and $||X_1 - X_2||$ is congruent to the triangle with sides of length $||x_1||$, $||x_2||$, and $||x_1 - x_2||$. Hence, the angle between the vectors x_1 and x_2 must be the same as the angle between X_1 and X_2. In general $\mathbf{x} \cdot \mathbf{y} = ||\mathbf{x}||\,||\mathbf{y}||\cos\theta$ where θ is the angle between $\mathbf{x}$ and $\mathbf{y}$. Since $||x_i|| = ||X_i||$ in our case, it follows that

$$x_1 \cdot x_2 = X_1 \cdot X_2$$

Observing that $\mathbf{x} \cdot \mathbf{y} = \mathbf{x}^T \mathbf{y}$ and $x_i = AX_i$, we see that

$$(AX_1)^T(AX_2) = X_1^T X_2 \tag{2.1}$$

Moving everything to the left side of the equation, and using the transpose rule for matrix vector multiplication, Equation (2.1) is rewritten as

$$X_1^T(A^T A - \mathbb{1})X_2 = 0$$

where $\mathbb{1}$ is the 3×3 identity matrix.

Since X_1 and X_2 were arbitrary points to begin with, this holds *for all* possible choices. The only way this can hold is if

$$A^T A = \mathbb{1} \tag{2.2}$$

An easy way to see this is to choose $X_1 = e_i$ and $X_2 = e_j$ for $i, j \in \{1, 2, 3\}$. This forces all the components of the matrix $A^T A - \mathbb{1}$ to be zero.

Equation (2.2) says that a rotation matrix is one whose inverse is its transpose. Taking the determinant of both sides of this equation yields $(\det A)^2 = 1$. There are two possibilities: $\det A = \pm 1$. The case $\det A = -1$ is a reflection and is not physically realizable in the sense that a rigid body cannot be reflected (only its image can be). A rotation is what remains:

$$\det A = +1 \tag{2.3}$$

Thus, a rotation matrix A is one which satisfies both Equation (2.2) and Equation (2.3). The set of all real matrices satisfying both Equation (2.2) and Equation (2.3) is called the set of *special orthogonal*[1] matrices. In general, the set of all $N \times N$ special orthogonal matrices is called $SO(N)$, and the set of all rotations in three-dimensional space is referred to as $SO(3)$.

In the special case of rotation about a fixed axis by an angle ϕ, the rotation has only one degree of freedom. In particular, for counterclockwise rotations about the e_3, e_2, and e_1 axes:

$$R_3(\phi) = \begin{pmatrix} \cos\phi & -\sin\phi & 0 \\ \sin\phi & \cos\phi & 0 \\ 0 & 0 & 1 \end{pmatrix} \tag{2.4}$$

$$R_2(\phi) = \begin{pmatrix} \cos\phi & 0 & \sin\phi \\ 0 & 1 & 0 \\ -\sin\phi & 0 & \cos\phi \end{pmatrix} \tag{2.5}$$

$$R_1(\phi) = \begin{pmatrix} 1 & 0 & 0 \\ 0 & \cos\phi & -\sin\phi \\ 0 & \sin\phi & \cos\phi \end{pmatrix} \tag{2.6}$$

---

[1] Also called proper orthogonal.

2.1.1 Rules for Composing Rotations

Consider three frames of reference A, B, and C, all of which have the same origin. The vectors $\mathbf{x}^A, \mathbf{x}^B, \mathbf{x}^C$ represent *the same* arbitrary point in space, $\mathbf{x}$, as it is viewed in the three different frames. With respect to some common frame fixed in space with axes defined by $\{\mathbf{e}_1, \mathbf{e}_2, \mathbf{e}_3\}$ where $(\mathbf{e}_i)_j = \delta_{ij}$, the rotation matrices describing the basis vectors of the frames A, B, and C are

$$R_A = \begin{bmatrix} \mathbf{e}_1^A, \mathbf{e}_2^A, \mathbf{e}_3^A \end{bmatrix} \quad R_B = \begin{bmatrix} \mathbf{e}_1^B, \mathbf{e}_2^B, \mathbf{e}_3^B \end{bmatrix} \quad R_C = \begin{bmatrix} \mathbf{e}_1^C, \mathbf{e}_2^C, \mathbf{e}_3^C \end{bmatrix}$$

where the vectors $\mathbf{e}_i^A$, $\mathbf{e}_i^B$, and $\mathbf{e}_i^C$ are unit vectors along the ith axis of frame A, B, or C. The "absolute" coordinates of the vector $\mathbf{x}$ are then given by

$$\mathbf{x} = R_A \mathbf{x}^A = R_B \mathbf{x}^B = R_C \mathbf{x}^C$$

In this notation, which is often used in the field of robotics (see e.g., [1, 2]), there is effectively a "cancellation" of indices along the upper right to lower left diagonal.

Given the rotation matrices R_A, R_B, and R_C, it is possible to define rotations of one frame *relative* to another by observing that, for instance, $R_A \mathbf{x}^A = R_B \mathbf{x}^B$ implies $\mathbf{x}^A = (R_A)^{-1} R_B \mathbf{x}^B$. Therefore, given any vector $\mathbf{x}^B$ as it looks in B, we can find how it looks in A, $\mathbf{x}^A$, by performing the transformation:

$$\mathbf{x}^A = R_B^A \mathbf{x}^B \quad \text{where} \quad R_B^A = (R_A)^{-1} R_B \tag{2.7}$$

It follows from substituting the analogous expression $\mathbf{x}^B = R_C^B \mathbf{x}^C$ into $\mathbf{x}^A = R_B^A \mathbf{x}^B$ that concatenation of rotations is calculated as

$$\mathbf{x}^A = R_C^A \mathbf{x}^C \quad \text{where} \quad R_C^A = R_B^A R_C^B \tag{2.8}$$

Again there is effectively a cancellation of indices, and this propagates through for any number of relative rotations. Note that the order of multiplication is critical.

In addition to changes of basis, rotation matrices can be viewed as descriptions of motion. Multiplication of a rotation matrix Q (which represents a frame of reference) by a rotation matrix R (representing motion) on the left, RQ, has the effect of moving Q by R relative to the *base frame*. Multiplying by the same rotation matrix on the right, QR, has the effect of moving by R relative to the the frame Q.

To demonstrate the difference, consider a frame of reference $Q = [\mathbf{a}, \mathbf{b}, \mathbf{n}]$ where $\mathbf{a}$ and $\mathbf{b}$ are unit vectors orthogonal to each other, and $\mathbf{a} \times \mathbf{b} = \mathbf{n}$. First rotating from the identity $\mathbb{1} = [\mathbf{e}_1, \mathbf{e}_2, \mathbf{e}_3]$ fixed in space to Q and then rotating relative to Q by $R_3(\theta)$ results in $QR_3(\theta)$. On the other hand, a rotation about the vector $\mathbf{e}_3^Q = \mathbf{n}$ as viewed in the fixed frame is a rotation $A(\theta, \mathbf{n})$. Hence, shifting the frame of reference Q by multiplying on the left by $A(\theta, \mathbf{n})$ has the same effect as $QR_3(\theta)$, and so we write

$$A(\theta, \mathbf{n}) Q = QR_3(\theta) \quad \text{or} \quad A(\theta, \mathbf{n}) = QR_3(\theta) Q^T \tag{2.9}$$

This is one way to define the matrix

$$A(\theta, \mathbf{n}) = \begin{pmatrix} n_1^2 v\theta + c\theta & n_2 n_1 v\theta - n_3 s\theta & n_3 n_1 v\theta + n_2 s\theta \\ n_1 n_2 v\theta + n_3 s\theta & n_2^2 v\theta + c\theta & n_3 n_2 v\theta - n_1 s\theta \\ n_1 n_3 v\theta - n_2 s\theta & n_2 n_3 v\theta + n_1 s\theta & n_3^2 v\theta + c\theta \end{pmatrix}$$

where $s\theta = \sin\theta$, $c\theta = \cos\theta$, and $v\theta = 1 - \cos\theta$. This expresses a rotation in terms of its axis and angle, and is a mathematical statement of *Euler's Theorem*.

Note that $\mathbf{a}$ and $\mathbf{b}$ do not appear in the final expression. There is nothing magical about $\mathbf{e}_3$, and we could have used the same construction using any other basis vector, $\mathbf{e}_i$, and we would get the same result so long as $\mathbf{n}$ is in the ith column of Q.

2.1.2 Euler Angles

Euler angles are by far the most widely known parametrization of rotation. They are generated by three successive rotations about independent axes. Three of the most common choices are the ZXZ, ZYZ, and ZYX Euler angles. We will denote these as

$$A_{ZXZ}(\alpha, \beta, \gamma) = R_3(\alpha) R_1(\beta) R_3(\gamma) \tag{2.10}$$

$$A_{ZYZ}(\alpha, \beta, \gamma) = R_3(\alpha) R_2(\beta) R_3(\gamma) \tag{2.11}$$

$$A_{ZYX}(\alpha, \beta, \gamma) = R_3(\alpha) R_2(\beta) R_1(\gamma) \tag{2.12}$$

Of these, the ZXZ and ZYZ Euler angles are the most common, and the corresponding matrices are explicitly

$$A_{ZXZ}(\alpha, \beta, \gamma) = \begin{pmatrix} \cos\gamma\cos\alpha - \sin\gamma\sin\alpha\cos\beta & -\sin\gamma\cos\alpha - \cos\gamma\sin\alpha\cos\beta & \sin\beta\sin\alpha \\ \cos\gamma\sin\alpha + \sin\gamma\cos\alpha\cos\beta & -\sin\gamma\sin\alpha + \cos\gamma\cos\alpha\cos\beta & -\sin\beta\cos\alpha \\ \sin\beta\sin\gamma & \sin\beta\cos\gamma & \cos\beta \end{pmatrix}$$

and

$$A_{ZYZ}(\alpha, \beta, \gamma) = \begin{pmatrix} \cos\gamma\cos\alpha\cos\beta - \sin\gamma\sin\alpha & -\sin\gamma\cos\alpha\cos\beta - \cos\gamma\sin\alpha & \sin\beta\cos\alpha \\ \sin\alpha\cos\gamma\cos\beta + \sin\gamma\cos\alpha & -\sin\gamma\sin\alpha\cos\beta + \cos\gamma\cos\alpha & \sin\beta\sin\alpha \\ -\sin\beta\cos\gamma & \sin\beta\sin\gamma & \cos\beta \end{pmatrix}$$

The ranges of angles for these choices are $0 \leq \alpha \leq 2\pi$, $0 \leq \beta \leq \pi$, and $0 \leq \gamma \leq 2\pi$. When ZYZ Euler angles are used,

$$R_3(\alpha) R_2(\beta) R_3(\gamma) = R_3(\alpha)(R_3(\pi/2) R_1(\beta) R_3(-\pi/2)) R_3(\gamma)$$
$$= R_3(\alpha + \pi/2) R_1(\beta) R_3(-\pi/2 + \gamma)$$

and so

$$R_{ZYZ}(\alpha, \beta, \gamma) = R_{ZXZ}(\alpha + \pi/2, \beta, \gamma - \pi/2)$$

2.1.3 The Matrix Exponential

The result of Euler's theorem discussed earlier can be viewed in another way using the concept of a matrix exponential. Recall that the Taylor series expansion of the scalar exponential function is

$$e^x = 1 + \sum_{k=1}^{\infty} \frac{x^k}{k!}$$

The matrix exponential is the same formula evaluated at a square matrix:

$$e^X = \mathbb{1} + \sum_{k=1}^{\infty} \frac{X^k}{k!}$$

Let $N\mathbf{y} = \mathbf{n} \times \mathbf{y}$ for the unit vector $\mathbf{n}$ and any $\mathbf{y} \in \mathbb{R}^3$, and let this relationship be denoted as $\mathbf{n} = \text{vect}(N)$. It may be shown that $N^2 = \mathbf{n}\mathbf{n}^T - \mathbb{1}$.

All higher powers of N can be related to either N or N^2 as

$$N^{2k+1} = (-1)^k N \quad \text{and} \quad N^{2k} = (-1)^{k+1} N^2 \tag{2.13}$$

The first few terms in the Taylor series of $e^{\theta N}$ are then expressed as

$$e^{\theta N} = \mathbb{1} + (\theta - \theta^3/3! + \cdots) N + (\theta^2/2! - \theta^4/4! + \cdots) N^2$$

Hence for any rotational displacement, we can write

$$A(\theta, \mathbf{n}) = e^{\theta N} = \mathbb{1} + \sin\theta\, N + (1 - \cos\theta) N^2$$

This form clearly illustrates that $(\theta, \mathbf{n})$ and $(-\theta, -\mathbf{n})$ correspond to the same rotation.
Since $\theta = \|\mathbf{x}\|$ where $\mathbf{x} = \text{vect}(X)$ and $N = X/\|\mathbf{x}\|$, one sometimes writes the alternative form

$$e^X = \mathbb{1} + \frac{\sin\|\mathbf{x}\|}{\|\mathbf{x}\|} X + \frac{(1 - \cos\|\mathbf{x}\|)}{\|\mathbf{x}\|^2} X^2$$

2.2 Full Rigid-Body Motion

The following statements address what comprises the complete set of rigid-body motions.

Chasles' Theorem [12]: (1) Every motion of a rigid body can be considered as a translation in space and a rotation about a point; (2) Every spatial displacement of a rigid body can be equivalently affected by a single rotation about an axis and translation along the same axis.

In modern notation, (1) is expressed by saying that every point $\mathbf{x}$ in a rigid body may be moved as

$$\mathbf{x}' = R\mathbf{x} + \mathbf{b} \tag{2.14}$$

where $R \in SO(3)$ is a rotation matrix, and $\mathbf{b} \in \mathbb{R}^3$ is a translation vector.

The pair $g = (R, \mathbf{b}) \in SO(3) \times \mathbb{R}^3$ describes both motion of a rigid body *and* the relationship between frames fixed in space and in the body. Furthermore, motions characterized by a pair $(R, \mathbf{b})$ could describe the behavior either of a rigid body or of a deformable object undergoing a rigid-body motion during the time interval for which this description is valid.

2.2.1 Composition of Motions

Consider a rigid-body motion which moves a frame originally coincident with the "natural" frame $(\mathbb{1}, \mathbf{0})$ to $(R_1, \mathbf{b}_1)$. Now consider a relative motion of the frame $(R_2, \mathbf{b}_2)$ with respect to the frame $(R_1, \mathbf{b}_1)$. That is, given any vector $\mathbf{x}$ defined in the terminal frame, it will look like $\mathbf{x}' = R_2\mathbf{x} + \mathbf{b}_2$ in the frame $(R_1, \mathbf{b}_1)$. Then the same vector will appear in the natural frame as

$$\mathbf{x}'' = R_1(R_2\mathbf{x} + \mathbf{b}_2) + \mathbf{b}_1 = R_1 R_2 \mathbf{x} + R_1 \mathbf{b}_2 + \mathbf{b}_1$$

The net effect of composing the two motions (or changes of reference frame) is equivalent to the definition

$$(R_3, \mathbf{b}_3) = (R_1, \mathbf{b}_1) \circ (R_2, \mathbf{b}_2) \triangleq (R_1 R_2, R_1 \mathbf{b}_2 + \mathbf{b}_1) \tag{2.15}$$

From this expression, we can calculate the motion $(R_2, \mathbf{b}_2)$ that for any $(R_1, \mathbf{b}_1)$ will return the floating frame to the natural frame. All that is required is to solve $R_1 R_2 = \mathbb{1}$ and $R_1 \mathbf{b}_2 + \mathbf{b}_1 = \mathbf{0}$ for the variables R_2 and $\mathbf{b}_2$, given R_1 and $\mathbf{b}_1$. The result is $R_2 = R_1^T$ and $\mathbf{b}_2 = -R_1^T \mathbf{b}_1$. Thus, we denote the inverse of a motion as

$$(R, \mathbf{b})^{-1} = (R^T, -R^T \mathbf{b}) \tag{2.16}$$

This inverse, when composed either on the left or right side of $(R, \mathbf{b})$, yields $(\mathbb{1}, \mathbf{0})$.

The set of all pairs $(R, \mathbf{b})$ together with the operation $\circ$ is denoted as $SE(3)$ for "Special Euclidean" group.

Note that every rigid-body motion (element of $SE(3)$) can be decomposed into a pure translation followed by a pure rotation as

$$(R, \mathbf{b}) = (\mathbb{1}, \mathbf{b}) \circ (R, \mathbf{0})$$

and every translation *conjugated*[2] by a rotation yields a translation:

$$(R, \mathbf{0}) \circ (\mathbb{1}, \mathbf{b}) \circ (R^T, \mathbf{0}) = (\mathbb{1}, R\mathbf{b}) \tag{2.17}$$

2.2.2 Screw Motions

The axis in the second part of Chasles' theorem is called the *screw axis*. It is a line in space about which a rotation is performed and along which a translation is performed.[3] As with any line in space, its direction is specified completely by a unit vector, $\mathbf{n}$, and the position of any point $\mathbf{p}$ on the line. Hence, a line is parametrized as

$$\mathbf{L}(t) = \mathbf{p} + t\mathbf{n}, \quad \forall t \in \mathbb{R}$$

Since there are an infinite number of vectors $\mathbf{p}$ on the line that can be chosen, the one which is "most natural" is that which has the smallest magnitude. This is the vector originating at the origin of the coordinate system and terminating at the line which it intersects orthogonally.

Hence the condition $\mathbf{p} \cdot \mathbf{n} = 0$ is satisfied. Since $\mathbf{n}$ is a unit vector and $\mathbf{p}$ satisfies a constraint equation, a line is uniquely specified by only four parameters. Often instead of the pair of line coordinates $(\mathbf{n}, \mathbf{p})$, the pair $(\mathbf{n}, \mathbf{p} \times \mathbf{n})$ is used to describe a line because this implicitly incorporates the constraint $\mathbf{p} \cdot \mathbf{n} = 0$. That is, when $\mathbf{p} \cdot \mathbf{n} = 0$, $\mathbf{p}$ can be reconstructed as $\mathbf{p} = \mathbf{n} \times (\mathbf{p} \times \mathbf{n})$, and it is clear that for unit $\mathbf{n}$, the pair $(\mathbf{n}, \mathbf{p} \times \mathbf{n})$ has four degrees of freedom. Such a description of lines is called the Plücker coordinates.

Given an arbitrary point $\mathbf{x}$ in a rigid body, the transformed position of the same point after translation by d units along a screw axis with direction specified by $\mathbf{n}$ is $\mathbf{x}' = \mathbf{x} + d\mathbf{n}$. Rotation about the same screw axis is given as $\mathbf{x}'' = \mathbf{p} + e^{\theta N}(\mathbf{x}' - \mathbf{p})$.

Since $e^{\theta N}\mathbf{n} = \mathbf{n}$, it does not matter if translation along a screw axis is performed before or after rotation. Either way, $\mathbf{x}'' = \mathbf{p} + e^{\theta N}(\mathbf{x} - \mathbf{p}) + d\mathbf{n}$.

It may be shown that the screw axis parameters $(\mathbf{n}, \mathbf{p})$ and motion parameters (θ, d) always can be extracted from a given rigid displacement $(R, \mathbf{b})$.

2.3 Homogeneous Transforms and the Denavit-Hartenberg Parameters

It is of great convenience in many fields, including robotics, to represent each rigid-body motion with a transformation matrix instead of a pair of the form $(R, \mathbf{b})$ and to use matrix multiplication in place of a composition rule.

2.3.1 Homogeneous Transformation Matrices

One can assign to each pair $(R, \mathbf{b})$ a unique 4×4 matrix

$$H(R, \mathbf{b}) = \begin{pmatrix} R & \mathbf{b} \\ \mathbf{0}^T & 1 \end{pmatrix} \tag{2.18}$$

This is called a homogeneous transformation matrix, or simply a *homogeneous transform*. It is easy to see by the rules of matrix multiplication and the composition rule for rigid-body motions that

$$H((R_1, \mathbf{b}_1) \circ (R_2, \mathbf{b}_2)) = H(R_1, \mathbf{b}_1) H(R_2, \mathbf{b}_2)$$

---

[2]Conjugation of a motion $g = (R, \mathbf{x})$ by a motion $h = (Q, \mathbf{y})$ is defined as $h \circ g \circ h^{-1}$.
[3]The theory of screws was developed by Sir Robert Stawell Ball (1840–1913) [13].

Likewise, the inverse of a homogeneous transformation matrix represents the inverse of a motion:

$$H((R,\mathbf{b})^{-1}) = [H(R,\mathbf{b})]^{-1}$$

In this notation, vectors in $\mathbb{R}^3$ are augmented by appending a "1" to form a vector

$$\mathbf{X} = \begin{pmatrix} \mathbf{x} \\ 1 \end{pmatrix}$$

The following are then equivalent expressions:

$$\mathbf{Y} = H(R,\mathbf{b})\mathbf{X} \leftrightarrow \mathbf{y} = R\mathbf{x} + \mathbf{b}$$

Homogeneous transforms representing pure translations and rotations are respectively written in the form

$$\text{trans}(\mathbf{n},d) = \begin{pmatrix} \mathbb{I} & d\mathbf{n} \\ \mathbf{0}^T & 1 \end{pmatrix}$$

and

$$\text{rot}(\mathbf{n},\mathbf{p},\theta) = \begin{pmatrix} e^{\theta N} & (\mathbb{I} - e^{\theta N})\mathbf{p} \\ \mathbf{0}^T & 1 \end{pmatrix}$$

Rotations around, and translations along, the same axis commute, and the homogeneous transform for a general rigid-body motion along screw axis $(\mathbf{n},\mathbf{p})$ is given as

$$\text{rot}(\mathbf{n},\mathbf{p},\theta)\text{trans}(\mathbf{n},d) = \text{trans}(\mathbf{n},d)\text{rot}(\mathbf{n},\mathbf{p},\theta) = \begin{pmatrix} e^{\theta N} & (\mathbb{I} - e^{\theta N})\mathbf{p} + d\mathbf{n} \\ \mathbf{0}^T & 1 \end{pmatrix} \quad (2.19)$$

2.3.2 The Denavit-Hartenberg Parameters in Robotics

The Denavit-Hartenberg (D-H) framework is a method for assigning frames of reference to a serial robot arm constructed of sequential rotary (and/or translational) joints connected with rigid links [15]. If the robot arm is imagined at any fixed time, the axes about which the joints turn are viewed as lines in space. In the most general case, these lines will be skew, and in degenerate cases, they can be parallel or intersect.

In the D-H framework, a frame of reference is assigned to each link of the robot at the joint where it meets the previous link. The z-axis of the ith D-H frame points along the ith joint axis. Since a robot arm is usually attached to a base, there is no ambiguity in terms of which of the two ($\pm$) directions along the joint axis should be chosen, i.e., the "up" direction for the first joint is chosen. Since the $(i+1)$st joint axis in space will generally be skew relative to axis i, a unique x-axis is assigned to frame i, by defining it to be the unit vector pointing in the direction of the shortest line segment from axis i to axis $i+1$. This segment intersects both axes orthogonally. In addition to completely defining the relative orientation of the ith frame relative to the $(i-1)$st, it also provides the relative position of the origin of this frame.

The D-H parameters, which completely specify this model, are [1]:

- The distance from joint axis i to axis $i+1$ as measured along their mutual normal. This distance is denoted as a_i.
- The angle between the projection of joint axes i and $i+1$ in the plane of their common normal. The sense of this angle is measured counterclockwise around their mutual normal originating at axis i and terminating at axis $i+1$. This angle is denoted as α_i.
- The distance between where the common normal of joint axes $i-1$ and i, and that of joint axes i and $i+1$ intersect joint axis i, as measured along joint axis i. This is denoted as d_i.

- The angle between the common normal of joint axes $i-1$ and i, and the common normal of joint axes i and $i+1$. This is denoted as θ_i, and has positive sense when rotation about axis i is counterclockwise.

Hence, given all the parameters $\{a_i, \alpha_i, d_i, \theta_i\}$ for all the links in the robot, together with how the base of the robot is situated in space, one can completely specify the geometry of the arm at any fixed time. Generally, θ_i is the only parameter that depends on time.

In order to solve the *forward kinematics* problem, which is to find the position and orientation of the distal end of the arm relative to the base, the homogeneous transformations of the relative displacements from one D-H frame to another are multiplied sequentially. This is written as

$$H_N^0 = H_1^0 H_2^1 \cdots H_N^{N-1}$$

The relative transformation, H_i^{i-1} from frame $i-1$ to frame i is performed by first rotating about the x-axis of frame $i-1$ by α_{i-1}, then translating along this same axis by a_{i-1}. Next we rotate about the z-axis of frame i by θ_i and translate along the same axis by d_i. Since all these transformations are relative, they are multiplied sequentially on the right as rotations (and translations) about (and along) natural basis vectors. Furthermore, since the rotations and translations appear as two screw motions (translations and rotations along the same axis), we write

$$H_i^{i-1} = \text{Screw}(\mathbf{e}_1, a_{i-1}, \alpha_{i-1})\text{Screw}(\mathbf{e}_3, d_i, \theta_i)$$

where in this context

$$\text{Screw}(\mathbf{v}, c, \gamma) = \text{rot}(\mathbf{v}, \mathbf{0}, \gamma)\text{trans}(\mathbf{v}, c)$$

Explicitly,

$$H_i^{i-1}(a_{i-1}, \alpha_{i-1}, d_i, \theta_i) = \begin{pmatrix} \cos\theta_i & -\sin\theta_i & 0 & a_{i-1} \\ \sin\theta_i \cos\alpha_{i-1} & \cos\theta_i \cos\alpha_{i-1} & -\sin\alpha_{i-1} & -d_i \sin\alpha_{i-1} \\ \sin\theta_i \sin\alpha_{i-1} & \cos\theta_i \sin\alpha_{i-1} & \cos\alpha_{i-1} & d_i \cos\alpha_{i-1} \\ 0 & 0 & 0 & 1 \end{pmatrix}$$

2.4 Infinitesimal Motions and Associated Jacobian Matrices

A "small motion" is one which describes the relative displacement of a rigid body at two times differing by only an instant. Small rigid-body motions, whether pure rotations or combinations of rotation and translation, differ from large displacements in both their properties and the way they are described. This section explores these small motions in detail.

2.4.1 Angular Velocity and Jacobians Associated with Parametrized Rotations

It is clear from Euler's theorem that when $|\theta| \ll 0$, a rotation matrix reduces to the form

$$A_E(\theta, \mathbf{n}) = \mathbb{1} + \theta N$$

This means that for small rotation angles θ_1 and θ_2, rotations commute:

$$A_E(\theta_1, \mathbf{n}_1)A_E(\theta_2, \mathbf{n}_2) = \mathbb{1} + \theta_1 N_1 + \theta_2 N_2 = A_E(\theta_2, \mathbf{n}_2)A_E(\theta_1, \mathbf{n}_1)$$

Given two frames of reference, one of which is fixed in space and the other of which is rotating relative to it, a rotation matrix describing the orientation of the rotating frame as seen in the fixed frame is written as $R(t)$ at each time t. One connects the concepts of small rotations and angular velocity by observing that if $\mathbf{x}_0$ is a fixed (constant) position vector in the rotating frame of reference, then the position of the same

Rigid-Body Kinematics

point as seen in a frame of reference fixed in space with the same origin as the rotating frame is related to this as

$$\mathbf{x} = R(t)\mathbf{x}_0 \leftrightarrow \mathbf{x}_0 = R^T(t)\mathbf{x}$$

The velocity as seen in the frame fixed in space is then

$$\mathbf{v} = \dot{\mathbf{x}} = \dot{R}\mathbf{x}_0 = \dot{R}R^T \mathbf{x} \qquad (2.20)$$

Observing that since R is a rotation matrix,

$$\frac{d}{dt}(RR^T) = \frac{d}{dt}(\mathbf{I}) = 0$$

and one writes

$$\dot{R}R^T = -R\dot{R}^T = -(\dot{R}R^T)^T$$

Due to the skew-symmetry of this matrix, we can rewrite Equation (2.20) in the form most familiar to engineers:

$$\mathbf{v} = \boldsymbol{\omega}_L \times \mathbf{x}$$

and the notation $\boldsymbol{\omega}_L = \text{vect}(\dot{R}R^T)$ is used. The vector $\boldsymbol{\omega}_L$ is the angular velocity as seen in the space-fixed frame of reference (i.e., the frame in which the moving frame appears to have orientation given by R). The subscript L is not conventional but serves as a reminder that the $\dot{R}$ appears on the "left" of the R^T inside the $\text{vect}(\cdot)$. In contrast, the angular velocity as seen in the rotating frame of reference (where the orientation of the moving frame appears to have orientation given by the identity rotation) is the dual vector of $R^T \dot{R}$, which is also a skew-symmetric matrix. This is denoted as $\boldsymbol{\omega}_R$ (where the subscript R stands for "right"). Therefore we have

$$\boldsymbol{\omega}_R = \text{vect}(R^T \dot{R}) = \text{vect}(R^T(\dot{R}R^T)R) = R^T \boldsymbol{\omega}_L$$

This is because for any skew-symmetric matrix X and any rotation matrix R, $\text{vect}(R^T X R) = R^T \text{vect}(X)$. Another way to write this is

$$\boldsymbol{\omega}_L = R\boldsymbol{\omega}_R$$

In other words, the angular velocity as seen in the frame of reference fixed in space is obtained from the angular velocity as seen in the rotating frame in the same way in which the absolute position is obtained from the relative position.

When a time-varying rotation matrix is parametrized as

$$R(t) = A(q_1(t), q_2(t), q_3(t)) = A(\mathbf{q}(t))$$

then by the chain rule from calculus, one has

$$\dot{R} = \frac{\partial A}{\partial q_1}\dot{q}_1 + \frac{\partial A}{\partial q_2}\dot{q}_2 + \frac{\partial A}{\partial q_3}\dot{q}_3$$

Multiplying on the right by R^T and extracting the dual vector from both sides, one finds that

$$\boldsymbol{\omega}_L = J_L(A(\mathbf{q}))\dot{\mathbf{q}} \qquad (2.21)$$

where one can define [6]:

$$J_L(A(\mathbf{q})) = \left[\text{vect}\left(\frac{\partial A}{\partial q_1}A^T\right), \text{vect}\left(\frac{\partial A}{\partial q_2}A^T\right), \text{vect}\left(\frac{\partial A}{\partial q_3}A^T\right)\right]$$

Similarly,

$$\omega_R = J_R(A(\mathbf{q}))\dot{\mathbf{q}} \qquad (2.22)$$

where

$$J_R(A(\mathbf{q})) = \left[\text{vect}\left(A^T \frac{\partial A}{\partial q_1}\right), \text{vect}\left(A^T \frac{\partial A}{\partial q_2}\right), \text{vect}\left(A^T \frac{\partial A}{\partial q_3}\right)\right]$$

These two Jacobian matrices are related as

$$J_L = A J_R \qquad (2.23)$$

It is easy to verify from the above expressions that for an arbitrary constant rotation $R_0 \in SO(3)$:

$$J_L(R_0 A(\mathbf{q})) = R_0 J_L(A(\mathbf{q})), \quad J_L(A(\mathbf{q}) R_0) = J_L(A(\mathbf{q}))$$
$$J_R(R_0 A(\mathbf{q})) = J_R(A(\mathbf{q})), \quad J_R(A(\mathbf{q}) R_0) = R_0^T J_R(A(\mathbf{q}))$$

In the following subsection we provide the explicit forms for the left and right Jacobians for the ZXZ Euler-angle parametrization. Analogous computations for the other parametrizations discussed earlier in this chapter can be found in [6].

2.4.2 The Jacobians for ZXZ Euler Angles

In this subsection we explicitly calculate the Jacobian matrices J_L and J_R for the ZXZ Euler angles. In this case, $A(\alpha, \beta, \gamma) = R_3(\alpha) R_1(\beta) R_3(\gamma)$, and the skew-symmetric matrices whose dual vectors form the columns of the Jacobian matrix J_L are given as[4]:

$$\frac{\partial A}{\partial \alpha} A^T = (R_3'(\alpha) R_1(\beta) R_3(\gamma))(R_3(-\gamma) R_1(-\beta) R_3(-\alpha)) = R_3'(\alpha) R_3(-\alpha)$$

$$\frac{\partial A}{\partial \beta} A^T = (R_3(\alpha) R_1'(\beta) R_3(\gamma))(R_3(-\gamma) R_1(-\beta) R_3(-\alpha))$$
$$= R_3(\alpha)(R_1'(\beta) R_1(-\beta)) R_3(-\alpha)$$

$$\frac{\partial A}{\partial \gamma} A^T = R_3(\alpha) R_1(\beta)(R_3'(\gamma)(R_3(-\gamma)) R_1(-\beta) R_3(-\alpha)$$

Noting that $\text{vect}(R_i' R_i^T) = \mathbf{e}_i$ regardless of the value of the parameter, and using the rule $\text{vect}(RXR^T) = R\text{vect}(X)$, one finds

$$J_L(\alpha, \beta, \gamma) = [\mathbf{e}_3, R_3(\alpha)\mathbf{e}_1, R_3(\alpha) R_1(\beta) \mathbf{e}_3]$$

This is written explicitly as

$$J_L(\alpha, \beta, \gamma) = \begin{pmatrix} 0 & \cos\alpha & \sin\alpha \sin\beta \\ 0 & \sin\alpha & -\cos\alpha \sin\beta \\ 1 & 0 & \cos\beta \end{pmatrix}$$

The Jacobian J_R can be derived similarly, or one can calculate it easily from J_L as [6]

$$J_R = A^T J_L = [R_3(-\gamma) R_1(-\beta) \mathbf{e}_3, R_3(-\gamma)\mathbf{e}_1, \mathbf{e}_3]$$

---

[4] For one-parameter rotations we use the notation $'$ to denote differentiation with respect to the parameter.

Rigid-Body Kinematics

Explicitly, this is

$$J_R = \begin{pmatrix} \sin\beta\sin\gamma & \cos\gamma & 0 \\ \sin\beta\cos\gamma & -\sin\gamma & 0 \\ \cos\beta & 0 & 1 \end{pmatrix}$$

These Jacobian matrices are important because they relate rates of change of Euler angles to angular velocities. When the determinants of these matrices become singular, the Euler-angle description of rotation becomes degenerate.

2.4.3 Infinitesimal Rigid-Body Motions

As with pure rotations, the matrix exponential can be used to describe general rigid-body motions [2,14]. For "small" motions the matrix exponential description is approximated well when truncated at the first two terms:

$$\exp\left[\begin{pmatrix} \Omega & \mathbf{v} \\ \mathbf{0}^T & 0 \end{pmatrix} \Delta t\right] \approx \mathbb{1} + \begin{pmatrix} \Omega & \mathbf{v} \\ \mathbf{0}^T & 0 \end{pmatrix} \Delta t \quad (2.24)$$

Here $\Omega = -\Omega^T$ and $\text{vect}(\Omega) = \boldsymbol{\omega}$ describe the rotational part of the displacement. Since the second term in Equation (2.24) consists mostly of zeros, it is common to extract the information necessary to describe the motion as

$$\begin{pmatrix} \Omega & \mathbf{v} \\ \mathbf{0}^T & 0 \end{pmatrix}^\vee = \begin{pmatrix} \boldsymbol{\omega} \\ \mathbf{v} \end{pmatrix}$$

This six-dimensional vector is called an *infinitesimal* screw motion or *infinitesimal twist*.

Given a homogeneous transform

$$H(\mathbf{q}) = \begin{pmatrix} R(\mathbf{q}) & \mathbf{b}(\mathbf{q}) \\ \mathbf{0}^T & 0 \end{pmatrix}$$

parametrized with $(q_1, \ldots, q_6)$, which we write as a vector $\mathbf{q} \in \mathbb{R}^6$, one can express the homogeneous transform corresponding to a slightly changed set of parameters as the truncated Taylor series

$$H(\mathbf{q} + \Delta\mathbf{q}) = H(\mathbf{q}) + \sum_{i=1}^{6} \Delta q_i \frac{\partial H}{\partial q_i}(\mathbf{q})$$

This result can be shifted to the identity transformation by multiplying on the right or left by H^{-1} to define an equivalent relative infinitesimal motion. In this case we write [6]

$$\begin{pmatrix} \boldsymbol{\omega}_L \\ \mathbf{v}_L \end{pmatrix} = \mathcal{J}_L(\mathbf{q})\dot{\mathbf{q}} \quad \text{where} \quad \mathcal{J}_L(\mathbf{q}) = \left[\left(\frac{\partial H}{\partial q_1} H^{-1}\right)^\vee, \ldots, \left(\frac{\partial H}{\partial q_6} H^{-1}\right)^\vee\right] \quad (2.25)$$

where $\boldsymbol{\omega}_L$ is defined as before in the case of pure rotation and

$$\mathbf{v}_L = -\boldsymbol{\omega}_L \times \mathbf{b} + \dot{\mathbf{b}} \quad (2.26)$$

Similarly,

$$\begin{pmatrix} \boldsymbol{\omega}_R \\ \mathbf{v}_R \end{pmatrix} = \mathcal{J}_R(\mathbf{q})\dot{\mathbf{q}} \quad \text{where} \quad \mathcal{J}_R(\mathbf{q}) = \left[\left(H^{-1}\frac{\partial H}{\partial q_1}\right)^\vee, \ldots, \left(H^{-1}\frac{\partial H}{\partial q_6}\right)^\vee\right] \quad (2.27)$$

Here

$$\mathbf{v}_R = R^T \dot{\mathbf{b}}$$

The left and right Jacobian matrices are related as

$$\mathcal{J}_L(H) = [Ad(H)]\mathcal{J}_R(H) \qquad (2.28)$$

where the matrix $[Ad(H)]$ is called the *adjoint* and is written as [2, 6, 8]:

$$[Ad(H)] = \begin{pmatrix} R & 0 \\ BR & R \end{pmatrix} \qquad (2.29)$$

The matrix B is skew-symmetric, and vect$(B) = \mathbf{b}$.

Note that when the rotations are parametrized as $R = R(q_1, q_2, q_3)$ and the translations are parametrized using Cartesian coordinates $\mathbf{b}(q_4, q_5, q_6) = [q_4, q_5, q_6]^T$, one finds that

$$\mathcal{J}_R = \begin{pmatrix} J_R & 0 \\ 0 & R^T \end{pmatrix} \quad \text{and} \quad \mathcal{J}_L = \begin{pmatrix} J_L & 0 \\ BJ_L & \mathbb{1} \end{pmatrix} \qquad (2.30)$$

where J_L and J_R are the left and right Jacobians for the case of rotation.

Given

$$H_0 = \begin{pmatrix} R_0 & \mathbf{b}_0 \\ \mathbf{0}^T & 1 \end{pmatrix}$$

$$\mathcal{J}_L(HH_0) = \left[\left(\frac{\partial H}{\partial q_1} H_0 (HH_0)^{-1} \right)^{\vee} \cdots \left(\frac{\partial H}{\partial q_6} H_0 (HH_0)^{-1} \right)^{\vee} \right]$$

Since $(HH_0)^{-1} = H_0^{-1} H^{-1}$, and $H_0 H_0^{-1} = 1$, we have that $\mathcal{J}_L(HH_0) = \mathcal{J}_L(H)$.

Similarly,

$$\mathcal{J}_L(H_0 H) = \left[\left(H_0 \frac{\partial H}{\partial q_1} H^{-1} H_0^{-1} \right)^{\vee} \cdots \left(H_0 \frac{\partial H}{\partial q_6} H^{-1} H_0^{-1} \right)^{\vee} \right]$$

where

$$\left(H_0 \frac{\partial H}{\partial q_i} H^{-1} H_0^{-1} \right)^{\vee} = [Ad(H_0)] \left(\frac{\partial H}{\partial q_i} H^{-1} \right)^{\vee}$$

Therefore,

$$\mathcal{J}_L(H_0 H) = [Ad(H_0)] \mathcal{J}_L(H).$$

Analogous expressions can be written for $\mathcal{J}_R$, which is left invariant.

Further Reading

This chapter has provided a brief overview of rigid-body kinematics. A number of excellent textbooks on kinematics and robotics including [1–4,8,16–18] treat this material in greater depth. Other classic works from a number of fields in which rotations are described include [5,7,9–11]. The interested reader is encouraged to review these materials.

References

[1] Craig, J.J., *Introduction to Robotics, Mechanics and Control*, Addison-Wesley, Reading, MA, 1986.
[2] Murray, R.M., Li, Z., and Sastry, S.S., *A Mathematical Introduction to Robotic Manipulation*, CRC Press, Boca Raton, 1994.

[3] Angeles, J., *Rational Kinematics,* Springer-Verlag, New York, 1988.
[4] Bottema, O. and Roth, B., *Theoretical Kinematics,* Dover Publications, New York, reprinted 1990.
[5] Cayley, A., On the motion of rotation of a solid Body, *Cam. Math. J.,* 3, 224–232, 1843.
[6] Chirikjian, G.S. and Kyatkin, A.B., *Engineering Applications of Noncommutative Harmonic Analysis,* CRC Press, Boca Raton, 2001.
[7] Goldstein, H., *Classical Mechanics,* 2nd ed., Addison-Wesley, Reading, MA, 1980.
[8] McCarthy, J.M., *Introduction to Theoretical Kinematics,* MIT Press, Cambridge, MA, 1990.
[9] Rooney, J., A survey of representations of spatial rotation about a fixed point, *Environ. Plann.,* B4, 185–210, 1977.
[10] Shuster, M.D., A survey of attitude representations, *J. Astron. Sci.,* 41, 4, 439–517, 1993.
[11] Stuelpnagel, J.H., On the parameterization of the three-dimensional rotation group, *SIAM Rev.,* 6, 422–430, 1964.
[12] Chasles, M., Note sur les propriétés générales du systéme de deux corps semblables entr'eux et placés d'une manière quelconque dans l'espace; et sur le désplacement fini ou infiniment petit d'un corps solids libre. *Férussac, Bulletin des Sciences Mathématiques,* 14, 321–326, 1830.
[13] Ball, R.S., *A Treatise on the Theory of Screws,* Cambridge University Press, Cambridge, England, 1900.
[14] Brockett, R.W., Robotic manipulators and the product of exponentials formula, in *Mathematical Theory of Networks and Systems* (A. Fuhrman, ed.), 120–129, Springer-Verlag, New York, 1984.
[15] Denavit, J. and Hartenberg, R.S., A kinematic notation for lower-pair mechanisms based on matrices, *J. Appl. Mech.,* 22, 215–221, June 1955.
[16] Tsai, L.-W., *Robot Analysis: The Mechanics of Serial and Parallel Manipulators,* John Wiley & Sons, New York, 1999.
[17] Karger, A. and Novák, J., *Space Kinematics and Lie Groups,* Gordon and Breach Science Publishers, New York, 1985.
[18] Selig, J.M., *Geometrical Methods in Robotics,* Springer-Verlag, New York, 1996.

3
Inverse Kinematics

3.1 Introduction 3-1
3.2 Preliminaries 3-1
    Existence and Uniqueness of Solutions • Notation and Nomenclature
3.3 Analytical Approaches 3-4
    Reduction of Inverse Kinematics to Subproblems • Pieper's Solution • Example • Other Approaches
3.4 Numerical Techniques 3-13
    Newton's Method • Inverse Kinematics Solution Using Newton's Method
3.5 Conclusions 3-17

Bill Goodwine
University of Notre Dame

3.1 Introduction

This chapter presents results related to the *inverse kinematics problem* for robotic manipulators. As presented elsewhere, the forward kinematics problem of a manipulator is to determine the configuration (position and orientation) of the end effector of the manipulator as a function of the manipulator's joint angles. The inverse problem of that, i.e., determining the joint angles given a desired end effector configuration, is the inverse kinematics problem and the subject of this chapter. This chapter will outline and provide examples for two main categories of approaches to this problem; namely, closed-form analytical methods and numerical approaches.

The main difficulty of the inverse kinematics problem in general is that for some desired end effector configuration, there may be no solutions, there may be a unique solution, or there may be multiple solutions. The advantage of a numerical approach is that it is relatively easy to implement. As illustrated subsequently, however, one drawback is that the method only leads to one solution for one set of starting values for what is fundamentally an iterative method. Also, if no solutions exist, a numerical approach will simply fail to converge, so care must be taken to distinguish between an attempted solution that will never converge and one that is simply slow to converge. The advantage of analytical approaches is that all solutions can be found and if no solutions exist, it will be evident from the computations. The disadvantage is that they are generally algebraically cumbersome and involve many steps and computations. Also, closed form solutions only exist for certain categories of manipulators, but fortunately, the kinematics associated with the most common manipulators generally seem to belong to the class of solvable systems.

3.2 Preliminaries

This section will elaborate upon the nature of the inherent difficulties associated with the inverse kinematics problem and also provide a summary of the nomenclature and notation used in this chapter. The first part of this section provides simple examples illustrating the fact that a various number of solutions may exist

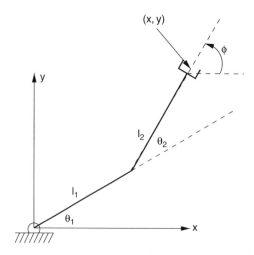

FIGURE 3.1 Simple two link manipulator.

for a given desired end effector configuration of a robot. The second section provides a summary of the notation and nomenclature used subsequently in this chapter.

3.2.1 Existence and Uniqueness of Solutions

Consider the very simple planar two link robot illustrated in Figure 3.1. Assume that the first link has a length of l_1, the second link has a length of l_2 and that θ_1 and θ_2 denote the angles of links one and two, respectively, as illustrated in the figure. The variables x, y, and ϕ denote the position and orientation of the end effector. Since this planar robot has only two joints, only two variables may be specified for the desired end effector configuration. For the purposes of this example, the (x, y) location of the end effector is utilized; however, any two variables, i.e., (x, y), (x, ϕ), or (y, ϕ) may be specified.

For simplicity, if we assume that $l_1 = l_2$, then Figure 3.2 illustrates a configuration in which two inverse kinematic solutions exist, which is the case when $(x, y) = (l_1, l_2)$. The two solutions are obviously, $(\theta_1, \theta_2) = (0°, 90°)$ and $(\theta_1, \theta_2) = (90°, 0°)$. Figure 3.3 illustrates a configuration in which an infinite number of kinematic solutions exist, which is the case where $(x, y) = (0, 0)$. In this case, the manipulator is "folded back" upon itself, which is typically physically unrealizable, but certainly is mathematically feasible. Since this configuration will allow the robot to rotate arbitrarily about the origin, there are an infinite number of configurations that place the end effector at the origin. Figure 3.4 illustrates a configuration in which only one inverse kinematics solution exists, which is the case when $(x, y) = (l_1 + l_2, 0)$. Finally, Figure 3.5 illustrates a case where no inverse kinematic solutions exist, which is the case when $(x, y) = (l_1 + l_2 + 1, 0)$.

While the preceding example was very simple, it illustrates the fundamental point that the inverse kinematics problem is complicated by the fact that there may be zero, one, multiple, or an infinite number of

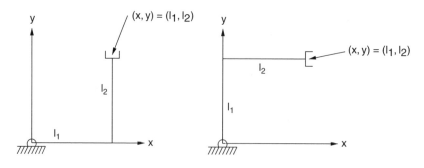

FIGURE 3.2 A configuration with two inverse kinematic solutions.

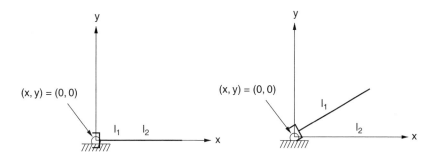

FIGURE 3.3 A configuration with an infinite number of inverse kinematic solutions.

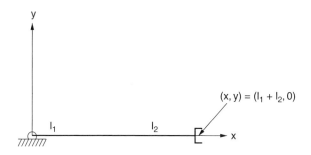

FIGURE 3.4 A configuration with one inverse kinematic solution.

solutions to the problem for a specified configuration. This phenomenon extends to the more complicated kinematics of six degree of freedom manipulators as well. As will become apparent in Section 3.3, certain assumptions must be made regarding the kinematics of the manipulator to make the inverse kinematics problem more feasible.

3.2.2 Notation and Nomenclature

This chapter will utilize the popular notation from Craig [1]. In particular:

$^A P$	is a position vector, P, referenced to the coordinate frame A;
$\hat{X}_A, \hat{Y}_A, \hat{Z}_A$	are coordinate axes for frame A;
$^B\hat{X}_A, {}^B\hat{Y}_A, {}^B\hat{Z}_A$	are coordinate axes for frame A expressed in frame B;
$^A_B R$	is the rotation matrix describing the orientation of frame B relative to frame A;

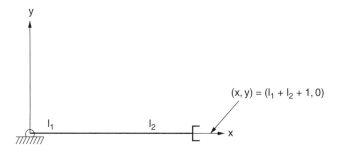

FIGURE 3.5 A configuration with no inverse kinematic solutions.

$^A P_{BORG}$ is the origin of frame B expressed in coordinate of frame A;
$^A_B T$ is the homogeneous transformation relating frame B to frame A; and,
$J(\theta)$ is the Jacobian which maps joint angle velocities to the rigid body velocity of the nth coordinate frame.

Also, unless otherwise indicated, this chapter will assume that coordinate frames are assigned to axes of the manipulator in accordance with the Denavit-Hartenberg [2] procedure presented in Craig [1]. This particular frame assignment procedure is implicit in some of the equations that are part of the algorithms presented. In particular, frames i and $i+1$ are affixed to the manipulator in accordance with the following rules:

1. At the point of intersection between the joint axes i and $i+1$, or the point where the common perpendicular between axes i and $i+1$ intersects axis i, assign the link frame origin for frame i.
2. Assign $\hat{Z}_i$ to point along the ith joint axis.
3. Assign $\hat{X}_i$ to point along the common perpendicular with axis $i+1$, or if axes i and $i+1$ intersect, normal to the plane defined by the axes i and $i+1$.
4. Assign frame 0 to match frame 1 when the first joint variable is zero, and for the nth frame (the last frame), the origin and $\hat{X}_n$ axis can be assigned arbitrarily.

If this procedure is followed, then the following *link parameters* are well-defined:

α_i is the angle between $\hat{Z}_i$ and $\hat{Z}_{i+1}$ measured about $\hat{X}_i$;
a_i is the distance from $\hat{Z}_i$ to $\hat{Z}_{i+1}$ measured along $\hat{X}_i$;
d_i is the distance from $\hat{X}_{i-1}$ to $\hat{X}_i$ measured along Z_i; and,
θ_i is the angle between $\hat{X}_{i-1}$ and $\hat{X}_i$ measured about $\hat{Z}_i$.

A detailed, but straightforward derivation shows that

$$^{i-1}_i T = \begin{bmatrix} \cos\theta_i & -\sin\theta_i & 0 & a_{i-1} \\ \sin\theta_i \cos\alpha_{i-1} & \cos\theta_i \cos\alpha_{i-1} & -\sin\alpha_{i-1} & -\sin\alpha_{i-1} d_i \\ \sin\theta_i \sin\alpha_{i-1} & \cos\theta_i \sin\alpha_{i-1} & \cos\alpha_{i-1} & \cos\alpha_{i-1} d_i \\ 0 & 0 & 0 & 1 \end{bmatrix} \quad (3.1)$$

3.3 Analytical Approaches

This section outlines various analytical solution techniques that lead to closed form solutions to the inverse kinematics problem. This section is not completely comprehensive because specific manipulators may have kinematic features that allow for unique approaches. However, the primary procedures are outlined. First, Section 3.3.1 outlines the general approach of decoupling the manipulator kinematics so that the inverse kinematics problem can be decomposed into a set of subproblems. Section 3.3.2 presents the so-called "Pieper's solution," which is applicable to six degree of freedom manipulators in which the last three axes are rotational axes which mutually intersect. This approach essentially decomposes the inverse kinematics problem into two subproblems, which are solved separately. Finally, Section 3.3.4 outlines two other relatively recently developed althernative approaches.

3.3.1 Reduction of Inverse Kinematics to Subproblems

The basic approach of many analytical approaches is to decompose the complete inverse kinematics problem into a series of decoupled subproblems. This approach will mirror that presented in [6], but will be presented in a manner consistent with the Denavit-Hartenberg approach rather than the product of exponentials approach in [6]. First, two relatively simple motivational example problems will be presented, followed by the characterization of some more general results.

Inverse Kinematics

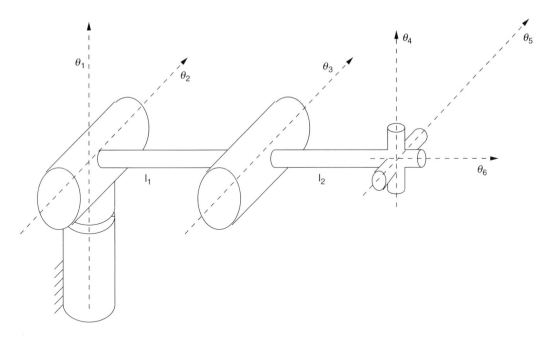

FIGURE 3.6 Elbow manipulator.

3.3.1.1 Inverse Kinematics for Two Examples *via* Subproblems

Consider the schematic illustration of the "Elbow Manipulator" in Figure 3.6. The link frame attachments are illustrated in Figure 3.7. With respect to the elbow manipulator, we can make the following observations:

- If $_6^0 T_{des}$ is specified for the manipulator in Figure 3.6, generally two values for θ_3 may be determined since the location of the common origin of frames 4, 5, and 6 is given by $_6^0 T_{des}$ and the distance from

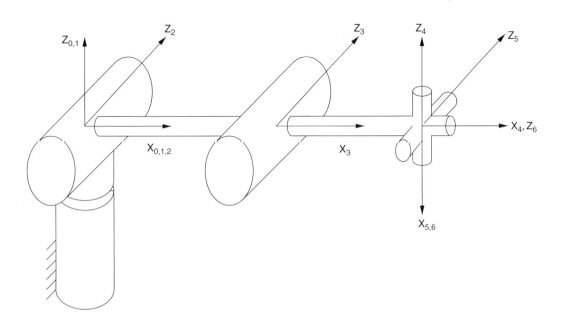

FIGURE 3.7 Link frame attachments for the elbow manipulator.

- the common origin of each of the 0, 1, and 2 frames to the origins of any of the 4, 5, and 6 frames is only affected by θ_3.
- Once θ_3 is determined, the height of the origins of frames 4, 5, and 6, i.e., $^0P_{iORG}$, $i = 4, 5, 6$, is only affected by θ_2. Again, generally two values of θ_2 can be determined.
- For each pair of (θ_2, θ_3) values, the x and y components of $^0P_{iORG}$, $i = 4, 5, 6$, determines one unique θ_1 value.
- Once θ_1, θ_2, and θ_3 are known, ^0_3T can be computed. Using this, $^3_6T_{des}$ can be computed from

$$^3_6T_{des} = {^0_3T}^{-1}\,{^0_6T_{des}}$$

- Since axes 4, 5, and 6 intersect, they will share a common origin. Therefore,

$$a_4 = a_5 = d_5 = d_6 = 0$$

Hence,

$$^3_6T_{des} = \begin{bmatrix} c_4c_5c_6 - s_4s_6 & c_6s_4 - c_4c_5s_6 & c_4s_5 & l_2 \\ c_6s_5 & -s_5s_6 & -c_5 & 0 \\ c_5c_6s_4 + c_4s_6 & c_4c_6 - c_5s_4s_6 & s_4s_5 & 0 \\ 0 & 0 & 0 & 1 \end{bmatrix} \quad (3.2)$$

where c_i and s_i are shorthand for $\cos\theta_i$ and $\sin\theta_i$ respectively. Hence, two values for θ_5 can be computed from the third element of the second row.

- Once θ_5 is computed, one value of θ_4 can be computed from the first and third elements of the third column.
- Finally, two remaining elements of $^3_6T_{des}$, such as the first two elements of the second row, can be used to compute θ_6.

Generically, this procedure utilized the two following subproblems:

1. Determining a rotation that produced a specified distance. In the example, θ_3 determined the distance from the origins of the 1, 2, and 3 frames to the origins of the 4, 5, and 6 frames and subsequently, θ_2 was determined by the height of the origins of the 4, 5, and 6 frames.
2. Determining a rotation about a single axis that specified a particular point to be located in a desired position. In the example, θ_1 was determined in this manner.

Other subproblems are possible as well, such as determining two rotations, which, concatenated together, specify a particular point to be located in a particular position. These concepts are presented with full mathematical rigor in [6] and are further elaborated in [8].

As an additional example, consider the variation on the Stanford manipulator illustrated in Figure 3.8. The assumed frame configurations are illustrated in Figure 3.9. By analogous reasoning, we can compute the inverse kinematic solutions using the following procedure:

- Determine d_3 (the prismatic joint variable) from the distance between the location of $^0P_{6ORG}$ and $^0P_{0ORG}$.
- Determine θ_2 from the height of $^0P_{6ORG}$.
- Determine θ_1 from the location of $^0P_{6ORG}$.
- Determine θ_4, θ_5, and θ_6 in the same manner as for the elbow manipulator.

The presentation of these subproblems is intended to be a motivational conceptual introduction rather than a complete exposition. As is clear from the examples, for some manipulators, the inverse kinematics problem may be solved using only one or two of the subproblems. In contrast, some inverse kinematics problems cannot be solved in this manner. The following section presents Pieper's solution, which is a more mathematically complete solution technique, but one based fundamentally on such subproblems.

Inverse Kinematics

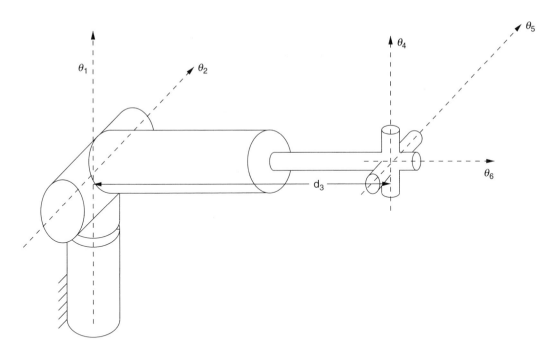

FIGURE 3.8 Variation of Stanford manipulator.

3.3.2 Pieper's Solution

The general formulation of Pieper's solution will closely parallel that presented in Craig [1], but more computational details will be provided. As mentioned previously, the approach presented only applied to manipulators in which the last three axes intersect and where all six joints are revolute. Fortunately,

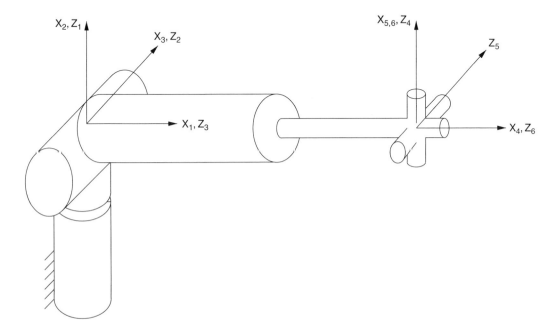

FIGURE 3.9 Link frame attachments for the variation of Stanford manipulator.

however, this happens to be the case when the manipulator is equipped with a three-axis spherical wrist for joints four through six.

Assuming a six degree of freedom manipulator and assuming that axes four, five, and six intersect, then the point of intersection will be the origins of frames 4, 5, and 6. Thus, the problem of solving for $\theta_1, \theta_2,$ and θ_3 simplifies to a three-link position problem, since $\theta_4, \theta_5,$ and θ_6 do not affect the position of their common origins, $^0P_{4ORG} = {}^0P_{5ORG} = {}^0P_{6ORG}$.

Recall that the first three elements in the fourth column of Equation (3.1) give the position of the origin of frame i expressed in frame $i-1$. Thus, from ^3_4T,

$$^3P_{4ORG} = \begin{bmatrix} a_3 \\ -\sin\alpha_3 d_4 \\ \cos\alpha_3 d_4 \end{bmatrix}$$

Expressing this in the 0 frame,

$$^0P_{4ORG} = {}^0_1T\,{}^1_2T\,{}^2_3T \begin{bmatrix} a_3 \\ -\sin\alpha_3 d_4 \\ \cos\alpha_3 d_4 \\ 1 \end{bmatrix}$$

Since

$$^2_3T = \begin{bmatrix} \cos\theta_3 & -\sin\theta_3 & 0 & a_2 \\ \sin\theta_3\cos\alpha_2 & \cos\theta_3\cos\alpha_2 & -\sin\alpha_2 & -\sin\alpha_2 d_3 \\ \sin\theta_3\sin\alpha_2 & \cos\theta_3\sin\alpha_2 & \cos\alpha_2 & \cos\alpha_2 d_3 \\ 0 & 0 & 0 & 1 \end{bmatrix}$$

we can define three functions that are a function only of the joint angle θ_3,

$$\begin{bmatrix} f_1(\theta_3) \\ f_2(\theta_3) \\ f_3(\theta_3) \\ 1 \end{bmatrix} = {}^2_3T \begin{bmatrix} a_3 \\ -\sin\alpha_3 d_4 \\ \cos\alpha_3 d_4 \\ 1 \end{bmatrix}$$

where

$$f_1(\theta_3) = a_2\cos\theta_3 + d_4\sin\alpha_3\sin\theta_3 + a_2$$
$$f_2(\theta_3) = a_3\sin\theta_3\cos\alpha_2 - d_4\sin\alpha_3\cos\alpha_2\cos\theta_3 - d_4\sin\alpha_2\cos\alpha_3 - d_3\sin\alpha_2$$
$$f_3(\theta_3) = a_3\sin\alpha_2\sin\theta_3 - d_4\sin\alpha_3\sin\alpha_2\cos\theta_3 + d4\cos\alpha_2\cos\alpha_3 + d_3\cos\alpha_2$$

and thus

$$^0P_{4ORG} = {}^0_1T\,{}^1_2T \begin{bmatrix} f_1(\theta_3) \\ f_2(\theta_3) \\ f_2(\theta_3) \\ 1 \end{bmatrix} \quad (3.3)$$

If frame 0 is specified in accordance with step four of the Denavit-Hartenberg frame assignment outlined in Section 3.2.2, then $\hat{Z}_0$ will be parallel to $\hat{Z}_1$, and hence $\alpha_0 = 0$. Also, since the origins of frames 0 and 1 will be coincident, $a_0 = 0$ and $d_1 = 0$. Thus, substituting these values into ^0_1T and expanding Equation (3.3),

Inverse Kinematics

we have

$$^0P_{4ORG} = \begin{bmatrix} \cos\theta_1 & -\sin\theta_1 & 0 & 0 \\ \sin\theta_1 & \cos\theta_i & 0 & 0 \\ 0 & 0 & 1 & 0 \\ 0 & 0 & 0 & 1 \end{bmatrix}$$
$$\times \begin{bmatrix} \cos\theta_2 & -\sin\theta_2 & 0 & a_1 \\ \sin\theta_2\cos\alpha_1 & \cos\theta_i\cos\alpha_1 & -\sin\alpha_1 & -\sin\alpha_1 d_i \\ \sin\theta_2\sin\alpha_1 & \cos\theta_2\sin\alpha_1 & \cos\alpha_1 & \cos\alpha_1 d_2 \\ 0 & 0 & 0 & 1 \end{bmatrix} \begin{bmatrix} f_1(\theta_3) \\ f_2(\theta_3) \\ f_3(\theta_3) \\ 1 \end{bmatrix}$$

Since the second matrix is only a function of θ_2, we can write

$$^0P_{4ORG} = \begin{bmatrix} \cos\theta_1 & -\sin\theta_1 & 0 & 0 \\ \sin\theta_1 & \cos\theta_i & 0 & 0 \\ 0 & 0 & 1 & 0 \\ 0 & 0 & 0 & 1 \end{bmatrix} \begin{bmatrix} g_1(\theta_1,\theta_2) \\ g_2(\theta_1,\theta_2) \\ g_3(\theta_1,\theta_2) \\ 1 \end{bmatrix} \tag{3.4}$$

where

$$g_1(\theta_2,\theta_3) = \cos\theta_2 f_1(\theta_3) - \sin\theta_2 f_2(\theta_3) + a_1 \tag{3.5}$$
$$g_2(\theta_2,\theta_3) = \sin\theta_2\cos\alpha_1 f_1(\theta_3) + \cos\theta_2\cos\alpha_1 f_2(\theta_3) - \sin\alpha_1 f_3 - d_2\sin\alpha_1 \tag{3.6}$$
$$g_3(\theta_2,\theta_3) = \sin\theta_2\sin\alpha_1 f_1(\theta_3) + \cos\theta_2\sin\alpha_1 f_2(\theta_3) + \cos\alpha_1 f_3(\theta_3) + d_2\cos\alpha_1. \tag{3.7}$$

Hence, multiplying Equation (3.4),

$$^0P_{4ORG} = \begin{bmatrix} \cos\theta_1 g_1(\theta_2,\theta_3) - \sin\theta_1 g_2(\theta_2,\theta_3) \\ \sin\theta_1 g_1(\theta_2,\theta_3) + \cos\theta_1 g_2(\theta_2,\theta_3) \\ g_3(\theta_2,\theta_3) \\ 1 \end{bmatrix} \tag{3.8}$$

It is critical to note that the "height" (more specifically, the z coordinate of the center of the spherical wrist expressed in frame 0) is the third element of the vector in Equation (3.8) and is independent of θ_1. Specifically,

$$z = \sin\theta_2\sin\alpha_1 f_1(\theta_3) + \cos\theta_2\sin\alpha_1 f_2(\theta_3) + \cos\alpha_1 f_3(\theta_3) + d_2\cos\alpha_1 \tag{3.9}$$

Furthermore, note that the distance from the origin of the 0 and 1 frames to the center of the spherical wrist will also be independent of θ_1. The square of this distance, denoted by r^2 is simply the sum of the squares of the first three elements of the vector in Equation (3.8); namely,

$$\begin{aligned} r^2 &= g_1^2(\theta_2,\theta_3) + g_2^2(\theta_2,\theta_3) + g_3^2(\theta_2,\theta_3) \\ &= f_1^2(\theta_3) + f_2^2(\theta_3) + f_3^2(\theta_3) + a_1^2 + d_2^2 + 2d_2 f_3(\theta_3) \\ &\quad + 2a_1(\cos\theta_2 f_1(\theta_3) - \sin\theta_2 f_2(\theta_3)) \end{aligned} \tag{3.10}$$

We now consider three cases, the first two of which simplify the problem considerably, and the third of which is the most general approach.

3.3.2.1 Simplifying Case Number 1: $a_1 = 0$

Note that if $a_1 = 0$ (this will be the case when axes 1 and 2 intersect), then from Equation (3.10), the distance from the origins of the 0 and 1 frames to the center of the spherical wrist (which is the origin of frames 4, 5, and 6) is a function of θ_3 only; namely,

$$r^2 = f_1^2(\theta_3) + f_2^2(\theta_3) + f_3^2(\theta_3) + a_1^2 + d_2^2 + 2d_2 f_3(\theta_3) \quad (3.11)$$

Since it is much simpler in a numerical example, the details of the expansion of this expression will be explored in the specific examples subsequently; however, at this point note that since the f_i's contain trigonometric function of θ_3, there will typically be two values of θ_3 that satisfy Equation (3.11). Thus, given a desired configuration,

$$^0_6T = \begin{bmatrix} t_{11} & t_{12} & t_{13} & t_{14} \\ t_{21} & t_{22} & t_{23} & t_{24} \\ t_{31} & t_{32} & t_{33} & t_{34} \\ 0 & 0 & 0 & 1 \end{bmatrix}$$

the distance from the origins of frames 0 and 1 to the origins of frames 4, 5, and 6 is simply given by

$$r^2 = t_{14}^2 + t_{24}^2 + t_{34}^2 = f_1^2(\theta_3) + f_2^2(\theta_3) + f_3^2(\theta_3) + a_1^2 + d_2^2 + 2d_2 f_3(\theta_3) \quad (3.12)$$

Once one or more values of θ_3 that satisfy Equation (3.12) are determined, then the height of the center of the spherical wrist in the 0 frame is given by

$$t_{34} = g_3(\theta_2, \theta_3) \quad (3.13)$$

Since one or more values of θ_3 are known, Equation (3.13) will yield one value for θ_2 for each value of θ_3. Finally, returning to Equation (3.8), one value of θ_1 can be computed for each pair of (θ_2, θ_3) which have already been determined.

Finding a solution for joints 4, 5, and 6 is much more straightforward. First note that ^3_6R is determined by

$$^0_6R = ^0_3R\,^3_6R \implies ^3_6R = ^0_3R^T\,^0_6R$$

where ^0_6R is specified by the desired configuration, and ^0_3R can be computed since $(\theta_1, \theta_2, \theta_3)$ have already been computed. This was outlined previously in Equation (3.2) and the corresponding text.

3.3.2.2 Simplifying Case Number 2: $\alpha_1 = 0$

Note that if $\alpha_1 = 0$, then, by Equation (3.5), the height of the spherical wrist center in the 0 frame will be

$$g_3(\theta_2, \theta_3) = \sin\theta_2 \sin\alpha_1 f_1(\theta_3) + \cos\theta_2 \sin\alpha_1 f_2(\theta_3) + \cos\alpha_1 f_3(\theta_3) + d_2 \cos\alpha_1$$
$$g_3(\theta_3) = f_3(\theta_3) + d_2,$$

so typically, two values can be determined for θ_3. Then Equation (3.10), which represents the distance from the origin of the 0 and 1 frames to the spherical wrist center, is used to determine one value for θ_2. Finally, returning to Equation (3.8) and considering the first two equations expressed in the system, one value of θ_1 can be computed for each pair of (θ_2, θ_3) which have already been determined.

3.3.2.3 General Case when $a_1 \neq 0$ and $\alpha_1 \neq 0$

This case is slightly more difficult and less intuitive, but it is possible to combine Equation (3.7) and Equation (3.11) to eliminate the θ_2 dependence and obtain a fourth degree equation in θ_3. For a few more details regarding this more complicated case, the reader is referred to [1].

Inverse Kinematics

TABLE 3.1 Example PUMA 560 Denavit-Hartenberg Parameters

i	α_{i-1}	a_{i-1}	d_i	θ_i
1	0	0	0	θ_1
2	$-90°$	0	0	θ_2
3	0	2 ft	0.5 ft	θ_3
4	$-90°$	0.1666 ft	2 ft	θ_4
5	$90°$	0	0	θ_5
6	$90°$	0	0	θ_6

3.3.3 Example

Consider a PUMA 560 manipulator with

$$
{}^0_6 T_{des} = \begin{bmatrix} -\frac{1}{\sqrt{2}} & 0 & \frac{1}{\sqrt{2}} & 1 \\ 0 & -1 & 0 & 1 \\ \frac{1}{\sqrt{2}} & 0 & \frac{1}{\sqrt{2}} & -1 \\ 0 & 0 & 0 & 1 \end{bmatrix}
$$

and assume the Denavit-Hartenberg parameters are as listed in Table 3.1. Note that $a_1 = 0$, so the procedure to follow is that outlined in Section 3.3.2.1.

First, substituting the link parameter values and the value for r^2 (the desired distance from the origin of frame 0 to the origin of frame 6) into Equation (3.11) and rearranging gives

$$-8 \sin \theta_3 + 0.667 \cos \theta_3 = -5.2778 \tag{3.14}$$

Using the two trigonometric identities

$$\sin \theta = \frac{2 \tan \frac{\theta}{2}}{1 + \tan^2 \left(\frac{\theta}{2}\right)} \quad \text{and} \quad \cos \theta = \frac{1 - \tan^2 \left(\frac{\theta}{2}\right)}{1 + \tan^2 \left(\frac{\theta}{2}\right)} \tag{3.15}$$

and substituting into Equation (3.14) and rearranging gives

$$-4.611 \tan^2 \left(\frac{\theta_3}{2}\right) + 16 \tan \left(\frac{\theta_3}{2}\right) - 5.944 = 0 \tag{3.16}$$

Using the quadratic formula gives

$$\tan \frac{\theta_3}{2} = \frac{-16 \pm \sqrt{(16)^2 - 4(-4.611)(-5.944)}}{2(-4.611)} \tag{3.17}$$

$$\implies \theta_3 = 45.87°, 143.66° \tag{3.18}$$

Considering the $\theta_3 = 45.87°$ solution, substituting the link parameter values and the z-coordinate for ${}^0 P_{6ORG}$ into Equation (3.8) and rearranging gives

$$0.6805 \sin \theta_2 + 1.5122 \cos \theta_2 = 1$$

$$\implies \theta_2 = -28.68° \quad \text{or} \quad 77.14°$$

Considering the $\theta_3 = 143.66°$ solution, substituting the link parameter values and the z-coordinate for ${}^0 P_{6ORG}$ into Equation (3.8) and rearranging gives

$$-0.6805 \sin \theta_2 + 1.5122 \cos \theta_2 = -1$$

$$\implies \theta_2 = 102.85° \quad \text{or} \quad -151.31°$$

At this point, we have four combination of (θ_2, θ_3) solutions; namely

$$(\theta_2, \theta_3) = (-28.68°, 45.87°)$$
$$(\theta_2, \theta_3) = (77.14°, 45.87°)$$
$$(\theta_2, \theta_3) = (102.85°, 143.66°)$$
$$(\theta_2, \theta_3) = (-151.31°, 143.66°)$$

Now, considering the x and y elements of $^0P_{6ORG}$ in Equation (3.8) and substituting a pair of values for (θ_2, θ_3) into $g_1(\theta_2, \theta_3)$ and $g_1(\theta_2, \theta_3)$ in the first two lines and rearranging we have

$$\begin{bmatrix} g_1(\theta_2, \theta_3) & -g_2(\theta_2, \theta_3) \\ g_2(\theta_2, \theta_3) & g_1(\theta_2, \theta_3) \end{bmatrix} \begin{bmatrix} \cos \theta_1 \\ \sin \theta_1 \end{bmatrix} = \begin{bmatrix} 1 \\ 1 \end{bmatrix}$$

or

$$\begin{bmatrix} \cos \theta_1 \\ \sin \theta_1 \end{bmatrix} = \begin{bmatrix} g_1(\theta_2, \theta_3) & -g_2(\theta_2, \theta_3) \\ g_2(\theta_2, \theta_3) & g_1(\theta_2, \theta_3) \end{bmatrix}^{-1} \begin{bmatrix} 1 \\ 1 \end{bmatrix}$$

Substituting the $(\theta_2, \theta_3) = (-151.31°, 143.66°)$ solution for the g_i's gives

$$\begin{bmatrix} \cos \theta_1 \\ \sin \theta_1 \end{bmatrix} = \begin{bmatrix} -1.3231 & -0.5 \\ 0.5 & -1.3231 \end{bmatrix}^{-1} \begin{bmatrix} 1 \\ 1 \end{bmatrix} = \begin{bmatrix} -0.4114 \\ -0.9113 \end{bmatrix} \implies \theta_1 = 245.7°$$

Substituting the four pairs of (θ_2, θ_3) into the g_i's gives the following four sets of solutions including θ_1:

1. $(\theta_1, \theta_2, \theta_3) = (24.3°, -28.7°, 45.9°)$
2. $(\theta_1, \theta_2, \theta_3) = (24.3°, 102.9°, 143.7°)$
3. $(\theta_1, \theta_2, \theta_3) = (-114.3°, 77.14°, 45.9°)$
4. $(\theta_1, \theta_2, \theta_3) = (-114.3°, -151.32°, 143.7°)$

Now that θ_1, θ_2, and θ_3 are known, ^0_3T can be computed. Using this, $^3_6T_{des}$ can be computed from

$$^3_6T_{des} = {^0_3T}^{-1} {^0_6T_{des}}$$

Since axes 4, 5, and 6 intersect, they will share a common origin. Therefore,

$$a_4 = a_5 = d_5 = d_6 = 0$$

Hence,

$$^3_6T_{des} = \begin{bmatrix} c_4c_5c_6 - s_4s_6 & c_6s_4 - c_4c_5s_6 & c_4s_5 & l_2 \\ c_6s_5 & -s_5s_6 & -c_5 & 0 \\ c_5c_6s_4 + c_4s_6 & c_4c_6 - c_5s_4s_6 & s_4s_5 & 0 \\ 0 & 0 & 0 & 1 \end{bmatrix} \quad (3.19)$$

where c_i and s_i are shorthand for $\cos \theta_i$ and $\sin \theta_i$ respectively. Hence, two values for θ_5 can be computed from the third element of the second row.

Considering only the case where $(\theta_1, \theta_2, \theta_3) = (-114.29°, -151.31°, 143.65°)$,

$$^3_6T_{des} = \begin{bmatrix} 0.38256 & 0.90331 & -0.194112 & 2. \\ -0.662034 & 0.121451 & -0.739568 & 24. \\ -0.644484 & 0.411438 & 0.644484 & -2.922\,0.0 \\ 0. & 0. & 0. & 1. \end{bmatrix}$$

Inverse Kinematics

TABLE 3.2 Complete List of Solutions for the PUMA 560 Example

θ_1	θ_2	θ_3	θ_4	θ_5	θ_6
−114.29°	−151.31°	143.65°	−106.76°	−137.69°	10.39°
−114.29°	−151.31°	143.65°	73.23°	137.69°	−169.60°
−114.29°	77.14°	45.86°	−123.98°	−51.00°	−100.47°
−114.29°	77.14°	45.86°	56.01°	51.00°	79.52°
24.29°	−28.68°	45.86°	−144.42°	149.99°	−165.93°
24.29°	−28.68°	45.86°	35.57°	−149.99°	14.06°
24.29°	102.85°	143.65°	−143.39°	29.20°	129.34°
24.29°	102.85°	143.65°	36.60°	−29.20°	−50.65°

using the third element of the second row, $\theta_5 = \pm 137.7°$. Using the first and second elements of the second row, and following the procedure presented in Equation (3.14) through Equation (3.17), if $\theta_5 = 137.7°$, then $\theta_6 = -169.6°$ and using the first and third elements of the third column, $\theta_4 = 73.23°$.

Following this procedure for the other combinations of $(\theta_1, \theta_2, \theta_3)$ produces the complete set of solutions presented in Table 3.2.

3.3.4 Other Approaches

This section very briefly outlines two other approaches and provides references for the interested reader.

3.3.4.1 Dialytical Elimination

One of the most general approaches is to reduce the inverse kinematics problem to a system of polynomial equations through rather elaborate manipulation and then use results from elimination theory from algebraic geometry to solve the equations. The procedure has the benefit of being very general. A consequence of this fact, however, is that it therefore does not exploit any specific geometric kinematic properties of the manipulator, since such specific properties naturally are limited to a subclass of manipulators. Details can be found in [5, 9].

3.3.4.2 Zero Reference Position Method

This approach is similar in procedure to Pieper's method, but is distinct in that only a single, base coordinate system is utilized in the computations. In particular, in the case where the last three link frames share a common origin, the position of the wrist center can be used to determine θ_1, θ_2, and θ_3. This is in contrast to Pieper's method which uses the distance from the origin of the wrist frame to the origin of the 0 frame to determine θ_3. An interested reader is referred to [4] for a complete exposition.

3.4 Numerical Techniques

This section presents an outline of the inverse kinematics problem utilizing numerical techniques. This approach is rather straightforward in that it utilizes the well-known and simple Newton's root finding technique. The two main, somewhat minor, complications are that generally, for a six degree of freedom manipulator, the desired configuration is specified as a homogeneous transformation, which contains 12 elements, but the total number of degrees of freedom is only six. Also, Jacobian singularities are also potentially problematics. As mentioned previously, however, one drawback is that for a given set of initial values for Newton's method, the algorithm converges to only one solution when perhaps multiple solutions may exist. The rest of this section assumes that a six degree of freedom manipulator is under consideration; however, the results presented are easily extended to more or fewer degrees of freedom systems.

3.4.1 Newton's Method

Newton's method is directed toward finding a solution to the system of equations

$$f_1(\theta_1, \theta_2, \theta_3, \theta_4, \theta_5, \theta_6) = a_1 \quad (3.20)$$
$$f_2(\theta_1, \theta_2, \theta_3, \theta_4, \theta_5, \theta_6) = a_2$$
$$f_3(\theta_1, \theta_2, \theta_3, \theta_4, \theta_5, \theta_6) = a_3$$
$$f_4(\theta_1, \theta_2, \theta_3, \theta_4, \theta_5, \theta_6) = a_4$$
$$f_5(\theta_1, \theta_2, \theta_3, \theta_4, \theta_5, \theta_6) = a_5$$
$$f_6(\theta_1, \theta_2, \theta_3, \theta_4, \theta_5, \theta_6) = a_6$$

which may be nonlinear. In matrix form, these equations can be expressed as

$$\mathbf{f}(\theta) = \mathbf{a} \quad \text{or} \quad \mathbf{f}(\theta) - \mathbf{a} = 0 \quad (3.21)$$

where $\mathbf{f}$, θ, and $\mathbf{a}$ are vectors in $\mathbb{R}^6$. Newton's method is an iterative technique to find the roots of Equation (3.21), which is given concisely by

$$\theta_{n+1} = \theta_n - \mathbf{J}^{-1}(\theta_n)\mathbf{f}(\theta_n) \quad (3.22)$$

where

$$\mathbf{J}(\theta_n) = \left[\frac{\partial f_i(\theta)}{\partial \theta_j}\right] = \begin{bmatrix} \frac{\partial f_1(\theta)}{\partial \theta_1} & \frac{\partial f_1(\theta)}{\partial \theta_2} & \frac{\partial f_1(\theta)}{\partial \theta_3} & \cdots & \frac{\partial f_1(\theta)}{\partial \theta_{n-1}} & \frac{\partial f_1(\theta)}{\partial \theta_n} \\ \frac{\partial f_2(\theta)}{\partial \theta_1} & \frac{f_2(\theta)}{\theta_2} & & \cdots & & \frac{\partial f_2(\theta)}{\partial \theta_n} \\ \vdots & \vdots & & \ddots & & \vdots \\ \frac{\partial f_n(\theta)}{\partial \theta_1} & \frac{f_n(\theta)}{\theta_2} & & \cdots & & \frac{\partial f_n(\theta)}{\partial \theta_n} \end{bmatrix}$$

Two theorems regarding the conditions for convergence of Newton's method are presented in Appendix A.

3.4.2 Inverse Kinematics Solution Using Newton's Method

This section elaborates upon the application of Newton's method to the inverse kinematics problem. Two approaches are presented. The first approach considers a six degree of freedom system utilizing a 6×6 Jacobian by choosing from the 12 elements of the desired T, 6 elements. The second approach is to consider all 12 equations relating the 12 components of ${}^0_6T_{des}$ which results in an overdetermined system which must then utilize a pseudo-inverse in implementing Newton's method.

For the six by six approach, the three position elements and three *independent* elements of the rotation submatrix matrix of ${}^0_6T_{des}$. These six elements, as a function of the joint variables, set equal to the corresponding values of ${}^0_6T_{des}$ then constitute all the elements of Equation (3.20). Let

$$ {}^0_6T(\theta_1, \theta_2, \theta_3, \theta_4, \theta_5, \theta_6) = {}^0_6T_\theta $$

denote the forward kinematics as a function of the joint variables, and let ${}^0_6T_\theta(a, b)$ denote the element of ${}^0_6T_\theta$ that is in the ath row and bth column. Similarly, let ${}^0_6T_{des}(a, b)$ denote the element of ${}^0_6T_{des}$ that is in the ath row and bth column.

Inverse Kinematics

Then the six equations in Equation (3.20) may be, for example

$${}_6^0T_\theta(1,4) = {}_6^0T_{des}(1,4)$$
$${}_6^0T_\theta(1,5) = {}_6^0T_{des}(1,5)$$
$${}_6^0T_\theta(1,6) = {}_6^0T_{des}(1,6)$$
$${}_6^0T_\theta(2,3) = {}_6^0T_{des}(2,3)$$
$${}_6^0T_\theta(3,3) = {}_6^0T_{des}(3,3)$$
$${}_6^0T_\theta(3,2) = {}_6^0T_{des}(3,2).$$

One must take care, however, that the last three equations are actually independent. For example, in the case of

$${}_6^0T_{des} = \begin{bmatrix} -\frac{1}{\sqrt{2}} & 0 & \frac{1}{\sqrt{2}} & 2 \\ 0 & -1 & 0 & 1 \\ \frac{1}{\sqrt{2}} & 0 & \frac{1}{\sqrt{2}} & -1 \\ 0 & 0 & 0 & 1 \end{bmatrix}$$

these six equation will **not** be independent because both ${}_6^0T_{des}(2,3)$ and ${}_6^0T_{des}(3,2)$ are equal to zero and the interation may converge to

$$T = \begin{bmatrix} -\frac{1}{\sqrt{2}} & 0 & -\frac{1}{\sqrt{2}} & 2 \\ 0 & -1 & 0 & 1 \\ -\frac{1}{\sqrt{2}} & 0 & \frac{1}{\sqrt{2}} & -1 \\ 0 & 0 & 0 & 1 \end{bmatrix} \neq \begin{bmatrix} -\frac{1}{\sqrt{2}} & 0 & \frac{1}{\sqrt{2}} & 2 \\ 0 & -1 & 0 & 1 \\ \frac{1}{\sqrt{2}} & 0 & \frac{1}{\sqrt{2}} & -1 \\ 0 & 0 & 0 & 1 \end{bmatrix}$$

A more robust approach is to construct a system of 12 equations and six unknowns; in particular,

$${}_6^0T_\theta(1,4) = {}_6^0T_{des}(1,4)$$
$${}_6^0T_\theta(1,5) = {}_6^0T_{des}(1,5)$$
$${}_6^0T_\theta(1,6) = {}_6^0T_{des}(1,6)$$
$${}_6^0T_\theta(2,3) = {}_6^0T_{des}(2,3)$$
$${}_6^0T_\theta(3,3) = {}_6^0T_{des}(3,3)$$
$${}_6^0T_\theta(3,2) = {}_6^0T_{des}(3,2)$$
$${}_6^0T_\theta(1,1) = {}_6^0T_{des}(1,1)$$
$${}_6^0T_\theta(1,2) = {}_6^0T_{des}(1,2)$$
$${}_6^0T_\theta(1,3) = {}_6^0T_{des}(1,3)$$
$${}_6^0T_\theta(2,1) = {}_6^0T_{des}(2,1)$$
$${}_6^0T_\theta(2,2) = {}_6^0T_{des}(2,2)$$
$${}_6^0T_\theta(3,1) = {}_6^0T_{des}(3,1),$$

which are all 12 elements of the position vector (three elements) plus the entire rotation matrix (nine elements). In this case, however, the Jacobian will not be invertible since it will not be square; however, the standard pseudo-inverse which minimizes the norm of the error of the solution of an overdetermined system can be utilized. In this case

$$J(\theta_n) = \left[\frac{\partial f_i(\theta)}{\partial \theta_j}\right] = \begin{bmatrix} \frac{\partial f_1(\theta)}{\partial \theta_1} & \frac{\partial f_1(\theta)}{\partial \theta_2} & \frac{\partial f_1(\theta)}{\partial \theta_3} & \cdots & \frac{\partial f_1(\theta)}{\partial \theta_5} & \frac{\partial f_1(\theta)}{\partial \theta_6} \\ \frac{\partial f_2(\theta)}{\partial \theta_1} & \frac{f_2(\theta)}{\partial \theta_2} & & \cdots & & \frac{\partial f_2(\theta)}{\partial \theta_6} \\ \vdots & \vdots & & \ddots & & \vdots \\ \frac{\partial f_{12}(\theta)}{\partial \theta_1} & \frac{f_{12}(\theta)}{\partial \theta_2} & & \cdots & & \frac{\partial f_1 2(\theta)}{\partial \theta_6} \end{bmatrix}$$

which clearly is not square with six columns and 12 rows and, hence, not invertible. The analog of Equation (3.22) which utilizes a pseudo-inverse is

$$\theta_{n+1} = \theta_n - (\mathbf{J}^T\mathbf{J})^{-1}\mathbf{J}^T(\theta_n)\mathbf{f}(\theta_n) \tag{3.23}$$

Appendix B presents C code that implements Newton's method for a six degree of freedom manipulator. Note, that the values for each are constrained such that $-\pi \leq \theta_i \leq \pi$ by adding or subtracting π from the value of θ_i if the value of θ_i becomes less than $-\pi$ or greater than π, respectively. Finally, implementing the program for the same PUMA 560 example from Section 3.3.3, the following represents the iterative evolution of the θ values for a typical program run where the initial conditions were picked randomly:

```
Iteration   theta1    theta2    theta3    theta4    theta5    theta6
    0      -136.23   -139.17    88.74    102.89    137.92    -23.59
    1        -2.20    -27.61   178.80     34.10     14.86     21.61
    2        44.32     69.98    51.25    -38.03     97.29     20.31
    3        69.22    -32.48   114.17    -51.25    109.17    -68.76
    4       120.32     56.05    29.90    -11.28     -3.81    -80.29
    5        63.89    -20.66    91.33   -115.55     33.74      4.29
    6       -70.59     32.21   169.15    -87.01    135.05    161.33
    7       -50.98    108.14    54.72   -143.86    141.17    107.12
    8       -88.86     53.58    81.08   -156.97    135.54     54.00
    9      -173.40    127.80    38.88   -174.21      4.49    -27.35
    1      -130.68     72.18    81.86   -104.16     14.69     55.74
   11       -76.12     15.76    58.71    106.37    -27.14    -40.86
   12       164.40   -155.70    58.08     13.01     43.00   -176.37
   13       -18.18   -153.18   148.99     74.92     92.84    -32.89
   14       -47.83   -160.55    99.49     45.87    112.59     25.23
   15       -99.11     16.87    21.47   -177.35    -30.07   -115.73
   16       -75.66     14.52    84.43   -132.83     -9.84   -122.40
   17      -162.83     83.18    34.29    100.41    113.24    -93.66
   18      -121.64     63.91    57.62     77.15    151.31    -97.00
   19      -110.76     77.23    47.40     54.07    160.48   -117.69
   20      -114.32     77.03    45.94     47.39    158.26   -129.10
   21      -114.30     77.14    45.87     17.15    154.29   -165.67
   22      -114.30     77.14    45.87    -12.91    134.74    -28.24
   23      -114.30     77.14    45.87     35.38    140.57     26.89
   24      -114.30     77.14    45.87     88.80    101.19     89.46
   25      -114.30     77.14    45.87     57.85     63.60     69.39
   26      -114.30     77.14    45.87     57.38     51.07     78.59
   27      -114.30     77.14    45.87     56.02     51.00     79.53
   28      -114.30     77.14    45.87     56.02     51.01     79.53
   29      -114.30     77.14    45.87     56.02     51.01     79.53
```

3.5 Conclusions

This chapter outlined various approaches to the robotic manipulator inverse kinematics problem. The foundation for many analytical approaches was presented in Section 3.3.1, and one particular method, namely, Pieper's solution, was presented in complete detail. Numerical approaches were also presented with particular emphasis on Newton's iteration method utilizing a pseudo-inverse approach so that convergence to all twelve desired elements of $^0_6T_{des}$ is achieved. Sample C code is presented in Appendix B.

Appendix A: Theorems Relating to Newton's Method

Two theorems regarding convergence of Newton's method are presented here. See [4] for proofs. Since Newton's method is

$$\theta_{n+1} = \theta_n - \mathbf{J}^{-1}(\theta_n)\mathbf{f}(\theta_n)$$

if $\mathbf{g}(\theta) = \theta - \mathbf{J}^{-1}(\theta)\mathbf{f}(\theta)$, convergence occurs when

$$\theta = \mathbf{g}(\theta) \tag{3.24}$$

Theorem A.1 *Let Equation (3.24) have a root $\theta = \alpha$ and $\theta \in \mathbb{R}^n$. Let ρ be the interval*

$$\|\theta - \alpha\| < \rho \tag{3.25}$$

in which the components, $g_i(\theta)$ have continuous first partial derivatives that satisfy

$$\left|\frac{\partial g_i(\theta)}{\partial \theta_j}\right| \leq \frac{\lambda}{n}, \quad \lambda < 1$$

Then

1. *For any θ_0 satisfying Equation (3.25), all iterates θ_n of Equation (3.22) also satisfy Equation (3.25).*
2. *For any θ_0 satisfying Equation (3.25), all iterates θ_n of Equation (3.22) converge to the root α of Equation (3.24) which is unique in Equation (3.25).*

This theorem basically ensures convergence if the initial value, θ_0 is close enough to the root α, where "close enough" means that $\|\theta_0 - \alpha\| < \rho$. A more constructive approach, which gives a test to ensure that θ_0 is "close enough" to α is given by Theorem A.3. First, define the vector and matrix norms as follows:

Definition A.2 Let $\theta \in \mathbb{R}^n$. Define

$$\|\theta\| = \max_i |\theta_i|$$

Let $\mathbf{J} \in \mathbb{R}^{n \times n}$. Define

$$\|\mathbf{J}\| = \max_i \left\{ \sum_j = 1^m |j_{ij}| \right\}$$

Note that other definitions of vector and matrix norms are possible as well [7].

Theorem A.3 *Let $\theta_0 \in \mathbb{R}^n$ be such that*

$$\|\mathbf{J}(\theta_0)\| \leq a$$

let
$$\|\theta_1 - \theta_0\| \leq b$$

and let
$$\sum_{k=1}^{n} \left| \frac{\partial^2 f_i(\theta)}{\partial \theta_j \partial \theta_k} \right| \leq \frac{c}{n}$$

for all $\theta \in \|\theta - \theta\| \leq 2b$, where $i, j \in \{1, \ldots, n\}$. If
$$abc \leq \frac{1}{2}$$

then

1. θ_i defined by Equation (3.22) are uniquely defined as
$$\|\theta_n - \theta_0\| \leq 2b$$

and

2. the iterates converge to some vector, α for which $\mathbf{g}(\alpha) = \mathbf{0}$ and
$$\|\theta_n - \theta_0\| \leq \frac{2b}{2^n}$$

Appendix B: Implementation of Newton's Method

This appendix presents C code that implements Newton's method for a six degree of freedom manipulator. It assumes that all the joints are revolute, i.e., the joint angles, θ_i, are the variables to be determined. It has been written not with the goal of complete robustness or efficiency, but rather for a (hopefully) optimal combination of *readability*, robustness, and efficiency, with emphasis on readability. Most of the variables that one may need to tweak are contained in the header file.

The main file is "inversekinematics.c" which utilizes Newton's method to numerically find a solution to the inverse kinematics problem. This file reads the Denavit-Hartenberg parameters for the manipulator from the file "dh.dat," reads the desired configuration for the sixth frame from the file "Tdes.dat," reads the initial values from the file "theta.dat." The other files are:

- "computejacobian.c" which numerically approximates the Jacobian by individually varying the joint angles and computing a finite approximation of each partial derivative;
- "forwardkinematics.c" which computes the forward homogeneous transformation matrix using the Denavit-Hartenberg parameters (including the joint angles, θ_i);
- "homogeneoustransformation.c" which multiplies the six forward transformation matrices to determine the overall homogeneous transformation;
- "matrixinverse.c" which inverts a matrix;
- "matrixproduct.c" which multiplies two matrices;
- "dh.dat" which contains the Denavit-Hartenberg parameters α_{i-1}, a_{i-1}, and d_i;
- "theta.dat" which contains the initial values for the last Denavit-Hartenberg parameter, θ_i; and,
- "inversekinematics.h" which is a header file for the various C files.

On a Unix machine, move all the files to the same directory and compile the program, by typing

```
> gcc *.c -lm
```

at a command prompt. To execute the program type

Inverse Kinematics

```
./a.out
```

at a command prompt.

To compile these programs on a PC running Windows using Microsoft Visual C++, it must be incorporated into a project (and essentially embedded into a C++ project) to be compiled.

B.1 File "inversekinematics.c"

```
/* This file is inversekinematics.c
 *
 * This program numerically solves the inverse kinematics problem for
 * an n degree of freedom robotic manipulator.
 *
 * It reads the Denavit-Hartenberg parameters stored in a file named
 * "dh.dat". The format of "dh.dat" is:
 *
 * alpha_0    a_0    d_1
 * alpha_1    a_1    d_2
 * ...
 *
 * The number of degrees of freedom is determined by the number of
 * rows in "dh.dat". For this program, it is assumed that the number
 * of degrees of freedom is six.
 *
 * This program reads the desired configuration from the file
 * "Tdes.dat" and stores it in the matrix Tdes[][].
 *
 * This program reads the initial values for Newton's iteration from
 * the file "theta.dat" and stores them in the array theta[]. The
 * initial values are in degrees.
 *
 * The convergence criterion is that the sum of the squares of the
 * change in joint variables between two successive iterations is less
 * that EPS, which is set in "inversekinematics.h".
 *
 * This program assumes that the d_i are fixed and the joint variables
 * are the theta_i.
 *
 * This program assumes a 6-by-6 Jacobian, where n is the number of
 * degrees of freedom for the system. The six elements utilized in
 * the homogeneous transformation are the first three elements of the
 * fourth column (the position of the origin of the sixth frame and
 * elements (2,3), (3,3), and (3,2) of the rotation submatrix of the
 * homogeneous transformation).
 *
 * Copyright (C) 2003 Bill Goodwine.
 *
 */

#include "inversekinematics.h"

int main() {
```

```c
double **J,**T, **Tdes,**Jinv, *f;
double *alpha,*a,*d,*theta;
double sum;
int i,j,k,l,n;
FILE *fp;

/* Allocate memory for the first line of DH parameters. */
alpha = malloc(sizeof(double));
a = malloc(sizeof(double));
d = malloc(2*sizeof(double));
theta = malloc(2*sizeof(double));

n = 1;
fp = fopen("dh.dat","r");

/* Read the DH parameters and reallocate memory accordingly.
   After reading all the data, n will be the number of DOF+1. */
while(fscanf(fp,"%lf %lf %lf",&alpha[n-1],&a[n-1],&d[n]) != EOF) {
  n++;
  alpha = realloc(alpha,n*sizeof(double));
  a = realloc(a,n*sizeof(double));
  d = realloc(d,(n+1)*sizeof(double));
  theta = realloc(theta,(n+1)*sizeof(double));
}
fclose(fp);

if(n-1 != N) {
  printf("Warning, this code is only written for 6 DOF manipulators!\n");
  exit(1);
}

/* Allocate memory for the actual homogeneous transformation and
   the desired final transformation. */
T=(double **) malloc((unsigned) 4*sizeof(double*));
Tdes=(double **) malloc((unsigned) 4*sizeof(double*));
for(i=0;i<4;i++) {
  T[i]=(double *) malloc((unsigned) 4*sizeof(double));
  Tdes[i] = (double *) malloc((unsigned) 4*sizeof(double));
}


/* Read the desired configuration. */
fp = fopen("Tdes.dat","r");
for(i=0;i<4;i++)
  fscanf(fp,"%lf%lf%lf%lf",&Tdes[i][0],&Tdes[i][1],
          &Tdes[i][2],&Tdes[i][3]);
fclose(fp);

printf("Desired T = \n");
for(i=0;i<4;i++) {
  for(j=0;j<4;j++) {
```

Inverse Kinematics   3-21

```c
      printf("%f\t",Tdes[i][j]);
    }
    printf("\n");
  }

  /* Allocate memory for the Jacobian, its inverse and a homogeneous
     transformation matrix. */

  f=(double *) malloc((unsigned) 2*N*sizeof(double));

  J=(double **) malloc((unsigned) 2*N*sizeof(double*));
  for(i=0;i<2*N;i++) {
    J[i]=(double *) malloc((unsigned) N*sizeof(double));
  }

  Jinv=(double **) malloc((unsigned) N*sizeof(double*));
  for(i=0;i<N;i++) {
    Jinv[i]=(double *) malloc((unsigned) 2*N*sizeof(double));
  }

  /* Read the starting values for the iteration. */
  fp = fopen("theta.dat","r");
  for(i=1;i<=N;i++)
    fscanf(fp,"%lf",&theta[i]);

  /* Change the angle values from degrees to radians. */
  for(i=1;i<=N;i++)
    theta[i] *= M_PI/180.0;

  for(i=1;i<=N;i++) {
    theta[i] = 2.0*(double)rand()/(double)RAND_MAX*M_PI - M_PI;
    printf("%.2f\n",theta[i]*180/M_PI);
  }

  /* Begin the iteration. The variable i is the number of iterations.
     MAX_ITERATIONS is set in the file "inversekinematics.h". */
  i = 0;

  while(i<MAX_ITERATIONS) {
    T = forwardkinematics(alpha,a,d,theta,T);
    J = computejacobian(alpha,a,d,theta,T,J);

    Jinv = pseudoinverse(J,Jinv);

    f[0] = T[0][3]-Tdes[0][3];
    f[1] = T[1][3]-Tdes[1][3];
    f[2] = T[2][3]-Tdes[2][3];
    f[3] = T[1][2]-Tdes[1][2];
    f[4] = T[2][2]-Tdes[2][2];
    f[5] = T[2][1]-Tdes[2][1];
    f[6] = T[0][0]-Tdes[0][0];
```

```c
      f[7] = T[0][1]-Tdes[0][1];
      f[8] = T[0][2]-Tdes[0][2];
      f[9] = T[1][0]-Tdes[1][0];
      f[10] = T[1][1]-Tdes[1][1];
      f[11] = T[2][0]-Tdes[2][0];

      for(k=0;k<N;k++) {
        for(l=0;l<2*N;l++) {
          theta[k+1] -= Jinv[k][l]*f[l];
        }
        while(fabs(theta[k+1]) > M_PI)
          theta[k+1] -= fabs(theta[k+1])/theta[k+1]*M_PI;
      }

      printf("%d\t",i);
      for(k=0;k<6;k++)
        printf("%.2f\t",theta[k+1]*180/M_PI);
      printf("\n");

      sum = 0.0;
      for(k=0;k<2*N;k++) {
        sum += pow(f[k],2.0);
      }

      if(sum < EPS)
        break;
      i++;
    }

    printf("Iteration ended in %d iterations.\n",i);
    if(i>=MAX_ITERATIONS) {
      printf("Warning!\n");
      printf("The system failed to converge in %d iterations.\n",MAX_ITERATION
      printf("The solution is suspect!\n");
      exit(1);
    }


    printf("Final T = \n");
    for(i=0;i<4;i++) {
      for(j=0;j<4;j++) {
        printf("%f\t",T[i][j]);
      }
      printf("\n");
    }
    printf("Converged ${\\bf \\theta}:$\n");
    for(i=1;i<n;i++)
      printf("& $%.2f^\\circ$\n",theta[i]*180/M_PI);

    return(0);
}
```

B.2 File "computejacobian.c"

```c
/* This file is "computejacobian.c".
 *
 * It constructs an approximate * Jacobain by numerically
 * approximating the partial derivatives with * respect to each joint
 * variable. It returns a pointer to a * six-by-six Jacobain.
 *
 * The magnitude of the perturbations is determined by the
 * PERTURBATION, which is set in "inversekinematics.h".
 *
 * Copyright (C) 2003 Bill Goodwine.
 *
 */

#include "inversekinematics.h"

double** computejacobian(double* alpha,double* a, double* d,
 double* theta, double** T, double** J) {
  double **T1, **T2;
  int i;

  /* Allocate memories for the perturbed homogeneous transformations.
   * T1 is the forward perturbation and T2 is the backward perturbation.
   */
  T1=(double **) malloc((unsigned) (4)*sizeof(double*));
  T2=(double **) malloc((unsigned) (4)*sizeof(double*));
  for(i=0;i<4;i++) {
    T1[i]=(double *) malloc((unsigned) (4)*sizeof(double));
    T2[i]=(double *) malloc((unsigned) (4)*sizeof(double));
  }

    for(i=1;i<=N;i++) {

      /* Compute the actual homogeneous transformation. */
      T = forwardkinematics(alpha,a,d,theta,T);
      theta[i] += PERTURBATION;

      /* Compute the forward perturbation. */
      T1 = forwardkinematics(alpha,a,d,theta,T1);
      theta[i] -= 2.0*PERTURBATION;

      /* Compute the backward perturbation. */
      T2 = forwardkinematics(alpha,a,d,theta,T2);
      theta[i] += PERTURBATION;

      /* Let the Jacobain elements be the average of the forward
       * and backward perturbations.
       */
      J[0][i-1] = ((T1[0][3]-T[0][3])/(PERTURBATION) +
   (T2[0][3]-T[0][3])/(-PERTURBATION))/2.0;
```

```
      J[1][i-1] = ((T1[1][3]-T[1][3])/(PERTURBATION) +
   (T2[1][3]-T[1][3])/(-PERTURBATION))/2.0;

      J[2][i-1] = ((T1[2][3]-T[2][3])/(PERTURBATION) +
   (T2[2][3]-T[2][3])/(-PERTURBATION))/2.0;

      J[3][i-1] = ((T1[1][2]-T[1][2])/(PERTURBATION) +
   (T2[1][2]-T[1][2])/(-PERTURBATION))/2.0;

      J[4][i-1] = ((T1[2][2]-T[2][2])/(PERTURBATION) +
   (T2[2][2]-T[2][2])/(-PERTURBATION))/2.0;

      J[5][i-1] = ((T1[2][1]-T[2][1])/(PERTURBATION) +
   (T2[2][1]-T[2][1])/(-PERTURBATION))/2.0;

      J[6][i-1] = ((T1[0][0]-T[0][0])/(PERTURBATION) +
   (T2[0][0]-T[0][0])/(-PERTURBATION))/2.0;

      J[7][i-1] = ((T1[0][1]-T[0][1])/(PERTURBATION) +
   (T2[0][1]-T[0][1])/(-PERTURBATION))/2.0;

      J[8][i-1] = ((T1[0][2]-T[0][2])/(PERTURBATION) +
   (T2[0][2]-T[0][2])/(-PERTURBATION))/2.0;

      J[9][i-1] = ((T1[1][0]-T[1][0])/(PERTURBATION) +
   (T2[1][0]-T[1][0])/(-PERTURBATION))/2.0;

      J[10][i-1] = ((T1[1][1]-T[1][1])/(PERTURBATION) +
   (T2[1][1]-T[1][1])/(-PERTURBATION))/2.0;

      J[11][i-1] = ((T1[2][0]-T[2][0])/(PERTURBATION) +
   (T2[2][0]-T[2][0])/(-PERTURBATION))/2.0;
    }

    free(T1);
    free(T2);
    return J;
}
```

B.3 File "forwardkinematics.c"

```
/* This file is "forwardkinematics.c".
 *
 * It computes the homogeneous transformation as a function of the
 * Denavit-Hartenberg parameters as presented in Craig, Introduction
 * to Robotics Mechanics and Control, Second Ed.
 *
 * Copyright (C) 2003 Bill Goodwine.
 *
 */
```

Inverse Kinematics

```c
#include "inversekinematics.h"

double** forwardkinematics(double *alpha, double *a, double *d,
   double *theta, double **T) {

  double **T1,**T2;
  int i,j,k;

  /* Allocate memory for two additional homogeneous transformations
   * which are necessary to multiply all six transformations.
   */
  T1=(double **) malloc((unsigned) 4*sizeof(double*));
  T2=(double **) malloc((unsigned) 4*sizeof(double*));
  for(i=0;i<4;i++) {
    T1[i]=(double *) malloc((unsigned) 4*sizeof(double));
    T2[i]=(double *) malloc((unsigned) 4*sizeof(double));
  }


  T1 = homogeneoustransformation(alpha[0],a[0],d[1],theta[1],T1);

  /* This loop multiplies all six transformations. The final
   * homogeneous transformation is stored in T and a pointer is
   * returned to T.
   */
  for(i=2;i<=N;i++) {
    T2 = homogeneoustransformation(alpha[i-1],a[i-1],d[i],theta[i],T2);

    T = matrixproduct(T1,T2,T,4);

    for(j=0;j<4;j++)
      for(k=0;k<4;k++)
        T1[j][k] = T[j][k];
  }

  free(T1);
  free(T2);

  return T;
}
```

B.4 File "homogeneoustransformation.c"

```c
/* This file is "homogeneoustransformation.c".
 *
 * This function computes a homogeneous transformation as a function
 * of the Denavit-Hartenberg parameters as presented in Craig,
 * Introduction to Robotics Mechanics and Control, Second Ed.
 *
 * The homogeneous transformation is stored in T and a pointer to T is
 * returned.
```

```
 *
 * Copyright (C) 2003 Bill Goodwine.
 *
 */

#include "inversekinematics.h"

double** homogeneoustransformation(double alpha, double a, double d,
        double theta, double **T) {

  T[0][0] = cos(theta);
  T[0][1] = -sin(theta);
  T[0][2] = 0;
  T[0][3] = a;

  T[1][0] = sin(theta)*cos(alpha*M_PI/180.0);
  T[1][1] = cos(theta)*cos(alpha*M_PI/180.0);
  T[1][2] = -sin(alpha*M_PI/180.0);
  T[1][3] = -d*sin(alpha*M_PI/180.0);

  T[2][0] = sin(theta)*sin(alpha*M_PI/180.0);
  T[2][1] = cos(theta)*sin(alpha*M_PI/180.0);
  T[2][2] = cos(alpha*M_PI/180.0);
  T[2][3] = d*cos(alpha*M_PI/180.0);

  T[3][0] = 0;
  T[3][1] = 0;
  T[3][2] = 0;
  T[3][3] = 1;

  return T;
}
```

B.5 File "matrixinverse.c"

```
/* This file is "matrixinverse.c".
 *
 * This program computes the inverse of a matrix using Gauss-Jordan
 * elimination. Row shifting is only utilized if a diagonal element
 * of the original matrix to be inverted has a magnitude less than
 * DIAGONAL_EPS, which is set in "inversekinematics.h".
 *
 * The inverse matrix is stored in y[][], and a pointer to y is
 * returned.
 *
 * Copyright (C) 2003 Bill Goodwine.
 *
 */

#include "inversekinematics.h"
```

```c
double **matrixinverse(double **a, double **y, int n) {
  double temp,coef;
  double max;
  int max_row;
  int i,j,k;

  /* Initialize y[][] to be the identity element. */
  for(i=0;i<n;i++) {
    for(j=0;j<n;j++) {
      if(i==j)
        y[i][j] = 1;
      else
        y[i][j] = 0;
    }
  }

  /* Gauss-Jordan elimination with selective initial pivoting */

  /* Check the magnitude of the diagonal elements, and if one is less
   *  than DIAGONAL_EPS, then search for an element lower in the same
   *  column with a larger magnitude.
   */
  for(i=0;i<n;i++) {
    if(fabs(a[i][i]) < DIAGONAL_EPS) {
      max = a[i][i];
      max_row = i;
      for(j=i;j<n;j++) {
        if(fabs(a[j][i]) > max) {
          max = fabs(a[j][i]);
          max_row = j;
        }
      }

      if(max < DIAGONAL_EPS) {
        printf("Ill-conditioned matrix encountered. Exiting...\n");
        exit(1);
      }

      /* This loop switches rows if needed. */
      for(k=0;k<n;k++) {
        temp = a[max_row][k];
        a[max_row][k] = a[i][k];
        a[i][k] = temp;

        temp = y[max_row][k];
        y[max_row][k] = y[i][k];
        y[i][k] = temp;
      }
    }
  }
```

```c
  /* This is the forward reduction. */
  for(i=0;i<n;i++) {

    coef = a[i][i];
    for(j=n-1;j>=0;j--) {
      y[i][j] /= coef;
      a[i][j] /= coef;
    }

    for(k=i+1;k<n;k++) {
      coef = a[k][i]/a[i][i];
      for(j=n-1;j>=0;j--) {
        y[k][j] -= coef*y[i][j];
        a[k][j] -= coef*a[i][j];
      }
    }
  }

  /* This is the back substitution. */
  for(i=n-1;i>=0;i--) {

    for(k=i-1;k>=0;k--) {
      coef = a[k][i]/a[i][i];
      for(j=0;j<n;j++) {
        y[k][j] -= coef*y[i][j];
        a[k][j] -= coef*a[i][j];
      }
    }
  }

  return y;
}
```

B.6 File "matrixproduct.c"

```c
/* This file is "matrix product.c".
 *
 * This file multiples a and b (both square, n by n matrices) and
 * returns a pointer to the matrix c.
 *
 * Copyright (C) 2003 Bill Goodwine.
 *
 */

double** matrixproduct(double **a,double **b,double **c,int n) {
  int i,j,k;

  for(i=0;i<n;i++)
    for(j=0;j<n;j++)
      c[i][j] = 0.0;
```

```
  for(i=0;i<n;i++)
    for(j=0;j<n;j++)
      for(k=0;k<n;k++)
        c[i][j] += a[i][k]*b[k][j];

  return c;

}
```

B.7 File "inversekinematics.h"

```
/*
 * Copyright (C) 2003 Bill Goodwine.
 */

#include<stdio.h>
#include<stdlib.h>
#include<math.h>

#define MAX_ITERATIONS 1000
#define EPS 0.0000000001
#define PERTURBATION 0.001
#define DIAGONAL_EPS 0.0001
#define N 6

double** forwardkinematics(double *alpha, double *a, double *d,
   double *theta, double **T);

double** matrixinverse(double **J, double **Jinv, int n);

double** pseudoinverse(double **J, double **Jinv);

double** computejacobian(double* alpha, double* a, double* d,
 double* theta, double** T, double** J);

double** matrixproduct(double **a,double **b, double **c, int n);

double** homogeneoustransformation(double alpha, double a, double d,
   double theta, double **T);
```

B.8 File "dh.dat"

```
0 0 0
-90 0 0
0 24 6
-90 2 24
90 0 0
-90 0 0
```

B.9 File "Tdes.dat"

```
-0.707106 0 0.707106 12
0 -1 0 12
0.707106 0 0.707106 -12
0 0 0 1
```

B.10 File "theta.dat"

```
24.0
103.0
145.0
215.0
29.2
30.0
```

References

[1] Craig, J.J. (1989). *Introduction to Robotics Mechanics and Control*. Addison-Wesley, Reading, MA.
[2] Denavit, J. and Hartenberg, R.S. (1955). A kinematic notation for lower-pair mechanisms based on matrices, *J. Appl. Mech.*, 215–221.
[3] Murray, R.M., Zexiang, L.I., and Sastry, S.S. (1994). *A Mathematical Introduction to Robotic Manipulation*, CRC Press, Boca Raton, FL.
[4] Gupta, K.C. (1997). *Mechanics and Control of Robots*. Springer-Verlag, Heidelberg.
[5] Paden, B. and Sastry, S. (1988). Optimal kinematic design of 6 manipulators. *Int. J. Robotics Res.*, 7(2), 43–61.
[6] Lee, H.Y. and Liang, C.G. (1988). A new vector theory for the analysis of spatial mechanisms. *Mechanisms and Machine Theory*, 23(3), 209–217.
[7] Raghavan, M. (1990). Manipulator kinematics. In Roger Brockett (ed.), *Robotics: Proceedings of Symposia in Applied Mathematics*, Vol. 41, American Mathematical Society, 21–48.
[8] Isaacson, E. and Keller, H.B. (1966). *Analysis of Numerical Methods*. Wiley, New York.
[9] Naylor, A.W. and Sell, G.R. (1982). *Linear Operator Theory in Engineering and Science*. Springer-Verlag, Heidelberg.

4
Newton-Euler Dynamics of Robots

4.1	Introduction	4-1
	Scope • Background	
4.2	Theoretical Foundations	4-2
	Newton-Euler Equations • Force and Torque Balance on an Isolated Link • Two-Link Robot Example • Closed-Form Equations	
4.3	Additional Considerations.............................	4-8
	Computational Issues • Moment of Inertia • Spatial Dynamics	
4.4	Closing	4-9

Mark L. Nagurka
Marquette University
Ben-Gurion University of the Negev

4.1 Introduction

4.1.1 Scope

The subject of this chapter is the dynamic analysis of robots modeled using the Newton-Euler method, one of several approaches for deriving the governing equations of motion. It is assumed that the robots are characterized as open kinematic chains with actuators at the joints that can be controlled. The kinematic chain consists of discrete rigid members or links, with each link attached to neighboring links by means of revolute or prismatic joints, i.e., lower-pairs (Denavit and Hartenberg, 1955). One end of the chain is fixed to a base (inertial) reference frame and the other end of the chain, with an attached end-effector, tool, or gripper, can move freely in the robotic workspace and/or be used to apply forces and torques to objects being manipulated to accomplish a wide range of tasks.

Robot dynamics is predicated on an understanding of the associated kinematics, covered in a separate chapter in this handbook. An approach for robot kinematics commonly adopted is that of a 4×4 homogeneous transformation, sometimes referred to as the Denavit-Hartenberg (D-H) transformation (Denavit and Hartenberg, 1955). The D-H transformations essentially determine the position of the origin and the rotation of one link coordinate frame with respect to another link coordinate frame. The D-H transformations can also be used in deriving the dynamic equations of robots.

4.1.2 Background

The field of robot dynamics has a rich history with many important developments. There is a wide literature base of reported work, with myriad articles in professional journals and established textbooks (Featherstone, 1987; Shahinpoor, 1987; Spong and Vidyasager, 1989; Craig, 1989; Yoshikawa, 1990; McKerrow, 1991;

Sciavicco and Siciliano, 2000; Mason, 2001; Niku, 2001) as well as research monographs (Robinett et al., 2001; Vukobratovic et al., 2003). A review of robot dynamic equations and computational issues is presented in Featherstone and Orin (2000). A recent article (Swain and Morris, 2003) formulates a unified dynamic approach for different robotic systems derived from first principles of mechanics.

The control of robot motions, forces, and torques requires a firm grasp of robot dynamics, and plays an important role given the demand to move robots as fast as possible. Robot dynamic equations of motion can be highly nonlinear due to the configuration of links in the workspace and to the presence of Coriolis and centrifugal acceleration terms. In slow-moving and low-inertia applications, the nonlinear coupling terms are sometimes neglected making the robot joints independent and simplifying the control problem.

Several different approaches are available for the derivation of the governing equations pertaining to robot arm dynamics. These include the Newton-Euler (N-E) method, the Lagrange-Euler (L-E) method, Kane's method, bond graph modeling, as well as recursive formulations for both Newton-Euler and Lagrange-Euler methods. The N-E formulation is based on Newton's second law, and investigators have developed various forms of the N-E equations for open kinematic chains (Orin et al., 1979; Luh et al., 1980; Walker and Orin, 1982).

The N-E equations can be applied to a robot link-by-link and joint-by-joint either from the base to the end-effector, called forward recursion, or vice versa, called backward recursion. The forward recursive N-E equations transfer kinematic information, such as the linear and angular velocities and the linear and angular accelerations, as well as the kinetic information of the forces and torques applied to the center of mass of each link, from the base reference frame to the end-effector frame. The backward recursive equations transfer the essential information from the end-effector frame to the base frame. An advantage of the forward and backward recursive equations is that they can be applied to the robot links from one end of the arm to the other providing an efficient means to determine the necessary forces and torques for real-time or near real-time control.

4.2 Theoretical Foundations

4.2.1 Newton-Euler Equations

The Newton-Euler (N-E) equations relate forces and torques to the velocities and accelerations of the center of masses of the links. Consider an intermediate, isolated link n in a multi-body model of a robot, with forces and torques acting on it. Fixed to link n is a coordinate system with its origin at the center of mass, denoted as centroidal frame C_n, that moves with respect to an inertial reference frame, R, as shown in Figure 4.1. In accordance with Newton's second law,

$$\mathbf{F}_n = \frac{d\mathbf{P}_n}{dt}, \quad \mathbf{T}_n = \frac{d\mathbf{H}_n}{dt} \quad (4.1)$$

where $\mathbf{F}_n$ is the net external force acting on link n, $\mathbf{T}_n$ is the net external torque about the center of mass of link n, and $\mathbf{P}_n$ and $\mathbf{H}_n$ are, respectively, the linear and angular momenta of link n,

$$\mathbf{P}_n = m_n{}^R\mathbf{v}_n, \quad \mathbf{H}_n = {}^C\mathbf{I}_n{}^R\boldsymbol{\omega}_n \quad (4.2)$$

In Equation (4.2) ${}^R\mathbf{v}_n$ is the linear velocity of the center of mass of link n as seen by an observer in R, m_n is the mass of link n, ${}^R\boldsymbol{\omega}_n$ is the angular velocity of link n (or equivalently C_n) as seen by an observer in R, and ${}^C\mathbf{I}_n$ is the mass moment of inertia matrix about the center of mass of link n with respect to C_n.

Two important equations result from substituting the momentum expressions (4.2) into (4.1) and taking the time derivatives (with respect to R):

$$^R\mathbf{F}_n = m_n{}^R\mathbf{a}_n \quad (4.3)$$

$$^R\mathbf{T}_n = {}^C\mathbf{I}_n{}^R\boldsymbol{\alpha}_n + {}^R\boldsymbol{\omega}_n \times \left({}^C\mathbf{I}_n{}^R\boldsymbol{\omega}_n\right) \quad (4.4)$$

Newton-Euler Dynamics of Robots

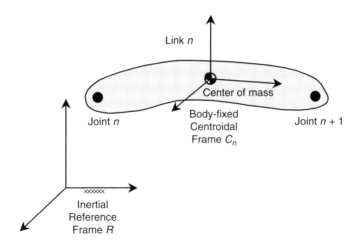

FIGURE 4.1 Coordinate systems associated with link n.

where $^R\mathbf{a}_n$ is the linear acceleration of the center of mass of link n as seen by an observer in R, and $^R\alpha_n$ is the angular acceleration of link n (or equivalently C_n) as seen by an observer in R. In Equation (4.3) and Equation (4.4) superscript R has been appended to the external force and torque to indicate that these vectors are with respect to R.

Equation (4.3) is commonly known as Newton's equation of motion, or Newton's second law. It relates the linear acceleration of the link center of mass to the external force acting on the link. Equation (4.4) is the general form of Euler's equation of motion, and it relates the angular velocity and angular acceleration of the link to the external torque acting on the link. It contains two terms: an inertial torque term due to angular acceleration, and a gyroscopic torque term due to changes in the inertia as the orientation of the link changes. The dynamic equations of a link can be represented by these two equations: Equation (4.3) describing the translational motion of the center of mass and Equation (4.4) describing the rotational motion about the center of mass.

4.2.2 Force and Torque Balance on an Isolated Link

Using a "free-body" approach, as depicted in Figure 4.2, a single link in a kinematic chain can be isolated and the effect from its neighboring links can be accounted for by considering the forces and torques they apply. In general, three external forces act on link n: (i) link $n-1$ applies a force through joint n, (ii) link

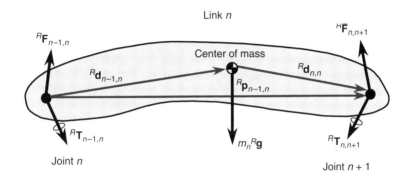

FIGURE 4.2 Forces and torques acting on link n.

$n + 1$ applies a force through joint $n + 1$, and (*iii*) the force due to gravity acts through the center of mass. The dynamic equation of motion for link n is then

$$^R\mathbf{F}_n = {}^R\mathbf{F}_{n-1,n} - {}^R\mathbf{F}_{n,n+1} + m_n{}^R\mathbf{g} = m_n{}^R\mathbf{a}_n \tag{4.5}$$

where $^R\mathbf{F}_{n-1,n}$ is the force applied to link n by link $n - 1$, $^R\mathbf{F}_{n,n+1}$ is the force applied to link $n + 1$ by link n, $^R\mathbf{g}$ is the acceleration due to gravity, and all forces are expressed with respect to the reference frame R. The negative sign before the second term in Equation (4.5) is used since the interest is in the force exerted on link n by link $n + 1$.

For a torque balance, the forces from the adjacent links result in moments about the center of mass. External torques act at the joints, due to actuation and possibly friction. As the force of gravity acts through the center of mass, it does not contribute a torque effect. The torque equation of motion for link n is then

$$\begin{aligned}{}^R\mathbf{T}_n &= {}^R\mathbf{T}_{n-1,n} - {}^R\mathbf{T}_{n,n+1} - {}^R\mathbf{d}_{n-1,n} \times {}^R\mathbf{F}_{n-1,n} + {}^R\mathbf{d}_{n,n} \times {}^R\mathbf{F}_{n,n+1} \\ &= {}^C\mathbf{I}_n{}^R\boldsymbol{\alpha}_n + {}^R\boldsymbol{\omega}_n \times \left({}^C\mathbf{I}_n{}^R\boldsymbol{\omega}_n\right)\end{aligned} \tag{4.6}$$

where $^R\mathbf{T}_{n-1,n}$ is the torque applied to link n by link $n - 1$ as seen by an observer in R, and $^R\mathbf{d}_{n-1,n}$ is the position vector from the origin of the frame $n - 1$ at joint n to the center of mass of link n as seen by an observer in R.

In dynamic equilibrium, the forces and torques on the link are balanced, giving a zero net force and torque. For example, for a robot moving in free space, a torque balance may be achieved when the actuator torques match the torques due to inertia and gravitation. If an external force or torque is applied to the tip of the robot, changes in the actuator torques may be required to regain torque balance. By rearranging Equation (4.5) and Equation (4.6) it is possible to write expressions for the joint force and torque corresponding to dynamic equilibrium.

$$^R\mathbf{F}_{n-1,n} = m_n{}^R\mathbf{a}_n + {}^R\mathbf{F}_{n,n+1} - m_n{}^R\mathbf{g} \tag{4.7}$$

$$^R\mathbf{T}_{n-1,n} = {}^R\mathbf{T}_{n,n+1} + {}^R\mathbf{d}_{n-1,n} \times {}^R\mathbf{F}_{n-1,n} - {}^R\mathbf{d}_{n,n} \times {}^R\mathbf{F}_{n,n+1} + {}^C\mathbf{I}_n{}^R\boldsymbol{\alpha}_n + {}^R\boldsymbol{\omega}_n \times \left({}^C\mathbf{I}_n{}^R\boldsymbol{\omega}_n\right) \tag{4.8}$$

Substituting Equation (4.7) into Equation (4.8) gives an equation for the torque at the joint with respect to R in terms of center of mass velocities and accelerations.

$$\begin{aligned}{}^R\mathbf{T}_{n-1,n} &= {}^R\mathbf{T}_{n,n+1} + {}^R\mathbf{d}_{n-1,n} \times m_n{}^R\mathbf{a}_n + {}^R\mathbf{p}_{n-1,n} \times {}^R\mathbf{F}_{n,n+1} \\ &\quad - {}^R\mathbf{d}_{n-1,n} \times m_n{}^R\mathbf{g} + {}^C\mathbf{I}_n{}^R\boldsymbol{\alpha}_n + {}^R\boldsymbol{\omega}_n \times \left({}^C\mathbf{I}_n{}^R\boldsymbol{\omega}_n\right)\end{aligned} \tag{4.9}$$

where $^R\mathbf{p}_{n-1,n}$ is the position vector from the origin of frame $n - 1$ at joint n to the origin of frame n as seen from by an observer in R. Equation (4.7) and Equation (4.9) represent one form of the Newton-Euler (N-E) equations. They describe the applied force and torque at joint n, supplied for example by an actuator, in terms of other forces and torques, both active and inertial, acting on the link.

4.2.3 Two-Link Robot Example

In this example, the dynamics of a two-link robot are derived using the N-E equations. From Newton's Equation (4.3) the net dynamic forces acting at the center of mass of each link are related to the mass center accelerations,

$$^0\mathbf{F}_1 = m_1{}^0\mathbf{a}_1, \quad {}^0\mathbf{F}_2 = m_2{}^0\mathbf{a}_2 \tag{4.10}$$

and from Euler's Equation (4.4) the net dynamic torques acting around the center of mass of each link are related to the angular velocities and angular accelerations,

$$^0\mathbf{T}_1 = \mathbf{I}_1{}^0\boldsymbol{\alpha}_1 + {}^0\boldsymbol{\omega}_1 \times \left(\mathbf{I}_1{}^0\boldsymbol{\omega}_1\right), \quad {}^0\mathbf{T}_2 = \mathbf{I}_2{}^0\boldsymbol{\alpha}_2 + {}^0\boldsymbol{\omega}_2 \times \left(\mathbf{I}_2{}^0\boldsymbol{\omega}_2\right) \tag{4.11}$$

where the mass moments of inertia are with respect to each link's centroidal frame. Superscript 0 is used to denote the inertial reference frame R.

Equation (4.9) can be used to find the torques at the joints, as follows:

$$^0T_{1,2} = {}^0T_{2,3} + {}^0\mathbf{d}_{1,2} \times m_2 {}^0\mathbf{a}_2 + {}^0\mathbf{p}_{1,2} \times {}^0\mathbf{F}_{2,3} \\ - {}^0\mathbf{d}_{1,2} \times m_2 {}^0\mathbf{g} + \mathbf{I}_2 {}^0\boldsymbol{\alpha}_2 + {}^0\boldsymbol{\omega}_2 \times \left(\mathbf{I}_2 {}^0\boldsymbol{\omega}_2\right) \tag{4.12}$$

$$^0T_{0,1} = {}^0T_{1,2} + {}^0\mathbf{d}_{0,1} \times m_1 {}^0\mathbf{a}_1 + {}^0\mathbf{p}_{0,1} \times {}^0\mathbf{F}_{1,2} \\ - {}^0\mathbf{d}_{0,1} \times m_1 {}^0\mathbf{g} + \mathbf{I}_1 {}^0\boldsymbol{\alpha}_1 + {}^0\boldsymbol{\omega}_1 \times \left(\mathbf{I}_1 {}^0\boldsymbol{\omega}_1\right) \tag{4.13}$$

An expression for ${}^0\mathbf{F}_{1,2}$ can be found from Equation (4.7) and substituted into Equation (4.13) to give

$$^0T_{0,1} = {}^0T_{1,2} + {}^0\mathbf{d}_{0,1} \times m_1 {}^0\mathbf{a}_1 + {}^0\mathbf{p}_{0,1} \times m_2 {}^0\mathbf{a}_2 + {}^0\mathbf{p}_{0,1} \times {}^0\mathbf{F}_{2,3} - {}^0\mathbf{p}_{0,1} \times m_2 {}^0\mathbf{g} \\ - {}^0\mathbf{d}_{0,1} \times m_1 {}^0\mathbf{g} + \mathbf{I}_1 {}^0\boldsymbol{\alpha}_1 + {}^0\boldsymbol{\omega}_1 \times \left(\mathbf{I}_1 {}^0\boldsymbol{\omega}_1\right) \tag{4.14}$$

For a planar two-link robot with two revolute joints, as shown in Figure 4.3, moving in a horizontal plane, i.e., perpendicular to gravity, several simplifications can be made in Equation (4.11) to Equation (4.14): (i) the gyroscopic terms ${}^0\boldsymbol{\omega}_n \times ({}^C\mathbf{I}_n {}^0\boldsymbol{\omega}_n)$ for $n = 1, 2$ can be eliminated since the mass moment of inertia matrix is a scalar and the cross product of a vector with itself is zero, (ii) the gravity terms ${}^0\mathbf{d}_{n-1,n} \times m_n {}^0\mathbf{g}$ disappear since the robot moves in a horizontal plane, and (iii) the torques can be written as scalars and act perpendicular to the plane of motion. Equation (4.12) and Equation (4.14) can then be written in scalar form,

$$^0T_{1,2} = {}^0T_{2,3} + \left({}^0\mathbf{d}_{1,2} \times m_2 {}^0\mathbf{a}_2\right) \cdot \hat{\mathbf{u}}_z + \left({}^0\mathbf{p}_{1,2} \times {}^0\mathbf{F}_{2,3}\right) \cdot \hat{\mathbf{u}}_z + I_2 {}^0\alpha_2 \tag{4.15}$$

$$^0T_{0,1} = {}^0T_{1,2} + \left({}^0\mathbf{d}_{0,1} \times m_1 {}^0\mathbf{a}_1\right) \cdot \hat{\mathbf{u}}_z + \left({}^0\mathbf{p}_{0,1} \times m_2 {}^0\mathbf{a}_2\right) \cdot \hat{\mathbf{u}}_z + \left({}^0\mathbf{p}_{0,1} \times {}^0\mathbf{F}_{2,3}\right) \cdot \hat{\mathbf{u}}_z + I_1 {}^0\alpha_1 \tag{4.16}$$

where the dot product is taken (with the terms in parentheses) with the unit vector $\hat{\mathbf{u}}_z$ perpendicular to the plane of motion.

A further simplification can be introduced by assuming the links have uniform cross-sections and are made of homogeneous (constant density) material. The center of mass then coincides with the geometric center and the distance to the center of mass is half the link length. The position vectors can be

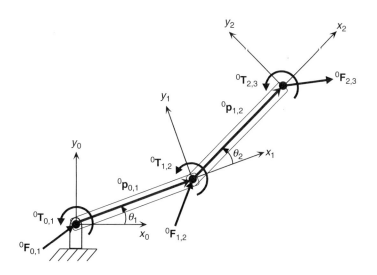

FIGURE 4.3 Two-link robot with two revolute joints.

written as,

$$^0\mathbf{p}_{0,1} = 2\,^0\mathbf{d}_{0,1}, \quad ^0\mathbf{p}_{1,2} = 2\,^0\mathbf{d}_{1,2} \tag{4.17}$$

with the components,

$$\begin{bmatrix} ^0p_{0,1x} \\ ^0p_{0,1y} \end{bmatrix} = 2 \begin{bmatrix} ^0d_{0,1x} \\ ^0d_{0,1y} \end{bmatrix} = \begin{bmatrix} l_1 C_1 \\ l_1 S_1 \end{bmatrix}, \quad \begin{bmatrix} ^0p_{1,2x} \\ ^0p_{1,2y} \end{bmatrix} = 2 \begin{bmatrix} ^0d_{1,2x} \\ ^0d_{1,2y} \end{bmatrix} = \begin{bmatrix} l_2 C_{12} \\ l_2 S_{12} \end{bmatrix} \tag{4.18}$$

where the notation $S_1 = \sin\theta_1, C_1 = \cos\theta_1, S_{12} = \sin(\theta_1 + \theta_2); C_{12} = \cos(\theta_1 + \theta_2)$ is introduced; and l_1, l_2 are the lengths of links 1, 2, respectively.

The accelerations of the center of mass of each link are needed in Equation (4.15) and Equation (4.16). These accelerations can be written as,

$$^0\mathbf{a}_1 = \,^0\boldsymbol{\alpha}_1 \times \,^0\mathbf{d}_{0,1} + \,^0\boldsymbol{\omega}_1 \times \left(^0\boldsymbol{\omega}_1 \times \,^0\mathbf{d}_{0,1}\right) \tag{4.19}$$

$$^0\mathbf{a}_2 = 2\,^0\mathbf{a}_1 + \,^0\boldsymbol{\alpha}_2 \times \,^0\mathbf{d}_{1,2} + \,^0\boldsymbol{\omega}_2 \times \left(^0\boldsymbol{\omega}_2 \times \,^0\mathbf{d}_{1,2}\right) \tag{4.20}$$

Expanding Equation (4.19) and Equation (4.20) gives the acceleration components:

$$\begin{bmatrix} ^0a_{1x} \\ ^0a_{1y} \end{bmatrix} = \begin{bmatrix} -\frac{1}{2}l_1 S_1 \ddot\theta_1 - \frac{1}{2}l_1 C_1 \dot\theta_1^2 \\ \frac{1}{2}l_1 C_1 \ddot\theta_1 - \frac{1}{2}l_1 S_1 \dot\theta_1^2 \end{bmatrix} \tag{4.21}$$

$$\begin{bmatrix} ^0a_{2x} \\ ^0a_{2y} \end{bmatrix} = \begin{bmatrix} -\left(l_1 S_1 + \frac{1}{2}l_2 S_{12}\right)\ddot\theta_1 - \frac{1}{2}l_2 S_{12}\ddot\theta_2 - l_1 C_1 \dot\theta_1^2 - \frac{1}{2}l_2 C_{12}(\dot\theta_1+\dot\theta_2)^2 \\ \left(l_1 C_1 + \frac{1}{2}l_2 C_{12}\right)\ddot\theta_1 + \frac{1}{2}l_2 C_{12}\ddot\theta_2 - l_1 S_1 \dot\theta_1^2 - \frac{1}{2}l_2 S_{12}(\dot\theta_1+\dot\theta_2)^2 \end{bmatrix} \tag{4.22}$$

Equation (4.15) gives the torque at joint 2:

$$^0T_{1,2} = \,^0T_{2,3} + m_2 \left[\left(^0d_{1,2x}\hat{\mathbf{u}}_{x0} + \,^0d_{1,2y}\hat{\mathbf{u}}_{y0}\right) \times \left(^0a_{2x}\hat{\mathbf{u}}_{x0} + \,^0a_{2y}\hat{\mathbf{u}}_{y0}\right)\right] \cdot \hat{\mathbf{u}}_z$$
$$+ \left[\left(^0p_{1,2x}\hat{\mathbf{u}}_{x0} + \,^0p_{1,2y}\hat{\mathbf{u}}_{y0}\right) \times \left(^0F_{2,3x}\hat{\mathbf{u}}_{x0} + \,^0F_{2,3y}\hat{\mathbf{u}}_{y0}\right)\right] \cdot \hat{\mathbf{u}}_z + I_2\,^0\alpha_2$$

where $\hat{\mathbf{u}}_{x0}, \hat{\mathbf{u}}_{y0}$ are the unit vectors along the x_0, y_0 axes, respectively. Substituting and evaluating yields,

$$^0T_{1,2} = \,^0T_{2,3} + \left(\tfrac{1}{2}m_2 l_1 l_2 C_2 + \tfrac{1}{4}m_2 l_2^2 + I_2\right)\ddot\theta_1 + \left(\tfrac{1}{4}m_2 l_2^2 + I_2\right)\ddot\theta_2 + \tfrac{1}{2}m_2 l_1 l_2 S_2 \dot\theta_1^2$$
$$+ l_2 C_{12}\,^0F_{2,3y} - l_2 S_{12}\,^0F_{2,3x} \tag{4.23}$$

Similarly, Equation (4.16) gives the torque at joint 1:

$$^0T_{0,1} = \,^0T_{1,2} + m_1 \left[\left(^0d_{0,1x}\hat{\mathbf{u}}_{x0} + \,^0d_{0,1y}\hat{\mathbf{u}}_{y0}\right) \times \left(^0a_{1x}\hat{\mathbf{u}}_{x0} + \,^0a_{1y}\hat{\mathbf{u}}_{y0}\right)\right] \cdot \hat{\mathbf{u}}_z$$
$$+ m_2 \left[\left(^0p_{0,1x}\hat{\mathbf{u}}_{x0} + \,^0p_{0,1y}\hat{\mathbf{u}}_{y0}\right) \times \left(^0a_{2x}\hat{\mathbf{u}}_{x0} + \,^0a_{2y}\hat{\mathbf{u}}_{y0}\right)\right] \cdot \hat{\mathbf{u}}_z$$
$$+ \left[\left(^0p_{0,1x}\hat{\mathbf{u}}_{x0} + \,^0p_{0,1y}\hat{\mathbf{u}}_{y0}\right) \times \left(^0F_{2,3x}\hat{\mathbf{u}}_{x0} + \,^0F_{2,3y}\hat{\mathbf{u}}_{y0}\right)\right] \cdot \hat{\mathbf{u}}_z + I_1\,^0\alpha_1$$

which yields after substituting,

$$^0T_{0,1} = \,^0T_{1,2} + \left(\tfrac{1}{4}m_1 l_1^2 + m_2 l_1^2 + \tfrac{1}{2}m_2 l_1 l_2 C_2 + I_1\right)\ddot\theta_1 + \left(\tfrac{1}{2}m_2 l_1 l_2 C_2\right)\ddot\theta_2$$
$$- \tfrac{1}{2}m_2 l_1 l_2 S_2 \dot\theta_1^2 - \tfrac{1}{2}m_2 l_1 l_2 S_2 \dot\theta_2^2 - m_2 l_1 l_2 S_2 \dot\theta_1 \dot\theta_2$$
$$+ l_1 C_1\,^0F_{2,3y} - l_1 S_1\,^0F_{2,3x} \tag{4.24}$$

The first term on the right hand side of Equation (4.24) has been found in Equation (4.23) as the torque at joint 2. Substituting this torque into Equation (4.24) gives:

$$
\begin{aligned}
{}^0T_{0,1} = {}^0T_{2,3} &+ \left(\tfrac{1}{4}m_1 l_1^2 + \tfrac{1}{4}m_2 l_2^2 + m_2 l_1^2 + m_2 l_1 l_2 C_2 + I_1 + I_2\right)\ddot{\theta}_1 \\
&+ \left(\tfrac{1}{4}m_2 l_2^2 + \tfrac{1}{2}m_2 l_1 l_2 C_2 + I_2\right)\ddot{\theta}_2 - \tfrac{1}{2}m_2 l_1 l_2 S_2 \dot{\theta}_2^2 \\
&- m_2 l_1 l_2 S_2 \dot{\theta}_1 \dot{\theta}_2 + (l_1 C_1 + l_2 C_{12}){}^0F_{2,3y} - (l_1 S_1 + l_2 S_{12}){}^0F_{2,3x}
\end{aligned}
\quad (4.25)
$$

4.2.4 Closed-Form Equations

The N-E equations, Equation (4.7) and Equation (4.9), can be expressed in an alternative form more suitable for use in controlling the motion of a robot. In this recast form they represent input-output relationships in terms of independent variables, sometimes referred to as the generalized coordinates, such as the joint position variables. For direct dynamics, the inputs can be taken as the joint torques and the outputs as the joint position variables, i.e., joint angles. For inverse dynamics, the inputs are the joint position variables and the outputs are the joint torques. The inverse dynamics form is well suited for robot control and programming, since it can be used to determine the appropriate inputs necessary to achieve the desired outputs.

The N-E equations can be written in scalar closed-form,

$$
T_{n-1,n} = \sum_{j=1}^{N} J_{nj} \ddot{\theta}_j + \sum_{j=1}^{N} \sum_{k=1}^{N} D_{njk} \dot{\theta}_j \dot{\theta}_k + T_{ext,n}, \quad n = 1, \ldots, N \quad (4.26)
$$

where J_{nj} is the mass moment of inertia of link j reflected to joint n assuming N total links, D_{njk} is a coefficient representing the centrifugal effect when $j = k$ and the Coriolis effect when $j \neq k$ at joint n, and $T_{ext,n}$ is the torque arising from external forces and torques at joint n, including the effect of gravity and end-effector forces and torques.

For the two-link planar robot example, Equation (4.26) expands to:

$$
T_{0,1} = J_{11} \ddot{\theta}_1 + J_{12} \ddot{\theta}_2 + \left(D_{112} + D_{121}\right) \dot{\theta}_1 \dot{\theta}_2 + D_{122} \dot{\theta}_2^2 + T_{ext,1} \quad (4.27)
$$

$$
T_{1,2} = J_{21} \ddot{\theta}_1 + J_{22} \ddot{\theta}_2 + D_{211} \dot{\theta}_1^2 + T_{ext,2} \quad (4.28)
$$

Expressions for the coefficients in Equation (4.27) and Equation (4.28) can be found by comparison to Equation (4.23) and Equation (4.25). For example,

$$
J_{11} = \tfrac{1}{4} m_1 l_1^2 + \tfrac{1}{4} m_2 l_2^2 + m_2 l_1^2 + m_2 l_1 l_2 C_2 + I_1 + I_2 \quad (4.29)
$$

is the total effective moment of inertia of both links reflected to the axis of the first joint, where, from the parallel axis theorem, the inertia of link 1 about joint 1 is $\tfrac{1}{4} m_1 l_1^2 + I_1$ and the inertia of link 2 about joint 1 is $m_2(l_1^2 + \tfrac{1}{4} l_2^2 + l_1 l_2 C_2) + I_2$. Note that the inertia reflected from link 2 to joint 1 is a function of the configuration, being greatest when the arm is straight (when the cosine of θ_2 is 1).

The first term in Equation (4.27) is the torque due to the angular acceleration of link 1, whereas the second term is the torque at joint 1 due to the angular acceleration of link 2. The latter arises from the contribution of the coupling force and torque across joint 2 on link 1. By comparison, the total effective moment of inertia of link 2 reflected to the first joint is

$$
J_{12} = \tfrac{1}{4} m_2 l_2^2 + m_2 l_1 l_2 C_2 + I_2
$$

which is again configuration dependent.

The third term in Equation (4.27) is the torque due to the Coriolis force,

$$T_{0,1\,Coriolis} = (D_{112} + D_{121})\dot{\theta}_1\dot{\theta}_2 = -m_2 l_1 l_2 S_2 \dot{\theta}_1 \dot{\theta}_2$$

resulting from the interaction of the two rotating frames. The fourth term in Equation (4.27) is the torque due to the centrifugal effect on joint 1 of link 2 rotating at angular velocity $\dot{\theta}_2$,

$$T_{0,1\,centrifugal} = D_{122}\dot{\theta}_2^2 = -\tfrac{1}{2} m_2 l_1 l_2 S_2 \dot{\theta}_2^2$$

The Coriolis and centrifugal torque terms are also configuration dependent and vanish when the links are co-linear. For moving links, these terms arise from Coriolis and centrifugal forces that can be viewed as acting through the center of mass of the second link, producing a torque around the second joint which is reflected to the first joint.

Similarly, physical meaning can be associated with the terms in Equation (4.28) for the torque at the second joint.

4.3 Additional Considerations

4.3.1 Computational Issues

The form of Equation (4.26) matches the inverse dynamics representation, with the inputs being the desired trajectories, typically described as time functions of $\theta_1(t)$ through $\theta_N(t)$, and known external loading. The outputs are the joint torques applied by the actuators needed to follow the specific trajectories. These joint torques are found by evaluating Equation (4.26) at each instant after computing the joint angular velocities and angular accelerations from the given time functions. In the inverse dynamics approach, the joint positions, velocities, and accelerations are assumed known, and the joint torques required to cause these known time-dependent motions are sought.

The computation can be intensive, posing a challenge for real time implementation. Nonetheless, the inverse dynamics approach is particularly important for robot control, since it provides a means to account for the highly coupled, nonlinear dynamics of the links. To alleviate the computational burden, efficient inverse dynamic algorithms have been developed that formulate the dynamic equations in recursive forms allowing the computation to be accomplished sequentially from one link to another.

4.3.2 Moment of Inertia

The mass distribution properties of a link with respect to rotations about the center of mass are described by its mass moment of inertia tensor. The diagonal components in the inertia tensor are the moments of inertia, and the off-diagonal components are the products of inertia with respect to the link axes. If these link axes are chosen along the principal axes, for which the link is symmetrical with respect to each axis, the moments of inertia are called the principal moments of inertia and the products of inertia vanish. Expressions for the moments of inertia can be derived and those for common shapes are readily available in tabulated form. For the planar robot example above, if each link is a thin-walled cylinder of radius r and length l the moment of inertia about its centroidal axis perpendicular to the plane is $I = \tfrac{1}{2}mr^2 + \tfrac{1}{12}ml^2$. Other geometries can be evaluated. If each link is a solid cylinder of radius r and length l the moment of inertia about its centroidal axis perpendicular to the plane is $I = \tfrac{1}{4}mr^2 + \tfrac{1}{12}ml^2$. If each link is a solid rectangle of width w and length l, the moment of inertia is $I = \tfrac{1}{12}mw^2 + \tfrac{1}{12}ml^2$.

4.3.3 Spatial Dynamics

Despite their seeming complexity, the equations of motion for the two-link planar robot in the example (Equation (4.23) and Equation (4.25)) are relatively simple due to the assumptions, in particular, restricting the motion to two-dimensions. The expansion of the torque balance equation for robot motions in three-dimensions results in expressions far more involved.

In the example, gravity is neglected since the motion is assumed in a horizontal plane. The Euler equation is reduced to a scalar as all rotations are about axes perpendicular to the plane. In two-dimensions, the inertia tensor is reduced to a scalar moment of inertia, and the gyroscopic torque is zero since it involves the cross product of a vector with itself.

In the three-dimensional case, the inertia tensor has six independent components in general. The vectors for translational displacement, velocity, and acceleration each have three components, in comparison with two for the planar case, adding to the complexity of Newton's equations. The angular displacement, velocity, and acceleration of each link are no longer scalars but have three components. Transformation of these angular vectors from frame to frame requires the use of rotation transforms, as joint axes may not be parallel. Angular displacements in three dimensions are not commutative, so the order of the rotation transforms is critical. Gyroscopic torque terms are present, as are terms accounting for the effect of gravity that must be considered in the general case. For these reasons, determining the spatial equations of motion for a general robot configuration can be a challenging and cumbersome task, with considerable complexity in the derivation.

4.4 Closing

Robot dynamics is the study of the relation between the forces and motions in a robotic system. The chapter explores one specific method of derivation, namely the Newton-Euler approach, for rigid open kinematic chain configurations. Many topics in robot dynamics are not covered in this chapter. These include algorithmic and implementation issues (recursive formulations), the dynamics of robotic systems with closed chains and with different joints (e.g., higher-pairs), as well as the dynamics of flexible robots where structural deformations must be taken into account. Although some forms of non-rigid behavior, such as compliance in the joint bearings, are relatively straightforward to incorporate into a rigid body model, the inclusion of dynamic deformation due to link (arm) flexibility greatly complicates the dynamic equations of motion.

References

Craig, J.J., *Introduction to Robotics: Mechanics and Control,* 2nd ed., Addison-Wesley, Reading, MA, 1989.
Denavit, J. and Hartenberg, R.S., A kinematics notation for lower pair mechanisms based on matrices, *ASME J. Appl. Mech.,* 22, 215–221, 1995.
Featherstone, R., *Robot Dynamics Algorithms,* Kluwer, Dordrecht, 1987.
Featherstone, R. and Orin, D.E., Robot dynamics: equations and algorithms, *IEEE Int. Conf. Robotics & Automation,* San Francisco, CA, 826–834, 2000.
Luh, J.Y.S., Walker, M.W., and Paul, R.P., On-line computational scheme for mechanical manipulators, *Trans. ASME, J. Dyn. Syst., Meas. Control,* 120, 69–76, 1980.
Mason, M.T., *Mechanics of Robotic Manipulation,* MIT Press, Cambridge, MA, 2001.
McKerrow, P.J., *Introduction to Robotics,* Addison-Wesley, Reading, MA, 1991.
Niku, S., *Introduction to Robotics: Analysis, Systems, Applications,* Prentice Hall, Upper Saddle River, NJ, 2001.
Orin, D.E., McGhee, R.B., Vukobratovic, M., and Hartoch, G., Kinematic and kinetic analysis of open-chain linkages utilizing Newton-Euler methods, *Math. Biosci.,* 43, 107–130, 1979.
Robinett, R.D. III (ed.), Feddema, J., Eisler, G.R., Dohrmann, C., Parker, G.G., Wilson, D.G., and Stokes, D., *Flexible Robot Dynamics and Controls,* Kluwer, Dordrecht, 2001.
Sciavicco, L. and Siciliano, B., *Modelling and Control of Robot Manipulators,* 2nd ed., Springer-Verlag, Heidelberg, 2000.
Shahinpoor, M., *A Robot Engineering Textbook,* Harper and Row, New York, 1987.

Spong, M.W. and Vidyasager, M., *Robot Dynamics and Control*, John Wiley & Sons, New York, 1989.

Swain, A.K. and Morris, A.S., A unified dynamic model formulation for robotic manipulator systems, *J. Robotic Syst.*, 20, No. 10, 601–620, 2003.

Vukobratovic, M., Potkonjak, V., and Matijevi, V., *Dynamics of Robots with Contact Tasks*, Kluwer, Dordrecht, 2003.

Walker, M.W. and Orin, D.E., Efficient dyamic computer simulation of robotic mechanisms, *Trans. ASME, J. Syst., Meas. Control*, 104, 205–211, 1982.

Yoshikawa, T., *Foundations of Robotics: Analysis and Control*, MIT Press, Cambridge, MA, 1990.

5
Lagrangian Dynamics

	5.1	Introduction 5-1
	5.2	Preliminaries 5-1
		Velocities and Forces • Kinematics of Serial Linkages
	5.3	Dynamic Equations 5-6
		Inertial Properties of a Rigid Body • Euler-Lagrange Equations for Rigid Linkages • Generalized Force Computation • Geometric Interpretation
Miloš Žefran	5.4	Constrained Systems 5-11
University of Illinois at Chicago		Geometric Interpretation
Francesco Bullo	5.5	Impact Equations 5-13
University of Illinois at	5.6	Bibliographic Remarks 5-14
Urbana-Champaign	5.7	Conclusion 5-14

5.1 Introduction

The motion of a mechanical system is related via a set of *dynamic equations* to the forces and torques the system is subject to. In this work we will be primarily interested in robots consisting of a collection of rigid links connected through joints that constrain the relative motion between the links. There are two main formalisms for deriving the dynamic equations for such mechanical systems: Newton-Euler equations that are directly based on Newton's laws and Euler-Lagrange equations that have their root in the classical work of d'Alembert and Lagrange on analytical mechanics and the work of Euler and Hamilton on variational calculus. The main difference between the two approaches is in dealing with constraints. While Newton's equations treat each rigid body separately and explicitly model the constraints through the forces required to enforce them, Lagrange and d'Alembert provided systematic procedures for eliminating the constraints from the dynamic equations, typically yielding a simpler system of equations. Constraints imposed by joints and other mechanical components are one of the defining features of robots so it is not surprising that the Lagrange's formalism is often the method of choice in robotics literature.

5.2 Preliminaries

The approach and the notation in this section follow [21], and we refer the reader to that text for additional details. A starting point in describing a physical system is the formalism for describing its motion. Since we will be concerned with robots consisting of rigid links, we start by describing rigid body motion. Formally, a rigid body $\mathcal{O}$ is a subset of $\mathbb{R}^3$ where each element in $\mathcal{O}$ corresponds to a point on the rigid body. The defining property of a rigid body is that the distance between arbitrary two points on the rigid body remains unchanged as the rigid body moves. If a *body-fixed* coordinate frame B is attached to $\mathcal{O}$, an arbitrary point $p \in \mathcal{O}$ can be described by a fixed vector p_B. As a result, the position of any point in $\mathcal{O}$ is uniquely determined by the location of the frame B. To describe the location of B in space we choose a global coordinate frame S. The position and orientation of the frame B in the frame S is called the *configuration* of $\mathcal{O}$ and can be described by a 4×4 *homogeneous matrix* g_{SB}:

$$g_{SB} = \begin{bmatrix} R_{SB} & d_{SB} \\ 0 & 1 \end{bmatrix}, \quad R_{SB} \in \mathbb{R}^{3\times 3}, d_{SB} \in \mathbb{R}^3, R_{SB}^T R_{SB} = I_3, \det(R_{SB}) = 1 \tag{5.1}$$

Here I_n denotes the identity matrix in $\mathbb{R}^{n \times n}$. The set of all possible configurations of $\mathcal{O}$ is known as $SE(3)$, the special Euclidean group of rigid body transformations in three dimensions. By the above argument, $SE(3)$ is equivalent to the set of homogeneous matrices. It can be shown that it is a matrix group as well as a smooth manifold and, therefore, a *Lie group*; for more details we refer to [11, 21]. It is convenient to denote matrices $g \in SE(3)$ by the pair (R, d) for $R \in SO(3)$ and $d \in \mathbb{R}^3$ (recall that $SO(3) = \{R \in \mathbb{R}^{3 \times 3} | R^T R = I_3, \det(R) = 1\}$).

Given a point $p \in \mathcal{O}$ described by a vector p_B in the frame B, it is natural to ask what is the corresponding vector in the frame S. From the definition of g_{SB} we have

$$\overline{p}_S = g_{SB} \overline{p}_B$$

where for a vector $p \in \mathbb{R}^3$, the corresponding *homogeneous vector* $\overline{p}$ is defined as

$$\overline{p} = \begin{bmatrix} p \\ 1 \end{bmatrix}$$

The tangent space of $SE(3)$ at $g_0 \in SE(3)$ is the vector space of matrices of the form $\dot{g}(0)$, where $g(t)$ is a curve in $SE(3)$ such that $g(0) = g_0$. The tangent space of a Lie group at the group identity is called the *Lie algebra* of the Lie group. The Lie algebra of $SE(3)$, denoted by $se(3)$, is

$$se(3) = \left\{ \begin{bmatrix} \Omega & v \\ 0 & 0 \end{bmatrix} \Big| \Omega \in \mathbb{R}^{3 \times 3}, v \in \mathbb{R}^3, \Omega^T = -\Omega \right\} \tag{5.2}$$

Elements of $se(3)$ are called *twists*. Recall that a 3×3 skew-symmetric matrix Ω can be uniquely identified with a vector $\omega \in \mathbb{R}^3$ so that for an arbitrary vector $x \in \mathbb{R}^3$, $\Omega x = \omega \times x$, where $\times$ is the vector cross product operation in $\mathbb{R}^3$. Each element $T \in se(3)$ can be thus identified with a 6×1 vector

$$\xi = \begin{bmatrix} v \\ \omega \end{bmatrix}$$

called the *twist coordinates* of T. We also write $\Omega = \widehat{\omega}$ and $T = \widehat{\xi}$ to denote these transitions between vectors and matrices.

An important relation between the Lie group and its Lie algebra is provided by the *exponential map*. It can be shown that for $\widehat{\xi} \in se(3)$, $\exp(\widehat{\xi}) \in SE(3)$, where $\exp : \mathbb{R}^{n \times n} \to \mathbb{R}^{n \times n}$ is the usual matrix exponential. Using the properties of $SE(3)$, it can be shown that, for $\xi^T = [v^T \omega^T]$,

$$\exp(\widehat{\xi}) = \begin{cases} \begin{bmatrix} I_3 & \emptyset v \\ 0 & 1 \end{bmatrix}, & \omega = 0, \\ \begin{bmatrix} \exp(\widehat{\omega}) & \frac{1}{\|\omega\|^2}[(I_3 - \exp(\widehat{\omega}))(\omega \times v) + \omega \omega^T v] \\ 0 & 1 \end{bmatrix}, & \omega \neq 0 \end{cases} \tag{5.3}$$

where the relation

$$\exp(\widehat{\omega}) = I_3 + \frac{\sin \|\omega\|}{\|\omega\|} \widehat{\omega} + \frac{1 - \cos \|\omega\|}{\|\omega\|^2} \widehat{\omega}^2$$

is known as *Rodrigues' formula*. From the last formula it is easy to see that the exponential map exp : $se(3) \to SE(3)$ is many to one. It can be shown that the map is in fact onto. In other words, every matrix $g \in SE(3)$ can be written as $g = \exp(\widehat{\xi})$ for some $\widehat{\xi} \in se(3)$. The components of ξ are also called *exponential coordinates* of g.

To every twist $\xi^T = [v^T \omega^T]$, $\omega \neq 0$, we can associate a triple (l, h, M) called the *screw associated with* $\widehat{\xi}$, where we define $l = \{p + \lambda \omega \in \mathbb{R}^3 | p \in \mathbb{R}^3, \lambda \in \mathbb{R}\}$, $h \in \mathbb{R}$, and $M \in \mathbb{R}$ so that the following relations hold:

$$M = \|\omega\|, \qquad v = -\omega \times p + h\omega$$

Note that l is the line in $\mathbb{R}^3$ in the direction of ω passing through the point $p \in \mathbb{R}^3$. If $\omega = 0$, the corresponding screw is $(l, \infty, \|v\|)$, where $l = \{\lambda v \in \mathbb{R}^3 | \lambda \in \mathbb{R}\}$. In this way, the exponential map can be given an interesting geometric interpretation: if $g = \exp(\widehat{\xi})$ with $\omega \neq 0$, then the rigid body transformation represented by g can be realized as a rotation around the line l by the angle M followed by a translation in the direction of this line for the distance hM. If $\omega = 0$ and $v \neq 0$, the rigid body transformation is simply a translation in the direction of $\frac{v}{\|v\|}$ for a distance M. The line l is the *axis*, h is the *pitch*, and M is the *magnitude* of the screw. This geometric interpretation and the fact that every element $g \in SE(3)$ can be written as the exponential of a twist is the essence of *Chasles Theorem*.

5.2.1 Velocities and Forces

We have seen that $SE(3)$ is the configuration space of a rigid body $\mathcal{O}$. By considering a rigid body moving in space, a more intuitive interpretation can be also given to $se(3)$. At every instant t, the configuration of $\mathcal{O}$ is given by $g_{SB}(t) \in SE(3)$. The map $t \mapsto g_{SB}(t)$ is therefore a curve representing the motion of the rigid body. The time derivative $\dot{g}_{SB} = \frac{d}{dt} g_{SB}$ corresponds to the velocity of the rigid body motion. However, the matrix curve $t \mapsto \dot{g}_{SB}$ does not allow an easy geometric interpretation. Instead, it is not difficult to show that the matrices $\widehat{V}^b_{SB} = g_{SB}^{-1} \dot{g}_{SB}$ and $\widehat{V}^s_{SB} = \dot{g}_{SB} g_{SB}^{-1}$ take values in $se(3)$. The matrices $\widehat{V}^b_{SB}$ and $\widehat{V}^s_{SB}$ are called the *body* and *spatial velocity*, respectively, of the rigid body motion. Their twist coordinates will be denoted by V^b_{SB} and V^s_{SB}. Assume a point $p \in \mathcal{O}$ has coordinates p_S and p_B in the spatial frame and in the body frame, respectively. As the rigid body moves along the trajectory $t \mapsto g_{SB}(t)$, a direct computation shows that the velocity of the point can be computed by $\dot{\overline{p}}_S = \widehat{V}^s_{SB} \overline{p}_S$ or, alternatively, $\dot{\overline{p}}_B = \widehat{V}^b_{SB} \overline{p}_B$.

The body and spatial velocities are related by

$$V^s_{SB} = \mathrm{Ad}_{g_{SB}} V^b_{SB}$$

where for $g = (R, d) \in SE(3)$, the *adjoint transformation* $\mathrm{Ad}_g : \mathbb{R}^6 \to \mathbb{R}^6$ is

$$\mathrm{Ad}_g = \begin{bmatrix} R & \widehat{d} R \\ 0 & R \end{bmatrix}$$

In general, if $\widehat{\xi} \in se(3)$ is a twist with twist coordinates $\xi \in \mathbb{R}^6$, then for any $g \in SE(3)$, the twist $g \widehat{\xi} g^{-1}$ has twist coordinates $\mathrm{Ad}_g \xi \in \mathbb{R}^6$.

Quite often it is necessary to relate the velocity computed in one set of frames to that computed with respect to a different set of frames. Let X, Y, and Z be three coordinate frames. Let g_{XY} and g_{YZ} be the homogeneous matrices relating the Y to the X frame and the Z to the Y frame, respectively. Let $V^s_{XY} = \dot{g}_{XY} g_{XY}^{-1}$ and $V^b_{XY} = g_{XY}^{-1} \dot{g}_{XY}$ be the spatial velocity and the body velocity of the frame Y with respect to the frame X, respectively; define V^s_{XZ}, V^s_{YZ}, V^b_{XZ}, and V^b_{YZ} analogously. The following relations can be verified by a direct calculation:

$$\begin{aligned} V^s_{XZ} &= V^s_{XY} + \mathrm{Ad}_{g_{XY}} V^s_{YZ} \\ V^b_{XZ} &= \mathrm{Ad}_{g_{YZ}^{-1}} V^b_{XY} + V^b_{YZ} \end{aligned} \qquad (5.4)$$

In the last equation, Ad^{-1} is the inverse of the adjoint transformation and can be computed using the formula $\text{Ad}_g^{-1} = \text{Ad}_{g^{-1}}$, for $g \in SE(3)$.

The body and spatial velocities are, in general, time dependent. Take the spatial velocity $V_{SB}^s(t)$ at time t and consider the screw (l, h, M) associated with it. We can show that at this time instant t, the rigid body is moving with the same velocity as if it were rotating around the screw axis l with a constant rotational velocity of magnitude M, and translating along this axis with a constant translational velocity of magnitude hM. If $h = \infty$, then the motion is a pure translation in the direction of l with velocity M. A similar interpretation can be given to $V_{SB}^b(t)$.

The above arguments show that elements of the Lie algebra $se(3)$ can be interpreted as generalized velocities. It tuns out that the elements of its dual $se^*(3)$, known as *wrenches*, can be interpreted as generalized forces. The fact that the names *twist* and *wrench* which originally derived from the screw calculus [5] are used for elements of the Lie algebra and its dual is not just a coincidence; the present treatment can be viewed as an alternative interpretation of the screw calculus.

Using coordinates dual to the twist coordinates, a wrench can be written as a pair $F = [f \ \tau]$, where f is the force component and τ the torque component. Given a twist $V^T = [v^T \ \omega^T]$, for $v, \omega \in \mathbb{R}^{3\times 1}$, and a wrench $F = [f \ \tau]$, for $f, \tau \in \mathbb{R}^{1\times 3}$, the *natural pairing* $\langle W; F \rangle = fv + \tau\omega$ is a scalar representing the *instantaneous work* performed by the generalized force F on a rigid body that is instantaneously moving with the generalized velocity V. In computing the instantaneous work, it is important to consider generalized force and velocity with respect to the same coordinate frame.

As was the case with the twists, we can also associate a screw to a wrench. Given a wrench $F = [f \ \tau]$, $f \neq 0$, the associated screw (l, h, M) is given by $l = \{p + \lambda f \in \mathbb{R}^3 | \ p \in \mathbb{R}^3, \lambda \in \mathbb{R}\}$ and by:

$$M = \|f\|, \quad \tau = -f \times p + hf$$

For $f = 0$, the screw is $(l, \infty, \|\tau\|)$, where $l = \{\lambda\tau \in \mathbb{R}^3 | \ \lambda \in \mathbb{R}\}$. Geometrically, a wrench with the associated screw (l, h, M) corresponds to a force with magnitude M applied in the direction of the screw axis l and a torque with magnitude hM applied about this axis. If $f = 0$ and $\tau \neq 0$, then the wrench is a pure torque of magnitude M applied around l. The fact that every wrench can be interpreted in this way is known as *Poinsot Theorem*. Note the similarity with Chasles Theorem.

5.2.2 Kinematics of Serial Linkages

A serial linkage consists of a sequence of rigid bodies connected through one-degree-of-freedom (DOF) revolute or prismatic joints; a multi-DOF joint can be modeled as a sequence of individual 1-DOF joints. To uniquely describe the position of such a serial linkage it is only necessary to know the *joint variables*, i.e., the angle of rotation for a revolute joint or the linear displacement for a prismatic joint. Typically it is then of interest to know the configuration of the links as a function of these joint variables. One of the links is typically designated as the *end-effector*, the link that carries the tool involved in the robot task. For serial manipulators, the end-effector is the last link of the mechanism. The map from the joint space to the end-effector configuration space is known as the *forward kinematics* map. We will use the term *extended forward kinematics map* to denote the map from the joint space to the Cartesian product of the configuration spaces of each of the links.

The configuration spaces of a revolute and of a prismatic joint are connected subsets of the unit circle S^1 and of the real line $\mathbb{R}$, respectively. For simplicity, we shall assume that these connected subsets are S^1 and $\mathbb{R}$, respectively. If a mechanism with n joints has r revolute and $p = n - r$ prismatic joints, its configuration space is $Q = S^r \times \mathbb{R}^p$, where $S^r = \underbrace{S^1 \times \cdots \times S^1}_{r}$. Assume the coordinate frames attached to the individual links are $B_1, \ldots, B_n$. The configuration space of each of the links is $SE(3)$, so the extended forward kinematics map is

$$\kappa: Q \to SE^n(3)$$
$$q \mapsto (g_{SB_1}(q), \ldots, g_{SB_n}(q))$$
(5.5)

We shall call $SE^n(3)$ the Cartesian space.

For serial linkages the forward kinematics map has a particularly revealing form. Choose a *reference configuration* of the manipulator, i.e. the configuration where all the joint variables are 0, and let $\xi_1, \ldots, \xi_n$ be the *joint twists* in this configuration expressed in the global coordinate frame S. If the ith joint is revolute and $l = \{p + \lambda \omega \in \mathbb{R}^3 \mid \lambda \in \mathbb{R}, \|\omega\| = 1\}$ is the axis of rotation, then the joint twist corresponds to the screw $(l, 0, 1)$ and is

$$\xi_i = \begin{bmatrix} -\omega \times p \\ \omega \end{bmatrix} \tag{5.6}$$

If the ith joint is prismatic and $v \in \mathbb{R}^3$, $\|v\| = 1$, is the direction in which it moves, then the joint twist corresponds to the screw $(\{\lambda v \in \mathbb{R}^3 \mid \lambda \in \mathbb{R}\}, \infty, 1)$ and is

$$\xi_i = \begin{bmatrix} v \\ 0 \end{bmatrix}$$

One can show that

$$g_{SB_i}(q) = \exp(\xi_1 q^1) \cdots \exp(\xi_i q^i) g_{SB_i,0} \tag{5.7}$$

where $g_{SB_i,0} \in SE(3)$ is the reference configuration of the ith link. This is known as the *Product of Exponentials Formula* and was introduced in [7].

Another important relation is that between the joint rates and the link body and spatial velocities. A direct computation shows that along a curve $t \mapsto q(t)$,

$$V^s_{SB_i}(t) = J^s_{SB_i}(q(t)) \dot{q}(t)$$

where $J^s_{SB_i}$ is a configuration-dependent $6 \times n$ matrix, called the *spatial manipulator Jacobian*, that for serial linkages equals

$$J^s_{SB_i}(q) = [\xi_{S,1}(q) \ \ldots \ \xi_{S,i}(q) \ 0 \ \ldots \ 0]$$

Here the jth column $\xi_{S,j}(q)$ of the spatial manipulator Jacobian is the jth joint twist expressed in the global reference frame S after the manipulator has moved to the configuration described by q. It can be computed from the forward kinematics map using the formula:

$$\xi_{S,j}(q) = \mathrm{Ad}_{(\exp(\xi_1 q^1) \cdots \exp(\xi_{j-1} q^{j-1}))} \xi_j$$

Similar expressions can be obtained for the body velocity

$$V^b_{SB_i}(t) = J^b_{SB_i}(q(t)) \dot{q}(t) \tag{5.8}$$

where $J^b_{SB_i}$ is a configuration-dependent $6 \times n$ *body manipulator Jacobian* matrix. For serial linkages,

$$J^b_{SB_i}(q) = [\xi_{B_i,1}(q) \ \ldots \ \xi_{B_i,i}(q) \ 0 \ \ldots \ 0] \tag{5.9}$$

where the jth column $\xi_{B_i,j}(q)$, $j \leq i$ is the jth joint twist expressed in the link frame B_i after the manipulator has moved to the configuration described by q and is given by

$$\xi_{B_i,j}(q) = \mathrm{Ad}^{-1}_{(\exp(\xi_j q^j) \cdots \exp(\xi_i q^i) g_{SB_i,0})} \xi_j$$

5.3 Dynamic Equations

In this section we continue our study of mechanisms composed of rigid links connected through prismatic or revolute joints. One way to describe a system of interconnected rigid bodies is to describe each of the bodies independently and then explicitly model the joints between them through the constraint forces. Since the configuration space of a rigid body is $SE(3)$, a robot manipulator with n links would be described with $6n$ parameters. Newton's equations can be then directly used to describe robot dynamics. An alternative is to use *generalized coordinates* and describe just the degrees of freedom of the mechanism. As discussed earlier, for a robot manipulator composed of n links connected through 1-DOF joints the generalized coordinates can be chosen to be the joint variables. In this case n parameters are therefore sufficient. The Lagrange-d'Alembert Principle can then be used to derive the Euler-Lagrange equations describing the dynamics of the mechanism in generalized coordinates. Because of the dramatic reduction in the number of parameters describing the system this approach is often preferable. We shall describe this approach in some detail.

The Euler-Lagrange equations are derived directly from the energy expressed in the generalized coordinates. Let q be the vector of generalized coordinates. We first form the system *Lagrangian* as the difference between the kinetic and the potential energies of the system. For typical manipulators the Lagrangian function is

$$L(q,\dot{q}) = T(q,\dot{q}) - V(q)$$

where $T(q,\dot{q})$ is the kinetic energy and $V(q)$ the potential energy of the system. The *Euler-Lagrange* equations describing the dynamics for each of the generalized coordinates are then

$$\frac{d}{dt}\frac{\partial L}{\partial \dot{q}^i} - \frac{\partial L}{\partial q^i} = Y_i \quad (5.10)$$

where Y_i is the *generalized force* corresponding to the generalized coordinate q^i. The following sections discuss these equations in the context of a single rigid body and of a rigid linkage.

5.3.1 Inertial Properties of a Rigid Body

In order to apply the Lagrange's formalism to a rigid body $\mathcal{O}$, we need to compute its Lagrangian function. Assume that the body-fixed frame B is attached to the center of mass of the rigid body. The kinetic energy of the rigid body is the sum of the kinetic energy of all its particles. Therefore, it can be computed as

$$T = \int_\mathcal{O} \frac{1}{2}\|\dot{p}\|^2 dm = \int_\mathcal{O} \frac{1}{2}\|\dot{p}_S\|^2 \rho(p_S) dV$$

where ρ is the body density and dV is the volume element in $\mathbb{R}^3$. If $(R_{SB}, d_{SB}) \in SE(3)$ is the rigid body configuration, the equality $p_S = R_{SB} p_B + d_{SB}$ and some manipulations lead to

$$T = \frac{1}{2}m\|\dot{d}_{SB}\|^2 + \frac{1}{2}(\omega_{SB}^b)^T \left(-\int_\mathcal{O} \widehat{p}_B^2 \rho(p_B) dV\right) \omega_{SB}^b \quad (5.11)$$

where m is the total mass of the rigid body and $\widehat{p}_B$ is the skew-symmetric matrix corresponding to the vector p_B. The formula shows that the kinetic energy is the sum of two terms referred to as the *translational* and the *rotational components*. The quantity $\mathcal{I} = -\int_\mathcal{O} \widehat{p}_B^2 \rho(p_B) dV$ is the *inertia tensor* of the rigid body; one can show that $\mathcal{I}$ is a symmetric positive-definite 3×3 matrix. By defining the *generalized inertia matrix*

$$\mathcal{M} = \begin{bmatrix} mI_3 & 0 \\ 0 & \mathcal{I} \end{bmatrix}$$

Lagrangian Dynamics

the kinetic energy can be written as

$$T = \frac{1}{2}(V_{SB}^b)^T \mathcal{M} V_{SB}^b \tag{5.12}$$

where V_{SB}^b is the body velocity of the rigid body.

Equation (5.12) can be used to obtain the expression for the generalized inertia matrix when the body-fixed frame is not at the center of mass. Assume that $g_{AB} = (R_A, d_A) \in SE(3)$ is the transformation between the frame A attached to the center of mass of the rigid body and the body-fixed frame B. According to Equation (5.4), $V_{SA}^b = \text{Ad}_{g_{AB}} V_{SB}^b$. We thus have

$$T = \frac{1}{2}(V_{SA}^b)^T \mathcal{M}_A V_{SA}^b = \frac{1}{2}(\text{Ad}_{g_{AB}} V_{SB}^b)^T \mathcal{M}(\text{Ad}_{g_{AB}} V_{SB}^b) = (V_{SB}^b)^T \mathcal{M}_B V_{SB}^b$$

where $\mathcal{M}_B$ is the generalized inertia matrix with respect to the body-fixed frame B:

$$\mathcal{M}_B = (\text{Ad}_{g_{AB}})^T \mathcal{M}_A \text{Ad}_{g_{AB}} = \begin{bmatrix} mI_3 & mR_A^T \hat{d}_A R_A \\ -mR_A^T \hat{d}_A R_A & R_A^T(\mathcal{I} - m\hat{d}_A^2)R_A \end{bmatrix}$$

By observing that $g_{BA} = g_{AB}^{-1} = (R_B, d_B)$, where $R_B = R_A^T$ and $d_B = -R_A^T d_A$, $\mathcal{M}_B$ can be written as:

$$\mathcal{M}_B = \begin{bmatrix} mI_3 & -m\hat{d}_B \\ m\hat{d}_B & R_B \mathcal{I} R_B^T - \hat{d}_B^2 \end{bmatrix}$$

5.3.2 Euler-Lagrange Equations for Rigid Linkages

To obtain the expression for the kinetic energy of a linkage composed of n rigid bodies, we need to add the kinetic energy of each of the links:

$$T = \sum_{i=1}^n T_i = \frac{1}{2} \sum_{i=1}^n (V_{SB_i}^b)^T \mathcal{M}_{B_i} V_{SB_i}^b$$

Using the relation (5.8), this becomes

$$T = \sum_{i=1}^n \frac{1}{2}(J_{SB_i}^b \dot{q})^T \mathcal{M}_{B_i} J_{SB_i}^b \dot{q} = \frac{1}{2} \dot{q}^T M(q) \dot{q}$$

where

$$M(q) = \sum_{i=1}^n (J_{SB_i}^b)^T \mathcal{M}_{B_i} J_{SB_i}^b$$

is the *manipulator inertia matrix*. For serial manipulators, the body manipulator Jacobian $J_{SB_i}^b$ is given by Equation (5.9).

The potential energy of the linkage typically consists of the sum of the gravitational potential energies of each of the links. Let $h_i(q)$ denote the height of the center of mass of the ith link. The potential energy of the link is then $V_i(q) = m_i g h_i(q)$, and the potential energy of the linkage is

$$V(q) = \sum_{i=1}^n m_i g h_i(q)$$

If other conservative forces act on the manipulator, the corresponding potential energy can be simply added to V.

The Lagrangian for the manipulator is the difference between the kinetic and potential energies, that is,

$$L(q,\dot{q}) = \frac{1}{2}\dot{q}^T M(q)\dot{q} - V(q) = \frac{1}{2}\sum_{i,j=1}^{n} M_{ij}(q)\dot{q}^i \dot{q}^j - V(q)$$

where $M_{ij}(q)$ is the component (i,j) of the manipulator inertia matrix at q. Substituting these expressions into the Euler-Lagrange Equations (5.10) we obtain

$$\sum_{j=1}^{n} M_{ij}(q)\ddot{q}^j + \sum_{j,k=1}^{n} \Gamma_{ijk}(q)\dot{q}^j\dot{q}^k + \frac{\partial V}{\partial q^i}(q) = Y_i, \quad i \in \{1,\ldots,n\}$$

where the functions Γ_{ijk} are the *Christoffel symbols (of the first kind)* of the inertia matrix M and are defined by

$$\Gamma_{ijk}(q) = \frac{1}{2}\left(\frac{\partial M_{ij}(q)}{\partial q^k} + \frac{\partial M_{ik}(q)}{\partial q^j} - \frac{\partial M_{jk}(q)}{\partial q^i}\right) \tag{5.13}$$

Collecting the equations in a vector format, we obtain

$$M(q)\ddot{q} + C(q,\dot{q})\dot{q} + G(q) = Y \tag{5.14}$$

where $C(q,\dot{q})$ is the *Coriolis matrix* for the manipulator with components

$$C_{ij}(q,\dot{q}) = \sum_{k=1}^{n} \Gamma_{ijk}(q)\dot{q}^k$$

and where $G_i(q) = \frac{\partial V}{\partial q^i}(q)$. Equation (5.14) suggests that the robot dynamics consists of four components: the inertial forces $M(q)\ddot{q}$, the *Coriolis and centrifugal forces* $C(q,\dot{q})\dot{q}$, the conservative forces $G(q)$, and the generalized force Y composed of all the non-conservative external forces acting on the manipulator. The Coriolis and centrifugal forces depend quadratically on the generalized velocities $\dot{q}^i$ and reflect the inter-link dynamic interactions. Traditionally, terms involving products $\dot{q}^i\dot{q}^j, i \neq j$ are called Coriolis forces, while centrifugal forces have terms of the form $(\dot{q}^i)^2$.

A direct calculation can be used to show that the robot dynamics has the following two properties [29]. For all $q \in Q$

1. the manipulator inertia matrix $M(q)$ is symmetric and positive-definite, and
2. the matrix $(\dot{M}(q) - 2C(q,\dot{q})) \in \mathbb{R}^{n \times n}$ is skew-symmetric.

The first property is a mathematical statement of the following fact: the kinetic energy of a system is a quadratic form which is positive unless the system is at rest. The second property is referred to as the *passivity property* of rigid linkages; this property implies that the total energy of the system is conserved in the absence of friction. The property plays an important role in stability analysis of many robot control schemes.

5.3.3 Generalized Force Computation

Given a configuration manifold Q, the tangent bundle is the set $TQ = \{(q,\dot{q}) | q \in Q\}$. The Lagrangian is formally a real-valued map on the tangent bundle, $L : TQ \to \mathbb{R}$. Recall that given a scalar function on a vector space, its partial derivative is a map from the dual space to the reals. Similarly, it is possible to interpret the partial derivatives $\frac{\partial L}{\partial q}$ and $\frac{\partial L}{\partial \dot{q}}$ as functions taking values in the dual of the tangent bundle, T^*Q. Accordingly, the Euler-Lagrange equations in vector form

$$\frac{d}{dt}\frac{\partial L}{\partial \dot{q}} - \frac{\partial L}{\partial q} = Y$$

Lagrangian Dynamics

can be interpreted as an equality on T^*Q. The external force Y is thus formally a one-form, i.e., a map $Y : Q \to T^*Q$. We let Y_i denote the ith component of Y. Roughly speaking, Y_i is the component of generalized force Y that directly affects the coordinate q^i. In what follows we derive an expression for the generalized force Y_i as a function of the wrenches acting on the individual links.

Let us start by introducing two useful concepts. Recall that given two manifolds Q_1 and Q_2 and a smooth map $\phi : Q_1 \to Q_2$, the *tangent map* $T\phi : TQ_1 \to TQ_2$ is a linear function that maps tangent vectors from $T_q Q_1$ to $T_{\phi(q)} Q_2$. If $X \in T_q Q_1$ is a tangent vector and $\gamma : \mathbb{R} \to Q_1$ is a smooth curve tangent to X at q, then $T\phi(X)$ is the vector tangent to the curve $\phi \circ \gamma$ at $\phi(q)$. In coordinates this linear map is the Jacobian matrix of ϕ. Given a one-form ω on Q_2, the *pull-back* of ω is the one-form $(\phi^*\omega)$ on Q_1 defined by

$$\langle (\phi^*\omega)(q); X \rangle = \langle \omega; T_q\phi(X) \rangle$$

for all $q \in Q_1$ and $X \in T_q Q_1$.

Now consider a linkage consisting of n rigid bodies with the configuration space Q. Recall that the extended forward kinematics function $\kappa : Q \to SE^n(3)$ from Equation (5.5) maps a configuration $q \in Q$ to the configuration of each of the links, i.e., to a vector $(g_{SB_1}(q), \ldots, g_{SB_n}(q))$ in the Cartesian space $SE^n(3)$. Forces and velocities in the Cartesian space are described as twists and wrenches; they are thus elements of $se^n(3)$ and $se^{*n}(3)$, respectively. Let W_i be the wrench acting on the ith link. The total force in the Cartesian space is thus $W = (W_1, \ldots, W_n)$. The generalized force Y_i is the component of the total configuration space force Y in the direction of q^i. Formally, $Y_i = \langle Y; \frac{\partial}{\partial q^i} \rangle$, where $\frac{\partial}{\partial q^i}$ is the ith coordinate vector field.

It turns out that the total configuration space force Y is the pull-back of the total Cartesian space force W to T^*Q through the extended forward kinematics map, i.e.,

$$Y = \kappa^*(W)$$

Using this relation and the definition of the pull-back we thus have

$$Y_i(q) = \left\langle Y(q); \frac{\partial}{\partial q^i} \right\rangle = \left\langle W(\kappa(q)); T_q\kappa\left(\frac{\partial}{\partial q^i}\right) \right\rangle \tag{5.15}$$

It is well established that in the absence of external forces, the generalized force Y_i is the torque applied at the ith joint if the joint is revolute and the force applied at the ith joint if the joint is prismatic; let us formally derive this result using Equation (5.15). Consider a serial linkage with the forward kinematics map given by (5.7). Assume that the ith joint connecting $(i-1)$th and ith link is revolute and let τ_i be the torque applied at the joint. At $q \in Q$, we compute

$$T_q\kappa\left(\frac{\partial}{\partial q^j}\right) = (\underbrace{0, \ldots, 0}_{j-1}, \xi_{S,j}(q), \ldots, \xi_{S,j}(q))$$

where $\xi_{S,j}(q)$ is defined by Equation (5.6) and it is the unit twist corresponding to a rotation about the jth joint axis. The wrench in the Cartesian space resulting from the torque τ_i acting along the ith joint axis is $W = \tau_i (\underbrace{0, \ldots, 0}_{i-2}, -\sigma_{S,i}(q), \sigma_{S,i}(q), 0, \ldots, 0)$, where $\sigma_{S,i}(q)$ is the unit wrench corresponding to a torque about the ith joint axis. It is then easy to check that W results in

$$Y_j = \begin{cases} 0, & j \neq i, \\ \tau_i, & j = i \end{cases}$$

5.3.4 Geometric Interpretation

Additional differential geometric insight into the dynamics can be gained by observing that the kinetic energy provides a *Riemannian metric* on the configuration manifold Q, i.e., a smoothly changing rule for computing the inner product between tangent vectors. Recall that given $(q, \dot{q}) \in TQ$, the kinetic energy of a manipulator is $T(q, \dot{q}) = \frac{1}{2} \dot{q}^T M(q) \dot{q}$, where $M(q)$ is a positive-definite matrix. If we have two vectors $v_1, v_2 \in T_q Q$, we can thus define their inner product by $\langle\!\langle v^1, v^2 \rangle\!\rangle = \frac{1}{2}(v^1)^T M(q) v^2$. Physical properties of a manipulator guarantee that $M(q)$ and thus the rule for computing the inner product changes smoothly over Q. For additional material on this section we refer to [11].

Given two smooth vector fields X and Y on Q, the *covariant derivative* of Y with respect to X is the vector field $\nabla_X Y$ with coordinates

$$(\nabla_X Y)^i = \frac{\partial Y^i}{\partial q^j} X^j + \widehat{\Gamma}^i_{jk} X^j Y^k$$

where X^i and Y^j are the ith and jth component of X and Y, respectively. The operator ∇ is called an *affine connection* and it is determined by the n^3 functions $\widehat{\Gamma}^i_{jk}$. A direct calculation shows that for real valued functions f and g defined over Q, and vector fields X, Y, and Z:

$$\nabla_{fX+gY} Z = f \nabla_X Z + g \nabla_Y Z$$
$$\nabla_X(Y + Z) = \nabla_X Y + \nabla_X Z$$
$$\nabla_X(fY) = f \nabla_X Y + X(f) Y$$

where in coordinates, $X(f) = \sum_{i=1}^n \frac{\partial f}{\partial q^i} X^i$.

When the functions $\widehat{\Gamma}^i_{jk}$ are computed according to

$$\widehat{\Gamma}^i_{jk}(q) = \frac{1}{2} \sum_{l=1}^n M^{li} \left(\frac{\partial M_{lj}(q)}{\partial q^k} + \frac{\partial M_{lk}(q)}{\partial q^j} - \frac{\partial M_{jk}(q)}{\partial q^l} \right) \tag{5.16}$$

where $M^{li}(q)$ are the components of $M^{-1}(q)$, they are called the *Christoffel symbols (of the second kind)* of the Riemannian metric M, and the affine connection is called the *Levi-Civita connection*. From Equation (5.13) and Equation (5.16), it is easy to see that for a Levi-Civita connection,

$$\Gamma_{ijk} = \sum_{l=1}^n M_{li} \widehat{\Gamma}^l_{jk}$$

Assume that the potential forces are not present so that $L(q, \dot{q}) = T(q, \dot{q})$. The manipulator dynamics equations (5.14) can be thus written as:

$$\ddot{q} + M^{-1}(q) C(q, \dot{q}) \dot{q} = M^{-1}(q) Y$$

It can be seen that the last equation can be written in coordinates as

$$\ddot{q}^i + \widehat{\Gamma}^i_{jk}(q) \dot{q}^j \dot{q}^k = F^i$$

and in vector format as

$$\nabla_{\dot{q}} \dot{q} = F \tag{5.17}$$

where ∇ is the Levi-Civita connection corresponding to $M(q)$ and $F = M^{-1}(q) Y$ is the vector field obtained from the one-form Y through the identity $\langle Y; X \rangle = \langle\!\langle F, X \rangle\!\rangle$. In the absence of the external forces, Equation (5.17) becomes the so-called *geodesic equation* for the metric M. It is a second order differential equation whose solutions are curves of locally minimal length – like straight lines in $\mathbb{R}^n$ or great circles on a sphere. It can also be shown that among all the curves $\gamma : [0, 1] \to Q$ that connect two given points, a geodesic is

Lagrangian Dynamics

the curve that minimizes the energy integral:

$$E(\gamma) = \int_0^1 T(\gamma(t), \dot{\gamma}(t)) dt$$

This geometric insight implies that if a mechanical system moves freely with no external forces present, it will move along the minimum energy paths.

When the potential energy is present, Equation (5.17) still applies if the resulting conservative forces are included in F. The one-form Y describing the conservative forces associated with the potential energy V is the differential of V, that is, $Y = -dV$, where the *differential* dV is defined by $\langle dV; X \rangle = X(V)$. The corresponding vector field $F = -M^{-1}(q)dV$ is related to the notion of *gradient* of V, specifically, $F = -\operatorname{grad} V$. In coordinates, differential and gradient of V have components:

$$(dV)_i = \frac{\partial V}{\partial q^i}, \quad \text{and} \quad (\operatorname{grad} V)^i = M^{ij} \frac{\partial V}{\partial q^j}$$

5.4 Constrained Systems

One of the advantages of Lagrange's formalism is a systematic procedure to deal with constraints. Generalized coordinates in themselves are a way of dealing with constraints. For example, for robot manipulators, joints limit the relative motion between the links and thus represent constraints. We can avoid modeling these constraints explicitly by choosing joint variables as generalized coordinates. However, in robotics it is often necessary to constrain the motion of the end-effector in some way, model rolling of the wheels of a mobile robot without slippage or, for example, taking into account various conserved quantities for space robots. Constraints of this nature are external to the robot itself so it is desirable to model them explicitly. Mathematically, a constraint can be described by an equation of the form $\varphi(q, \dot{q}) = 0$, where $q \in \mathbb{R}^n$ are the generalized coordinates and $\varphi : TQ \to \mathbb{R}^m$. However, there is a fundamental difference between constraints of the form $\varphi(q) = 0$, called *holonomic constraints*, and those where the dependence on the generalized velocities cannot be eliminated through the integration, known as *nonholonomic constraints*. Also, among the nonholonomic constraints, only those that are linear in the generalized velocities turn out to be interesting in practice. In other words, typical nonholonomic constraints are of the form $\varphi(q, \dot{q}) = A(q)\dot{q}$.

Holonomic constraints restrict the motion of the system to a submanifold of the configuration manifold Q. For nonholonomic constraints such a constraint manifold does not exist. However, formally the holonomic and nonholonomic constraints can be treated in the same way by observing that a holonomic constraint $\varphi(q) = 0$ can be differentiated to obtain $A(q)\dot{q} = 0$, where $A(q) = \frac{\partial \varphi}{\partial q} \in \mathbb{R}^{m \times n}$. The general form of constraints can be thus assumed to be

$$A(q)\dot{q} = 0 \tag{5.18}$$

We will assume that $A(q)$ has full rank everywhere.

Constraints are enforced on the mechanical system via a constraint force. A typical assumption is that the constraint force does no work on the system. This excludes examples such as the end-effector sliding with friction along a constraint surface. With this assumption, the constraint force must be at all times perpendicular to the velocity of the system (strictly speaking, it must be in the annihilator of the set of admissible velocities). Since the velocity satisfies Equation (5.18), the constraint force, say Λ, will be of the form

$$\Lambda = A^T(q)\lambda \tag{5.19}$$

where $\lambda \in \mathbb{R}^m$ is a set of *Lagrange multipliers*. The constrained Euler-Lagrange equations thus take the form:

$$\frac{d}{dt}\frac{\partial L}{\partial \dot{q}} - \frac{\partial L}{\partial q} = Y + A^T(q)\lambda \tag{5.20}$$

These equations, together with the constraints (5.18), determine all the unknown quantities (either Y and λ, if the motion of the system is given, or trajectories for q and λ, if Y is given).

Several methods can be used to eliminate the Lagrange multipliers and simplify the set of equations (5.18)–(5.20). For example, let $S(q)$ be a matrix whose columns are the basis of the null-space of $A(q)$. Thus we can write

$$\dot{q} = S(q)\eta \tag{5.21}$$

for some vector $\eta \in \mathbb{R}^{n-m}$. The components of η are called *pseudo-velocities*. Now multiply (5.20) on the left with $S^T(q)$ and use Equation (5.21) to eliminate $\dot{q}$. The dynamic equations then become

$$S^T(q)M(q)S(q)\dot{\eta} + S^T(q)M(q)\frac{\partial S(q)}{\partial q}S(q)\eta + S^T(q)C(q, S(q)\eta) + S^T(q)G(q) = S^T(q)Y \tag{5.22}$$

The last equation together with Equation (5.21) then completely describes the system subject to constraints. Sometimes, such procedures are also known as *embedding of constraints* into the dynamic equations.

5.4.1 Geometric Interpretation

The constrained Euler-Lagrange equation can also be given an interesting geometric interpretation. As we did in Section 5.3.4, consider a manifold Q with the Riemannian metric M. Equation (5.18) allows the system to move only in the directions given by the null-space of $A(q)$. Geometrically, the velocity of the system at each point $q \in Q$ must lie in the subset $\mathcal{D}(q) = \emptyset A(q)$ of the tangent space $T_q Q$. Formally, $\mathcal{D} = \cup_{q \in Q}\mathcal{D}(q)$ is a *distribution* on Q. For obvious reasons, this distribution will be called the *constraint distribution*. Note that given a constraint distribution $\mathcal{D}$, Equation (5.18) simply becomes $\dot{q} \in \mathcal{D}(q)$.

Let $P : TQ \to \mathcal{D}$ denote the orthogonal (with respect to the metric M) projection onto the distribution of feasible velocities $\mathcal{D}$. In other words, at each $q \in Q$, $P(q)$ maps $T_q Q$ into $\mathcal{D}(q) \subset T_q Q$. Let $\mathcal{D}^\perp$ denote the orthogonal complement to $\mathcal{D}$ with respect to the metric M and let $P^\perp = I - P$, where I is the identity map on TQ. Geometrically, the Equation (5.18) and Equation (5.20) can be written as:

$$\nabla_{\dot{q}}\dot{q} = \Lambda(t) + F \tag{5.23}$$
$$P^\perp(\dot{q}) = 0 \tag{5.24}$$

where $\Lambda(t) \in \mathcal{D}^\perp$ is the same as that in Equation (5.19) and is the Lagrange multiplier enforcing the constraint.

It is known, e.g., see [9, 17], that Equation (5.23) can be written as:

$$\widetilde{\nabla}_{\dot{q}}\dot{q} = P(Y) \tag{5.25}$$

where $\widetilde{\nabla}$ is the affine connection given by

$$\widetilde{\nabla}_X Y = \nabla_X Y + \nabla_X(P^\perp(Y)) - P^\perp(\nabla_X Y) \tag{5.26}$$

We refer to the affine connection $\widetilde{\nabla}$ as the *constraint connection*. Typically, the connection $\widetilde{\nabla}$ is only applied to the vector fields that belong to $\mathcal{D}$. A short computation shows that for $Y \in \mathcal{D}$,

$$\widetilde{\nabla}_X Y = P(\nabla_X Y) \tag{5.27}$$

Lagrangian Dynamics

The last expression allows us to evaluate $\tilde{\nabla}_X Y$ directly and thus avoid significant amounts of computation needed to explicitly compute $\tilde{\nabla}$ from Equation (5.26). Also, by choosing a basis for $\mathcal{D}$, we can directly derive Equation (5.22) from Equation (5.25).

The important implication of this result is that even when a system is subject to holonomic or nonholonomic constraints, it is indeed possible to write the equations of motion in the form (5.17). In every case the systems evolve over the same manifold Q, but they are described with different affine connections. This observation has important implications for control of mechanical systems in general and methods developed for the systems in the form (5.17) can be quite often directly applied to systems with constraints.

5.5 Impact Equations

Quite often, robot systems undergo impacts as they move. A typical example is walking robots, but impacts also commonly occur during manipulation and assembly. In order to analyze and effectively control such systems, it is necessary to have a systematic procedure for deriving the impact equations. We describe a model for elastic and plastic impacts.

In order to derive the impact equations, recall that the Euler-Lagrange Equations (5.10) are derived from the *Lagrange-d'Alembert principle*. First, for a given C^2 curve $\gamma : [a, b] \to Q$, define a *variation* $\sigma : (-\epsilon, \epsilon) \times [a, b] \to Q$, a C^2 map with the properties:

1. $\sigma(0, t) = \gamma(t)$;
2. $\sigma(s, a) = \gamma(a), \sigma(s, b) = \gamma(b)$

Let $\delta q = \frac{d}{ds}|_{s=0} \sigma(s, t)$. Note that δq is a vector field along γ. A curve $\gamma(t)$ satisfies the Lagrange-d'Alembert Principle for the force Y and the Lagrangian $L(q, \dot{q})$ if for every variation $\sigma(s, t)$:

$$\frac{d}{ds}\bigg|_{s=0} \int_a^b L\left(\sigma(s,t), \frac{d}{dt}\sigma(s,t)\right) dt + \int_a^b \langle Y; \delta q \rangle dt = 0 \tag{5.28}$$

Now assume a robot moving on a submanifold $M_1 \subseteq Q$ before the impact, on $M_2 \subseteq Q$ after the impact, with the impact occurring on a $M_3 \subseteq Q$, where $M_3 \subseteq M_1$ and $M_3 \subseteq M_2$. Typically, we would have $M_1 = M_2$ when the system bounces off M_3 (impact with some degree of restitution) or $M_2 = M_3 \subseteq M_1$ for a plastic impact. Assume that $M_i = \varphi_i^{-1}(0)$, where $\varphi_i : Q \to \mathbb{R}^{p_i}$. In other words, we assume that 0 is a regular value of φ_i and that the submanifold M_i corresponds to the zero set of φ_i. The value p_i is the co-dimension of the submanifold M_i. Let τ be the time at which the impact occurs. A direct application of (5.28) leads to the following equation:

$$\int_a^b \left(\frac{\partial L}{\partial q} - \frac{d}{dt}\frac{\partial L}{\partial \dot{q}}\right) \delta q \, dt + \int_a^b \langle Y; \delta q \rangle \, dt + \left(\frac{\partial L}{\partial \dot{q}}\bigg|_{t=\tau^-} - \frac{\partial L}{\partial \dot{q}}\bigg|_{t=\tau^+}\right) \delta q|_{t=\tau} = 0$$

Given that δq is arbitrary, the first two terms would yield the constrained Euler-Lagrange Equations (5.20). The impact equations can be derived from the last term:

$$\left(\frac{\partial L}{\partial \dot{q}}\bigg|_{t=\tau^-} - \frac{\partial L}{\partial \dot{q}}\bigg|_{t=\tau^+}\right) \delta q|_{t=\tau} = 0 \tag{5.29}$$

Since $\gamma(s, \tau) \in M_3$, we have $d\varphi_3 \cdot \delta q|_{t=\tau} = 0$. Equation (5.29) thus implies

$$\left(\frac{\partial L}{\partial \dot{q}}\bigg|_{t=\tau^-} - \frac{\partial L}{\partial \dot{q}}\bigg|_{t=\tau^+}\right) \in \text{span } d\varphi_3 \tag{5.30}$$

The expression $\frac{\partial L}{\partial \dot{q}}$ can be recognized as the momentum. Equation (5.30) thus leads to the expected geometric interpretation: the momentum of the system during the impact can change only in the directions "perpendicular" to the surface on which the impact occurs.

Let $\varphi_3^1, \ldots, \varphi_3^{p_3}$ be the components of φ_3, and let grad φ_3^j be the gradient of φ_3^j. Observe that $\langle \frac{\partial L(q,\dot{q})}{\partial \dot{q}} ; V \rangle = \langle\!\langle \dot{q}, V \rangle\!\rangle$. Equation (5.30) can be therefore written as:

$$(\dot{q}|_{t=\tau^-} - \dot{q}|_{t=\tau^+}) = \sum_{j=1}^{p_3} \lambda_{3j} \operatorname{grad} \varphi_3^j \qquad (5.31)$$

In the case of a bounce, the equation suggests that the velocity in the direction orthogonal (with respect to the Riemannian metric M) to the impact surface undergoes the change; the amount of change depends on the coefficient of restitution. In the case of plastic impact, the velocity should be orthogonally (again, with respect to the metric M) projected to the impact surface. In other words, using appropriate geometric tools, the impact can indeed be given the interpretation that agrees with the simple setting typically taught in the introductory physics courses.

5.6 Bibliographic Remarks

The literature on robot dynamics is vast and the following list is just a small sample. A good starting point to learn about robot kinematics and dynamics is many excellent robotics textbooks [3, 10, 21, 26, 29]. The earliest formulations of the Lagrange equations of motions for robot manipulators are usually considered to be [32] and [15]. Computationally efficient procedures for numerically formulating dynamic equations of serial manipulators using Newton-Euler formalism are described in [19, 22, 30], and in the Lagrange's formalism in [14]. A comparison between the two approaches is given in [28]. These algorithms have been generalized to more complex structures in [4, 12, 18]. A good review of robot dynamics algorithms is [13].

Differential geometric aspects of modeling and control of mechanical systems are the subjects of [1, 6, 9, 20, 27].

Classical dynamics algorithms are recast using differential geometric tools in [31] and [24] for serial linkages and in [25] for linkages containing closed kinematic chains.

Control and passivity of robot manipulators and electromechanical systems are discussed in [2, 23].

Impacts are extensively studied in [8]. An interesting analysis is also presented in [16].

5.7 Conclusion

The chapter gives an overview of the Lagrange's formalism for deriving the dynamic equations for robotic systems. The emphasis is on the geometric interpretation of the classical results as such interpretation leads to a very intuitive and compact treatment of constraints and impacts. We start by providing a brief overview of kinematics of serial chains. Rigid-body transformations are described with exponential coordinates; the product of exponentials formula is the basis for deriving forward kinematics map and for the velocity analysis. The chosen formalism allows us to provide a direct mathematical interpretation to many classical notions from screw calculus. Euler-Lagrange equations for serial linkages are presented next. First, they are stated in the traditional coordinate form. The equations are then rewritten in the Riemannian setting using the concept of affine connections. This form of Euler-Lagrange equations enables us to highlight their variational nature. Next, we study systems with constraints. Again, the constrained Euler-Lagrange equations are first presented in their coordinate form. We then show how they can be rewritten using the concept of a constraint connection. The remarkable outcome of this procedure is that the equations describing systems with constraints have exactly the same form as those for unconstrained systems; what changes is the affine connection that is used. We conclude the chapter with a derivation of equations for systems that undergo impacts. Using geometric tools we show that the impact equations have a natural interpretation as a linear velocity projection on the tangent space of the impact manifold.

References

[1] Abraham, R. and Marsden, J.E., *Foundations of Mechanics*, Addison-Wesley, Reading, MA, 2nd ed. 1987.

[2] Arimoto, S., *Control Theory of Non-linear Mechanical Systems: A Passivity-Based and Circuit-Theoretic Approach*, Volume 49 of *OESS*, Oxford University Press, Oxford, 1996.

[3] Asada, H. and Slotine, J.-J.E., *Robot Analysis and Control*, John Wiley & Sons, New York, 1986.

[4] Balafoutis, C.A. and Patel, R.V., *Dynamic Analysis of Robot Manipulators: A Cartesian Tensor Approach*, Number 131 in *Kluwer International Series in Engineering and Computer Science*, Kluwer, Dordrecht, 1991.

[5] Ball, R.S., *The Theory of Screws*, Hodges & Foster, Dublin, 1876.

[6] Bloch, A.M., *Nonholonomic Mechanics and Control*, Volume 24 of *Interdisciplinary Texts in Mathematics*, Springer-Verlag, Heidelberg, 2003.

[7] Brockett, R.W., Robotic manipulators and the product of exponentials formula, In: *Mathematical Theory of Networks and Systems (MTNS)*, 120–129, Beer Sheba, Israel, 1983.

[8] Brogliato, B., *Nonsmooth Mechanics: Models, Dynamics and Control*, Springer-Verlag, New York, 2nd ed., 1999.

[9] Bullo, F. and Lewis, A.D., *Geometric Control of Mechanical Systems*, Texts in Applied Mathematics, Springer-Verlag, New York, December 2002 (to be published).

[10] Craig, J.J., *Introduction to Robotics: Mechanics and Control*, Addison-Wesley, Reading, MA, 2nd ed., 1989.

[11] do Carmo, M.P., *Riemannian Geometry*, Birkhauser, Boston, 1992.

[12] Featherstone, R., *Robot Dynamics Algorithms*, Volume 22 of *Kluwer International Series in Engineering and Computer Science*, Kluwer, Dordrecht, 1987.

[13] Featherstone, R. and Orin, D.E., Robot dynamics: equations and algorithms, In: *IEEE Conference on Robotics and Automation*, 826–834, San Francisco, CA, April 2000.

[14] Hollerbach, J.M., A recursive Lagrangian formulation of manipulator dynamics and a comparative study of dynamics formulation, *IEEE Trans. Syst., Man, & Cybernetics*, 10(11):730–736, 1980.

[15] Kahn, M.E. and Roth, B., The near-minimum-time control of open-loop articulated kinematic chains, *ASME J. Dynamic Syst., Meas., Control*, 93:164–172, September 1971.

[16] Kozlov, V.V. and Treshchev, D.V., *Billiards: A Generic Introduction to the Dynamics of Systems with Impacts*, Volume 89 of *Translations of Mathematical Monographs*, American Mathematical Society, Providence, RI, June 1991.

[17] Lewis, A.D., Simple mechanical control systems with constraints, *IEEE Trans. Automatic Control*, 45(8):1420–1436, 2000.

[18] Lilly, K.W., *Efficient Dynamic Simulation of Robotic Mechanisms*, Kluwer, Dordrecht, 1992.

[19] Luh, J.Y.S., Walker, M.W., and Paul, R.P.C., On-line computational scheme for mechanical manipulators, *ASME J. Dynamic Syst., Meas., Control*, 102(2):69–76, 1980.

[20] Marsden, J.E. and Ratiu, T.S., *Introduction to Mechanics and Symmetry*, Springer-Verlag, New York, 1994.

[21] Murray, R.M., Li, Z., and Sastry, S.S., *A Mathematical Introduction to Robotic Manipulation*, CRC Press, Boca Raton, 1994.

[22] Orin, D.E., McGhee, R.B., Vukobratović, M., and Hartoch, G., Kinematic and kinetic analysis of open-chain linkages utilizing Newton-Euler methods, *Mathe. Biosci.*, 43:107–130, 1979.

[23] Ortega, R., Loria, A., Nicklasson, P.J., and Sira-Ramirez, H., Passivity-based control of Euler-Lagrange systems: mechanical, electrical and electromechanical applications, In: *Communications and Control Engineering*, Springer-Verlag, New York, 1998.

[24] Park, F.C., Bobrow, J.E., and Ploen, S.R., A Lie group formulation of robot dynamics, *Int. J. Robotics Res.*, 14(6):609–618, 1995.

[25] Park, F.C., Choi, J., and Ploen, S.R., Symbolic formulation of closed chain dynamics in independent coordinates, *Mechanism and Machine Theory*, 34(5):731–751, 1999.

[26] Sciavicco, L. and Siciliano, B., *Modeling and Control of Robot Manipulators,* Springer-Verlag, New York, 2nd ed., 2000.
[27] Selig, J.M., *Geometrical Methods in Robotics,* Springer-Verlag, Heidelberg, 1996.
[28] Silver, W.M., On the equivalence of Lagrangian and Newton-Euler dynamics for manipulators, *Int. J. Robotics Res.,* 1(2):60–70, 1982.
[29] Spong, M.W. and Vidyasagar, M., *Robot Dynamics and Control,* John Wiley & Sons, New York, 1989.
[30] Stepanenko, Y. and Vukobratović, M., Dynamics of articulated open-chain active mechanisms, *Math. Biosci.,* 28:137–170, 1976.
[31] Stokes, A. and Brockett, R.W., Dynamics of kinematic chains, *Int. J. Robotics Res.,* 15(4):393–405, 1996.
[32] Uicker, J.J., Dynamic force analysis of spatial linkages, *ASME J. Appl. Mech.,* 34:418–424, 1967.

6
Kane's Method in Robotics

	6.1	Introduction	**6**-1
	6.2	The Essence of Kane's Method	**6**-3
	6.3	Two DOF Planar Robot with Two Revolute Joints	**6**-4
		Preliminaries • Generalized Coordinates and Speeds • Velocities • Partial Velocities • Accelerations • Generalized Inertia Forces • Generalized Active Forces • Equations of Motion • Additional Considerations	
	6.4	Two-DOF Planar Robot with One Revolute Joint and One Prismatic Joint	**6**-8
		Preliminaries • Generalized Coordinates and Speeds • Velocities • Partial Velocities • Accelerations • Generalized Inertia Forces • Generalized Active Forces • Equations of Motion	
	6.5	Special Issues in Kane's Method................	**6**-13
		Linearized Equations • Systems Subject to Constraints • Continuous Systems	
	6.6	Kane's Equations in the Robotics Literature	**6**-22
	6.7	Commercially Available Software Packages Related to Kane's Method	**6**-25
Keith W. Buffinton		SD/FAST • Pro/ENGINEER Simulation (formerly Pro/MECHANICA) • AUTOLEV • ADAMS • Working Model	
Bucknell University	6.8	Summary	**6**-29

6.1 Introduction

In 1961, Professor Thomas Kane of Stanford University published a paper entitled "Dynamics of Nonholonomic Systems" [1] that described a method for formulating equations of motion for complex dynamical systems that was equally applicable to either holonomic or nonholonomic systems. As opposed to the use of Lagrange's equations, this new method allowed dynamical equations to be generated without the differentiation of kinetic and potential energy functions and, for nonholonomic systems, without the need to introduce Lagrange multipliers. This method was later complemented by the publication in 1965 of a paper by Kane and Wang entitled "On the Derivation of Equations of Motion" [2] that described the use of motion variables (later called generalized speeds) that could be any combinations of the time derivatives of the generalized coordinates that describe the configuration of a system. These two papers laid the foundation for what has since become known as Kane's method for formulating dynamical equations.

The origins of Kane's method can be found in Kane's undergraduate dynamics texts entitled *Analytical Elements of Mechanics* volumes 1 and 2 [3, 4] published in 1959 and 1961, respectively. In particular, in Section 4.5.6 of [4], Kane states a "law of motion" containing a term referred to as the *activity in R* (a reference frame) *of the gravitational and contact forces on P* (a particular particle of interest). Kane's focus on the *activity* of a set of forces was a significant step in the development of his more general dynamical method, as is elaborated in Section 6.2. Also important to Kane's approach to formulating dynamical equations was his desire to avoid what he viewed as the vagaries of the Principle of Virtual Work, particularly when applied to the analysis of systems undergoing three-dimensional rotational motions. Kane's response to the need to clarify the process of formulating equations of motion using the Principle of Virtual Work was one of the key factors that led to the development of his own approach to the generation of dynamical equations.

Although the application of Kane's method has clear advantages over other methods of formulating dynamical equations [5], the importance of Kane's method only became widely recognized as the space industry of the 1960s and 1970s drove the need to model and simulate ever more complex dynamical systems and as the capabilities of digital computers increased geometrically while computational costs concomitantantly decreased. In the 1980s and early 1990s, a number of algorithms were developed for the dynamic analysis of multibody systems (references [6–9] provide comprehensive overviews of various forms of these dynamical methods), based on variations of the dynamical principles developed by Newton, Euler, Lagrange, and Kane. During this same time, a number of algorithms lead to commercially successful computer programs [such as ADAMS (Automatic Dynamic Analysis of Mechanisms) [10], DADS (Dynamic Analysis and Design of Systems) [11], NEWEUL [12], SD/FAST [13], AUTOLEV [14], Pro/MECHANICA MOTION, and Working Model [15], to name just a few], many of which are still on the market today. As elaborated in Section 6.7, many of the most successful of these programs were either directly or indirectly influenced by Kane and his approach to dynamics.

The widespread attention given to efficient dynamical methods and the development of commercially successful multibody dynamics programs set the stage for the application of Kane's method to complex robotic mechanisms. Since the early 1980s, numerous papers have been written on the use of Kane's method in analyzing the dynamics of various robots and robotic devices (see Section 6.6 for brief summaries of selected articles). These robots have incorporated revolute joints, prismatic joints, closed-loops, flexible links, transmission mechanisms, gear backlash, joint clearance, nonholonomic constraints, and other characteristics of mechanical devices that have important dynamical consequences. As evidenced by the range of articles described in Section 6.6, Kane's method is often the method of choice when analyzing robots with various forms and functions.

The broad goal of this chapter is to provide an introduction to the application of Kane's method to robots and robotic devices. It is essentially tutorial while also providing a limited survey of articles that address robot analysis using Kane's method as well as descriptions of multipurpose dynamical analysis software packages that are either directly or indirectly related to Kane's approach to dynamics. Although a brief description of the fundamental basis for Kane's method and its relationship to Lagrange's equations is given in Section 6.2, the purpose of this chapter is not to enter into a prolonged discussion of the relationship between Kane's method and other similar dynamical methods, such as the "orthogonal complement method" (the interested reader is referred to references [16,17] for detailed commentary) or Jourdain's principle (see "Kane's equations or Jourdain's principle?" by Piedboeuf [18] for further information and a discussion of Jourdain's original 1909 work entitled "Note on an analogue of Gauss' principle of least constraint" in which he established the principle of virtual power) or Gibbs-Appell equations [interested readers are referred to a lively debate on the subject that appeared in volumes 10 (numbers 1 and 6), 12(1), and 13(2) of the *Journal of Guidance, Control, and Dynamics* from 1987 to 1990]. The majority of this chapter (Section 6.3, Section 6.4, and Section 6.5) is in fact devoted to providing a tutorial illustration of the application of Kane's method to the dynamic analysis of two relatively simple robots: a two-degree-of-freedom planar robot with two revolute joints and a two-degree-of-freedom planar robot with one revolute joint and one prismatic joint. Extentions and modifications of these analyses that are facilitated by the use of Kane's method are also discussed as are special issues in the use of Kane's method, such as formulating linearized equations, generating equations of motion for systems subject to constraints, and developing equations of motion for systems with continuous elastic elements, leading to a detailed analysis of the two-degree-of-freedom planar robot with one revolute joint and one prismatic joint when the element traversing the prismatic joint is regarded as elastic.

Following these tutorial sections, a brief summary of the range of applications of Kane's method in robotics is presented. Although not meant to be an exhaustive list of publications involving the use of Kane's method in robotics, an indication of the popularity and widespread use of Kane's method in robotics is provided. The evolution of modern commercially available dynamical analysis computer software is also briefly described as is the relationship that various programs have to either Kane's method or more general work that Kane has contributed to the dynamics and robotics literature. Through this chapter, it is hoped that readers previously unfamiliar with Kane's method will gain at least a flavor of its application. Readers already acquainted with Kane's method will hopefully gain new insights into the method as well as have the opportunity to recognize the large number of robotic problems in which Kane's method can be used.

6.2 The Essence of Kane's Method

Kane's contributions to dynamics have been not only to the development of "Kane's method" and "Kane's equations" but also to the clarity with which one can deal with basic kinematical principles (including the explicit and careful accounting for the reference frames in which kinematical and dynamical relationships are developed), the definition of basic kinematical, and dynamical quantities (see, for example, Kane's paper entitled "Teaching of Mechanics to Undergraduates" [19]), the careful deductive way in which he derives all equations from basic principles, and the algorithmic approach he prescribes for the development of dynamical equations of motion for complex systems. These are in addition to the fundamental elements inherent in Kane's method, which allow for a clear and convenient separation of kinematical and dynamical considerations, the exclusion of nonworking forces, the use of generalized speeds to describe motion, the systematic way in which constraints can be incorporated into an analysis, and the ease and confidence with which linearized of equations of motion can be developed.

Before considering examples of the use of Kane's method in robotics, a simple consideration of the essential basis for the method may be illuminating. Those who have read and studied *DYNAMICS: Theory and Applications* [20] will recognize that the details of Kane's approach to dynamics and to Kane's method can obscure the fundamental concepts on which Kane's method is based. In Section 5.8 of the first edition of *DYNAMICS* [21] (note that an equivalent section is not contained in *DYNAMICS: Theory and Applications*), a brief discussion is given of the basis for "Lagrange's form of D'Alembert's principle" [Equation (6.1) in [21] and Equation (6.1) of Chapter 6 in [20] where it is referred to as *Kane's dynamical equations*]. This section of *DYNAMICS* offers comments that are meant to "shed light" on Kane's equations "by reference to analogies between these equations and other, perhaps more familiar, relationships." Section 5.8 of *DYNAMICS* is entitled "The Activity and Activity-Energy Principles" and considers the development of equations of motion for a single particle. While the analysis of a single particle does not give full insight into the advantages (and potential disadvantages) inherent in the use of Kane's method, it does provide at least a starting point for further discussion and for understanding the origins of Kane's method.

For a single particle P for which $\mathbf{F}$ is the resultant of all contact and body forces acting on P and for which $\mathbf{F}^*$ is the inertia force for P in an inertial reference frame R (note that for a single particle, $\mathbf{F}^*$ is simply equal to $-m\mathbf{a}$, where m is the mass of P and $\mathbf{a}$ is the acceleration of P in R), D'Alembert's principle states that

$$\mathbf{F} + \mathbf{F}^* = 0 \tag{6.1}$$

When this equation is dot-multiplied with the velocity $\mathbf{v}$ of P in R, one obtains

$$\mathbf{v} \cdot \mathbf{F} + \mathbf{v} \cdot \mathbf{F}^* = 0 \tag{6.2}$$

Kane goes on in Section 5.8 of *DYNAMICS* to define two scalar quantities A and A^* such that

$$A = \mathbf{v} \cdot \mathbf{F} \tag{6.3}$$

$$A^* = \mathbf{v} \cdot \mathbf{F}^* \tag{6.4}$$

and then presents

$$A + A^* = 0 \tag{6.5}$$

as a statement of the *activity principle* for a single particle P for which A and A^* are called the *activity* of force $\mathbf{F}$ and the *inertia activity* of the inertia force $\mathbf{F}^*$, respectively (note that Kane refers to A^* as the *activity* of the force $\mathbf{F}^*$; here A^* is referred to the *inertia activity* to distinguish it from the *activity A*).

Kane points out that Equation (6.5) is a scalar equation, and thus it cannot "furnish sufficient information for the solution in which P has more than one degree of freedom." He continues by noting that Equation (6.5) is weaker than Equation (6.1), which is equivalent to three scalar equations. Equation (6.5) does, however, possess one advantage over Equation (6.1). If $\mathbf{F}$ contains contributions from (unknown) constraint forces, these forces will appear in Equation (6.1) and then need to be eliminated from the final dynamical equation(s) of motion; whereas, in cases in which the components of $\mathbf{F}$ corresponding to constraint directions are ultimately not of interest, they are automatically eliminated from Equation (6.5) by the dot multiplication needed to produce A and A^* as given in Equation (6.3) and Equation (6.4).

The essence of Kane's method is thus to arrive at a procedure for formulating dynamical equations of motion that, on the one hand, contain sufficient information for the solution of problems in which P has more than one degree of freedom, and on the other hand, automatically eliminate unknown constraint forces. To that end, Kane noted that one may replace Equation (6.2) with

$$\mathbf{v}_r \cdot \mathbf{F} + \mathbf{v}_r \cdot \mathbf{F}^* = 0 \tag{6.6}$$

where $\mathbf{v}_r (r = 1, \ldots, n)$ are the *partial velocities* (see Section 6.3 for a definition of partial velocities) of P in R and n is the number of degrees of freedom of P in R (note that the $\mathbf{v}_r$ form a set of independent quantities). Furthermore, if F_r and F_r^* are defined as

$$F_r = \mathbf{v}_r \cdot \mathbf{F}, \qquad F_r^* = \mathbf{v}_r \cdot \mathbf{F}^* \tag{6.7}$$

one can then write

$$F_r + F_r^* = 0 \quad (r = 1, \ldots, n) \tag{6.8}$$

where F_r and F_r^* are referred to as the rth *generalized active force* and the rth *generalized inertia force* for P in R. Although referred to in Kane's earlier works, including [21], as *Lagrange's form of D'Alembert's principle*, Equation (6.9) has in recent years come to be known as *Kane's equations*.

Using the expression for the generalized inertia force given in Equation (6.8) as a point of departure, the relationship between Kane's equations and Lagrange's equations can also be investigated. From Equation (6.8),

$$F_r^* = \mathbf{v}_r \cdot \mathbf{F}^* = \mathbf{v}_r \cdot (-m\mathbf{a}) = -m\mathbf{v}_r \cdot \frac{d\mathbf{v}}{dt} = -\frac{m}{2}\left(\frac{d}{dt}\frac{\partial \mathbf{v}^2}{\partial \dot{q}_r} - \frac{\partial \mathbf{v}^2}{\partial q_r}\right) \quad \text{[20, p 50]} \tag{6.9}$$

$$= -\frac{d}{dt}\frac{\partial}{\partial \dot{q}_r}\left(\frac{m\mathbf{v}^2}{2}\right) + \frac{\partial}{\partial q_r}\left(\frac{m\mathbf{v}^2}{2}\right) = -\frac{d}{dt}\frac{\partial K}{\partial \dot{q}_r}\frac{\partial K}{\partial q_r} \tag{6.10}$$

where K is the kinetic energy of P in R. Substituting Equation (6.10) into Equation (6.8) gives

$$\frac{d}{dt}\frac{\partial K}{\partial \dot{q}_r} + \frac{\partial K}{\partial q_r} = F_r \quad (r = 1, \ldots, n) \tag{6.11}$$

which can be recognized as *Lagrange's equations of motion of the first kind*.

6.3 Two DOF Planar Robot with Two Revolute Joints

In order to provide brief tutorials on the use of Kane's method in deriving equations of motion and to illustrate the steps that make up the application of Kane's method, in this and the following section, the dynamical equations of motion for two simple robotic systems are developed. The first system is a

Kane's Method in Robotics

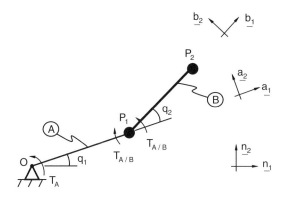

FIGURE 6.1 Two DOF planar robot with two revolute joints.

two-degree-of-freedom robot with two revolute joints moving in a vertical plane. The second is a two-degree-of-freedom robot with one revolute and one prismatic joint moving in a horizontal plane. Both of these robots have been chosen so as to be simple enough to give a clear illustration of the details of Kane's method without obscuring key points with excessive complexity.

As mentioned at the beginning of Section 6.2, Kane's method is very algorithmic and as a result is easily broken down into discrete general steps. These steps are listed below:

- Definition of preliminary information
- Introduction of generalized coordinates and speeds
- Development of requisite velocities and angular velocities
- Determination of partial velocities and partial angular velocities
- Development of requisite accelerations and angular accelerations
- Formulation of generalized inertia forces
- Formulation of generalized active forces
- Formulation of dynamical equations of motion by means of Kane's equations

These steps will now be applied to the system shown in Figure 6.1. This system represents a very simple model of a two-degree-of-freedom robot moving in a vertical plane. To simplify the system as much as possible, the mass of each of the links of the robot has been modeled as being lumped into a single particle.

6.3.1 Preliminaries

The first step in formulating equations of motion for any system is to introduce symbols for bodies, points, constants, variables, unit vectors, and generalized coordinates. The robot of Figure 6.1 consists of two massless rigid link A and B, whose motions are confined to parallel vertical planes, and two particles P_1 and P_2, each modeled as being of mass m. Particle P_1 is located at the distal end of body A and P_2 is located at the distal end of body B. Body A rotates about a fixed horizontal axis through point O, while body B rotates about a horizonal axis fixed in A and passing through P_1. A constant needed in the description of the robot is L, which represents the lengths of both links A and B. Variables for the system are the torque T_A, applied to link A by an inertially fixed actuator, and the torque $T_{A/B}$, applied to link B by an actuator attached to A. Unit vectors needed for the description of the motion of the system are $\mathbf{n}_i$, $\mathbf{a}_i$, and $\mathbf{b}_i$ ($i = 1, 2, 3$). Unit vectors $\mathbf{n}_1$ and $\mathbf{n}_2$ are fixed in an inertial reference frame N, $\mathbf{a}_1$, and $\mathbf{a}_2$ are fixed in A, and $\mathbf{b}_1$ and $\mathbf{b}_2$ are fixed in B, as shown. The third vector of each triad is perpendicular to the plane formed by the other two such that each triad forms a right-handed set.

6.3.2 Generalized Coordinates and Speeds

While even for simple systems there is an infinite number of possible choices for the generalized coordinates that describe a system's configuration, generalized coordinates are usually selected based on physical relevance and analytical convenience. Generalized coordinates for the system of Figure 6.1 that are both relevant and convenient are the angles q_1 and q_2. The quantity q_1 measures the angle between an inertially fixed horizontal line and a line fixed in A, and q_2 measures the angle between a line fixed in A and a line fixed in B, both as shown.

Within the context of Kane's method, a complete specification of the kinematics of a system requires the introduction of quantities known as *generalized speeds*. Generalized speeds are defined as any (invertible) linear combination of the time derivatives of the generalized coordinates and describe the *motion* of a system in a way analogous to the way that generalized coordinates describe the *configuration* of a system. While, as for the generalized coordinates, there is an infinite number of possibilities for the generalized speeds describing the motion of a system, for the system at hand, reasonable generalized speeds (that will ultimately lead to equations of motion based on a "joint space" description of the robot) are defined as

$$u_1 = \dot{q}_1 \tag{6.12}$$

$$u_2 = \dot{q}_2 \tag{6.13}$$

An alternate, and equally acceptable, choice for generalized speeds could have been the $\mathbf{n}_1$ and $\mathbf{n}_2$ components of the velocity of P_2 (this choice would lead to equations of motion in "operational space"). For a comprehensive discussion of guidelines for the selection of generalized speeds that lead to "exceptionally efficient" dynamical equations for a large class of systems frequently encountered in robotics, see [22].

6.3.3 Velocities

The angular and translational velocities required for the development of the equations of motion for the robot of Figure 6.1 are the angular velocities of bodies A and B as measured in reference frame N and the translational velocities of particles P_1 and P_2 in N. With the choice of generalized speeds given above, expressions for the angular velocities are

$$\boldsymbol{\omega}^A = u_1 \mathbf{a}_3 \tag{6.14}$$

$$\boldsymbol{\omega}^B = (u_1 + u_2)\mathbf{b}_3 \tag{6.15}$$

Expressions for the translational velocities can be developed either directly from Figure 6.1 or from a straightforward application of the kinematical formula for relating the velocities of two points fixed on a single rigid body [20, p. 30]. From inspection of Figure 6.1, the velocities of P_1 and P_2 are

$$\mathbf{v}^{P_1} = L u_1 \mathbf{a}_2 \tag{6.16}$$

$$\mathbf{v}^{P_2} = L u_1 \mathbf{a}_2 + L(u_1 + u_2)\mathbf{b}_2 \tag{6.17}$$

6.3.4 Partial Velocities

With all the requisite velocity expressions in hand, Kane's method requires the identification of *partial velocities*. Partial velocities must be identified from the angular velocities of all nonmassless bodies and of bodies acted upon by torques that ultimately contribute to the equations of motion (i.e., bodies acted upon by nonworking torques and nonworking sets of torques need not be considered) as well as from the translational velocities of all nonmassless particles and of points acted upon by forces that ultimately contribute to the equations of motion. These partial velocities are easily identified and are simply the coefficients of the generalized speeds in expressions for the angular and translational velocities. For the system of Figure 6.1, the partial velocities are determined by inspection from Equation (6.14) through (6.17). The resulting partial velocities are listed in Table 6.1, where $\boldsymbol{\omega}_r^A$ is the rth partial angular velocity of A in N, $\mathbf{v}_r^{P_1}$ is the rth partial translational velocity of P_1 in N, etc.

TABLE 6.1 Partial Velocities for Robot of Figure 6.1

	$r = 1$	$r = 2$
ω_r^A	$\mathbf{a}_3$	0
ω_r^B	$\mathbf{b}_3$	$\mathbf{b}_3$
$\mathbf{v}_r^{P_1}$	$L\mathbf{a}_2$	0
$\mathbf{v}_r^{P_2}$	$L(\mathbf{a}_2 + \mathbf{b}_2)$	$L\mathbf{b}_2$

6.3.5 Accelerations

In order to complete the development of kinematical quantities governing the motion of the robot, one must develop expressions for the translational accelerations of particles P_1 and P_2 in N. The translational accelerations can be determined either by direct differentiation or by applying the basic kinematical formula for relating the accelerations of two points fixed on the same rigid body [20, p. 30]. For the problem at hand, the latter approach is more convenient and yields

$$\mathbf{a}^{P_1} = -Lu_1^2\mathbf{a}_1 + L\dot{u}_1\mathbf{a}_2 \qquad (6.18)$$

$$\mathbf{a}^{P_2} = -Lu_1^2\mathbf{a}_1 + L\dot{u}_1\mathbf{a}_2 - L(u_1 + u_2)^2\mathbf{b}_1 + L(\dot{u}_1 + \dot{u}_2)\mathbf{b}_2 \qquad (6.19)$$

where $\mathbf{a}^{P_1}$ and $\mathbf{a}^{P_2}$ are the translational accelerations of P_1 and P_2 in N.

6.3.6 Generalized Inertia Forces

In general, the generalized inertia forces F_r^* $(i = 1, \ldots, n)$, for a system of particles with n degrees of freedom in a reference frame N can be expressed as

$$F_r^* = -\sum_{i=1}^{n} \mathbf{v}_r^{P_i} \cdot m_i \mathbf{a}^{P_i} \qquad (r = 1, \ldots, n) \qquad (6.20)$$

where $\mathbf{v}_r^{P_i}$ is the rth partial velocity of particle P_i in N, $\mathbf{a}^{P_i}$ is the acceleration of P_i in N, and m_i is the mass of particle P_i. For the robot of Figure 6.1, by making use of the partial velocities in Table 6.1 and the accelerations in Equation (6.18) and Equation (6.19), the general expression of Equation (6.20) yields

$$F_1^* = -mL^2\left\{(3 + 2c_2)\dot{u}_1 + (1 + c_2)\dot{u}_2 + s_2\left[u_1^2 - (u_1 + u_2)^2\right]\right\} \qquad (6.21)$$

$$F_2^* = -mL^2\left[(1 + c_2)\dot{u}_1 + \dot{u}_2 + s_2 u_2^2\right] \qquad (6.22)$$

where s_2 and c_2 are defined as the sine and cosine of q_2, respectively.

6.3.7 Generalized Active Forces

Nonworking forces (or torques), or sets of forces (or sets of torques), make no net contribution to the generalized active forces. To determine the generalized active forces for the robot of Figure 6.1, therefore, one need only consider the gravitational force acting on each of the two particles and the two torques T_A and $T_{A/B}$. The fact that only these forces and torques contribute to the generalized active forces highlights one of the principal advantages of Kane's method. Forces such as contact forces exerted on parts of a system across smooth surfaces of rigid bodies, contact and body forces exerted by parts of a rigid body on one another, and forces that are exerted between two bodies that are in rolling contact make no net contribution to generalized active forces. Indeed, the fact that all such forces do not contribute to the generalized active forces is one of the principal motivations for introducing the concept of generalized active forces.

In general, generalized active forces are constructed by dot multiplying all contributing forces and torques with the partial translational velocities and partial angular velocities of the points and bodies to which they are applied. For the system at hand, therefore, one can write the generalized active forces F_r as

$$F_r = \mathbf{v}_r^{P_1} \cdot (-mg\mathbf{n}_2) + \mathbf{v}_r^{P_2} \cdot (-mg\mathbf{n}_2) + \boldsymbol{\omega}_r^A \cdot (T_A \mathbf{a}_3 - T_{A/B}\mathbf{b}_3) + \boldsymbol{\omega}_r^B \cdot T_{A/B}\mathbf{b}_3 \quad (r = 1, 2) \quad (6.23)$$

Substituting from Table 6.1 into Equation (6.23) produces

$$F_1 = T_A + mgL(2s_2 + s_{23}) \quad (6.24)$$
$$F_2 = T_{A/B} + mgLs_{23} \quad (6.25)$$

where s_{23} is the sine of $q_2 + q_3$.

6.3.8 Equations of Motion

Finally, now that all generalized active and inertia forces have been determined, the equations of motion for the robot can be formed by substituting from Equation (6.21), Equation (6.22), Equation (6.24), and Equation (6.25) into Kane's equations:

$$F_r + F_r^* = 0 \quad (r = 1, 2) \quad (6.26)$$

Equation (6.26) provides a complete description of the dynamics of the simple robotic system of Figure 6.1.

6.3.9 Additional Considerations

Although two of the primary advantages of Kane's method are the ability to introduce motion variables (generalized speeds) as freely as configuration variables (generalized coordinates) and the elimination of nonworking forces, Kane's method also facilitates modifications to a system once equations of motion have already been formulated. For example, to consider the consequence to the equations of motion of applying an external force $F_x^E \mathbf{n}_1 + F_y^E \mathbf{n}_2$ to the distal end of the robot (at the location of P_2), one simply determines additional contributions F_r^{Ext} to the generalized active forces given by

$$F_r^{Ext} = \mathbf{v}_r^{P_2} \cdot \left(F_x^E \mathbf{n}_1 + F_y^E \mathbf{n}_2 \right) \quad (r = 1, 2) \quad (6.27)$$

and adds these contributions to the equations of motion in Equation (6.26). One could similarly consider the effect of viscous damping torques at the joints by adding contributions to the generalized active forces given by

$$F_r^{Damp} = \boldsymbol{\omega}_r^A \cdot [-b_{t1}u_1\mathbf{a}_3 + b_{t2}(u_1 + u_2)\mathbf{b}_3] + \boldsymbol{\omega}_r^B \cdot [-b_{t2}(u_1 + u_2)\mathbf{b}_3] \quad (r = 1, 2) \quad (6.28)$$

where b_{t1} and b_{t2} are viscous damping coefficients at the first and second joints.

Another consideration that often arises in robotics is the relationship between formulations of equations of motion in joint space and operational space. As mentioned at the point at which generalized speeds were defined, the above derivation could easily have produced equations corresponding to operational space simply by defining the generalized speeds to be the $\mathbf{n}_1$ and $\mathbf{n}_2$ components of the velocity of P_2 and then using this other set of generalized speeds to define partial velocities analogous to those appearing in Table 6.1. A systematic approach to directly converting between the joint space equations in Equation (6.26) and corresponding operational space equations is described in Section 6.5.2.

6.4 Two-DOF Planar Robot with One Revolute Joint and One Prismatic Joint

The steps outlined at the beginning of the previous section will now be applied to the system shown in Figure 6.2. This system represents a simple model of a two-degree-of-freedom robot with one revolute and one prismatic joint moving in a horizontal plane. While still relatively simple, this system is significantly more complicated than the one analyzed in the preceding section and gives a fuller understanding of issues

Kane's Method in Robotics

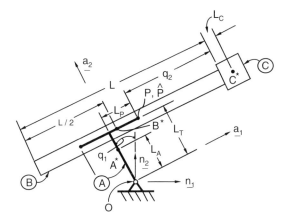

FIGURE 6.2 Two DOF robot with one revolute joint and one prismatic joint.

related to the application of Kane's method. Moreover, as is discussed in greater detail in Section 6.5.3, this robot represents a restricted model of the Stanford Arm [23] and thus provides insights into the formulation of equations of motion for Stanford-like manipulators.

6.4.1 Preliminaries

As in the preceding section, the first step in formulating equations of motion for the system now under consideration is to introduce symbols for bodies, points, constants, variables, unit vectors, and generalized coordinates. The robot of Figure 6.2 consists of three rigid bodies A, B, and C, whose motions are confined to parallel horizontal planes. Body A rotates about a fixed vertical axis while body B translates along an axis fixed in A. Body C is rigidly attached to one end of B, as shown. The masses of A, B, and C are denoted m_A, m_B, and m_C, respectively, and the moments of inertia of the three bodies about vertical lines passing through their mass centers are denoted I_A, I_B, and I_C, respectively. Points of interest are O, P, $\hat{P}$, $\overline{A}^{B^*}$, and the mass centers of bodies A, B, and C. Point O lies on the axis about which body A rotates, P is a point of body A at that location where the force that drives the motion of B relative to A is applied, $\hat{P}$ is a point of body B whose location instantaneously coincides with P, and $\overline{A}^{B^*}$ is a point regarded as fixed on body A but whose location is instantaneously coincident with the mass center of B. The mass centers of bodies A, B, and C are denoted A^*, B^*, and C^* and have locations as indicated in Figure 6.2. Constants needed in the description of the robot are L_A, L_T, L_P, L, and L_C. These constants all denote lengths as indicated in the figure. Variables in the analysis are a torque T_A applied to body A by an actuator fixed in an inertial reference frame N and a force $F_{A/B}$ applied to B at P by an actuator fixed in A. Unit vectors needed for the description of the motion of the system are $\mathbf{n}_i$ and $\mathbf{a}_i$ ($i = 1, 2, 3$). Unit vectors $\mathbf{n}_1$ and $\mathbf{n}_2$ are fixed in N, and unit vectors $\mathbf{a}_1$ and $\mathbf{a}_2$ are fixed in body A, as shown. The third vector of each triad is perpendicular to the plane formed by the other two such that each triad forms a right-handed set.

6.4.2 Generalized Coordinates and Speeds

A reasonable choice for generalized coordinates for this robot are q_1 and q_2, as shown in Figure 6.2. The coordinate q_1 measures the angle between an inertially fixed line and a line fixed in body A, while q_2 measures the distance between point P and the end of B attached to C. For the problem at hand, generalized speeds that are physically relevant and analytically convenient can be defined as

$$u_1 = \boldsymbol{\omega}^A \cdot \mathbf{a}_3 \tag{6.29}$$

$$u_2 = {}^A\mathbf{v}^{C^*} \cdot \mathbf{a}_1 \tag{6.30}$$

where ω^A is the angular velocity of A in reference frame N and $^A\mathbf{v}^{C^*}$ is the velocity of C^* as measured in body A.

6.4.3 Velocities

The angular and translational velocities required for the development of equations of motion for the robot are the angular velocities of A, B, and C in N, and the velocities of points A^*, P, $\widehat{P}$, B^*, and C^* in N. With the choice of generalized speeds given above, expressions for the angular velocities are

$$\omega^A = u_1 \mathbf{a}_3 \tag{6.31}$$
$$\omega^B = u_1 \mathbf{a}_3 \tag{6.32}$$
$$\omega^C = u_1 \mathbf{a}_3 \tag{6.33}$$

An expression for the translational velocity of A^* in N can be developed directly by inspection of Figure 6.2, which yields

$$\mathbf{v}^{A^*} = -L_A u_1 \mathbf{a}_1 \tag{6.34}$$

The velocity of point P is determined by making use of the facts that the velocity of point O, fixed on the axis of A, which is zero and the formula for the velocities of two points fixed on a rigid body. Specifically,

$$\mathbf{v}^P = \mathbf{v}^O + \omega^A \times \mathbf{p}^{OP} \tag{6.35}$$

where $\mathbf{p}^{OP}$ is the position vector from O to P given by

$$\mathbf{p}^{OP} = L_P \mathbf{a}_1 + L_T \mathbf{a}_2 \tag{6.36}$$

Evaluating Equation (6.35) with the aid of Equation (6.31) and Equation (6.36) yields

$$\mathbf{v}^P = -L_T u_1 \mathbf{a}_1 + L_P u_1 \mathbf{a}_2 \tag{6.37}$$

The velocity of $\widehat{P}$ is determined from the formula for the velocity of a single point moving on a rigid body [20, p. 32]. For $\widehat{P}$, this formula is expressed as

$$\mathbf{v}^{\widehat{P}} = \mathbf{v}^{\overline{A}^{\widehat{P}}} + {^A}\mathbf{v}^{\widehat{P}} \tag{6.38}$$

where $\mathbf{v}^{\overline{A}^{\widehat{P}}}$ is the velocity of that point of body A whose location is instantaneously coincident with $\widehat{P}$, and $^A\mathbf{v}^{\widehat{P}}$ is the velocity of $\widehat{P}$ in A. The velocity of $\mathbf{v}^{\overline{A}^{\widehat{P}}}$, for the problem at hand, is simply equal to $\mathbf{v}^P$. The velocity of $\widehat{P}$ in A is determined with reference to Figure 6.2, as well as the definition of u_2 given in Equation (6.30), and is

$$^A\mathbf{v}^{\widehat{P}} = u_2 \mathbf{a}_1 \tag{6.39}$$

Substituting from Equation (6.37) and Equation (6.39) into Equation (6.38), therefore, produces

$$\mathbf{v}^{\widehat{P}} = (-L_T u_1 + u_2)\mathbf{a}_1 - L_P u_1 \mathbf{a}_2 \tag{6.40}$$

The velocity of B^* is determined in a manner similar to that used to find the velocity of $\widehat{P}$, which produces

$$\mathbf{v}^{B^*} = (-L_T u_1 + u_2)\mathbf{a}_1 + (L_P + q_2 - L/2)u_1 \mathbf{a}_2 \tag{6.41}$$

Since C^* is rigidly connected to B, the velocity of C^* can be related to the velocity of B^* by making use of the formula for two points fixed on a rigid body, which yields

$$\mathbf{v}^{C^*} = (-L_T u_1 + u_2)\mathbf{a}_1 + (L_P + q_2 + L_C)u_1 \mathbf{a}_2 \tag{6.42}$$

Kane's Method in Robotics

TABLE 6.2 Partial Velocities for Robot of Figure 6.2

	$r = 1$	$r = 2$
ω_r^A	$\mathbf{a}_3$	0
ω_r^B	$\mathbf{a}_3$	0
ω_r^C	$\mathbf{a}_3$	0
$\mathbf{v}_r^{A^*}$	$-L_A \mathbf{a}_1$	0
$\mathbf{v}_r^P$	$-L_T \mathbf{a}_1 + L_P \mathbf{a}_2$	0
$\mathbf{v}_r^{\hat{P}}$	$-L_T \mathbf{a}_1 + L_P \mathbf{a}_2$	$\mathbf{a}_1$
$\mathbf{v}_r^{B^*}$	$-L_T \mathbf{a}_1 + (L_P + q_2 - L/2)\mathbf{a}_2$	$\mathbf{a}_1$
$\mathbf{v}_r^{C^*}$	$-L_T \mathbf{a}_1 + (L_P + q_2 + L_C)\mathbf{a}_2$	$\mathbf{a}_1$

6.4.4 Partial Velocities

As explained in the previous section, partial velocities are simply the coefficients of the generalized speeds in the expressions for the angular and translational velocities and here are determined by inspection from Equations (6.31) through (6.34), (6.37), and (6.40) through (6.42). The resulting partial velocities are listed in Table 6.2, where ω_r^A is the rth partial angular velocity of A in N, $\mathbf{v}_r^{A^*}$ is the rth partial linear velocity of A^* in N, etc.

6.4.5 Accelerations

In order to complete the development of requisite kinematical quantities governing the motion of the robot, one must develop expressions for the angular accelerations of bodies A, B, and C in N as well as for the translational accelerations of A^*, B^*, and C^* in N. The angular accelerations can be determined by differentiating Equations (6.31) through (6.33) in N. Since the unit vector $\mathbf{a}_3$ is fixed in N, this is straightforward and produces

$$\boldsymbol{\alpha}^A = \dot{u}_1 \mathbf{a}_3 \qquad (6.43)$$

$$\boldsymbol{\alpha}^B = \dot{u}_1 \mathbf{a}_3 \qquad (6.44)$$

$$\boldsymbol{\alpha}^C = \dot{u}_1 \mathbf{a}_3 \qquad (6.45)$$

where $\boldsymbol{\alpha}^A$, $\boldsymbol{\alpha}^B$, and $\boldsymbol{\alpha}^C$ are the angular acceleration of A, B, and C in N.

The translational accelerations can be determined by direct differentiation of Equation (6.34), Equation (6.41), and Equation (6.42). The acceleration of A^* in N is obtained from

$$\mathbf{a}^{A^*} = \frac{{}^N d\mathbf{v}^{A^*}}{dt} = \frac{{}^A d\mathbf{v}^{A^*}}{dt} + \boldsymbol{\omega}^A \times \mathbf{v}^{A^*} \quad [20, \text{ p. 23}] \qquad (6.46)$$

where $\frac{{}^N d\mathbf{v}^{A^*}}{dt}$ and $\frac{{}^A d\mathbf{v}^{A^*}}{dt}$ are the derivatives of $\mathbf{v}^{A^*}$ in reference frames A and N, respectively. This equation takes advantage of the fact the velocity of A^* is written in terms of unit vectors fixed in A and expresses the derivative of $\mathbf{v}^{A^*}$ in N in terms of its derivative in A plus terms that account for the rotation of A relative to N. Evaluation of Equation (6.46) produces

$$\mathbf{a}^{A^*} = -L_A \dot{u}_1 \mathbf{a}_1 - L_A u_1^2 \mathbf{a}_2 \qquad (6.47)$$

The accelerations of B^* and C^* can be obtained in a similar manner and are

$$\mathbf{a}^{B^*} = \left[-L_T \dot{u}_1 + \dot{u}_2 - (L_P + q_2 - L/2)u_1^2 \right] \mathbf{a}_1 + \left[(L_P + q_2 - L/2)\dot{u}_1 - L_T u_1^2 + 2u_1 u_2 \right] \mathbf{a}_2 \qquad (6.48)$$

$$\mathbf{a}^{C^*} = \left[-L_T \dot{u}_1 + \dot{u}_2 - (L_P + q_2 + L_C)u_1^2 \right] \mathbf{a}_1 + \left[(L_P + q_2 + L_C)\dot{u}_1 + -L_T u_1^2 + 2u_1 u_2 \right] \mathbf{a}_2 \qquad (6.49)$$

6.4.6 Generalized Inertia Forces

In general, the generalized inertia forces F_r^* in a reference frame N for a rigid body B that is part of a system with n degrees of freedom are given by

$$(F_r^*)_B = \mathbf{v}_r^* \cdot \mathbf{R}^* + \boldsymbol{\omega}_r \cdot \mathbf{T}^* \qquad (r = 1, \ldots, n) \tag{6.50}$$

where $\mathbf{v}_r^*$ is the rth partial velocity of the mass center of B in N, $\boldsymbol{\omega}_r$ is the rth partial angular velocity of B in N, and $\mathbf{R}^*$ and $\mathbf{T}^*$ are the inertia force for B in N and the inertia torque for B in N, respectively. The inertia force for a body B is simply

$$\mathbf{R}^* = -M\mathbf{a}^* \tag{6.51}$$

where M is the total mass of B and $\mathbf{a}^*$ is the acceleration of the mass center of B in N. In its most general form, the inertia torque for B is given by

$$\mathbf{T}^* = -\boldsymbol{\alpha} \cdot \mathbf{I} - \boldsymbol{\omega} \times \mathbf{I} \cdot \boldsymbol{\omega} \tag{6.52}$$

where $\boldsymbol{\alpha}$ and $\boldsymbol{\omega}$ are, respectively, the angular acceleration of B in N and the angular velocity of B in N, and $\mathbf{I}$ is the central inertia dyadic of B.

For the problem at hand, generalized inertia forces are most easily formed by first formulating them individually for each of the bodies A, B, and C and then substituting the individual results into

$$F_r^* = (F_r^*)_A + (F_r^*)_B + (F_r^*)_C \qquad (r = 1, 2) \tag{6.53}$$

where $(F_r^*)_A$, $(F_r^*)_B$, and $(F_r^*)_C$ are the generalized inertia forces for bodies A, B, and C, respectively. To generate the generalized inertia forces for A, one must first develop expressions for its inertia force and inertia torque. Making use of Equations (6.31), Equation (6.43), and Equation (6.47), in accordance with Equation (6.51) and Equation (6.52), one obtains

$$\mathbf{R}_A^* = -m_A \left(-L_A \dot{u}_1 \mathbf{a}_1 - L_A u_1^2 \mathbf{a}_2 \right) \tag{6.54}$$
$$\mathbf{T}_A^* = -I_A \dot{u}_1 \mathbf{a}_3 \tag{6.55}$$

The resulting generalized inertia forces for A, formulated with reference to Equation (6.50), Equation (6.54), and Equation (6.55), as well as the partial velocities of Table 6.2, are

$$(F_1^*)_A = -\left(m_A L_A^2 + I_A\right)\dot{u}_1 \tag{6.56}$$
$$(F_2^*)_A = 0 \tag{6.57}$$

Similarly, for bodies B and C,

$$(F_1^*)_B = -\left\{m_B \left[(L_P + q_2 - L/2)^2 + L_T^2\right] + I_B\right\}\dot{u}_1 + m_B L_T \dot{u}_2 - 2m_B u_1 u_2 \tag{6.58}$$
$$(F_2^*)_B = m_B L_T \dot{u}_1 - m_B \dot{u}_2 + m_B (L_P + q_2 - L/2)u_1^2 \tag{6.59}$$

and

$$(F_1^*)_C = -\left\{m_C \left[(L_P + q_2 + L_C)^2 + L_T^2\right] + I_C\right\}\dot{u}_1 + m_C L_T \dot{u}_2 - 2m_C u_1 u_2 \tag{6.60}$$
$$(F_2^*)_C = m_C L_T \dot{u}_1 - m_C \dot{u}_2 + m_C (L_P + q_2 - L_C)u_1^2 \tag{6.61}$$

Substituting from Equations (6.56) through (6.61) into Equation (6.53) yields the generalized inertia forces for the entire system.

6.4.7 Generalized Active Forces

Since nonworking forces, or sets of forces, make no net contribution to the generalized active forces, one need only consider the torque T_A and the force $F_{A/B}$ to determine the generalized active forces for the robot of Figure 6.2. One can, therefore, write the generalized active forces F_r as

$$F_r = \omega_r^A \cdot T_A \mathbf{a}_3 + \mathbf{v}_r^{\hat{P}} \cdot F_{A/B} \mathbf{a}_1 + \mathbf{v}_r^P \cdot (-F_{A/B} \mathbf{a}_1) \qquad (r = 1, 2) \tag{6.62}$$

Substituting from Table 6.2 into Equation (6.62) produces

$$F_1 = T_A \tag{6.63}$$
$$F_2 = F_{A/B} \tag{6.64}$$

6.4.8 Equations of Motion

The equations of motion for the robot can now be formulated by substituting from Equation (6.53), Equation (6.63), and Equation (6.64) into Kane's equations:

$$F_r + F_r^* = 0 \qquad (r = 1, 2) \tag{6.65}$$

Kane's method can, of course, be applied to much more complicated robotic systems than the two simple illustrative systems analyzed in this and the preceding section. Section 6.6 describes a number of analyses of robotic devices that have been performed using Kane's method over the last two decades. These studies and the commercially available software packages related to the use of Kane's method described in Section 6.7, as well as studies of the efficacy of Kane's method in the analysis of robotic mechanisms such as in [24, 25], have shown that Kane's method is both analytically convenient for hand analyses as well as computationally efficient when used as the basis for general purpose or specialized robotic system simulation programs.

6.5 Special Issues in Kane's Method

Kane's method is applicable to a wide range of robotic and nonrobotic systems. In this section, attention is focused on ways in which Kane's method can be applied to systems that have specific characteristics or for which equations of motion in a particular form are desired. Specifically, described below are approaches that can be utilized when linearized dynamical equations of motion are to be developed, when equations are sought for systems that are subject to kinematical constraints, and when systems that have continuous elastic elements are analyzed.

6.5.1 Linearized Equations

As discussed in Section 6.4 of [20], dynamical equations of motion that have been linearized in all or some of the configuration or motion variables (i.e., generalized coordinates or generalized speeds) are often useful either for the study of the stability of motion or for the development of linear control schemes. Moreover, linearized differential equations have the advantage of being easier to solve than nonlinear ones while still yielding information that may be useful for restricted classes of motion of a system. In situations in which fully nonlinear equations for a system are already in hand, one develops linearized equations simply by expanding in a Taylor series all terms containing the variables in which linearization is to be performed and then eliminating all nonlinear contributions. However, in situations in which linearized dynamical equations are to be formulated directly without first developing fully nonlinear ones, or in situations in which fully nonlinear equations *cannot* be formulated, one can efficiently generate linear dynamical equations with Kane's method by proceeding as follows: First, as was done for the illustrative systems of Section 6.3 and Section 6.4, develop fully nonlinear expressions for the requisite angular and translational velocities of the particles and rigid bodies comprising the system under consideration.

These nonlinear expressions are then used to determine nonlinear partial angular velocities and partial translational velocities by inspection. Once the nonlinear partial velocities have been identified, however, they are no longer needed in their nonlinear form and these partial velocities can be linearized. Moreover, with the correct linearized partial velocities available, the previously determined nonlinear angular and translational velocities can also be linearized and then used to construct linearized angular and translational accelerations. These linearized expressions can then be used in the procedure outlined in Section 6.3 for formulating Kane's equations of motion while only retaining linearized terms in each expression throughout the process. The significant advantage of this approach is that the transition from nonlinear expressions to completely linearized ones can be made at the very early stages of an analysis, thus avoiding the need to retain terms that ultimately make no contribution to linearized equations of motion. While this is important for any system for which linearized equations are desired, it is particularly relevant to continuous systems for which fully nonlinear equations cannot be formulated in closed form (such as for the system described later in Section 6.5.3).

A specific example of the process of developing linearized equations of motion using Kane's method is given in Section 6.4 of [20]. Another example of issues associated with developing linearized dynamical equations using Kane's method is given below in Section 6.5.3 on continuous systems. Although a relatively complicated example, it demonstrates the systematic approach of Kane's method that guarantees that all the terms that should appear in linearized equations actually do. The ability to confidently and efficiently formulate linearized equations of motion for continuous systems is essential to ensure (as discussed at length in works such as [26,27]) that linear phenomena, such as "centrifugal stiffening" in rotating beam systems, are corrected and taken into account.

6.5.2 Systems Subject to Constraints

Another special case in which Kane's method can be used to particular advantage is in systems subject to constraints. This case is useful when equations of motion have already been formulated, and new equations of motion reflecting the presence of additional constraints are needed, and allows the new equations to be written as a recombination of terms comprising the original equations. This approach avoids the need to introduce the constraints as kinematical equations at an early stage of the analysis or to increase the number of equations through the introduction of unknown forces. Introducing unknown constraint forces is disadvantageous unless the constraint forces themselves are of interest, and the early introduction of kinematical constraint equations typically unnecessarily complicates the development of the dynamical equations. This approach is also useful in situations after equations of motion have been formulated and additional constraints are applied to a system, for example, when design objectives change, when a system's topology changes during its motion, or when a system is replaced with a simpler one as a means of checking a numerical simulation. In such situations, premature introduction of constraints deprives one of the opportunity to make maximum use of expressions developed in connection with the unconstrained system. The approach described below, which provides a general statement of how dynamical equations governing constrained systems can be generated, is based on the work of Wampler et al. [28].

In general, if a system described by Equation (6.9) is subjected to m independent constraints such that the number of degrees of freedom decreases from n to $n - m$, the independent generalized speeds for the system $u_1, \ldots, u_n$ must be replaced by a new set of independent generalized speeds $u_1, \ldots, u_{n-m}$. The equations of motion for the constrained system can then be generated by considering the problem as a completely new one, or alternatively, by making use of the following, one can make use of many of the expressions that were generated in forming the original set of equations.

Given an n degree-of-freedom system possessing n independent partial velocities, n generalized inertia forces F_r^*, and n generalized active forces F_r, each associated with the n independent generalized speeds $u_1, \ldots, u_n$ that are subject to m linearly independent constraints that can be written in the form

$$u_k = \sum_{l=1}^{n-m} \alpha_{kl} u_l + \beta_k \qquad (k = n - m + 1, \ldots, n) \tag{6.66}$$

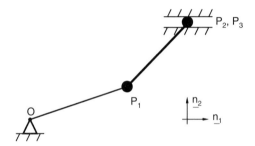

FIGURE 6.3 Two-DOF planar robot "grasping" an object.

where the u_l ($l = 1, \ldots, n - m$) are a set of independent generalized speeds governing the constrained system [α_{kl} and β_k ($l = 1, \ldots, n - m; k = n - m + 1, \ldots, n$) are functions solely of the generalized coordinates and time], the equations of motion for the constrained system are written as

$$F_r + F_r^* + \sum_{k=n-m+1}^{n} \alpha_{kr}(F_k + F_k^*) = 0 \qquad (r = 1, \ldots, n - m) \tag{6.67}$$

As can be readily observed from the above two equations, formulating equations of motion for the constrained system simply involves identifying the α_{kl} coefficients appearing in constraint equations expressed in the form of Equation (6.66) and then recombining the terms appearing in the unconstrained equations in accordance with Equation (6.67).

As an example of the above procedure, consider again the two-degree-of-freedom system of Figure 6.1. If this robot were to grasp a particle P_3 that slides in a frictionless horizontal slot, as shown in Figure 6.3, the system of the robot and particle, which when unconnected would have a total of three degrees of freedom, is reduced to one having only one degree of freedom. The two constraints arise as a result of the fact that when P_3 is grasped, the velocity of P_3 is equal to the velocity of P_1. If the velocity of P_3 before being grasped is given by

$$\mathbf{v}^{P_3} = u_3 \mathbf{n}_1 \tag{6.68}$$

where u_3 is a generalized speed chosen to describe the motion of P_3, then the two constraint equations that are in force after grasping can be expressed as

$$-Ls_1 u_1 - Ls_{12}(u_1 + u_2) = u_3 \tag{6.69}$$

$$Lc_1 u_1 + Lc_{12}(u_1 + u_2) = 0 \tag{6.70}$$

where s_1, c_1, s_{12}, and c_{12} are equal to $\sin q_1$, $\cos q_1$, $\sin(q_1 + q_2)$, and $\cos(q_1 + q_2)$, respectively. Choosing u_1 as the independent generalized speed for the constrained system, expressions for the dependent generalized speeds u_2 and u_3 are written as

$$u_2 = -\frac{c_1 + c_{12}}{c_{12}} u_1 \tag{6.71}$$

$$u_3 = L \frac{s_2}{c_{12}} u_1 \tag{6.72}$$

where s_2 is equal to $\sin q_2$ and where the coefficients of u_1 in Equation (6.71) and Equation (6.72) correspond to the terms α_{21} and α_{31} defined in the general form for constraint equations in Equation (6.66). Noting that the generalized inertia and active forces for P_3 before the constraints are applied are given by

$$F_3^* = -m_3 \dot{u}_3 \tag{6.73}$$

$$F_3 = 0 \tag{6.74}$$

where m_3 is the mass of particle P_3, the single equation of motion governing the constrained system is

$$F_1 + F_1^* + \alpha_{21}(F_2 + F_2^*) + \alpha_{31}(F_3 + F_3^*) = 0 \tag{6.75}$$

where F_r and F_r^* ($r = 1, 2$) are the generalized inertia and active forces appearing in Equation (6.21), Equation (6.22), Equation (6.24), and Equation (6.25).

There are several observations about this approach to formulating equations of motion for constrained systems that are worthy of consideration. The first is that it makes maximal use of terms that were developed for the *unconstrained* system in producing equations of motion for the *constrained* system. Second, this approach retains all the advantages inherent in the use of Kane's method, most significantly the abilities to use generalized speeds to describe motion and to disregard in an analysis nonworking forces that ultimately do not contribute to the final equations. Additionally, this approach can be applied to either literal or numerical developments of the equations of motion, and while particularly useful when constraints are added to a system once equations have been developed for the unconstrained system, it is also often a convenient approach to formulating equations of motion for constrained system from the outset of an analysis. Three particular categories of systems for which this approach is particularly appropriate from the outset of an analysis are (1) those whose topologies and number of degrees of freedom change during motion, (2) systems that are so complex that the early introduction of constraints unnecessarily encumbers the formulation of dynamical equations, and (3) those for which the introduction of "fictitious constraints" affords a means of checking numerical solutions of dynamical equations (detailed descriptions of these three categories are given in [28]).

6.5.3 Continuous Systems

Another particular class of problems to which Kane's method can be applied is nonrigid body problems. There have been many such studies in the literature (for example, see [17,26,29–32]), but to give a specific example of the process by which this is done, outlined below are the steps taken to construct by means of Kane's method the equations of motion for the system of Figure 6.2 (considered previously in Section 6.3) when the translating link is regarded as elastic rather than rigid. The system is modeled as consisting of a uniform elastic beam connected at one end to a rigid block and capable of moving longitudinally over supports attached to a rotating base. Deformations of the beam are assumed to be both "small" and adequately described by Bernoulli-Euler beam theory (i.e., shear deformations and rotatory inertia are neglected), axial deformations are neglected, and all motions are confined to a single horizontal plane. The equations are formulated by treating the supports of the translating link as kinematical constraints imposed on an unrestrained elastic beam and by discretizing the beam by means of the assumed modes method (details of this approach can be found in [31,33]). This example, though relatively complex, thus represents not only an illustration of the application of Kane's method to continuous systems but also an opportunity to consider issues associated with developing linearized dynamical equations (discussed in Section 6.5.1) as well as issues arising when formulating equations of motion for complex systems regarded from the outset as subject to kinematical constraints (following the procedure outlined in Section 6.5.2).

6.5.3.1 Preliminaries

For the purposes at hand, the system can be represented schematically as shown in Figure 6.4. This system consists of a rigid T-shaped base A that rotates about a vertical axis and that supports a nonrigid beam B at two distinct points $\widehat{P}$ and $\widehat{Q}$. The beam is capable of longitudinal motions over the supports and is connected at one end to a rigid block C. Mutually perpendicular lines N_1 and N_2, fixed in an inertial reference frame N and intersecting at point O, serve as references to which the position and orientation of the components of the system can be related. The base A rotates about an axis fixed in N and passing through O so that the orientation of A in N can be described by the single angle θ_1 between N_2 and a line fixed in A, as shown. The mass center of A, denoted A^*, is a distance L_A from O, and the distance from O to the line A_1, fixed in A, is L_T. The two points $\widehat{P}$ and $\widehat{Q}$ at which B is supported are fixed in A. The distance from the line connecting O and A^* to $\widehat{P}$ is L_P, and the distance between $\widehat{P}$ and $\widehat{Q}$ is L_D. The

Kane's Method in Robotics

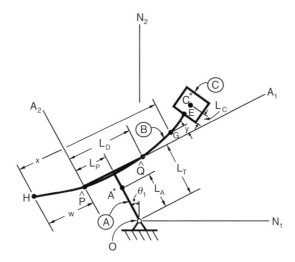

FIGURE 6.4 Robot of Figure 6.2 with continuously elastic translating link.

distance from $\widehat{P}$ to H, one endpoint of B, is w, and the distance from H to a generic point G of B, as measured along A_1, is x. The displacement of G from A_1 is y, as measured along the line A_2 that is fixed in A and intersects A_1 at $\widehat{P}$. The block C is attached to the endpoint E of B. The distance from E to the mass center C^* of C is L_C.

To take full advantage of the methods of Section 6.5.2, the system is again depicted in Figure 6.5, in which the beam B is now shown released from its supports and in a general configuration in the A_1-A_2 plane. As is described in [33], this representation facilitates the formulation of equations of motion and permits the formulation of a set of equations that can be numerically integrated in a particularly efficient manner. To this end, one can introduce an auxiliary reference frame R defined by the mutually perpendicular lines R_1 and R_2 intersecting at R^* and lying in the A_1-A_2 plane. The position of R^* relative to $\widehat{P}$, as measured

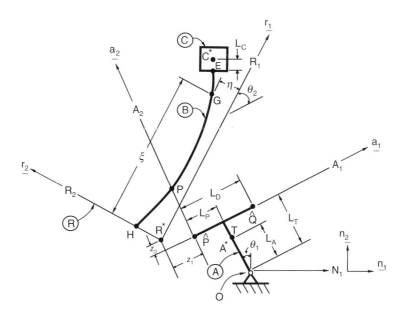

FIGURE 6.5 Robot of Figure 6.4 with translating link released from supports.

along A_1 and A_2, is characterized by z_1 and z_2, respectively, and the orientation of R in A is described by the angle θ_2 between A_1 and R_1. The R_1-coordinate and R_2-coordinate of G are ξ and η, respectively, while the angle (not shown) between R_1 and the tangent to B at G is α. Finally, P is a point on B lying on A_2, and a similar point Q (not shown) is on B lying on a line parallel to A_2 passing through $\hat{Q}$.

6.5.3.2 Kinematics

By making use of the assumed modes method, one can express η in terms of modal functions ϕ_i as

$$\eta(s, t) = \sum_{i=1}^{\nu} \phi_i(s) q_i(t) \tag{6.76}$$

where the ϕ_i ($i = 1, \ldots, \nu$) are functions of s, the arc length as measured along B from H to G, and ν is a positive integer indicating the number of modes to be used in the analysis. The quantities ξ and α can be directly related to ϕ_i and q_i, and thus the $\nu + 4$ quantities q_i ($i = 1, \ldots, \nu$), θ_1, z_1, z_2, and θ_2 suffice to fully describe the configuration of the complete system (the system has $\nu + 4$ degrees of freedom). Generalized speeds u_i ($i = 1, \ldots, \nu + 4$) describing the motion of the system can be introduced as

$$\begin{aligned}
u_i &= \dot{q}_i \quad (i = 1, \ldots, \nu) \\
u_{\nu+1} &= \boldsymbol{\omega}^A \cdot \mathbf{a}_3 = \dot{\theta}_1 \\
u_{\nu+2} &= {}^A\mathbf{v}^{R^*} \cdot \mathbf{a}_1 = -\dot{z}_1 \\
u_{\nu+3} &= {}^A\mathbf{v}^{R^*} \cdot \mathbf{a}_2 = \dot{z}_2 \\
u_{\nu+4} &= {}^A\boldsymbol{\omega}^R \cdot \mathbf{a}_3 = \dot{\theta}_2
\end{aligned} \tag{6.77}$$

where $\boldsymbol{\omega}^A$ is the angular velocity of A in N, ${}^A\mathbf{v}^{R^*}$ is the velocity of R^* in A, ${}^A\boldsymbol{\omega}^R$ is the angular velocity of R in A, and $\mathbf{a}_1, \mathbf{a}_2$, and $\mathbf{a}_3$ are elements of a dextral set of unit vectors such that $\mathbf{a}_1$ and $\mathbf{a}_2$ are directed as shown in Figure 6.5 and $\mathbf{a}_3 = \mathbf{a}_1 \times \mathbf{a}_2$.

The angular and translational velocities essential to the development of equations of motion are the angular velocity of A in N, the velocity of A^* in N, the velocity of G in N, the angular velocity of C in N, and the velocity of C^* in N. These are given by

$$\boldsymbol{\omega}^A = u_{\nu+1} \mathbf{a}_3 \tag{6.78}$$

$$\mathbf{v}^{A^*} = -L_A u_{\nu+1} \mathbf{a}_1 \tag{6.79}$$

$$\begin{aligned}
\mathbf{v}^G = &\left\{ -\sum_{i=1}^{\nu} \left[\int_0^s \phi_i'(\sigma) \tan\alpha(\sigma, t) d\sigma \right] u_i - [(L_T + z_2) c_2 + (L_P + z_1) s_2 + \eta] u_{\nu+1} \right. \\
&\left. + c_2 u_{\nu+2} + s_2 u_{\nu+3} - \eta u_{\nu+4} \right\} \mathbf{r}_1 \\
&+ \left\{ \sum_{i=1}^{\nu} \phi_i u_i + [(L_T + z_2) s_2 - (L_P + z_1) c_2 + \xi] u_{\nu+1} - s_2 u_{\nu+2} + c_2 u_{\nu+3} + \xi u_{\nu+4} \right\} \mathbf{r}_2
\end{aligned} \tag{6.80}$$

$$\boldsymbol{\omega}^C = \left[\frac{1}{\cos \alpha^E} \sum_{i=1}^{\nu} \phi_i'(L) u_i + u_{\nu+1} + u_{\nu+4} \right] \mathbf{r}_3 \tag{6.81}$$

$$\begin{aligned}
\mathbf{v}^{C^*} = \mathbf{v}^G \bigg|_{s=L} &- L_C \sin\alpha^E \left[\frac{1}{\cos \alpha^E} \sum_{i=1}^{\nu} \phi_i'(L) u_i + u_{\nu+1} + u_{\nu+4} \right] \mathbf{r}_1 \\
&+ L_C \cos\alpha^E \left[\frac{1}{\cos \alpha^E} \sum_{i=1}^{\nu} \phi_i'(L) u_i + u_{\nu+1} + u_{\nu+4} \right] \mathbf{r}_2
\end{aligned} \tag{6.82}$$

where s_2 and c_2 are defined as $\sin\theta_2$ and $\cos\theta_2$, respectively, $\mathbf{r}_1$ and $\mathbf{r}_2$ are unit vectors directed as shown in Figure 6.5, α^E is the value of α corresponding to the position of point E, and $\mathbf{v}^G|_{s=L}$ is the velocity of G evaluated at $s = L$, where L is the undeformed length of B.

It should be noted that, so far, all expressions have been written in their full nonlinear form. This is in accordance with the requirements of Kane's method described in Section 6.5.2 for developing linearized equations of motion. These requirements mandate that completely nonlinear expressions for angular and translational velocities be formulated, and thereafter completely nonlinear partial angular velocities and partial translational velocities, before any kinematical expressions can be linearized. From Equation (6.78) through Equation (6.82), the requisite nonlinear partial velocities are identified as simply the coefficients of the generalized speeds u_i ($i = 1, \ldots, \nu + 4$).

With all nonlinear partial velocities identified, Equation (6.76) and Equation (6.80) through Equation (6.82) may now be linearized in the quantities q_i ($i = 1, \ldots, \nu$), z_2, θ_2, u_i ($i = 1, \ldots, \nu$), $u_{\nu+3}$, and $u_{\nu+4}$ to produce equations governing "small" displacements of B from the A_1 axis. Linearization yields

$$\eta(\xi, t) = \sum_{i=1}^{\nu} \phi_i(\xi) q_i(t) \tag{6.83}$$

$$\mathbf{v}^G = \left\{ -\left[L_T + z_2 + (L_P + z_1)\theta_2 + \sum_{i=1}^{\nu} \phi_i q_i \right] u_{\nu+1} + u_{\nu+2} \right\} \mathbf{r}_1$$
$$+ \left[\sum_{i=1}^{\nu} \phi_i u_i + (L_T \theta_2 - L_P - z_1 + \xi) u_{\nu+1} - \theta_2 u_{\nu+2} + u_{\nu+3} + \xi u_{\nu+4} \right] \mathbf{r}_2 \tag{6.84}$$

$$\boldsymbol{\omega}^C = \left[\sum_{i=1}^{\nu} \phi_i'(L) u_i + u_{\nu+1} + u_{\nu+4} \right] \mathbf{r}_3 \tag{6.85}$$

$$\mathbf{v}^{C^*} = \mathbf{v}^G \bigg|_{\xi=L} - L_C u_{\nu+1} \sum_{i=1}^{\nu} \phi_i'(L) q_i \, \mathbf{r}_1 + L_C \left[\sum_{i=1}^{\nu} \phi_i'(L) u_i + u_{\nu+1} + u_{\nu+4} \right] \mathbf{r}_2 \tag{6.86}$$

while leaving Equation (6.78) and Equation (6.79) unchanged. Linearization of the partial velocities obtained from Equation (6.78) through Equation (6.82) yields the expressions recorded in Table 6.3, where

$$\epsilon_{ij}(\xi) = \int_0^\xi \phi_i'(\sigma) \phi_j'(\sigma) d\sigma \qquad (i, j = 1, \ldots, \nu) \tag{6.87}$$

Note that if linearization had been performed prematurely, and partial velocities had been obtained from the linearized velocity expressions given in Equation (6.84) through Equation (6.86), terms involving ϵ_{ij} would have been lost. These terms must appear in the equations of motion in order to correctly account for the coupling between longitudinal accelerations and transverse deformations of the beam.

As was done in Section 6.4, the generalized inertia forces F_r^* for the system depicted in Figure 6.5 can be expressed as

$$F_r^* = (F_r^*)_A + (F_r^*)_B + (F_r^*)_C \qquad (r = 1, \ldots, \nu + 4) \tag{6.88}$$

where $(F_r^*)_A$, $(F_r^*)_B$, and $(F_r^*)_C$ are the generalized inertia forces for bodies A, B, and C, respectively. The generalized inertia forces for A are generated by summing the dot product between the partial angular velocity of A and the inertia torque of A with the dot product between the partial translational velocity of A^* and the inertia force of A. This can be written as

$$(F_r^*)_A = \boldsymbol{\omega}_r^A \cdot \left(-I_3^A \dot{u}_{\nu+1} \mathbf{a}_3 \right) + \mathbf{v}_r^{A^*} \cdot (-m_A \mathbf{a}^{A^*}) \qquad (r = 1, \ldots, \nu + 4) \tag{6.89}$$

where $\boldsymbol{\omega}_r^A$ and $\mathbf{v}_r^{A^*}$ are the partial velocities for body A given in Table 6.4, I_3^A is the moment of inertia of A about a line parallel to $\mathbf{a}_3$ and passing through A^*, m_A is the mass of A, and $\mathbf{a}^{A^*}$ is the acceleration of A^*.

TABLE 6.3 Linearized Partial Velocities for Robot of Figure 6.5

r	ω_r^A	$\mathbf{v}_r^{A*}$	$\mathbf{v}_r^G$	ω_r^C	$\mathbf{v}_r^{C*}$
$j = 1, \ldots, \nu$	0	0	$-\sum_{i=1}^{\nu} \epsilon_{ij} q_i\, \mathbf{r}_1 + \phi_j\, \mathbf{r}_2$	$\phi_j'(L)\mathbf{r}_3$	$\mathbf{v}_j^G\big|_{\xi=L}$ $-L_C \sum_{i=1}^{\nu} \phi_i'(L)\phi_j'(L) q_i\, \mathbf{r}_1$ $+ L_C \phi_j'(L)\, \mathbf{r}_2$
$\nu + 1$	$\mathbf{a}_3$	$-L_A\, \mathbf{a}_1$	$-[L_T + z_2 + (L_P + z_1)\theta_2$ $+ \sum_{i=1}^{\nu} \phi_i q_i]\mathbf{r}_1 + (L_T\,\theta_2$ $- L_P - z_1 + \xi)\mathbf{r}_2$	$\mathbf{r}_3$	$\mathbf{v}_{\nu+1}^G\big|_{\xi=L}$ $-L_C \sum_{i=1}^{\nu} \phi_i'(L) q_i\, \mathbf{r}_1$ $+ L_C\, \mathbf{r}_2$
$\nu + 2$	0	0	$\mathbf{r}_1 - \theta_2\, \mathbf{r}_2$	0	$\mathbf{v}_{\nu+2}^G\big|_{\xi=L}$
$\nu + 3$	0	0	$\theta_2\, \mathbf{r}_1 + \mathbf{r}_2$	0	$\mathbf{v}_{\nu+3}^G\big|_{\xi=L}$
$\nu + 4$	0	0	$-\sum_{i=1}^{\nu} \phi_i q_i\, \mathbf{r}_1 + \xi\, \mathbf{r}_2$	$\mathbf{r}_3$	$\mathbf{v}_{\nu+4}^G\big|_{\xi=L}$ $-L_C \sum_{i=1}^{\nu} \phi_i'(L) q_i\, \mathbf{r}_1$ $+ L_C\, \mathbf{r}_2$

The generalized inertia forces for B are developed from

$$(F_r^*)_B = \int_0^L \mathbf{v}_r^G \cdot (-\rho\, \mathbf{a}^G)\, d\xi \qquad (r = 1, \ldots, \nu + 4) \tag{6.90}$$

where $\mathbf{v}_r^G$ ($r = 1, \ldots, \nu + 4$) and $\mathbf{a}^G$ are the partial velocities and the acceleration of a differential element of B at G, respectively, and ρ is the mass per unit length of B. An expression for the generalized inertia forces for the block C is produced in a manner similar to that used to generate Equation (6.89) and has the form

$$\mathbf{F}_C^* = \omega_r^C \cdot \left\{ I_3^C \left[\sum_{i=1}^{\nu} \phi_i'(L)\dot{u}_i + \dot{u}_{\nu+1} + \dot{u}_{\nu+4} \right] \mathbf{r}_3 \right\}$$
$$+ \mathbf{v}_r^{C*} \cdot (-m_C\, \mathbf{a}^{C*}) \qquad (r = 1, \ldots, \nu + 4) \tag{6.91}$$

where ω_r^C and $\mathbf{v}_r^{C*}$ are the partial velocities for body C, I_3^C is the moment of inertia of C about a line parallel to $\mathbf{a}_3$ and passing through C^*, m_C is the mass of C, and $\mathbf{a}^{C*}$ is the acceleration of C^*.

For the sake of brevity, explicit expressions for the generalized inertia forces F_r^* ($r = 1, \ldots, \nu + 4$) are not listed here. Once linearized expressions for the accelerations $\mathbf{a}^{A*}$, $\mathbf{a}^G$, and $\mathbf{a}^{C*}$ are constructed by differentiating Equation (6.79), Equation (6.84), and Equation (6.86) with respect to time in N, however, all the necessary information is available to develop the generalized inertia forces by performing the operations indicated above in Equations (6.88) through (6.91). The interested reader is referred to [33] for complete details.

As in the development of the generalized inertia forces, it is convenient to consider contributions to the generalized active forces F_r from bodies A, B, and C separately and write

$$F_r = (F_r)_A + (F_r)_B + (F_r)_C \qquad (r = 1, \ldots, \nu + 4) \tag{6.92}$$

The only nonzero contribution to the generalized active forces for A is due to an actuator torque T_A so that the generalized active forces for A are given by

$$(F_r)_A = \omega_r^A \cdot (T_A\, \mathbf{a}_3) \qquad (r = 1, \ldots, \nu + 4) \tag{6.93}$$

Using a Bernoulli-Euler beam model to characterize deformations, the generalized active forces for B are determined from

$$(F_r)_B = \int_0^L \mathbf{v}_r^G \cdot \left(-EI\frac{\partial^4 \eta}{\partial \xi^4}\mathbf{r}_2\right) d\xi \qquad (r = 1, \ldots, \nu+4) \qquad (6.94)$$

where E and I are the modulus of elasticity and the cross-sectional area moment of inertia of B, respectively. Nonzero contributions to the generalized active forces for C arise from the bending moment and shear force exerted by B on C so that $(F_r)_C$ is given by

$$(F_r)_C = \boldsymbol{\omega}_r^C \cdot \left(-EI\frac{\partial^2 \eta}{\partial \xi^2}\bigg|_{\xi=L}\mathbf{r}_3\right) + \mathbf{v}_r^G\bigg|_{\xi=L} \cdot \left(EI\frac{\partial^3 \eta}{\partial \xi^3}\bigg|_{\xi=L}\mathbf{r}_2\right) \qquad (r = 1, \ldots, \nu+4) \quad (6.95)$$

Final expressions for the generalized active forces for the entire system are produced by substituting Equations (6.93) through (6.95) into Equation (6.92).

It is important to remember at this point that the generalized inertia and active forces of Equation (6.88) and Equation (6.92) govern *unrestrained* motions of the beam B. The equations of interest, however, are those that govern motions of B when it is constrained to remain in contact with the supports at $\widehat{P}$ and $\widehat{Q}$. These equations can be formulated by first developing constraint equations expressed explicitly in terms of the generalized speeds of Equation (6.77) and by then making use of the procedure for formulating equations of motion for constrained systems described in Section 6.5.2.

One begins the process by identifying the constraints on the position of the points P and Q of the beam B. These are

$$\frac{d}{dt}(\mathbf{p}^{\widehat{P}P} \cdot \mathbf{a}_2) = 0, \qquad \frac{d}{dt}(\mathbf{p}^{\widehat{Q}Q} \cdot \mathbf{a}_2) = 0 \qquad (6.96)$$

where $\mathbf{p}^{\widehat{P}P}$ and $\mathbf{p}^{\widehat{Q}Q}$ are the position vectors of P and Q relative to $\widehat{P}$ and $\widehat{Q}$, respectively. These equations must next be expressed in terms of the generalized speeds of Equation (6.77). Doing so yields the constraint equations

$$u_{\nu+3} - (s_2\eta^P - c_2\xi^P)u_{\nu+4} + s_2\frac{d\xi^P}{dt} + c_2\frac{d\eta^P}{dt} = 0 \qquad (6.97)$$

$$u_{\nu+3} - (s_2\eta^Q - c_2\xi^Q)u_{\nu+4} + s_2\frac{d\xi^Q}{dt} + c_2\frac{d\eta^Q}{dt} = 0 \qquad (6.98)$$

where s^P and s^Q are the arc lengths, measured along B, from H to P and from H to Q, respectively. Note that since P and Q are not fixed on B, s^P and s^Q are not independent variables, but rather are functions of time.

One is now in a position to express Equation (6.97) and Equation (6.98) in a form such that the generalized speeds u_i ($i = 1, \ldots, \nu+4$) appear explicitly. Doing so produces

$$\sum_{i=1}^{\nu}\left[\sin\alpha^P \int_0^{s^P}\phi_i'(\sigma)\tan\alpha(\sigma,t)d\sigma + \phi_i(s^P)\cos\alpha^P\right]u_i - (s_2\cos\alpha^P + c_2\sin\alpha^P)u_{\nu+2}$$
$$+ (c_2\cos\alpha^P - s_2\sin\alpha^P)u_{\nu+3} + (\eta^P\sin\alpha^P + \xi^P\cos\alpha^P)u_{\nu+4} = 0 \qquad (6.99)$$

and

$$\sum_{i=1}^{\nu}\left[\sin\alpha^Q \int_0^{s^Q}\phi_i'(\sigma)\tan\alpha(\sigma,t)d\sigma + \phi_i(s^Q)\cos\alpha^Q\right]u_i - (s_2\cos\alpha^Q + c_2\sin\alpha^Q)u_{\nu+2}$$
$$+ (c_2\cos\alpha^Q - s_2\sin\alpha^Q)u_{\nu+3} + (\eta^Q\sin\alpha^Q + \xi^Q\cos\alpha^Q)u_{\nu+4} = 0 \qquad (6.100)$$

where α^P and α^Q denote the values of $\alpha(s,t)$ evaluated with $s = s^P$ and $s = s^Q$, respectively.

Since the ultimate goal is to develop equations of motion governing only small deformations of B, it is again appropriate to now linearize the coefficients of u_i ($i = 1, \ldots, \nu + 4$) in Equation (6.99) and Equation (6.100) in the quantities q_i ($i = 1, \ldots, \nu$), z_2, θ_2, u_i ($i = 1, \ldots, \nu$), $u_{\nu+3}$, and $u_{\nu+4}$. This is analogous to the linearization procedure used previously in developing linearized partial velocities. Note that linearization may not be performed prior to the formulation of Equation (6.99) and Equation (6.100) in their full nonlinear form. The consequence of premature linearization would be the loss of *linear* terms in the equations of motion reflecting the coupling between longitudinal and transverse motions of B.

Before the beam B is constrained to remain in contact with the supports at $\widehat{P}$ and $\widehat{Q}$, contributions to the generalized active forces from contact forces acting between bodies A and B cannot be accommodated in a consistent manner. Once the constraints are introduced, however, these forces must be taken into account. The only such forces that make nonzero contributions are $\widetilde{\mathbf{F}}_{\overline{P}}$ and $\widetilde{\mathbf{F}}_{\hat{P}}$. The force $\widetilde{\mathbf{F}}_{\overline{P}}$ drives B longitudinally over $\widehat{P}$ and $\widehat{Q}$ and is regarded as being applied tangentially to B at $\overline{P}$, where $\overline{P}$ is that point fixed on B whose location instantaneously coincides with that of P. The force $\widetilde{\mathbf{F}}_{\hat{P}}$ is equal in magnitude and opposite in direction to $\widetilde{\mathbf{F}}_{\overline{P}}$ and is applied to A at $\widehat{P}$. Taking advantage of the facts that $\widetilde{\mathbf{F}}_{\hat{P}}$ equals $-\widetilde{\mathbf{F}}_{\overline{P}}$ and that the velocity of $\overline{P}$ equals $\mathbf{v}^G|_{\xi=z_1}$, one can express the additional generalized forces $\widetilde{F}_r$ as

$$\widetilde{F}_r = \left(\mathbf{v}_r^G \bigg|_{\xi=z_1} - \mathbf{v}_r^{\hat{P}} \right) \cdot \left\{ F_{A/B} \left[\mathbf{r}_1 + \sum_{i=1}^{\nu} \phi_i'(z_1) q_i \, \mathbf{r}_2 \right] \right\} \qquad (r = 1, \ldots, \nu + 4) \qquad (6.101)$$

where the $\mathbf{v}_r^{\hat{P}}$ are the linearized partial velocities associated with the velocity of $\widehat{P}$ in N and $F_{A/B}$ is the component of $\widetilde{F}_{\overline{P}}$ tangent to B at P. The partial velocities needed in Equation (16.101) can be obtained from the expression for the velocity of $\widehat{P}$ in N given by

$$\mathbf{v}^{\hat{P}} = -u_{\nu+1}[(L_T c_2 + L_P s_2) \mathbf{r}_1 + (L_P c_2 - L_T s_2) \mathbf{r}_2] \qquad (6.102)$$

With all the necessary generalized inertia forces, generalized active forces, and constraint equations in hand, one is in a position to construct the complete dynamical equations governing the motion of the system of Figure 6.4. Without stating them explicitly, these are given by

$$\underline{\alpha}^T (\underline{F}^* + \underline{F} + \underline{\widetilde{F}}) = 0 \qquad (6.103)$$

where $\underline{F}^*$ is a $(\nu + 4) \times 1$ matrix whose elements are the generalized inertial forces of Equation (6.88), $\underline{F}$ and $\underline{\widetilde{F}}$ are $(\nu + 4) \times 1$ matrices that contain the generalized active forces of Equation (6.92) and Equation (6.101), respectively, and $\underline{\alpha}^T$ is the transpose of a matrix of coefficients of the independent generalized speeds appearing in the constraint equations of Equation (6.99) and Equation (6.100).

One particular point to reinforce about the use of Kane's method in formulating linearized equations of motion for this and other systems with nonrigid elements is the fact that it automatically and correctly accounts for the phenomenon known as "centrifugal stiffening." This is guaranteed when using Kane's method so long as fully nonlinear kinematical expressions are used up through the formulation of partial velocities. This fact ensures that terms resulting from the coupling between longitudinal inertia forces, due either to longitudinal motions of the beam or to rotational motions of the base, and transverse deflections are correctly taken into account. The benefit of having these terms in the resulting equations of motion is that the equations are valid for any magnitude of longitudinal accelerations of the beam or of rotational speeds of the base.

6.6 Kane's Equations in the Robotics Literature

The application of Kane's method in robotics has been cited in numerous articles in the robotics literature since approximately 1982. For example, a search of the COMPENDEX database using the words "Kane" AND "robot" OR "robotic" produced 79 unique references. Moreover, a search of citations of the "The Use of Kane's Dynamical Equations in Robotics" by Kane and Levinson [24] in the Web of Science database produced 76 citations (some, but not all, of which are the same as those found in the COMPENDEX database). Kane's method has also been addressed in recent dynamics textbooks such as *Analytical Dynamics* by Baruh [34] and *Dynamics of Mechanical Systems* by Josephs and Huston [35]. These articles and books discuss not only the development of equations of motion for specific robotic devices but also general methods of formulating equations for robotic-like systems.

Although a comprehensive survey of all articles in the robotics literature that focus on the use of Kane's method is not possible within the limits of this chapter, an attempt to indicate the range of applications of Kane's method is given below. The listed articles were published between 1983 and the present. In some cases, the articles represent seminal works that are widely known in the dynamics and robotics literature. The others, although perhaps less well known, give an indication of the many individuals around the world who are knowledgeable about the use of Kane's method and the many types of systems to which it can be applied. In each case, the title, authors, and year of publication are given (a full citation can be found in the list of references at the end of this chapter) as well as an abbreviated version of the abstract of the article.

- "The use of Kane's dynamical equations in robotics" by Kane and Levinson (1983) [24]: Focuses on the improvements in computational efficiency that can be achieved using Kane's dynamical equations to formulate explicit equations of motion for robotic devices as compared to the use of general-purpose, multibody-dynamics computer programs for the numerical formulation and solution of equations of motion. Presents a detailed analysis of the Stanford Arm so that each step in the analysis serves as an illustrative example for a general method of analysis of problems in robot dynamics. Simulation results are reported and form the basis for discussing questions of computational efficiency.
- "Simulation der dynamik von robotern nach dem verfahren von Kane" by Wloka and Blug (1985) [36]: Discusses the need for modeling robot dynamics when simulating the trajectory of a robot arm. Highlights the inefficiency of implementing the well-known methods of Newton-Euler or Lagrange. Develops recursive methods, based on Kane's equations, that reduce computational costs dramatically and provide an efficient way to calculate the equations of motion of mechanical systems.
- "Dynamique de mecanismes flexibles" by Gallay et al. (1986) [37]: Describes an approximate method suitable for the automatic generation of equation of motion of complex flexible mechanical systems, such as large space structures, robotic manipulators, etc. Equations are constructed based on Kane's method. Rigid-body kinematics use relative coordinates and a method of modal synthesis is employed for the treatment of deformable bodies. The method minimizes the number of equations, while allowing complex topologies (closed loops), and leads to the development of a computer code, called ADAM, for dynamic analysis of mechanical assemblies.
- "Algorithm for the inverse dynamics of n-axis general manipulators using Kane's equations" by Angeles et al. (1989) [38]: Presents an algorithm for the numerical solution of the inverse dynamics of serial-type, but otherwise arbitrary, robotic manipulators. The algorithm is applicable to manipulators containing n rotational or prismatic joints. For a given set of Denavit-Hartenberg and inertial parameters, as well as for a given trajectory in the space of joint coordinates, the algorithm produces the time histories of the n torques or forces required to drive the manipulator through the prescribed trajectory. The algorithm is based on Kane's dynamical equations. Moreover, the complexity of the presented algorithm is lower than that of the most efficient inverse-dynamics algorithm reported in the literature. The applicability of the algorithm is illustrated with two fully solved examples.
- "The use of Kane's method in the modeling and simulation of robotic systems" by Huston (1989) [39]: Discusses the use of Euler parameters, partial velocities, partial angular velocities, and generalized speeds. Explicit expressions for the coefficients of the governing equations are obtained using these methods in forms ideally suited for conversion into computer algorithms. Two applications are discussed: flexible systems and redundant systems. For flexible systems, it is shown that Kane's method using generalized speeds leads to equations that are decoupled in the joint force and

moment components. For redundant systems, with the desired end-effector motion, it is shown that Kane's method uses generalized constraint forces that may be directly related to the constraint equation coefficients, leading directly to matrix methods for solving the governing equations.

- "Dynamics of elastic manipulators with prismatic joints" by Buffinton (1992) [31]: Investigates the formulation of equations of motion for flexible robots containing translationally moving elastic members that traverse a finite number of distinct support points. Specifically studies a two-degree-of-freedom manipulator whose configuration is similar to that of the Stanford Arm and whose translational member is regarded as an elastic beam. Equations of motion are formulated by treating the beam's supports as kinematical constraints imposed on an unrestrained beam, by discretizing the beam by means of the assumed modes technique, and by applying an alternative form of Kane's method which is particularly well suited for systems subject to constraints. The resulting equations are programmed and are used to simulate the system's response when it performs tracking maneuvers. Results provide insights into some of the issues and problems involved in the dynamics and control of manipulators containing highly elastic members connected by prismatic joints. (The formulation of equations of motion presented in this article is summarized in Section 6.5.3 above.)

- "Explicit matrix formulation of the dynamical equations for flexible multibody systems: a recursive approach" by Amirouche and Xie (1993) [32]: Develops a recursive formulation based on the finite element method where all terms are presented in a matrix form. The methodology permits one to identify the coupling between rigid and flexible body motion and build the necessary arrays for the application at hand. The equations of motion are based on Kane's equation and the general matrix representation for n bodies of its partial velocities and partial angular velocities. The algorithm developed is applied to a single two-link robot manipulator and the subsequent explicit equations of motion are presented.

- "Developing algorithms for efficient simulation of flexible space manipulator operations" by Van Woerkom et al. (1995) [17]: Briefly describes the European robot arm (ERA) and its intended application for assembly, maintenance, and servicing of the Russian segment of the International Space Station Alpha. A short review is presented of efficient Order-N**3 and Order-N algorithms for the simulation of manipulator dynamics. The link between Kane's method, the natural orthogonal complement method, and Jourdain's principle is described as indicated. An algorithm of the Order-N type is then developed for use in the ERA simulation facility. Numerical results are presented, displaying accuracy as well as computational efficiency.

- "Dynamic model of an underwater vehicle with a robotic manipulator using Kane's method" by Tarn et al. (1996) [40]: Develops a dynamic model for an underwater vehicle with an n-axis robot arm based on Kane's method. The presented technique provides a direct method for incorporating external environmental forces into the model. The model includes four major hydrodynamic forces: added mass, profile drag, fluid acceleration, and buoyancy. The derived model provides a closed form solution that is easily utilized in modern model-based control schemes.

- "Ease of dynamic modeling of wheeled mobile robots (WMRs) using Kane's approach" by Thanjavur (1997) [41]: Illustrates the ease of modeling the dynamics of WMRs using Kane's approach for nonholonomic systems. For a control engineer, Kane's method is shown to offer several unique advantages over the Newton-Euler and Lagrangian approaches used in the available literature. Kane's method provides physical insight into the nature of nonholonomic systems by incorporating the motion constraints directly into the derivation of equations of motion. The presented approach focuses on the number of degrees of freedom and not on the configuration, and thus eliminates redundancy. Explicit expressions to compute the dynamic wheel loads needed for tire friction models are derived, and a procedure to deduce the dynamics of a differentially driven WMR with suspended loads and operating on various terrains is developed. Kane's approach provides a systematic modeling scheme, and the method proposed in the paper is easily generalized to model WMRs with various wheel types and configurations and for various loading conditions. The resulting dynamic model is mathematically simple and is suited for real time control applications.

- "Haptic feedback for virtual assembly" by Luecke and Zafer (1998) [42]: Uses a commercial assembly robot as a source of haptic feedback for assembly tasks, such as a peg-hole insertion task. Kane's method is used to derive the dynamics of the peg and the contact motions between the peg and the hole. A handle modeled as a cylindrical peg is attached to the end-effector of a PUMA 560 robotic arm equipped with a six-axis force/torque transducer. The user grabs the handle and the user-applied forces are recorded. Computed torque control is employed to feed the full dynamics of the task back to the user's hand. Visual feedback is also provided. Experimental results are presented to show several contact configurations.
- "Method for the analysis of robot dynamics" by Xu et al. (1998) [43]: Presents a recursive algorithm for robot dynamics based on Kane's dynamical equations. Differs from other algorithms, such as those based on Lagrange's equations or the Newton-Euler formulation, in that it can be used for solving the dynamics of robots containing closed-chains without cutting the closed-chain open. It is also well suited for computer implementation because of its recursive form.
- "Dynamic modeling and motion planning for a hybrid legged-wheeled mobile vehicle" by Lee and Dissanayake (2001) [44]: Presents the dynamic modeling and energy-optimal motion planning of a hybrid legged-wheeled mobile vehicle. Kane's method and the AUTOLEV symbolic manipulation package are used in the dynamic modeling, and a state parameterization method is used to synthesize the leg trajectories. The resulting energy-optimal gait is experimentally evaluated to show the effectiveness and feasibility of the path planned.
- "Kane's approach to modeling mobile manipulators" by Tanner and Kyriakopoulos (2002) [45]: Develops a detailed methodology for dynamic modeling of wheeled nonholonomic mobile manipulators using Kane's dynamic equations. The resulting model is obtained in closed form, is computationally efficient, and provides physical insight as to what forces really influence the system dynamics. Allows nonholonomic constraints to be directly incorporated into the model without introducing multipliers. The specific model constructed in the paper includes constraints for no slipping, no skidding, and no tipover. The dynamic model and the constraint relations provide a framework for developing control strategies for mobile manipulators.

Beyond the applications of Kane's method to specific problems in robotics, there are a large number of papers that either apply Kane's method to completely general multibody dynamic systems or develop from it related algorithms applicable to various subclasses of multibody systems. A small sampling of these articles is [13–15,46,47]. The algorithms described in [13–15] have, moreover, lead directly to the commercially successful software packages discussed below. A discussion of these packages is presented in the context of this chapter to provide those working in the field of robotics with an indication of the options available for simulating the motion of complex robotic devices without the need to develop the equations of motion and associated simulation programs on their own.

6.7 Commercially Available Software Packages Related to Kane's Method

There are many commercially available software packages that are based, either directly or indirectly, on Kane's method. The most successful and widely used are listed below:

- SD/FAST
- Pro/ENGINEER Simulation (formerly Pro/MECHANICA)
- AUTOLEV
- ADAMS
- Working Model and VisualNASTRAN

The general characteristics of each of these packages, as well as some of their advantages and disadvantages, are discussed below.

6.7.1 SD/FAST

SD/FAST is based on an extended version of Kane's method and was originally developed by Dan Rosenthal and Michael Sherman (founders of Symbolic Dynamics, Inc, and currently chief technical officer and executive vice president of Research and Development, respectively, at Protein Mechanics <www.proteinmechanics.com>) while students at Stanford University [13]. It was first offered commercially directly by Rosenthal and Sherman and later through Mitchell-Gauthier. SD/FAST is still available and since January 2001 has been distributed by Parametric Technologies Corporation, which also markets Pro/ENGINEER Simulation software (formerly Pro/MECHANICA; see below).

As described on the SD/FAST website <www.sdfast.com>:

> SD/FAST provides physically based simulation of mechanical systems by taking a short description of an articulated system of rigid bodies (bodies connected by joints) and deriving the full nonlinear equations of motion for that system. The equations are then output as C or Fortran source code, which can be compiled and linked into any simulation or animation environment. The symbolic derivation of the equations provides the fastest possible simulations. Many substantial systems can be executed in real time on modest computers.

The SD/FAST website also states that SD/FAST is "in daily use at thousands of sites worldwide." Typical areas of application mentioned include physically based animation, mechanism simulation, aerospace simulation, robotics, real-time hardware-in-the-loop and man-in-the-loop simulation, biomechanical simulation, and games. Also mentioned is that SD/FAST is the "behind-the-scenes dynamics engine" for both Pro/ENGINEER Mechanism Dynamics from Parametric Technologies Corporation and SIMM (Software for Interactive Musculoskeletal Modeling) from MusculoGraphics, Inc.

6.7.2 Pro/ENGINEER Simulation (formerly Pro/MECHANICA)

Pro/ENGINEER Simulation is a suite of packages "powered by MECHANICA technology" <www.ptc.com> for structural, thermal, durability, and dynamic analysis of mechanical systems. These packages are fully integrated with Pro/ENGINEER and the entire suite currently consists of

- Pro/ENGINEER Advanced Structural & Thermal Simulation
- Pro/ENGINEER Structural and Thermal Simulation
- Pro/ENGINEER Fatigue Advisor
- Pro/ENGINEER Mechanism Dynamics
- Pro/ENGINEER Behavioral Modeling

As mentioned above, Pro/ENGINEER Mechanism Dynamics is based on the dynamics engine originally developed by Rosenthal and Sherman. The evolution of the Mechanism Dynamics package progressed from SD/FAST to a package offered by Rasna Corporation and then to Pro/MECHANICA MOTION after Rasna was purchased in 1996 by Parametric Technology Corporation. In its latest version, Pro/ENGINEER Mechanism Dynamics allows dynamics analyses to be fully integrated with models constructed in the solids modeling portion of Pro/ENGINEER. Mechanisms can be created using a variety of joint and constraint conditions, input loads and drivers, contact regions, coordinate systems, two-dimensional cam and slot connections, and initial conditions. Various analyses can be performed on the resulting model including static, velocity, assembly, and full motion analyses for the determination of the position, velocity, and acceleration of any point or body of interest and of the reaction forces and torques exerted between bodies and ground. Mechanism Dynamics can also be used to perform design studies to optimize dimensions and other design parameters as well as to check for interferences. Results can be viewed as data reports or graphs or as a complete motion animation.

6.7.3 AUTOLEV

AUTOLEV was originally developed by David Schaechter and David Levinson at Lockheed Palo Alto Research Laboratory and was first released for commercial distribution in the summer of 1988 [14]. Levinson and Schaechter were later joined by Prof. Kane himself and Paul Mitiguy (a former Ph.D. student of Kane's) in forming a company built around AUTOLEV, OnLine Dynamics, Inc <www.autolev.com>. AUTOLEV is based on a symbolic construction of the equations of motion, and permits a user to formulate exact, literal motion equations. It is not restricted to any particular system topology and can be applied to the analysis of *any* holonomic or nonholonomic discrete dynamic system of particles and rigid bodies and allows a user to choose any convenient set of variables for the description of configuration and motion. Within the context of robotics, robots made up of a collection of rigid bodies containing revolute joints, prismatic joints, and closed kinematical loops can be analyzed in the most physically meaningful and computationally efficient terms possible.

While the use of AUTOLEV requires a user who is well-versed in dynamical and kinematical principles, the philosophy that has governed its development and evolution seeks to achieve a favorable balance between the tasks performed by an analyst and those relegated to a computer. Human analysts are best at creative activities that involve the integration of an individual's skills and experiences. A computer, on the other hand, is best suited to well-defined, repetitive operations. With AUTOLEV, an analyst is required to define only the number of degrees of freedom, the symbols with which quantities of interest are to be represented, and the *steps* that he would ordinarily follow in a manual derivation of the equations of motion. The computer is left to perform the mathematical operations, bookkeeping, and computer coding required to actually produce equations of motion and a computer program for their numerical solution. This approach necessitates that a user be skilled in dynamical principles but allows virtually any rigid multibody problem to be addressed. AUTOLEV thus does not necessarily address the needs of the widest possible class of users but rather the widest possible class of problems.

Beyond simply providing a convenient means of implementing Kane's method, AUTOLEV can also be used to symbolically evaluate mathematical expressions of many types, thus allowing it to be used to formulate equations of motion based on any variant of Newton's second law ($\mathbf{F} = m\mathbf{a}$). AUTOLEV is similar to other symbolic manipulation programs, such as MATLAB, Mathematica, and Maple, although it was specifically designed for motion analysis. It is thus a useful tool for solving problems in mechanics dealing with kinematics, mass and inertia properties, statics, kinetic energies, linear, or angular momentum, etc. and can perform either symbolic or numerical operations on scalars, matrices, vectors, and dyadics. It can also be used to automatically formulate linearized dynamical equations of motion, for example, for control system design, and can be used to produce ready-to-compile-link-and-run C and FORTRAN programs for numerical solving equations of motion. A complete list of the features and properties can be found at <www.autolev.com>.

6.7.4 ADAMS

The theory upon which ADAMS is based was originally developed by Dr. Nick Orlandea [10] at the University of Michigan and was subsequently transformed into the ADAMS computer program, marketed, and supported by Mechanical Dynamics, Inc., which was purchased in 2002 by MSC.Software. Although ADAMS was not originally developed using Kane's method, the evolution of ADAMS has been strongly influenced by Kane's approach to dynamics. Dr. Robert Ryan, a former Ph.D. student of Prof. Kane as well as a former professor at the University of Michigan, was directly involved in the development of ADAMS at Mechanical Dynamics, starting in 1988 as Vice President of Product Development, later as chief operating officer and executive vice president, and finally as president. Since the acquisition of Mechanical Dynamics by MSC.Software in 2002, Dr. Ryan has been executive vice president-products at MSC. One of Kane's method on ADAMS was through Dr. Ryan's early work on formulating dynamical equations for systems containing flexible members [29].

The broad features of the ADAMS "MSC.ADAMS suite of software packages" <www.mscsoftware.com> are described as including a number of modules that focus on various areas of application, such as ADAMS/3D Road (for the creation of roadways such as parking structures, racetracks, and highways), ADAMS/Aircraft (for the creation of aircraft virtual prototypes), ADAMS/Car (for the creation of virtual prototypes of complete vehicles combining mathematical models of the chassis subsystems, engine and driveline subsystems, steering subsystems, controls, and vehicle body), ADAMS/Controls (for integrating motion simulation and control system design), ADAMS/Exchange (for transferring data between MSC.ADAMS products and other CAD/CAM/CAE software), ADAMS/Flex (for studying the influence of flexible parts interacting in a fullly mechanical system), ADAMS/Linear (for linearizing or simplifying nonlinear MSC.ADAMS equations), ADAMS/Tire (for simulating virtual vehicles with the ADAMS/Tire product portfolio), ADAMS/Vibration (for the study of forced vibrations using frequency domain analysis), and MECHANISM/Pro (for studying the performance of mechanical simulations with Pro/ENGINEER data and geometry).

6.7.5 Working Model

As with ADAMS, Working Model is not directly based on Kane's method but was strongly influenced by Kane's overall approach to dynamics through the fact that one of the primary developers of Working Model is a former student of Kane, Dr. Paul Mitiguy (currently Director of Educational Products at MSC.Software). The dynamics engine behind Working Model is based on the use of the *constraint force algorithm* [15]. The constraint force algorithm circumvents the large set of equations produced with traditional Newton-Euler formulations by breaking the problem into two parts. The first part "uses linear algebra techniques to isolate and solve a relatively small set of equations for the constraint forces only" and then "with all the forces on each body known, it solves for the accelerations, without solving *any* additional linear equations" [15]. The constraint force algorithm is described by Dr. Mitiguy as the "inverse of Kane's method" in that Kane's method solves first for accelerations and then separately for forces. Although originally developed as an outgrowth of the software package Interactive Physics developed at Knowledge Revolution <www.krev.com>, Knowledge Revolution was acquired by MSC.Software in 1998 and Working Model is now sold and supported through MSC.Software (as is ADAMS, as described above). Working Model is described at <www.krev.com> as the "best selling motion simulation product in the world," and features of the program include the ability to

- analyze designs by measuring force, torque, acceleration, or other mechanical parameters acting on an object or set of objects;
- view output as vectors or in numbers and graphs in English or metric units;
- import 2D CAD drawings in DXF format;
- simulate nonlinear or user events using a built-in formula language;
- design linkages with pin joints, slots, motors, springs, and dampers;
- create bodies and specify properties such as mass properties, initial velocity, electrostatic charge, etc;
- simulate contact, collisions, and friction;
- analyze structures with flexible beams and shear and bending moment diagrams; and
- record simulation data and create graphs or AVI video files.

Working Model is a particularly attractive simulation package in an educational environment due to its graphical user interface, which allows it to be easily learned and mechanisms to be easily animated. Another positive aspect of Working Model is its approach to modeling. Modeling mechanisms in Working Model is performed "free-hand" and is noncoordinate based, thus allowing models to be created without the difficulties associated with developing complex geometries. The graphical user interface facilitates visualization of the motion of mechanisms and thus output is not restricted to simply numerical data or the analysis of a small number of mechanism configurations. Working model also allows the analysis of an array of relatively sophisticated problems, including those dealing with impacts, Coulomb friction, and gearing.

A three-dimensional version of working model was developed and marketed by Knowledge Revolution in the 1990s, but after the acquisition of Knowledge Revolution by MSC.Software, working model 3D became the basis for the creation of MSC.VisualNASTRAN 4D <www.krev.com>. VisualNASTRAN 4D integrates computer-aided-design, motion analysis, finite-element-analysis, and controls technologies into a single modeling system. Data can be imported from a number of different graphics packages, including SolidWorks, Solid Edge, Mechanical Desktop, Inventor, and Pro/ENGINEER, and then used to perform both motion simulations and finite element analyses. The interface of VisualNASTRAN 4D is similar to that of Working Model, and thus relatively complex models can be created easily and then tested and refined. In addition to the features of Working Model, VisualNASTRAN allows stress and strain to be calculated at every time step of a simulation using a linear finite element solver. Normal modes and buckling finite element analyses are also possible, as is the possibility of adding animation effects, such as photorealistic rendering, texture mapping, lights, cameras, and keyframing.

6.8 Summary

From all of the foregoing, the wide range of applications for which Kane's method is suitable is clear. It has been applied over the years to a variety of robotic and other mechanical systems and has been found to be both analytically convenient and computationally efficient. Beyond the systematic approach inherent in the use of Kane's method for formulating dynamical equations, there are two particular features of the method that deserve to be reiterated. The first is that Kane's method automatically eliminates "nonworking" forces in such a way that they may be neglected from an analysis at the outset. This feature was one of the fundamental bases for Kane's development of his approach and saves an analyst the need to introduce, and subsequently eliminate, forces that ultimately make no contribution to the equations of motion. The second is that Kane's method allows the use of motion variables (generalized speeds) that permit the selection of these variables as not only individual time-derivatives of the generalized coordinates but in fact any convenient linear combinations of them. These two features taken together mean that Kane's method not only allows dynamical equations of motion to be formulated with a minimum of effort but also enables one to express the equations in the simplest possible form.

The intent of this chapter has been to illustrate a number of issues associated with Kane's method. Brief historical information on the development of Kane's method has been presented, the fundamental essence of Kane's method and its relationship to Lagrange's equations has been discussed, and two simple examples have been given in a step-by-step format to illustrate the application of Kane's method to typical robotic systems. In addition, the special features and advantages of Kane's method have been described for cases related to formulating linearized dynamical equations, equations of motion for systems subject to constraints, and equations of motion for systems with continuous elastic elements. Brief summaries of a number of articles in the robotics literature have also been given to illustrate the popularity of Kane's method in robotics and the types of problems to which it can be applied. Finally, the influence of Kane and Kane's method on commercially available dynamics software packages has been described and some of the features of the most widely used of those packages presented. This chapter has hopefully provided those not familiar with Kane's method, as well as those who are, with a broad understanding of the fundamental issues related to Kane's method as well as a clear indication of the ways in which it has become the basis for formulating, either by hand or with readily available software, a vast array of dynamics problems associated with robotic devices.

References

[1] Kane, T.R., Dynamics of nonholonomic systems, *J. App. Mech.*, 28, 574, 1961.
[2] Kane, T.R. and Wang, C.F., On the derivation of equations of motion, *J. Soc. Ind. App. Math.*, 13, 487, 1965.

[3] Kane, T.R., *Analytical Elements of Mechanics*, vol. 1, Academic Press, New York, 1959.
[4] Kane, T.R., *Analytical Elements of Mechanics*, vol. 2, Academic Press, New York, 1961.
[5] Kane, T.R. and Levinson, D.A., Formulation of equations of motion for complex spacecraft, *J. Guid. Contr.*, 3, 99, 1980.
[6] Roberson, R.E. and Schwertassek, R., *Dynamics of Multibody Systems*, Springer-Verlag, Berlin, 1988.
[7] Shabana, A.A., *Dynamics of Multibody Systems*, John Wiley & Sons, New York, 1989.
[8] Schiehlen, W. (ed.), *Multibody Systems Handbook*, Springer-Verlag, Berlin, 1990.
[9] Amirouche, F.M.L., *Computational Methods in Multibody Dynamics*, Prentice Hall, Englewood Cliffs, New Jersey, 1992.
[10] Orlandea, N., Chace, M.A., and Calahan, D.A., A sparsity-oriented approach to the dynamic analysis and design of mechanical systems, part I and II, *J. Eng. Ind.*, 99, 773, 1977.
[11] Haug, E.J., Nikravesh, P.E., Sohoni, V.N., and Wehage, R.A., Computer aided analysis of large scale, constrained, mechanical systems, *Proc. 4th Int. Symp. on Large Engrg. Systems*, Calgary, Alberta, 1982, 3.
[12] Schiehlen, W.O. and Kreuzer, E.J., Symbolic computational derivations of equations of motion, *Proc. IUTAM Symp. on Dyn. of Multibody Sys.*, Munich, 1977, 290.
[13] Rosenthal, D.E. and Sherman, M.A., High performance multibody simulations via symbolic equation manipulation and Kane's method, *J. Astro. Sci.*, 34, 223, 1986.
[14] Schaechter, D.B. and Levinson, D.A., Interactive computerized symbolic dynamics for the dynamicist, *Proc. Amer. Contr. Conf.*, 1988, 177.
[15] Mitiguy, P.C. and Banerjee, A.K., A constraint force algorithm for formulating equations of motion, *Proc. First Asian Conf. on Multibody Dynamics*, 2002, 606.
[16] Lesser, M., A geometrical interpretation of Kane equations, *Proc. Royal Soc. London Ser. A - Math., Phy., Eng. Sci.*, Royal Soc. London, London, 1992, 69.
[17] Van Woerkom, P.Th.L.M., De Boer, A., Ellenbroek, M.H.M., and Wijker, J.J., Developing algorithms for efficient simulation of flexible space manipulator operations, *Acta Astronautica*, 36, 297, 1995.
[18] Piedboeuf, J.-C., Kane's equations or Jourdain's principle? *Proc. 36th Midwest Symp. on Circuits and Sys.*, IEEE, 1993, 1471.
[19] Kane, T.R., Teaching of mechanics to undergraduates, *Int. J. Mech. Eng. Ed.*, 6, 286, 1978.
[20] Kane, T.R. and Levinson, D.A., *DYNAMICS: Theory and Applications*, McGraw-Hill, New York, 1985.
[21] Kane, T.R., *DYNAMICS*, Holt, Rinehart and Winston, New York, 1968.
[22] Mitiguy, P.C and Kane, T.R., Motion variables leading to efficient equations of motion, *Int. J. Robotics Res.*, 15, 522, 1996.
[23] Scheinman, V.D., Design of a computer controlled manipulator, Engineer thesis, Stanford University, Stanford, 1969.
[24] Kane, T.R. and Levinson, D.A., The use of Kane's dynamical equations in robotics, *Int. J. Robotics Res.*, 2, 3, 1983.
[25] Wampler, C.W., A comparative study of computer methods in manipulator kinematics, dynamics, and control, Ph.D. thesis, Stanford University, Stanford, 1985.
[26] Kane, T.R., Ryan, R.R., and Banerjee, A.K., Dynamics of a cantilever beam attached to a moving base, *J. Guid., Contr., Dyn.*, 10, 139, 1987.
[27] Buffinton, K.W., Dynamics of beams moving over supports, Ph.D. thesis, Stanford University, Stanford, 1985.
[28] Wampler, C.W., Buffinton, K.W., and Jia, S.H., Formulation of equations of motion for systems subject to constraints, *J. Appl. Mech.*, 52, 465, 1985.
[29] Ryan, R.R., Flexibility modeling methods in multibody dynamics, Ph.D. thesis, Stanford University, Stanford, 1986.

[30] Huston, R.L., Multibody dynamics formulations via Kane's equations, in *Mechanics and Control of Large Flexible Structures,* Jenkins, J.L., ed., American Institute of Aeronautics and Astronautics (AIAA), Reston, Virginia, 1990, 71.

[31] Buffinton, K.W., Dynamics of elastic manipulators with prismatic joints, *J. Dyn. Sys., Meas., Contr.,* 114, 41, 1992.

[32] Amirouche, F.M.L. and Xie, M., Explicit matrix formulation of the dynamical equations for flexible multibody systems: a recursive approach, *Comp. and Struc.,* 46, 311, 1993.

[33] Buffinton, K.W. and Kane, T.R., Dynamics of a beam moving over supports, *Int. J. Solids Structures,* 21, 617, 1985.

[34] Baruh, H., *Analytical Dynamics,* McGraw-Hill, Boston, 1999.

[35] Josephs, H. and Huston, R.L., *Dynamics of Mechanical Systems,* CRC Press, Boca Raton, FL, 2002.

[36] Wloka, D. and Blug, K., Simulation der dynamik von robotern nach dem verfahren von Kane, *Robotersysteme,* 1, 211, 1985.

[37] Gallay, G., Girard, F., Auclair, J.-M., and Gerbeaux, F.-X., Dynamique de mecanismes flexibles, *Mecanique-Materiaux-Electricite,* 416, 17, 1986.

[38] Angeles, J., Ma, O., and Rojas, A., Algorithm for the inverse dynamics of n-axis general manipulators using Kane's equations, *Comp. Math. App.,* 17, 1545, 1989.

[39] Huston, R.L., The use of Kane's method in the modeling and simulation of robotic systems, *Proc. IMACS Symp.,* North-Holland, Amsterdam, 1989, 221.

[40] Tarn, T.J., Shoults, G.A., and Yang, S.P., Dynamic model of an underwater vehicle with a robotic manipulator using Kane's method, *Autonomous Robots,* 3, 269, 1996.

[41] Thanjavur, K., Ease of dynamic modeling of wheeled mobile robots (WMRs) using Kane's approach, *Proc. 1997 Int. Conf. on Robotics and Automation,* IEEE, 1997, 2926.

[42] Luecke, G.R. and Zafer, N., Haptic feedback for virtual assembly, *Proc. 1998 Conf. on Telemanipulator and Telepresence Technologies,* SPIE, 1998, 115.

[43] Xu, X.-R., Chung, W.-J., Choi, Y.-H., and Ma, X.-F., Method for the analysis of robot dynamics, *Modelling, Meas. Contr.,* 66, 59, 1998.

[44] Lee, K.Y. and Dissanayake, M.W.M.G., Dynamic modeling and motion planning for a hybrid legged-wheeled mobile vehicle, *J. Mech. Eng. Sci.,* 215, 7, 2001.

[45] Tanner, H.G. and Kyriakopoulos, K.J., Kane's approach to modeling mobile manipulators, *Advanced Robotics,* 16, 57, 2002.

[46] Kane, T.R. and Levinson, D.A., Multibody Dynamics, *J. Appl. Mech.,* 50, 1071, 1983.

[47] Huston, R.L., Multibody dynamics: modeling and analysis methods, *Appl. Mech. Rev.,* 44, 109, 1991.

7
The Dynamics of Systems of Interacting Rigid Bodies

7.1	Introduction 7-1
7.2	Newton's Law and the Covariant Derivative 7-2
7.3	Newton's Law in a Constrained Space 7-5
7.4	Euler's Equations and the Covariant Derivative 7-8
7.5	Example 1: Euler's Equations for a Rigid Body 7-11
7.6	The Equations of Motion of a Rigid Body 7-13
7.7	Constraint Forces and Torques between Interacting Bodies................................... 7-15
7.8	Example 2: Double Pendulum in the Plane 7-16
7.9	Including Forces from Friction and from Nonholonomic Constraints 7-18
7.10	Example 3: The Dynamics of the Interaction of a Disk and a Link 7-19
7.11	Example 4: Including Friction in the Dynamics....... 7-21
7.12	Conclusions 7-22

Kenneth A. Loparo
Case Western Reserve University

Ioannis S. Vakalis
Institute for the Protection and Security of the Citizen (IPSC) European Commission

7.1 Introduction

In this chapter, we begin by examining the dynamics of rigid bodies that interact with other moving or stationary rigid bodies. All the bodies are components of a multibody system and are allowed to have a single point of interaction that can be realized through contact or some type of joint constraint. The kinematics for the case of point contact has been formulated in previous works [52–54]. On the other hand, the case of joint constraints can be easily handled because the type of joint clearly defines the degrees of freedom that are allowed for the rigid bodies that are connected through the joint. Then we will introduce a methodology for the description of the dynamics of a rigid body generally constrained by points of interaction. Our approach is to use the geometric properties of Newton's equations and Euler's equations to accomplish this objective. The methodology is developed in two parts, we first investigate the geometric properties of the basic equations of motion of a rigid body. Next we consider a multibody system that includes point interaction that can occur through contact or some type of joint constraint. Each body is considered initially as an independent unit and forces and torques are applied to the bodies through the interaction points. There is a classification of the applied forces with respect to their type (e.g., constraint, friction, external). From the independent dynamic equations of motion for each body we can derive a reduced model of the overall system, exploiting the geometric properties of the physical laws. The framework that is developed can also be used to solve control problems in a variety of settings. For example, in addition to each point contact constraint, we can impose holonomic and/or nonholonomic constraints on individual bodies. Holonomic and nonholonomic constraints have been studied extensively in the areas of mechanics and dynamics, see, for example [4,37]. Holonomic control problems have been studied recently in the context of robotics and manipulation, for example [33,34]. On the other hand, nonholonomic control problems, which are more difficult to solve, have recently attracted the attention of a number of researchers and an extensive published literature is developing. Specific classes of problems have been studied, such as mechanical systems sliding or rolling in the plane, see, for example [9–11]. Another category of nonholonomic control problems deals with mobile robots and wheeled vehicles, for example [7,8,15,28–31,36,42–48,50,51]. Spacecrafts and space robots are a further class of mechanical systems with holonomic constraints. The reason for this characterization is the existence of certain model symmetries that correspond to conserved quantities. If these quantities are not integrable, then we have a nonholonomic problem. A number of works have been published in this area, see, for example [16,21,26,35,38,49,55], and the literature in this area is still developing. Important techniques based on the concept of geometric phases have also been developed for the solution of holonomic control problems [9,19,27,32,38–40].

7.2 Newton's Law and the Covariant Derivative

Application of Newton's law is a very familiar technique for determining the equations of motion for an object or a particle in three space ($\mathbb{R}^3$). The fact that we use Newton's law in $\mathbb{R}^3$, which has some nice geometric properties, disguises the intrinsic geometric aspects of the law. A geometric interpretation of Newton's law becomes apparent when we use it in cases where the object or the particle is constrained to move on a subset Q, or a submanifold M, in $\mathbb{R}^3$. As we have seen from some of our previous work [53], such submanifolds result when the object is confined to move in contact with other surfaces in $\mathbb{R}^3$, which can be stationary or moving (other objects). Whenever an object is in contact with a surface in $\mathbb{R}^3$ then we have "reaction" forces as a result of the contact, and three reaction forces are applied to the object and on the surface. These forces in the classical formulation are vectors vertical to the common tangent plane between the objects or the object and the surface at the point of contact, and it is necessary to distinguish these forces from the friction forces that are introduced tangent to the contacting surfaces. The "reaction" forces are introduced in Newton's formulation of the equations of motion because the object is constrained to move on a given surface. We begin our development by considering the original form of Newton's law in $\mathbb{R}^3$. Assume that there is an object moving in $\mathbb{R}^3$, then Newton's law states that the acceleration multiplied by the mass of the object is equal to the applied forces on the object. We can elaborate more on this simple form of the law so that its geometric aspects become more obvious. This can be done using concepts from Riemannian geometry in $\mathbb{R}^3$. The kinetic energy metric in $\mathbb{R}^3$, which when expressed in local coordinates is given by $E = \frac{1}{2}\sigma_{ij}v^i v^j$.[1] The kinetic energy metric gives a Riemannian structure to $\mathbb{R}^3$, denoted by $\langle \cdot, \cdot \rangle : T\mathbb{R}^3 \times T\mathbb{R}^3 \longrightarrow \mathbb{R}$ and $E = \frac{1}{2}\sigma_{ij}v^i v^j = \frac{1}{2}\langle v_i, v_j \rangle$

Definition 7.1 A Riemannian structure on a manifold N is a covariant tensor k of type $T_2^0(N)$. The covariant tensor $k : TN \times TN \longrightarrow \mathbb{R}$, is nondegenerate and positive definite.

Nondegenerate and positive definite mean that $k(v_m, v_m) > 0, \forall v_m \in T_m N$ when $v_m \neq 0$. In this chapter we will consider only positive definite Riemannian metrics. Some of the following definitions and theorems are also valid for pseudo-Riemannian metrics [1]. The Riemannian structure is important because later in the development we will need to induce a metric from one manifold to another manifold

---

[1] Tensor notation requires superscripts for coordinates instead of subscripts.

through an immersion. This requires a positive definite Riemannian metric on the first manifold and the extension to a pseudo-Riemannian metric requires further investigation and will not be discussed here.

In order to establish a geometric form for the acceleration component of Newton's law, we need to introduce the notion of a *connection* on a general Riemannian manifold. The connection is used to describe the acceleration along curves in more general spaces like a Riemannian manifold, see Boothby [12].

Definition 7.2 A C^∞ *connection* ∇ on a manifold N is a mapping $\nabla : \mathcal{X}(N) \times \mathcal{X}(N) \longrightarrow \mathcal{X}(N)$ defined by $\nabla(X, Y) \longrightarrow \nabla_X Y, \nabla$ satisfies the linearity properties for all C^∞ functions f, g on N and $X, X', Y, Y' \in \mathcal{X}(N)$:

1. $\nabla_{fX+gX'} Y = f(\nabla_X Y) + g(\nabla_{X'} Y)$
2. $\nabla_X(fY + gY') = f\nabla_X Y + g\nabla_X Y' + (Xf)Y + (Xg)Y'$

Here, $\mathcal{X}(M)$ denotes the set of C^∞ vector fields on the manifold M.

Definition 7.3 A connection on a Riemannian manifold N is called a *Riemannian connection* if it has the additional properties:

1. $[X, Y] = \nabla_X Y - \nabla_Y X$
2. $X\langle Y, Y'\rangle = \langle \nabla_X Y, Y'\rangle + \langle Y, \nabla_X Y'\rangle$

Here, $[\cdot, \cdot]$ denotes the Lie bracket.

Theorem 7.1 *If N is a Riemannian manifold, then there exists a uniquely determined **Riemannian connection** on N.*

A comprehensive proof of this theorem can be found in W.M. Boothby [12]. The acceleration along a curve $c(t) \in N$ is given by the connection $\nabla_{\dot{c}(t)}\dot{c}(t)$. In this context the Riemannian connection denotes the derivative of a vector field along the direction of another vector field, at a point $m \in N$.

To understand how the covariant derivative, defined on a submanifold M of $\mathbb{R}^n$, leads to the abstract notion of a connection, we need to introduce the derivative of a vector field along a curve in $\mathbb{R}^n$. Consider a vector field X defined on $\mathbb{R}^n$ and a curve $c(t) \in \mathbb{R}^n$. Let $X(t) = X \mid_{c(t)}$, then the derivative of $X(t)$ denoted by $\dot{X}(t)$ is the rate of change of the vector field X along this curve. Consider a point $c(t_0) = p \in \mathbb{R}^n$ and the vectors $X(t_0) \in T_{c(t_0)}\mathbb{R}^n$ and $X(t_0 + \Delta t) \in T_{c(t_0+\Delta t)}\mathbb{R}^n$. We can use the natural identification of $T_{c(t_0)}\mathbb{R}^n$ with $T_{c(t_0+\Delta t)}\mathbb{R}^n$, and the difference $X(t_0 + \Delta t) - X(t_0)$ can be defined in $T_{c(t_0)}\mathbb{R}^n$. Consequently, the derivative $\dot{X}(t_0)$ can be defined by

$$\dot{X}(t_0) = \lim_{\Delta t \to 0} \frac{X(t_0 + \Delta t) - X(t_0)}{\Delta t}$$

Consider a submanifold M imbedded in $\mathbb{R}^n$, and a vector field X on M, not necessarily tangent to M. Then, the derivative of X along a curve $c(t) \in M$ is denoted by $\dot{X}(t) \in T_{c(t)}\mathbb{R}^n$. At a point $c(t_0) = p \in M$, the tangent space $T_p\mathbb{R}^n$ can be decomposed into two mutually exclusive subspaces $T_p\mathbb{R}^n = T_pM \oplus T_pM^\perp$. Consider the projection $p_1 : T_p\mathbb{R}^n \longrightarrow T_pM$, the covariant derivative of X along $c(t) \in M$ is defined as follows:

Definition 7.4 The *covariant* derivative of a vector field X on a submanifold M of $\mathbb{R}^n$ along a curve $c(t) \in M$ is the projection $p_1(\dot{X}(t))$ and is denoted by $\frac{DX}{dt}$.

An illustration of the covariant derivative is given in Figure 7.1. The covariant derivative gives rise to the notion of a connection through the following construction: Consider a curve $c(t) \in M$, the point $p = c(t_0)$, and the tangent vector to the curve $X_p = \dot{c}(t_0)$ at p. We can define the map $\nabla_{X_p} : T_pM \longrightarrow T_pM$ by $\nabla_{X_p} Y : X_p \longrightarrow \frac{DY}{dt}\mid_{t=t_0}$, along any curve $c(t) \in M$ such that $\dot{c}(t_0) = X_p$ and $Y \in M$. Along the curve $c(t)$ we have $\nabla_{\dot{c}(t)} Y = \frac{DY}{dt}$. The connection can be defined as a map $\nabla : \mathcal{X}(M) \times \mathcal{X}(M) \longrightarrow \mathcal{X}(M)$, where

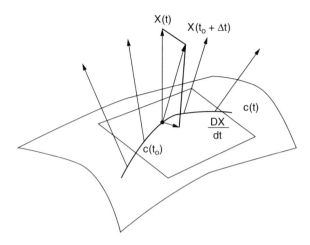

FIGURE 7.1 The covariant derivative of a vector field X.

$\nabla : (X, Y) \longrightarrow \nabla_X Y$. In a general Riemannian manifold, the connection is defined first and then the covariant derivative is introduced by $\frac{DY}{dt} = \nabla_{\dot{c}(t)} Y$. In general, the acceleration on a Riemannian manifold N, along a path $c(t)$, is given by

$$\nabla_{\dot{c}(t)} \dot{c}(t)$$

It then follows that the equation for a geodesic curve is given by $\nabla_{\dot{c}(t)} \dot{c}(t) = 0$.

Consider now a curve $c : I \longrightarrow \mathbb{R}^3$, which represents the motion of a body or a particle in $\mathbb{R}^3$. Computing the acceleration of the body along this curve provides only one part of the expression needed for Newton's law. Because $\mathbb{R}^3$ is a Riemannian manifold, the acceleration along a curve is given by the connection $\nabla_{\dot{c}(t)} \dot{c}(t) \in T\mathbb{R}^3$.

Consequently, we need to examine the geometric properties of the forces applied to the object. As mentioned previously, the forces in the classical approach are considered as vectors in $\mathbb{R}^3$, but this is not sufficient for a geometric interpretation of Newton's law. Geometrically the forces are considered to be 1-forms on the cotangent space associated with the state space of the object, which in our case is $T^*\mathbb{R}^3$. The main reason that the forces are considered to be 1-forms is because of the concept of "work." Actually, the work done by the forces along $c(t)$ is given by the integral of the 1-forms representing the forces that are acting along $c(t)$.

Although we have provided a geometric interpretation for the acceleration and the forces, there is still a missing link because the acceleration is a vector field on the tangent space $T\mathbb{R}^3$ and the forces are 1-forms on the cotangent space $T^*\mathbb{R}^3$. Thus, we cannot just multiply the acceleration by the mass and equate the product with the forces. The missing link is provided by the kinetic energy metric, using its properties as a covariant tensor. More specifically, if we consider a covariant tensor k on a manifold N of the type $T_2^0(N)$, then there is a map associated with the tensor called the *flat map* of k (refer to R. Abraham and J.E. Marsden [1] for more details). Let $k^\flat$ denote the flat map, defined by

$$\forall v \in TN \quad k^\flat(v) = k(v, \cdot) \quad \text{and} \quad k^\flat : TN \longrightarrow T^*N$$

thus $k^\flat(v)$ is a 1-form on T^*N. Consider the kinetic energy tensor $E = \frac{1}{2}\sigma(u, u)$ on $\mathbb{R}^3$, then $\sigma^\flat : T\mathbb{R}^3 \longrightarrow T^*\mathbb{R}^3$ and $\forall v \in T\mathbb{R}^3$, $\sigma^\flat(v)$ is a 1-form in $T^*\mathbb{R}^3$.

If we let the flat map of the kinetic energy tensor $\sigma^\flat$ act on the covariant derivative $\nabla_{\dot{c}(t)} \dot{c}(t)$, we obtain a 1-form in $T^*\mathbb{R}^3$. The geometric form of Newton's law is then given by

$$\sigma^\flat(\nabla_{\dot{c}(t)} \dot{c}(t)) = F \tag{7.1}$$

where F denotes the 1-forms corresponding to the forces applied on the body.

If the body is constrained to move on a submanifold M of $\mathbb{R}^3$, the state space of the object is no longer $\mathbb{R}^3$ but the submanifold M of $\mathbb{R}^3$.

7.3 Newton's Law in a Constrained Space

Next, we need to examine how to describe the dynamics of the constrained object on the submanifold M, instead of $\mathbb{R}^3$. More specifically we want to find a description of Newton's law on the constrained submanifold M, to develop dynamic models for systems of interacting rigid bodies, with the ultimate objective of developing control strategies for this class of systems. In order to constrain the dynamic equation $\sigma^\flat(\nabla_{\dot{c}(t)}\dot{c}(t)) = F$ on the submanifold M, we must constrain the different quantities involved in Newton's law; we begin with the restriction of the kinetic energy tensor σ to M.

Consider a smooth map $f : N \longrightarrow N'$, where N, N' are C^∞ manifolds. Let $T_k^0(N')$ denote the set of covariant tensors of order k on N'. Then if $\Phi \in T_k^0(N')$, we can define a tensor on N in the following way:

$$f^* : T_k^0(N') \longrightarrow T_k^0(N)$$
$$(f^*\Phi)_p(u_1, \ldots, u_k) = \Phi_{f(p)}(f_*u_1, \ldots, f_*u_k)$$

where

$$u_i \in T_pN \quad i = 1, \ldots, k \quad \text{and} \quad f_{*p} : T_pN \longrightarrow T_{f(p)}N'$$

denotes the derived map of f. The map f^* is called the *pull-back*. Earlier we denoted the dual map of f_*, by f^*. This notation is consistent with that of a pull-back because the dual $f^* : T^*_{f(p)}N' \longrightarrow T^*_pN$, pulls back 1-forms which are tensors of type $T_1^0(N')$ and belong to $T^*_{f(p)}N'$, to 1-forms in T^*_pN, which are tensors of type $T_1^0(N)$.

The pull-back in the algebraic sense can be defined without the further requirement of differentiability, and the manifold N need not be a submanifold of N'. Next, we examine under what conditions we can pull-back a covariant tensor on a manifold N and find a coordinate expression in the local coordinates of N. The answer is given by the following theorem from R. Abraham, J. E. Marsden, and T. Ratiu [2]:

Theorem 7.2 *Let N and N' be two finite dimensional manifolds and let $f : N \longrightarrow N'$ be a C^∞ map with a local expression[2] $y^i = f^i(x^1, \ldots, x^m)$ $i = 1, \ldots, n$ in the appropriate charts, where $\dim(N) = m$ and $\dim(N') = n$. Then, if t is a tensor in $T_r^0(N')$, the pull-back f^*t is a tensor in $T_r^0(N)$ and has the coordinate expression:*

$$(f^*t)_{j_1,\ldots,j_s} = \frac{\partial y^{k_1}}{\partial x^{j_1}} \cdots \frac{\partial y^{k_s}}{\partial x^{j_s}} t_{k_1,\ldots,k_s} \circ f$$

Here $(f^*t)_{j_1,\ldots,j_s}$ and $t_{k_1,\ldots,k_s}$, denote the components of the tensors f^*t and t, respectively. To pull-back a tensor from a manifold N' to a manifold N requires a C^∞ map $f : N \longrightarrow N'$, with a coordinate representation $y^j = f^j(x^1, \ldots, x_m)$. There is a certain condition that guarantees that a map between two manifolds can have a coordinate representation, and thus can be used to pull-back a covariant tensor. The notion of an *immersion* is important for the further developement of these ideas (see F. Brickell and R.S. Clark [13] for more details).

Definition 7.5 *Let N' and N be two finite dimensional differentiable manifolds and $f : N \longrightarrow N'$ a C^∞ map, then f is called immersion if for every point $a \in N$ $\text{rank}(f) = \dim(N)$.*

---

[2] Here we also use tensor notation for the coordinates.

When studying the dynamics of rigid bodies, we deal with Riemannian metrics, which are tensors of type $T_2^0(N')$. Using the general theorem for pulling back covariant tensors and the definition of an immersion, the following theorem results, see F. Brickell and R.S. Clark [13]:

Theorem 7.3 *If a manifold N' has a given positive definite Riemannian metric then any global immersion $f : N \longrightarrow N'$ induces a positive definite Riemannian metric on N.*

A manifold N is a submanifold of a manifold N' if N is a subset of N' and the natural injection $i : N \longrightarrow N'$ is an immersion. Submanifolds of this type are called immersed submanifolds. If in addition, we restrict the natural injection to be one to one, then we have an imbedded submanifold [13]. There is an alternative definition for a submanifold where the subset property is not included along with the C^∞ structure, see W.M. Boothby [12]. In our case, because we are dealing with the constrained configuration space of a rigid body which is subset of $\mathbb{E}(3)$, it seems more natural to follow the definition of a submanifold given by F. Brickell and R.S. Clark [13].

From the previous theorem we conclude that we can induce a Riemannian metric on a manifold N if there is a Riemannian manifold N' and a map $f : N \longrightarrow N'$, which is an immersion. The manifold N need not be a submanifold of N'. In general, N will be a subset of N' and the map $f : N \longrightarrow N'$ is needed to induce a metric. We consider a system of rigid bodies with each body of the system as a separate unit moving in a constraint state space because of the presence of some joint or point contact beween it and the rest of the bodies. The case of a joint constraint is easier to deal with because the type of joint also defines the degrees of freedom of the constrained body. The degrees of freedom then define the map that describes the constrained configuration space of motion as a submanifold of $\mathbb{E}(3)$. Accordingly we can then induce a Riemannian metric on the constrained configuration space. The case of point contact is more complicated. We have studied the problem of point contact, where the body moves on a smooth surface B in [53]. In this work we have shown that the resulting constrained configuration space M is a subset of $\mathbb{E}(3)$. M is also a submanifold of $\mathbb{E}(3) \times B$ and there is a map $\mu_1 : M \longrightarrow \mathbb{E}(3)$. This map is not necessarily the natural injection $i : M \longrightarrow \mathbb{E}(3)$. The map μ_1 therefore should be an immersion so that it induces a Riemannian metric on M from $\mathbb{E}(3)$. This is not generally the case when we deal with the dynamics of constrained rigid body motions defined by point contact.

Using this analysis we can now study the geometric form of Newton's law for bodies involved in constrained motion. To begin with we need to endow the submanifold M of $\mathbb{R}^3$ with a Riemannian structure. According to the discussion above this is possible using the pull-back of the Riemannian metric σ on $\mathbb{R}^3$ through the natural injection $j : M \longrightarrow \mathbb{R}^3$, where j is by definition an immersion. We let $\bar{\sigma} = j^*\sigma$ denote the induced Riemannian structure on the submanifold M by the pull-back j^*. We can explicitly compute the coordinate representation of the pull-back of the Riemannian metric $\bar{\sigma}$ if we use the fact that positive definite tensors of order $T_2^0(N')$ on a manifold N' can be represented as positive definite symmetric matrices. In our case, the Riemannian metric in $\mathbb{R}^3$ can be represented as a 3×3 positive definite symmetric matrix given by

$$\sigma = \begin{pmatrix} \sigma_{11} & \sigma_{12} & \sigma_{13} \\ \sigma_{12} & \sigma_{22} & \sigma_{23} \\ \sigma_{13} & \sigma_{23} & \sigma_{33} \end{pmatrix}$$

Assume that the submanifold M is two-dimensional and has coordinates $\{x^1, x^2\}$, and $\{y^1, y^2, y^3\}$ are the coordinates of $\mathbb{R}^3$. Then, the derived map $j_* : TM \longrightarrow T\mathbb{R}^3$ has the coordinate representation:

$$j_* = \begin{pmatrix} \frac{\partial j^1}{\partial x^1} & \frac{\partial j^1}{\partial x^2} \\ \frac{\partial j^2}{\partial x^1} & \frac{\partial j^2}{\partial x^2} \\ \frac{\partial j^3}{\partial x^1} & \frac{\partial j^3}{\partial x^2} \end{pmatrix}$$

The induced Riemannian metric $\bar{\sigma}$ on M (by using the pull-back) has the following coordinate representation:

$$\bar{\sigma} = j^*\sigma = \begin{pmatrix} \frac{\partial j^1}{\partial x^1} & \frac{\partial j^2}{\partial x^1} & \frac{\partial j^3}{\partial x^1} \\ \frac{\partial j^1}{\partial x^2} & \frac{\partial j^2}{\partial x^2} & \frac{\partial j^3}{\partial x^2} \end{pmatrix} \cdot \begin{pmatrix} \sigma_{11} & \sigma_{12} & \sigma_{13} \\ \sigma_{12} & \sigma_{22} & \sigma_{23} \\ \sigma_{13} & \sigma_{23} & \sigma_{33} \end{pmatrix} \cdot \begin{pmatrix} \frac{\partial j^1}{\partial x^1} & \frac{\partial j^1}{\partial x^2} \\ \frac{\partial j^2}{\partial x^1} & \frac{\partial j^2}{\partial x^2} \\ \frac{\partial j^3}{\partial x^1} & \frac{\partial j^3}{\partial x^2} \end{pmatrix}$$

In a similar way we can pull-back 1-forms from T^*N' to T^*N, considering that 1-forms are actually tensors of order $T_1^0(N')$. We can represent a 1-form ω in the manifold N', with coordinates $\{y^1, \ldots, y^n\}$, as $\omega = a_1 dy^1 + \cdots + a_n dy^n$. Then using the general form for a tensor of type $T_r^0(N')$, the pull-back of ω denoted by $\bar{\omega} = f^*\omega$, has the coordinate representation:

$$\bar{\omega} = \begin{pmatrix} \frac{\partial f^1}{\partial x^1} & \cdots & \frac{\partial f^n}{\partial x^1} \\ \vdots & & \vdots \\ \frac{\partial f^1}{\partial x^m} & \cdots & \frac{\partial f^n}{\partial x^m} \end{pmatrix} \begin{pmatrix} a_1 \\ \vdots \\ a_n \end{pmatrix}$$

where $f = f^1, \ldots, f^n$ is the coordinate representation of f and $\{x^1, \ldots, x^m\}$ are the coordinates of N. Using this formula we can "pull-back" the 1-forms that represent the forces acting on the object. Thus on the submanifold M, the forces acting on the object are described by $\bar{F} = j^*F$. If, for example, we consider the case where M is a two-dimensional submanifold of $\mathbb{R}^3$, then the 1-form representing the forces applied on the object $F = F_1 dy^1 + F_2 dy^2 + F_3 dy^3$ is pulled back to

$$\bar{F} = j^*F = \begin{pmatrix} \frac{\partial j^1}{\partial x^1} & \frac{\partial j^2}{\partial x^1} & \frac{\partial j^3}{\partial x^1} \\ \frac{\partial j^1}{\partial x^2} & \frac{\partial j^2}{\partial x^2} & \frac{\partial j^3}{\partial x^2} \end{pmatrix} \cdot \begin{pmatrix} F_1 \\ F_2 \\ F_3 \end{pmatrix}$$

Next we use the results from tensor analysis and Newton's law to obtain the dynamic equations of a rigid body on the submanifold M. In Newton's law the flat map $\sigma^\flat$, resulting from the Riemannian metric σ, is used. A complete description of the various relations is given by the commutative diagram

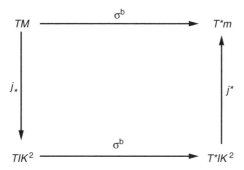

Let $\bar{c} = c \mid_M$ denote the restriction of the curve $c(t)$ to M. Let $\bar{\nabla}$ denote the connection on M associated with the Riemannian metric $\bar{\sigma}$. Then we can describe Newton's law on M by

$$\bar{\sigma}^\flat(\bar{\nabla}_{\dot{\bar{c}}(t)} \dot{\bar{c}}(t)) = \bar{F} \qquad (7.2)$$

We have to be careful how we interpret the different quantities involved in the constrained dynamic equation above, because the notation can cause some confusion. Actually, the way we have written

$\bar{\sigma}^{\flat} = j^{*}(\sigma^{\flat} \circ j_{*})$, j^{*} is the pull-back of the 1-form $\sigma^{\flat} \circ j_{*}$. In contrast with the formula, $\bar{\sigma} = j^{*}\sigma$, where j^{*} is the pull-back of the tensor σ. Actually it is true that $\bar{\sigma}^{\flat} = (j^{*}\sigma)^{\flat} = j^{*}(\sigma^{\flat} \circ j_{*})$.

The dynamic Equations (6.2) describe the motion of the object on the submanifold M. The main advantage of the description of the dynamics of the body on M is that we obtain a more convenient model of the system for analysis and control purposes. The "reaction" forces that were introduced as a consequence of the constraint do not appear explicitly in the model. In the modified 1-forms, that represent the forces on the constraint submanifold M denoted by $\bar{F} = j^{*}F$, the "reaction" forces belong to the null space of j^{*}. The "reaction" forces act in directions that tend to separate the rigid body from the constraint, and they do not belong to $T^{*}M$. The classical argument that the "reaction" forces do no work along the constrained path $\bar{c}(t)$ gives another justification for this result [6]. In the case of contact between a rigid body and a surface, the constraint submanifold M of the rigid body is a submanifold of $\mathbb{E}(3) \times B$. The argument that no work is done along the path $\bar{c}(t)$ holds for the case where the surface that the body moves in contact with is stationary. In the case of a dynamic system that consists of two contacting bodies where the contact surface is moving, the argument is that the "reaction" forces do exactly the same amount of work, but with opposite signs. In this case, of course, we have to consider the dynamic equations of both bodies as one system [20]. As we are going to see in the next section, the reduction of the dynamic equations to describe motion on the submanifold M, in the case of contact, also involves the description of Euler's equations written in geodesic form.

7.4 Euler's Equations and the Covariant Derivative

In order to have a complete description of the dynamic equations that determine the behavior of a rigid body in $\mathbb{R}^{3}$, we have to consider Euler's equations also. This set of differential equations was discovered by Euler in the 18th century and describes the rotational motion of a rigid body. The original form of the equations is given by

$$I_{1}\dot{\omega}_{1} + (I_{2} - I_{3})\omega_{2}\omega_{3} = M_{x}$$
$$I_{2}\dot{\omega}_{2} + (I_{3} - I_{1})\omega_{1}\omega_{3} = M_{y}$$
$$I_{3}\dot{\omega}_{3} + (I_{1} - I_{2})\omega_{1}\omega_{2} = M_{z}$$

where $\omega_{1}, \omega_{2}, \omega_{3}$ are the rotational velocities around each axis of a coordinate frame attached to the body. This coordinate frame is moving along with the body and is chosen such that the axes are coincident with the principal axes of the body. Thus I_{1}, I_{2}, and I_{3} are the principle moments of inertia. Finally M_{x}, M_{y}, and M_{z} are the external torques applied around each of these axes.

Euler's representation of the rotational dynamics, although it has a very simple form, comes with two major disadvantages. The first is that the equations are described in body coordinates and if we try to solve the equations for ω_{1}, ω_{2}, and ω_{3} we have no information about the orientation of the body. This is a consequence of the fact that there is no reference frame, with respect to which the angular orientation of the bodies can be measured. The second disadvantage is that this form of Euler's equations obscures the fact that they have the form of a geodesic on SO(3); SO(3) is the space of rotations of a rigid body in ($\mathbb{R}^{3}$). The geodesic form of Euler's equations was discussed by V.I. Arnold in [5]. In this work V.I. Arnold shows that Euler's equations in their original form are actually described using the local coordinates of $so(3)$; here $so(3)$ is the Lie algebra corresponding to SO(3) and is also the tangent space of SO(3) at the identity element. The main idea is to use the exponential map in order to identify the tangent space of a group G at the identity $T_{e}G$, with the group (G) itself. The following definition of the exponential map is based on notions of the flow of a vector field and left invariant vector fields on a group, see, for example, Boothby [12] or Brickell and Clark [13]:

Definition 7.6 Consider a group G and a left invariant vector field X on G. We denote by $\Phi : \mathbb{R} \times G \longrightarrow G$ the flow of X which is a differentiable function. For every $v \in T_{e}G$, there is a left invariant vector field X on G such that $X_{e} = v$. Then the exponential map $exp : T_{e}G \longrightarrow G$ is defined by $v \mapsto \Phi(1, e)$.

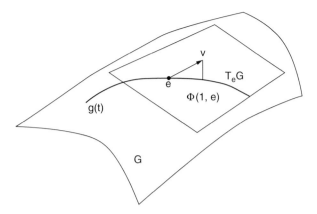

FIGURE 7.2 The action of the exponential map on a group.

Where $\Phi(1, e)$ is the result of the flow of a vector field X applied on the identity element e for $t = 1$. Figure 7.2 illustrates how the exponential map operates on elements of a group G.

Consider the tangent space of the Lie group G at the identity element e, denoted by $T_e G$. The Lie algebra of the Lie group G is identified with $T_e G$, and it can be denoted by $\mathcal{V}$. The main idea underlying this analysis is the fact that we can identify a chart in $\mathcal{V}$, in the neighborhood of $0 \in \mathcal{V}$, with a chart in the neighborhood of $e \in G$. More specifically, what V.I. Arnold proves in his work is that Euler's equations for a rigid body are equations that describe geodesic curves in the group $SO(3)$, but they are described in terms of coordinates of the corresponding Lie algebra $so(3)$. This is done by the identification mentioned above using the exponential map $exp : so(3) \longrightarrow SO(3)$. We are going to outline the main concepts and theorems involved in this construction. A complete presentation of the subject is included in [5]. A Lie group acts on itself by left and right translations. Thus for every element $g \in G$, we have the following diffeomorphisms:

$$L_g : G \longrightarrow G, \qquad L_g h = gh$$
$$R_g : G \longrightarrow G, \qquad R_g h = hg$$

As a result, we also have the following maps on the tangent spaces:

$$L_{g*} : T_g G \longrightarrow T_{gh} G, \qquad R_{g*} : T_g G \longrightarrow T_{gh} G$$

Next we consider the map $R_{g^{-1}} L_g : G \longrightarrow G$, which is a diffeomorphism on the group. Actually, it is an automorphism because it leaves the identity element of the group fixed. The derived map of $R_{g^{-1}} L_g$ is going to be very useful in the construction which follows:

$$Ad_g = (R_{g^{-1}} L_g)_{e*} : T_e G = \mathcal{V} \longrightarrow T_e G = \mathcal{V}$$

Thus, Ad_g is a linear map of the Lie algebra $\mathcal{V}$ to itself. The map Ad_g has certain properties related to the Lie bracket $[\cdot, \cdot]$ of the Lie algebra $\mathcal{V}$. Thus, if $exp : \mathcal{V} \longrightarrow G$ and $g(t) = exp(f(t))$ is a curve on G, then we have the following relations:

$$Ad_{exp(f(t))} \xi = \xi + t[f, \xi] + o(t^2) \qquad (t \rightarrow 0)$$
$$Ad_g [\xi, n] = [Ad_g \xi, Ad_g n]$$

There are two linear maps induced by the left and right translations on the cotangent space of G, T^*G. These maps are the duals to L_{g*} and R_{g*} and are known as *pull-backs* of L_g and R_g, respectively:

$$L_g^* : T_{gh}^* G \longrightarrow T_h^* G, \qquad R_g^* : T_{hg}^* G \longrightarrow T_h^* G$$

We have the following properties for the dual maps:

$$(L_g^*\xi, n) = (\xi, L_{g*}n)$$
$$(R_g^*\xi, n) = (\xi, R_{g*}n)$$

for $\xi \in T_g^*G, n \in T_gG$. Here, $(\xi, n) \in \mathbb{R}$ is the value of the linear map ξ applied on the vector field n, both evaluated at $g \in G$. In general we can define a Euclidean structure on the Lie algebra $\mathcal{V}$ via a symmetric positive definite linear operator $A : \mathcal{V} \longrightarrow \mathcal{V}^*$ ($A : T_eG \longrightarrow T_e^*G$), $(A\xi, n) = (An, \xi)$ for all $\xi, n \in \mathcal{V}$. Using the left translation we can define a symmetric operator $A_g : T_gG \longrightarrow T_g^*G$ according to the relation $A_g\xi = L_{g^{-1}}^* A L_{g^{-1}*}\xi$. Finally, via this symmetric operator we can define a metric on T_gG by

$$\langle \xi, n \rangle_g = (A_g n, \xi) = (A_g \xi, n) = \langle n, \xi \rangle_g$$

for all $\xi, n \in T_gG$. The metric $\langle \cdot, \cdot \rangle_g$ is a Riemannian metric that is invariant under left translations. At the identity element e, we denote the metric on $T_eG = \mathcal{V}$ by $\langle \cdot, \cdot \rangle$. We can define an operator $B : \mathcal{V} \times \mathcal{V} \longrightarrow \mathcal{V}$ using the relation $\langle [a, b], c \rangle = \langle B(c, a), b \rangle$ for all $b \in \mathcal{V}$. B is a bilinear operator, and if we fix the first argument, B is skew symmetric with respect to the second argument:

$$\langle B(c, a), b \rangle + \langle B(c, b), a \rangle = 0$$

In the case of rigid body motions the group is SO(3) and the Lie algebra is $so(3)$, but we are going to keep the general notation in order to emphasize the fact that this construction is more general and can be applied to a variety of problems. The rotational part of the motion of a body can be represented by a trajectory $g(t) \in G$. Thus $\dot{g}$ represents the velocity along the trajectory $\dot{g} \in T_{g(t)}G$. The rotational velocity with respect to the body coordinate frame is the left translation of the vector $\dot{g} \in G$ to $T_eG = \mathcal{V}$. Thus, if we denote the rotational velocity of the body with respect to the coordinate frame attached to the body by ω_c, then we have $\omega_c = L_{g^{-1}*}\dot{g} \in \mathcal{V}$. In a similar manner the rotational velocity of the body with respect to an inertial (stationary) frame is the right translation of the vector $\dot{g} \in T_{g(t)}G$ to $T_eG = \mathcal{V}$, which we denote by $\omega_s = R_{g^{-1}*}\dot{g} \in \mathcal{V}$.

In this case we have that $A_g : T_gG \longrightarrow T_g^*G$ is an inertia operator. Next, we denote the inertia operator at the identity by $A : T_eG \longrightarrow T_e^*G$. The angular momentum $M = A_g\dot{g} \in \mathcal{V}$ can be expressed with respect to the body coordinate frame as $M_c = L_g^*M = A\omega_c$ and with respect to an inertial frame $M_s = R_g^*M = Ad_{g^{-1}}^*M$ ($Ad_{g^{-1}}^*$ is the dual of $Ad_{g^{-1}*}$).

Euler's equations according to the notation established above are given by

$$\frac{d\omega_c}{dt} = B(\omega_c, \omega_c), \quad \omega_c = L_{g^{-1}*}\dot{g}$$

This form of Euler's equations can be derived in two steps. Consider first a geodesic $g(t) \in G$ such that $g(0) = e$ and $\dot{g}(0) = \omega_c$. Because the metric is left invariant, the left translation of a geodesic is also a geodesic. Thus the derivative $\frac{d\omega_c}{dt}$ depends only on ω_c and not on g.

Using the exponential map, we can consider a neighborhood of $0 \in \mathcal{V}$ as a chart of a neighborhood of the identity element $e \in G$. As a consequence, the tangent space at a point $a \in \mathcal{V}$, namely $T_a\mathcal{V}$, is identified naturally with $\mathcal{V}$. Thus the following lemma can be stated.

Lemma 7.1 Consider the left translation $L_{exp(a)}$ for $a \in \mathcal{V}$. This map can be identified with L_a for $\|a\| \to 0$. The corresponding derived map is denoted by L_{*a}, and $L_{*a} : \mathcal{V} = T_0\mathcal{V} \to \mathcal{V} = T_a\mathcal{V}$. If $\xi \in \mathcal{V}$ then

$$L_{a*}\xi = \xi + \frac{1}{2}[a, \xi] + o(a^2)$$

Because geodesics can be translated to the origin using coordinates of the algebra $\mathcal{V}$, the derivative $\frac{d\omega_c}{dt}$ gives the Euler equations. The proof of the lemma as well as more details on the rest of the arguments can be found in [5].

Thus, if $g(t)$ is a geodesic of the group G, then the equations that describe the rotational motion of a rigid body are given by

$$\nabla_{\dot{g}(t)}\dot{g}(t) = 0$$

and these are Euler's equations for the case that there are no torques applied on the body. Here, ∇ is the Riemannian connection on the group G, with the Riemannian metric given by the kinetic energy $K = \frac{1}{2}\langle \dot{g},\dot{g}\rangle_g = \frac{1}{2}\rho(\dot{g},\dot{g})$, and $\langle \cdot,\cdot\rangle_g$ [3] as defined above. In the case that there are external torques applied on the body, we have to modify the equations. We consider the torques T as 1-forms such that $T \in T^*G$. If we denote the kinetic energy metric by $\rho: T_gG \times T_gG \to \mathbb{R}$, then we can write Euler's equations in the form

$$\rho^{\flat}(\nabla_{\dot{g}(t)}\dot{g}(t)) = T$$

At this point we can further study the rotational dynamics of a rigid body in the case where the body is constrained to move on a submanifold G' of G. The equations of motion are still going to be geodesics of the constrained kinetic energy metric. Consider next a submanifold G' of G and let $j: G' \hookrightarrow G$ be the natural injection. In general, the submanifold G' will not have the property of a subgroup of G. For example, this would happen if the body was constrained to move in $\mathbb{R}^2$. In this case the subgroup is the one-dimensional subgroup of plane rotations. The same arguments that occur in the case of Newton's law apply here also. Thus if G' is a submanifold of G, it means that the natural injection is an immersion. Consequently, the Riemannian metric can be pulled back on the submanifold G' using the natural injection j. We can write the constrained Euler equations on the submanifold G' in the form

$$\bar{\rho}^{\flat}(\bar{\nabla}_{\dot{\bar{g}}(t)}\dot{\bar{g}}(t)) = j^*(T)$$

where $\bar{\rho}^{\flat} = j^*(\rho^{\flat} \circ j_*)$ and $\bar{g}(t) = g(t)|_{G'}$. $\bar{\nabla}$ denotes the Riemannian connection on the submanifold G', induced by the Riemannian metric $\bar{\rho}$. Because we are interested in studying the general case of a rigid body and the result of the presence of certain constraints, we have to consider the Euclidean manifold $\mathbb{R}^3 \times SO(3)$. In the next example we investigate the representation of Euler's equations in the form of geodesics in terms of the coordinates of the group $SO(3)$. This example will illustrate how the arguments developed up to this point can be applied to this case.

7.5 Example 1: Euler's Equations for a Rigid Body

Consider a coordinate chart of the Lie group $SO(3)$. We can use as coordinates any convention of the Euler angles. In this example, we use the Y convention (Goldstein [17]) with ϕ, θ, ψ as coordinates, where $0 \leq \phi \leq 2\pi$, $0 \leq \theta \leq \pi$, $0 \leq \psi \leq 2\pi$. Consider a trajectory $g(t) \in SO(3)$ representing the rotational motion of a rigid body in $\mathbb{R}^3$. The coordinates of the trajectory are given by $\phi(t), \theta(t), \psi(t)$, and thus the coordinates of the velocity are $\dot{g}(t) = \{\dot{\phi}(t), \dot{\theta}(t), \dot{\psi}(t)\}$. The velocity with respect to the body coordinate system is $\omega_c = L_{g^{-1}*}\dot{g}$, which when given in matrix form yields

$$\begin{pmatrix} \omega_{c1} \\ \omega_{c2} \\ \omega_{c3} \end{pmatrix} = \begin{pmatrix} -\sin(\theta)\cos(\psi) & \cos(\psi) & 0 \\ \sin(\theta)\sin(\psi) & \cos(\psi) & 0 \\ \cos(\theta) & 0 & 1 \end{pmatrix} \begin{pmatrix} \dot{\phi} \\ \dot{\theta} \\ \dot{\psi} \end{pmatrix}$$

---

[3] The subscript g emphasizes the local character of the metric because the velocities are measured with respect to a moving coordinate frame.

With respect to an inertial coordinate frame, the velocity of the body is $\omega_s = R_{g^{-1}*}\dot{g}$, which when given in matrix form yields

$$\begin{pmatrix} \omega_{s1} \\ \omega_{s2} \\ \omega_{s3} \end{pmatrix} = \begin{pmatrix} 0 & -\sin(\phi) & \sin(\theta)\cos(\phi) \\ 0 & \cos(\phi) & \sin(\theta)\sin(\phi) \\ 1 & 0 & \cos(\theta) \end{pmatrix} \begin{pmatrix} \dot{\phi} \\ \dot{\theta} \\ \dot{\psi} \end{pmatrix}$$

The kinetic energy metric is given by $K = \frac{1}{2}\langle \dot{g}, \dot{g}, \rangle_g = \frac{1}{2}\langle \omega_c, \omega_c \rangle$. We can choose a coordinate frame attached to the body such that the three axes are the principle axes of the body; thus,

$$A = \begin{pmatrix} I_1 & 0 & 0 \\ 0 & I_2 & 0 \\ 0 & 0 & I_3 \end{pmatrix}$$

We can calculate the quantity $A_g = L^*_{g^{-1}} A L_{g^{-1}*}$, which when given in matrix form yields

$$A_g = \begin{pmatrix} -\sin(\theta)\cos(\psi) & \sin(\theta)\sin(\psi) & \cos(\theta) \\ \sin(\psi) & \cos(\psi) & 0 \\ 0 & 0 & 1 \end{pmatrix}$$

$$\begin{pmatrix} I_1 & 0 & 0 \\ 0 & I_2 & 0 \\ 0 & 0 & I_3 \end{pmatrix} \begin{pmatrix} -\sin(\theta)\cos(\psi) & \sin(\psi) & 0 \\ \sin(\theta)\sin(\psi) & \cos(\psi) & 0 \\ \cos(\theta) & 0 & 1 \end{pmatrix}$$

We can write Euler's equations in group coordinates using the formula

$$\rho^\flat(\nabla_{\dot{g}}\dot{g}) = T$$

The complete system of Euler's equations, using the coordinates given by the three Euler angles, is given by

$$\ddot{\phi} I_2 \sin(\psi)^2 \sin(\theta)^2 + 2\dot{\phi}\dot{\psi} I_2 \cos(\psi)\sin(\psi)\sin(\theta)^2 - 2\dot{\phi}\dot{\psi} I_1 \cos(\psi)\sin(\psi)\sin(\theta)^2 + \ddot{\phi} I_1 \cos(\psi)^2 \sin(\theta)^2$$
$$+ 2\dot{\phi}\dot{\theta} I_2 \sin(\psi)^2 \cos(\theta)\sin(\theta) + 2\dot{\phi}\dot{\theta} I_1 \cos(\psi)^2 \cos(\theta)\sin(\theta) - 2\dot{\phi}\dot{\theta} I_3 \cos(\theta)\sin(\theta)$$
$$- \dot{\psi}\dot{\theta} I_2 \sin(\psi)^2 \sin(\theta) + \dot{\psi}\dot{\theta} I_1 \sin(\psi)^2 \sin(\theta) + \ddot{\theta} I_2 \cos(\psi)\sin(\psi)\sin(\theta) - \ddot{\theta} I_1 \cos(\psi)\sin(\psi)\sin(\theta)$$
$$+ \dot{\psi}\dot{\theta} I_2 \cos(\psi)^2 \sin(\theta) - \dot{\psi}\dot{\theta} I_1 \cos(\psi)^2 \sin(\theta) - \dot{\psi}\dot{\theta} I_3 \sin(\theta) + \ddot{\phi} I_3 \cos(\theta)^2$$
$$+ \dot{\theta}^2 I_2 \cos(\psi)\sin(\psi)\cos(\theta) - \dot{\theta}^2 I_1 \cos(\psi)\sin(\psi)\cos(\theta) + \ddot{\psi} I_3 \cos(\theta) = T_\phi$$

$$-\dot{\phi}^2 I_2 \sin(\psi)^2 \cos(\theta)\sin(\theta) - \dot{\phi}^2 I_1 \cos(\psi)^2 \cos(\theta)\sin(\theta) + \dot{\phi}^2 I_3 \cos(\theta)\sin(\theta) - \dot{\phi}\dot{\psi} I_2 \sin(\psi)^2 \sin(\theta)$$
$$+ \dot{\phi}\dot{\psi} I_1 \sin(\psi)^2 \sin(\theta) + \ddot{\phi} I_2 \cos(\psi)\sin(\psi)\sin(\theta) - \ddot{\phi} I_1 \cos(\psi)\sin(\psi)\sin(\theta) + \dot{\phi}\dot{\psi} I_2 \cos(\psi)^2 \sin(\theta)$$
$$- \dot{\phi}\dot{\psi} I_1 \cos(\psi)^2 \sin(\theta) + \dot{\phi}\dot{\psi} I_3 \sin(\theta) + \ddot{\theta} I_1 \sin(\psi)^2 - 2\dot{\psi}\dot{\theta} I_2 \cos(\psi)\sin(\psi)$$
$$+ 2\dot{\psi}\dot{\theta} I_1 \cos(\psi)\sin(\psi) + \ddot{\theta} I_2 \cos(\psi)^2 = T_\theta$$

$$-\dot{\phi}^2 I_2 \cos(\psi)\sin(\psi)\sin(\theta)^2 + \dot{\phi}^2 I_1 \cos(\psi)\sin(\psi)\sin(\theta)^2 + \dot{\phi}\dot{\theta} I_2 \sin(\psi)^2 \sin(\theta) - \dot{\phi}\dot{\theta} I_1 \sin(\psi)^2 \sin(\theta)$$
$$- \dot{\phi}\dot{\theta} I_2 \cos(\psi)^2 \sin(\theta) + \dot{\phi}\dot{\theta} I_1 \cos(\psi)^2 \sin(\theta) - \dot{\phi}\dot{\theta} I_3 \sin(\theta) + \ddot{\phi} I_3 \cos(\theta) + \dot{\theta}^2 I_2 \cos(\psi)\sin(\psi)$$
$$- \dot{\theta}^2 I_1 \cos(\psi)\sin(\psi) + \ddot{\psi} I_3 = T_\psi$$

Observe that the torques T are given in terms of the axes of the Euler angles and, thus, cannot be measured directly. This happens because the axis of rotation for an Euler angle is moving independent of the body during the motion. This can be circumvented by using the torques around the principle axes, and then expressing them via an appropriate transformation to torques measured with respect to the axes of the Euler angles. Because the torques are considered as 1-forms, if we want to translate the torques with respect to the group coordinates to torques with respect to body coordinates, we have to use a covariant transformation. The relationship between the corresponding velocities is given by $\omega_c = L_{g^{-1}*}\dot{g}$, and for the torques we have $T_c = L^{-1}_{g^{-1}*}T$, where T_c are the torques with respect to the body coordinate system, and T are the torques with respect to the group coordinate system. Thus, T can be replaced by $(L^*_{g^{-1}})^{-1}T_c$ according to the formula

$$\begin{pmatrix} T_\phi \\ T_\theta \\ T_\psi \end{pmatrix} = \begin{pmatrix} \frac{\cos(\psi)}{\sin(\theta)} & \sin(\psi) & \frac{\cos(\psi)\cos(\theta)}{\sin(\theta)} \\ \frac{\sin(\psi)}{\sin(\theta)} & \cos(\psi) & \frac{-\sin(\psi)\cos(\theta)}{\sin(\theta)} \\ 0 & 0 & 1 \end{pmatrix} \begin{pmatrix} T_x \\ T_y \\ T_z \end{pmatrix}$$

Up to this point we have investigated the geometric properties of the basic equations of motion of a rigid body. Next we will use these properties to model multibody systems where we assume that the rigid bodies interact through point contact or joints. We have used results from some of our previous work and techniques from Differential Geometry to study the geometric properties of Newton's and Euler's equations. The final goal is to extend the modeling method for dynamical systems introduced by Kron, which is presented in Hoffmann [20], to interacting mulibody systems.

7.6 The Equations of Motion of a Rigid Body

According to the analysis presented so far, we have derived a representation for the dynamics of a rigid body which suits our needs. The geodesic representation of the dynamics is the most appropriate to handle constrained problems. We have described Newton's and Euler's equations in their geodesic form on the manifolds $\mathbb{R}^3$ and SO(3), respectively, with the Riemannian kinetic energy metrics

$$\begin{aligned} \sigma^\flat(\nabla_{\dot{c}(t)}\dot{c}(t)) &= F \\ \rho^\flat(\nabla_{\dot{g}(t)}\dot{g}(t)) &= T \end{aligned} \quad (7.3)$$

These metrics can be written in tensor form as $E = \frac{1}{2}\sigma_{ij}v^i v^j$ and $K = \frac{1}{2}\rho_{ij}u^i u^j$. Where E and K are the translational and the rotational kinetic energies, respectively. Also $v^i, i = 1, 2, 3$ are the components of the velocity $\dot{c}(t)$ and $u^i, i = 1, 2, 3$ are the components of the velocity $\dot{g}(t)$. Next, we want to combine the two sets of equations in order to treat them geometrically as a unified set of dynamical equations for a rigid body. The configuration manifold where the position and the orientation of the rigid body in the Euclidean space are described is $\mathbb{E}(3) = \mathbb{R}^3 \times SO(3)$. It is obvious from the analysis so far that we can use the same process of transforming the translational and rotational dynamics of a rigid body, when the body is constrained to move on a subset of the configuration space, instead of the total configuration space. Combining the two sets of dynamics of a rigid body does not affect the geometric properties of Newton's and Euler's equations, as we described them in the previous sections of this chapter. To extend the reduction process of the dynamics of rigid body motion on the combined configuration manifold $\mathbb{E}(3)$, we have to introduce a combined kinetic energy metric. More precisely, the sum of the two metrics associated with the translational and rotational motions is appropriate for a combined kinetic energy metric, i.e., $E + K = \frac{1}{2}\sigma_{ij}v^i v^j + \frac{1}{2}\rho_{ij}u^i u^j$. We can then define the equations of motion of a free rigid body.

Definition 7.7 The dynamic equations which define the motion of a free rigid body are given by the equations for a geodesic on the manifold $\mathbb{E}(3)$ with Riemannian metric given by the combined metric $\varpi = \sigma + \rho$.

When there are forces and torques acting on the body, then we have to modify the equations of the geodesics in the form of Equations (7.3). In the case that the body is constrained to move on a subset M of $\mathbb{E}(3)$, we can describe the dynamics of the body on M under the same conditions we established in the previous sections for Newton's and Euler's equations. Thus, if M is a submanifold of $\mathbb{E}(3)$, then we can pull-back the Riemannian metric on M from $\mathbb{E}(3)$ using the natural injection from M to $\mathbb{E}(3)$, which is by definition an immersion. Actually this is the case that occurs when the rigid bodies interact through a joint. Depending on the type of joint, we can create a submanifold of $\mathbb{E}(3)$ by considering only the degrees of freedom allowed by the joint. The other case that we might encounter is when M is a subset of $\mathbb{E}(3)$, which is also a manifold (with differential structure established independent of the fact that is a subset) and there is a map $l : M \longrightarrow \mathbb{E}(3)$. This is what happens in the case where the rigid bodies interact with contact. As we proved in [53], the subset M is also a manifold with a differential structure established by the fact that M is a submanifold of $\mathbb{E}(3) \times B$. Also we proved in [53] that there exists a map $\mu_1 : M \longrightarrow \mathbb{E}(3)$. In the case that this map is an immersion, we can describe the dynamics of the rigid body on M.

According to our assumptions about the point interaction between rigid bodies, joints and contact are the two types of interactions that are being considered. Our intention is to develop a general model for the dynamics of a set of rigid interacting bodies; therefore, both cases of contact and joint interaction might appear simultaneously in a system of equations describing the dynamics of each rigid body in the rigid body system. Thus we will consider the more general case where the dynamics are constrained to a subset M of $\mathbb{E}(3)$, which is a manifold, and that there exists a map $l : M \longrightarrow \mathbb{E}(3)$ which is an immersion. We can use as before the pull-back $l^* : T^*\mathbb{E}(3) \longrightarrow T^*M$ to induce a metric on M. If we denote the Riemannian metric on $\mathbb{E}(3)$ by $\varpi = \sigma + \rho$, then the induced Riemannian metric on M is $\bar{\varpi}^\flat = l^*\varpi$. Consider next a trajectory on $\mathbb{E}(3)$, $r(t) : (c(t), g(t))$. If this curve is a solution for the dynamics of a rigid body, it satisfies Equations (7.3) which can be written in condensed form

$$\varpi^\flat(\nabla_{\dot{r}(t)}\dot{r}(t)) = F + T \tag{7.4}$$

The term $\nabla_{\dot{r}(t)}\dot{r}(t)$ has the same form as the RHS of the geodesic equations and depends only on the corresponding Riemannian metric ϖ. When we describe the dynamics on the manifold M, with the induced Riemannian metric $\bar{\varpi}$, the corresponding curve $\bar{r}(t) \in M$ should satisfy the following equations:

$$\bar{\varpi}^\flat(\bar{\nabla}_{\dot{\bar{r}}(t)}\dot{\bar{r}}(t)) = l^*(F + T) \tag{7.5}$$

where $F + T$ is the combined 1-form of the forces and the torques applied to the unconstrained body. The term $\bar{\nabla}_{\dot{\bar{r}}(t)}\dot{\bar{r}}(t)$ is again the same as the RHS of the equations for a geodesic curve on M and depends only on the induced Riemannian metric $\bar{\varpi}$. Thus in general, the resulting curve $\bar{r}(t) \in M$ will not be related to the solution of the unconstrained dynamics except in the case where M is a submanifold of $\mathbb{E}(3)$, and in this case we have $\bar{r}(t) = r(t) |_M$. It is natural to conclude that the 1-form $l^*(F + T)$ includes all forces and torques that allow the rigid body to move on the constraint manifold M. In the initial configuration space $\mathbb{E}(3)$, the RHS of the dynamic equations includes all the forces and torques that make the body move on the subset M of $\mathbb{E}(3)$; these are sometimes referred to as "generalized constraint forces." Let F_M and T_M denote, respectively, these forces and torques. We can separate the forces and torques into two mutually exclusive sets, one set consisting of the forces and torques that constrain the rigid body motion on M and the other set of the remaining forces and torques which we denote by F_S, T_S. Thus we have $F + T = (F_M + T_M) \oplus (F_S + T_S)$. We can define *a priori* the generalized constraint forces as those that satisfy the condition

$$l^*(F_M + T_M) = 0 \tag{7.6}$$

This is actually what B. Hoffmann described in his work [20] as Kron's method of subspaces. Indeed in this work Kron's ideas about the use of tensor mathematics in circuit analysis are applied to mechanical systems. As Hoffman writes, Kron himself gave some examples of the application of his method in mechanical systems in an unpublished manuscript. In this chapter we actually develop a methodology within Kron's framework, using the covariant derivative, for the modeling of multibody systems. From the analysis that we have presented thus far it appears as if this is necessary in order to include not only joints between the bodies of the system but also point contact interaction.

7.7 Constraint Forces and Torques between Interacting Bodies

Because one of the main objectives is to develop dynamic models of multibody systems suitable for control studies, our first task is to investigate some of the ways in which rigid bobies can interact. By interaction between rigid bodies here we do not mean dynamic interaction in the sense of gravity. Instead, we mean kinematic interaction through points on the surfaces of the rigid bodies. More precisely, in this work we say that two rigid bodies are interacting when for a given position of the rigid bodies, the set of points belonging to the surface of one rigid body coincides with a set of points belonging to the surface of another rigid body. To put this definition in a dynamic context, we can say that if $\gamma_1(t)$ and $\gamma_2(t)$ are two curves in $\mathbb{E}(3)$ describing the motion of two rigid bodies in $\mathbb{R}^3$, then we have interaction if at least two points of these two curves coincide for some instant of time $t = t_0$. In our construction we restrict the modeling of interaction by introducing the following assumptions:

1. There is only one point on each body that coincides with one point on the other body at any instant of time.
2. During the motion of the interacting bodies, there is a nonempty interval of time with positive measure for which interaction occurs.
3. The interaction points follow continuous paths on the surfaces of the interacting bodies.

The second condition avoids impulsive interaction, which requires special investigation. The third condition is necessary to avoid cases where the interaction might be interrupted for some time and then continue again. In this later case our model fails unless special care is taken to include such transitions in the modeling. With these three assumptions for the interaction between rigid bodies, the possible types of interaction are illustrated in Figure 7.3.

Precisely the possible interactions include contact, joint connection, and rigid connection. The last case is of no interest because the two bodies behave dynamically as one. In the other two cases, the interaction results in a kinematic constraint for one of the bodies. We can replace this constraint in the study of dynamics with a force acting on each body. In the case of a joint between the interacting bodies, the constraint can be replaced by forces of equal and opposite signs acting on each body. The direction of the force is not specified, see Figure 7.3. In the case of contact, the constraint is replaced by a force in the direction vertical to the common tangent plane at the point of contact. Forces can act in the tangential direction, but they have nothing to do with the constraint; they are friction forces. In both cases the forces are such that the constraints are satisfied. This is an *a priori* consideration for the constraints. In this way,

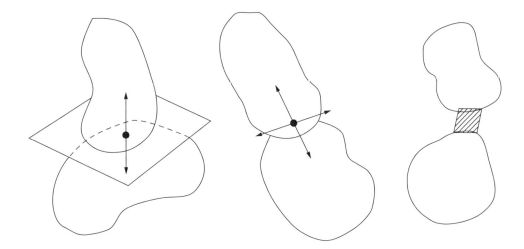

FIGURE 7.3 Possible types of interaction.

a constraint resulting from the interaction between two rigid bodies can be replaced by certain forces that act on the unconstrained bodies. The type of forces depends on the type of constraints. To make this point more clear, we present several examples.

7.8 Example 2: Double Pendulum in the Plane

In this example we will develop a model for the double pendulum with masses concentrated at the ends of the links. We regard initially each body separately (Figure 7.4). Thus, the first link has one degree of freedom θ, and the second has three degrees of freedom $\{\phi, x, y\}$. For the second body (link) x, y are the coordinates of the center of the mass attached to the end of the link, and ϕ is the rotational degree of freedom of the link of the second body. The joint between the two links acts as a constraint imposed on the second link that is kinematically expressed by

$$x = l_1 \sin(\theta) + l_2 \sin(\phi)$$
$$y = -l_1 \cos(\theta) - l_2 \cos(\phi)$$

Thus, in order that the point at one end of the second link always coincides with the point P of the other link, a force with components F_x, F_y is required to act on the unconstrained second link. These forces are exerted by the tip of the first link, and there is an equal but opposite force $-F_x, -F_y$ acting on the first link. We could have considered the first body as a free body in two dimensions constrained by the first joint, but this does not provide any additional insight for the double pendulum problem. We can separate the rotational and translational equations of motion for the system consisting of the two bodies in the form

$$\rho^\flat(\nabla_{\dot{r}(t)} \dot{r}(t)) = F + T$$

In our case

$$\rho = \begin{pmatrix} m_1 l_1^2 & 0 & 0 & 0 \\ 0 & 0 & 0 & 0 \\ 0 & 0 & m_2 & 0 \\ 0 & 0 & 0 & m_2 \end{pmatrix}$$

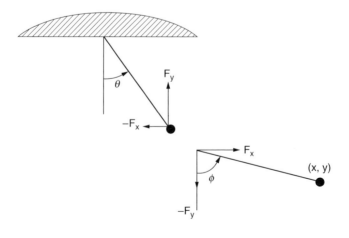

FIGURE 7.4 Double pendulum and associated interaction forces.

The degrees of freedom of the system are θ, ϕ, x and y, and the matrix expression for $\nabla_{\dot{r}(t)}\dot{r}(t)$ is

$$\begin{pmatrix} \ddot{\theta} \\ \ddot{\phi} \\ \ddot{x} \\ \ddot{y} \end{pmatrix}$$

The matrix expression of the 1-form of the forces and the torques $F + T$ is

$$F + T = \begin{pmatrix} -m_1 l_1 g \sin(\theta) - F_x l_1 \cos(\theta) + F_y l_1 \sin(\theta) \\ -F_x l_2 \cos(\phi) + F_y l_2 \sin(\phi) \\ F_x \\ -F_y - m_2 g \end{pmatrix}$$

At this point we have described the equations of motion for the two free bodies that make up the system. Next we proceed with the reduction of the model according to the methodology described above. The degrees of freedom of the constrained system are $\tilde{\theta}, \tilde{\phi}$, and we can construct the map ξ according to

$$\xi : \{\tilde{\theta}, \tilde{\phi}\} \longrightarrow \{\tilde{\theta}, \tilde{\phi}, l_1 \sin(\tilde{\theta}) + l_2 \sin(\tilde{\phi}), -l_1 \cos(\tilde{\theta}) - l_2 \cos(\tilde{\phi})\}$$

Next ξ_* is calculated:

$$\xi_* = \begin{pmatrix} 1 & 0 \\ 0 & 1 \\ l_1 \cos(\tilde{\theta}) & l_2 \cos(\tilde{\phi}) \\ l_1 \sin(\tilde{\theta}) & l_2 \sin(\tilde{\phi}) \end{pmatrix}$$

It is obvious that the map ξ is an immersion, because the rank$(\xi_*) = 2$. Consequently, we compute $\tilde{\rho} = \xi^* \rho$:

$$\tilde{\rho} = \begin{pmatrix} l_1^2 m_1 + l_1^2 m_2 & l_1 l_2 m_2 \cos(\tilde{\theta} - \tilde{\phi}) \\ l_1 l_2 m_2 \cos(\tilde{\theta} - \tilde{\phi}) & l_2^2 m_2 \end{pmatrix}$$

Applying ξ^* on the 1-forms of the forces and the torques on the RHS of the dynamic equations of motion we obtain

$$\xi^*(F + T) = \begin{pmatrix} -m_1 l_1 g \sin(\tilde{\theta}) - l_1 m_2 g \sin(\tilde{\theta}) \\ -l_2 m_2 g \sin(\tilde{\phi}) \end{pmatrix}$$

We observe that all the reaction forces have disappeared from the model because we are on the constrained submanifold. Thus, we could have neglected them in the initial free body modeling scheme. We continue with the calculation of $\tilde{\nabla}_{\dot{r}(t)}\dot{r}(t)$, which defines a geodesic on the constraint manifold. The general form of the geodesic is [12,14]

$$\frac{d^2 x^i}{dt^2} + \Gamma^i_{jk} \frac{dx^j}{dt} \frac{dx^k}{dt}$$

where x^i are the degrees of freedom of the constrained manifold and Γ^i_{jk} are the Christoffel symbols. We have that

$$\Gamma^1_{12} = 0$$
$$\Gamma^2_{12} = 0$$
$$\Gamma^1_{11} = \frac{l_1^2 l_2^2 m_2^2 \cos(\tilde{\theta} - \tilde{\phi}) \sin(\tilde{\theta} - \tilde{\phi})}{l_1^2 l_2^2 m_1 m_2 + l_1^2 l_2^2 m_2^2 - l_1^2 l_2^2 m_2^2 \cos(\tilde{\theta} - \tilde{\phi})^2}$$
$$\Gamma^2_{11} = \frac{l_1 l_2 m_2 (l_1^2 m_1 + l_1^2 m_2) \sin(\tilde{\theta} - \tilde{\phi})}{l_1^2 l_2^2 m_1 m_2 + l_1^2 l_2^2 m_2^2 - l_1^2 l_2^2 m_2^2 \cos(\tilde{\theta} - \tilde{\phi})^2}$$
$$\Gamma^1_{22} = \frac{l_1 l_2^3 m_2^3 \sin(\tilde{\theta} - \tilde{\phi})}{l_1^2 l_2^2 m_1 m_2 + l_1^2 l_2^2 m_2^2 - l_1^2 l_2^2 m_2^2 \cos(\tilde{\theta} - \tilde{\phi})^2}$$
$$\Gamma^2_{22} = \frac{l_1^2 l_2^2 m_2^2 \cos(\tilde{\theta} - \tilde{\phi}) \sin(\tilde{\theta} - \tilde{\phi})}{l_1^2 l_2^2 m_1 m_2 + l_1^2 l_2^2 m_2^2 - l_1^2 l_2^2 m_2^2 \cos(\tilde{\theta} - \tilde{\phi})^2}$$

$$\frac{d^2 x^i}{dt^2} + \Gamma^i_{jk} \frac{dx^j}{dt} \frac{dx^k}{dt} = \begin{pmatrix} \ddot{\tilde{\theta}} + 2\Gamma^1_{11} \dot{\tilde{\theta}}^2 + 2\Gamma^1_{11} \dot{\tilde{\phi}}^2 \\ \ddot{\tilde{\phi}} + 2\Gamma^1_{22} \dot{\tilde{\theta}}^2 + 2\Gamma^2_{22} \dot{\tilde{\phi}}^2 \end{pmatrix}$$

We apply $\bar{\rho}^b$ in the form of a matrix operating on $\bar{\nabla}_{\dot{r}(t)} \dot{r}(t)$, and equating the result with $\xi^*(F + T)$ yields the following equations of motion:

$$\begin{pmatrix} (m_1 + m_2) l_1^2 \ddot{\tilde{\theta}} + m_2 l_1 l_2 \cos(\tilde{\theta} - \tilde{\phi}) \ddot{\tilde{\phi}} + l_1 l_2 m_2 \sin(\tilde{\theta} - \phi) \dot{\phi}^2 \\ m_2 l_2^2 \ddot{\tilde{\phi}} + m_2 l_1 l_2 \cos(\tilde{\theta} - \tilde{\phi}) \ddot{\tilde{\theta}} - m_2 l_1 l_2 \sin(\tilde{\theta} - \tilde{\phi}) \dot{\tilde{\theta}}^2 \end{pmatrix} = \begin{pmatrix} -m_1 g l_1 \sin(\tilde{\theta}) - m_2 g l_1 \sin(\tilde{\theta}) \\ m_2 g l_2 \sin(\tilde{\phi}) \end{pmatrix}$$

which are exactly the equations that are obtained using the classical Lagrangian formulation [41].

7.9 Including Forces from Friction and from Nonholonomic Constraints

Friction forces, and the resulting torques, can be added naturally within the modeling formalism presented. Frictional forces are introduced when a body moves in contact with a nonideal surface and the friction forces have a direction which opposes the motion. There is a significant literature on the subject of friction, which actually constitutes the separate scientific field of triboblogy. What is important for our formulation is that the direction of the friction forces is opposite to the direction of the motion. Modeling of the friction forces is also a complicated subject and depends on additional information about the nature of contacting surfaces. The magnitude of the friction forces can depend on the velocity or the square power of the velocity [3], for example. At this point we will not investigate the inclusion of friction forces and torques as functions of the velocities and other parameters. Because a complete model requires this information, we will leave this matter for future investigation. Nonholonomic constraints are a special case of constraints that appear in a certain class of problems in dynamics. These constraints are not integrable (e.g., rolling) and can be represented as 1-forms. Integrability is a property defined for a set of equations describing kinematic constraints with the general form $J(q)dq = 0$ (1-forms), where $J(q)$ is $n \times (n - m)$ matrix with rank equal to $(n - m)$. If the initial dynamic system can be described on a n-dimensional manifold M, then the constraint equation $J(q)dq = 0$ is said to be *integrable* on M if there exists a smooth $(n - m)$-dimensional foliation whose leaves are $(n - m)$-dimensional submanifolds with all of their points tangent to the planes defined by the constraint equations [4]. Alternatively, integrability of the constraints can be defined with respect to the associated vector fields, vanishing when the 1-forms of constraints operate on them. In this context integrability of the vanishing vector fields is defined by the Frobenious theorem [12].

From the existing literature it is obvious that these types of constraints are more difficult to deal with and a lot of attention is required in their use. Many times they lead to falacies [41] and they have been the cause of many discussions. It has also been shown that using the nonholonomic constraints with different variational principles (D' Alembert, Hamilton, etc.) can lead to different solutions (Vakonomic mechanics [22–25]). In our formalism we deal with elementary (not generalized) entities like forces and torques applied on each body. Initially it appears as if there is no better way to introduce a force or a torque from a nonholonomic constraint in any modeling formalism. The forces and the torques from a nonholonomic constraint are frictional, but they produce no work. The virtual work produced by these forces and torques is of third order (δx^3) in the related displacement [18], and these forces and torques are acting on the body to keep the constraint valid. This is equivalent to requiring that the forces and the torques act in the direction of the 1-form representing the nonholonomic constraint (Burke [14]). This interpretation allows the introduction of Lagrange multipliers in a nonvariational framework, and a force or torque in the direction of the 1-form ω should be of the form $F = \lambda \omega$.

Next, consider a procedure where the objective is to model the dynamics of a set of interacting bodies. We begin by considering each body separately and all the forces and torques that are acting on the body, including applied and reaction forces and torques. Then, because of the existence of constraints, each body is actually moving on a submanifold of the original configuration space defined by a natural injection j. The resulting forces and torques are, as we pointed out earlier, $l^*(F+T)$. The constraint forces and torques disappear from the dynamics on the resulting submanifold because $l^*(F_M + T_M) = 0$. Thus, in considering the dynamics on the submanifold, we should not include the torques and forces that result because of the constraints at the initial stage of the modeling. What is important, though, is being able to distinguish between constraint forces and torques and the other exerted forces and torques caused from friction, for example. These other exerted forces and torques do have an effect on the dynamics on the submanifold. In the case of contact between two bodies for example, friction forces can be introduced in the direction of the common tangent plane. In the reduction process of the initial configuration space, this friction force will remain present. The effect of these types of forces can dramatically alter the behavior of a rigid body, producing interesting phenomena; for example, in the case of contact interaction, they can cause rolling phenomenon. To control a system using a model that includes these types of forces and torques, we need to introduce them as measurable quantities in the model. Certain assumptions can be used *a priori* for special cases, like pure sliding motion. However, there are other situations where the cause of a change from one type of motion to another, with little or no *a priori* information about the frictional forces and torques, is unknown. Neglecting information on how the forces and torques can affect the qualitative properties of the motion can be disastrous from a modeling perspective. On the other hand, trying to describe the dynamics using principles like the D'Alembert-Lagrange or Hamiltonian formulations can result in an ambiguous model, as in the case of nonholonomic constraints, see, for example [4,22–25]. Next we present a systematic procedure for modelling the dynamics of a set of interacting bodies using the configuration space reduction method. This is done in the context of the above discussion, regarding the forces and the torques that are involved as a result of the interaction. Several examples are used to illustrate the approach.

7.10 Example 3: The Dynamics of the Interaction of a Disk and a Link

The constrained configuration space of the disk is M_1, a two-dimensional submanifold of $\mathbb{E}(3) \times B_1$. Let B_1 denote the surface of the disk and let B_2 denote the surface of the link. The coordinates of $\mathbb{E}(2) \times B_1$ are $\{z_1, z_2, \theta, \phi\}$, and for the submanifold they are $\{\tilde{z}_1, \tilde{\theta}\}$. In a similar way the constrained configuration space of the link is M_2, a two-dimensional submanifold of $\mathbb{E}(2) \times B_2$. The coordinates of M_2 are $\{\tilde{y}_1, \tilde{\theta}_2\}$ and for $\mathbb{E}(2) \times B_2$ they are $\{y_1, y_2, \theta_2, \tau_1\}$.

The modeling methodology requires deriving Newton's and Euler's equations for each body separately, with the external and the reaction forces and torques included as shown in Figure 7.5. At this point, we

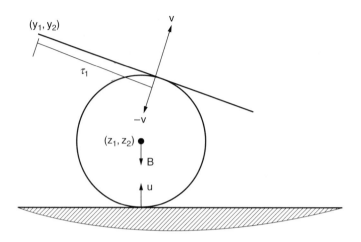

FIGURE 7.5 Disk in contact with a link.

assume that there are no friction forces generated by the contact. The set of the dynamic equations is

$$m\ddot{z}_1 = -v\sin(\theta_2)$$
$$m\ddot{z}_2 = u - v\cos(\theta_2) - mg$$
$$I\ddot{\theta} = 0$$
$$m_2\ddot{y}_1 = v\sin(\theta_2)$$
$$m_2\ddot{y}_2 = v\cos(\theta_2) - m_2 g$$
$$I_2\ddot{\theta}_2 = u_2 - v\tau_1$$

Where m, I are the mass and the moment of inertia of the disk, m_2, I_2 are the mass and the moment of inertia of the link, and u_2 is the input (applied) torque. As we mentioned previously, the dynamic equations that describe the motion for each body are described initially on $\mathbb{E}(2)$. The torques and the forces are 1-forms in $T^*\mathbb{E}(2)$. The dynamic equations of the combined system are defined on $\mathbb{E}(2) \times \mathbb{E}(2)$, but the constrained system actually evolves on $M_1 \times M_2$. We can restrict the original system of dynamic equations to $M_1 \times M_2$ using the projection method developed earlier. Thus we need to construct the projection function $\xi : M_1 \times M_2 \longrightarrow \mathbb{E}(2) \times \mathbb{E}(2)$. We already have a description of the projection mappings for each constrained submanifold, $\mu_1 : M_1 \longrightarrow \mathbb{E}(2)$ and $\nu_1 : M_2 \longrightarrow \mathbb{E}(2)$. We can use the combined projection function $\xi = (\mu_1, \nu_1)$ for the projection method. As stated previously, the forces and torques are 1-forms in $T^*\mathbb{E}(2)$, and for the link, for example, they should have the form $a(y_1, y_2, \theta_2)dy_1 + b(y_1, y_2, \theta_2)dy_2 + c(y_1, y_2, \theta_2)d\theta_2$. However, in the equations for the 1-forms we have the variables τ_1 and u_2, which are different from $\{y_1, y_2, \theta_2\}$. The variable u_2 is the input (torque) to the link, and τ_1 is the point of interaction between the link and the disk. Thus τ_1 can be viewed as another input to the link system. In other words, when we have contact between objects, the surfaces of the objects in contact must be included in the input space. When we project the dynamics on the constraint submanifold M_2, then τ_1 is assigned a special value because in M_2, τ_1 is a function of $\{\tilde{y}_1, \tilde{\theta}_2\}$ (the coordinates of M_2). We can compute the quantities ξ_* and ξ^* that are needed for the modeling process. The projection map ξ is defined as

$$\xi : \{\tilde{z}_1, \tilde{\theta}, \tilde{y}_1, \tilde{\theta}_2\} \longrightarrow \left\{\tilde{z}_1, 1, \tilde{\theta}, \tilde{y}_1, \frac{1 + \sin(\tilde{\theta}_2)(\tilde{z}_1 - \tilde{y}_1)}{\cos(\tilde{\theta}_2)} + 1, \tilde{\theta}_2\right\}$$

Next ξ_* is calculated:

$$\xi_* = \begin{pmatrix} 1 & 0 & 0 & 0 \\ 0 & 0 & 0 & 0 \\ 0 & 1 & 0 & 0 \\ 0 & 0 & 1 & 0 \\ \tan(\tilde{\theta}_2) & 0 & -\tan(\tilde{\theta}_2) & \frac{\tilde{z}_1 - \tilde{y}_1 + \sin(\tilde{\theta}_2)}{\cos(\tilde{\theta}_2)} \\ 0 & 0 & 0 & 1 \end{pmatrix}$$

We can easily see that the map ξ is an immersion, because the rank$(\xi_*) = 4$. Applying ξ^* to the 1-forms of the forces and the torques on the RHS of the dynamic equations, we obtain

$$\xi^*(F+T) = \begin{pmatrix} -m_2 g \tan(\tilde{\theta}_2) \\ 0 \\ m_2 g \tan(\tilde{\theta}_2) \\ -\frac{\tilde{z}_1 - \tilde{y}_1 + \sin(\tilde{\theta}_2)}{\cos(\tilde{\theta}_2)^2} m_2 g + u_2 \end{pmatrix}$$

We observe that all the reaction forces have disappeared from the model because the complete model is defined on the constraint submanifold $M_1 \times M_2$. Thus, we could have neglected them in the initial modeling of the individual interacting bodies. Special attention should be paid to include the frictional forces and torques that might occur at the points of interaction (see Example 4).

The kinetic energy tensor ρ has a matrix representation:

$$\rho = \begin{pmatrix} m & 0 & 0 & 0 & 0 & 0 \\ 0 & m & 0 & 0 & 0 & 0 \\ 0 & 0 & I & 0 & 0 & 0 \\ 0 & 0 & 0 & m_2 & 0 & 0 \\ 0 & 0 & 0 & 0 & m_2 & 0 \\ 0 & 0 & 0 & 0 & 0 & I_2 \end{pmatrix}$$

and we can compute $\bar{\rho} = \xi^* \rho$:

$$\bar{\rho} = \begin{pmatrix} m_2 \tan(\tilde{\theta}_2)^2 + m & 0 & -m_2 \tan(\tilde{\theta}_2)^2 & \frac{\tilde{z}_1 - \tilde{y}_1 + \sin(\tilde{\theta}_2)}{\cos(\tilde{\theta}_2)^2} m_2 \tan(\tilde{\theta}_2) \\ 0 & I & 0 & 0 \\ -m_2 \tan(\tilde{\theta}_2)^2 & 0 & m_2 \tan(\tilde{\theta}_2)^2 + m_2 & -\frac{\tilde{z}_1 - \tilde{y}_1 + \sin(\tilde{\theta}_2)}{\cos(\tilde{\theta}_2)^2} m_2 \tan(\tilde{\theta}_2) \\ \frac{\tilde{z}_1 - \tilde{y}_1 + \sin(\tilde{\theta}_2)}{\cos(\tilde{\theta}_2)^2} m_2 \tan(\tilde{\theta}_2) & 0 & -\frac{\tilde{z}_1 - \tilde{y}_1 + \sin(\tilde{\theta}_2)}{\cos(\tilde{\theta}_2)^2} m_2 \tan(\tilde{\theta}_2) & \frac{(\tilde{z}_1 - \tilde{y}_1 + \sin(\tilde{\theta}_2))^2}{\cos(\tilde{\theta}_2)^4} m_2 + I_2 \end{pmatrix}$$

7.11 Example 4: Including Friction in the Dynamics

In Example 2 we did not include the friction forces that can occur at the points of contact. If we introduce friction at the points of contact (Figure 7.6), we have to include these forces in the dynamics as additional components of the 1-forms representing the forces and torques. When we describe the dynamics of the disk and the link on the constraint manifolds M_1 and M_2, respectively, the resulting forces and torques must include the friction, which affects only the RHS of the dynamic equations.

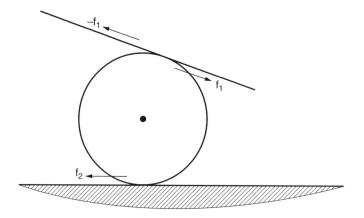

FIGURE 7.6 Friction forces as a result of contact.

Thus the friction forces f_1 and f_2 are included in the 1-forms representing the forces and the torques, and the RHS of Newton's and Euler's equations is given by

$$\begin{pmatrix} -v\sin(\theta_2) + f_1\cos(\theta_2) - f_2 \\ u - v\cos(\theta_2) - mg - f_1\sin(\theta_2) \\ -f_1 - f_2 \\ v\sin(\theta_2) - f_1\cos(\theta_2) \\ v\cos(\theta_2) - m_2 g + f_1\sin(\theta_2) \\ 0 \end{pmatrix}$$

Here the friction forces contribute to both the forces and torques, as we mentioned above. If we want to describe the dynamics on the product manifold, $M_1 \times M_2$, we pull-back this 1-form to obtain

$$\xi^*(F+T) = \begin{pmatrix} -m_2 g \tan(\tilde{\theta}_2) + f_1\sin(\tilde{\theta}_2)\tan(\tilde{\theta}_2) + f_1\cos(\tilde{\theta}_2) - f_2 \\ -f_2 - f_1 \\ m_2 g \tan(\tilde{\theta}_2) - f_1\sin(\tilde{\theta}_2)\tan(\tilde{\theta}_2) - f_1\cos(\tilde{\theta}_2) \\ -\frac{\tilde{z}_1 - \tilde{y}_1 + \sin(\tilde{\theta}_2)}{\cos(\tilde{\theta})^2} m_2 g + u_2 + \frac{f_1\sin(\tilde{\theta}_2)(\tilde{z}_1 - \tilde{y}_1 + \sin(\tilde{\theta}_2))}{\cos^2(\tilde{\theta}_2)} \end{pmatrix}$$

As we can see, the friction forces affect the dynamics of the system of interacting bodies on the constraint submanifold $M_1 \times M_2$, in contrast to the reaction forces which disappear. The LHS of the dynamic equations on $M_1 \times M_2$ is unchanged from before.

7.12 Conclusions

In this chapter we have presented a method for the description of the dynamics of a multibody system where individual elements of the system interact through point contact or through a joint. The modeling methodology uses results from Differential Geometry to develop a systemtic procedure for deriving a reduced configuration space and the equations for the system dynamics. The method can be viewed as an extention to the method proposed by Kron described in B. Hoffmann [20]. Based on this modeling method, we can restrict the configuration space of a multibody system and, at the same time, exclude the interaction forces and torques. The resulting constrained model is appropriate for the investigation of simulation and control of multibody systems, and other related problems.

References

[1] Abraham, R. and Marsden, J.E. (1985). *Foundations of Mechanics.* Addison-Wesley, Reading, MA.
[2] Abraham, R., Marsden, J.E., and Ratiu, T. (1988). Manifolds, Tensor Analysis and Applications. *Applied Mathematical Sciences.* Springer-Verlag, Heidelberg.
[3] Armstrong-Helouvry, B., Dupont, P., and Canudas de Wit, C. (1994). A survey of models, analysis tools and compensation methods for the control of machines with friction. *Automatica,* 30, 7, 1083–1038.
[4] Arnold, V.I., Kozlov, V.V., and Neishtadt, A.I. (1988). *Mathematical Aspects of Classical and Celestial Mechanics.* Springer-Verlag, Vol. 3.
[5] Arnold, V.I. (1966). Sur la Geometrie Differentielle des Groupes de Lie de Dimension Infinie et ses Applications a l' Hydrodynamique des Fluides Parfait. *Ann. Inst. Fourier.* Grenoble 16, 1, 319–361.
[6] Arnold, V.I. (1978). *Mathematical Methods of Classical Mechanics.* Springer-Verlag, Heidelberg.
[7] Barraquand, J. and Lacombe, J.C. (1989). On nonholonomic mobile robots and optimal maneuvering. *Proceedings of the IEEE International Symposium on Intelligent Control,* 25–26 September 1989, pp. 340–347.
[8] Barraquand, J. and Lacombe, J.C. (1991). Nonholonomic multibody mobile robots: Controllability and motion planning in the presence of obstacles. *Proceedings of the IEEE International Conference on Robotics and Automation,* 2328–2335.
[9] Bloch, A. , Reyhanoglu, M, and McClamroch, N.H. (1992). Control and stabilization of nonholonomic dynamic systems." *IEEE Transactions on Automatic Control,* 37, Il, 1746–1757.
[10] Bloch, A. and McClamroch, N.H. (1989). Control of mechanical systems with classical nonholonomic constraints. *Proceedings of the 28th IEEE Conference on Decision and Control,* 201–205.
[11] Bloch, A., McClamroch, N.H., and Reyhanoglu M. (1990). Controllability and stabilizability of a nonholonomic control system. *Proceedings of the 29th IEEE Coriference on Decision and Control,* 1312–1314.
[12] Boothby, W.M. (1986). *An Introduction to Differentiable Manifolds and Riemannian Geometry.* Academic Press, New York.
[13] Brickell, F. and Clark, R.S. (1970). *Differentiable Manifolds.* Van Nostrand Reinhold.
[14] Burke, W. (1976). *Applied Differential Geometry.* Cambridge University Press, Cambridge.
[15] Canudas de Wit, C. and Sordalen, O.J. (1992). Exponential stabilization of mobile robots with nonholonomic constraints. *IEEE Transactions on Automatic Control,* 13, 11, 1791–1797.
[16] Chen, C.K. and Sreenath , N. (1993). Control of coupled spatial two-body systems with nonholonomic constraints. *Proceedings of the 32nd IEEE Conference on Decision and Control,* 949–954.
[17] Goldstein, B. (1944). Mechanics. *Quart. Appl. Math.,* 2, 3, 218–223.
[18] Greenwood, D.T. (1965). *Principles of Dynamics.* Prentice Hall, NJ, 229–274.
[19] Gurvits, L. and Li, Z. (1993). Smooth time-periodic feedback solutions for nonholonomic motion planning. In: *Nonholonomic Motion Planning,* Z. Li and J. Canny, eds, Kluwer, Dordretch, 53–108.
[20] Hoffmann, B. (1944). Kron's Method of Subspaces. *Quart. Appl. Math.,* 2, 3, 218–223.
[21] Hussein, N.M. and Kane, T.R. (1994). Three dimensional reorientation maneuvers of a system of interconnected rigid bodies. *J. Astron. Sci.,* 42, 1, 1–25.
[22] Kozlov, V.V. (1982). Dynamical systems with nonintegrable restrictions I. (Engls. Transl.). *Mosc. Univ. Mech. Bull.,* 37, 3–4, 27–34.
[23] Kozlov, V.V. (1982). Dynamical systems with nonintegrable restrictions II. (Engls. Transl.). *Mosc. Univ. Mech. Bull.,* 37, 3–4, 78–80.
[24] Kozlov, V.V. (1983). Dynamical systems with nonintegrable restrictions III. (Engls. Transl.). *Mosc. Univ. Mech. Bull.,* 38, 3, 27–34.
[25] Kozlov, V.V. (1983). Realization of nonintegrable constraints in classical mechanics. (Engls. Transl.). *Soviet Phys. Dokl.,* 28, 9, 735–737.
[26] Krishnaprasad, P.S. and Yang, R. (1991). Geometric phases, anholonomy, and optimal movement. *Proceedings of the IEEE International Conference on Robotics and Automation,* 2185–2189.

[27] Krishnaprasad, P. S., Yang, R., and Dayawansa, W.P. (1991). Control problems on principle bundles and nonholonomic mechanics. *Proceedings of the 30th IEEE Conference on Decision and Control,* 1133–1137.

[28] Latombe, J.-C. (1991). *Robot Motion Planning.* Kluwer, Boston.

[29] Laumond, J.-P. (1993). Controllability of a multibody mobile robot. *IEEE Transactions on Robotics and Automation,* 9, 6, 755–763.

[30] Laumond, J.-P., Jacobs, P.E., Taix, M., and Murray, R.M. (1994). A motion planner for nonholonomic systems. *IEEE Transactions on Robotics and Automation,* 10, 5, 577–593.

[31] Laumond, J.-P., Sekhavat, S., and Vaisset, M. (1994). Collision-free motion planning for a nonholonomic mobile robot with trailers. *Proceedings of the IFAC Symposium on Robot Control,* Capri, Italy.

[32] Marsden, J.E. (1992). Lectures on mechanics. *London Mathematical Society Lecture Notes Series,* 174, Cambridge University Press, Cambridge.

[33] McClamroch, N.H. and Bloch, A.M. (1988). Control of constrained Hamiltonian systems and applications to control of constrained robots. In: *Dyanmical Systems Approaches to Nonlinear Problems in Systems and Control,* F.M.A. Salam and M.L. Levi, eds, SIAM, Philadelphia, 394–403.

[34] McClamroch, N.H. and Wang, D. (1988). Feedback stabilization and tracking of constrained robots. *IEEE Trans. Automat. Contr.,* 33, 419–426.

[35] McNally, P.J. and McClamroch, N.H. (1993). Space station attitude disturbance arising from internal motions. *Proceedings of the American Control Conference,* 2489–2493.

[36] Murray, R.M., Li, Z., and Sastry, S.S. (1994). *A Mathematical Introduction to Robotic Manipulation.* CRC Press, Bota Raton, FL.

[37] Neimark, Y. and Fufaev, N.A. (1972). Dynamics of Nonholonomic Systems. *Am. Math. Soc. Transl.,* vol. 33.

[38] Reyhanoglu, M. and McClamroch, N.H. (1992). Planar reorientation maneuvers of space multibody spacecraft using internal controls. *J. Guidance Control Dynamics,* 15, 6, 1475–1430.

[39] Reyhanoglu, M. and AI-Regib, E. (1994). Nonholonomic motion planning for wheeled mobile systems using geometric phases. *Proceedings of the 9th IEEE International Symposium on Intelligent Control,* Columbus, Ohio, 135–140.

[40] Reyhanoglu, M., McClamroch, N.H., and Bloch, A.M. (1993). Motion planning for nonholonomic dynamic systems. In: *Nonholonomic Motion Planning,* Z. Li and J. F. Canny, eds, Kluwer, Dordretch, 201–234.

[41] Rosenberg, M.R. (1977). *Analytical Dynamics of Discrete Systems.* Plenum Press, New York, 239–243.

[42] Sahai, A., Secor, M., and Bushnell, L. (1994). An obstacle avoidance algorithm for a car pulling many trailers with kingpin hitching. *Proceedings of the 33rd IEEE Conference on Decision and Control,* 2944–2949.

[43] Samson, C. (1991). Velocity and torque feedback control of a nonholonomic cart. *Advanced Robot Control,* C. Canudas de Wit, ed., *LNCIS 162,* Springer-Verlag, Germany, 125–151.

[44] Samson, C. (1993). Time-varying feedback stabilization of a car-like wheeled mobile robot. *Int. J. Robotics Res.,* 12, 1, 55–66.

[45] Samson, C. and Ait-Abderrahim, K. 1991. Feedback stabilization of a nonholonomic wheeled mobile robot. *Proceedings of International Conference on Intelligent Robots and Systems* (IROS).

[46] Samson, C. (1995). Control of chained system: Application to path following and time-varying point-stabilization of mobile robots. *IEEE Transactions on Automatic Control,* 40, 1, 64–77.

[47] Sordalen, O.J. and Wichlund, K.Y. (1993). Exponential stabilization of a car with n trailers. *Proceedings of the 32nd IEEE Control and Decision Conference,* 1, 978–983.

[48] Sordalen, O.J. (1993). Conversion of the kinematics of a car with n trailers into a chained form. *Proceedings of the IEEE International Conference on Robotics and Automation,* 382–387.

[49] Sreenath, N. (1992). Nonlinear control of planar multibody systems in shape space. *Math. Control Signals Syst.,* 5, 4, 343–363.

[50] Tilbury, D., Laumond, J.-P., Murray, R., Sastry, S.S., and Walsh, G.C. (1992). Steering, car-like systems with trailers using sinusoids. *Proceedings of the IEEE International Conference on Robotics and Automation*, 2, 1993–1998.

[51] Tilbury, D., Murray, R., and Sastry, S.S. (1993). Trajectory, generation for the N-trailer problem using Goursat normal form. *Proceedings of the 32nd IEEE Conference on Decision and Control*, 971–977.

[52] Vakalis, J.S. (1995). Conditions for point contact between robot arm and an object for manipulation modelling and control. *Stud. Informatics Control*, 7, 2, 107–127.

[53] Vakalis, J.S. and Loparo, K. (1996). Configuration space of a body moving on a surface and the kinematics of rolling. *Int. J. Engng. Sci.*, 34, 7, 831–850.

[54] Vakalis, J.S. (1990). "Kinematic and dynamic modelling of interacting manybody systems with applications to robotics." Ph.D Thesis, CWRU 1990.

[55] Walsh, G.C. and Sastry, S.S. (1995). On reorienting linked rigid bodies using internal motions. *IEEE Trans. Robotics Automation*, 11, 1, 139–145.

8
D-H Convention

Dr. Jaydev P. Desai
Drexel University

8.1 Introduction 8-1
8.2 D-H Parameters 8-1
8.3 Algorithm for Determining the Homogenous Transformation Matrix, A_n° 8-6
8.4 Examples 8-8

8.1 Introduction

Denavit-Hartenberg in 1955 developed a notation for assigning orthonormal coordinate frames to a pair of adjacent links in an open kinematic chain. The procedure involves finding the link coordinates and using them to find the 4 × 4 homogeneous transformation matrix composed of four separate submatrices to perform transformations from one coordinate frame to its adjacent coordinate frame. D-H notation is valuable to the area of robotics in which robot manipulators can be modeled as links of rigid bodies. Most industrial robot manipulators are open loop kinematic chains consisting of a base, joints, links, and an endeffector. The ability to control a robot endeffector in three-dimensional space requires the knowledge of a relationship between the robot's joints and the position and orientation of the endeffector. The relationship requires the use and an understanding of the rotation matrix and the translation vector.

8.2 D-H Parameters

Figure 8.1 shows a schematic of two orthonormal reference frames; one is the fixed (inertial) frame and the other is the moving (noninertial) frame. The fixed reference frame comprising the triad of unit vectors ($\vec{e_1}, \vec{e_2}, \vec{e_3}$) has the origin at point O. The moving frame comprising the triad of unit vectors ($\vec{e_1}', \vec{e_2}', \vec{e_3}'$) has the origin at point O'. The rotation matrix transforms coordinates from one reference frame to another. Since most robotic manipulators have individual joint reference frames, which are displaced by a finite distance and rotation, it is necessary to develop a uniform methodology for deriving the transformation from one reference frame to the next. The D-H parameters for a robot manipulator help to systematically derive the transformation from one joint to the next. As a result, it is possible to derive the transformation from the robot endeffector to the base coordinate frame of the robot arm.

In Figure 8.1(a), a point U in the fixed frame $\{F\}$ can be described by equation:

$$\vec{U} = U_1 \vec{e_1} + U_2 \vec{e_2} + U_3 \vec{e_3} \tag{8.1}$$

where U_1, U_2, and U_3 are the projections of the vector $\vec{U}$ along the x, y, and z axes, which have unit vectors, $\vec{e_1}$, $\vec{e_2}$, and $\vec{e_3}$, respectively. Similarly the same point U can also be described in the moving frame

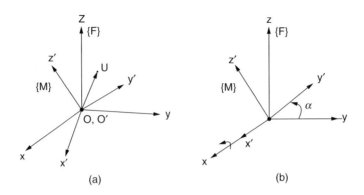

FIGURE 8.1 (a) Schematic of two orthonormal coordinates frames $\{F\}$ and $\{M\}$ and (b) rotation by fixed angle about the x-axis.

$\{M\}$, by the equation:

$$\vec{U} = U'_1 \vec{e_1}' + U'_2 \vec{e_2}' + U'_3 \vec{e_3}' \tag{8.2}$$

where U'_1, U'_2, and U'_3 are the projections of the vector $\vec{U}$ along the x', y', and z' axes, which have unit vectors, $\vec{e_1}'$, $\vec{e_2}'$, and $\vec{e_3}'$, respectively.

From Equation (8.1) and Equation (8.2), we can conclude that

$$U_1 \vec{e_1} + U_2 \vec{e_2} + U_3 \vec{e_3} = U'_1 \vec{e_1}' + U'_2 \vec{e_2}' + U'_3 \vec{e_3}' \tag{8.3}$$

By taking the dot product with $\vec{e_1}$ on both sides in Equation (8.3), we get

$$U_1 = (\vec{e_1}' \cdot \vec{e_1})U'_1 + (\vec{e_2}' \cdot \vec{e_1})U'_2 + (\vec{e_3}' \cdot \vec{e_1})U'_3 \tag{8.4}$$

Similarly taking the dot product with $\vec{e_2}$, and $\vec{e_3}$ separately in Equation (8.3), we get

$$U_2 = (\vec{e_1}' \cdot \vec{e_2})U'_1 + (\vec{e_2}' \cdot \vec{e_2})U'_2 + (\vec{e_3}' \cdot \vec{e_2})U'_3 \tag{8.5}$$
$$U_3 = (\vec{e_1}' \cdot \vec{e_3})U'_1 + (\vec{e_2}' \cdot \vec{e_3})U'_2 + (\vec{e_3}' \cdot \vec{e_3})U'_3 \tag{8.6}$$

From Equation (8.4), Equation (8.5), and Equation (8.6), we conclude that:

$$\begin{bmatrix} U_1 \\ U_2 \\ U_3 \end{bmatrix} = \begin{bmatrix} R^F_M \end{bmatrix} \begin{bmatrix} U'_1 \\ U'_2 \\ U'_3 \end{bmatrix} \tag{8.7}$$

where R^F_M is the general expression for rotation matrix transforming coordinates from the moving frame $\{M\}$ to the fixed frame $\{F\}$ and is given by

$$R^F_M = \begin{bmatrix} \vec{e_1} \cdot \vec{e_1}' & \vec{e_1} \cdot \vec{e_2}' & \vec{e_1} \cdot \vec{e_3}' \\ \vec{e_2} \cdot \vec{e_1}' & \vec{e_2} \cdot \vec{e_2}' & \vec{e_2} \cdot \vec{e_3}' \\ \vec{e_3} \cdot \vec{e_1}' & \vec{e_3} \cdot \vec{e_2}' & \vec{e_3} \cdot \vec{e_3}' \end{bmatrix} \tag{8.8}$$

For the specific case of rotation about the x-axis by an angle α, such as that shown in Figure 8.1(b), the rotation matrix is given by

$$R^F_M(x;\alpha) = \begin{bmatrix} 1 & 0 & 0 \\ 0 & e_2 \cdot e'_2 & e_2 \cdot e'_3 \\ 0 & e_3 \cdot e'_2 & e_3 \cdot e'_3 \end{bmatrix} = \begin{bmatrix} 1 & 0 & 0 \\ 0 & \cos(\alpha) & -\sin(\alpha) \\ 0 & \sin(\alpha) & \cos(\alpha) \end{bmatrix} \tag{8.9}$$

D-H Convention

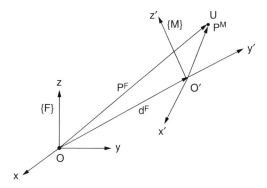

FIGURE 8.2 Schematic of two coordinates frames—one fixed frame {F} and the other moving frame {M} whose origins are displaced by a finite displacement.

Summarized below are the properties of the rotation matrix:

- The determinant of the rotation matrix is 1.
- The norm of any given row or column of the rotation matrix is 1.
- The dot product of any row or column vector with another row or column vector, (except itself), is zero.
- Inverse of the rotation matrix is a rotation matrix, and it is the transpose of the original matrix, i.e., $R^{-1} = R^T$.
- Product of rotation matrices is also a rotation matrix.

Let P^F be the vector representation in the fixed frame {F} of the point U shown in Figure 8.1(a). The same point can be expressed by vector p^M, in the moving frame {M}. Based on Equation (8.7), we can write

$$P^F = R_M^F p^M \tag{8.10}$$

Let us now analyze the second case of two coordinate frames whose origins O and O' are displaced by a finite displacement vector d^F (expressed in the fixed frame). As seen from Figure 8.2, the fixed reference frame {F} has its origin at O while the moving frame {M} has its origin at O'.

Based on previous analysis, if $\vec{d^F} = \vec{0}$, then we can write $P^F = R_M^F p^M$. However, since O and O' are a finite distance apart, $\vec{d^F} \neq \vec{0}$, we have

$$P^F = R_M^F p^M + d^F \tag{8.11}$$

where R_M^F is the rotation transformation from the moving frame {M} and the fixed frame {F}, and d^F is the displacement expressed in the fixed frame between the origin O' of the moving frame {M} and the origin O of the fixed frame {F}.

$$\begin{pmatrix} P_x \\ P_y \\ P_z \end{pmatrix}^F = R_M^F \begin{pmatrix} p_x \\ p_y \\ p_z \end{pmatrix}^M + \begin{pmatrix} d_x \\ d_y \\ d_z \end{pmatrix}^F \tag{8.12}$$

The rotation matrix and the displacement vector are submatrices of the 4×4 homogenous transformation matrix A_M^F. Equation (8.12) is the same as Equation (8.11) except that the point P^F has been written

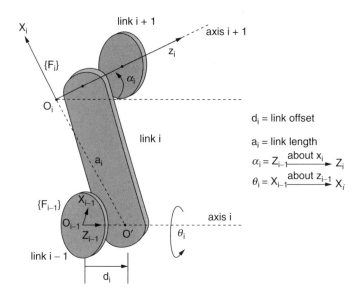

FIGURE 8.3 Schematic of the adjacent axes with the appropriately assigned reference frames for determining the Denavit-Hartenburg parameters.

in vector form. Equation (8.13) provides the expression for the 4 × 4 homogenous transformation matrix A_M^F. Thus, we can write

$$\begin{pmatrix} P_x \\ P_y \\ P_z \end{pmatrix}^F = \begin{bmatrix} R_{M3\times3}^F & d_{3\times1}^F \\ 0\ 0\ 0 & 1 \end{bmatrix} \begin{pmatrix} p_x \\ p_y \\ p_z \end{pmatrix} \Rightarrow A_M^F = \begin{bmatrix} R_{M3\times3}^F & d_{3\times1}^F \\ 0\ 0\ 0 & 1 \end{bmatrix} \quad (8.13)$$

Having described the general relationship for transforming coordinates from one reference frame to the other, we can now discuss the D-H parameters. In Figure 8.3, link i connects link $i-1$ to link $i+1$ through a_i. The figure is used to demonstrate the steps involved in determining the Denavit and Hartenberg parameters. D-H notation is a method of assigning coordinate frames to the different joints of a robotic manipulator. The method involves determining four parameters to build a complete homogeneous transformation matrix. These parameters are the twist angle α_i, link length a_i, link offset d_i, and joint angle θ_i. Parameters a_i and α_i are based on the geometry of the manipulator and are constant values based on the manipulator geometry, while parameters d_i and θ_i can be variable (depending on whether the joint is prismatic or revolute).

The following steps denote the systematic derivation of the D-H parameters.

1. Label each axis in the manipulator with a number starting from 1 as the base to n as the endeffector. O_o is the origin of the base coordinate frame. Every joint must have an axis assigned to it.
2. Set up a coordinate frame for each joint. Starting with the base joint, set up a right handed coordinate frame for each joint. For a rotational joint, the axis of rotation for axis i is always along Z_{i-1}. If the joint is a prismatic joint, Z_{i-1} should point in the direction of translation.
3. The X_i axis should always point away from the Z_{i-1} axis.
4. Y_i should be directed such that a right-handed orthonormal coordinate frame is created.
5. For the next joint, if it is not the endeffector frame, steps 2–4 should be repeated.
6. For the endeffector, the Z_n axis should point in the direction of the endeffector approach.
7. Joint angle θ_i is the rotation about Z_{i-1} to make X_{i-1} parallel to X_i (going from X_{i-1} to X_i).
8. Twist angle α_i is the rotation about X_i axis to make Z_{i-1} parallel to Z_i (going from Z_{i-1} to Z_i).

D-H Convention

9. Link length a_i is the perpendicular distance between axis i and axis $i+1$.
10. Link offset d_i is the offset along the Z_{i-1} axis as shown schematically in Figure 8.3. Thus d_i is the distance between O_{i-1} and O'_i along Z_{i-1} (axis i).

The above steps are shown schematically in the flowchart in Figure 8.4.

In a robot manipulator, there are commonly two types of joints: revolute and prismatic. The revolute joint allows for rotation between two links about an axis, and the prismatic joint allows for translation (sliding) motion along an axis. In a revolute joint, the link offset d is a constant while the joint angle θ is a variable, and in a prismatic joint, the link offset d is variable and the joint angle θ is normally zero. The link length a_i and the twist angle α_i are determined by the geometry of the manipulator and are therefore constant values.

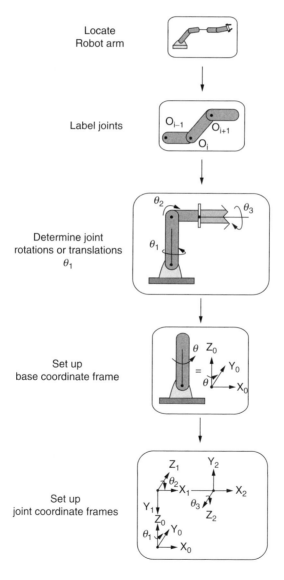

FIGURE 8.4 Flow chart for the process of determining the Denavit-Hartenburg parameters for a given robot manipulator.(*Continued*)

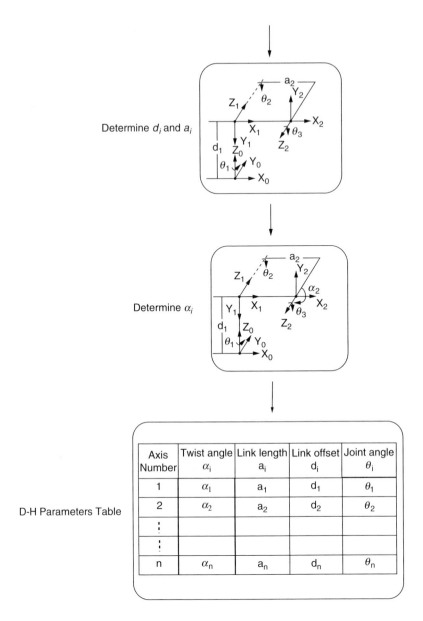

FIGURE 8.4 (Continued).

8.3 Algorithm for Determining the Homogenous Transformation Matrix, A_n^o

After determining the D-H parameters, the next step is to derive the homogeneous transformation matrix from one frame to the next. We will denote the $U_i(\theta_i)$ matrix as the rotation matrix about the Z_{i-1} axis while the $V_i(\alpha_i)$ matrix as the rotation matrix denoting the rotation about the X_i axis. To perform coordinate transformation from reference frame $\{F_i\}$ to reference frame $\{F_{i-1}\}$, we need to derive the rotation matrix and the displacement vector. The rotation matrix from $\{F_i\}$ to $\{F_{i-1}\}$ is given by

$$R_i^{i-1} = U_i V_i \qquad (8.14)$$

D-H Convention

where $U_i(\theta_i)$ is given by

$$U_i(\theta_i) = \begin{bmatrix} \cos(\theta_i) & -\sin(\theta_i) & 0 \\ \sin(\theta_i) & \cos(\theta_i) & 0 \\ 0 & 0 & 1 \end{bmatrix} \quad (8.15)$$

and $V_i(\alpha_i)$ is given by

$$V_i(\alpha_i) = \begin{bmatrix} 1 & 0 & 0 \\ 0 & \cos(\alpha_i) & -\sin(\alpha_i) \\ 0 & \sin(\alpha_i) & \cos(\alpha_i) \end{bmatrix} \quad (8.16)$$

The homogeneous transformation matrix transforming coordinates from $\{F_i\}$ to $\{F_{i-1}\}$ is given by

$$A_i^{i-1} = A_{i'}^{i-1} A_i^{i'} \quad (8.17)$$

where $A_i^{i'}$ is the screw displacement about the axis x_i through an angle α_i and distance a_i. Similarly, $A_{i'}^{i-1}$ is the screw displacement about the axis z_i through an angle θ_i and distance d_i. The expression for $A_{i'}^{i-1}$ and $A_i^{i'}$ is given by

$$A_{i'}^{i-1} = \begin{bmatrix} & & & 0 \\ & U_{i_{3\times3}} & & 0 \\ & & & d_i \\ 0 & 0 & 0 & 1 \end{bmatrix} \quad (8.18)$$

$$A_i^{i'} = \begin{bmatrix} & & & a_i \\ & V_{i_{3\times3}} & & 0 \\ & & & 0 \\ 0 & 0 & 0 & 1 \end{bmatrix} \quad (8.19)$$

Consequently,

$$A_i^{i-1} = \begin{bmatrix} & & & 0 \\ & U_i & & 0 \\ & & & d_i \\ 0 & 0 & 0 & 1 \end{bmatrix} \begin{bmatrix} & & & a_i \\ & V_i & & 0 \\ & & & 0 \\ 0 & 0 & 0 & 1 \end{bmatrix} = \begin{bmatrix} U_i V_i & U_i s_i \\ 0 & 1 \end{bmatrix} \quad (8.20)$$

Expanding the various terms in Equation (8.20) we get

$$U_i V_i = \begin{bmatrix} c\theta_i & -s\theta_i & 0 \\ s\theta_i & c\theta_i & 0 \\ 0 & 0 & 1 \end{bmatrix} \begin{bmatrix} 1 & 0 & 0 \\ 0 & c\alpha_i & -s\alpha_i \\ 0 & s\alpha_i & +c\alpha_i \end{bmatrix} = \begin{bmatrix} c\theta_i & -s\theta_i c\alpha_i & s\theta_i s\alpha_i \\ s\theta_i & c\theta_i c\alpha_i & -c\theta_i s\alpha_i \\ 0 & s\alpha_i & +c\alpha_i \end{bmatrix}$$

$$U_i s_i = \begin{bmatrix} c\theta_i & -s\theta_i & 0 \\ s\theta_i & c\theta_i & 0 \\ 0 & 0 & 1 \end{bmatrix} \begin{bmatrix} a_i \\ 0 \\ 0 \end{bmatrix} + \begin{bmatrix} 0 \\ 0 \\ d_i \end{bmatrix} = \begin{bmatrix} c\theta_i a_i \\ s\theta_i a_i \\ d_i \end{bmatrix} \quad (8.21)$$

where

$$s_i = \begin{bmatrix} a_i \\ 0 \\ d_i \end{bmatrix} \quad (8.22)$$

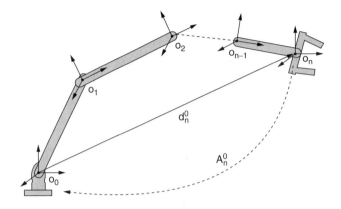

FIGURE 8.5 Transformation from the endeffector frame to the base frame.

Performing the composition from the nth frame to the base frame (see Figure 8.5), we get

$$A_n^0 = A_1^0 A_2^1 \ldots A_n^{n-1} = \begin{bmatrix} R_n^0 & d_n^0 \\ 0 & 1 \end{bmatrix} \quad (8.23)$$

Based on Equations (8.20) and (8.23) we can write

$$R_n^0 = (U_1 V_1)(U_2 V_2) \cdots (U_n V_n) \quad (8.24)$$

and

$$d_n^0 = U_1 s_1 + U_1 V_1 U_2 s_2 + U_1 V_1 U_2 V_2 U_3 s_3 \ldots + U_1 V_1 U_2 V_2 \ldots U_{n-1} V_{n-1} U_n s_n \quad (8.25)$$

8.4 Examples

Example 8.1

Figure 8.6 is an example of a robotic arm manipulator with five joints. In this manipulator, three joints are rotational and two joints are prismatic. Thus, the robot has five degrees of freedom.

No.	Twist Angle α_i	Link Length a_i	Joint Offset d_i	Joint Angle θ_i
$1_{(0-1)}$	0	0	0	$\theta_1^{\text{Variable}}$
$2_{(1-2)}$	$\pi/2$	0	d_2^{Variable}	0
$3_{(2-3)}$	$\pi/2$	0	0	$\theta_3^{\text{Variable}}$
$4_{(3-4)}$	0	0	d_4^{Variable}	0
$5_{(4-5)}$	0	0	0	$\theta_5^{\text{Variable}}$

The first step in determining the D-H parameters is to locate the joints of the robot arm manipulator and determine if the joint is prismatic or revolute. For the five-degrees-of-freedom manipulator, starting from the base joint, joints 1, 3, and 5 are revolute joints, while joints 2 and 4 are prismatic joints. Because

D-H Convention

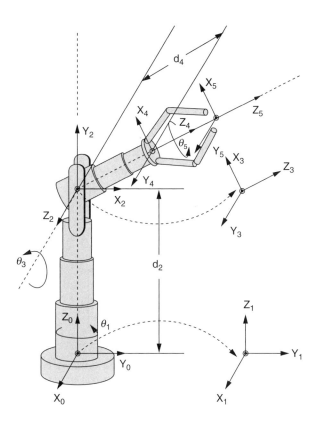

FIGURE 8.6 D-H notation for a five-degrees-of-freedom robot arm manipulator.

joints 1, 3, and 5 are revolute, the θ_i values are variable, i.e., $\theta_1^{\text{Variable}}$, $\theta_3^{\text{Variable}}$, and $\theta_5^{\text{Variable}}$, respectively. Since there is no rotation about prismatic joints, the θ_i values for joints 2 and 4 are zero. Similarly for prismatic joints 2 and 4, the d_i values are variable, i.e., d_2^{Variable} and d_4^{Variable} while θ_2 and θ_4 are zero.

As per the flow chart in Figure 8.4, the next step is to set up the joint coordinate frames starting with the base joint 1. The Z_0 axis points in the direction such that θ_1 is about the Z_0 axis. X_0 and Y_0 are set up for a right-hand coordinate frame. Joint 2 is a prismatic joint. Thus Z_1 points in the direction of the translation of the prismatic joint, X_1 is in the plane perpendicular to Z_0 and the direction of Y_1 completes the right-handed coordinate system. Joint 3 is a revolute joint so the same system follows for assigning the coordinate frame as the first joint. X_2 lies in the plane perpendicular to Z_2 (and also away from Z_1). Similarly the coordinate system for joints 4 and 5 can be assigned as shown in Figure 8.6.

Having established the coordinate frames, the next step is to determine the D-H parameters. We begin by first determining that α_i. α_i is the rotation about X_i to make Z_{i-1} parallel with Z_i (starting from Z_{i-1}). Starting with axis 1, the rotation about X_1 to make Z_0 parallel with Z_1 is zero because the Z axes for both are parallel. For axis 2, the rotation required about X_2 to take Z_1 parallel to Z_2 is 90° or $\pi/2$. Similarly α_3 is also $\pi/2$. Both α_4 and α_5 are zero because the Z axes are parallel for the last two joints.

The next step is to determine a_i and d_i. a_i is the link length and always points away from the Z_{i-1} axis. d_i is the offset and is always along the Z_{i-1} axis. For axis 1, there is no offset in the Z_0 direction from joint 1 to joint 2, so d_1 is equal to zero. Also, the distance between axes 1 and 2 is zero, so a_1 is zero. As seen from the schematic of the five-degrees-of-freedom manipulator in Figure 8.6, the only nonzero a_i and d_i D-H parameters are d_2^{Variable} and d_4^{Variable}. Each of these distances is in the Z_{i-1} direction, so the respective d values are equal to these distances.

Having determined all the D-H parameters, the transformation matrix A_5^0 can now be computed. The transformation matrix consists of the rotation matrix R_5^0 and the displacement vector d_5^0. Using the expression in Equation (8.24) and Equation (8.25) we get

$$R_5^0 = (U_1 V_1)(U_2 V_2)(U_3 V_3)(U_4 V_4)(U_5 V_5) \tag{8.26}$$

$$d_5^0 = U_1 s_1 + (U_1 V_1) U_2 s_2 + (U_1 V_1)(U_2 V_2) U_3 s_3 + (U_1 V_1)(U_2 V_2)(U_3 V_3) U_4 s_4$$
$$+ (U_1 V_1)(U_2 V_2)(U_3 V_3)(U_4 V_4) U_5 s_5 \tag{8.27}$$

The following matrices are the individual U and V matrices for each axis of the manipulator.

$$U_1 = \begin{bmatrix} \cos(\theta_1) & -\sin(\theta_1) & 0 \\ \sin(\theta_1) & \cos(\theta_1) & 0 \\ 0 & 0 & 1 \end{bmatrix}, \quad V_1 = \begin{bmatrix} 1 & 0 & 0 \\ 0 & 1 & 0 \\ 0 & 0 & 1 \end{bmatrix}$$

$$U_2 = \begin{bmatrix} 1 & 0 & 0 \\ 0 & 1 & 0 \\ 0 & 0 & 0 \end{bmatrix}, \quad V_2 = \begin{bmatrix} 1 & 0 & 0 \\ 0 & 0 & -1 \\ 0 & 1 & 0 \end{bmatrix}$$

$$U_3 = \begin{bmatrix} \cos(\theta_3) & -\sin(\theta_3) & 0 \\ \sin(\theta_3) & \cos(\theta_3) & 0 \\ 0 & 0 & 1 \end{bmatrix}, \quad V_3 = \begin{bmatrix} 1 & 0 & 0 \\ 0 & 0 & -1 \\ 0 & 1 & 0 \end{bmatrix}$$

$$U_4 = \begin{bmatrix} 1 & 0 & 0 \\ 0 & 1 & 0 \\ 0 & 0 & 1 \end{bmatrix}, \quad V_4 = \begin{bmatrix} 1 & 0 & 0 \\ 0 & 1 & 0 \\ 0 & 0 & 1 \end{bmatrix}$$

$$U_5 = \begin{bmatrix} \cos(\theta_5) & -\sin(\theta_5) & 0 \\ \sin(\theta_5) & \cos(\theta_5) & 0 \\ 0 & 0 & 1 \end{bmatrix}, \quad V_5 = \begin{bmatrix} 1 & 0 & 0 \\ 0 & 1 & 0 \\ 0 & 0 & 1 \end{bmatrix}$$

Similarly the s_i vector is given by

$$s_1 = \begin{bmatrix} 0 \\ 0 \\ 0 \end{bmatrix}, \quad s_2 = \begin{bmatrix} 0 \\ 0 \\ d_2 \end{bmatrix}, \quad s_3 = \begin{bmatrix} 0 \\ 0 \\ 0 \end{bmatrix}, \quad s_4 = \begin{bmatrix} 0 \\ 0 \\ d_4 \end{bmatrix}, \quad s_5 = \begin{bmatrix} 0 \\ 0 \\ 0 \end{bmatrix}$$

Substituting the above U_i, V_i, and s_i into Equation (8.26) and Equation (8.27) we get

$$R_5^0 = \begin{bmatrix} C_1 C_3 C_5 + S_1 S_5 & -C_1 C_3 S_5 + S_1 C_5 & C_1 S_3 \\ S_1 C_3 C_5 - C_1 S_5 & -S_1 C_3 S_5 - C_1 C_5 & S_1 S_3 \\ S_3 C_5 & -S_3 S_5 & -C_3 \end{bmatrix} \tag{8.28}$$

$$d_5^0 = \begin{bmatrix} C_1 S_3 d_4 \\ S_1 S_3 d_4 \\ d_2 - C_3 d_4 \end{bmatrix} \tag{8.29}$$

where $C_a = \cos(a)$ and $S_a = \sin(a)$.

The final transformation matrix is thus given by

$$A_5^0 = \begin{bmatrix} C_1 C_3 C_5 + S_1 S_5 & -C_1 C_3 S_5 + S_1 C_5 & C_1 S_3 & C_1 S_3 d_4 \\ S_1 C_3 C_5 - C_1 S_5 & -S_1 C_3 S_5 - C_1 C_5 & S_1 S_3 & S_1 S_3 d_4 \\ S_3 C_5 & -S_3 S_5 & -C_3 & d_2 - C_3 d_4 \\ 0 & 0 & 0 & 1 \end{bmatrix} \tag{8.30}$$

D-H Convention

TABLE 8.1 D-H Parameters for the SCARA

No.	Twist Angle α_i	Link Length a_i	Link Offset d_i	Joint Angle θ_i
$1_{(0-1)}$	0	a_1	d_1	$\theta_1^{\text{Variable}}$
$2_{(1-2)}$	π	a_2	0	$\theta_2^{\text{Variable}}$
$3_{(2-3)}$	0	0	d_3^{Variable}	0
$4_{(3-4)}$	0	0	d_4	$\theta_4^{\text{Variable}}$

Example 8.2: The SCARA Robot

The SCARA robot is a four axis robot arm manipulator with three revolute joints and one prismatic joint (see Figure 8.7 and Table 8.1). The name SCARA is an acronym for *selective compliance assembly robot arm*.

The first step in determining D-H parameters for the SCARA is to locate the joints and determine the rotation or translation of each joint. The SCARA has four joints with three of them being revolute joints and one prismatic joint. Each of the revolute joints, 1, 2, and 4, have a θ value of $\theta_i^{\text{variable}}$ with i being the joint number. The prismatic joint (joint 3) has a θ value of zero and a d value of d_3^{variable}. Starting from the base, the joint coordinate frames are assigned based on the algorithm outlined before.

Having established the coordinate frames, the next step is the determination of the D-H parameters. We begin by first determining α_i. α_i is the rotation about X_i to make Z_{i-1} parallel with Z_i (starting from Z_{i-1}). Starting with axis 1, the rotation about X_1 to make Z_0 parallel to Z_1 is 0 because they are parallel. For axis 2, the α is π or 180° because Z_2 is pointing down along the translation of the prismatic joint. α_3 and α_4 are zero because Z_3 and Z_4 are parallel with Z_2 and Z_3, respectively.

The next step is to determine a_i and d_i. a_i is the link length and always points away from the Z_{i-1} axis. d_i is the offset and is always along the Z_{i-1} axis. For axis 1, there is an offset between axes 1 and 2 in the

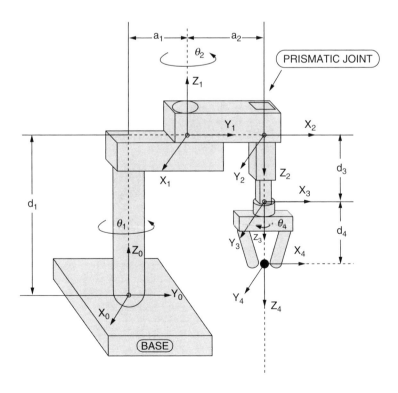

FIGURE 8.7 Schematic of the SCARA robot with the appropriate coordinate frames.

Z_0 direction so the offset is equal to d_1. There is also a distance between the axes, which is equal to a_1. For axis 2, there is a distance between axes 2 and 3 away from the Z_1 axis equal to a_2. d_2 is equal to zero. Axis 3 is a prismatic joint, so d is variable, i.e., d_3^{Variable}. Between axes 3 and 4, there is an offset only in the Z direction, so d_4 is equal to this distance. a_3 and a_4 are both zero.

Having determined all the D-H parameters, the transformation matrix A_4^0 can now be computed. The transformation matrix consists of the rotation matrix R_4^0 and the displacement vector d_4^0. Using the expression in Equation (8.24) and Equation (8.25) we get

$$R_4^0 = (U_1 V_1)(U_2 V_2)(U_3 V_3)(U_4 V_4) \quad (8.31)$$

$$d_4^0 = U_1 s_1 + (U_1 V_1) U_2 s_2 + (U_1 V_1)(U_2 V_2) U_3 s_3 + (U_1 V_1)(U_2 V_2)(U_3 V_3) U_4 s_4 \quad (8.32)$$

The following matrices are the individual U and V matrices for each axis of the manipulator.

$$U_1 = \begin{bmatrix} \cos(\theta_1) & -\sin(\theta_1) & 0 \\ \sin(\theta_1) & \cos(\theta_1) & 0 \\ 0 & 0 & 1 \end{bmatrix}, \quad V_1 = \begin{bmatrix} 1 & 0 & 0 \\ 0 & 1 & 0 \\ 0 & 0 & 1 \end{bmatrix}$$

$$U_2 = \begin{bmatrix} \cos(\theta_2) & -\sin(\theta_2) & 0 \\ \sin(\theta_2) & \cos(\theta_2) & 0 \\ 0 & 0 & 1 \end{bmatrix}, \quad V_2 = \begin{bmatrix} 1 & 0 & 0 \\ 0 & -1 & 0 \\ 0 & 0 & -1 \end{bmatrix}$$

$$U_3 = \begin{bmatrix} 1 & 0 & 0 \\ 0 & 1 & 0 \\ 0 & 0 & 1 \end{bmatrix}, \quad V_3 = \begin{bmatrix} 1 & 0 & 0 \\ 0 & 1 & 0 \\ 0 & 0 & 1 \end{bmatrix}$$

$$U_4 = \begin{bmatrix} \cos(\theta_4) & -\sin(\theta_4) & 0 \\ \sin(\theta_4) & \cos(\theta_4) & 0 \\ 0 & 0 & 1 \end{bmatrix}, \quad V_4 = \begin{bmatrix} 1 & 0 & 0 \\ 0 & 1 & 0 \\ 0 & 0 & 1 \end{bmatrix}$$

Similarly the s_i vector is given by

$$s_1 = \begin{bmatrix} a_1 \\ 0 \\ d_1 \end{bmatrix}, \quad s_2 = \begin{bmatrix} a_2 \\ 0 \\ 0 \end{bmatrix}, \quad s_3 = \begin{bmatrix} 0 \\ 0 \\ d_3 \end{bmatrix}, \quad s_4 = \begin{bmatrix} 0 \\ 0 \\ d_4 \end{bmatrix}$$

Substituting the above U_i, V_i, and s_i into Equation (8.31) and Equation (8.32) we get

$$R_4^0 = \begin{bmatrix} C_{12}C_4 + S_{12}S_4 & -C_{12}S_4 + S_{12}C_4 & 0 \\ S_{12}C_4 - C_{12}S_4 & -S_{12}S_4 - C_{12}C_4 & 0 \\ 0 & 0 & -1 \end{bmatrix} \quad (8.33)$$

$$d_4^0 = \begin{bmatrix} C_1 a_1 + C_{12} a_2 \\ S_1 a_1 + S_{12} a_2 \\ d_1 - d_3 - d_4 \end{bmatrix} \quad (8.34)$$

where $C_a = \cos(a)$, $S_a = \sin(a)$, $C_{ab} = \cos(a+b)$ and $S_{ab} = \sin(a+b)$.

The final transformation matrix is

$$A_4^0 = \begin{bmatrix} C_{12}C_4 + S_{12}S_4 & -C_{12}S_4 + S_{12}C_4 & 0 & C_1 a_1 + C_{12} a_2 \\ S_{12}C_4 - C_{12}S_4 & -S_{12}S_4 - C_{12}C_4 & 0 & S_1 a_1 + S_{12} a_2 \\ 0 & 0 & -1 & d_1 - d_3 - d_4 \\ 0 & 0 & 0 & 1 \end{bmatrix} \quad (8.35)$$

D-H Convention

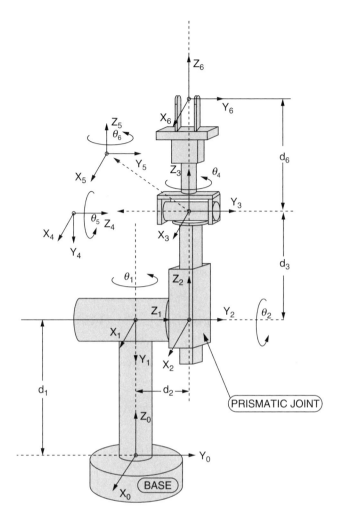

FIGURE 8.8 Schematic of the Stanford arm.

Example 8.3: The Stanford Arm

Designed in 1969 by Victor Scheinman, the Stanford arm was one of the first robots exclusively designed for computer control. [1] The Stanford arm is a 6-axis robot manipulator with five revolute joints and one prismatic joint, giving it six degrees of freedom

The first step in determining the D-H parameters for the Stanford arm is to locate the joints and determine if the joint is prismatic or revolute (Table 8.2). Starting from the base joint, joints 1, 2, 4, 5, and 6 are all revolute joints. The θ_i values are $\theta_1^{Variable}$, $\theta_2^{Variable}$, $\theta_4^{Variable}$, $\theta_5^{Variable}$, and $\theta_6^{Variable}$, respectively. Since there is no rotation about prismatic joints, θ_3 is equal to zero but d_3 is equal to $d_3^{Variable}$. Starting from the base, the joint coordinate frames are assigned based on the algorithm outlined before.

Having established the coordinate frames, the next step is to determine the D-H parameters. We begin by first determining α_i. α_i is the rotation about X_i to make Z_{i-1} parallel with Z_i. For axis 1, the rotation required about X_1 to take Z_0 parallel to Z_1 is $-90°$ or $-\pi/2$. Similarly α_2 and α_5 are equal to $90°$ or $\pi/2$. For axes 3 and 6, Z_{i-1} and Z_i are parallel, so α_3 and α_6 are equal to zero.

---

[1] http://www-db.stanford.edu/pub/voy/museum/pictures/display/1-Robot.htm

TABLE 8.2 D-H Parameters for the Stanford Arm

No.	Twist Angle α_i	Link Length a_i	Link Offset d_i	Joint Angle θ_i
$1_{(0-1)}$	$-\pi/2$	0	d_1	$\theta_1^{\text{Variable}}$
$2_{(1-2)}$	$\pi/2$	0	d_2	$\theta_2^{\text{Variable}}$
$3_{(2-3)}$	0	0	d_3^{Variable}	0
$4_{(3-4)}$	$-\pi/2$	0	0	$\theta_4^{\text{Variable}}$
$5_{(4-5)}$	$\pi/2$	0	0	$\theta_5^{\text{Variable}}$
$6_{(5-6)}$	0	0	d_6	$\theta_6^{\text{Variable}}$

The next step is to determine a_i and d_i. a_i is the link length and always points away from the Z_{i-1} axis. d_i is the offset and is along the Z_{i-1} axis. As seen from the schematic of the Stanford arm, there is no distance between axes in any other direction besides the Z direction, thus the a_i for each link is zero.

Having determined all the D-H parameters, the transformation matrix A_6^0 can now be computed. The transformation matrix consists of the rotation matrix R_6^0 and the displacement vector d_6^0. Using the expression in Equation (8.24) and Equation (8.25) we get

$$R_6^0 = (U_1 V_1)(U_2 V_2)(U_3 V_3)(U_4 V_4)(U_5 V_5)(U_6 V_6) \qquad (8.36)$$

$$d_6^0 = U_1 s_1 + (U_1 V_1)U_2 s_2 + (U_1 V_1)(U_2 V_2)U_3 s_3 + (U_1 V_1)(U_2 V_2)(U_3 V_3)U_4 s_4$$
$$+ (U_1 V_1)(U_2 V_2)(U_3 V_3)(U_4 V_4)U_5 s_5 + (U_1 V_1)(U_2 V_2)(U_3 V_3)(U_4 V_4)(U_5 V_5)U_6 s_6 \qquad (8.37)$$

The following matrices are the individual U and V matrices for each axis of the manipulator.

$$U_1 = \begin{bmatrix} \cos(\theta_1) & -\sin(\theta_1) & 0 \\ \sin(\theta_1) & \cos(\theta_1) & 0 \\ 0 & 0 & 1 \end{bmatrix}, \quad V_1 = \begin{bmatrix} 1 & 0 & 0 \\ 0 & 0 & 1 \\ 0 & -1 & 0 \end{bmatrix}$$

$$U_2 = \begin{bmatrix} \cos(\theta_2) & -\sin(\theta_2) & 0 \\ \sin(\theta_2) & \cos(\theta_2) & 0 \\ 0 & 0 & 1 \end{bmatrix}, \quad V_2 = \begin{bmatrix} 1 & 0 & 0 \\ 0 & 0 & -1 \\ 0 & 1 & 0 \end{bmatrix}$$

$$U_3 = \begin{bmatrix} 1 & 0 & 0 \\ 0 & 1 & 0 \\ 0 & 0 & 1 \end{bmatrix}, \quad V_3 = \begin{bmatrix} 1 & 0 & 0 \\ 0 & 1 & 0 \\ 0 & 0 & 1 \end{bmatrix}$$

$$U_4 = \begin{bmatrix} \cos(\theta_4) & -\sin(\theta_4) & 0 \\ \sin(\theta_4) & \cos(\theta_4) & 0 \\ 0 & 0 & 1 \end{bmatrix}, \quad V_4 = \begin{bmatrix} 1 & 0 & 0 \\ 0 & 0 & 1 \\ 0 & -1 & 0 \end{bmatrix}$$

$$U_5 = \begin{bmatrix} \cos(\theta_5) & -\sin(\theta_5) & 0 \\ \sin(\theta_5) & \cos(\theta_5) & 0 \\ 0 & 0 & 1 \end{bmatrix}, \quad V_5 = \begin{bmatrix} 1 & 0 & 0 \\ 0 & 0 & -1 \\ 0 & 1 & 0 \end{bmatrix}$$

$$U_6 = \begin{bmatrix} \cos(\theta_6) & -\sin(\theta_6) & 0 \\ \sin(\theta_6) & \cos(\theta_6) & 0 \\ 0 & 0 & 1 \end{bmatrix}, \quad V_6 = \begin{bmatrix} 1 & 0 & 0 \\ 0 & 1 & 0 \\ 0 & 0 & 1 \end{bmatrix}$$

Similarly the s_i vector is given by

$$s_1 = \begin{bmatrix} 0 \\ 0 \\ d_1 \end{bmatrix}, \quad s_2 = \begin{bmatrix} 0 \\ 0 \\ d_2 \end{bmatrix}, \quad s_3 = \begin{bmatrix} 0 \\ 0 \\ d_3 \end{bmatrix}, \quad s_4 = \begin{bmatrix} 0 \\ 0 \\ 0 \end{bmatrix}, \quad s_5 = \begin{bmatrix} 0 \\ 0 \\ 0 \end{bmatrix}, \quad s_6 = \begin{bmatrix} 0 \\ 0 \\ d_6 \end{bmatrix}$$

Substituting the above U_i, V_i, and s_i into Equation (8.36) and Equation (8.37), we get the following, where $C_i = \cos(i)$ and $S_i = \sin(i)$:

$$R_6^0 = \begin{bmatrix} r_{11} & r_{12} & r_{13} \\ r_{21} & r_{22} & r_{23} \\ r_{31} & r_{32} & r_{33} \end{bmatrix} \tag{8.38}$$

$r_{11} = ((C_1C_2C_4 - S_1S_4)C_5 - C_1S_2S_5)C_6 + (-C_1C_2S_4 - S_1C_4)S_6$
$r_{12} = -((C_1C_2C_4 - S_1S_4)C_5 - C_1S_2S_5)S_6 + (-C_1C_2S_4 - S_1C_4)C_6$
$r_{13} = (C_1C_2C_4 - S_1S_4)S_5 + C_1S_2C_5$
$r_{21} = ((S_1C_2C_4 + C_1S_4)C_5 - S_1S_2S_5)C_6 + (-S_1C_2S_4 + C_1C_4)S_6$
$r_{22} = -((S_1C_2C_4 + C_1S_4)C_5 - S_1S_2S_5)S_6 + (-S_1C_2S_4 + C_1C_4)C_6$
$r_{23} = (S_1C_2C_4 + C_1S_4)S_5 + S_1S_2C_5$
$r_{31} = (-S_2C_4C_5 - C_2S_5)C_6 + S_2S_4S_6$
$r_{32} = -(-S_2C_4C_5 - C_2S_5)S_6 + S_2S_4C_6$
$r_{33} = -S_2C_4S_5 + C_2C_5$

$$d_6^0 = \begin{bmatrix} d_{11} \\ d_{21} \\ d_{31} \end{bmatrix} \tag{8.39}$$

$d_{11} = -S_1d_2 + C_1S_2d_3 + ((C_1C_2C_4 - S_1S_4)S_5 + C_1S_2C_5)d_6$
$d_{21} = C_1d_2 + S_1S_2d_3 + ((S_1C_2C_4 + C_1S_4)S_5 + S_1S_2C_5)d_6$
$d_{31} = d_1 + C_2d_3 + (-S_2C_4S_5 + C_2C_5)d_6$

The final transformation matrix is

$$A_6^0 = \begin{bmatrix} R_6^0 & d_6^0 \\ 0 \ 0 \ 0 & 1 \end{bmatrix} \tag{8.40}$$

Example 8.4: The Mitsubishi PA-10 Robot Arm

The Mitsubishi PA-10 robot arm is a 7-joint robot manipulator. All seven of the joints are revolute, and thus the manipulator has seven degrees of freedom. The first step in determining the D-H parameters of the Mitsubishi PA-10 robot arm is to locate each joint and determine whether it is rotational or prismatic (Table 8.3 and Figure 8.9). As seen in Figure 8.9, all the joints are rotational joints. Because all the joints are rotational the θ_i values are variable. Starting from the base, the joint coordinate frames are assigned based on the algorithm outlined before.

TABLE 8.3 D-H Parameters of the Mitsubishi PA-10 Robot Arm

No.	Twist Angle α_i	Link Length a_i	Joint Offset d_i	Joint Angle θ_i
$1_{(0-1)}$	$-\pi/2$	0	0	$\theta_1^{\text{Variable}}$
$2_{(1-2)}$	$-\pi/2$	0	0	$\theta_2^{\text{Variable}}$
$3_{(2-3)}$	$\pi/2$	0	d_3	$\theta_3^{\text{Variable}}$
$4_{(3-4)}$	$\pi/2$	0	0	$\theta_4^{\text{Variable}}$
$5_{(4-5)}$	$\pi/2$	0	d_5	$\theta_5^{\text{Variable}}$
$6_{(5-6)}$	$-\pi/2$	0	0	$\theta_6^{\text{Variable}}$
$7_{(6-7)}$	0	0	d_7	$\theta_7^{\text{Variable}}$

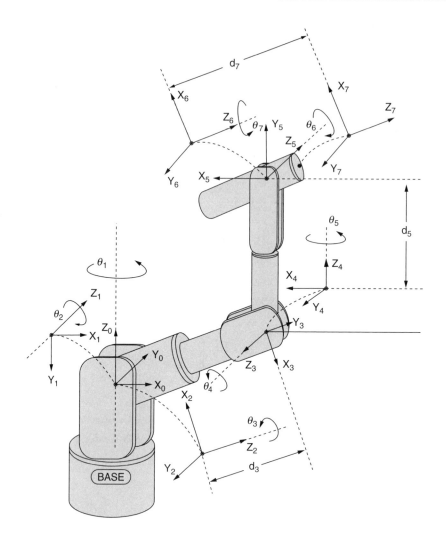

FIGURE 8.9 Schematic of the Mitsubishi PA-10 robot arm.

Having established the coordinate frames, the next step is to determine the D-H parameters. We begin by first determining α_i. α_i is the rotation about X_i to make Z_{i-1} parallel with Z_i. For axis 1, the rotation about X_1 to make Z_0 parallel with Z_1 is $-90°$ or $-\pi/2$ radians. For axis 2, the rotation about X_2 to make Z_1 parallel with Z_2 is $-90°$ or $-\pi/2$ radians. α_3 through α_6 are determined in the same manner. α_7 is zero because Z_7 is parallel with Z_6.

The next step is to determine a_i and d_i. a_i is the link length and always points away from the Z_{i-1} axis. d_i is the offset and is always along the Z_{i-1} axis. In the case of the Mitsubishi PA-10 robot arm, there are no link lengths so a_1 through a_7 are equal to zero. As seen from the schematic of the Mitsubishi arm, the only nonzero link offsets, d_i's D-H parameters are d_3, d_5, and d_7.

Having determined all the D-H parameters, the transformation matrix A_7^0 can now be computed. The transformation matrix consists of the rotation matrix R_7^0 and the displacement vector d_7^0. Using the expression in Equation (8.24) and Equation (8.25) we get:

$$R_7^0 = (U_1 V_1)(U_2 V_2)(U_3 V_3)(U_4 V_4)(U_5 V_5)(U_6 V_6)(U_7 V_7) \tag{8.41}$$

$$\begin{aligned} d_7^0 = &\, U_1 s_1 + (U_1 V_1) U_2 s_2 + (U_1 V_1)(U_2 V_2) U_3 s_3 + (U_1 V_1)(U_2 V_2)(U_3 V_3) U_4 s_4 \\ &+ (U_1 V_1)(U_2 V_2)(U_3 V_3)(U_4 V_4) U_5 s_5 + (U_1 V_1)(U_2 V_2)(U_3 V_3)(U_4 V_4)(U_5 V_5) U_6 s_6 \\ &+ (U_1 V_1)(U_2 V_2)(U_3 V_3)(U_4 V_4)(U_5 V_5)(U_6 V_6) U_7 s_7 \end{aligned} \tag{8.42}$$

D-H Convention

The following matrices are the individual U and V matrices for each axis of the manipulator.

$$U_1 = \begin{bmatrix} \cos(\theta_1) & -\sin(\theta_1) & 0 \\ \sin(\theta_1) & \cos(\theta_1) & 0 \\ 0 & 0 & 1 \end{bmatrix}, \quad V_1 = \begin{bmatrix} 1 & 0 & 0 \\ 0 & 0 & 1 \\ 0 & -1 & 0 \end{bmatrix}$$

$$U_2 = \begin{bmatrix} \cos(\theta_2) & -\sin(\theta_2) & 0 \\ \sin(\theta_2) & \cos(\theta_2) & 0 \\ 0 & 0 & 1 \end{bmatrix}, \quad V_2 = \begin{bmatrix} 1 & 0 & 0 \\ 0 & 0 & 1 \\ 0 & -1 & 0 \end{bmatrix}$$

$$U_3 = \begin{bmatrix} \cos(\theta_3) & -\sin(\theta_3) & 0 \\ \sin(\theta_3) & \cos(\theta_3) & 0 \\ 0 & 0 & 1 \end{bmatrix}, \quad V_3 = \begin{bmatrix} 1 & 0 & 0 \\ 0 & 0 & -1 \\ 0 & 1 & 0 \end{bmatrix}$$

$$U_4 = \begin{bmatrix} \cos(\theta_4) & -\sin(\theta_4) & 0 \\ \sin(\theta_4) & \cos(\theta_4) & 0 \\ 0 & 0 & 1 \end{bmatrix}, \quad V_4 = \begin{bmatrix} 1 & 0 & 0 \\ 0 & 0 & -1 \\ 0 & 1 & 0 \end{bmatrix}$$

$$U_5 = \begin{bmatrix} \cos(\theta_5) & -\sin(\theta_5) & 0 \\ \sin(\theta_5) & \cos(\theta_5) & 0 \\ 0 & 0 & 1 \end{bmatrix}, \quad V_5 = \begin{bmatrix} 1 & 0 & 0 \\ 0 & 0 & -1 \\ 0 & 1 & 0 \end{bmatrix}$$

$$U_6 = \begin{bmatrix} \cos(\theta_6) & -\sin(\theta_6) & 0 \\ \sin(\theta_6) & \cos(\theta_6) & 0 \\ 0 & 0 & 1 \end{bmatrix}, \quad V_6 = \begin{bmatrix} 1 & 0 & 0 \\ 0 & 0 & 1 \\ 0 & -1 & 0 \end{bmatrix}$$

$$U_7 = \begin{bmatrix} \cos(\theta_7) & -\sin(\theta_7) & 0 \\ \sin(\theta_7) & \cos(\theta_7) & 0 \\ 0 & 0 & 1 \end{bmatrix}, \quad V_7 = \begin{bmatrix} 1 & 0 & 0 \\ 0 & 1 & 0 \\ 0 & 0 & 1 \end{bmatrix}$$

Similarly the s_i vector is given by

$$s_1 = \begin{bmatrix} 0 \\ 0 \\ 0 \end{bmatrix}, \quad s_2 = \begin{bmatrix} 0 \\ 0 \\ 0 \end{bmatrix}, \quad s_3 = \begin{bmatrix} 0 \\ 0 \\ d_3 \end{bmatrix}, \quad s_4 = \begin{bmatrix} 0 \\ 0 \\ 0 \end{bmatrix}, \quad s_5 = \begin{bmatrix} 0 \\ 0 \\ d_5 \end{bmatrix}, \quad s_6 = \begin{bmatrix} 0 \\ 0 \\ 0 \end{bmatrix}, \quad s_7 = \begin{bmatrix} 0 \\ 0 \\ d_7 \end{bmatrix}$$

Substituting the above U_i, V_i, and s_i into Equation (8.41) and Equation (8.42) we get the following, where $C_i = \cos(i)$ and $S_i = \sin(i)$:

$$R_7^0 = \begin{bmatrix} r_{11} & r_{12} & r_{13} \\ r_{21} & r_{22} & r_{23} \\ r_{31} & r_{32} & r_{33} \end{bmatrix} \quad (8.43)$$

$r_{11} = ((((C_1C_2C_3 + S_1S_3)C_4 - C_1S_2S_4)C_5 + (C_1C_2S_3 - S_1C_3)S_5)C_6 + ((C_1C_2C_3 + S_1S_3)S_4 + C_1S_2C_4)S_6)C_7 + (-((C_1C_2C_3 + S_1S_3)C_4 - C_1S_2S_4)S_5 + (C_1C_2S_3 - S_1C_3)C_5)S_7$

$r_{12} = -((((C_1C_2C_3 + S_1S_3)C_4 - C_1S_2S_4)C_5 + (C_1C_2S_3 - S_1C_3)S_5)C_6 + ((C_1C_2C_3 + S_1S_3)S_4 + C_1S_2C_4)S_6)S_7 + (-((C_1C_2C_3 + S_1S_3)C_4 - C_1S_2S_4)S_5 + (C_1C_2S_3 - S_1C_3)C_5)C_7$

$r_{13} = -(((C_1C_2C_3 + S_1S_3)C_4 - C_1S_2S_4)C_5 + (C_1C_2S_3 - S_1C_3)S_5)S_6 + ((C_1C_2C_3 + S_1S_3)S_4 + C_1S_2C_4)C_6$

$r_{21} = ((((S_1C_2C_3 - C_1S_3)C_4 - S_1S_2S_4)C_5 + (S_1C_2S_3 + C_1C_3)S_5)C_6 + ((S_1C_2C_3 - C_1S_3)S_4 + S_1S_2C_4)S_6)C_7 + (-((S_1C_2C_3 - C_1S_3)C_4 - S_1S_2S_4)S_5 + (S_1C_2S_3 + C_1C_3)C_5)S_7$

$$r_{22} = -((((S_1C_2C_3 - C_1S_3)C_4 - S_1S_2S_4)C_5 + (S_1C_2S_3 + C_1C_3)S_5)C_6 + ((S_1C_2C_3 - C_1S_3)S_4$$
$$+ S_1S_2C_4)S_6)S_7 + (-((S_1C_2C_3 - C_1S_3)C_4 - S_1S_2S_4)S_5 + (S_1C_2S_3 + C_1C_3)C_5)C_7$$

$$r_{23} = -(((S_1C_2C_3 - C_1S_3)C_4 - S_1S_2S_4)C_5 + (S_1C_2S_3 + C_1C_3)S_5)S_6 + ((S_1C_2C_3 - C_1S_3)S_4$$
$$+ S_1S_2C_4)C_6$$

$$r_{31} = (((-S_2C_3C_4 - C_2S_4)C_5 - S_2S_3S_5)C_6 + (-S_2C_3S_4 + C_2C_4)S_6)C_7 + (-(-S_2C_3C_4 - C_2S_4)S_5$$
$$- S_2S_3C_5)S_7$$

$$r_{32} = -(((-S_2C_3C_4 - C_2S_4)C_5 - S_2S_3S_5)C_6 + (-S_2C_3S_4 + C_2C_4)S_6)S_7 + (-(-S_2C_3C_4 - C_2S_4)S_5$$
$$- S_2S_3C_5)C_7$$

$$r_{33} = -((-S_2C_3C_4 - C_2S_4)C_5 - S_2S_3S_5)S_6 + (-S_2C_3S_4 + C_2C_4)C_6$$

$$d_7^0 = \begin{bmatrix} d_{11} \\ d_{21} \\ d_{31} \end{bmatrix} \tag{8.44}$$

$$d_{11} = -C_1S_2d_3 + ((C_1C_2C_3 + S_1S_3)S_4 + C_1S_2C_4)d_5 + (-(((C_1C_2C_3 + S_1S_3)C_4 - C_1S_2S_4)C_5$$
$$+ (C_1C_2S_3 - S_1C_3)S_5)S_6 + ((C_1C_2C_3 + S_1S_3)S_4 + C_1S_2C_4)C_6)d_7$$

$$d_{21} = -S_1S_2d_3 + ((S_1C_2C_3 - C_1S_3)S_4 + S_1S_2C_4)d_5 + (-(((S_1C_2C_3 - C_1S_3)C_4 - S_1S_2S_4)C_5$$
$$+ (S_1C_2S_3 + C_1C_3)S_5)S_6 + ((S_1C_2C_3 - C_1S_3)S_4 + S_1S_2C_4)C_6)d_7$$

$$d_{31} = -C_2d_3 + (-S_2C_3S_4 + C_2C_4)d_5 + (-((-S_2C_3C_4 - C_2S_4)C_5 - S_2S_3S_5)S_6$$
$$+ (-S_2C_3S_4 + C_2C_4)C_6)d_7$$

The transformation matrix is now given by

$$A_7^0 = \begin{bmatrix} R_7^0 & d_7^0 \\ 0\ 0\ 0 & 1 \end{bmatrix} \tag{8.45}$$

Example 8.5: The PUMA 600 Robot Arm

The PUMA 600 has six revolute joints giving it six degrees of freedom. The motion of the arm is controlled by six brushed DC servo motors (see Figure 8.10 and Table 8.4). The first step in determining the D-H parameters is to locate the joints of the robot arm manipulator and determine if the joint is prismatic or revolute. There are six joints in the PUMA 600. All of them are revolute joints so θ_i of each joint is variable. Starting from the base, the joint coordinate frames are assigned based on the algorithm outlined before.

Having established the coordinate frames, the next step is to determine the D-H parameters. We begin by first determining α_i. α_i is the rotation about X_i to make Z_{i-1} parallel with Z_i. The rotation

TABLE 8.4 D-H Parameters for the PUMA 600

No.	Twist Angle α_i	Link Length a_i	Link Offset d_i	Joint Angle θ_i
$1_{(0-1)}$	$\pi/2$	0	0	$\theta_1^{\text{Variable}}$
$2_{(1-2)}$	0	a_2	d_2	$\theta_2^{\text{Variable}}$
$3_{(2-3)}$	$-\pi/2$	0	0	$\theta_3^{\text{Variable}}$
$4_{(3-4)}$	$\pi/2$	0	d_4	$\theta_4^{\text{Variable}}$
$5_{(4-5)}$	$\pi/2$	0	0	$\theta_5^{\text{Variable}}$
$6_{(5-6)}$	0	0	d_6	$\theta_6^{\text{Variable}}$

D-H Convention

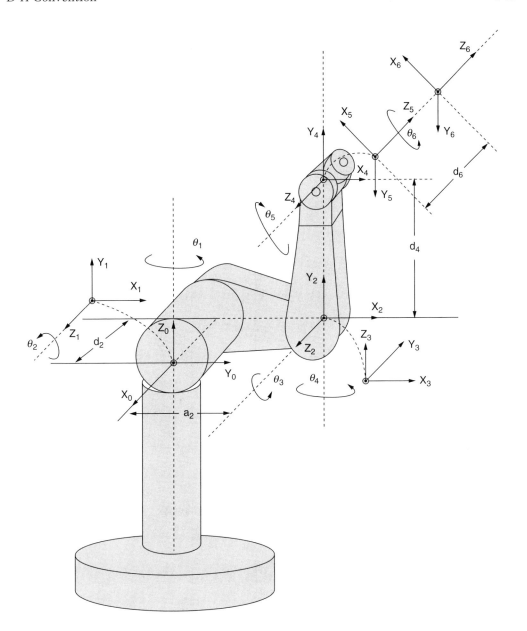

FIGURE 8.10 Schematic of the PUMA 600 robot arm.

about X_1 to take Z_0 parallel to Z_1 is $\pi/2$. α_2 through α_6 can be determined by following the same procedure.

The next step is to determine a_i and d_i. a_i is the link length and always points away from the Z_{i-1} axis. d_i is the offset and is always along the Z_{i-1} axis. For axis 1, there is no offset between joint 1 and joint 2 and no distance between axes 1 and 2, so d_1 and a_1 are zero. There is no offset between joint 2 and joint 3 as well as joint 3 and joint 4, so d_i is equal to zero. The only link length is between axes 2 and 3 and the distance is equal to a_2. The other offsets between joints are between axes 3 and 4, and axes 5 and 6. The offsets are equal to d_4 and d_6, respectively.

Having determined all the D-H parameters, the transformation matrix A_6^0 can now be computed. The transformation matrix consists of the rotation matrix R_6^0 and the displacement vector d_6^0. Using the expression in Equation (8.24) and Equation (8.25) we get

$$R_6^0 = (U_1 V_1)(U_2 V_2)(U_3 V_3)(U_4 V_4)(U_5 V_5)(U_6 V_6) \tag{8.46}$$

$$d_6^0 = U_1 s_1 + (U_1 V_1) U_2 s_2 + (U_1 V_1)(U_2 V_2) U_3 s_3 + (U_1 V_1)(U_2 V_2)(U_3 V_3) U_4 s_4 \\
+ (U_1 V_1)(U_2 V_2)(U_3 V_3)(U_4 V_4) U_5 s_5 + (U_1 V_1)(U_2 V_2)(U_3 V_3)(U_4 V_4)(U_5 V_5) U_6 s_6 \tag{8.47}$$

The following matrices are the individual U and V matrices for each axis of the manipulator.

$$U_1 = \begin{bmatrix} \cos(\theta_1) & -\sin(\theta_1) & 0 \\ \sin(\theta_1) & \cos(\theta_1) & 0 \\ 0 & 0 & 1 \end{bmatrix}, \quad V_1 = \begin{bmatrix} 1 & 0 & 0 \\ 0 & 0 & -1 \\ 0 & 1 & 0 \end{bmatrix}$$

$$U_2 = \begin{bmatrix} \cos(\theta_2) & -\sin(\theta_2) & 0 \\ \sin(\theta_2) & \cos(\theta_2) & 0 \\ 0 & 0 & 1 \end{bmatrix}, \quad V_2 = \begin{bmatrix} 1 & 0 & 0 \\ 0 & 1 & 0 \\ 0 & 0 & 1 \end{bmatrix}$$

$$U_3 = \begin{bmatrix} \cos(\theta_3) & -\sin(\theta_3) & 0 \\ \sin(\theta_3) & \cos(\theta_3) & 0 \\ 0 & 0 & 1 \end{bmatrix}, \quad V_3 = \begin{bmatrix} 1 & 0 & 0 \\ 0 & 0 & 1 \\ 0 & -1 & 0 \end{bmatrix}$$

$$U_4 = \begin{bmatrix} \cos(\theta_4) & -\sin(\theta_4) & 0 \\ \sin(\theta_4) & \cos(\theta_4) & 0 \\ 0 & 0 & 1 \end{bmatrix}, \quad V_4 = \begin{bmatrix} 1 & 0 & 0 \\ 0 & 0 & -1 \\ 0 & 1 & 0 \end{bmatrix}$$

$$U_5 = \begin{bmatrix} \cos(\theta_5) & -\sin(\theta_5) & 0 \\ \sin(\theta_5) & \cos(\theta_5) & 0 \\ 0 & 0 & 1 \end{bmatrix}, \quad V_5 = \begin{bmatrix} 1 & 0 & 0 \\ 0 & 0 & -1 \\ 0 & 1 & 0 \end{bmatrix}$$

$$U_6 = \begin{bmatrix} \cos(\theta_6) & -\sin(\theta_6) & 0 \\ \sin(\theta_6) & \cos(\theta_6) & 0 \\ 0 & 0 & 1 \end{bmatrix}, \quad V_6 = \begin{bmatrix} 1 & 0 & 0 \\ 0 & 1 & 0 \\ 0 & 0 & 1 \end{bmatrix}$$

Similarly the s_i vector is given by

$$s_1 = \begin{bmatrix} 0 \\ 0 \\ 0 \end{bmatrix}, \quad s_2 = \begin{bmatrix} a_2 \\ 0 \\ d_2 \end{bmatrix}, \quad s_3 = \begin{bmatrix} 0 \\ 0 \\ 0 \end{bmatrix}, \quad s_4 = \begin{bmatrix} 0 \\ 0 \\ d_4 \end{bmatrix}, \quad s_5 = \begin{bmatrix} 0 \\ 0 \\ 0 \end{bmatrix}, \quad s_6 = \begin{bmatrix} 0 \\ 0 \\ d_6 \end{bmatrix}$$

Substituting the above U_i, V_i, and s_i into Equation (8.46) and Equation (8.47), we get the following where $C_a = \cos(a)$ and $S_a = \sin(a)$.

$$R_6^0 = \begin{bmatrix} r_{11} & r_{12} & r_{13} \\ r_{21} & r_{22} & r_{23} \\ r_{31} & r_{32} & r_{33} \end{bmatrix} \tag{8.48}$$

D-H Convention

$$r_{11} = (((C_1C_2C_3 - C_1S_2S_3)C_4 - S_1S_4)C_5 + (-C_1C_2S_3 - C_1S_2C_3)S_5)C_6 + (((C_1C_2C_3 - C_1S_2S_3)S_4 + S_1C_4)S_6$$

$$r_{12} = -(((C_1C_2C_3 - C_1S_2S_3)C_4 - S_1S_4)C_5 + (-C_1C_2S_3 - C_1S_2C_3)S_5)S_6 + (((C_1C_2C_3 - C_1S_2S_3)S_4 + S_1C_4)C_6$$

$$r_{13} = ((C_1C_2C_3 - C_1S_2S_3)C_4 - S_1S_4)S_5 - (-C_1C_2S_3 - C_1S_2C_3)C_5$$

$$r_{21} = (((S_1C_2C_3 - S_1S_2S_3)C_4 + C_1S_4)C_5 + (-S_1C_2S_3 - S_1S_2C_3)S_5)C_6 + (((S_1C_2C_3 - S_1S_2S_3)S_4 - C_1C_4)S_6$$

$$r_{22} = -(((S_1C_2C_3 - S_1S_2S_3)C_4 + C_1S_4)C_5 + (-S_1C_2S_3 - S_1S_2C_3)S_5)S_6 + (((S_1C_2C_3 - S_1S_2S_3)S_4 - C_1C_4)C_6$$

$$r_{23} = ((S_1C_2C_3 - S_1S_2S_3)C_4 + C_1S_4)S_5 - (-S_1C_2S_3 - S_1S_2C_3)C_5$$

$$r_{31} = ((S_2C_3 + C_2S_3)C_4C_5 + (-S_2S_3 + C_2C_3)S_5)C_6 + (S_2C_3 + C_2S_3)S_4S_6$$

$$r_{32} = -((S_2C_3 + C_2S_3)C_4C_5 + (-S_2S_3 + C_2C_3)S_5)S_6 + (S_2C_3 + C_2S_3)S_4C_6$$

$$r_{33} = (S_2C_3 + C_2S_3)C_4S_5 - (-S_2S_3 + C_2C_3)C_5$$

$$d_6^0 = \begin{bmatrix} d_{11} \\ d_{21} \\ d_{31} \end{bmatrix} \tag{8.49}$$

$$d_{11} = C_1C_2a_2 + S_1d_2 + (-C_1C_2S_3 - C_1S_2C_3)d_4 + (((C_1C_2C_3 - C_1S_2S_3)C_4 - S_1S_4)S_5 - (-C_1C_2S_3 - C_1S_2C_3)C_5)d_6$$

$$d_{21} = S_1C_2a_2 - C_1d_2 + (-S_1C_2S_3 - S_1S_2C_3)d_4 + (((S_1C_2C_3 - S_1S_2S_3)C_4 + C_1S_4)S_5 - (-S_1C_2S_3 - S_1S_2C_3)C_5)d_6$$

$$d_{31} = S_2a_2 + (-S_2S_3 + C_2C_3)d_4 + ((S_2C_3 + C_2S_3)C_4S_5 - (-S_2S_3 + C_2C_3)C_5)d_6$$

The final transformation matrix is

$$A_6^0 = \begin{bmatrix} R_6^0 & d_6^0 \\ 0 \quad 0 \quad 0 & 1 \end{bmatrix} \tag{8.50}$$

9
Trajectory Planning for Flexible Robots

9.1	Introduction	9-1
9.2	Command Generation	9-4
	Bridge Crane Example • Generating Zero Vibration Commands • Using Zero-Vibration Impulse Sequences to Generate Zero-Vibration Commands • Robustness to Modeling Errors • Multi-Mode Input Shaping • Real-Time Implementation • Applications • Extensions Beyond Vibration Reduction	
9.3	Feedforward Control Action	9-15
	Feedforward Control of a Simple System with Time Delay • Feedforward Control of a System with Nonlinear Friction • Zero Phase Error Tracking Control • Conversion of Feedforward Control to Command Shaping	
9.4	Summary	9-24

William E. Singhose
Georgia Institute of Technology

9.1 Introduction

When a robotic system is pushed to its performance limits in terms of motion velocity and throughput, the problem of flexibility usually arises. Flexibility comes from physical deformation of the structure or compliance introduced by the feedback control system. For example, implementing a proportional and derivative (PD) controller is analogous to adding a spring and damper to the system. The controller "spring" can lead to problematic flexibility in the system. Deformation of the structure can occur in the links, cables, and joints. This can lead to problems with positioning accuracy, trajectory following, settling time, component wear, and stability and may also introduce nonlinear dynamics if the deflections are large.

An example of a robot whose flexibility is obviously detrimental to its positioning accuracy is the long-reach manipulator sketched in Figure 9.1. This robotic arm was built to test methods for cleaning nuclear waste storage tanks [1]. The arm needs to enter a tank through an access hole and then reach long distances to clean the tank walls. These types of robots have mechanical flexibility in the links, the joints, and possibly the base to which they are attached. Both the gross motion of the arm and the cleaning motion of the end effector can induce vibration. An example of a less conventional flexible positioning system is the cable-driven "Hexaglide" shown in Figure 9.2 [2]. This system is able to perform six-degrees-of-freedom positioning by sliding the cable support points along overhead rails.

The cables present in the Hexaglide make its flexibility obvious. However, all robotic systems will deflect if they are moved rapidly enough. Consider the moving-bridge coordinate measuring machine (CMM)

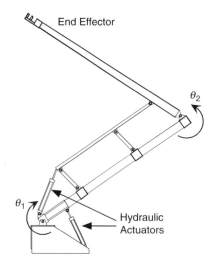

FIGURE 9.1 Long reach manipulator RALF.

sketched in Figure 9.3. The machine is composed of stiff components including a granite base and large cross-sectional structural members. The goal of the CMM is to move a probe throughout its workspace so that it can contact the surface of manufactured parts that are fixed to the granite base. In this way it can accurately determine the dimensions of the part. The position of the probe is measured by optical encoders that are attached to the granite base and the moving bridge. However, if a laser interferometer is used to measure the probe, its location will differ from that indicated by the encoders. This difference arises because the physical structure deflects between the encoders and the probe endpoint. Figure 9.4 shows the micron-level deflection for a typical move where the machine rapidly approaches a part's surface and then slows down just before making contact with the part [3]. If the machine is attempting to measure with micron resolution, then the 20–25 μm vibration during the approach phase is problematic.

Flexibility introduced by the control system is also commonplace. Feedback control works by detecting a difference between the actual response and the desired response. This difference is then used to generate a corrective action that mimics a restoring force. In many cases this restoring force acts something like a spring. Increasing the gains may have the effect of stiffening the system to combat excessive compliance, but it also increases actuator demands and noise problems and can cause instability.

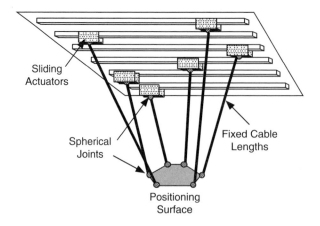

FIGURE 9.2 Hexaglide mechanism.

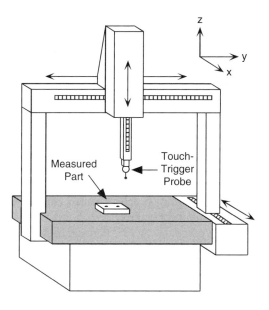

FIGURE 9.3 Moving-bridge coordinate measuring machine.

For robotic systems attempting to track trajectories, flexible dynamics from either the physical structure or the control system cause many difficulties. If a system can be represented by the block diagram shown in Figure 9.5, then four main components are obvious: feedback control, feedforward control, command generation, and the physical hardware. Each of the four blocks provides unique opportunities for improving the system performance. For example, the feedback control can be designed to reject random disturbances, while the feedforward block cannot. On the other hand, the feedforward block can compensate for unwanted trajectory deviations before they show up in the output, while the feedback controller cannot.

Trajectory planning for flexible robots centers on the design of the command generator and the feedforward controller. However, adequate feedback control must be in place to achieve additional performance requirements such as disturbance rejection and steady-state positioning. Although both command generation and feedforward control can greatly aid in trajectory following, they work in fundamentally different ways. The command generator creates a specially shaped reference command that is fed to the feedback

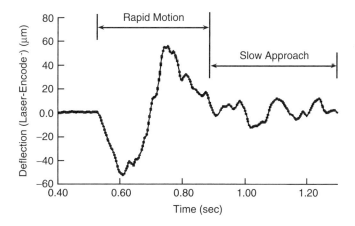

FIGURE 9.4 Deflection of coordinate measuring machine.

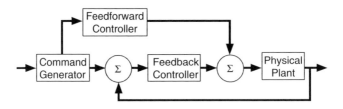

FIGURE 9.5   Block diagram of generic system.

control loop. The actual actuator effort is then generated by the feedback controller and applied to the plant. On the other hand, a feedforward controller injects control effort directly into the plant, thereby aiding, or overruling, the action of the feedback control.

There is a wide range of command generators and feedforward controllers, so globally characterizing their strengths and weaknesses is difficult. However, in general, command generators are less aggressive, rely less on an accurate system model, and are consequently more robust to uncertainty and plant variation. Feedforward control action can produce better trajectory tracking than command generation, but it is usually less robust. There are also control techniques that can be implemented as command generation or as feedforward control, so it is not always obvious at first how a control action should be characterized.

The desired motion of a robotic system is often a rapid change in position or velocity without residual oscillation at the new setpoint. However, tracking trajectories that are complex functions of time and space is also of prime importance for many robotic systems. For *rigid* multi-link serial and parallel robots, the process of determining appropriate commands to achieve a desired endpoint trajectory can be quite challenging and is addressed elsewhere. This chapter assumes that some baseline commands, possibly generated from kinematic requirements, already exist. The challenge discussed here is how to modify the commands to accommodate the *flexible* nature of the robotic system.

9.2   Command Generation

Creating specially shaped reference commands that move flexible systems in a desired fashion is an old idea [4, 5, 51]. Commands can be created such that the system's motion will cancel its own vibration. Some of these techniques require that the commands be pre-computed using boundary conditions before the move is initiated. Others can be implemented in real time. Another significant difference between the various control methods is robustness to modeling errors. Some techniques require a very good system model to work effectively, while others need only rough estimates of the system parameters.

9.2.1   Bridge Crane Example

A simple, but difficult to accurately control, flexible robotic system is an automated overhead bridge crane like the one shown schematically in Figure 9.6. The payload is hoisted up by an overhead suspension cable. The upper end of the cable is attached to a trolley that travels along a bridge to position the payload. Furthermore, the bridge on which the trolley travels can also move perpendicular to the trolley motion, thereby providing three-dimensional positioning. These cranes are usually controlled by a human operator who presses buttons to cause the trolley to move, but there are also automated versions where a control computer drives the motors. If the control button is depressed for a finite time period, then the trolley will move a finite distance and come to rest. The payload on the other hand, will usually oscillate about the new trolley position. The planar motion of the payload for a typical trolley movement is shown in Figure 9.7.

The residual payload motion is usually undesirable because the crane needs to place the payload at a desired location. Furthermore, the payload may need to be transported through a cluttered work environment containing obstacles and human workers. Oscillations in the payload make collision-free transport through complex trajectories much more difficult. Figure 9.8 shows an overhead view of the position of

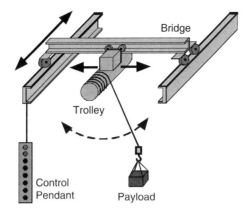

FIGURE 9.6 Overhead bridge crane.

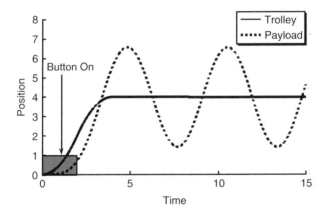

FIGURE 9.7 Crane response when operator presses move button.

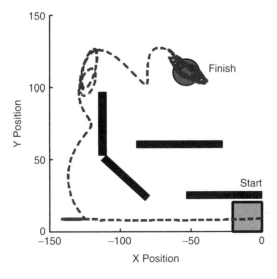

FIGURE 9.8 Payload response moving through an obstacle field.

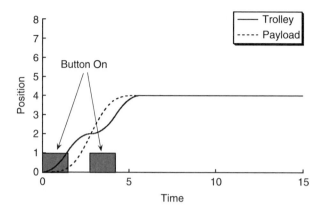

FIGURE 9.9 Crane response when operator presses move button two times.

a bridge crane payload while it is being driven through an obstacle field by a novice operator. There is considerable payload sway both during the transport and at the final position. These data were obtained via an overhead camera that tracked the payload motion but was not capable of measuring the position for feedback control purposes. An experienced crane operator can often produce the desired payload motion with much less vibration by pressing the control buttons multiple times at the proper time instances. This type of operator command and payload response for planar motion is shown in Figure 9.9. When compared with the response shown in Figure 9.7, the benefits of properly choosing the reference command are obvious. This is the type of effect that the command generator block in Figure 9.5 strives to achieve.

9.2.2 Generating Zero Vibration Commands

As a first step to understanding how to generate commands that move flexible robots without vibration, it is helpful to start with the *simplest* such command. A fundamental building block for all commands is an impulse. This theoretical command is often a good approximation of a short burst of force such as that from a hammer blow or from momentarily turning the actuator full on. Applying an impulse to a flexible robot will cause it to vibrate. However, if we apply a second impulse to the robot, we can cancel the vibration induced by the first impulse. This concept is demonstrated in Figure 9.10. Impulse A_1 induces the vibration indicated by the dashed line, while A_2 induces the dotted response. Combining the two responses using superposition results in zero residual vibration. The second impulse must be applied at the correct time and must have the appropriate magnitude. Note that this two-impulse sequence is analogous to the two-pulse crane command shown in Figure 9.9.

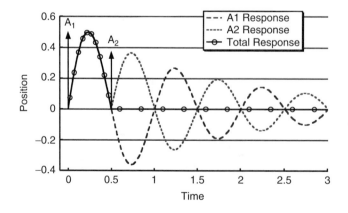

FIGURE 9.10 Two impulses can cancel vibration.

In order to derive the amplitudes and time locations of the two-impulse command shown in Figure 9.10, a mathematical description of the residual vibration that results from a series of impulses must be utilized. If the system's natural frequency is ω and the damping ratio is ζ, then the residual vibration that results from a sequence of impulses applied to a second-order system can be described by [5–7]

$$V(\omega,\zeta) = e^{-\zeta\omega t_n}\sqrt{[C(\omega,\zeta)]^2 + [S(\omega,\zeta)]^2} \qquad (9.1)$$

where,

$$C(\omega,\zeta) = \sum_{i=1}^{n} A_i e^{\zeta\omega t_i}\cos(\omega_d t_i) \quad \text{and} \quad S(\omega,\zeta) = \sum_{i=1}^{n} A_i e^{\zeta\omega t_i}\sin(\omega_d t_i) \qquad (9.2)$$

A_i and t_i are the amplitudes and time locations of the impulses, n is the number of impulses in the impulse sequence, and

$$\omega_d = \omega\sqrt{1 - \zeta^2} \qquad (9.3)$$

Note that Equation (9.1) is expressed in a nondimensional form. It is generated by taking the absolute amplitude of residual vibration from an impulse series and then dividing by the vibration amplitude from a single, unity-magnitude impulse. This expression predicts the percentage of residual vibration that will remain after input shaping has been implemented. For example, if Equation (9.1) has a value of 0.05 when the impulses and system parameters are entering into the expression, then input shaping with the impulse sequence should reduce the residual vibration to 5% of the amplitude that occurs without input shaping. Of course, this result applies to only underdamped systems, because overdamped systems do not have residual vibration.

To generate an impulse sequence that causes no residual vibration, we set Equation (9.1) equal to zero and solve for the impulse amplitudes and time locations. However, we must place a few more restrictions on the impulses, because the solution can converge to zero-valued or infinitely-valued impulses. To avoid the trivial solution of all zero-valued impulses and to obtain a normalized result, we require the impulse amplitudes to sum to one:

$$\sum A_j = 1 \qquad (9.4)$$

At this point, the impulses could satisfy Equation (9.1) by taking on very large positive and negative values. These large impulses would saturate the actuators. One way to obtain a bounded solution is to limit the impulse amplitudes to positive values:

$$A_i > 0, \quad i = 1,\ldots,n \qquad (9.5)$$

Limiting the impulses to positive values provides a good solution. However, performance can be pushed even further by allowing a limited amount of negative impulses [8].

The problem we want to solve can now be stated explicitly: find a sequence of impulses that makes Equation (9.1) equal to zero, while also satisfying Equation (9.4) and Equation (9.5).[1] Because we are looking for the two-impulse sequence shown in Figure 9.10 that satisfies the above specifications, the problem has four unknowns — the two impulse amplitudes (A_1, A_2) and the two impulse time locations (t_1, t_2).

Without loss of generality, we can set the time location of the first impulse equal to zero:

$$t_1 = 0 \qquad (9.6)$$

The problem is now reduced to finding three unknowns (A_1, A_2, t_2). In order for Equation (9.1) to equal

---

[1] This problem statement and solution are similar to one first published by Smith [4] in 1957.

zero, the expressions in Equation (9.2) must both equal zero independently because they are squared in Equation (9.1). Therefore, the impulses must satisfy

$$0 = \sum_{i=1}^{2} A_i e^{\zeta \omega t_i} \cos(\omega_d t_i) = A_1 e^{\zeta \omega t_1} \cos(\omega_d t_1) + A_2 e^{\zeta \omega t_2} \cos(\omega_d t_2) \qquad (9.7)$$

$$0 = \sum_{i=1}^{2} A_i e^{\zeta \omega t_i} \sin(\omega_d t_i) = A_1 e^{\zeta \omega t_1} \sin(\omega_d t_1) + A_2 e^{\zeta \omega t_2} \sin(\omega_d t_2) \qquad (9.8)$$

Substituting Equation (9.6) into Equation (9.7) and Equation (9.8) reduces the equations to

$$0 = A_1 + A_2 e^{\zeta \omega t_2} \cos(\omega_d t_2) \qquad (9.9)$$

$$0 = A_2 e^{\zeta \omega t_2} \sin(\omega_d t_2) \qquad (9.10)$$

In order for Equation (9.10) to be satisfied in a nontrivial manner, the sine term must equal zero. This occurs when its argument is a multiple of π:

$$\omega_d t_2 = p\pi, \quad p = 1, 2, \ldots \qquad (9.11)$$

In other words

$$t_2 = \frac{p\pi}{\omega_d} = \frac{p T_d}{2}, \quad p = 1, 2, \ldots \qquad (9.12)$$

where T_d is the damped period of vibration. This result tells us that there are an infinite number of possible values for the location of the second impulse—they occur at multiples of the half period of vibration. The solutions at odd multiples of the half period correspond to solutions with positive impulses, while those at even multiples would require a negative impulse. Time considerations require using the smallest value for t_2, that is

$$t_2 = \frac{T_d}{2} \qquad (9.13)$$

The impulse time locations can be described very simply; the first impulse is at time zero and the second impulse is located at half the period of vibration. For this simple two-impulse case, the amplitude constraint given in Equation (9.4) reduces to:

$$A_1 + A_2 = 1 \qquad (9.14)$$

Using the expression for the damped natural frequency given in Equation (9.3) and substituting Equation (9.13) and Equation (9.14) into Equation (9.9) gives

$$0 = A_1 - (1 - A_1) e^{\left(\frac{\zeta \pi}{\sqrt{1-\zeta^2}}\right)} \qquad (9.15)$$

Rearranging Equation (9.15) and solving for A_1 gives

$$A_1 = \frac{e^{\left(\frac{\zeta \pi}{\sqrt{1-\zeta^2}}\right)}}{1 + e^{\left(\frac{\zeta \pi}{\sqrt{1-\zeta^2}}\right)}} \qquad (9.16)$$

To simplify the expression, multiply top and bottom of the right-hand side by the inverse of the exponential term to get

$$A_1 = \frac{1}{1 + K} \qquad (9.17)$$

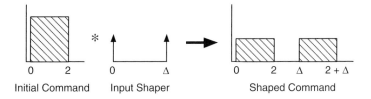

FIGURE 9.11 Input shaping a short pulse command.

where the inverse of the exponential term is

$$K = e^{\left(\frac{-\zeta\pi}{\sqrt{1-\zeta^2}}\right)} \quad (9.18)$$

substituting Equation (9.17) back into Equation (9.14), we get

$$A_2 = \frac{K}{1+K} \quad (9.19)$$

The sequence of two impulses that leads to zero vibration (ZV) can now be summarized in matrix form as

$$\begin{bmatrix} A_i \\ t_o \end{bmatrix} = \begin{bmatrix} \frac{1}{1+K} & \frac{K}{1+K} \\ 0 & 0.5T_d \end{bmatrix} \quad (9.20)$$

9.2.3 Using Zero-Vibration Impulse Sequences to Generate Zero-Vibration Commands

Real systems cannot be moved around with impulses, so we need to convert the properties of the impulse sequence given in Equation (9.20) into a usable command. This can be done by simply convolving the impulse sequence with any desired command. The convolution product is then used as the command to the system. If the impulse sequence, also known as an input shaper, causes no vibration, then the convolution product will also cause no vibration [7, 9]. This command generation process, called input shaping, is demonstrated in Figure 9.11 for an initial pulse function. This particular input shaper was designed for an undamped system, so both impulses have the same amplitude. Note that the convolution product in this case is the two-pulse command shown in Figure 9.9, which moved the crane with no residual vibration. In this case, the shaper is longer than the initial command, but in most cases the impulse sequence will be much shorter than the command profile. This is especially true when the baseline command is generated to move a robot through a complex trajectory and the periods of the system vibration are small compared with the duration of the trajectories. When this is the case, the components of the shaped command that arise from the individual impulses run together to form a smooth continuous function as shown in Figure 9.12.

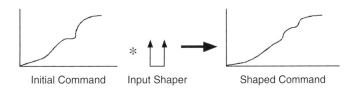

FIGURE 9.12 Input shaping a generic trajectory command.

9.2.4 Robustness to Modeling Errors

The amplitudes and time locations of the impulses depend on the system parameters (ω and ζ). If there are errors in these values (and there always are), then the input shaper will not result in zero vibration. In fact, when using the two-impulse sequence discussed above, there can be a noticeable amount of vibration for a relatively small modeling error. This lack of robustness was a major stumbling block for the original formulation of this idea that was developed in the 1950s [10].

This problem can be visualized by plotting a sensitivity curve for the input shaper. These curves show the amplitude of residual vibration caused by the shaper as a function of the system frequency and/or damping ratio. One such sensitivity curve for the zero-vibration (ZV) shaper given in Equation (9.20) is shown in Figure 9.13 with a normalized frequency on the horizontal axis and the percentage vibration on the vertical axis. Note that as the actual frequency ω deviates from the modeling frequency ω_m, the amount of vibration increases rapidly.

The first input shaper designed to have robustness to modeling errors was developed by Singer and Seering in the late 1980s [7, 11]. This shaper was designed by requiring the derivative of the residual vibration, with respect to the frequency, to be equal to zero at the modeling frequency. Mathematically, this can be stated as

$$\frac{\partial V(\omega, \zeta)}{\partial \omega} = 0 \tag{9.21}$$

Including this constraint has the effect of keeping the vibration near zero as the actual frequency starts to deviate from the modeling frequency. The sensitivity curve for this zero vibration and derivative (ZVD) shaper is also shown in Figure 9.13. Note that this shaper keeps the vibration at a low level over a much wider range of frequencies than the ZV shaper.

Since the development of the ZVD shaper, several other robust shapers have been developed. In fact, shapers can now be designed to have any amount of robustness to modeling errors [12]. Any real robotic system will have some amount of tolerable vibration; real machines are always vibrating at some level. Using this tolerance, a shaper can be designed to suppress any frequency range. The sensitivity curve for a very robust shaper is included in Figure 9.13 [12–14]. This shaper is created by establishing a tolerable vibration limit V_{tol} and then restricting the vibration to below this value over any desired range of frequency errors.

Input shaping robustness is not restricted to errors in the frequency value. Figure 9.14 shows a three-dimensional sensitivity curve for a shaper that was designed to suppress vibration between 0.7 and 1.3 Hz and also over the range of damping ratios between 0 and 0.2. Notice that the shaper is very robust to changes in the damping ratio.

To achieve greater robustness, input shapers generally must contain more than two impulses and their durations must increase. For example, the ZVD shaper [7] obtained by satisfying Equation (9.21) contains

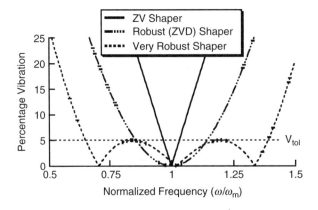

FIGURE 9.13 Sensitivity curves of several input shapers.

Trajectory Planning for Flexible Robots

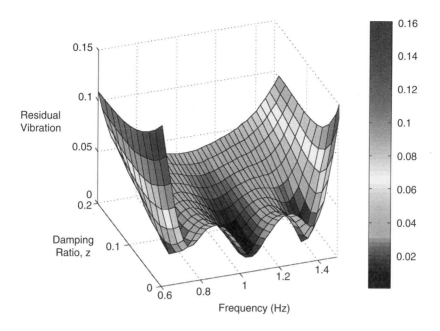

FIGURE 9.14 Three-dimensional sensitivity curve.

three impulses given by

$$\begin{bmatrix} A_i \\ t_i \end{bmatrix} = \begin{bmatrix} \frac{1}{(1+K)^2} & \frac{2K}{(1+K)^2} & \frac{K^2}{(1+K)^2} \\ 0 & 0.5T_d & T_d \end{bmatrix} \quad (9.22)$$

The increase in shaper duration means that the shaped command will also increase in duration. Fortunately, even very robust shapers have fairly short durations. For example, the ZVD shaper has a duration of only one period of the natural frequency. This time penalty is often a small cost in exchange for the improved robustness to modeling errors. To demonstrate this tradeoff, Figure 9.15 shows the response of a spring-mass system to step commands shaped with the three shapers used to generate Figure 9.13. Figure 9.15a shows the response when the model is perfect, and Figure 9.15b shows the case when there is a 30% error in the estimated system frequency. The increase in rise time caused by the robust shapers is apparent in Figure 9.15a, while Figure 9.15b shows the vast improvement in vibration reduction that the robust shapers provide in the presence of modeling errors. In this case the very robust shaper yields essentially zero vibration, even with the 30% frequency error.

9.2.5 Multi-Mode Input Shaping

Many robotic systems will have more than one flexible mode that can degrade trajectory tracking. There have been several methods developed for generating input shapers to suppress multiple modes of vibration [15–18]. These techniques can be used to solve the multiple vibration constraint equations sequentially, or concurrently. Suppressing multiple modes can lead to a large increase in the number of impulses in the input shaper, so methods have been developed to limit the impulses to a small number [16]. Furthermore, the nature of multi-input systems can be exploited to reduce the complexity and duration of a shaper for a multi-mode system [18].

A simple, yet slightly suboptimal, method is to design an input shaper independently for each mode of problematic vibration. Then, the individual input shapers can simply be convolved together. This straightforward process is shown in Figure 9.16.

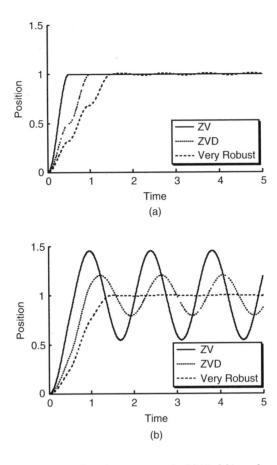

FIGURE 9.15 Spring-mass response to shaped step commands: (a) Model is perfect; (b) 30% frequency error.

9.2.6 Real-Time Implementation

One of the strengths of command shaping is that it can often be implemented in a real-time control system. Many command shaping methods can be thought of as filtering the reference signal before it is fed to the closed-loop control. For example, the input shaping technique requires only a simple convolution that can usually be implemented with just a few multiplication and addition operations each time through the control loop.

Many motion control boards and DSP chips have built in algorithms for performing the real-time convolution that is necessary for input shaping. If these features are not available, then a very simple algorithm can be added to the control system. The algorithm starts by creating a buffer, just a vector variable of a finite length. This buffer is used to store the command values for each time step. For example, the first

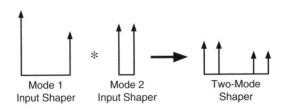

FIGURE 9.16 Forming a two-mode shaper through convolution.

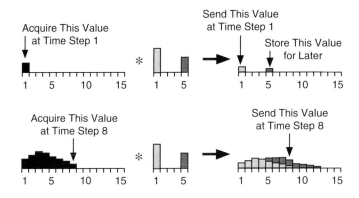

FIGURE 9.17   Real-time input shaping.

value in the buffer would be the shaped command at the first time instance. A graphical representation of such a buffer is shown in the upper right-hand corner of Figure 9.17. The upper left-hand portion of the figure shows the unshaped baseline command in the digital domain. This baseline command can be created in real time, for example, by reading a joystick position.

In order to fill the buffer with the input-shaped command, the algorithm determines the baseline command each time through the control loop. The algorithm multiplies the value of the baseline command by the amplitude of the first impulse in the input shaper. This value then gets added to the current time location in the command buffer. The amplitude of the second impulse is then multiplied by the baseline command value. However, this value is not sent directly out to the feedback control loop. Rather, it is added to the future buffer slot that corresponds to the impulse's location in time. For example, assuming a 10 Hz sampling rate, if the time location of the second impulse were at 0.5 sec, then this second value would be added to the buffer five slots ahead of the current position. This real-time process will build up the shaped command as demonstrated in Figure 9.17. The figure indicates what the value of the shaped command would be at the first time step and at the eighth time step and how the index to the current command value advances through the buffer. To avoid having the index exceed the size of the buffer, a circular buffer is used where the index goes back to the beginning when it reaches the end of the buffer.

9.2.7   Applications

The robustness and ease of use of input shaping has enabled its implementation on a large variety of systems. Shaping has been successfully implemented on a number of cranes and crane-like structures [19–22]. The performance of both long-reach manipulators [23, 24] and coordinate measuring machines [3, 25, 26] improves considerably with input shaping. The multiple modes of a silicon-handling robot were eliminated with input shaping [27]. Shaping was an important part of a control system developed for a wafer stepper [28, 29]. The throughput of a disk-drive-head tester was significantly improved with shaping [8]. A series of input-shaping experiments was even performed on board the space shuttle Endeavor [13, 30].

Although input shaping may not be specifically designed to optimize a desired trajectory, its ability to reduce vibration allows the system to track a trajectory without continually oscillating around the trajectory path. In this respect, input shaping relies on a good baseline command for the desired trajectory. This trajectory needs to be nominally of the correct shape and needs to account for physical limitations of the hardware such as workspace boundaries and actuator limits. This baseline trajectory would normally be derived from these physical and kinematic requirements and this topic is addressed elsewhere.

As a demonstration of input shaping trajectories, consider the painting robot shown in Figure 9.18. The machine has orthogonal decoupled modes arising from a two-stage beam attached vertically to an

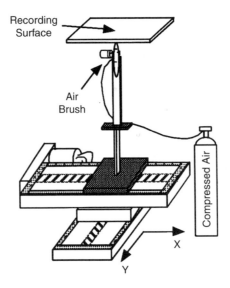

FIGURE 9.18 Painting robot.

XY positioning stage moving in a horizontal plane. The end effector is a compressed-air paint brush that paints on paper suspended above the airbrush. The stages are positioned by PD controllers. No information about the paintbrush position is utilized in the control system. Experiments were conducted by turning on the flow of air to the paint brush, commencing the desired trajectory, and then shutting off the flow of paint at the end of the move. Figure 9.19 shows the result when a very fast 3 in. × 3 in. square trajectory was commanded and input shaping was not used. When input shaping was enabled, the system followed the desired trajectory much closer, as shown in Figure 9.20.

9.2.8 Extensions Beyond Vibration Reduction

At the core of command generation is usually the desire to eliminate vibration. However, many other types of performance specifications can also be satisfied by properly constructing the reference command. Methods for constructing commands that limit actuator effort have been developed [8, 31, 32]. If the actuator limits are well known and the fastest move times are desired, then command generation can be used to develop the time-optimal command profiles [33–37]. When the system is Multi-Input Multi-Output, the shaped commands can be constructed to take this effect into account [18, 31, 38]. When

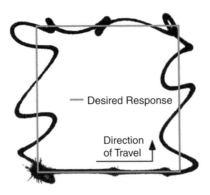

FIGURE 9.19 Response to unshaped square trajectory.

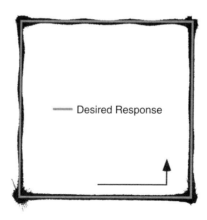

FIGURE 9.20 Response to shaped square trajectory.

constructing reaction jet commands, the amount of fuel can be limited or set to a specific amount [39–44]. It is also possible to limit the transient deflection [45, 46]. Furthermore, vibration reduction and slewing can be completed simultaneously with momentum dumping operations [47].

The sections above are not intended to be complete, but they are an attempt to give an introduction to and a reference for using command generation in the area of trajectory following. The list of successful applications and extensions of command generation will undoubtedly increase substantially in the years to come.

9.3 Feedforward Control Action

Feedforward control is concerned with directly generating a control action (force, torque, voltage, etc.), rather than generating a reference command. By including an anticipatory corrective action before an error shows up in the response, a feedforward controller can provide much better trajectory tracking than with feedback control alone. It can be used for a variety of cases such as systems with time delays, nonlinear friction [48], or systems performing repeated motions [49]. Most feedforward control methods require an accurate system model, so robustness is an important issue to consider. Feedforward control can also be used to compensate for a disturbance if the disturbance itself can be measured before the effect of the disturbance shows up in the system response. A block diagram for such a case is shown in Figure 9.21.

The generic control system diagram that was first shown in Figure 9.5 shows the feedforward block injecting control effort directly into the plant, as an auxiliary to the effort of the feedback controller. Decoupling this type of control from the action of the command generator, which creates an appropriate reference command, makes analysis and design of the overall control system simpler. However, this nomenclature is not universal. There are numerous papers and books that refer to command generation

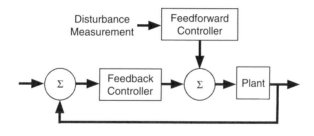

FIGURE 9.21 Feedforward compensation of disturbances.

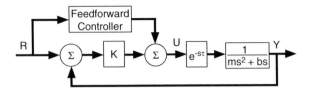

FIGURE 9.22 Feedforward compensation of a system with a time delay.

as feedforward control. In order to establish clarity between these two fundamentally different control techniques, the following nomenclature will be used here:

> *Command Generation* attempts to produce an appropriate command signal to a system. The system could be open or closed loop. In an open loop system, the command would be a force acting directly on the plant. In a closed-loop system, the command would be a reference signal to the feedback controller.
>
> *Feedforward Control* produces a force acting directly on the plant that is auxiliary to the feedback control force. Without a feedback control loop, there cannot be feedforward control action.

One reason for the inconsistent use of the term *feedforward* is because some techniques can be employed in either a feedforward manner or in the role of a command generator. However, the strengths and weakness of the techniques change when their role changes, so this effect should be noted in the nomenclature.

9.3.1 Feedforward Control of a Simple System with Time Delay

To demonstrate a very simple feedforward control scheme, consider a system under proportional feedback control that can be modeled as a mass-damper system with a time delay. The block diagram for this case is shown in Figure 9.22. The desired output is represented by the reference signal **R**. The actions of the feedforward control and the feedback control combine to produce the actuator effort **U** that then produces the actual output **Y**.

Let us first examine the response of the system without feedforward compensation. Suppose that the feedback control system is running at 10 Hz, the time delay is 0.1 sec, $m = 1$, and $b = 0.5$. The dynamic response of the system can be adjusted to some degree by varying the proportional gain K. Figure 9.23 shows the oscillatory step response to a variety of K values. This is a case where the system flexibility results from the feedback control rather than from the physical plant. Note that for low values of K, the system is

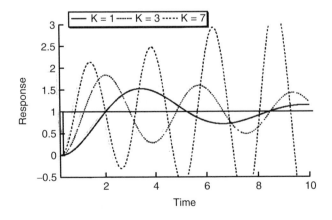

FIGURE 9.23 Step response of time-delay system without feedforward compensation.

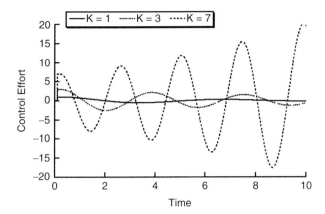

FIGURE 9.24 Control effort without feedforward compensation.

sluggish. The system rise time can be improved by increasing K, but that strategy soon drives the system unstable. The corresponding control effort is shown in Figure 9.24.

Rather than performing a step motion, suppose the desired motion was a smooth trajectory function such as:

$$r(t) = 1 - \cos(\omega t) \tag{9.23}$$

The simple proportional feedback controller might be able to provide adequate tracking if the frequency of the desired trajectory was very low. However, if the trajectory is demanding (relative to system frequency), then the feedback controller will provide poor tracking as shown in Figure 9.25. To improve performance, we have several options ranging from redesigning the physical system to improve the dynamics and reduce the time delay, adding additional sensors, improving the feedback controller, using command shaping, or adding feedforward compensation. Given the system delay, feedforward compensation is a natural choice.

A simple feedforward control strategy would place the inverse of the plant in the feedforward block shown in Figure 9.22. If this controller can be implemented, then the overall transfer function from desired response to system response would be unity. That is, the plant would respond exactly in the desired manner. There are, or course, limitations to what can be requested of the system. But, let us proceed with this example and discuss the limitations after the basic concept is demonstrated. In this case, the plant

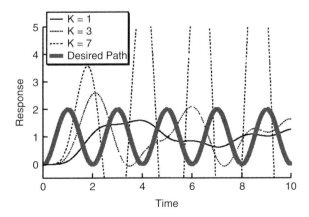

FIGURE 9.25 Tracking a smooth function without feedforward compensation.

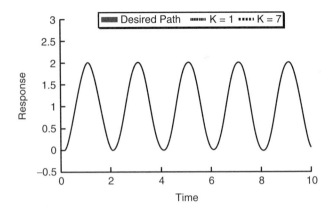

FIGURE 9.26 Tracking a smooth function with feedforward compensation.

transfer function including the time delay is

$$G_p = \frac{e^{-s\tau}}{ms^2 + bs} \tag{9.24}$$

The feedforward controller would then be

$$G_{FF} = \frac{ms^2 + bs}{e^{-s\tau}} \tag{9.25}$$

Note that this would be implemented in the digital domain, so the time delay in the denominator becomes a time shift in the numerator. This time shift would be accomplished by essentially looking ahead at the desired trajectory. Without knowing the future desired trajectory for at least the amount of time corresponding to the delay, this process cannot be implemented.

Figure 9.26 shows that under feedforward compensation, the system perfectly tracks the desired trajectory for various values of the feedback gain. This perfect result will not apply to real systems because there will always be modeling errors. Figure 9.27 shows the responses when there is a 5% error in the system mass and damping parameters. With a low proportional gain, the tracking is still fairly good, but the system goes unstable for the higher gain. If the error is increased to 10%, then the tracking performance with the low gain controller also starts to degrade as shown in Figure 9.28.

One important issue to always consider with feedforward control is the resulting control effort. Given that the feedback controller generates some effort and the feedforward adds to this effort, the result might

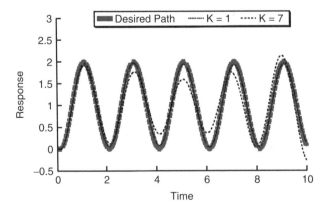

FIGURE 9.27 Effect of 5% model errors on feedforward compensation.

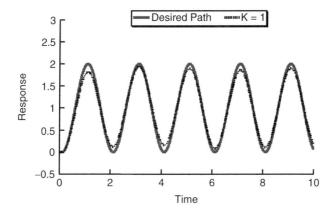

FIGURE 9.28 Effect of 10% model errors on feedforward compensation.

be unrealistically high effort that saturates the actuators. Furthermore, the control effort can be highly dependent on the desired trajectory. Consider again the smooth trajectory given in Equation (9.23). The Laplace transform is

$$R(s) = \frac{\omega^2}{s(s^2 + \omega^2)} \tag{9.26}$$

Sending the desired trajectory through the feedforward compensator, results in a feedforward control effort of

$$G_{FF}(s)R(s) = e^{s\tau} \frac{\omega^2(ms + b)}{s^2 + \omega^2} \tag{9.27}$$

Converting this into the time domain yields

$$m\omega^2 \cos(\omega[t + \tau]) + b\omega\sin(\omega[t + \tau]) \tag{9.28}$$

or

$$C_1 \sin(\omega[t + \tau] + C_2) \tag{9.29}$$

where

$$C_1 = \omega\sqrt{m^2\omega^2 + b} \quad \text{and} \quad C_2 = \tan^{-1}\left(\frac{m\omega}{b}\right) \tag{9.30}$$

Note that the control effort increases with the desired frequency of response. Therefore, requesting very fast response will lead to a large control effort that will saturate the actuators. This effect is demonstrated in Figure 9.29. Finally, if the desired trajectory has discontinuous derivatives that cannot physically be realized, such as step and ramp commands for systems that have an inertia, then the feedforward control effort would also be unrealizable and saturate the actuators.

9.3.2 Feedforward Control of a System with Nonlinear Friction

The above example showed the possibility of using feedforward compensation to deal with a time delay and flexibility induced by the feedback controller. Another good use of feedforward control is to help compensate for nonlinear friction. Consider the system shown in Figure 9.30. An applied force moves a mass subject to Coulomb friction. Attached to the base unit is a flexible appendage modeled as a mass-spring-damper system.

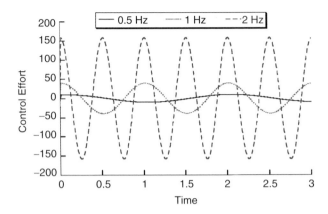

FIGURE 9.29 Control effort tracking various frequencies with feedforward compensation.

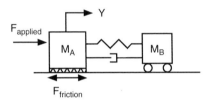

FIGURE 9.30 Model of system with Coulomb friction.

If a reasonable model of the friction dynamics exists, then a feedforward compensator could be useful to improve trajectory tracking. A block diagram of such a control system is shown in Figure 9.31. Note again that the feedforward compensator contains an inverse of the plant dynamics.

Let us first examine the performance of the system without the feedforward compensation. To do this, we start with a baseline system where Mass A is 1, Mass B is 0.5, the spring constant is 15, the damping constant is 0.3, and the coefficient of friction is 0.3. Figure 9.32 shows the response of Mass A for various values of the feedback control gain when the desired path is a cosine function in Equation (9.23). For a gain of 1, the control effort is too small to overcome the friction, so the system does not move. Larger values of gain are able to break the system free, but the trajectory following is very poor. Figure 9.33 shows the response of Mass B for the same range of feedback gains. This part of the system responds with additional flexible dynamics. The corresponding control effort is shown in Figure 9.34.

When the feedforward compensator is turned on, the tracking improves greatly, as shown in Figure 9.35. The control effort necessary to achieve this trajectory following is shown in Figure 9.36. The control effort

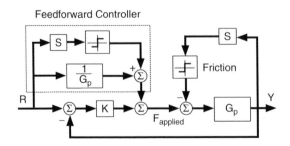

FIGURE 9.31 Block diagram of friction system with feedforward compensation.

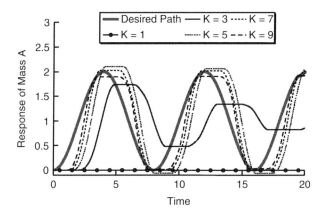

FIGURE 9.32 Response of mass A in friction system without feedforward compensation.

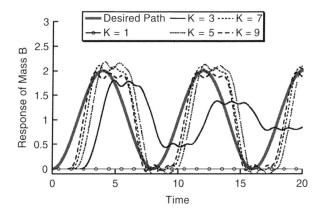

FIGURE 9.33 Response of mass B in friction system without feedforward compensation.

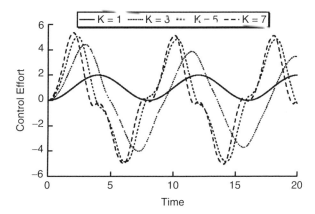

FIGURE 9.34 Control effort in friction system without feedforward compensation.

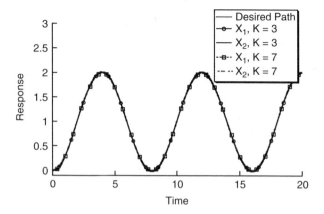

FIGURE 9.35 Response of friction system with feedforward compensation.

contains discontinuous jumps that will not be strictly realizable with real actuators, so the actual trajectory tracking will not be perfect, as predicted through simulation. As with any feedforward control technique, robustness to modeling errors is an issue. Figure 9.37 shows how the tracking degrades as modeling errors are introduced. Recall that feedforward control is dependent not only on the system model but also on the desired trajectory. This is apparent in Figure 9.38 which shows the same modeling errors cause considerably more tracking problems when the period of the desired trajectory is changed from 8 sec, as in Figure 9.37, to 10 sec.

9.3.3 Zero Phase Error Tracking Control

The above feedforward control scheme is subject to the same limitations as for the system with a time delay. For example, only physically realizable desired trajectories can be utilized. Another important limitation exists with model-inverting feedforward control schemes. If the model contains non-minimum phases zeros, then the model cannot be inverted. This would require implementing unstable poles in the feedforward control path. A way around this problem is to invert only the portion of the plant that yields a stable feedforward path.

To improve the performance of such partial-plant-inversion schemes, various extensions and scaling techniques have been developed. A good example of a feedforward controller that only inverts the acceptable parts of the plant is the zero phase error tracking controller (ZPETC) [50]. In the digital formulation of

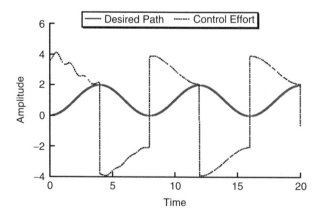

FIGURE 9.36 Control effort for friction system with feedforward compensation.

Trajectory Planning for Flexible Robots

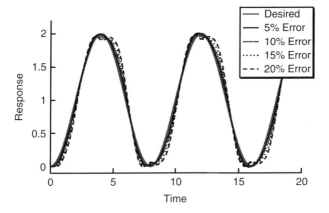

FIGURE 9.37 Response of mass B with modeling errors.

this controller, the plant transfer function is written in the following form:

$$G(z^{-1}) = \frac{z^{-d} B^a(z^{-1}) B^u(z^{-1})}{A(z^{-1})} \qquad (9.31)$$

The numerator is broken down into three parts—a pure time delay z^{-d}, a part that is acceptable for inversion B^a, and a part that should not be inverted B^u.

9.3.4 Conversion of Feedforward Control to Command Shaping

Consider the control structure using a ZPETC controller shown in Figure 9.39. The physical plant and the feedback controller have been reduced to a single transfer function described by

$$G_c(z^{-1}) = \frac{z^{-d} B_c^a(z^{-1}) B_c^u(z^{-1})}{A_c(z^{-1})} \qquad (9.32)$$

where $B_c^u(z^{-1}) = b_{c0}^u + b_{c1}^u z^{-1} + \cdots + b_{cs}^u z^{-s}$. The c subscripts have been added to denote the transfer function now represents the entire closed-loop system dynamics. The superscripts on the numerator again refer to parts that are acceptable and unacceptable for inversion. In this case, the output of the ZPETC is a reference signal for the closed-loop controller. It does not directly apply a force to the plant. It therefore

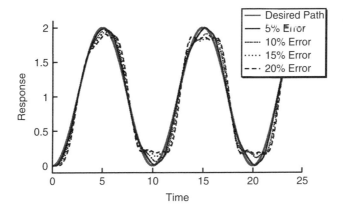

FIGURE 9.38 Response of mass B with modeling errors and slower tr-ajectory.

FIGURE 9.39 ZPETC as a command generator.

does not try to overrule or add to the efforts of the feedback controller. Given this structure, the ZPETC should be considered a command generator, rather than a feedforward compensator.

To cancel the acceptable portion of the plant model, create the ZPETC so that the reference signal is given by

$$r(k) = \frac{A_c(z^{-1})}{B_c^a(z^{-1})B_c^u(1)} y_d^*(k+d) \tag{9.33}$$

Rather than invert the unacceptable zeros, the reference signal is formed using the unacceptable portion evaluated at 1. $B_c^u(1)$ is a scaling term to compensate for the portion of the system that is not inverted. Note that the ZPETC operates on a term related to the desired trajectory denoted by y_d^*. Choosing this function carefully can also help compensate for the incomplete model inversion. Note that for perfect trajectory tracking, we would choose

$$y_d^*(k) = \frac{B_c^u(1)}{B_c^u(z^{-1})} y_d(k) \tag{9.34}$$

However, the term in the denominator would cause unacceptable oscillation or instability in the calculation of y_d^*. The natural choice would then be to simply choose

$$y_d^*(k) = y_d(k) \tag{9.35}$$

However, this would lead to a phase lag between the desired response and the actual response. In order to reduce the phase lag, the ZPETC uses

$$y_d^*(k) = \frac{B_c^u(z)}{B_c^u(1)} y_d(k) \tag{9.36}$$

The total effect of the ZPETC can then be summarized as

$$r(k) = \frac{A_c(z^{-1})B_c^{u*}(z^{-1})}{B_c^a(z^{-1})\left[B_c^u(1)\right]^2} y_d(k+d+s) \tag{9.37}$$

where s is the number of unacceptable zeros and $B_c^{u*}(z^{-1}) = b_{cs}^u + b_{c(s-1)}^u z^{-1} + \cdots + b_{c0}^u z^{-s}$.

9.4 Summary

Trajectory following with flexible robotic systems presents many challenges. For these applications, it is obviously very important to design good mechanical hardware and a good feedback controller. Furthermore, a baseline command trajectory must be generated using knowledge of the physical limitations of the system. Once that is accomplished, the detrimental effects of the system flexibility on trajectory following can be reduced through use of command shaping and feedforward control.

Command shaping operates by taking the baseline command and changing its shape slightly so that it will not excite the flexible modes in the system. Given that this process changes the shape of the reference command, there is the possibility that the system will not exactly follow the intended trajectory. However, the minor deviations produced by command shaping are usually less than the trajectory deviations caused by the deflection that would occur without command shaping. Many command shaping methods have good robustness to modeling errors. Therefore, they can be used on a wide variety of systems, even if their dynamics are somewhat uncertain or change over time.

Feedforward control uses a system model to create auxiliary forces that are added to the force generated by the feedback controller. In this way, it can produce a corrective action before the feedback controller has sensed the problem. With knowledge of the intended trajectory, feedforward control can drive the system along the trajectory better than when only feedback control is utilized. Because feedforward control can dominate the forces from the feedback control, unexpected disturbances and errors in the system model may be problematic when an aggressive feedforward control system is operating.

A highly successful robotic system for trajectory following would likely be composed of four well-designed components: hardware, feedback control, feedforward control, and command shaping. Each of the components has its strengths and weaknesses. Luckily, they are all very compatible with each other and, thus, a good solution will make use of all of them when necessary.

References

[1] Magee, D.P. and Book, W.J., Eliminating multiple modes of vibration in a flexible manipulator, presented at IEEE International Conference on Robotics and Automation, Atlanta, GA, 1993.

[2] Honegger, M., Codourey, A., and Burdet, E., Adaptive control of the hexaglide, a 6 DOF parallel manipulator, presented at IEEE International Conference on Robotics and Automation, Albuquerque, 1997.

[3] Singhose, W., Singer, N., and Seering, W., Improving repeatability of coordinate measuring machines with shaped command signals, *Precision Eng.*, 18, 138–146, 1996.

[4] Smith, O.J.M., Posicast control of damped oscillatory systems, *Proceedings of the IRE*, 45, 1249–1255, 1957.

[5] Smith, O.J.M., *Feedback Control Systems*, McGraw-Hill, New York, 1958.

[6] Bolz, R.E. and Tuve, G.L., *CRC Handbook of Tables for Applied Engineering Science*, CRC Press, Boca Raton, FL, 1973.

[7] Singer, N.C. and Seering, W.P., Preshaping command inputs to reduce system vibration, *J. Dynamic Syst., Meas., Control*, 112, 76–82, 1990.

[8] Singhose, W., Singer, N., and Seering, W., Time-optimal negative input shapers, *ASME J. Dynamic Syst., Meas., Control*, 119, 198–205, 1997.

[9] Bhat, S.P. and Miu, D.K., Precise point-to-point positioning control of flexible structures, *J. Dynamic Syst., Meas., Control*, 112, 667–674, 1990.

[10] Tallman, G.H. and Smith, O.J.M., Analog study of dead-beat posicast control, *IRE Trans. Autom. Control*, 14–21, 1958.

[11] Singer, N.C., Seering, W.P., and Pasch, K.A., Shaping command inputs to minimize unwanted dynamics, MIT, Ed.: U.S. Patent 4,916,635, 1990.

[12] Singhose, W.E., Seering, W.P., and Singer, N.C., Input shaping for vibration reduction with specified insensitivity to modeling errors, presented at Japan-U.S.A. Symposium on Flexible Automation, Boston, MA, 1996.

[13] Singhose, W.E., Porter, L.J., Tuttle, T.D., and Singer, N.C., Vibration reduction using multi-hump input shapers, *ASME J. Dynamic Syst., Meas., Control*, 119, 320–326, 1997.

[14] Singhose, W., Singer, N., Rappole, W., Derezinski, S., and Pasch, K., Methods and apparatus for minimizing unwanted dynamics in a physical system, June 10: U.S. Patent 5,638,267, 1997.

[15] Hyde, J.M. and Seering, W.P., Using input command pre-shaping to suppress multiple mode vibration, presented at IEEE International Conference on Robotics and Automation, Sacramento, CA, 1991.

[16] Singh, T. and Heppler, G.R., Shaped input control of a system with multiple modes, *ASME J. Dynamic Syst., Meas., Control*, 115, 341–347, 1993.

[17] Singhose, W.E., Crain, E.A., and Seering, W.P., Convolved and simultaneous two-mode input shapers, *IEE Control Theory and Applications*, 515–520, 1997.

[18] Pao, L.Y., Multi-input shaping design for vibration reduction, *Automatica*, 35, 81–89, 1999.

[19] Feddema, J.T., Digital filter control of remotely operated flexible robotic structures, presented at American Control Conference, San Francisco, CA, 1993.
[20] Kress, R.L., Jansen, J.F., and Noakes, M.W., Experimental implementation of a robust damped-oscillation control algorithm on a full sized, two-DOF, AC induction motor-driven crane, presented at 5th ISRAM, Maui, HA, 1994.
[21] Singer, N., Singhose, W., and Kriikku, E., An input shaping controller enabling cranes to move without sway, presented at ANS 7th Topical Meeting on Robotics and Remote Systems, Augusta, GA, 1997.
[22] Singhose, W., Porter, L., Kenison, M., and Kriikku, E., Effects of hoisting on the input shaping control of gantry cranes, *Control Eng. Pract.*, 8, 1159–1165, 2000.
[23] Jansen, J.F., Control and analysis of a single-link flexible beam with experimental verification, Oak Ridge National Laboratory ORNL/TM-12198, December 1992.
[24] Magee, D.P. and Book, W.J., Filtering micro-manipulator wrist commands to prevent flexible base motion, presented at American Control Conference, Seattle, WA, 1995.
[25] Seth, N., Rattan, K., and Brandstetter, R., Vibration, control of a coordinate measuring machine, presented at IEEE Conference on Control Apps., Dayton, OH, 1993.
[26] Jones, S. and Ulsoy, A.G., An approach to control input shaping with application to coordinate measuring machines, *J. Dynamics, Meas., Control*, 121, 242–247, 1999.
[27] Rappole, B.W., Singer, N.C., and Seering, W.P., Multiple-mode impulse shaping sequences for reducing residual vibrations, presented at 23rd Biennial Mechanisms Conference, Minneapolis, MN, 1994.
[28] deRoover, D., Sperling, F.B., and Bosgra, O.H., Point-to-point control of a MIMO servomechanism, presented at American Control Conference, Philadelphia, PA, 1998.
[29] deRoover, D., Bosgra, O.H., Sperling, F.B., and Steinbuch, M., High-performance motion control of a wafer stage, presented at Philips Conference on Applications of Control Technology, Epe, The Netherlands, 1996.
[30] Tuttle, T.D. and Seering, W.P., Vibration reduction in flexible space structures using input shaping on MACE: Mission Results, presented at IFAC World Congress, San Francisco, CA, 1996.
[31] Lim, S., Stevens, H.D., and How, J.P., Input shaping design for multi-input flexible systems, *J. Dynamic Sys., Meas., Control*, 121, 443–447, 1999.
[32] Pao, L.Y. and Singhose, W.E., Robust minimum time control of flexible structures, *Automatica*, 34, 229–236, 1998.
[33] Pao, L.Y., Minimum-time control characteristics of flexible structures, *J. Guidance, Control, Dynamics*, 19, 123–29, 1996.
[34] Pao, L.Y. and Singhose, W.E., Verifying robust time-optimal commands for multi-mode flexible spacecraft, *AIAA J. Guidance, Control, Dynamics*, 20, 831–833, 1997.
[35] Tuttle, T. and Seering, W., Creating time optimal commands with practical constraints, *J. Guidance, Control, Dynamics*, 22, 241–250, 1999.
[36] Liu, Q. and Wie, B., Robust time-optimal control of uncertain flexible spacecraft, *J. Guidance, Control, Dynamics*, 15, 597–604, 1992.
[37] Singh, T. and Vadali, S.R., Robust time-optimal control: a frequency domain approach, *J. Guidance, Control, Dynamics*, 17, 346–353, 1994.
[38] Cutforth, C.F. and Pao, L.Y., A modified method for multiple actuator input shaping, presented at American Control Conference, San Diego, CA, 1999.
[39] Meyer, J.L. and Silverberg, L., Fuel optimal propulsive maneuver of an experimental structure exhibiting spacelike dynamics, *J. Guidance, Control, Dynamics*, 19, 141–149, 1996.
[40] Singhose, W., Bohlke, K., and Seering, W., Fuel-efficient pulse command profiles for flexible spacecraft, *AIAA J. Guidance, Control, Dynamics*, 19, 954–960, 1996.
[41] Singhose, W., Singh, T., and Seering, W., On-off control with specified fuel usage, *J. Dynamic Syst., Meas., Control*, 121, 206–212, 1999.

[42] Wie, B., Sinha, R., Sunkel, J., and Cox, K., Robust fuel- and time-optimal control of uncertain flexible space structures, presented at AIAA Guidance, Navigation, and Control Conference, Monterey, CA, 1993.

[43] Lau, M. and Pao, L., Characteristics of time-optimal commands for flexible structures with limited fuel usage, *J. Guidance, Control, Dynamics,* 25, 2002.

[44] Singh, T., Fuel/time optimal control of the benchmark problem, *J. Guidance, Control, Dynamics,* 18, 1225–31, 1995.

[45] Singhose, W., Banerjee, A., and Seering, W., Slewing flexible spacecraft with deflection-limiting input shaping, *AIAA J. Guidance, Control, Dynamics,* 20, 291–298, 1997.

[46] Kojima, H. and Nakajima, N., Multi-objective trajectory optimization by a hierarchical gradient algorithm with fuzzy decision Logic, presented at AIAA Guidance, Navigation, and Control Conference, Austin, TX, 2003.

[47] Banerjee, A., Pedreiro, N., and Singhose, W., Vibration reduction for flexible spacecraft following momentum dumping with/without slewing, *AIAA J. Guidance, Control, Dynamics,* 24, 417–428, 2001.

[48] Tung, E.D. and Tomizuka, M., Feedforward tracking controller design based on the identification of low frequency dynamics, *ASME J. Dynamic Syst., Meas., Control,* 115, 348–356, 1993.

[49] Sadegh, N., Synthesis of a stable discrete-time repetitive controller for MIMO systems, *J. Dynamic Sys., Meas., Control,* 117, 92–97, 1995.

[50] Tomizuka, M., Zero phase error tracking algorithm for digital control, *ASME J. Dynamic Syst., Meas., Control,* 109, 65–68, 1987.

[51] Gimpel, D.J. and Calvert, J.F., Signal component control, *AIEE Transactions,* 339–343, 1952.

10
Error Budgeting

10.1	Introduction **10**-1
10.2	Probability in Error Budgets **10**-2
10.3	Tolerances **10**-3
10.4	Error Sources **10**-5
10.5	Kinematic Modeling **10**-7
10.6	Assessing Accuracy and Process Capability .... **10**-12 Sensitive Directions • Correlation among Multiple Criteria • Interactions among Processing Steps • Spatial Distribution of Errors
10.7	Modeling Material Removal Processes **10**-15
10.8	Summary **10**-19

Daniel D. Frey
Massachusetts Institute of Technology

10.1 Introduction

An error budget is a tool for predicting and managing variability in an engineering system. Error budgets are particularly important for systems with stringent accuracy requirements such as robots, machine tools, coordinate measuring machines, and industrial automation systems. To provide reasonable predictions of accuracy, an error budget must include a systematic account of error sources that may affect a machine. Error sources include structural compliance, thermally induced deflections, and imperfections in machine components. An error budget must transmit these error sources through the system and determine the effects on the position of end effectors, work-pieces, and/or cutting tools. Ultimately, an error budget should combine the effects of multiple error sources to determine overall system performance which is often determined by the tolerances held on finished goods.

Error budgeting was first comprehensively applied to the design of an industrial machine by Donaldson [1]. Over the past two decades error budgeting methods have been improved and extended by several researchers [2–7] and applied to a wide variety of manufacturing-related systems [1, 8–11]. Error budgets are also frequently applied to optical systems, weapons systems, satellites, and so on, but this article will focus on robotics, manufacturing, and industrial automation.

An appropriately formulated error budget allows the designer to make better decisions throughout the design cycle. When bidding on a machine contract, an error budget provides evidence of feasibility. During conceptual design, an error budget aids in selection among different machine configurations. During system design the error budget can be used to allocate allowable error among subsystems to balance the level of difficulty across design teams. At the detailed design stage, an error budget can inform selection of components, materials, and manufacturing process steps. Error models of machine tools can also be used to enhance machine accuracy [12]. During operation, an error model can be used for diagnosis if system accuracy degrades [13] or to help schedule routine maintenance [5]. Error budgets can also improve acceptance procedures because tolerances can be evaluated more effectively using a statistical model of expected manufacturing errors [14].

Error budgeting is a multidisciplinary topic. Several fields of engineering science and mathematics must be applied to achieve good results. The subsections below provide an introduction to the required foundations. The first topic is probability, which enables more rigorous development throughout the subsequent chapters. Next is a discussion of tolerancing and process capability, which defines the required outputs of an error budget. The remaining sections concern the construction of an error budget including characterization of error sources, kinematic modeling, combination of multiple error effects, and modeling of the material removal processes.

10.2 Probability in Error Budgets

Error budgets codify our knowledge about variability and uncertainty of an engineering system. This article must therefore rely substantially on probabilistic and statistical concepts. This section provides an overview of basic concepts including random variables, probability density functions, mean, standard deviation, and sums of random variables.

A random variable is a quantity that may take different values at different times or under different conditions. Random variables may be used to describe quantities about which we have imperfect knowledge. For example, the ambient temperature in a factory may vary from morning to night, from one day to another, and from season to season. There may be bounds on our uncertainty. For example, we may be confident that an HVAC system will maintain the ambient temperature within a few degrees of a specified value.

A probability density function (pdf) can codify one's knowledge and uncertainty about a random variable. A pdf represents the probability that a random variable will be within any given range of values. As depicted in Figure 10.1, given a random variable x, the pdf $f_x(x)$ is defined so that the probability that x lies in a range between lower limit L and upper limit U is

$$P\{L < x \le U\} = \int_L^U f_x(x)dx \qquad (10.1)$$

Given that any random variable must take *some* value, it follows that $\int_{-\infty}^{\infty} f_x(x)dx = 1$. And given that the probability of x taking any particular value cannot be negative, it follows that $0 \le f_x(x)$ for all x.

In error budgeting, we often need to determine whether two random variables are independent of one another, that is, if one variable carries information about the other. The concept of a probability density function (Equation 10.1) allows us to rigorously define the concept of statistical independence. Random variables x and y are said to be statistically independent if and only if

$$f_{xy}(x, y) = f_x(x) f_y(y) \qquad (10.2)$$

A consequence of independence is that the probability that x and y both lie within ranges defined over each variable is simply the product of the probability that either one lies within its specified range.

In error budgets we often wish to keep our description of random variables simple. One way to do that is to focus on a measure of central tendency of the variable and a measure of spread of the variable. The most natural measure of central tendency for use in error budgeting is the expected value, $E(x)$, which is also known as the mean, $\mu(x)$.

$$E(x) = \mu(x) = \int_a^b x \cdot f_x(x)dx \qquad (10.3)$$

Error Budgeting

The most common measure of spread for use in error budgeting is the standard deviation $\sigma(x)$.

$$\sigma(x) = \sqrt{E((x - E(x))^2)} \tag{10.4}$$

It is often reasonable to assume that random variables are distributed according to a normal distribution. In this case, the probability density function is fully defined by just the mean and standard deviation

$$f_x(x) = \frac{1}{\sigma\sqrt{2\pi}} e^{-\frac{(x-\mu)^2}{2\sigma^2}} \tag{10.5}$$

In some cases, it may be more practical to characterize a random variable by stating a range of possible values and assuming that any value within that range is equally probable (i.e., that the random variable is uniformly distributed). The probability density function for a uniformly distributed variable x that ranges from L to U is equal to $1/(U - L)$ everywhere within its range and is zero everywhere outside that range. The mean of the variable is the center of the range $(U + L)/2$, and the standard deviation is $(U - L)/2\sqrt{3}$.

In error budgeting, it is necessary to combine the effects of many random variables. This combination can often be accurately modeled as a sum. Fortunately, there are fairly simple rules for summation of random variables. The means of random variables can be summed in a straightforward way

$$E(x + y) = E(x) + E(y) \tag{10.6}$$

Equation (10.6) holds regardless of distributions of the random variables or whether the variables are independent.

The combination rule for standard deviations is sometimes known as the "root sum of squares" or "RSS" rule

$$\sigma(x + y) = \sqrt{\sigma^2(x) + \sigma^2(y)} \tag{10.7}$$

Equation (10.7) has one very important restriction in its use. It applies only if the variables x and y are probabilistically independent as defined in Equation (10.2) [15]. If the variables are not independent, then the variation in the sum could be much higher than that given by the RSS rule but not higher than a sum of the standard deviations. The standard deviation of the sum could also be lower than that given by the RSS rule, especially if the correlations are induced deliberately as in selective assembly techniques or in machines designed to compensate for existing errors.

10.3 Tolerances

Error budgets are intimately tied to concepts of product tolerance and manufacturing process capability. An error budget is a tool for managing variation and uncertainty, tolerances are ways of placing limits on variation, and process capability is a measure of a system's ability to manufacture products that meet their associated tolerances.

A widely used standard for tolerancing is ANSI Y14.5M [16]. According to this standard, each *feature* of a part has *dimensions* with associated *tolerances*. A feature is "the general term applied to a physical portion of a part, such as a surface, hole, or slot." A dimension is "a numerical value expressed in appropriate units

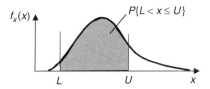

FIGURE 10.1 Probability density functions.

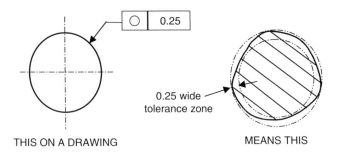

FIGURE 10.2 ANSI definition of circularity for a sphere.

of measure ... to define the size or geometric characteristic, or both, of a part or feature." A tolerance is the amount by which a dimension is permitted to vary. Tolerances are defined in different ways depending on the type of dimension and can be classified according to their spatial frequency:

- Tolerances on surface finish (highest spatial frequencies)
- Tolerances of size and location (zero spatial frequency or "DC" component of error) — Tolerances of size and location are usually defined by a basic dimension (the ideal size or location), and upper and lower limits.
- Tolerances of form (intermediate spatial frequencies) — A form tolerance defines a zone in which all the points on the surface of a part must lie. For example, a tolerance of circularity specifies that each element of the surface in a plane perpendicular to an axis must lie between two concentric circles (Figure 10.2).

Specialized standards exist for many different manufacturing processes. In defense electronics, the acceptance criteria for lead alignment with pads is defined by MIL-STD 2000A. Figure 10.3 depicts the military standard for positioning of a lead from an electronic component with respect to a land or pad on a printed wiring board.

Given any set of specifications for a product, a manufacturing system can be rated in terms of its ability to meet that tolerance. The most common measures are the process capability index (C_p) and the performance index (C_{pk}) [17]. The process capability index (C_p) is defined as

$$C_p \equiv \frac{(U - L)/2}{3\sigma} \qquad (10.8)$$

where U and L are upper and lower limits of the tolerance and σ is the standard deviation of the dimension in the population of parts made by that system. The process capability index is a dimensionless ratio of the amount of variation that can be tolerated and the amount of variation present. It is a useful measure

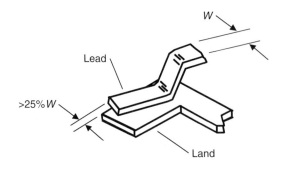

FIGURE 10.3 Specification for foot side overhang of electronic leads.

of the capability of a manufacturing system, but it does not account for *bias*, the difference between the mean value of the dimension and target value. The process capability index can be modified to include the effect of bias which leads to the definition of a performance index C_{pk}.

$$C_{pk} \equiv C_p(1-k) \tag{10.9}$$

where k is a dimensionless ratio of the absolute value of the bias and tolerance width.

$$k \equiv \frac{\left|\mu - \frac{U+L}{2}\right|}{(U-L)/2} \tag{10.10}$$

The performance index (C_{pk}) is closely correlated to the yield of a manufacturing system under the influence of both variation and bias. C_{pk} is often used as a measure of success in six sigma programs [18].

Many products have multiple features, each with an associated tolerance. For products with multiple tolerances, engineering specifications generally require *simultaneous* conformance to *all* criteria. Therefore, a measure of manufacturing system capability is rolled throughput yield (Y_{RT}) [18], which is defined as the probability that *all* the quality characteristics of any product will meet their associated tolerance limits.

As defined by American National Standards Institute (ANSI), product acceptance is binary — either a part is accurate enough or it is defective. Furthermore, variations within the tolerance limits are defined as permissible. However, most deviations from perfect form, even those within tolerance, can contribute degraded performance and lower value in the eyes of the customer. To reflect this reality, quality loss functions were proposed by Taguchi as a means to quantify the economic impacts of variation. The simplest form of quality loss function is a quadratic loss function

$$\text{Quality Loss} = \frac{A_o}{[(U-L)/2]^2}\left(d - \frac{U+L}{2}\right)^2 \tag{10.11}$$

where d is the dimension of the part and A_o is the cost to scrap or rework the part [19]. It can be shown that, for a system which exhibits both variation and bias, the expected value of quality loss is

$$E(\text{Quality Loss}) = \frac{A_o}{[(U-L)/2]^2}\left(\left(\mu - \frac{U+L}{2}\right)^2 + \sigma^2\right) = A_o\left(k^2 + \frac{1}{9C_p^2}\right) \tag{10.12}$$

Equation (10.12) shows that expected quality loss is equally influenced by the variance and the squared bias. Equation (10.12) also shows a link between expected quality loss and process capability. Expected quality loss is similar to the performance index C_{pk} in that both are functions of k and C_p, but expected quality loss more clearly shows the effects of variation on the value delivered to the customer.

10.4 Error Sources

An error source is the physical cause of inaccuracy. An essential step in building an error budget is to list and characterize the most important error sources affecting the system. Table 10.1 lists error sources commonly included in an error budget. The table is not exhaustive but it does provide a representative set of the main error sources in robots, machine tools, and coordinate measuring machines.

The error sources for linear and rotary joints listed at the top of Table 10.1 are considered to be fundamental in forming an error budget. For example, a prismatic three-axis robot, CMM, or machine tool is said to have 21 fundamental measurable errors [3]. For each axis there are three rotary errors (roll, pitch, and yaw), two straightness errors, and one drive-related error motion. That makes six error motions per axis. With three axes, that makes 18 errors associated with the axes individually. In addition, there are three squareness errors which describe the imperfection in assembling the axes into a mutually perpendicular set.

TABLE 10.1 Sources of Error Typically Included in an Error Budget

Error Source	Description
Errors for linear axes	
Straightness	Linear axes exhibit imperfections in straightness in the two directions perpendicular to the direction of motion.
Squareness	When two linear axes are intended to be mutually perpendicular there is an error in their squareness. One error is associated with each pair of axes.
Angular error motions (roll, pitch, and yaw)	Every carriage may exhibit three angular error motions. Roll is the rotational error motion about the axis of intended linear motion. Pitch and yaw are the other two error motions.
Drive related errors	Lead screws can have error in the advance per revolution (cumulative lead error) and can also exhibit periodic errors that repeat every revolution (once per revoulition lead error). Drives for a stage can also exhibit backlash (lost motion) which occurs when the direction of movement is reversed.
Errors for rotary axes	
Parallelism	When two rotary axes are intended to be parallel, there are two errors required to describe the lack of parallelism between two axes.
Linear error motions	Rotory axes may exhibit unintended linear motions as they rotate. The motions along the axis of rotation are known as face motions. The motions in the other two directions are radial error motions which contribute to run-out.
Angular error motions	Rotary axes may exhibit random error motion about either of the two axes perpendicular to the axis of rotation.
Drive related errors	Rotary axes each have one error associated with the difference between the commanded and the realized angular position. These errors are often associated with the finite resolution of the encoder.
Load induced deformation	The weight of the payload, arms, and carriages or the forces applied by the machine to the workpiece result in deflections of the machine's components due to their limited stiffness.
Dynamic effects	Acceleration due to external influences (vibration of the floor) or internal sources (vibration of motors or acceleration of the machine's carriages) lead to deflections due to the machine's limited stiffness and damping.
Thermal deformation	Heating of the machine due to internal or external sources results in deformation of the machine's components due to either differences in thermal expansion coefficients or non-uniform temperature distributions.
Fixturing errors	The workpiece is held by the machine imperfectly due to the variable dimensions of the part itself or the inherent non-repeatability of the fixturing process.
Tool related errors	In material removal processes, cutting tools contribute to error due to wear, scalloping, chatter, deflection, etc.

There are many types of errors that may be included in an error budget such as thermally induced deflections, dynamic effects, and so on. However, error sources should only be characterized as fundamentally as is useful for the purpose at hand [1]. For example, Table 10.1 lists thermal deformation as a key error source. It is not always necessary to characterize temperature variation directly. In some cases it is more convenient to include thermal error in the distribution function for squareness, straightness, etc.

Once a list of error sources has been made, each error source should be characterized with measures that enable probabilistic analysis. Typically, error sources are described by means and standard deviations or upper and lower limits. In many cases it is also necessary to describe the frequency content or spatial distribution properties of the errors. For example, many linear drives are based on a lead screw. These drives exhibit periodic errors that repeat each time the screw completes a full revolution. They also exhibit a cumulative error that increases as the lead screw turns over and over (Figure 10.4).

An important step in forming an error budget is to gather needed information about the error sources. In some cases the needed information is available from component suppliers. For example, the cumulative lead error and once per revolution lead error are normally quantified by the lead screw manufacturer. In other cases, the errors will have to be estimated based on engineering science. For example, the load induced deformation can usually be estimated early in the design stage using beam theory and can be refined later in the design using finite element analysis.

Error Budgeting

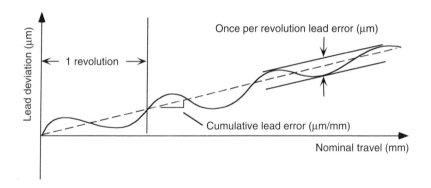

FIGURE 10.4 Cumulative and once per revolution lead error associated with a lead screw drive for a linear actuator.

Once the error sources have been listed and characterized, it is necessary to infer the effects of the errors on the machine's motions. The next section will show how to do this via kinematic modeling.

10.5 Kinematic Modeling

Kinematics is a subset of dynamics concerned with the study of motion without reference to mass or force. A kinematic model of a robot or machine tool describes how the components of the machine are arranged in space and how the components move with respect to one another. In error budgeting, a kinematic model includes both the intended motions and any unintended motions due to error sources. The value of kinematic models in error budgeting is widely recognized [1, 2].

An important first step in building a kinematic model is to assign local coordinate systems (LCSs) to each of the rigid bodies of the machine. These local coordinate systems will serve as reference points for describing the motions of the machine's components. All angular motions will be made about the axes of these LCSs, so it is good practice to assign the LCS to a rigid body at the center of stiffness of the joint that attaches it to the rest of the machine. As an example, LCSs are assigned to a surface grinder in Figure 10.5.

The local coordinate systems are usually configured in two serial kinematic chains. One chain extends from the base of the machine out to an end effector or cutting tool. The other chain extends from the base out to the workpiece. The serial chains describe the topology of the machine — how each component is connected to other components and eventually connected to the base. This topology is critical to the kinematic model because whenever a body moves, all the other bodies farther out on the same chain also execute the same motion.

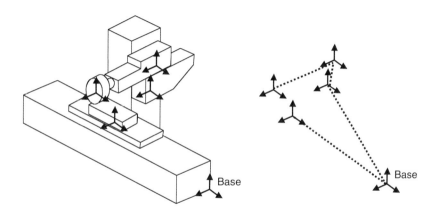

FIGURE 10.5 A surface grinder with local coordinate systems and the resulting serial kinematic chains.

There are many formal ways to mathematically model the motions of a robot or machine tool. Chao and Yang [4] persuasively argued for use of homogeneous transformation matrices (HTMs) because error budgets must account for all six possible error motions of a rigid body. HTMs include six degrees of freedom while Denavit-Hartenberg [20] notation includes only four degrees of freedom. This article will, therefore, describe the use of HTMs for kinematic modeling of machine tools. A general form of HTM is

$$\begin{bmatrix} 1 & 0 & 0 & X \\ 0 & 1 & 0 & Y \\ 0 & 0 & 1 & Z \\ 0 & 0 & 0 & 1 \end{bmatrix} \cdot \begin{bmatrix} 1 & 0 & 0 & 0 \\ 0 & \cos\Theta_x & -\sin\Theta_x & 0 \\ 0 & \sin\Theta_x & \cos\Theta_x & 0 \\ 0 & 0 & 0 & 1 \end{bmatrix} \cdot \begin{bmatrix} \cos\Theta_y & 0 & \sin\Theta_y & 0 \\ 0 & 1 & 0 & 0 \\ -\sin\Theta_y & 0 & \cos\Theta_y & 0 \\ 0 & 0 & 0 & 1 \end{bmatrix} \cdot \begin{bmatrix} \cos\Theta_z & -\sin\Theta_z & 0 & 0 \\ \sin\Theta_z & \cos\Theta_z & 0 & 0 \\ 0 & 0 & 1 & 0 \\ 0 & 0 & 0 & 1 \end{bmatrix}$$

(10.13)

where X, Y, and Z are displacements in the x, y, and z directions and Θ_x, Θ_y, and Θ_z are rotations about the x, y, and z axes. This form of HTM allows for the large rotations required to model the commanded motions of the robot or machine tool.

HTMs can be used to model the joints in robots and machine tools. For a rotary joint, let Θ_z represent the commanded motion of the joint while δ_x, δ_y, δ_z and ε_x, ε_y, ε_z represent the translational and rotational error motions of the joint, respectively. Entering these symbols into the general form of HTM (Equation (10.13)) and employing small angle approximations to simplify yields

$$\begin{bmatrix} \varepsilon_z + \cos\Theta_z & -\sin\Theta_z & \varepsilon_y & \delta_x \\ \sin\Theta_z & \varepsilon_z + \cos\Theta_z & -\varepsilon_x & \delta_y \\ \varepsilon_x \sin\Theta_z - \varepsilon_y \cos\Theta_z & \varepsilon_x \cos\Theta_z + \varepsilon_y \sin\Theta_z & 1 & \delta_z \\ 0 & 0 & 0 & 1 \end{bmatrix}$$

(10.14)

For a linear joint, let X represent the commanded motion of the joint in the x direction while δ_x, δ_y, δ_z and ε_x, ε_y, ε_z represent the translational and rotational error motions of the joint, respectively. Entering these symbols into the general form of HTM (Equation (10.13)) and employing small angle approximations to simplify yields

$$\begin{bmatrix} 1 & -\varepsilon_z & \varepsilon_y & X+\delta_x \\ \varepsilon_z & 1 & -\varepsilon_x & \delta_y \\ -\varepsilon_y & \varepsilon_x & 1 & \delta_z \\ 0 & 0 & 0 & 1 \end{bmatrix}$$

(10.15)

If the linear axis is arranged so that the bearings do not move with the rigid body they support but are instead fixed with respect to the previous body in the kinematic chain, then the ordering of the error motions and the commanded translations must be reversed. This kind of joint is, therefore, modeled by the equation below. Note the difference in Figure 10.6 and Figure 10.7. When the commanded motion X is zero, the effect of an error motion ε_y is the same with either Equation (10.15) or Equation (10.16). But when X is non-zero, the error motion ε_y causes a much larger z displacement using Equation (10.16). Equation (10.15) is most often used in machine tool error budgets, but some linear joints are arranged the other way, so Equation (10.16) is sometimes needed.

$$\begin{bmatrix} 1 & -\varepsilon_z & \varepsilon_y & X+\delta_x \\ \varepsilon_z & 1 & -\varepsilon_x & \delta_y + X\varepsilon_z \\ -\varepsilon_y & \varepsilon_x & 1 & \delta_z + X\varepsilon_y \\ 0 & 0 & 0 & 1 \end{bmatrix}$$

(10.16)

Error Budgeting

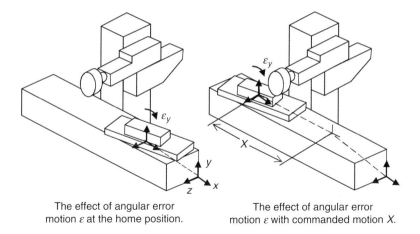

The effect of angular error motion ε at the home position.

The effect of angular error motion ε with commanded motion X.

FIGURE 10.6 When the bearings are attached to the component whose motion is being driven by the joint, use Equation (10.15) to model the commanded linear motion, X, and angular error motion, ε.

It is straightforward to modify Equation (10.14) to Equation (10.16) to model rotary and linear joints oriented along different axes. The choice of orientation of the joints with respect to the local coordinate systems can be made according to the context in which they are used.

To apply HTMs to construct a complete kinematic model of a robot or machine tool, one must model the relationship of each local coordinate system in each serial chain with respect to the adjacent local coordinate systems. Following Soons et al. [3], one may consider two kinds of HTMs in the kinematic model, one for the joints themselves and one for the shapes of the components which are between the joints and determine the relative location of the joints. The HTMs for the joints have already been described in Equation (10.14), Equation (10.15), and Equation (10.16). They are labeled according to the joint number so that the HTM for the ith joint is $\mathbf{J}_i$.

The HTMs for the shapes describe the relative position and orientation of the LCSs at the "home" position (when the commanded motions X, Θ_z, etc. are zero). They are labeled according to the two LCSs they relate to one another so that the HTM relating the ith LCS to the LCS before it in the kinematic chain is $^{i-1}\mathbf{S}_i$.

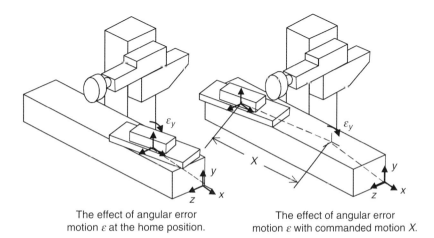

The effect of angular error motion ε at the home position.

The effect of angular error motion ε with commanded motion X.

FIGURE 10.7 When the bearings are attached to the fixed component, use Equation (10.16) to model the commanded linear motion, X, and angular error motion, ε.

The complete kinematic model is formed by multiplying all the joint and shape HTMs in sequence. Thus, an HTM that models a kinematic chain of n LCSs from the base to the end effector is

$$^0\mathbf{T}_n = {}^0\mathbf{S}_1 \cdot \mathbf{J}_1 \cdot {}^1\mathbf{S}_2 \cdot \mathbf{J}_2 \cdots \mathbf{J}_{n-1}{}^{n-1}\mathbf{S}_n \tag{10.17}$$

With this HTM in Equation (10.17), one can carry out analysis of the position and orientation of any object held by the end effector by mapping local coordinates into the base (global) coordinate system. If the coordinates of any point on with respect to the nth LCS are (p_x, p_y, p_z), then the coordinates with respect to the base (global) LCS (p'_x, p'_y, p'_z) are given by

$$\begin{Bmatrix} p'_x \\ p'_y \\ p'_z \\ 1 \end{Bmatrix} = {}^0\mathbf{T}_n \begin{Bmatrix} p_x \\ p_y \\ p_z \\ 1 \end{Bmatrix} \tag{10.18}$$

As an example of kinematic modeling, consider the SCARA type robot in Figure 10.8. Assume that the pose depicted in the figure is the home or zero angle pose. Local coordinate systems have been assigned to each of four major components of the robot at the location of the bearing centers of stiffness.

Joints 1 and 2 are modeled using Equation (10.14). Joint 3 is best modeled using Equation (10.16) assuming the quill is supported by a set of bearings fixed to the arm of the robot rather than to the quill

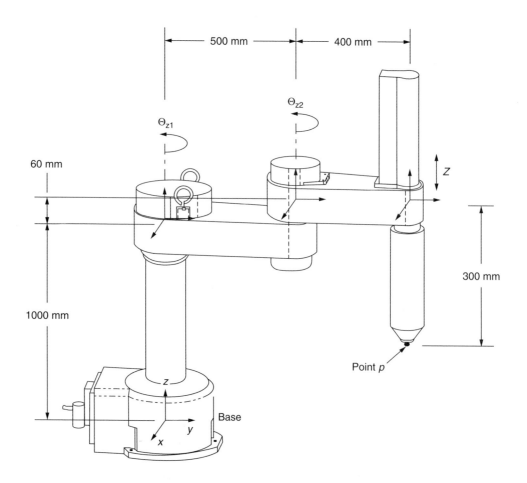

FIGURE 10.8 A robot to be modeled kinematically.

Error Budgeting

TABLE 10.2 Error motions for the Example HTM Model of a SCARA Robot

Error	Description	μ	σ
ε_{z1}	Drive error of joint #1	0 rad	0.0001 rad
ε_{z2}	Drive error of joint #2	0 rad	0.0001 rad
δ_{z3}	Drive error of joint #3	$Z \cdot 0.0001$	0.01 mm
ε_{x3}	Pitch of joint #3	0 rad	0.00005 rad
ε_{y3}	Yaw of joint #3	0 rad	0.00005 rad
xp_2	Parallelism of joint #2 in the x direction	0.0002 rad	0.0001 rad

itself. The shape transformations require translations only in this case and the translations required can be inferred from the dimensions in Figure 10.8. Let us include only six error motions in this model as described in Table 10.2. The resulting HTM follows the form of Equation (10.17) with alternating shape and joint transformations

$$^0T_3 = \begin{bmatrix} 1 & 0 & 0 & 0\text{ mm} \\ 0 & 1 & 0 & 0\text{ mm} \\ 0 & 0 & 1 & 1000\text{ mm} \\ 0 & 0 & 0 & 1 \end{bmatrix} \begin{bmatrix} \cos(\Theta_{z1}+\varepsilon_{z1}) & -\sin\Theta_{z1} & 0 & 0\text{ mm} \\ \sin\Theta_{z1} & \cos(\Theta_{z1}+\varepsilon_{z1}) & 0 & 0\text{ mm} \\ 0 & 0 & 1 & 0\text{ mm} \\ 0 & 0 & 0 & 1 \end{bmatrix}$$

$$\cdot \begin{bmatrix} 1 & 0 & 0 & 500\text{ mm} \\ 0 & 1 & -xp_2 & 0\text{ mm} \\ 0 & xp_2 & 1 & 60\text{ mm} \\ 0 & 0 & 0 & 1 \end{bmatrix} \begin{bmatrix} \cos(\Theta_{z2}+\varepsilon_{z2}) & -\sin\Theta_{z2} & 0 & 0\text{ mm} \\ \sin\Theta_{z2} & \cos(\Theta_{z2}+\varepsilon_{z2}) & 0 & 0\text{ mm} \\ 0 & 0 & 1 & 0\text{ mm} \\ 0 & 0 & 0 & 1 \end{bmatrix}$$

$$\cdot \begin{bmatrix} 1 & 0 & 0 & 400\text{ mm} \\ 0 & 1 & 0 & 0\text{ mm} \\ 0 & 0 & 1 & 0\text{ mm} \\ 0 & 0 & 0 & 1 \end{bmatrix} \begin{bmatrix} 1 & 0 & \varepsilon_{y3} & 0\text{ mm} \\ 0 & 1 & -\varepsilon_{x3} & 0\text{ mm} \\ -\varepsilon_{y3} & \varepsilon_{x3} & 1 & -Z-\delta_{z3} \\ 0 & 0 & 0 & 1 \end{bmatrix} \quad (10.19)$$

The HTM can be used to compute the location of the tip of the end effector at any location in the working volume as a function of the joint motions and error motions.

$$\begin{Bmatrix} p'_x \\ p'_y \\ p'_z \\ 1 \end{Bmatrix} = {}^0T_3 \begin{Bmatrix} 0 \\ 0 \\ 300 \\ 1 \end{Bmatrix} \quad (10.20)$$

The resulting expressions are generally very complex. In this case the z position of the end effector is the simplest term. It is generally safe to eliminate the second order terms — those that contain products of two or more error motions. With this simplification, the z position of the end effector is

$$p'_z = 760 - Z - \delta_{z3} + [400\sin\Theta_{z2}]xp_2 \quad (10.21)$$

Examining Equation (10.21), one can observe that the position of a point in space varies from the commanded position according to a sum of error terms. The error terms are either linear error motions or angular error motions multiplied by the perpendicular distance of the point of interest from the center of rotation of each error motion.

The fact that angular error motions are amplified by perpendicular distances was formally codified by Ernst Abbé who was concerned with the effect of parallax in measurement tasks. The concept was later extended to machine design by Bryan [21]. As a recognition of this, large displacements caused by small angular error motions are sometimes called "Abbé errors." One value of the kinematic model is that it automatically accounts for the changing perpendicular distances as the machine moves through its working volume. In effect, the kinematic model is an automatic accounting system for Abbé errors.

Models for use in error budgeting can vary widely in complexity. The example shown in this chapter is simple. To make the model more complete, one might also model the changing static and dynamic loads on the machine throughout the working volume. This can be accomplished by expressing the error motions as functions of machine stiffness, payload weight, payload position, etc.

This section has shown how to develop a formal kinematic model. The inputs to the model are the dimensions of the machine, the commanded motions, and the error motions. The model allows one to express the position of any point on the machine as a function of the model inputs. The next section describes how the errors in the kinematic model combine to affect the machine's accuracy and the overall process capability.

10.6 Assessing Accuracy and Process Capability

A common way to characterize the accuracy of a machine is by defining one's uncertainty about the location of a point. The point may be the tip of an end effector, a probe, or cutting tool. The confidence can be separately assessed at multiple points in space throughout the working volume of the machine [22]. It is standard practice to separately characterize systematic deviations, random deviations, and hysteresis [23]. These three types of error may be described in terms of probability theory. The systematic deviations are related to the mean of the distribution. The random deviation is related to the spread of the distribution and, therefore, the standard deviation. The hysteresis effects can be modeled as random variation or can be treated as systematic deviations depending on the context. A comparison of different approaches to combining errors is provided by Shen and Duffie [24].

As an example of machine accuracy assessemnt using error combination formulae, consider the SCARA robot modeled in the previous section. Equation (10.21) is the expression for z position of the end effector. The commanded z position is $760 - Z$. The deviation from the commanded position is $-\delta_{z3} + 400 \sin \Theta_{z2} x p_2$. The deviation from the commanded position is a function of two random variables and the commanded joint angle which is known exactly. To assess the overall effect of these two random variables on the deviation, we may employ the combination rules for mean and standard deviation. The mean deviation is simply the sum of the mean deviations of each term (as stated in Equation (10.6)).

$$-E(\delta_{z3}) + E([400 \sin \Theta_{z2}] x p_2) = -0.0001 Z + 400 \sin \Theta_{z2} \cdot 0.0002 \tag{10.22}$$

The standard deviation of the z position can be estimated using the root sum of squares rule (Equation (10.7)) if we assume that the error motions are uncorrelated.

$$\sqrt{\sigma^2(\delta_{z3}) + \sigma^2([400 \sin \Theta_{z2}] x p_2)} = \sqrt{(0.01 \text{ mm})^2 + (400 \sin \Theta_{z2} \cdot 0.0001)^2} \tag{10.23}$$

We have just described a machine's accuracy as the mean and standard deviation of a point. In many cases, this is not a sufficient basis for evaluating accuracy. A more general definition is **"the degree of conformance of the finished part to dimensional and geometric specifications"** [25]. This definition ties the accuracy of the machine to its fitness for use in manufacturing, making it a powerful tool for decision making. However, to use this definition, several factors must be considered, including:

- Sensitive directions for the task
- Correlation among multiple criteria
- Interactions among processing steps
- Spatial distribution of errors

Error Budgeting

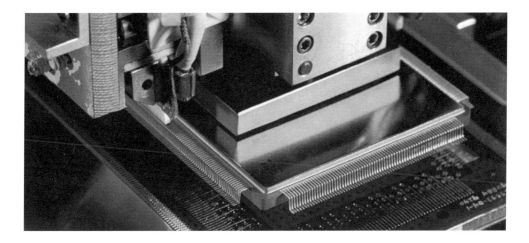

FIGURE 10.9 A robot assembles an electronic package onto a printed wiring board.

10.6.1 Sensitive Directions

The required accuracy for any particular task may be very high in one direction (a sensitive direction) as compared with other directions. Consider a robot which must place an electronic component on a printed wiring board (Figure 10.9). The leads extending from the package must be aligned with the pads on the printed wiring board. This places tight requirements on the x and y position of the electronic component. On the other hand, the flexibility of the leads may allow for significant error in the z direction. Therefore, for this assembly task, x and y are sensitive directions and z is an insensitive direction. A complete assessment of accuracy cannot generally be made without consideration of sensitive direction for a particular task.

10.6.2 Correlation among Multiple Criteria

The variations due to any individual processing step may induce correlation among the multiple dimensions of the product. For example, the lead forming operation depicted in Figure 10.10 generates the "foot" shape on the wire leads of an electronics package (in Figure 10.9). Errors in the lead forming process may result in variation of the position of the leads. Figure 10.11 shows that this variation is linearly dependent on

FIGURE 10.10 An automated system for forming the leads on electronic packages.

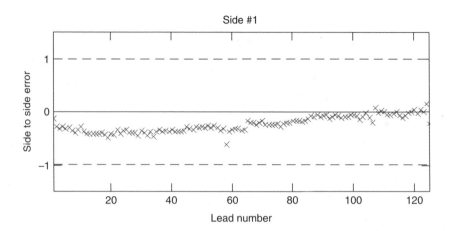

FIGURE 10.11 Data on the location of leads formed on an automated system.

the location of the lead along the edge of the package. Correlations such as this can have a significant impact on the process capability. Any package that has one lead out of tolerance is likely to have many failed leads on the same side. As a result, one cannot simply multiply the probability of failure for a lead by the number of leads to determine rolled throughput yield. Many sophisticated statistical techniques are available to estimate yield under these circumstances, but Monte Carlo simulation is adequate for most situations and is conceptually very simple. For each of many random trials, the error sources are sampled from their respective distributions, the manufacturing process is simulated, and the product is compared to the specifications. By this means, the rolled throughput yield can be estimated even if there is correlation among multiple acceptance criteria.

10.6.3 Interactions among Processing Steps

Most products undergo several processing steps in their manufacture. Errors are transmitted though multi-step processes in many ways. The errors introduced by one step may be added to by subsequent processing steps, for example, when two parts are assembled to deliver one dimension across both parts. The errors of one step may be amplified by another step, for example, when the material properties set in one step affect a subsequent forming operation. The errors of a previous step may be corrected to some degree by a subsequent step, for example, the errors in the position of wire leads on an electronic component can be compensated by robots with machine vision that place the electronic components so as to minimize the sum squared deviation of the leads from the pads. Addition rules for error are only adequate when the separate error terms are additive or nearly additive, so these rules should be used with care when forming an error budget for a manufacturing system with multiple processing steps. Specialized techniques for assessing the process capability of complex manufacturing systems, including error compensation, are discussed by Frey et al. [6].

10.6.4 Spatial Distribution of Errors

The errors in a machine's position at one pose may be correlated to the errors at another pose. These patterns of correlation may be smoothly distributed across the working envelope, for example, when the deflection due to static load rises as robot's arm extends farther from the base. Or the patterns of correlation may be periodic when they arise from the repeated rotations of a lead screw (see Figure 10.4). The periodic patterns can vary in frequency, content, and phase. As described in the section on tolerancing, form and surface finish tolerances are related to different spatial frequencies of error. One important consequence of these facts is that the effects of individual error sources on tolerances cannot always be linearly superposed.

Error Budgeting

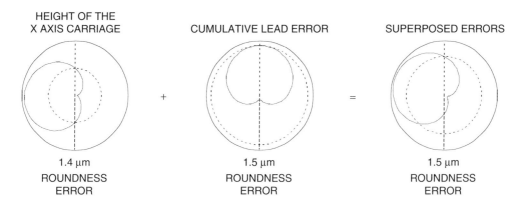

FIGURE 10.12 Superposition of two error sources and their effects on roundness.

Figure 10.12 shows the effect of two error sources applied separately in an error budget for a crankpin grinding operation. The polar plots indicate the deviation of the location of points on the surface as a function of angular position. Figure 10.12 shows that when the two errors are superposed, the effect on form tolerance is not additive. For tolerances of form, a root sum of squared combination rule may be used but could prove to be overly conservative. A more precise assessment of process capability requires direct simulation of the manufacturing process under a range of error source values.

This section demonstrated the application of error combination rules from probability theory to simple error budgets. It also pointed out some common scenarios in which these combination rules should not be applied. If the error in the final product is not a linear function of the error sources, a more general probabilistic simulation may be required.

10.7 Modeling Material Removal Processes

Material removal processes are a commercially important class of manufacturing operations which present special challenges for error budgeting. As discussed in the previous section, sensitive directions have a profound impact on the mapping between error motions and accuracy of the finished part. In machining, only motions normal to the surface of the part are sensitive. Error motions parallel to the surface of the part have little effect. But the sensitive directions change in complicated ways as the tool moves and even vary across the length of the cutting tool at a single point in time. This section presents a general solution to incorporating the geometry of material removal into an error budget.

The mapping from cutting tool positioning error into workpiece geometry error can be formulated as a problem in computational geometry. Machining can be modeled as the sweeping of a cutting tool through space which removed everything in its path. The mathematics of this process is described by swept envelope theory as illustrated by Figure 10.13. The motion of the tool is determined by the kinematic model. The swept volume of the cutting tool is the collection of the set of points in the interior of the cutting tool as it moved from its initial position to its final position (Figure 10.13a and Figure 10.13b). The machined surfaces are the subset of the swept envelope of the cutting tool that lie in the interior of the workpiece stock. The swept envelope of the cutting tool is the set of points on the surface of the swept volume. The swept envelope consists of two types of components: a subset of the boundary of the generator at the initial and final positions, and surfaces generated during the motion of the generator [26] (Figure 10.13c). For many machining operations (plunge grinding, horizontal milling, etc.), only the surfaces of type 2 are of interest. Therefore, only surfaces of type 2 will be considered here.

Wang and Wang [26] give a general method for developing an implicit formulation of the envelope surface of a swept volume from a description of a generator curve and a motion function. The key to the

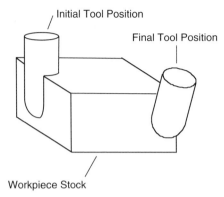

(a) The cutting tool in its initial and final positions with respect to the workpiece

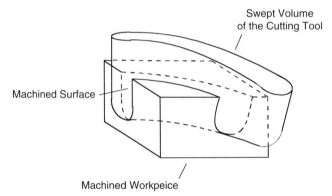

(b) The swept volume of the cutting tool as it moves between the initial and final positions.

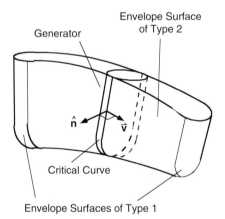

(c) The envelope surface of the cutting tool with normal and velocity vectors.

FIGURE 10.13 Swept envelopes of cutting tools. (a) The cutting tool in its initial and final positions with respect to the workpiece. (b) The swept volume of the cutting tool as it moves between the initial and final positions. (c) The envelope surface of the cutting tool with normal and velocity vectors.

method is the realization that, at any point in time, the generator curve is tangent to the envelope surface along a "critical curve" that lies on both the generator and envelope surface. Therefore at any point ($\vec{p}$) along the tangent curve, the unit normal vector to the envelope surface is identical to the unit normal vector ($\hat{n}$) of the generator at that point and at that instant in time. Further, the velocity ($\vec{v}$) of point $\vec{p}$

Error Budgeting

must be tangent to the envelope surface. Therefore

$$\mathbf{v}(\mathbf{p}, t) \cdot \hat{\mathbf{n}}(\mathbf{p}, t) = 0 \tag{10.24}$$

Equation (10.24) above is, in effect, a procedural definition of the envelope surface of a swept solid. If one can construct parametric formulations

$$\mathbf{v}(\mathbf{p}, t) = \mathbf{v}(u, v, t) \tag{10.25}$$

$$\text{and} \quad \hat{\mathbf{n}}(\mathbf{p}, t) = \hat{\mathbf{n}}(u, v, t) \tag{10.26}$$

then Equation (10.24) will result in a nonlinear equation in u, v, and t which all the points on the surface of the envelope must satisfy. Solution of Equation (10.24) is very complex in general, but with some restrictions on the geometry of the cutting tool, a closed form solution is available which is useful for error budgets. For example, grinding wheels and milling cutters can, under certain conditions, be modeled as surfaces of revolution.

Let us assume that the z axis of the tool local coordinate system lies on the cutting tool's axis of rotation (Figure 10.14). The shape of the cutting tool is defined by a function $r(u)$, equal to the perpendicular distance of the cutting edge (or grain) from the axis of rotation as a function of u, the component along the axis of rotation of the distance from the cutting point to the origin of the tool local coordinate system.

If the machine is modeled by two kinematic chains $\mathbf{T}$ (for the cutting tool) and $\mathbf{W}$ (for the workpiece), then the location of the origin of the tool LCS with respect to the workpiece is

$$\mathbf{c}(t) = {}_0\mathbf{W}_n^{-1} \cdot {}_0\mathbf{T}_m \cdot (0\ 0\ 0\ 1)^T \tag{10.27}$$

The unit vector pointing along the axis of rotation of the tool is

$$\hat{\mathbf{a}}(t) = {}_0\mathbf{W}_n^{-1} \cdot {}_0\mathbf{T}_m \cdot (0\ 0\ 1\ 0)^T \tag{10.28}$$

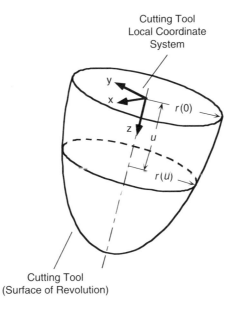

FIGURE 10.14 Representing a cutting tool as a surface of revolution.

Using $\hat{\mathbf{a}}$ and $\mathbf{c}$, we can define a local coordinate system on the tool defined by the set of mutually orthogonal unit vectors $\hat{\mathbf{e}}_1$, $\hat{\mathbf{e}}_2$, and $\hat{\mathbf{e}}_3$. The local coordinate system is defined as

$$\hat{\mathbf{e}}_1 = \hat{\mathbf{a}}$$
$$\text{if } |\dot{\hat{\mathbf{a}}}| \neq 0 \quad \text{then} \quad \hat{\mathbf{e}}_2 = \frac{\dot{\hat{\mathbf{a}}}}{|\dot{\hat{\mathbf{a}}}|} \quad \text{and} \quad \hat{\mathbf{e}}_3 = \hat{\mathbf{e}}_1 \times \hat{\mathbf{e}}_2 \qquad (10.29)$$
$$\text{if } |\dot{\hat{\mathbf{a}}}| = 0 \quad \text{and} \quad \dot{\hat{\mathbf{a}}} \cdot \dot{\mathbf{c}} \neq 0 \quad \text{then} \quad \hat{\mathbf{e}}_3 = \frac{\hat{\mathbf{a}} \times \dot{\mathbf{c}}}{|\dot{\mathbf{c}}|} \quad \text{and} \quad \hat{\mathbf{e}}_2 = \hat{\mathbf{e}}_3 \times \hat{\mathbf{e}}_1$$

Consider, as depicted in Figure 10.15, the circle defined by the intersection of a plane normal to the cutting tool's axis of rotation $\hat{\mathbf{a}}$, offset from the origin of the tool coordinate system $\mathbf{c}$, by the perpendicular distance u. A point $\mathbf{p}$ lies in that plane at an angle θ measured from the axis $\hat{\mathbf{e}}_2$. The velocity of point is

$$\mathbf{v} = \dot{\mathbf{c}} + u|\dot{\hat{\mathbf{a}}}|\hat{\mathbf{e}}_2 - r(u)|\dot{\hat{\mathbf{a}}}|\hat{\mathbf{e}}_1 \cos\theta \qquad (10.30)$$

The unit normal vector to the surface of revolution at point $\mathbf{p}$ is

$$\hat{\mathbf{n}} = \frac{-r'(u)\hat{\mathbf{e}}_1 + \hat{\mathbf{e}}_2 \cos\theta + \hat{\mathbf{e}}_3 \sin\theta}{\sqrt{1 + r'(u)^2}} \qquad (10.31)$$

where

$$r'(u) = \frac{dr}{dz}\bigg|_{z=u} \qquad (10.32)$$

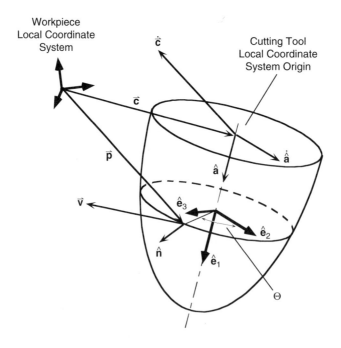

FIGURE 10.15 Notation for calculating points on the critical curve.

Combining Equation (10.30) and Equation (10.31) into Equation (10.24) yields an equation of the form

$$A \cos \Theta + B \sin \Theta + C = 0 \tag{10.33}$$

where

$$A = |\mathring{\mathbf{a}}| r'(u) r(u) + \dot{\mathbf{c}} \cdot \hat{\mathbf{e}}_2 + |\mathring{\mathbf{a}}| u \tag{10.34}$$

$$B = \dot{\mathbf{c}} \cdot \hat{\mathbf{e}}_3 \tag{10.35}$$

$$C = -r'(u) \dot{\mathbf{c}} \cdot \hat{\mathbf{e}}_1 \tag{10.36}$$

Equation (10.33) admits a closed form solution

$$D = \sqrt{B^2 + A^2 - C^2} \tag{10.37}$$

$$\cos \theta = \frac{C^2 - AC - B^2 + BD}{A^2 - AC + B^2 - BD} \tag{10.38}$$

$$\sin \theta = \frac{(A - C)(B - D)}{A^2 - AC + B^2 - BD} \tag{10.39}$$

The results of Equation (10.38) and Equation (10.39) can be used to compute the positions of the points $\mathbf{p}(t, u)$ on the surface of the working envelope

$$\mathbf{p}(t, u) = \mathbf{c} + u \hat{\mathbf{e}}_1 + r(u)(\hat{\mathbf{e}}_2 \cos \theta \pm \hat{\mathbf{e}}_3 \sin \theta) \tag{10.40}$$

The "$\pm$" symbol indicates that there are two solutions for $\mathbf{p}(t, u)$. If the tool is cutting a trough, then both solutions are valid. If the tool is contouring counterclockwise as viewed in the positive $\hat{\mathbf{e}}_3$ direction, then retain only the solution corresponding to the plus sign. If the tool is contouring clockwise as viewed in the positive $\hat{\mathbf{e}}_3$ direction, then retain only the solution corresponding to the minus sign.

As is evident from Equation (10.39) above, if A is equal to $-C$ then $\sin \theta$ is zero, $\cos \theta$ is one, and θ equals zero. For any values of u such that $A - C < 0$, that cross section of the tool is not in contact with the workpiece surface. Similarly, if the value of D is imaginary, the results of the calculation can be safely disregarded because the cutting tool is again not in contact with the workpiece. This occurs only when the tool is plunging at a steep angle so that every point cut by that section of the tool is immediately swept away by the cross sections following behind it.

The closed form solution described above can be used in error budgeting by performing three operations: (1) compute the swept envelope of the cutting tool using a kinematic model with **no error motions** to determine the nominal machined surface; (2) compute the swept envelope of the cutting tool **including error motions** to determine the perturbed machined surface; (3) evaluate the accuracy of the machining process by comparing the perturbed and nominal machined surfaces. The procedure described above has proved useful for modeling form grinding, centerless grinding, and "chasing the pin" cylindrical grinding. It should also be useful for ball end milling and flank milling with conical milling cutters. Due to the assumption that the cutter is a perfect surface or revolution, it is most useful for evaluating tolerances of form, profile, location, and size and is probably less useful for surface finish.

10.8 Summary

An error budget is a tool for predicting and managing variability in an engineering system. This chapter has reviewed basic theory of probability, tolerances, and kinematics and described a framework for error budgeting based upon those theoretical foundations. The framework presented here is particularly suited to manufacturing systems including robots, machine tools, and coordinate measuring machines.

An error budget must be developed with great care because small mistakes in the underlying assumptions or the mathematical implementation can lead to erroneous results. For this reason, error budgets should be kept as simple as possible, consistent with the needs of the task at hand. When error budgets are scoped appropriately, developed rigorously, and consistent with theoretical foundations (e.g., engineering science, mathematics, and probability), they are an indispensable tool for system design.

References

[1] Donaldson, R.R. (1980). Error Budgets. *Technology of Machine Tools,* Vol. 5, Machine Tool Task Force, Robert J. Hocken, Chairman, Lawrence Livermore National Laboratory.

[2] Slocum, A.H. (1992). *Precision Machine Design.* Prentice Hall, Englewood Cliffs, NJ.

[3] Soons, J.A., Theuws, F.C., and Schellekens, P.H. (1992). Modeling the errors of multi-axis machines: a general methodology. *Precision Eng.,* vol. 14, no. 1, pp. 5–19.

[4] Chao, L.M. and Yang, J.C.S. (1987). Implementation of a scheme to improve the positioning accuracy of an articulate robot by using laser distance-measuring interferometry, *Precision Eng.,* vol. 9, no. 4, pp. 210–217.

[5] Frey, D.D., Otto, K.N., and Pflager, W. (1997). Swept envelopes of cutting tools in integrated machine and workpiece error budgeting. *Ann. CIRP,* vol. 46, no. 1, pp. 475–480.

[6] Frey, D.D., Otto, K.N., and Taketani, S. (2001). Manufacturing system block diagrams and optimal adjustment procedures. *ASME J. Manuf. Sci. Eng.,* vol. 123, no. 1, pp. 119–127.

[7] Frey, D.D. and Hykes, T. (1997). *A Method for Virtual Machining.* U.S. Patent #5,691,909.

[8] Treib, T. (1987). Error budgeting — applied to the calculation and optimization of the volumetric error field of multiaxis systems. *Ann. CIRP,* vol. 36, no. 1, pp. 365–368.

[9] Portman, T. (1980). Error summation in the analytical calculation of lathe accuracy. *Machines and Tooling,* vol. 51, no. 1, pp. 7–10.

[10] Narawa, L., Kowalski, M., and Sladek, J. (1989). The influence of kinematic errors on the profile shapes by means of CMM. *Ann. CIRP,* vol. 38, no. 1, pp. 511–516.

[11] Whitney, D.E., Gilbert, O.L., and Jastrzebski, M. (1994). Representation of geometric variations using matrix transforms for statistical tolerance analysis in assemblies. *Res. Eng. Design,* vol. 6, pp. 191–210.

[12] Donmez, A. (1995). *A General Methodology for Machine Tool Accuracy Enhancement: Theory, Application, and Implementation,* Ph.D. thesis, Purdue University.

[13] Ceglarek, D. and Shi, J. (1996). Fixture failure diagnosis for the autobody assembly using pattern recognition. *ASME J. Eng. Ind.,* vol. 118, no. 1, pp. 55–66.

[14] Kurfess, T.R., Banks, D.L., and Wolfson, J.J. (1996). A multivariate statistical approach to metrology. *ASME J. Manuf. Sci. Eng.,* vol. 118, no. 1, pp. 652–657.

[15] Drake, A.W. (1967). *Fundamentals of Applied Probability Theory.* McGraw-Hill, New York.

[16] ASME (1983). *ANSI Y14.5M — Dimensioning and Tolerancing.* American Society of Mechanical Engineering, New York.

[17] Kane, V.E. (1986). Process capability indices. *J. Qual. Technol.,* vol. 18, no. 1, pp. 41–52.

[18] Harry, M.J. and Lawson, J.R. (1992). *Six Sigma Producibility Analysis and Process Characterization.* Addison-Wesley, Reading, MA.

[19] Phadke, M.S. (1989). *Quality Engineering Using Robust Design.* Prentice Hall, Englewood Cliffs, NJ.

[20] Denavit, J. and Hartenberg, R. (1955). A kinematic notation for lower pair mechanisms based on matrices. *J. Appl. Mech,* vol. 1, pp. 215–221.

[21] Bryan, J.B. (1989). The Abbé principle revisited — an updated interpretation. *Precision Eng.,* vol. 1, no. 3, pp. 129–132.

[22] Lin, P.D. and Ehmann, K.F. (1993). Direct volumetric error evaluation for multi-axis machines. *Int. J. Machine Tools Manuf.,* vol. 33, no. 5, pp. 675–693.

[23] CIRP (1978). A proposal for defining and specifying the dimensional uncertainty of multiaxis measuring machines. *Ann. CIRP,* vol. 27, no. 2, pp. 623–630.

[24] Shen, Y.L. and Duffie, N.A. (1993). Comparison of combinatorial rules for machine error budgets. *Ann. CIRP,* vol. 42, no. 1, pp. 619–622.
[25] Hocken, R.J. and Machine Tool Task Force (1980). *Technology of Machine Tools,* UCRL-52960-5, Lawrence Livermore National Laboratory, University of California, Livermore, CA.
[26] Wang, W.P. and Wang, K.K. (1986). Geometric modeling for swept volume of moving solids. *IEEE Comput. Graphics Appl.,* vol. 6, no. 12, pp. 8–17.

11
Design of Robotic End Effectors

	11.1	Introduction **11**-1
	11.2	Process and Environment **11**-2
		System Design • Special Environments
	11.3	Robot Attachment and Payload Capacity............ **11**-3
		Integrated End Effector Attachment • Attachment Precision • Special End Effector Locations • Wrist Compliance: Remote Compliance Centers • Payloads • Payload Force Analysis
	11.4	Power Sources **11**-7
		Compressed Air • Vacuum • Hydraulic Fluid Power • Electrical Power • Other Actuators
	11.5	Gripper Kinematics **11**-9
		Parallel Axis/Linear Motion Jaws • Pivoting/Rotary Action Jaws • Four-Bar Linkage Jaws • Multiple Jaw/Chuck Style • Articulating Fingers • Multi-Component End Effectors
	11.6	Grasping Modes, Forces, and Stability **11**-11
		Grasping Stability • Friction and Grasping Forces
	11.7	Design Guidelines for Grippers and Jaws............ **11**-13
		Gripper and Jaw Design Geometry • Gripper Design Procedure • Gripper Design: Case Study • Gripper Jaw Design Algorithms • Interchangeable End Effectors • Special Purpose End Effectors/Complementary Tools
	11.8	Sensors and Control Considerations **11**-17
		Proximity Sensors • Collision Sensors • Tactile Feedback/Force Sensing • Acceleration Control for Payload Limits • Tactile Force Control
Hodge Jenkins	11.9	Conclusion **11**-19
Mercer University		

11.1 Introduction

Aside from the robot itself, the most critical device in a robotic automation system is the end effector. Basic grasping end effector forms are referred to as grippers. Designs for end effectors are as numerous as the applications employing robots. End effectors can be part of the robot's integral design or added-on to the base robot. The design depends on the particular robot being implemented, objects to be grasped, tasks to be performed, and the robot work environment.

This chapter outlines many of the design and selection decisions of robotic end effectors. First, process and environment considerations are discussed. Robot considerations including power, joint compliance, payload capacity, and attachment are presented. Sections reviewing basic end effector styles and design guidelines follow this. Sensors and control issues are also presented. The chapter concludes with an end effector design.

11.2 Process and Environment

Robots vary in size and payload capacities for many diverse operations. Some robots are designed for specific, singular tasks such as materials handling, painting, welding, cutting, grinding, or deburing. These robots use specific tools as end effectors. Primary considerations for end effector designs, in these instances, are the tool, orientation, and control of the tool for effective processing, as well as the robot payload capacity (to be discussed later). Other robots are designed for general purposes and material handling. These robots require additional engineering detail in end effector design. In all cases the tasks and the robot environment must be considered when selecting or designing the appropriate robot and end effector. The end effector is seen as part of the overall system design subject to the same constraints as the entire system.

11.2.1 System Design

The end effector is but part of a system. As with all system designs, it is important to have a process flow diagram describing the tasks and how they are to be accomplished, before designing the end effector. This will clarify the process in terms of the objects to be handled and what functions are necessary for the end effector, as well as the entire system. A typical process flow chart defines the items to be handled, where it is grasped, what is done to the item, where it is placed, and special orientations. A sample chart is shown in Figure 11.1. This defines the roles of the robot, system controller, peripheral devices, end effector, and specialized tools. From the activities necessary to complete the process, the end effector design requirements and specifications are formed.

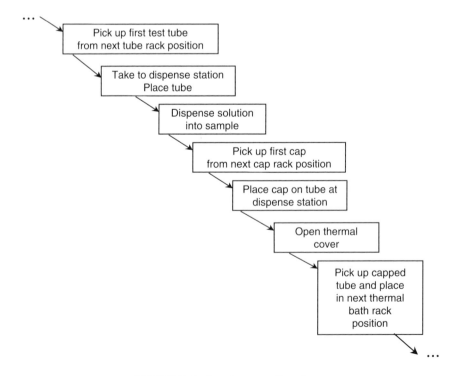

FIGURE 11.1 Sample process flow chart.

Design of Robotic End Effectors

Many other aspects affect the design of the end effectors. Even with robot, process, and system requirements and specifications available, not enough necessary information may be provided about a process. Many questions will need to be examined to complete the system and end effector design.

For example, answers to these questions and others will be necessary for designing end effectors.

1. What are the shape and weight of the grasped objects?
2. What is the uniformity or tolerance of handled objects?
3. Are the objects' surfaces smooth? What is the coefficient of friction?
4. What is the spacing of objects?
5. How will they be oriented and presented?
6. What placement and spacing dimensions can be modified?
7. What are placement tolerances?
8. What are grasping forces or pressure limits?
9. Are complementary tools needed?
10. Is compliance necessary for assembling parts?
11. Are there special environmental considerations?

11.2.2 Special Environments

Many industrial situations have special system environmental requirements that impact the robot and end effector design. A large area of operation for robots is semi-conductor manufacturing. Here robots must operate in various clean room conditions. End effectors must be selected for compatibility to the class of clean room in which the robot will operate. It is important that the end effector not generate many particles in this situation. End effectors must be designed with all surfaces to be either stainless steel or polymer or to be coated with a clean room acceptable material (such as an anodized aluminum surface or a baked-on powder coat). Polymer (oil-less) washers and bearings should be used in the end effector mechanisms to ensure that the surfaces do not contact, wear, or generate particles. In some circumstances, mechanisms must be enclosed with seals or bellows. To meet the requirements for Class 1 clean room operation, air bearings may be required. Many manufacturers sell end effectors designed for clean room operation.

A different problem arises in hazardous environments such as in the presence of highly reactive chemicals (e.g., chlorides, fluorides). Here the robot and end effector must be protected. Robot systems must be designed for chemical resistance. If using a custom end effector, component materials should be selected to be as inert as possible based on the chemicals present. Nickel or aluminum alloys, for example, work fairly well with oxidizers such as chlorine, fluorine, or high concentrations of oxygen. In many cases a covering (including a glove for the end effector) with a pressurized gas purge system must be provided to prevent gaseous vapors or aerosols from reacting with the end-effector or robot.

Another robotic environment requiring special treatments is food processing. In food handling tasks, the end effector and robot must be able to be cleaned and sterilized to kill germs and bacteria. High temperature water and disinfectant sprays are commonly used to clean equipment. Non-corrosive materials and sealed electrical connections for wash downs are required.

11.3 Robot Attachment and Payload Capacity

End effectors for general purpose robots are mounted at the end of the robot arm or wrist for articulating arm robots, or at the end of the last affixed stage or motion device for SCARA and Cartesian coordinate robots. Some robots have built-in grippers. Aside from these self-contained mechanisms, end effectors are selected from standard components, a custom design, or most likely a combination of both. End effectors can have multiple tools or grippers, but they must conform to the mounting attachment and payload limits of the individual robot.

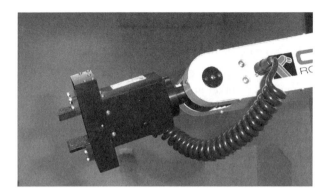

FIGURE 11.2 Robot end effector integrated into arm controller.

11.3.1 Integrated End Effector Attachment

Several types of robots have their end effectors highly integrated into the robot. In specialized applications such as welding, painting, gluing, and plasma cutting, the end effector is essentially a pre-engineered subsystem of the robot. For example, Fanuc offers arc welders as end effectors for their robots, as a complete system. Several smaller robots, such as those used in light manufacturing or bench-top applications, have the basic motion of the end effector built into the robot itself. These integral end effectors are basic gripper designs. The gripper integration feature is seen in robots such as the Microbot Alpha II (Questech, Inc.) and the A465 (Thermo-CRS, Ltd.) (Figure 11.2). The Microbot uses two parallel beam mechanisms for each side of the gripper, as seen in Figure 11.3. The control is provided by springs for opening the grippers and by a motor spooling a lightweight cable to close the grippers. The A654 robot has a servo controlled parallel jaw gripper. Control of the gripper opening is included in the robot controller. Even using the existing mechanisms, custom application jaws for the gripper can be designed and mounted to the gripper base. Note: these robots allow the removal of the standard gripper and provide a mounting base for other end effectors.

11.3.2 Attachment Precision

The attachment of the end effector to the robot must be secure and maintain the desired precision of the robot system. Location and tolerance are extremely critical for maintenance and end effector replacement. Precision dowel pins are used to locate and provide orientation for the end-effector while bolts secure it to the robot. Figure 11.2 and Figure 11.4 show mounting configurations at the end of industrial robot arms. Attachment means vary. Figure 11.2 shows an integral control attachment, while Figure 11.4 depicts an internal passageway provided for power connections.

It is still possible for a replacement end effector to lose positional accuracy, even with previously described design precautions. Thus, calibration of a newly placed end effector is necessary. This can be accomplished

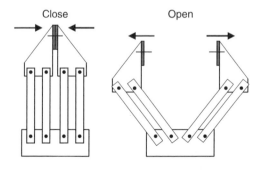

FIGURE 11.3 Four bar linkages gripper arms.

Design of Robotic End Effectors

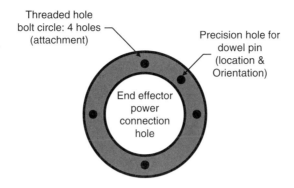

FIGURE 11.4 Typical robot arm end.

in several ways. The most common techniques are to place the robot in teach mode to redefine key geometry locations to hard stop locations, or to have specialized devices with proximity sensors to programmatically locate a position or grasped object. Since all components of a robot and the grasped objects may have varying tolerances, it is better to locate from a grasped artifact of calibrated dimensions, rather than a sample of the actual grasped object, unless the object meets the desired tolerance specifications.

11.3.3 Special End Effector Locations

In grasping large objects such as metal sheets, special attention must be given to the location of grippers (here multiple grippers are used in the end effector). Stability and preventing damage to the sheet are the primary concerns. The reader is referred to Ceglarek et al. [1] for additional information.

11.3.4 Wrist Compliance: Remote Compliance Centers

Robots, workpieces, and fixturing have limited and defined tolerances. Some variation in workpieces or orientation from fixturing may exist. In robotic or automation applications of part insertion into another component or assembly, the use of a remote center compliance device (RCC) (typically wrist mounted) is often required. The RCC allows an assembly robot or machine to compensate for positioning errors (misalignments) during insertion. These position errors are due to machine or workpiece fixturing inaccuracy, vibration, end effector path error, or tolerances. The RCC accomplishes this by lowering contact forces (perpendicular to the axis of insertion) via lower horizontal and rotational stiffnesses while retaining relatively higher axial insertion stiffness. This compliance can reduce or avoid the potential for part, robot, or end effector damage.

A typical remote compliance center design is seen in Figure 11.5. When the remote compliance center is near the contact point, the part will align with the hole automatically; correcting lateral and rotational misalignment. RCC are typically passive mechanical devices with high axial stiffness but low lateral stiffness. Many commercial designs use elastomer shear pads or springs or pneumatic pistons to create the compliance. Other custom designs use a simple beam with greater horizontal than axial compliance (Ciblak [2]). In all cases the RCC has limits on rotational or horizontal misalignments (typically less than $+/-0.25$ in. and $+/-10°$ [3]). Considerations for using an RCC are the added weight of the device and increased moment caused by the additional length of end effector. However, if accounted for properly in design, the RCC is a great aid for speedy part insertion.

11.3.5 Payloads

For all end effector designs the payload capacity of the robot must be carefully considered in the design. Manufacturers specify the payload for each robot. It is important to consider acceleration forces and moments as part of the load, if not already accounted for by the manufacturer.

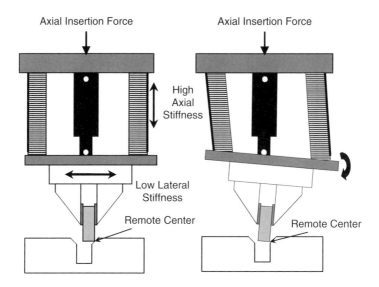

FIGURE 11.5 Remote compliance center device.

Where the end effector is provided or integrated into the robot by the manufacturer, the payload is specified as a weight or mass at a given distance on the gripper jaw of the end effector. These robots have lighter allowable payloads (e.g., 2 kg for the CRS A465 [4]) with the manufacturer provided end effector. If the standard end effector or rigid gripper attachment is replaced with a longer jaw, the allowable payload will be less. This is because a higher torque will be generated at the robot joints.

Most general-purpose, articulating robots have specified the payload capacity in force and moment load limits at the end of arm wrist attachment as well as other joint locations. The load and moment limits must be adhered to when designing the end effector and gripper attachments. In these cases the acceleration of the payload must also be included in addition to gravitational loading. For SCARA or Cartesian coordinate style robots, payload capacity and maximum final joint inertia are sufficient, given the robot kinematics.

11.3.6 Payload Force Analysis

Force calculations for end effector payloads must be determined at all critical locations including the gripper jaws and end of arm payload wrist reactions. Jaw designs and part-grasped orientation must be known or have a bounding estimate. Figure 11.6 and Figure 11.7 depict typical force and moment locations for end of arm geometry. The three-dimensional reaction force and moment at the end of arm location is given in Cartesian coordinates by Equation (11.1) and Equation (11.2). Note that in developing dynamical relationships for control design other coordinate systems, such as a polar or the generalized coordinate methods developed by Kane [5] may be more advantageous.

$$\vec{F}_R = F_X \hat{i} + F_Y \hat{j} + F_Z \hat{k} \tag{11.1}$$

$$\vec{M}_R = M_X \hat{i} + M_Y \hat{j} + M_Z \hat{k} \tag{11.2}$$

The apparent force of an accelerated, grasped object is determined by application of Newton's Second Law, Equation (11.3). Note: In the configuration examined here, gravity is assumed to be present only in the negative Y-direction. The subscript o denotes parameters associated with the object to be grasped, while the g subscript indicates those parameters associated with the gripper/end effector.

$$\vec{F}_R = (m_g + m_o)(A_X \hat{i} + |A_Y + g_Y|\hat{j} + A_Z \hat{k}) \tag{11.3}$$

Design of Robotic End Effectors

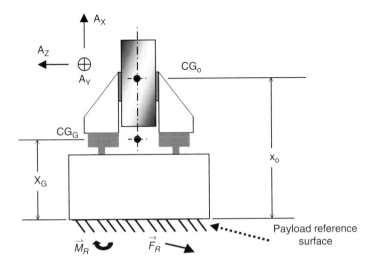

FIGURE 11.6 Payload force and moment: X-Z plane.

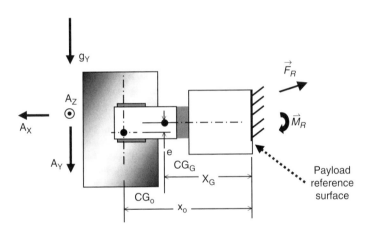

FIGURE 11.7 Payload force and moment: X-Y plane.

Equation (11.4) is the apparent moment at the end effector attachment for a object grasped as shown in Figure 11.6 and Figure 11.7, where the objects center of mass is located at the same X and Z coordinates as the gripper's center of mass. In this instance the centers of mass in the Y-direction are not coincident.

$$\vec{M}_R = \sum \vec{r} \times \vec{F} \tag{11.4}$$

$$\vec{M}_R = (x_g \hat{i}) \times m_g(A_X \hat{i} + |A_Y + g_Y|\hat{j} + A_Z \hat{k}) + (x_o \hat{i} + e\hat{j}) \times m_o(A_X \hat{i} + |A_Y + g_Y|\hat{j} + A_Z \hat{k}) \tag{11.5}$$

As can be seen from the equations the acceleration components may be more significant than the gravitational force alone.

11.4 Power Sources

Power sources for end effectors are generally electrical, pneumatic, vacuum, or hydraulic. Many robots have internal routings for a variety power and control options for end effectors. Common power sources and actuators are discussed with their application advantages.

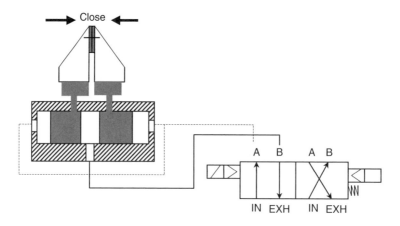

FIGURE 11.8 Pneumatic valve connections for safety.

11.4.1 Compressed Air

The most common end effector power source is compressed air. Compressed air is reliable, Readily available in most industrial settings, and can be adjusted to preset gripper clamping force. Pneumatic grippers and associated valve and piping components can be lightweight, low cost, and safe when designed properly. A variety of pneumatic grippers can be purchased from industrial manufacturers.

When using a pneumatically operated end effector, caution should be taken in the selection of the pneumatic valve connections so that that the gripper will retain a held object if electric power is interrupted. Pneumatic solenoid valves activate air operated grippers and have both normally open or normally closed valve ports. The normally open valve port of the valve should be used to supply air to close the gripper. Figure 11.8 shows a typical connection using a four-way valve. This will help prevent a held object from being released unintentionally, in the case of interrupted electric power. Compressed air supply is less susceptible to interruptions than electric power. Note that three-way solenoid valves may be used if the pneumatic cylinder has a spring return for opening. For added safety, use four-way solenoid valves with dual operators; thus, both the exhaust and supply will be isolated in the event of a power failure. If using pneumatic speed adjustments, they should be placed on the exhaust side of the air cylinder to allow full flow into the cylinder.

For clean room operation, the compressed air exhausts from the gripper air cylinder should be ported away from the clean room. However, if a clean, dry air source is used in an application, the exhausts can be released into the clean room if necessary. Care must be taken to use non-lubricated seals in the pneumatic cylinder, as there is no oil mist in a clean dry air supply.

11.4.2 Vacuum

Vacuum suction devices are used in many instances for lifting objects that have a smooth relatively flat surface. The vacuum either is obtained from a vacuum pump or, in most cases, is generated from compressed air blowing across a venturi. Using compressed air is the common and least expensive approach. Many commercial sources of these devices are available.

11.4.3 Hydraulic Fluid Power

In heavy lifting operations, hydraulic fluid is used to obtain higher pressures and resulting higher gripping forces than could be attained with compressed air. Many gripper devices are available as either pneumatic or hydraulic power equipped.

11.4.4 Electrical Power

Electrical devices can be heavy (electromagnets, DC motors, solenoids), limiting their use in end effectors of lighter payload robot arms. Many new end effectors are commercially available with electric motor actuation. Typically DC servomotors are used with electric power application. To reduce payload weight, lightweight and high strength cables can be used to connect the gripper mechanism to the drive motor remote from the wrist. For limited gripping force applications, stepper motor actuation is available.

For lifting of ferrous materials, electomagnets are potential end effector alternatives. This is a relatively simple design but has a high weight. Positional accuracy is also limited.

11.4.5 Other Actuators

While most industrial applications can adequately power end effectors with standard actuators, special applications and university research are exploring other novel power devices. Among the latest actuators are piezoelectrics, magnetostrictive materials, and shape memory alloys.

For high precision with nanoscale displacements, piezoelectric drives are of primary use. These high stiffness actuators change shape with a charge applied. Magnetostrictive actuators provide a similar scale and precision movement. Magnetostrictive materials, such as Terfenol-D, displace when a magnetic field is applied. Both types of actuators are commercially available.

A promising device under current study is shape memory metal alloys (Yang and Gu [6]). Controlled electrical heating is used to change shape between crystalline phases of alloys such as nickel-titanium. These actuator designs have a weight-to-strength ratio advantage over other alternative actuator strategies but are not commercially available.

11.5 Gripper Kinematics

Grippers are a basic element of end effectors. The geometric motion and kinematics of the end effector gripper are of great importance in the design of the gripper attachments, motion, object spacing, and auxiliary tooling. The design of end effectors or simple grippers is infinite. However, there are several simple functional kinematic designs widely used as gripper mechanisms. The majority of these designs have planar motion.

11.5.1 Parallel Axis/Linear Motion Jaws

The most common gripper design is a parallel axis design where two opposing parallel jaws (or fingers) move either toward or away from each other (Figure 11.9). Linear motion slides or slots are used to keep both gripper attachments parallel and collinear. This form of gripper is readily designed using a pneumatic cylinder as power for each jaw. For successful object centered grasping, it is important that the mechanism constrains the jaw movements to be synchronized. Otherwise the object may be pushed out of the grasp by

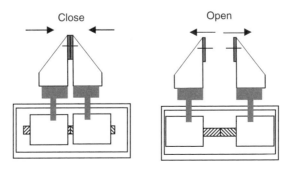

FIGURE 11.9 Parallel axes/linear jaws.

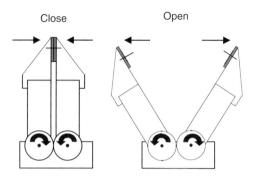

FIGURE 11.10 Rotating axes/pivoting jaws.

one jaw. An advantage of this style of gripper is that the center of the jaws does not move perpendicular to the axis of motion. Thus, once the gripper is centered on the object, it remains centered while the jaws close.

11.5.2 Pivoting/Rotary Action Jaws

Another basic gripper design is the pivoting or rotating jaws mechanism of Figure 11.10. Again this design is well suited to pneumatic cylinder power (Figure 11.11). Symmetric links to a center air cylinder or a rack with two pinions or two links may be used. While this design is simple and only requires one power source for activation, it has several disadvantages including jaws that are not parallel and a changing center of grasp while closing.

11.5.3 Four-Bar Linkage Jaws

A common integrated gripper design is a four-bar linkage, as previously depicted in Figure 11.3. Each jaw is a four-bar linkage that maintains the opposing jaws parallel while closing. A disadvantage of this design is the changing center of grasp while closing the jaws. Moving the robot end effector while closing will compensate for this problem. This requires more complicated and coordinated motions. A cable-linked mechanism allows one source of power to supply motion to both jaws. An implementation of this design is the Microbot Alpha II.

11.5.4 Multiple Jaw/Chuck Style

When objects are rod-like and can be grasped on an end, a chuck-style gripper with multiple jaws (typically three or four) can be used. Gripper jaws operated similarly to a machine tool multi-jaw chuck as seen in

FIGURE 11.11 Rotating axes pneumatic gripper.

Design of Robotic End Effectors 11-11

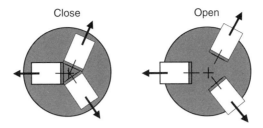

FIGURE 11.12 Multi-jaw chuck axes.

Figure 11.12. These devices are more complicated and heavier, but they can provide a significantly stronger grip on a rod-like object. Drawbacks to these multi-jaw designs are heavier weight and limited application.

11.5.5 Articulating Fingers

The current state-of-the-art design in robotic end effectors is the articulating fingered hand. The goal of these designs is to mimic human grasping and dexterous manipulation. These types of end effectors allow three-dimensional grasping. A minimum of three fingers is required to grasp in this manner. A distinct advantage of this design is the potential to grasp irregular and unknown objects [7]. These anthropomorphic end effectors require force sensing and significant computer control to operate effectively. Currently, the application of articulating fingers as end effector is limited. There are no commercially available end effectors of this type.

However, these designs remain an active university research area. Recent research on articulating finger grasping includes finger-tip force control mechanical drives mimicking tendons with Utah/MIT Dexterous Hand Project. Universities are exploring many new avenues of these end effectors [8–14].

11.5.6 Multi-Component End Effectors

In some industrial situations, it is desirable to have more than one type of grasping device on an end effector. A typical situation is in sheet metal fabrication. Thin sheets of sheet metal must be lifted from stacks or conveyors. Vacuum suction cups are used to lift and relocate the flat sheets, while mechanical grippers then provide the grasping for bending operations (Figure 11.19).

11.6 Grasping Modes, Forces, and Stability

11.6.1 Grasping Stability

Robotic grasping has been the topic of numerous research efforts. This is best summarized by Bicchi and Kumar [15]. Stable object grasping is the primary goal of gripper design. The most secure grasp is to enclose the gripper jaws or finger around the center of gravity of the object.

Six basic grasping patterns for human hands have been identified by Taylor and Schwarz [16] in the study of artificial limbs. Six grasping forms of (1) spherical, (2) cylindrical, (3) hook, (4) lateral, (5) palmar, and (6) tip are identified in Figure 11.13. For conventional two-dimensional grippers described in the previous sections, only the cylindrical, hook, and lateral grips apply. To securely grasp an object the cylindrical and lateral grips are effective for plane motion grippers. The two-jaw mechanisms of most end effectors/grippers most closely approximate the cylindrical grasp.

Kaneko [9] also discusses stability of grasps for articulating multi-fingered robots. Envelop or spherical grasping is the most robust, as it has a greater number of contact points for an object that is enclosed by articulating fingers.

For plane-motion grippers using a cylindrical style of grasp, stability can be similarly defined. Stability increases as the object has more points of contact on the gripper jaws and when the object's center of

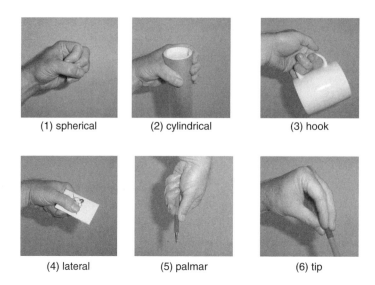

FIGURE 11.13 Grasp types for human hands.

gravity is most closely centered within the grasp. When using a cylindrical grip, it is desirable to grasp an object in a safe, non-slip manner if possible. For example, when grasping a cylindrical object with a flange, the jaws of the gripper should be just below the flange. This allows the flange to contact the top of the gripper jaws, to remove any possibility of slipping. If a vertical contact feature is not available, then a frictional grip must be used.

11.6.2 Friction and Grasping Forces

While a slip-proof grasp previously described is preferred for end effectors, in the majority of cases a frictional grip is all that can be attained. When designed properly, the frictional grip is very successful. Friction forces can be visualized on a gripper jaw in Figure 11.14. It is desirable to have the center of gravity for grasped objects and end effector coincident in the Z-direction, to not have moments at the jaw surfaces. To successfully grasp an object the applied gripping frictional force on each jaw of a two-jaw gripper must be equal to or greater than the half the vertical weight and acceleration payload, as defined by Equation (11.6). From this the applied gripping normal force is found by dividing the required friction force by the static coefficient of friction. Typically a factor of safety is applied [17].

$$F_{friction} = \mu F_{grip} \geq \frac{(1 + A_{vert}/g)w}{2} \qquad (11.6)$$

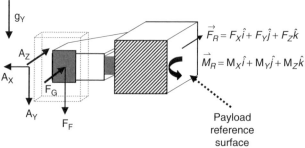

FIGURE 11.14 Gripper forces and moments.

Friction coefficients are a function of materials and surface geometries. Estimates can be found using standard references. Typically most surfaces will have a static coefficient of friction greater than 0.2. For metal to metal contacts, the static coefficient of friction is much higher (e.g., aluminum to mild steel has 0.6 and mild steel to hard steel has 0.78) [18].

In addition to an object's weight, surface texture, rigidity, and potential damage must also be considered in the selection or design of an end effector or the gripper. Pads are used on the jaws of the end effector to prevent surface damage to the object. Pads can also be used to increase the coefficient of friction between the object and the gripper jaws.

11.7 Design Guidelines for Grippers and Jaws

11.7.1 Gripper and Jaw Design Geometry

Gripper jaw design is best done graphically using computer aided design and drafting tools. The footprint of the gripper must be examined to determine how to approach an object and grasp it. Typically footprint is thought of in terms of only the vertical projection of the gripper. However, grasping is a three-dimensional issue, especially in a dense object environment. Gripper motion, either opening or closing, should be free of interferences from an adjacent object or a geometric feature of the grasped object. Smaller jaw footprints are more desirable for denser spacing of objects.

For example, a gripper designed to grasp a flanged object may require additional opening of the gripper as the end effector passes over the flange of the object. This may push neighboring parts out of the way of the object. Spacing of objects may need to be adjusted if possible.

Much has been said already on design considerations for end effectors. Causey and Quinn [19] have provided an excellent reference for design guidelines for grippers in modular manufacturing. Others [20] have also contributed to important design parameters for end effectors. Application of these general design considerations should result in a more robust and successful gripper design. A summary of general design guidelines for end effectors grippers follows.

11.7.2 Gripper Design Procedure

1. Determine objects, spacing, orientation, and weights for grasping. While objects are predetermined, based on the task to be accomplished, spacing and orientation may be adjustable. Note: All the grasping surfaces, support ledges, and other graspable features of each object must be identified.
2. Determine roughly how each object will be grasped (i.e., from top, from side, gripper opening), based on part geometry and presentation. If part presentation (orientation and spacing) is variable, choose the most stable grasp orientation, with the widest spacing around the object.
3. Calculate end effector payload reactions; estimate accelerations on controller information if none are specified. Establish necessary and safe grasping forces not to damage the object. Account for friction in frictional grasps.
4. Make preliminary selection of the gripper device based on weight and system design parameters.
5. Determine gripper maximum variable or fixed stroke. This may be a preliminary number if a variety of strokes is available. A trade-off between stroke and the physical dimensions and weight exists, so the minimum acceptable stroke is generally the best.
6. Design grippers to allow grasping of all objects. Objects must be grasped securely. Multiple grippers may be required to an individual end effector.
7. The grasp center of each gripper mechanism should be at the object center of gravity whenever possible. Also, keep the jaw lengths as short as possible and the grasp center close to robot wrist to minimize moment loading and to maximize stiffness.
8. Jaws must work reliably with the specific parts to being grasped and moved. The jaws should be self-centering to align parts, if desired. Ideally the jaws should conform to the surface of the held part. At a minimum, three-point or four-point contact is required. V-groove styles are often used

to accomplish this feature. It is also good practice to have chamfered edges on grippers to more closely match the part geometry and allow for greater clearance when grasping.

9. Determine the footprint of objects and presentation spacing with the gripper jaw design. Horizontal and vertical footprints must be checked with grippers open and closed so that there is no interference with other parts or features of the part being grasped. Minimize footprint of the gripper jaws.
10. Revise payload reaction calculations. To minimize gripper weight to reduce the payload, lightening holes are often added.
11. Gripper stiffness and strength must also be evaluated prior to completing the design. Note: Low end effector stiffness can cause positioning errors.
12. Iteration on the jaw design and part presentation may be necessary to converge on a good design.

11.7.3 Gripper Design: Case Study

The selection and design of an end effector for a small flexible automation system is worked through as an example. The objects to be grasped are cylindrical and have diameters of 6, 50, and 90 mm, with masses of 0.20 kg, 0.100 kg, and 0.350 kg, respectively. Surfaces of the objects are glass and the end effector was specified as aluminum. The objects must be presented on end, and the 90 mm object has a 100 mm top circular flange. The robot selected for the application has a payload capacity at the arm end of 3 kg, a maximum allowable attached inertia of 0.03 kg · m^2, and a maximum moment of 2.94 N · m [21].

For ease in grasp centering, a parallel gripper mechanism is selected for the design. While gripper jaw designs are infinite, two basic jaw designs evolved using footprints of the grasped objects: a single gripper jaw using a 70 mm minimum stroke or dual-jaw design with a 25 mm minimum stroke. Figure 11.15 and Figure 11.16 show the vertical footprints of jaw designs with the objects. Note the chamfering of the jaws to conform to the cylinders. Also notice the use of V-grooves for grasping the small cylinder. The dual jaw design requires a wrist rotation of 180° when alternating between object grasping and limits entry and exit ease of grasped objects. The longer stroke single jaw design has a higher payload moment. In this case the longer stroke is preferred for simplicity if the allowable payload is not exceeded. A three-dimensional rendering (Figure 11.17) shows an offset is needed near the gripper attachment point to prevent gripper interference while grasping under the object's flange. For strength and weight considerations, aluminum (e.g., 6061T6, 6063T6, or 5052H32) would be a good candidate material.

A standard industrial pneumatic gripper with parallel motion is commercially available with a total stroke of 90 mm. The cylinder bore is 14 mm. The total gripper mass (m_g) is 0.67 kg, without jaws. The gripper body center of gravity (X_g) is 17 mm from the attachment side. Each of 3 mm thick aluminum jaws weighs 40 g. The mass of the jaws will be lumped with the object as a conservative approximation. The

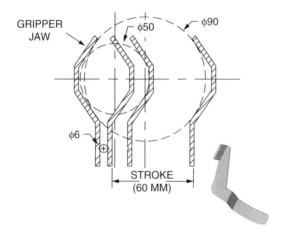

FIGURE 11.15 Single jaw gripper design.

Design of Robotic End Effectors 11-15

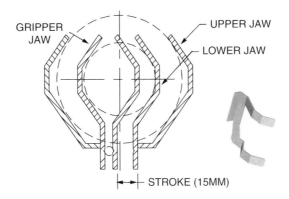

FIGURE 11.16 Multi-jaw gripper design.

total weight of the heaviest object plus jaws (m_o) is 0.43 kg, at location (X_o) of 96 mm from the end of arm attachment (refer to Figure 11.6 and Figure 11.7). The reaction forces and moments can be determined for robot limits load and deflection analysis using Equation (11.3) and Equation (11.5), after establishing acceleration limits.

Using 1g accelerations in all directions plus y-axis gravitational acceleration, reactions are evaluated. The maximum resulting magnitudes of force and moment are 26.4 N (2.69 kg equivalent static mass) and 1.34 N · m. Both are within the specified arm limitations of 3 kg and 2.94 N · m. A finite element analysis indicates the maximum loading; maximum stresses are less than 4% of material elastic yield. Deflections are under 0.010 mm for this loading, indicating a sufficient stiffness.

Frictional considerations may require jaw pads as the objects grasped are glass, and the gripper jaws will be anodized aluminum. Coefficient of friction is estimated at 0.20 between glass and various metals, while it is approximately 0.9 between rubber and glass. A vector magnitude of a 2 g vertical acceleration (including gravity) and 1 g horizontal acceleration on the 0.35 kg payload will cause a frictional force of slightly less than 3 N per gripper jaw, using Equation (11.6). With a 0.2 coefficient of friction and a factor of safety of 2, a minimum gripper pressure of 2 bar is required. Since this pressure is less than the readily available supply of air pressure of 6 bar, friction pads on the gripper are not required. The end effector with gripper jaw mechanical design is complete.

11.7.4 Gripper Jaw Design Algorithms

As indicated by the previous, most gripper jaw designs are ad hoc and follow rules of thumb. There is current research focusing on optimizing jaw features. A common problem is the orienting and fixturing of grasped parts. Typically parts are oriented for the end effector to grasp. Some part alignment can be accomplished for simple shapes such as cylinders and cubes.

FIGURE 11.17 Jaw with grasped object.

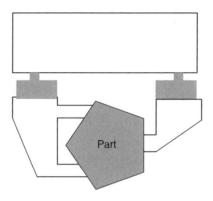

FIGURE 11.18 Part orienting gripper design.

Special algorithms have been developed for use with parallel axis grippers (the most common) to design trapezoid jaw shapes that will orient multi-sided parts for assembly (Zhang and Goldberg [22]). It has been shown that the jaw shapes need not be solid or continuous. In fact smaller contact areas are of value to allow parts to topple, slide, and rotate into alignment while being grasped. Figure 11.18 shows a simple example of this technique to orient parts. A limitation of this design procedure is that the gripper is optimized for one particular part geometry.

11.7.5 Interchangeable End Effectors

In many situations it is not feasible for a single end-effector or gripper to grasp all the objects necessary. An option is the use of interchangeable end effectors. While this will simplify the task performance, additional design constraints are placed on the system.

Key design considerations in developing interchangeable end effectors are attachment means, registration of connection points from the robot to the end effector, ease of change, and security of the tool connection. The system base is a coupling plate attached to the end of the arm and mating tool plates integrated into tools/end effectors. The most important feature of a design is the secure and fail-safe mounting of the end effector [23].

The additional end effectors require locating fixtures that occupy some of the robot work envelope. Aside from increased cost, the interchangeable end effectors also reduce repeatability, graspable payload, and range of movements. The flexibility of the additional end effectors can require significant hardware, engineering and programming costs. However, this option is significantly less expensive than purchasing an additional robot. The decision to use interchangeable end effectors must be carefully weighed.

If the decision to use interchangeable end effectors is made, it is important to note that systems are commercially available for robots.

11.7.6 Special Purpose End Effectors/Complementary Tools

Complementary tools can augment the usability of end effectors. These tools can assist in the dispensing or transfer of liquids or in the pick up of objects not easily grasped by fixed grippers. Tools can ease the constraints on end effectors. The shape of the tool handle will have a grasping center similar to the other grasped objects. The tools add function to the effector. Each tool will require a storing and locating rack. Festoon cable systems and constant force reel springs on overhead gantries are sometimes necessary to reduce weight and forces of the tool and attachments to not exceed the robot payload limits.

Tools will depend on the task to be performed. Some of the more common special purpose tools are liquid transfer tools including pipettes and dispensers, vacuum pick-ups, and electromagnets. Examples of vacuum and pipette tools are shown with their location rack in Figure 11.19. These tool elements can be off-the-shelf items with custom grips, such as the vacuum pick-up tool.

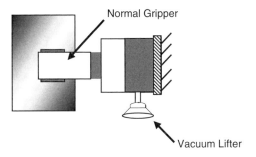

FIGURE 11.19 Multi-tool end effector.

11.8 Sensors and Control Considerations

A variety of sensors is available for inclusion in end effectors. The most common sensors employed in end effectors measure proximity, collision, and force.

11.8.1 Proximity Sensors

Through beam and reflective optical proximity sensors are widely used to verify the grasping of an object. Typically the sensors are used as presence indicators. The weight of the sensor on the end effector should be minimized because of payload considerations. Most optical proximity sensors are available with an optical fiber end connection so that sensor electronics will not necessarily be located on the end effector. Place as much sensor hardware as possible remote to the end effector. Grippers may be designed with built-in limit switches for grasping fixed dimensional objects. A common implementation of this is the use of Hall effect or reed switches with air cylinder pistons.

Inductive and capacitive proximity sensors may be used with compatible object materials such as carbon steels. Materials such as stainless steel, aluminum, brass, and copper may also be sensed but at a reduced detection distance.

If a non-contact distance measurement is required rather than a presence verification, laser diode-based analog sensors or a machine vision sensor can be used. For extremely close-range and precise measurements (such as part dimensional measurements) eddy current sensors may be used on metallic components.

11.8.2 Collision Sensors

Although compliance centers (discussed earlier) help align parts, ease assembly, and reduce potential damage, damage may still occur because of design limits (angle and position tolerances, as specified by manufacturers or design). Thus, a collision detection sensor can be a good end effector component to mitigate damage to the robot, inserted part, or workpiece where the part is being attached. Collision detection sensors detect the first onset of a collision in all directions. After detecting the start of a collision, the robot can be stopped or backed away before damage has occurred. (Note that while proximity sensors can also be used as collision avoidance devices, most are unidirectional with a limited field of view.) Collision detection devices are usually attached at the end of arm before the end effector.

There are several types of collision detection sensors available for wrist mounting. These include force sensors as an active control device. Force sensors are primarily strain gauge devices and require additional programming for monitoring and control.

Specific collision detection sensors are readily available in the market place. These are passive devices. Some utilize compressed air for variability of fault loading in three-dimensional space. Other collision sensor designs are based on mechanical springs at preset loads. Both types indicate a collision by the opening of a normally closed switch. This design can be hardwired into the E-stop circuit for safety or be wired to a digital input for programmatic monitoring.

11.8.3 Tactile Feedback/Force Sensing

Force sensing during grasping is important for successful grasping of varying parts, compliant part, and for reduction of part deformation. In the past strain gauge sensors were the primary device used. These, however, are of limited use in the multiple-point force sensing. Piezoelectric force sensors are also available but are better for measuring dynamic loads because of charge leakage.

Piezoresistor force sensors can be used for tactile force sensing using the principle that the resistance of silicon-implanted piezoresistors will increase when the resistors flex under any applied force. The sensor concentrates the applied force through a stainless steel plunger directly onto the silicon sensing element. The amount of resistance changes in proportion to the amount of force being applied. This change in circuit resistance results in a corresponding mV output level.

A relatively new development is the commercial availability of force sensing resistors (FSR). The FSR is a polymer thick film (0.20 to 1.25 mm PTF) device which produces a decrease in resistance with any increase in force applied to its surface. Although FSRs have properties similar to load cells, they are not well suited for precision measurement equipment (variation of force from grasping can be on the order of 10%). Its force sensitivity is optimized for use in human touch control switches of electronic devices with pressure sensitivity in the range of 0.5 to 150 psi (0.03 to 10 kg/cm^2) [24].

In general, the FSR's resistance vs. force curve approximately follows a power-law relationship (resistance decreasing with force). At the low force end of the force-resistance characteristic, a step response is evident. At this threshold force, the resistance drops substantially. At the high force end of the sensor range, the resistance deviates from the power-law behavior and ultimately levels to an output voltage saturation point. Between these extremes is a limited but usable region of a fairly linear response. The relatively low cost, small thickness, and reasonable accuracy make these sensors attractive for multi-sensor tactile control.

11.8.4 Acceleration Control for Payload Limits

Another important control aspect is the force caused during acceleration. Thus, the accelerations of the joints must be known and prescribed (or carefully estimated). In most instances the acceleration limits on each joint are known. Typically accelerations are defined by S-curves for velocity, with maximum accelerations limits.

11.8.5 Tactile Force Control

Traditional position and velocity control for trajectories are obviously important for painting, welding, and deburing robots. However, to create quality products, it is also necessary to have force control for material removal operations such as grinding and deburing. In these situation the compliance of the robot, end effector, and tool make up total compliance (flexibility). When a tool is moved against an object, the inverse of this compliance (stiffness) creates a force. To assure that dimensional accuracy is maintained, an additional position measurement of the tool tip must be made or position must be taken into account by the use of a compliance model. The stiffness of each joint is easily determined. However, each unique position of the joints creates a unique stiffness.

Generally fixed-gain controllers (such as PID-type controllers) are preferred for force applications because of their simplicity in implementation. These controllers are easily tuned using standard approaches.

Tuning the force control loop is best done using position control and the effective stiffness from the end effector-workpiece interaction. Many industrial controllers provide utilities for automatic tuning of position loops. Many are based on the Ziegler-Nichols PID tuning [25] or other heuristic techniques. Although this technique is based on continuous systems for very fast sampling rates (greater than 20 times the desire bandwidth), the results also apply well to discrete systems.

11.9 Conclusion

From the preceding text it is clear that the design, selection, control, and successful implementation of a robotic system relies heavily on the end effector subsystem. End effector designs and technology continue to evolve with new actuators, sensors, and devices.

References

[1] Ceglarek, D., Li, H.F., and Tang, Y., Modeling and optimization of end effector layout for handling compliant sheet metal parts, *J. Manuf. Sci. Eng.,* 123, 473, 2001.
[2] Ciblak, N., Analysis of Cartesian stiffness and compliance with applications, Ph.D. Thesis Defense, Georgia Institute of Technology, 1998.
[3] Robotic Accessories, Alignment Device 1718 Specifications, 2003.
[4] Thermo CRS, A465 Six Axis Robot Specification, 2003.
[5] Kane, T.R., *Dynamics: Theory and Applications,* McGraw-Hill, New York, 1985.
[6] Yang, K. and Gu, C.L., A novel robot hand with embedded shape memory alloy actuators, *J. Mech. Eng. Sci.,* 216, 737, 2002.
[7] Francois, C., Ikeuchi, K., and Herbert, M., A three finger gripper for manipulation in unstructured Environments, *IEEE Int. Conf. Robot. Autom.,* 3, 2261–2265, 1991.
[8] Foster, A., Akin, D., and Carignan, C., Development of a four-fingered dexterous robot end-effector for space operations, *IEEE Int. Conf. Robot. Autom.,* 3, 2302–2308, 2002.
[9] Kaneko, M. et al., Grasp and manipulation for multiple objects, *Adv. Robot,* 13, 353, 1999.
[10] Kumazaki, K. et al., A study of the stable grasping by a redundant multi-fingered robot hand, *SICE,* 631, 2002.
[11] Mason, M. and Salisbury, J., *Robotic Hands and the Mechanics of Manipulation,* MIT Press, Boston, 1985.
[12] Seguna, C.M., The design, construction and testing of a dexterous robotic end effector, *IEEE SPC,* 2001.
[13] Bicchi, A., Hands for dexterous manipulation and robust grasping: a difficult road toward simplicity, *IEEE Trans. Robot. Autom.,*16, 652, 2000.
[14] Caldwell, D.G. and Tsagarakis, N., Soft grasping using a dexterous hand, *Indust. Robot,* 3, 194, 2000.
[15] Bicchi, A. and Kumar, V., Robotic grasping and contact: a review, *IEEE Int. Conf. Robot. Autom.,* 200, 348, 2000.
[16] Taylor, C.L. and Schwarz, R.J., *The Anatomy and Mechanics of the Human Hand: Artificial Limbs,* vol. 2, 22–35, 1955.
[17] Zajac, T., Robotic gripper sizing: the science, technology and lore, *ZAYTRAN,* 2003.
[18] Baumeister, T. (ed.), *Marks' Standard Handbook for Mechanical Engineers,* 8th ed., McGraw-Hill, New York, 6–24, 1978.
[19] Causey, G.C. and Quinn, R.R., Gripper design guidelines for modular manufacturing, *IEEE Int. Conf. Robot. Autom.,* 1453, 1998.
[20] Walsh, S., Gripper design: guidelines for effective results, *Manuf. Eng.,* 93, 53, 1984.
[21] Adept Technologies, AdeptSix 300 Specifications Data Sheet, 2003.
[22] Zhang, T. and Goldberg, K., Design of robot gripper jaws based on trapezoidal modules, *IEEE Int. Conf. Robot. Autom.,* 29, 354, 2001.
[23] Derby, S. and McFadden, J., A high precision robotic docking end effector: the dockbot(TM), *Ind. Robot,* 29, 354, 2002.
[24] Interlink Electronics, FSR Data sheet, 2003.
[25] Franklin, G.F., Powell, J.D., and Workman, M.L., *Digital Control of Dynamic Systems,* Addison-Wesley, Reading, MA, 1990.

12
Sensors and Actuators

Jeanne Sullivan Falcon
National Instruments

12.1 Encoders **12**-1
   Rotary and Linear Incremental Encoders • Tachometer
   • Quadrature Encoders • Absolute Encoders
12.2 Analog Sensors **12**-4
   Analog Displacement Sensors • Strain • Force and Torque
   • Acceleration
12.3 Digital Sensors **12**-10
   Switches as Digital Sensors • Noncontact Digital Sensors
   • Solid State Output • Common Uses for Digital Sensors
12.4 Vision **12**-12
12.5 Actuators **12**-12
   Electromagnetic Actuators • Fluid Power Actuators

12.1 Encoders

12.1.1 Rotary and Linear Incremental Encoders

Incremental encoders are the most common feedback devices for robotic systems. They typically output digital pulses at TTL levels. Rotary encoders are used to measure the angular position and direction of a motor or mechanical drive shaft. Linear encoders measure linear position and direction. They are often used in linear stages or in linear motors. In addition to position and direction of motion, velocity can also be derived from either rotary or linear encoder signals.

In a rotary incremental encoder, a glass or metal disk is attached to a motor or mechanical drive shaft. The disk has a pattern of opaque and transparent sectors known as a code track. A light source is placed on one side of the disk and a photodetector is placed on the other side. As the disk rotates with the motor shaft, the code track interrupts the light emitted onto the photodetector, generating a digital signal output (Figure 12.1).

The number of opaque/transparent sector pairs, also known as line pairs, on the code track corresponds to the number of cycles the encoder will output per revolution. The number of cycles per revolution (CPR) defines the base resolution of the encoder.

12.1.2 Tachometer

An incremental encoder with a single photodetector is known as a tachometer encoder. The frequency of the output pulses is used to indicate the rotational speed of the shaft. However, the output of the single-channel encoder cannot give any indication of direction.

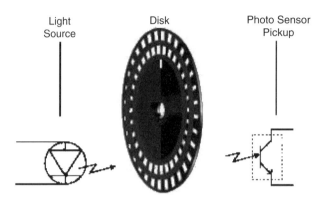

FIGURE 12.1 Typical encoder design.

12.1.3 Quadrature Encoders

Quadrature encoders are another type of incremental encoder. A two-channel quadrature encoder uses two photodetectors to sense both position and direction. The photodetectors are offset from each other by 90° relative to one line pair on the code track. Since the two output signals, A and B, are 90° out of phase, one signal will lead the other as the disk rotates. If A leads B, the disk is rotating in a clockwise direction, as shown in Figure 12.2. If B leads A, the disk is rotating in a counterclockwise direction, as shown in Figure 12.3.

Figure 12.2 and Figure 12.3 illustrate four separate pulse edges occurring during each cycle. A separate encoder count is generated with each rising and falling edge, which effectively quadruples the encoder resolution such that a 500 CPR (cycles per revolution) encoder provides 2000 counts per revolution with quadrature decoding.

The electrical complements of channels A and B may be included as differential signals to improve noise immunity. This is especially important in applications featuring long cable lengths between the encoder and the motion controller or electrical drive.

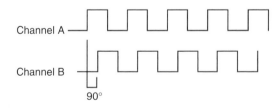

FIGURE 12.2 Clockwise motion.

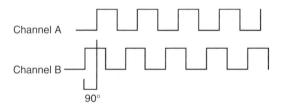

FIGURE 12.3 Counterclockwise motion.

Sensors and Actuators 12-3

Some quadrature encoders include a third output channel, known as an index or zero pulse. This signal supplies a single pulse per revolution and is used for referencing the position of the system. During power up of the system, the motor can be rotated until the index pulse occurs. This specifies the current position of the motor in relation to the revolution.

Moving to the index position is not enough to determine the position of the motor or mechanical system during system startup. The absolute position is usually determined through a homing routine where the system is moved to limit switches or sensors. Once the robot is commanded to move to the limits, then the encoder readings can be reset to zero or some other position value to define the absolute position. Subsequent motion is measured by the encoders to be relative to this absolute position.

Encoders can be attached to components other than motors throughout a mechanical system for measurement and control. For example, a rotary motor may be attached to a belt drive that is in turn attached to a payload under test. A rotary encoder is attached to the motor to provide feedback for control, but a second rotary encoder can also be attached to the payload to return additional feedback for improved positioning. This technique is known as dual-loop feedback control and can reduce the effects of backlash in the mechanical components of the motion system.

Some encoders output analog sine and cosine signals instead of digital pulses. These types of encoders are typically used in very high precision applications that require positioning accuracy at the submicron level. In this case, an interpolation module is necessary between the encoder output and the robot controller or intelligent drive. This functionality may be included in the robot controller. Interpolation modules increase the resolution of the encoder by an integer value and provide a digital quadrature output for use by the robot controller. Some encoder manufacturers offer interpolation modules in a wide range of multiplier values. Manufacturers of encoders include BEI, Renco, US Digital, Renishaw, Heidenhain, and Micro-E.

12.1.4 Absolute Encoders

A position can be read from an absolute encoder if the application requires knowledge of the position of the motors immediately upon system start-up. An absolute encoder is similar to an incremental encoder, except that the disk used has multiple concentric code tracks and a separate photodetector is used with each code track. The number of code tracks is equivalent to the binary resolution of the encoder, as shown in Figure 12.4.

An 8-bit absolute encoder has eight code tracks. The 8-bit output is read to form an 8-bit word indicating absolute position. While absolute encoders are available in a wide variety of resolutions, 8-, 10-, and 12-bit binary are the most common. Due to their complexity, absolute encoders are typically more expensive than quadrature encoders. Absolute encoders may output position in either parallel or serial format.

Because of the variety of output formats available for absolute encoders, it is important to ensure that the robot controller or intelligent drive is compatible with the particular model of the absolute encoder

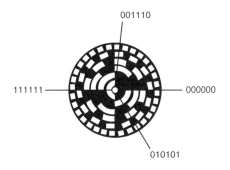

FIGURE 12.4 Absolute encoder example.

12.2 Analog Sensors

Analog sensors commonly used in robotic applications include displacement, force, torque, acceleration, and strain sensors. As in the case of encoders, these sensors may be used in either an open-loop or closed-loop fashion within the robot system. For example, a force sensor may be used to measure the weight of objects being assembled for quality control. Or, a force sensor may be added to the gripper in a robot end effector as feedback to the gripper actuator. The gripper control system would allow objects to be held with a constant force.

12.2.1 Analog Displacement Sensors

These sensors can include both angular and translation position measurement relative to a reference position. They provide a continuously varying output signal which is proportional to the position of the sensed object. The most common sensors and technologies used for displacement measurement include:

- Potentiometer
- LVDT — contact sensor
- Resolvers
- Inductive
- Capacitive
- Optical
- Ultrasonic
- Hall effect

12.2.1.1 Potentiometers

Potentiometers offer a low cost method of contact displacement measurement. Depending upon their design, they may be used to measure either rotary or linear motion. In either case, a movable slide or wiper is in contact with a resistive material or wire winding. The slide is attached to the target object in motion. A DC or an AC voltage is applied to the resistive material. When the slide moves relative to the material, the output voltage varies linearly with the total resistance included within the span of the slide. An advantage of potentiometers is that they can be used in applications with a large travel requirement. Some industrial potentiometers are offered in sealed configurations that can offer protection from environmental contamination.

12.2.1.2 Linear Variable Differential Transformers (LVDT)

Linear variable differential transformers are used for contact displacement measurement. They include a moving core which extends into the center of a hollow tube as shown in Figure 12.5. One primary and two secondary coils are wrapped around the hollow tube with the secondary coils symmetrically placed around the primary. The core is attached to the target object in motion.

An AC signal is applied to the primary coil and, because of coupling of magnetic flux between the coils, this induces AC voltages in the secondary coils. The magnitude and phase of the output signal of the LVDT

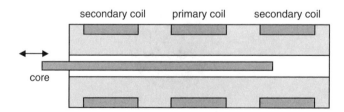

FIGURE 12.5 Linear variable differential transformer.

is determined by the position of the core within the tube. The magnitude of the output of the signal is a function of the distance of the core from the primary coil, and the phase of the output signal is a function of the direction of the core from the primary coil — towards one secondary coil or the other as shown in the figure.

LVDT sensors can be used in applications with large travel requirements. However, mechanical alignment along the direction of travel is important for this type of sensor.

12.2.1.3 Resolvers

A resolver is essentially a rotary transformer that can provide absolute position information over one revolution. The resolver consists of a primary winding located on the rotor shaft and a two secondary windings located on the stator. The secondary windings are oriented 90° relative to each other. Energy is applied to the primary winding on the rotor. As the rotor moves, the output energy of the secondary windings varies sinusoidally. Resolvers are an alternative to encoders for joint feedback in robotic applications.

12.2.1.4 Inductive (Eddy Current)

Inductive sensors are noncontact sensors and can sense the displacement of metallic (both ferrous and nonferrous) targets. The most common type of inductive sensor used in robotics is the eddy current sensor. The sensor typically consists of two coils of wire: an active coil and a balance coil, with both driven with a high frequency alternating current. When a metallic target is placed near the active sensor coil, the magnetic field from the active coil induces eddy currents in the target material. The eddy currents are closed loops of current and thus create their own magnetic field. This field causes the impedance of the active coil to change. The active coil and balance coil are both included in a bridge circuit. The impedance change of the active coil can be detected by measuring the imbalance in the bridge circuit. Thus, the output of the sensor is dependent upon the displacement of the target relative to the face of the sensor coil.

The effective depth of the eddy currents in the target material, δ, is given by

$$\delta = \frac{1}{\sqrt{\pi f \mu \sigma}}$$

where f is the excitation frequency of the coil, μ is the magnetic permeability of the target material, and σ is the conductivity of the target material. In order to ensure adequate measurement, the target material must be three times as thick as the effective depth of the eddy currents.

In general, the linear measurement range for inductive sensors is approximately 30% of the sensor active coil diameter. The target area must be at least as large as the surface area of the sensor probe. It is possible to use curved targets, but their diameter should be three to four times the diameter of the sensor probe. In addition, the sensor signal is weaker for ferrous target materials compared with nonferrous target materials. This can lead to a reduced measurement range and so this should be investigated with the sensor manufacturer.

Inductive sensors can sense through nonmetallic objects or nonmetallic contamination. However, if measurement of a nonmetallic target displacement is required, then a segment of metallic material must be attached to the target.

12.2.1.5 Capacitive

Capacitive displacement sensors are another type of noncontact sensor and are capable of directly sensing both metallic and nonmetallic targets. These sensors are designed using parallel plate capacitors. The capacitance is given by

$$C = \frac{\varepsilon_r \varepsilon_0 A}{d}$$

where ε_r is the relative permittivity (dielectric constant) of the material between the plates, ε_0 is the absolute permittivity of free space, A is the overlapping area between the plates, and d is the distance between the plates.

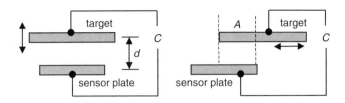

FIGURE 12.6 Distance and area variation in capacitive sensor measurement.

In displacement sensor designs, a capacitive sensor typically incorporates one of the capacitor plates within its housing and the target forms the other plate of the capacitor. The sensor then operates on the principle that the measured capacitance is affected by variation in the distance d or the overlapping area A between the plates. The capacitance is measured by detecting changes in an oscillatory circuit that includes the capacitor. Figure 12.6 shows the distance and area variation methods for capacitive displacement measurement.

For detection of nonmetallic targets, a stationary metallic reference is used as the external capacitor plate. The presence of the nonmetallic target in the gap between the sensor and the stationary metallic reference will change the permittivity and thus affect the measured capacitance. The capacitance will be determined by the thickness and location of the nonmetallic target. Figure 12.7 shows the dielectric variation approach for capacitive displacement measurement.

In general, the linear measurement range for capacitive sensors is approximately 25% of the sensor diameter. The target should be 30% larger than the sensor diameter for optimum performance. In addition, environmental contamination can change the dielectric constant between the sensor and the target, thus reducing measurement accuracy.

12.2.1.6 Optical Sensors

Optical sensors provide another means of noncontact displacement measurement. There are several types which are commonly used in robotics: optical triangulation, optical time-of-flight, and photoelectric.

12.2.1.6.1 Optical Triangulation

Optical triangulation sensors use a light emitter, either a laser or an LED, in combination with a light receiver to sense the position of objects. Both the emitter and receiver are contained in the same housing as shown in Figure 12.8. The emitter directs light waves toward a target. These are reflected off the target, through a lens, to the receiver. The location of the incident light on the receiver is used to determine the position of the target in relation to the sensor face. The type of receiver used may be a position sensitive detector (PSD) or a pixelized array device such as a charge coupled device (CCD). The PSD receiver generates a single analog output and has a faster response time than the output pixelized array device because less post-processing is required. It is also typically smaller so that the overall sensor size will be smaller. Pixelized array devices, however, are useful when the surface of the target is irregular or transparent.

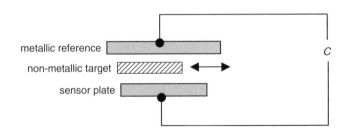

FIGURE 12.7 Dielectric variation in capacitive sensor measurement.

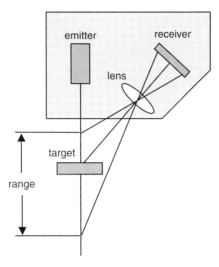

FIGURE 12.8 Optical triangulation displacement sensor.

Important specifications for this type of sensor are the working range and the standoff distance. The standoff distance is the distance from the sensor face to the center of the working range.

Both diffuse and specular designs are available for this type of sensor. A diffuse design is useful for targets with surfaces which scatter light such as anodized aluminum. A specular design is useful for targets with surfaces which reflect light well such as mirrors. In addition, the target color and transparency should also be considered when investigating this type of sensor because these properties affect the absorption of light by the target.

Optical triangulation sensors are high resolution and typically offer ranges up to one half meter. Higher ranges can be achieved, albeit with significantly increased cost.

12.2.1.6.2 Optical Time-Of-Flight

Optical time-of-flight sensors detect the position of objects by measuring the time it takes for light to travel to the object and back. As in the case of optical triangulation sensors, time-of-flight sensors also contain an emitter and a receiver. The emitter is a laser or an LED while the receiver is a photodiode. The emitter generates one or more pulses of light toward a target. Some of the light is reflected off the target and captured by the photodiode. The photodiode generates a pulse when it receives the light, and the time difference between the emitted pulse and the received pulse is determined by the sensor electronics. The distance to the target is then calculated based on the speed of light and the time difference.

Most time-of-flight sensors have measurement ranges of several meters. However, laser based time-of-flight sensors can have a range of several miles if a series of pulses from the emitter is used. The accuracy of these sensors is not as high as optical triangulation sensors but the range is typically greater.

12.2.1.6.3 Analog Photoelectric

Photoelectric sensors are noncontact displacement sensors which measure the position of objects by measuring the intensity of reflected light from the object. Again, an emitter and a receiver are used. An LED is used as the emitter and a phototransistor or photocell is used as the receiver. Photoelectric sensors are most often used as digital proximity sensors, but one category, diffuse-mode proximity, can be used as an analog sensor for measuring distance. In this design, light generated by the emitter is directed toward the target. Light reflected from the target is captured by the receiver. The analog output from the receiver is proportional to the intensity of the light received from the target. This output will be proportional to the distance of the target from the receiver.

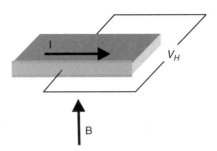

FIGURE 12.9 Hall effect sensor.

12.2.1.7 Ultrasonic Sensors

Ultrasonic displacement sensors use the same approach as optical time-of-flight sensors with an analog output proportional to the distance to target. Instead of a light pulse, however, an ultrasonic pulse is generated, reflected from a target, and then detected by a receiver. The response is dependent upon the density of the target. The frequency of the ultrasonic pulse is greater than 20 kHz and thus beyond the audible range for humans. Piezoelectric and electrostatic transducers are typically used to generate the ultrasonic pulse.

As compared with optical time-of-flight sensors, ultrasonic sensors are less sensitive to light disturbances. However, ultrasonic sensor output can be affected by large temperature gradients because the speed of sound is affected by air temperature.

12.2.1.8 Hall Effect Sensors

Hall effect sensors are noncontact sensors which output a signal proportional to input magnetic field strength. The Hall effect refers to the voltage generated when a current carrying conductor or semiconductor is exposed to magnetic flux in a direction perpendicular to the direction of the current. A voltage, the Hall voltage, is generated in a direction perpendicular to both the current, **I**, and the applied magnetic field, **B**, as shown in Figure 12.9. In order to use a Hall effect sensor as a displacement sensor, the sensor is typically matched with a moving permanent magnet. This magnet is applied to the target.

Since the output of the Hall effect sensor is directly proportional to the applied magnetic field strength, the output will be nonlinearly related to the distance from the sensor to the permanent magnet. This nonlinear relationship must be included in the post processing of the output signal. The sensor is most often made of a semiconductor material and therefore is available in standard integrated circuit packages with integrated electronics.

12.2.2 Strain

The most common strain sensor used in robotics is the strain gage. The electrical resistance of this sensor changes as a function of the input strain. The input strain may be either positive (tensile) or negative (compressive). An example of a bonded metallic strain gage is shown in Figure 12.10. The sensor consists of an insulating carrier, a foil or wire grid bonded to the insulating carrier, and an additional insulating layer applied to the top of the grid. The sensor is bonded a structure such that the grid legs are aligned with the desired direction of strain measurement. The design of the grid makes it most sensitive to strain along its length. The change in resistance of the gage is a function of the gage factor, G_L, multiplied by the strain

$$\frac{\Delta R}{R} = G_L \varepsilon_L$$

where $\Delta R/R$ is the relative resistance change, G_L is the gage factor along the sensor length, and ε_L is the strain along the sensor length. The gage factor for the strain gage is provided by the manufacturer.

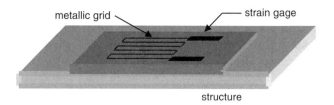

FIGURE 12.10 Strain gage sensor applied to structure.

A Wheatstone bridge circuit is used to measure the resistance change of the sensor. Strain gages are available with nominal resistance values between 30 and 3000 Ω. Temperature variation is the primary environmental concern for strain gages. Strain gage manufacturers have developed grid materials which can compensate for the thermal expansion of the test structure. Temperature compensation may also be achieved through the use of an additional strain gage applied in a direction normal to the measuring gage. This additional gage is then used as one of the legs in the Wheatstone bridge circuit for comparison.

12.2.3 Force and Torque

Force and torque sensors typically employ an elastic structural element with deflection proportional to the applied force or torque. Either strain gages or piezoelectric materials are used to measure the deflection of the elastic structure. Designs based on strain gages are useful for measuring static and low frequency forces and torques.

Piezoelectric materials develop an electrostatic charge when a mechanical load is applied. This charge is collected by electrodes and is converted to an output voltage signal through the use of a charge amplifier. Piezoelectric sensors are useful for dynamic measurements, but due to charge leakage across the electrodes, these sensors cannot measure static deflection.

Force sensors are frequently termed "load cells." Different load cell designs are capable of measuring compressive, tensile, or shear forces. With both force and torque sensors, the natural frequency of the elastic structure used in the sensor must be taken into consideration for the measurement.

12.2.4 Acceleration

Acceleration sensors are based on the measurement of the deflection of a proof mass suspended in the sensor housing. There are many types of accelerometers available for use in robotics. The various designs include piezoelectric, strain gage, piezoresistive, capacitive, and inductive technologies.

Figure 12.11 shows two simplified examples of two accelerometer designs. The first is that of a compressive piezoelectric design. The proof mass is suspended within the sensor housing using piezoelectric material. The output charge from the piezoelectric material, most often a PZT crystal, is proportional to the inertial force acting on the proof mass.

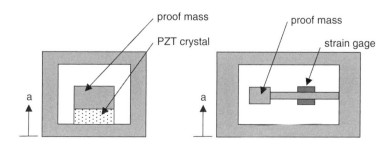

FIGURE 12.11 Piezoelectric and strain gage accelerometer designs.

The second design shown in Figure 12.11 is the basis for a strain gage accelerometer design. A cantilevered beam suspends the proof mass within the sensor housing. Strain gages are applied to the cantilevered beam and measure the deflection of the beam. MEMS accelerometers have found wide use in the automotive market and are suitable for many robotics applications. The most common designs in this category are based on capacitive displacement sensing.

The sensitivity, usable frequency range, and maximum input acceleration are limited by the mechanical dynamics of the accelerometer. For example, if the suspension is designed to be very compliant, the sensitivity of the accelerometer will increase with respect to input acceleration. However, the natural frequency of the suspension will be decreased and this will lead to a decreased frequency range for the measurement.

12.3 Digital Sensors

A digital sensor will output either an "on" or an "off" electrical state. Apart from encoders, the majority of digital sensors used in robotic applications are *static* digital sensors in that their value is solely based on the digital state of the output as opposed to the frequency of an output pulse train. Static digital sensors do not require counter electronics for acquisition.

12.3.1 Switches as Digital Sensors

A mechanical switch is the simplest and lowest cost type of digital sensor used in robotics. Switches may be purchased with "normally open" (NO) or "normally closed" (NC) contacts. The word "normally" refers to the inactive state of the switch. They also may be available with both options and can be wired in one configuration or the other. When a switch is configured as normally open, it will have zero continuity (or infinite resistance) when it is not activated. Activating the switch provides electrical continuity (zero resistance) between the contacts. Normally closed switches have continuity when they are not activated and zero continuity when they are activated.

Switches may also be designed to have multiple poles (P) and multiple throws (T) at each pole. A pole is a moving part of the switch and a throw is a potential contact position for the pole. Figure 12.12 shows an example of a single pole double throw switch (SPDT) and a double pole single throw (DPST) switch. Because mechanical switches are contact sensors, they can exhibit failures due to cyclical wear. In addition, the acquisition electronics used with switches must be able to filter out electrical noise at switch transitions due to contact bounce.

12.3.2 Noncontact Digital Sensors

In order to reduce problems of contact wear and switch bounce, noncontact digital sensors are frequently used in robotics. Digital sensor technology includes the transduction technologies discussed in the previous section on analog sensors:

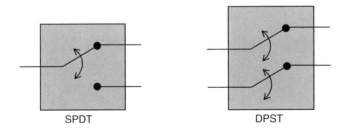

FIGURE 12.12 Single pole double throw (SPDT) and double pole single throw (DPST) switches.

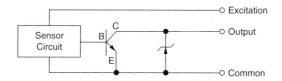

FIGURE 12.13 Digital sensor with NPN open collector output.

- Inductive
- Capacitive
- Optical
- Hall effect

An analog sensor design can be converted to a digital sensor design through the use of additional electronic circuitry. A comparator or Schmitt trigger circuit is frequently used to compare the output of the sensor amplifier with a predetermined reference. The Schmitt trigger will activate when the amplifier output is greater than the reference. If the amplifier output is less than the reference, then the Schmitt trigger will turn off.

12.3.3 Solid State Output

Digital sensors frequently use transistors as the output driver technology. These sensors are classified as having NPN or PNP outputs depending upon the type of transistor used for the output stage. Figure 12.13 shows a sensor with NPN open collector output. The base of the transistor is connected to the sensor circuit, the emitter to common, and the collector to the output. Since the output is not yet connected to any other signal, the NPN sensor has an "open collector" output. In the active state, the transistor will drive the output to common. In the inactive state, the transistor will leave the output floating. Typically, the output is connected to a data acquisition circuit with a pull-up resistor to a known voltage. In this way, the inactive state of the sensor will register at the known voltage and the active state will register as common or ground.

In the case of a PNP sensor, the base of the transistor is still connected to the sensor circuit, the collector is tied to the DC excitation source for the sensor, and the emitter is used as the output. In this case, the PNP sensor has an "open emitter" output. When active, the sensor output switches from floating to a positive voltage value. When inactive, the sensor output is left floating. In the inactive case, a pull-down resistor in the acquisition circuitry will bring the output to common or ground.

Any type of digital sensor can be classified as having a sinking or a sourcing output. A sinking output sensor means the sensor can sink current from a load by providing a path to a supply common or ground. This type of sensor must be wired to a sourcing input in a data acquisition system. A sourcing output sensor means the sensor can source current to a load by providing a path to a supply source. This type of sensor must be wired to a sinking input in a data acquisition system. NPN sensors are examples of sinking output devices while PNP sensors are examples of sourcing output devices. Switches and mechanical relays may be wired as either sinking or sourcing sensors.

12.3.4 Common Uses for Digital Sensors

Digital sensors can be used in a wide variety of applications within robotics. These include proximity sensors, limit sensors, and safety sensors such as light curtains.

12.3.4.1 Proximity Sensors

Proximity sensors are similar to analog displacement sensors, but they offer a static digital output as opposed to an analog output. Proximity sensors are used to determine the presence or absence of an object. They may be used as limit sensors, counting devices, or discrete positioning sensors. They are typically noncontact digital sensors and are based on inductive, capacitive, photoelectric, or Hall effect technology. These technologies are discussed in the previous section on analog sensors. Their design is frequently similar to that of analog position sensors but with threshold detecting electronics included so that their output is digital.

12.3.4.2 Limit Switches and Sensors

Limit switches or limit sensors are digital inputs to a robot controller that signal the end of travel for motors, actuators, or other mechanisms. The incorporation of limit sensors helps prevent mechanical failure caused by part of a mechanism hitting a hard stop in the system. The limit sensor itself can be a physical switch with mechanical contacts or a digital proximity sensor as described above. Limit sensors may be mounted to individual joints in the robot or to axes of motion in a robotic workcell. When the limit sensor is encountered for a particular joint or axis, the robot controller will bring the motion to a safe stop.

Both a forward and a reverse limit sensor can be connected to each joint or axis in a robotic system. Forward is defined as the direction of increasing position as measured by the encoder or analog feedback signal. Limit sensors can be used with both rotational and linear axes. A home switch or sensor can also be built into each axis and used to indicate a center position reference for the axis.

12.3.4.3 Light Curtains

Light curtains can automatically detect when an operator moves within the danger area for a robot or robotic operation. This danger area will usually include the entire workspace of the robot. Light curtains are typically based on photoelectric sensors which emit multiple beams of infrared light. When any of the beams of light is broken, the control circuit for the light curtain is activated, and the robotic system is immediately shut down in a safe manner.

12.4 Vision

Many robots use industrial cameras for part detection, inspection and, sometimes, guidance. The camera output may be analog or digital and may be acquired by a computer through several means. Often, a frame-grabber or image acquisition plug-in board is used and installed in a computer. More recently, computer bus technologies such as IEEE 1394, Camera Link®, Gigabit Ethernet, and USB have been used to transfer data between the camera and the computer. Machine vision software is installed on the computer and is used to examine the image so that image features can be determined and measurements can be made.

Smart cameras include embedded processing and machine vision software in their design. Smart cameras may be one integrated unit or they may use a tethered design with an electronic link between the processor and the camera.

12.5 Actuators

12.5.1 Electromagnetic Actuators

12.5.1.1 Solenoid

Solenoids are the most basic type of electromagnetic actuator. The linear solenoid concept is shown in Figure 12.14 and consists of a coil of wire, a fixed iron or steel frame, and a movable iron or steel plunger. The part of the plunger that extends from the frame is attached to the load.

When current flows in the coil, a magnetic field is generated around the coil. The frame serves to concentrate the magnetic field such that the maximum magnetic force is exerted on the plunger. The magnetic force causes the plunger to be attracted to the rear of the frame or the back stop and close the

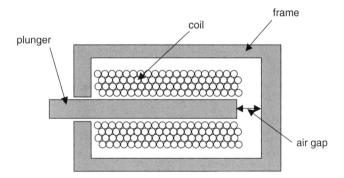

FIGURE 12.14 Linear solenoid concept.

initial air gap. This force is inversely proportional to the length of the air gap so that the solenoid exerts greater force as the air gap is closed. In addition, the magnetic force is increased by increasing the applied current and the number of turns of wire within the coil.

The magnetic field around the coil will only cause motion towards the center of the coil. When no current flows in the coil, the plunger is free to move to its initial position due to a restoring force caused by the load and, possibly, a return spring.

The diagram shown in the figure is an example of a pull type solenoid. All solenoids act to pull in the plunger. However, push type solenoids are designed such that the plunger extends out the rear of the frame and is then connected to the load. This causes a pushing action on the load when the solenoid is energized.

Solenoids are also specified based on whether they will be used continuously or only for intermittent duty. Continuous duty assumes that the solenoid will have a 100% duty cycle. Because the coil has limited heat dissipation, a 100% duty cycle means that the applied current will have to be reduced to prevent damage to the coil. This, in turn, results in a decreased applied force. Intermittent duty means that the duty cycle is less than 100%. Intermittent duty allows higher forces than continuous duty because more current can be applied.

Rotary solenoids are also available and are similar in design to linear solenoids. However, the linear motion is converted to rotary motion through the use of ball bearings moving down spiral grooves, called "ball races," to close the air gap.

12.5.1.2 Stepper Motors

Stepper motors convert input electrical pulses into discrete mechanical motion. They are available in either rotational or linear designs, but rotational is far more common. For rotary stepper motors, each step corresponds to a set number of degrees of rotation. The number of electrical pulses determines the number of steps moved while the timing of the pulses controls the speed of the stepper motor. Because no feedback sensors are required for their operation, they can be used open-loop and do not require any tuning. They are well-suited for use in positioning applications and are easy to use.

Permanent magnet stepper motors are common and include a stator (stationary part) with windings and a rotor (rotating part) made with permanent magnets. A simple example of a permanent magnet stepper motor is shown in Figure 12.15. The stator has four windings or poles and the rotor has one complete magnet or two poles. The stator itself is magnetically conductive. The windings on the stator are wired in pairs such that there are two phases: phase A and phase B. Power is applied to only one phase at a time and the power sequencing is known as "electronic commutation." When this occurs, one of the windings in the phase pair becomes a north pole and the other winding becomes a south pole. Magnetic attraction therefore causes the permanent magnet rotor to be aligned with the energized winding pair. The alignment of the rotor with a winding pair corresponds with a "full step."

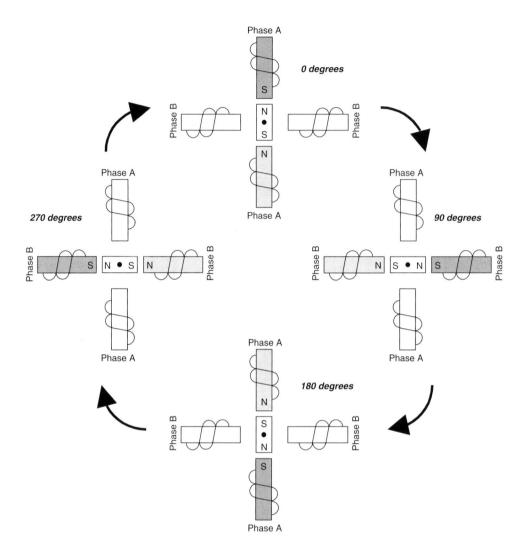

FIGURE 12.15 Two-phase stepper motor power sequence.

As shown in the diagram at the top of the figure, phase A is powered (indicated by shading) causing the rotor to be aligned at the 0° position. Then, going clockwise in the figure, phase B is powered while phase A is unpowered, causing the rotor to rotate to 90°. Then, phase A is again powered but with current of opposite polarity compared with the 0° position. This causes the rotor to rotate to 180°. The next step of 270° is achieved when phase B is powered again with opposite polarity compared with the 90° position. Thus, the two poles in the rotor result in four discrete steps of 90° within one complete rotation. The direction of rotation can be changed by reversing the pulse sequence applied to the phases.

The torque required to move the rotor when there is no power applied to either phase is known as the "detent torque." The detent torque causes the clicks that can be felt if an unpowered stepper motor is moved manually. The torque required to move the rotor when one of the phases is powered with DC current is known as the "holding torque."

Stepper motors may have higher numbers of poles in the rotor and the stator. These are created using a magnet with multiple teeth on the rotor and additional windings on the stator. This will increase the number of full steps per revolution. Many stepper motor manufacturers make stepper motors with 200 and 400 steps per revolution. This corresponds to step angles of 1.8° and 0.9°, respectively. Microstepping

Sensors and Actuators

is a common feature on stepper motor drives that adjusts the amount of current applied to the phases. This allows each full step to be subdivided into smaller steps and results in smoother motion.

Other types of stepper motors include variable reluctance and hybrid stepper motors. Variable reluctance stepper motors have a stator which is similar to the permanent magnet stepper motor, but the rotor is composed of soft iron rather than a permanent magnet and also has multiple teeth. Because there are no permanent magnets on the rotor, it is free to move with no detent torque if there is no power applied to either of the phases. Hybrid stepper motors combine the principles of permanent magnet and variable reluctance motors. The rotor has teeth like a variable reluctance motor but contains an axially magnetized cylindrical magnet which is concentric with the shaft.

12.5.1.3 DC Brush Motor

DC brush motors convert applied current into output mechanical motion. They are a type of servomotor, and the output torque is directly proportional to the applied current. As compared with stepper motors, DC brush motors require a feedback signal for stable operation. They must be used closed-loop, and this can make the system more complex than one using stepper motors. They also use conductive brushes for mechanical rather than electric commutation and thus can have higher maintenance costs due to wear. However, DC brush motors have smooth motion and higher peak torque and can be used at higher speeds than stepper motors.

DC brush motors incorporate a stator with either permanent magnets or windings and a rotor with windings. Windings in the stator are known as field windings or field coils. DC motors using permanent magnets in the stator are known as permanent magnet DC motors. DC motors with windings in the stator are known as series-wound or shunt-wound motors depending upon the connectivity of the windings in relation to rotor windings. Series-wound motors have the field windings in series with the rotor while shunt-wound motors have the field windings in parallel with the rotor.

All DC brush motors include conductive brushes which make sliding contact with a commutator as the rotor turns. The commutator is attached to the rotor shaft and the rotor windings are connected to the individual sections of the commutator. The commutator has as many sections as there are poles in the rotor.

A simple example of a permanent magnet DC motor is shown in Figure 12.16. This motor has two poles in the rotor. When voltage is applied to the brushes, current flows from the brushes to the commutator and, in turn, to the windings. This creates a magnetic field in the rotor. The interaction between this field

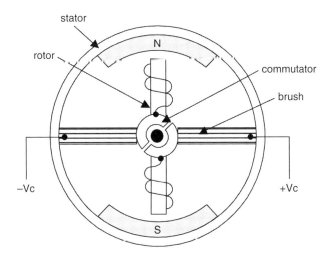

FIGURE 12.16 Permanent magnet DC brush motor with two poles.

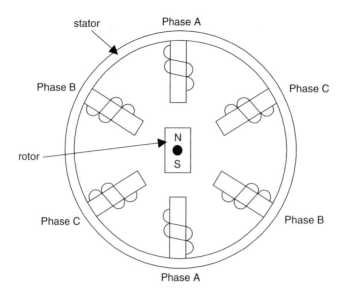

FIGURE 12.17 Three-phase DC brushless motor.

and that of the stator causes the rotor to move until it is aligned with the magnetic field of the stator. Just as the rotor is aligned, the brushes switch to the next section or contact of the commutator and the magnetic field of the rotor is reversed. This causes the rotor to continue moving until it is aligned with the stator again.

DC brush motors typically include multiple poles in the rotor to smooth out the motion and increase torque.

12.5.1.4 DC Brushless Motor

DC brushless motors are another type of servomotor in that feedback is required for stable operation. A DC brushless motor is like a DC brush motor turned inside out because the rotor contains a permanent magnet and the stator contains windings. The windings are electronically commutated so that the mechanical commutator and brushes are no longer required as compared with a DC brush motor. Figure 12.17 shows an example of a brushless motor with three phases (six poles connected in pairs).

DC brushless motors are commonly used in robotics applications because of their high speed capability, improved efficiency, and low maintenance in comparison with DC brush motors. They are capable of higher speeds because of the elimination of the mechanical commutator. They are more efficient because heat from the windings in the stator can be dissipated more quickly through the motor case. Finally, they require less maintenance because they do not have brushes that require periodic replacement. However, the total system cost for brushless motors is higher than that for DC brush motors due to the complexity of electronic commutation.

The position of the rotor must be known so that the polarity of current in the windings of the stator can be switched at the correct time. Two types of commutation are used with brushless motors. With trapezoidal commutation, the rotor position must only be known to within 60° so that only three digital Hall effect sensors are typically used. Sinusoidal commutation is employed instead of trapezoidal commutation when torque ripple must be reduced for the motor. In this case, the rotor position must be determined more accurately so that a resolver is used or an encoder is used in addition to Hall effect sensors. The Hall effect sensors are needed with the encoder to provide rotor shaft position at startup. The resolver provides absolute position information of the rotor shaft, so that the Hall effect sensors are not required for startup.

12.5.2 Fluid Power Actuators

12.5.2.1 Hydraulic Actuators

Hydraulic actuators are frequently used as joint or leg actuators in robotics applications requiring high payload lifting capability. Hydraulic actuators output mechanical motion through the control of incompressible fluid flow or pressure. Because incompressible fluid is used, these actuators are well suited for force, position, and velocity control. In addition, these actuators can be used to suspend a payload without significant power consumption. Another useful option when using hydraulics is that mechanical damping can be incorporated into the system design.

The primary components in a hydraulic actuation system include:

1. A pump — converts input electrical power to hydraulic pressure
2. Valves — to control fluid direction, flow, and pressure
3. An actuator — converts fluid power into output mechanical energy
4. Hoses or piping — used to transport fluids in the system
5. Incompressible fluid — transfers power within the system
6. Filters, accumulator, and reservoirs
7. Sensors and controls

Positive displacement pumps are used in hydraulic actuator systems and include gear, rotary vane, and piston pumps. The valves that are used include directional valves (also called distributors), on-off or check valves, pressure regulator valves, flow regulator valves, and proportional or servovalves.

Both linear and rotary hydraulic actuators have been developed to convert fluid power into output motion. A linear actuator is based on a rod connected to a piston which slides inside of a cylinder. The rod is connected to the mechanical load in motion. The cylinder may be single or double action. A single action cylinder can apply force in only one direction and makes use of a spring or external load to return the piston to its nominal position. A double action cylinder can be controlled to apply force in two directions. In this case, the hydraulic fluid is applied to both faces of the piston.

Rotary hydraulic actuators are similar to hydraulic pumps. Manufacturers offer gear, vane, and piston designs. Another type of rotary actuator makes use of a rack and pinion design where a piston is used to drive the rack and the pinion is used for the output motion.

Working pressures for hydraulic actuators vary between 150 and 300 bar. When using these actuators, typical concerns include hydraulic fluid leaking and system maintenance. However, these can be mitigated through intelligent engineering design.

Hydraulic actuators have been used in many factory automation problems and have also been used in mobile robotics. Figure 12.18 is a picture of the TITAN 3 servo-hydraulic manipulator system from Schilling Robotics. This is a remote manipulator that was originally developed for mobile underwater applications but is also being used in the nuclear industry.

12.5.2.2 Pneumatic Actuators

Pneumatic actuators are similar to hydraulic actuators in that they are also fluid powered. The difference is that a compressible fluid, pressurized air, is used to generate output mechanical motion. Pneumatic actuators have less load carrying capability than hydraulic actuators because they have lower working pressure. However, pneumatic actuators have advantages in lower system weight and relative size. They are also less complex in part because exhausted pressurized air in the actuator can be released to the environment through an outlet valve rather than sent through a return line.

Because compressed air is used, the governing dynamic equations of pneumatic actuators are nonlinear. In addition, compressed air adds passive compliance to the actuator. These two factors make these actuators more difficult to use for force, position, and velocity control. However, pneumatic actuators are frequently used in industry for discrete devices such as grippers on robotic end effectors.

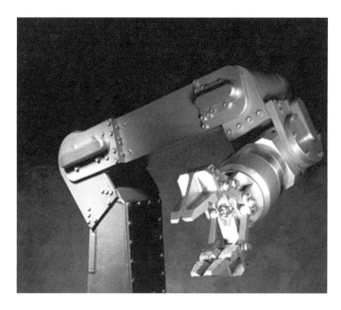

FIGURE 12.18 Titan 3 servo-hydraulic manipulator (Courtesy of Schilling Robotics).

The primary components in a pneumatic actuation system include:

1. A compressor — converts input electrical power to air pressure
2. Compressed air treatment unit — includes filters and pressure regulation
3. Valves — to control pneumatic power
4. An actuator — converts pneumatic power into output mechanical power
5. Hoses or piping — used to transport air in the system
6. Sensors and controls

There are many types of pump technologies used in pneumatic compressors. They include positive displacement pumps such as piston, diaphragm, and rotary vane types as well as non-positive displacement pumps such as centrifugal, axial, and regenerative blowers. The compressor may include a storage tank or it may output pressurized air directly to a regulator valve. The types of valves used are similar to those used in hydraulic actuation systems as described in the previous section. Both rotary and linear actuators are available and are also similar in design to those used in hydraulic actuation systems.

Working pressures for pneumatic actuators are typically less than 10 bar. Installation for these types of actuators is facilitated by the availability of compressed air on the factory floor.

References

Elbestawi, M.A., Force Measurement, in: *The Measurement, Instrumentation and Sensors Handbook*, Webster, J., ed., CRC Press, Boca Raton, FL, 1999.

Global Spec, 350 Jordan Rd., Troy, NY 12180, website: www.globalspec.com.

Honeywell Sensing and Control, 11 W. Spring St., Freeport, IL 61032, *Hall Effect Sensing and Application Book*.

Kennedy, W.P., The basics of triangulation sensors, *Sensors*, 15(5), 76, May 1998.

Lynch, K.M. and Peshkin, M.A., Linear and rotational sensors, in: *The Mechatronics Handbook*, Bishop, R.H., ed., CRC Press, Boca Raton, FL, 2002.

Mavroidis, C., Pfeiffer, C., and Mosley, M., Conventional actuators, shape memory alloys, and electrorheological fluids, Invited Chapter in *Automation, Miniature Robotics and Sensors for Non-Destructive Testing and Evaluation*, Bar-Cohen, Y., ed., The American Society for Nondestructive Testing, 2000.

Motion Control Course Manual, National Instruments, Part Number 323296A-01, March 2002.

National Instruments, 11500 N. Mopac Expwy, Austin, TX 78759, *Measuring Strain With Strain Gages,* website: www.zone.ni.com.

National Instruments, 11500 N. Mopac Expwy, Austin, TX 78759, *Choosing The Right Industrial Digital I/O Module for Your Digital Output Sensor,* website: www.zone.ni.com.

Prosser, S.J., The evolution of proximity, displacement and position sensing, *Sensors,* 15(4), April 1998.

Sorli, M. and Pastorelli, S., Hydraulic and pneumatic actuation systems, in: *The Mechatronics Handbook,* Bishop, R.H., ed., CRC Press, Boca Raton, FL, 2002.

Welsby, S.D., Capacitive and inductive noncontact measurement, *Sensors,* 20(3), March 2003.

13

Precision Positioning of Rotary and Linear Systems

13.1 Introduction **13**-1
13.2 Precision Machine Design Fundamentals **13**-2
   Definitions of Precision • Determinism • Alignment Errors (Abbé Principle) • Force and Metrology Loops • Constraint • Thermal Management • Cable Management • Environmental Considerations • Serviceability and Maintenance
13.3 Mechatronic Systems **13**-8
   Definition of Mechatronic Systems • Discrete-Time System Fundamentals • Precision Mechanics • Controller Implementation • Feedback Sensors • Control Algorithms
13.4 Conclusions **13**-21

Stephen Ludwick
Aerotech, Inc
Pittsburgh, Pennsylvania

13.1 Introduction

Precision positioning systems have historically been a key part of successful industrial societies. The need to *make something* requires the ability to *move something* with a very high level of accuracy. This has not changed in the Information Age, but instead has become even more important as global competition forces manufacturers to hold ever tighter specifications, with increased throughput and reduced costs. Automation is the equalizer that allows manufacturers in countries with high labor rates to compete globally with developing countries. The definition of a *precision* machine continues to evolve as technology advances. The components used to build machines become more precise, and so the machines themselves improve. As loosely defined here, precision machines are those that repeatably and reliably position to within a tolerance zone smaller than is possible in commonly available machines. Designing machines to best use the available components and manufacturing techniques requires specialized skills beyond those of general machine designers. It is interesting to note that many of the fundamental rules for designing precision machines have not changed for hundreds of years. Evans [9] tracks the evolution of precision machines and provides a historical context to the present state-of-the-art.

Modern precision-positioning systems are largely *mechatronic* in nature. Digital computers interface with electronic sensors and actuators to affect the motion of the system mechanics. Auslander and Kempf [2] review the basic elements required to interface between system mechanics and control electronics, while Åström and Wittenmark [24] and Franklin et al. [11] present the required discrete-time control theory and implementation. Kiong et al. [17] describe how these elements are specifically combined in the context of precision motion control. The performance of the overall system depends on all of its mechanical, electronic, and software components, and so a multi-disciplinary design team is usually needed to undertake the project development. Precision machines are increasingly being used as a part of an overall manufacturing system and, thus, are required to share process information over a network with many other systems. This information may be high- or low-speed and synchronous or asynchronous to the processes. Decisions on the control hardware and network structure are often very expensive to change later in the design and should be evaluated in parallel with the more-traditional electromechanical elements of the machine design.

This chapter focuses on issues important to engineers involved in the specification and evaluation of commercially available motion control systems. The emphasis is primarily on presenting precision machine design principles and comparing the uses and specifications of different existing technologies; a thorough review of bottom-up precision machine design is beyond the scope of this work, but the interested reader is referred to Shigley and Mischke [30], Slocum [34], Smith and Chetwynd [35], and Hale [15] for a more in-depth look at undertaking a precision machine design. The American Society for Precision Engineering [1] also publishes a journal and maintains a list of recommended readings.

We will limit the discussion of precision machines to those features most common in modern designs. These include discrete-time microprocessor-based control, ballscrew or direct-drive mechanisms, brushless motors, and quantized position feedback. This is not to suggest that all precision machines contain these elements or that any device lacking them is not precise, rather it is an observation that a designer looking to build his own motion control system, or select from available subsystems, will generally find these to be economical alternatives. Kurfess and Jenkins [18] detail precision control techniques as applied to systems with predominantly analog feedback sensors and compensation. The fundamental analog and continuous-time control design techniques still apply, but adding microprocessor control and quantized feedback adds an additional set of advantages and limitations that will be explored here.

13.2 Precision Machine Design Fundamentals

Precision machines vary widely in their design and application but follow several common principles. Machines are generally not accurate by accident but because of the effort taken in design throughout the process. Determinism is a proven machine design philosophy that guides engineers in these developments. Error motions and imperfections in the components used to build precision machines have different effects on the overall performance depending on their location in the system. Some errors are relatively inconsequential; other errors heavily influence the overall performance. One class of errors occurs when an angular error is multiplied by a lever arm into a linear error. These errors, referred to as Abbé errors, should be minimized in all precision machine designs. Applying forces to a machine strains the mechanical elements. These dimensional changes may affect the accuracy of the measurements. Machines should, therefore, be designed such that the load-bearing elements are as separate as possible from the elements used for metrology. Machines are all made of interconnected elements, and the method of connecting these elements affects both the precision of the design and the ease with which it can be analyzed. Exact-constraint designs are based on having contact at the minimum number of points necessary to constrain the required degrees-of-freedom of the free part. Elastic-averaging is the opposite and relies on contact over so many points that errors are averaged down to a low level. Problematic designs are those that occur somewhere in the middle of the two extremes. Finally, some often-neglected elements of a precision machine design include cable management, the environment that the system will operate in, analysis of heat generation and flow, and maintenance. Each of these items will be further detailed here.

13.2.1 Definitions of Precision

The terms *accuracy*, *repeatability*, and *resolution* are frequently misused when applied to precision machines. Accuracy is the nearness of a measurement to the standard calibrated value. Absolute accuracy is expensive, and there are few shortcuts to achieving it. The position readout of a system may indicate a 100 mm move,

Precision Positioning of Rotary and Linear Systems

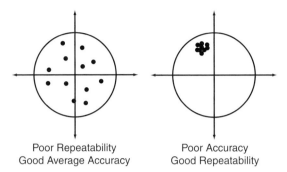

FIGURE 13.1 The figure on the left shows a measurement with poor repeatability and poor accuracy on any single measurement but good accuracy when the measurements are averaged together. The figure on the right shows a measurement with poor accuracy but excellent repeatability. Accuracy can usually be improved by mapping techniques, and so a machine with high repeatability is usually preferred.

but the *accuracy* of the system quantifies how closely this 100 mm compares with the standard measure of a meter. In applications that require absolute accuracy (e.g., semiconductor wafer mask writers, machine tools, and metrology instruments), the designer should be prepared to invest in a temperature-controlled environment, temperature-compensated mapping, and high attention to details of sensor location relative to the workpiece. In some applications, accuracy can be achieved statistically by taking many measurements and averaging the results together. In many other applications, absolute accuracy is less important than repeatability.

Repeatability is the nearness of successive samples to each other, and it is the goal of a deterministic machine design to produce as repeatable a machine as is possible. Figure 13.1 illustrates the difference between repeatability and accuracy. If the 100 mm move on our hypothetical machine is truly 99.95 mm, but is the same 99.95 mm each time it makes the move, then the machine is inaccurate but repeatable. Repeatability is a broad term that captures the overall uncertainty in both the positioning ability of a system and the measurement of this positioning. Repeatability can be measured as uni-directional (in which the same point is approached repeatably from the same direction) and bi-directional (in which the point is approached from both directions). Systems can have excellent repeatability when approached from a single direction, but terrible when approached from both (consider the case of a backlash in a leadscrew). Know which specification you are looking at. Depending on the level of modeling required, a statistical distribution of the measurements can be constructed to provide confidence intervals on the positioning ability of the system. The repeatability measurement itself can be separated into short-term and long-term repeatability. Short-term apparent nonrepeatability is generally caused by small variations in the size or roundness of the rolling elements in the bearings or ballscrew. Deterministic machine design holds that even these errors could be predicted with enough modeling, but the effort required to do so is forbidding. Temperature variations are usually the cause of longer-term apparent nonrepeatability. Nonuniform drag from any cables or hoses can also cause different performance over time. The goal of a precision machine design, therefore, becomes to design a system such that the level of nonrepeatability is small enough to be unimportant to the application.

The other frequently misused term regarding machine performance is resolution. Resolution (by itself) can be taken to mean mechanical resolution, which is the minimum usable mechanical displacement, or electrical resolution, which is the least significant digit of the displacement sensor. These two items are only loosely related, and the designer should again be clear about which value is given in the specification. Mechanical resolution is the ultimate requirement and is limited by characteristics of the control system, vibration in the environment, electrical noise in the sensors, and mechanical static friction effects. Fine mechanical resolution is required for good in-position stability, implementing error-correcting maps, and achieving tight velocity control.

13.2.2 Determinism

Researchers at the Lawrence Livermore National Laboratory developed and championed the philosophy of *determinism* in precision machine design. Much of this work was performed in the context of designing large-scale high-precision lathes, referred to as diamond-turning machines. The philosophy of determinism maintains that machine behavior follows and can be predicted by *familiar engineering principles* [4]. The emphasis here is on familiar, since esoteric models are generally not required to predict behavior even at the nanometer level. A careful and thorough accounting of all elements is often sufficient. Determinism maintains that machine behavior is highly repeatable. Apparent randomness in machine behavior is simply a result of inadequate modeling or lack of control over the environment. Machines must behave exactly as designed but may simply not have been designed adequately for the required task. This is not to suggest that creating a mathematical model that captures all the nuances of the machine performance is straightforward, or even practical, given limited resources. Rather, determinism guides the selection and design of systems in which the apparent randomness or nonrepeatability is below a certain level and for which models can be developed economically. Central to the deterministic machine design is the *error budget* in which all of the machine errors are tabulated and translated into a final error between a tool and workpiece through the use of transformation matrices. Slocum [34] and Hale [15] provide details of these calculations. A properly designed error budget allows the engineer to quickly assess the impact of component changes on the machine performance and to identify the elements that have the largest impact on overall error.

13.2.3 Alignment Errors (Abbé Principle)

Alignment errors occur when the axis of measurement is not precisely aligned with the part to be measured. There are two basic types of alignment errors, named *cosine* and *sine* based on their influence on the measurement. Examples of cosine errors include nonparallelism between a linear stage and a linear scale, or taking a measurement with a micrometer when the part is seated at an angle. Figure 13.2 illustrates a typical cosine error. The true displacement D and the measured value D_c are related by simple geometry through

$$D_c - D = D_c(1 - \cos\theta) \tag{13.1}$$

where θ is the misalignment angle. For small angles of θ, the error is well-approximated by

$$D_c - D \approx D_c \frac{\theta^2}{2} \tag{13.2}$$

Cosine errors are typically small but should not be dismissed as insignificant at sub-micron levels. They can usually be compensated for with a simple scale factor correction on the feedback device.

Sine errors, also known as Abbé errors, occur when an angular error is amplified by a lever arm into a linear error. Minimizing these errors invokes a principle derived by Dr. Ernst Abbé who noted that

> If errors in parallax are to be avoided, the measuring system must be placed coaxially with the axis along which displacement is to be measured on the workpiece [34].

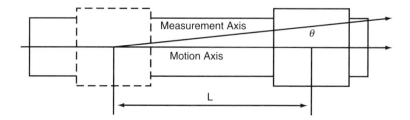

FIGURE 13.2 Example of cosine error in a typical machine setup.

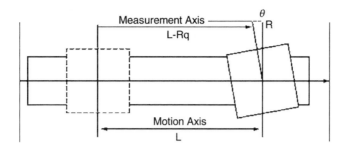

FIGURE 13.3 Illustration of Abbé error (sine error) in a typical machine setup.

Abbé errors typically occur on precision machines when a scale or measuring device is offset from the tool tip on a stage. The true displacement D and the measured value D_s in this case are related by

$$D_s - D = R \sin \theta \tag{13.3}$$

where R is the length of the lever arm, and θ is again the angular error (Figure 13.3). For small angles, the error is approximated by

$$D_s - D \approx R\theta \tag{13.4}$$

In contrast to cosine errors, Abbé errors are almost always significant in the overall performance of a machine and should be avoided to the greatest extent possible. As an example of the effect of sine errors, consider a linear slide having a linear accuracy of 1 μm at the center of the table and a yaw error of 20 μrad (approximately 4 arc-sec). At the edge of a 300 mm diameter payload ($R = 150$ mm), the yaw error contributes 3 μm to the linear error for a total of 4 μm. A secondary position sensor, usually a laser interferometer, can often be used to supplement the existing stage sensors and measure near to the point of interest. The Abbé Principle is perhaps the easiest to understand but most violated rule in precision machine design. Economy, serviceability, and size requirements often dictate that a multi-axis machine be made in a stacked arrangement of individual axes. It is not uncommon for angular errors on the lowest axis to be multiplied by a lever arm of several hundred millimeters before reaching the tool tip. In other cases, the geometry may offer only limited places to position the sensor, and this is often not at the ideal location. Deterministic machine design does not require strict adherence to the Abbé Principle, but it does require the designer to understand the consequences of violating it through the use of an error budget.

13.2.4 Force and Metrology Loops

Most precision machines are closed-systems, meaning that forces are applied and measurements are taken between machine elements. Force and metrology loops are conceptual tools used to guide a structural design and location of sensors and actuators. A metrology loop is the closed path containing all the elements between the sensor and the workpiece that affect the measurement. The metrology loop is not necessarily all mechanical, but can also consist of electrical or optical elements. These include sensor electronics or even the air in the beam path of a laser interferometer. Changes to the temperature, pressure, or humidity of the air change its index of refraction and thus also change the wavelength of the laser. This drift in the sensor readings is interpreted as stage motion. By definition, any change to elements in the metrology loop is impossible to distinguish from actual movement of the tool or workpiece. The force (or structural loop) is similarly defined as the closed path showing the conceptual flow of force in a loop around the structure. Applying force to structural elements generates stress, strain, and distortion of the elements. If any of the distorted elements is shared with the metrology loop, the result will be an inaccurate measurement. Precision machines should be designed to separate the force and metrology loops to the greatest possible extent. In extreme cases, this can be accomplished with a separate metrology frame with structural elements that are completely independent of load-bearing elements. The large optics diamond turning machine at

Lawrence Livermore is an example of a machine designed in this manner [7]. This separation may not be economical in all instances, but it demonstrates the general guideline of monitoring the flow of force in a machine, identifying the elements that influence the measurement, and separating these to the greatest extent practical.

13.2.5 Constraint

13.2.5.1 Exact-Constraint or Kinematic Design

Exact-constraint or *kinematic* design techniques create interfaces between machine elements in such a way that only the required number of the degrees of freedom are constrained leaving the element free to move in the remaining ones. A free-body has six degrees of freedom (three translation and three rotation), and each constraint reduces the available motion. For example, the ideal linear slide is kinematically constrained with only five points of contact leaving the stage with one remaining translational degree of freedom. Kinematic design is often used in precision engineering because of the relative ease with which geometric errors occurring at the points of constraint can be translated into errors at the workpiece through rigid-body coordinate transformations. There is nominally no structural deformation of the components. Blanding [3], Hale [15], and Schmiechen and Slocum [28] present quantitative analysis techniques for use with kinematic designs. Because the workpiece is exactly constrained, the structural elements behave fundamentally as rigid bodies. When a system is overconstrained, structural elements must deform in order to meet all of the constraints (consider the classic example of the three-legged vs. four-legged stool sitting on uneven ground), and this deformation often requires a finite-element model to analyze properly.

The design of kinematic couplings is a subset of exact-constraint design for items that are meant to be constrained in precisely six degrees of freedom. These fixtures contain exactly six points of contact and are often used for mounting components without distortion. Precision optical elements are often mounted using these techniques. Properly designed kinematic couplings are also highly repeatable, often allowing a component to be removed and replaced within micron or better tolerances. The classic kinematic couplings are the three-ball, three-groove coupling and the flat-vee-cone coupling (where the "cone" should more formally be a trihedral hole making three points of contact with the mating ball). Three-ball, three-groove couplings have the advantage of common elements in their construction with a well-defined symmetry about the center of the coupling. Any thermal expansion of the parts would cause growth about the center. The flat-vee-cone coupling has the advantage of allowing the entire coupling to pivot about the cone should any adjustments be necessary. The cone and mating ball are fixed together under thermal expansion. Slocum [32,33] presents analysis tools useful for designing kinematic couplings. In practice, kinematic couplings must be preloaded in order to achieve any meaningful level of stiffness. This stiffness can be determined quantitatively through an analysis of Hertzian contact stress. In general, larger radii on the contact points allow for the use of higher levels of preload, and greater stiffness. Under preload, however, the contact points become contact patches, and the coupling no longer has precisely six "points" of contact. Friction and wear at the contact patch decrease the repeatability of the coupling, and practical designs often include flexures that allow for this motion [14,29].

13.2.5.2 Elastic Averaging

Elastic averaging is the term used when contact occurs over such a large number of points that errors are averaged down to a low level. The classic hand-scraped machine way is one such example. All of the high points on the mating surfaces have been scraped down to lie in essentially the same plane, and contact between elements occurs uniformly over the entire area. Another example of elastic averaging in machine design is commonly used in precision rotary indexing tables. Two face gears are mated and lapped together. There is contact over so many teeth that individual pitch errors from one to the next are averaged out. This type of mechanism is often referred to as a *Hirth* coupling. A further example of elastic averaging occurs in recirculating-ball type linear bearings. A number of balls are constrained in a "racetrack" to maintain contact with the rail. Any individual ball will have size and roundness errors, but the average contact of many balls reduces the overall effect of the error.

13.2.6 Thermal Management

In all but the rarest cases, machines must be built with materials that expand and contract with temperature. The site where the machine is to be used should be prepared with appropriate HVAC systems to limit the temperature change in the environment, but it is rarely economical, or even possible, to control the temperature well enough that the machine accuracy will not be compromised to some level. The machine itself usually has some heat sources internal to it. Motors are essentially heating coils, friction in bearings and drivescrews creates heat, and even the process itself may lead to some heat generation. The precision machine design engineer must consider the flow of heat through the system and the effect that it will have on performance and accuracy. There are several techniques available for managing heat flow. One technique is to isolate the heat sources from the rest of the system and provide a well-defined path to transfer the heat away to a cooling sink. This could be as simple as a closed air or fluid path around a motor back out to a radiator, or strategic placement of materials with high thermal conductance. The entire system could also be placed in an air or fluid shower. A separate system keeps a tight control on the shower temperature. Note here that it is usually easier to heat and maintain a fluid or system above ambient temperature than it is to cool below. These techniques are not inexpensive but are often absolutely required for high-accuracy systems.

There are also mechanical design techniques that can be used to limit thermal-growth problems. Linear growth is usually much less of a problem than bending is, so the designer should try to maintain symmetry wherever possible in the design. Symmetric heat flow means equal temperatures and so equal growth about machine centerlines. Closed structural loops are preferable to open C-shapes. A C-shape will open under heating while a ring will expand uniformly. The designer must pay very careful attention to points where dissimilar materials are attached to each other. In each case, there is the possibility for a "bi-metallic strip" effect that can lead to bending. Where material mismatch and nonuniform growth is unavoidable, the designer should add specific compliant elements (i.e., expansion joints) so that the location of the displacement is known.

Finally, it is possible to map the errors induced by thermal growth of a machine. Given a series of measurements taken at different temperatures, and a set of temperature sensors properly located around the system, the control software can either move the axis to the correct position or at least present a corrected position measurement. As is generally the case with mapping and error correction techniques, there are usually a few key contributors to the overall error that are readily addressed, but identifying and correcting beyond these terms is challenging at best.

13.2.7 Cable Management

Cable management is a critical but often overlooked part of the overall machine design. It is often the most unreliable part of a precision machine. The cable management system (CMS) consists of the electrical cables themselves, pneumatic or hydraulic tubing, fittings and connectors, and often a carrier system. Common problems include conductor breakdown, insulation breakdown (shedding), connector reliability, and the influence of the drag force on the stage motion. The cable lifetime should be experimentally verified as early in the design as possible because the lifetime calculations are highly dependent on the actual implementation and mounting techniques. The magnitude of the cable drag force is generally position-dependent and can include a periodically varying component. This is particularly true when a chain-type cable carrier consisting of multiple links is used. In the highest-precision applications, a separate cable-carrier axis can be used to take up the main cable drag force while a short link connects it to the main stage with a relatively-constant force. There are two general techniques available for designing the CMS. The most-exercised portion of the CMS can be connectorized to allow for easy field replacement as part of a preventive maintenance schedule. This is appropriate when the duty cycle suggests that regular cable replacement will be required. However, the extra connectors required in this arrangement are themselves a possible failure point. In light-duty applications, it may be preferable to run continuous cable from the machine elements (motors, encoders, limits, and so on) through to a single junction block. In all cases, the CMS must be designed in parallel with the overall mechanics of the system.

13.2.8 Environmental Considerations

The design of a motion control system is often largely influenced by the environment in which it operates. Some systems, such as traditional metal-cutting machine tools, must contend with chips and coolant spray. Other forms of manufacturing, such as laser machining, form finer particulates that require additional sealing techniques. Different types of bellows and sliding seals are available for each application, and the disturbance forces that they impart can affect system repeatability and dynamic performance. Other systems must operate in clean-room environments where the emphasis is on preventing the stage from contaminating the process. In this case, special lubricants and seals must be used that are chosen specifically for low-levels of particulate generation.

Semiconductor processing technologies, such as extreme ultraviolet lithography (EUVL) and electron-beam inspection, often require precision motion control in a high-vacuum environment. The key considerations when designing or specifying a motion control system for operating in a vacuum are material selection and heat transfer. Most materials, when subjected to low pressures, will *outgass* into the vacuum environment. This outgassing rate will limit the achievable vacuum level, and the compounds released may even contaminate the process. In general, uncoated aluminum and stainless steel are acceptable materials for the main body of the structure. Polymers and epoxies used in connectors, cabling, and purchased components must be thoroughly reviewed. Care must be taken in the mechanical design to eliminate trapped volumes that can slowly outgas into the vacuum environment, increasing pumpdown times. This is generally done through the use of cross-drilled holes, machined channels, and vented fasteners. Systems that are designed for use in vacuum environments are generally cleaned and *baked out* at elevated temperature to eliminate as many contaminants as possible before use. Bakeout temperatures of over $100°C$ are not uncommon, and the motion system must be designed to survive this temperature and to maintain performance after a return to normal operating temperature. As will be described in the next section, serviceability is critical to the design of systems for use in a vacuum environment. Systems placed in a vacuum chamber generally are difficult (i.e., expensive) to remove after installation, and access to the stage may be limited to just a few windows through the chamber walls.

13.2.9 Serviceability and Maintenance

Machines must be designed to allow for routine preventive maintenance and for field-replacement of the items determined to be most-likely to fail. Many motion control systems are part of larger overall structures, and the designer should carefully monitor access points to the system after full integration. Preventive maintenance usually includes cleaning and bearing relubrication. It may also include replacing part of the cable management system in high-duty-cycle applications. It is important in the design to ensure that all required access is available in the fully assembled system, rather than just in the individual components.

13.3 Mechatronic Systems

13.3.1 Definition of Mechatronic Systems

Mechatronic systems are those systems whose performance relies on the interdependence between mechanical, electrical, and control components. This definition covers most modern precision linear and rotary motion systems. Designing in a mechatronic framework allows the designer to trade complexity between the various disciplines. Software-based error correction tables can be used to improve the accuracy of repeatable mechanics rather than requiring high accuracy of the components themselves. Torque ripple in motors and nonlinearities in feedback sensors can likewise be mapped. It is wrong to assume that all mechanics and electronics can be economically improved by the addition of software and control, but the mechatronic viewpoint shows that gains and tradeoffs are often possible. The emphasis in mechatronic design shifts heavily to the need for repeatability in designs, which agrees well with the deterministic machine design philosophy. Repeatability is the goal, and any non-repeatability or apparent randomness in the behavior is simply an inadequacy in the model. The philosophy is the same regardless of whether the system is fundamentally mechanical, electrical, or algorithmic in origin.

13.3.2 Discrete-Time System Fundamentals

Feedback compensation, and digital controllers in particular, are largely the element that distinguishes mechatronic systems from traditional electromechanical devices. Analog control continues to be used, and correctly so, in many applications. A compensator can often be built for just a few dollars in components. However, microprocessor control allows for greater flexibility in designing the controller, assures consistent operation, and may be the only method available for implementing some required functionality. Since microprocessor-based control will likely be encountered in some manner in precision system design, we will address it here. Two main features that distinguish analog from digital control are sampling (measurements taken only at fixed time intervals) with the associated aliasing problems and quantization (measurements have only a fixed number of values).

13.3.2.1 Sampling and Aliasing

Sampling refers to the process of taking measurements of system variables at a periodic rate. The system controller can then perform some action based on these measurements at that instant in time. The unavoidable problem is that no measurements of system behavior are available during the inter-sample period. If the frequency content of the signal to be sampled is low enough, then it can be proven that the samples alone are sufficient to exactly and uniquely reconstruct the original waveform. The criteria for this reconstruction is quantified with the *Nyquist Sampling Theorem* [22]. The Nyquist Sampling Theorem states that if $x_c(t)$ is a continuous-time bandlimited signal having a frequency content of

$$X_c(j\omega) = 0 \; for \, |\omega| > \omega_N \tag{13.5}$$

then $x_c(t)$ is *uniquely* determined by its samples $x[n] = x_c(nT), n = 0, \pm 1, \pm 2, \ldots$, if

$$\omega_s = \frac{2\pi}{T} > 2\omega_N \tag{13.6}$$

The frequency ω_N is referred to as the *Nyquist frequency*, and the sampling rate $(1/T)$ must be at least double this frequency in order for the samples to uniquely characterize the signal. More simply stated, a continuous-time signal can be *exactly* reconstructed from its samples if the signal contains no components at frequencies past half the sampling rate. In practice, it is usually preferable to sample several times faster than the Nyquist rate to ensure that several points are taken for each period of a sinusoid.

The consequence of sampling a signal at slower than the Nyquist rate is a phenomenon called *aliasing* in which the sampled waveform appears at a lower frequency than the original continuous-time signal. Figure 13.4 illustrates how any number of continuous-time frequencies can generate the same discrete-time sampled data points. Aliasing maps all continuous time frequencies onto the range from zero to the Nyquist frequency (i.e., half the sampling rate). One way to understand aliasing is by viewing it in the frequency domain. Oppenheim et al. [23] provide a detailed derivation of the process that is summarized here. Sampling a continuous-time signal can be modeled as convolving the signal with a train of impulses evenly spaced at the sample time. The resulting multiplication in the frequency domain places a copy of the frequency spectrum of the sampled signal at each integer multiple of the sampling frequency. If $x_c(t)$ is the original continuous-time waveform having a frequency content of $X_c(\omega)$, then the sampled signal will have a spectrum given by

$$X_P(\omega) = \frac{1}{T} \sum_{k=-\infty}^{+\infty} X_c(\omega - k\omega) \tag{13.7}$$

If there is any frequency content in the original signal past the Nyquist frequency, then there will be overlap between the repeated copies. This overlap, or mapping of higher-frequency signals into lower ones, is the phenomenon of *aliasing*. Figure 13.5 illustrates this operation. In the discrete-time frequency domain,

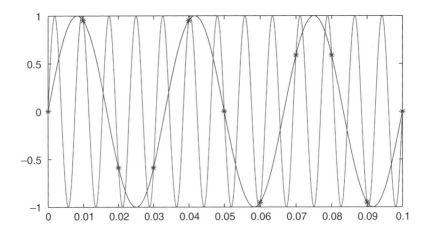

FIGURE 13.4 Multiple continuous-time frequencies can generate the same set of discrete-time samples. The lowest frequency sinusoid that fits all of the points is the fundamental frequency, and all higher-frequency sinusoids are aliased down to this lower frequency between zero and half the sampling frequency.

all frequencies are mapped into the region of $\pm\pi$ rad/sample. If the sampled signal is frequency-limited such that its highest component is less than half the sampling frequency, then the signal contained in the discrete-time frequency region of $\pm\pi$ rad/sample exactly captures the original continuous-time waveform. An ideal reconstruction filter passing only the initial copy of the components will reconstruct the original signal exactly. It is more common that the original waveform contains some components at frequencies above the Nyquist frequency, and the copies of these higher frequencies have wrapped down into the $\pm\pi$ band and added to the spectral content there. When reconstructed the higher-frequency components modify the original waveform.

Aliasing alters the apparent frequency of a signal but does not change the amplitude. This is an important distinction and is a reason why aliasing can be so problematic. As an example, consider measurements taken of a system variable at 1000 Hz. An apparent 300 Hz component in that measured waveform could be at 300 Hz, but could also be at 700, 1300, 1700 Hz, etc. There is no way to identify an aliased waveform, and so appropriate anti-aliasing precautions must be taken before sampling the waveform. A properly designed system will guard against aliasing, but as sampled-data systems are inherent in computer control (whether as part of the feedback system or simply diagnostic monitoring), the design engineer should be familiar with its effects.

Discussions of sampling are usually limited to time-based systems, but other indices can be used as well. For example, position-based sampling can be used to link sampling events in multi-axis systems to the

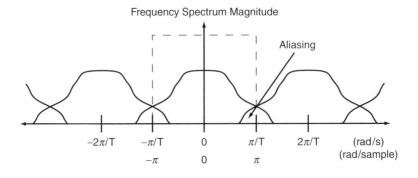

FIGURE 13.5 The frequency-domain view of aliasing shows that sampling creates repeated copies of the continuous-time frequency spectrum at each multiple of the sampling frequency. Aliasing occurs at the overlap between the copies.

position of a particular axis. This position-based sampling may be referred to as position-synchronized output (PSO), position event generation (PEG), or alternate terms. The fundamental idea is that actions are based in an external non-time-based signal. The same rules apply for sampling and aliasing whether the trigger signal is time-based, position-based, or otherwise. Sampling must be performed at least twice as fast as the highest frequency component (or shortest period for position-based systems) in the signal.

There are some instances, particularly in metrology, when aliasing can be used to simplify a measurement. This is because aliasing changes frequency but does not alter magnitude. As an example, consider an accuracy measurement that is to be taken of a ballscrew-based system. Mapping short-wave errors (occurring within one rotation of the screw) would seem to require taking many points per revolution of the screw. This can be time-consuming, and even counter-productive when longer term temperature drifts in the environment are considered. Sampling at exactly the period of the screw (once per revolution) will alias the short-wave errors into constant values. This measures only the long-wave accuracy errors of the screw. Sampling even slower, longer than the period of the screw, aliases the higher-frequency intercycle errors into lower-frequency (longer period) values. This measurement shows the long-wavelength accuracy errors and the short-wavelength intercycle errors both with the same economical test setup.

13.3.2.2 Quantization

The term *quantization* refers to the process of taking continuous signals (with infinite resolution) and expressing them as a finite resolution number for processing by digital systems. There is always some error inherent in the conversion, and at some level, the error can make a meaningful difference in the accuracy of the measurement or performance of the servo loop. The common sources of quantization in measurements are reviewed further here.

13.3.2.2.1 Analog-to-Digital Conversion

Most analog position sensors will be used with a digital control loop, and so the signal must be discretized with the use of an analog-to-digital (A/D) converter. To prevent aliasing, an analog low-pass filter should be placed before the A/D conversion that attenuates the signal content in all frequencies past one-half the sampling frequency. Typically the cutoff frequency of a practical analog filter is placed somewhat lower than this. The phase lag of the analog filter should always be included in the controller model. Analog-to-digital converters are rated in terms of *bits* of resolution where the minimum quanta size is given by

$$\Delta = \frac{Range}{2^{\#bits}} \qquad (13.8)$$

The two most common in precision control systems are 12-bit (having 4096 counts) and 16-bit (having 65536 counts). The total range of travel divided by the number of counts gives the fundamental electrical resolution. The true analog value is always somewhere between counts, and so the quantization process adds some error to the measurement. This error can be approximated as a random signal, uniformly distributed over a single quanta with a variance of $\Delta^2/12$.

The fundamental A/D resolution can be improved by oversampling and averaging the A/D converter signal. Typically an A/D converter can sample at several hundred kilohertz while the position control loop only updates at several thousand Hertz. Rather than taking one sample per servo cycle, the software can therefore average several hundred readings if sufficient processing power is available. Averaging N samples reduces the standard deviation of the signal by $\sqrt{N}$. This technique only works when the signal is sufficiently noisy to toggle between counts, otherwise a dither signal can be added to the sensor reading [5]. Typically however, the wiring in most systems absorbs enough electrical noise that adding anything additional is not required.

13.3.2.2.2 Encoders and Incremental Position Sensors

Some types of position sensors produce inherently quantized output. Digital encoders are one example of these. For a square-wave output encoder, the resolution is set simply by the period of the lines. Encoders used in precision machines are often supplied with electronics that interpolate the resolution much lower than the grating period. Typical quadrature decoders increase the resolution to four times the grating period, but other electronics are available that increase this to 4096 times or higher in compatible encoders. Encoders that have an amplified sinusoid output ideally have analog signals modeled by

$$v_a = V \sin\left(\frac{2\pi}{\lambda} \Delta x\right) \quad (13.9)$$

$$v_b = V \cos\left(\frac{2\pi}{\lambda} \Delta x\right) \quad (13.10)$$

where λ is the period of the encoder, and Δx is the displacement. When plotted vs. each other, they form a perfect circle,

$$v_a^2 + v_b^2 = V^2 \quad (13.11)$$

The displacement within a single period of the scale (i.e., angle around the circle) can be found by taking an inverse-tangent,

$$\Delta x = \tan^{-1}\left(\frac{v_a}{v_b}\right) \quad (13.12)$$

which is usually done through the use of a software lookup table parametric on v_a and v_b. The resolution of the A/D converter and size of the lookup table determine the new size of the position quanta.

There are errors in the practical application of encoder multiplication. The two analog sinusoid signals will have offsets in their DC level, gains, and phase. A more realistic representation of the encoder signals are

$$v_a = V_a \sin\left(\frac{2\pi}{\lambda} \Delta x\right) + a_0 + noise \quad (13.13)$$

$$v_b = V_b \cos\left(\frac{2\pi}{\lambda} \Delta x + \phi\right) + b_0 + noise \quad (13.14)$$

where each signal can have a DC offset, a different gain, and a nonorthogonal phase shift. Depending on the quality of the encoder, there may also be higher-order sinusoids present in the signal. When plotted against each other, these signals no longer form a perfect circle, but the shape is instead primarily a fuzzy off-axis ellipse. Higher-order sinusoids can create multi-lobed shapes. These signals are usually functions of alignment and analog circuit tuning and may change with position along the encoder scale. Higher-quality scales that are installed with careful attention to mounting details usually will have patterns that are more stable and circular, allowing higher levels of interpolation.

Interpolation errors are modeled as a sinusoidal noise source at speed-dependent frequencies. Offset errors in the multiplication occur at a frequency equal to the speed divided by the grating period, and errors in gain result in twice the frequency. For example, poor multiplier tuning in a stage with a 4 μm grating period traveling at 10 mm/s results in noise disturbances at 2500 and 5000 Hz. On typical systems, these disturbance frequencies can easily be several times the sampling frequency of the controller, meaning that they will alias down to below the Nyquist frequency. If this frequency falls well above the servo bandwidth of the stage, the noise source will appear in the frequency spectrum of the position error while scanning, if it falls within the servo bandwidth, it will not appear at all since the stage is tracking it. It will only show when the stage position is measured with a secondary sensor. There is no equivalent to the analog anti-aliasing filter used with A/D converters here. The best method to reduce this error is by carefully tuning the multiplier. Alternatively, the encoder signal can be sampled several times faster than the servo bandwidth and averaged for each update. This is more computationally intensive but has the effect of increasing the Nyquist frequency for this process.

13.3.2.2.3 Digital-to-Analog Conversion

A reconstruction filter must be used to generate a continuous-time waveform from discrete-time samples, and the characteristics of this filter affect the accuracy of the reconstruction. The ideal reconstruction filter is a low-pass filter that exactly passes all frequency components less than the Nyquist frequency and exactly blocks all frequencies higher. Such a filter is not realizable. The most common method to reconstruct a continuous-time waveform is through a digital-to-analog converter that holds an output at a discrete level for the duration of a sample period and changes at the next sample instant. This type of filter is referred to as a *zero-order-hold* (ZOH) reconstruction filter, and this filter adds some error to the reconstructed signal [22]. The ZOH reconstruction filter has an equivalent continuous-time frequency response of

$$H_{ZOH}(j\omega) = T e^{-j\frac{\omega T}{2}} \frac{\sin(\omega T/2)}{\omega T/2} \quad (13.15)$$

with magnitude and phase as shown in Figure 13.6. The sharp edges in the "staircase" output are the higher-frequency components in the signal at amplitudes indicated by the sidelobes of the frequency response.

The effect of the ZOH reconstruction filter is to include higher-frequency components in the spectrum of the output signal. As an example, consider the case of a 200 Hz sinusoid written through a D/A converter at a 1 kHz rate. Figure 13.7 shows the resulting waveform. The ideal reconstruction filter would create a waveform that is exactly

$$v_{ideal}(t) = \sin(2\pi 200 t) \quad (13.16)$$

However, according to Equation (13.15), the signal produced by the ZOH reconstruction filter is instead

$$v_{ZOH}(t) = 0.9355 \sin(2\pi 200 t - 36°) + 0.2339 \sin(2\pi 800 t - 144°) + H.F.T \quad (13.17)$$

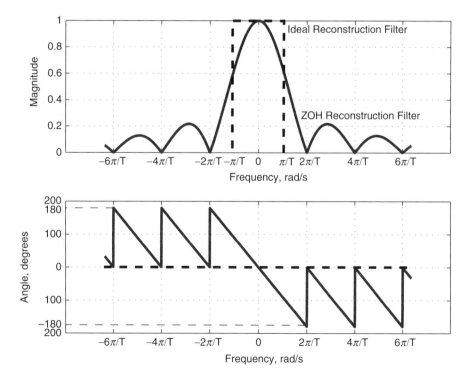

FIGURE 13.6 The magnitude and phase of the zero-order-hold reconstruction filter are significantly different from the ideal reconstruction filter, particularly as the frequency approaches the Nyquist frequency.

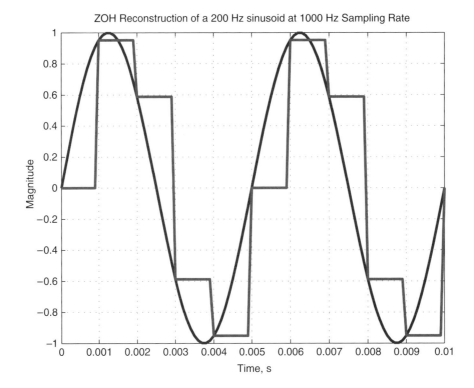

FIGURE 13.7 The zero-order-hold reconstruction filter creates a "stairstep" version of a signal that can be significantly distorted as compared with a continuous-time signal.

The difference between the target and actual outputs can be significant for precision machines, and a compensation filter may be required to pre-condition the discrete-time signal before sending it to the D/A converter. Note also in this case that the 800 Hz component of the output signal is above the Nyquist frequency. This command signal (presumably sent to the amplifiers) will generate motion at 800 Hz, but the 1000 Hz sampling frequency will alias 800 Hz down to 200 Hz, corrupting the measurement of actual motion of the stage at 200 Hz. A clear understanding of all sampling and reconstruction filters is usually needed to interpret the frequency content of signals that transfer between analog and digital domains.

13.3.3 Precision Mechanics

Precision mechatronic systems must begin with precision mechanics. Controls and electronics can be used to correct for some errors, but the controls problem is always easier when the mechanics are well-behaved (meaning repeatable and readily modeled). Some of the main components of the mechanics are the bearings, the machine structure itself, and the vibration isolation system.

13.3.3.1 Linear and Rotary Bearings

The bearings of a precision machine are the most critical element that defines the performance of a machine. Most bearings (or sets of bearings) are designed to constrain all but one degree of freedom of motion, and bearings of all types are used in precision machines. Many of the types of bearings are available in linear and rotary versions. The designer must choose a bearing based on its load carrying capability, stiffness, repeatability and resolution (the ability to move in small increments), friction, size, and cost. Slocum [34] provides a good overview of different bearing types, their advantages, limitations, and preferred uses.

13.3.3.2 Machine Structure

The overall machine structure and vibration isolation system must be considered as part of a precision machine design and installation. The primary concern in designing the structure of the machine is to provide a dimensionally-stable base with well-characterized, well-damped resonant modes. Vibrations of these modes can enter the feedback loop and either destabilize the system or, more typically, add an extra mode to the response that extends the settling time for a move. Achieving high stiffness is relatively quantitative, particular with the use of finite element models for analysis. However, achieving high damping is equally important in attenuating the influence of the structural modes. Most engineering materials (metals usually) have relatively little internal damping, and so damping must be explicitly added. Riven [26], Nayfeh [21], and Marsh and Slocum [20] detail methods for designing specific damping elements into structures. Jones [16] concentrates in a specific family of damping techniques using viscoelastic polymers. A common use of these materials is in *constrained layer dampers* in which a sheet of damping material is sandwiched between a structural member and a stiff constraining layer. Any vibration of the structural member shears and strains the viscoelastic material, thus creating a loss mechanism for the energy at the vibration frequency. These damping techniques are often applied in an attempt to fix existing problematic designs, but with mixed success. It is preferable to address damping in the mechanical elements at the earliest possible stage of the design.

13.3.3.3 Vibration Isolation

Vibration isolation systems are used primarily to attenuate the influence of ground-borne vibrations on the position stability of a precision machine. In other cases, the vibration isolation is to present movements of the machine itself from detrimentally influencing surrounding processes. Riven [25] and DeBra [6] provide detailed overviews of vibration isolation of precision machines. It should be noted that this is an active research area with significant publication activity. The two general types of isolation, passive and active, differ mainly on whether direct measurement of the vibration is used to help attenuate it. Passive isolation vibration isolation systems are usually chosen based on their natural frequency and damping level. Riven [27] details the design and application of passive isolation devices. Active isolation systems are generally more complex and require one of more vibration sensors (usually accelerometers or geophones) to measure payload and ground vibrations then apply forces to the payload to oppose this motion [12]. For cost reasons, passive systems are almost always preferable, but all-passive systems have a fundamental limitation that active systems can overcome.

The fundamentals of the problem can be seen with a conceptual single-degree-of-freedom model. Consider the case of a free mass, representing the machine base, attached to the ground with a combination spring-damper. To provide isolation from ground-based vibrations, the spring and damper should be made as soft as possible. However, there are also disturbance forces applied directly to the machine base. These forces are usually reaction forces created by motion of the devices attached to the base. Keeping the base stationary under these forces requires that the spring-damper system be as rigid as possible. The two requirements are in opposition and can be expressed mathematically as sensitivity and complementary sensitivity functions. This means they always add to unity, and any improvement in rejection of disturbance forces comes exactly at the cost of reduced rejection of ground vibrations. There is no way to adjust the impedance of a passive mount to improve the isolation from both sources. Isolating from ground vibrations requires a soft mount; isolating from disturbance forces requires a stiff mount. Active isolation systems do not have this same limitation since they are able to measure ground and payload vibration directly and (in-effect) adjust the instantaneous impedance of the mount as conditions require.

Most sources of vibration can be characterized as a summation of single-frequency sinusoids (from various pumps and motors in a facility) and random vibrations with a given spectral density. One common set of standards for characterizing the level of seismic vibration in a facility is the Bolt Beranek & Newman (BBN) criteria [13]. Their study presents vibration levels as a series of third-octave curves plotting RMS velocity levels vs. frequency. The curve levels range from VC–A, suitable for low-power optical microscopes, to VC–E, the highest level presumed to be adequate for the most-demanding applications. Most precision machines are designed to be placed in existing facilities, and so whenever possible, it is preferred to take a sample measurement of ground vibrations directly. Note that the ground does not just vibrate vertically, but laterally as well. Several measurements are usually necessary since vibration levels can vary greatly at different locations and at different times. These measured-vibration levels can be characterized and used as the input to a model of the isolation system. The response to the deterministic (single-frequency) elements of the disturbance can be predicted directly, but only a statistical characterization of the motion is possible given a random input. Wirsching et al. [38] detail techniques required to model the response of dynamic systems to random vibrations.

13.3.4 Controller Implementation

Dedicated electronics and hardware are required to implement the algorithms that control precision machines. The hardware usually consists of a top-level controller, power amplifiers, actuators, and position feedback sensors. Appropriate software and control algorithms complete the system design. These elements of a mechatronic design are as critical to the overall performance of a system as are the mechanics themselves.

13.3.4.1 Feedback Control Hardware

Microprocessor-based controllers implement the feedback compensation algorithms for most precision machines. Position is usually the controlled variable, but force, pressure, velocity, acceleration, or any other measured variable may also be controlled. The controller may be a standalone unit or may require a host PC. Commercially available systems differ in the number of axes they are able to control, the amount of additional I/O they can read and write to, their expandability, methods of communication to other electronic hardware, and as always, cost. In general, the analog circuitry in these controllers is being replaced by digital or microprocessor-based circuits. Analyzing and troubleshooting these systems require dedicated software routines rather than a simple oscilloscope, and the availability of such an interface should also factor into the choice of a controller. The selection of an appropriate motion controller may be driven by the need to be compatible with existing equipment, existing software, or existing expertise within a group. As is generally the case with microprocessor-based systems, the features available increase so rapidly that the system design engineer should always consult the controller manufacturers frequently. For unique applications, the best alternative may be to select a general-purpose processor and write the low level code required for implementing the controller, but this should usually be viewed as a last resort.

13.3.4.2 Power Amplifiers

Power amplifiers convert the low-level signals from the feedback compensation into high-power signals suitable for driving the actuators. Power amplifiers for electromagnetic actuators usually take the form of a current loop in which the command is a scaled representation of the desired actuator current. A proportional or proportional-integral control loop adjusts the voltage applied to the motor as required to maintain the commanded current. The voltage available to drive the current is limited by the amplifier bus rail and the speed-dependent back-emf voltage of the motor. Most higher-power actuators and electronics are multi-phase, typically three-phase, and so there will be three such compensators inside each amplifier. Power amplifiers have traditionally been designed with analog compensation for the current loop. Passive elements (resistors and capacitors) used to set the loop gains may be mounted to a removable board, also known as a personality module, to accommodate motors with different values of resistance and inductance. Potentiometers may also be included to allow fine-tuning of the control, but setting these can be difficult to quantify, and the setting may drift over time. Power amplifiers are increasingly moving to digital control loops, in which a microprocessor implements the control algorithm. A key advantage here is that the gain settings can be rapidly and repeatably matched for the desired performance of the current loop to the exact motor used.

Two basic types of power amplifiers are available for a machine designer to select from. *Linear* amplifiers are typically very low-noise, but larger and less power-efficient than comparable *switching* amplifiers. Switching amplifiers, also known as *pulse-width-modulation* (PWM) amplifiers operate by rapidly switching the voltage applied to the actuator between zero and full-scale. The duty cycle of the switching sets the average level of current flowing through the actuator, and the switching typically occurs at 20–50 kHz. PWM amplifiers rely on motor inductance to filter the rapidly changing voltage to a near-continuous current. Some level of ripple current will almost always be present in PWM systems.

In either case, the system designer should characterize the performance of the power amplifier. The amplifier should have a high-enough bandwidth on the current loop to track the required commands with an acceptable level of phase lag, but not so high that excessive high-frequency noise amplification occurs. The system designer should also monitor the fidelity of the amplified signal as it passes through zero amps. This is typically a handoff point in the amplifiers, where conduction passes from one set of transistors to another, and is a potential source of error.

13.3.4.3 Actuators

Precision systems use a variety of actuators to convert the amplifier commands into forces and motion. Brushless permanent-magnet motors, in linear and rotary forms, are the most common. These motors are usually three-phase systems and require a supervisory controller for *commutation* to ensure continuous motion. The system designer can select between *iron-core* and *ironless* motors in an application. Iron core motors offer higher force (or torque) density than do ironless models, but the magnetic lamination leads to cogging that affects the smoothness of the applied force. As usual, there is no best choice for all applications. Other types of actuators include voice-coil type motors, stepper motors, and piezoelectric drives.

13.3.5 Feedback Sensors

The selection, mounting, and location of the sensors are critical to the successful design of a motion control system. A controller with poor feedback devices will be unable to perform well regardless of the care taken in the mechanical design or the sophistication of the control algorithms. The most common sensors used in precision machines are encoders, laser interferometers, and a variety of analog feedback devices.

13.3.5.1 Rotary Encoders

Rotary encoders are used to measure the angular displacement of rotary tables and motors. They are inherently incremental devices and usually need some form of a reference marker to establish a home position. The primary source of errors in rotary encoders is misalignment between the optical center of the grating pattern and the geometric center of the axis of rotation. The angular errors caused by decentering can be estimated by

$$\epsilon = \pm 412 \frac{e}{D} \tag{13.18}$$

where ϵ is the angular error in arc-sec (1 arc-sec $\approx$ 4.85 μrad), e is the centering error in microns, and D is the diameter of the grating pattern in millimeters. A 5 μm centering error creates an angular error of 20 arc-sec on a 100 mm diameter encoder. This error is repeatable and can be mapped and corrected for in software. When used with a ballscrew, a rotary encoder can be used to indicate linear position, but allowances must be made in the error budget for changes to the lead of the screw over the range of operating temperatures. An alternative to software correction for the decentering error is to use additional encoder readheads. A second readhead, placed 180° opposite the first, reverses the decentering error. Averaging the signals from the two readheads cancels it. Reversal techniques abound in precision machine design [10], and the system design engineer should be familiar with them.

13.3.5.2 Linear Encoders

Linear encoders are somewhat easier to mount and align than are rotary encoders because the fundamental mounting error is to misalign the encoder with the direction of travel. This results in a cosine error and a resulting linear scale factor error. Once this factor is known, it is a simple scaling to correct in software. The more difficult problem is to reliably mount the scale to still perform properly under thermal changes. Scales made of zero-expansion glass are available, but they are typically mounted to stages that do expand and contract. In this case, the designer should compensate for the mismatch by fixing a point on the scale, preferably at some home signal, and allowing the scale to expand from there. The opposite case is to use a scale that adheres to the substrate firmly enough to expand and contract with the structure. Allowing for growth is not a problem, but it can be difficult to choose a fixed point about which growth occurs to use in thermal error mapping.

13.3.5.3 Laser Interferometers

Laser interferometers measure displacement by monitoring the interference between a reference beam, and one reflected from the target. Most interferometers use a stabilized Helium-Neon source with a wavelength of 632.2 nm as their basis. Linear displacement measuring interferometers are generally classified as either *homodyne* or *heterodyne* types. Homodyne interferometers use a single-frequency laser with optics that interfere a reference beam with one reflected from a moving target. The resulting constructive and destructive interference effectively creates a position-dependent periodic intensity that can be converted into an electrical signal with a set of photodiodes. To the end-user, the resulting sinusoidal signals (in quadrature) are treated no differently from the output of an analog encoder. Heterodyne interferometers use a two-frequency laser. One of the frequencies is diverted to a stationary reference, and the other frequency is reflected off the moving target. The frequency of the measurement beam is Doppler shifted by the velocity of the target on its return. This frequency shift is measured as a phase difference in the beat frequency produced when the measurement and reference beams are recombined. This phase difference can be accurately measured and converted to a velocity and displacement. Both categories, homodyne and heterodyne, have their fervent critics and supporters. In either case, sub-nanometer resolutions with sub-micron absolute accuracy are possible.

There are several major advantages to laser interferometers over linear encoders. The measurement is potentially very accurate due to its basis in the frequency of the laser beam. Calibration with a laser interferometer is generally used as the final confirmation of the accuracy of a machine. The measurement can also be made very close to the tool or workpiece, thus reducing or even eliminating Abbé errors. The noncontact nature of the measurement makes it easier to separate the force and metrology loops in the design.

The performance of a laser interferometer is usually limited by the environment that it operates in. Changes in air temperature, pressure, and humidity affect the index of refraction in the air and, thus, the wavelength of light and accuracy of measurement. Bulk changes can be compensated for with Edlens Equation [8,36], but localized changes (due to air turbulence) are nearly impossible to track. The best environment in which to operate a laser interferometer is a vacuum. When a vacuum is not practical, the beams should be routed through tubes that guard against air currents, and the free path of the beam kept as short as possible. The beam path of a laser interferometer is part of the metrology loop, and so errors in this measurement are not discernable in the feedback loop. In other cases, particularly when the stage is moving, uncertainty in the time of the measurement limits the overall accuracy. The interferometer can say where the stage was, but not precisely when it was there or where it may be now. Systems attempting for nanometer-level accuracy must incorporate this data age uncertainty into the error budget. Correctly applying a laser interferometer feedback system to a precision machine requires careful attention to detail, but in most cases, it provides the highest-resolution, highest-accuracy measurement possible.

13.3.5.4 Analog Sensors

Analog position sensors are frequently used in limited-travel applications and can resolve motion to less than 1 nm with appropriate electronics. Common types of sensors include capacitance gages, eddy-current sensors, and LVDTs. Their relatively small size and limited range make them a natural match to flexure-based mechanisms. Resolution is a function of the measurement bandwidth in analog systems, and the sensor noise floor should never be quoted without also noting the bandwidth at which it applies. Assuming that the noise floor of an analog sensor is relatively constant over all frequencies (i.e., white noise), then the standard deviation of this noise reduces with the square root of the bandwidth of the measurement. Figure 13.8 demonstrates the results of this operation. An analog sensor with a 3σ resolution of 10 nm at a 1 kHz bandwidth will have a resolution of 3.2 nm at 100 Hz, and 1 nm at 10 Hz. This increase in resolution increases the phase shift of the feedback system and possibly will require a reduction in

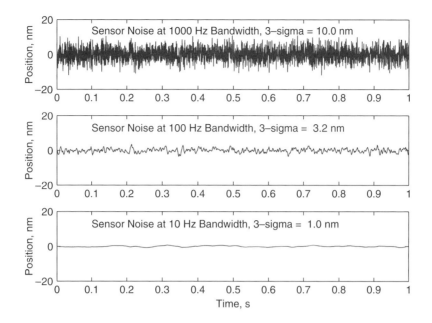

FIGURE 13.8 Effect of analog filtering on the resolution of an analog sensor.

controller loop bandwidth. If the noise present in the sensor measurement is essentially random, the sensor resolution can be expressed as a spectral density and quoted as nm/$\sqrt{\text{Hz}}$. The designer can then match the appropriate filter bandwidth to his resolution requirements. However, a caution here is that many sensors have inherently higher levels of noise at the lower frequencies. The amount of filtering required to achieve a resolution may be higher than expected. Analog sensors are usually supplied with signal conditioning electronics that convert the natural output of the sensor to a standard form (typically ± 10 V). These electronics should be placed as close as possible to the sensor because the low-level signals in the raw sensor output are often highly susceptible to electronic noise. The conditioned signal is usually somewhat less sensitive but should still be shielded and routed as far as possible from high-power conductors. Analog sensor accuracy and linearity is entirely distinct from the resolution specification. Sensors may be highly linear about an operating point but often become increasingly nonlinear over their full range of travel. This nonlinearity can be compensated for with a table or curve fit in software. Sensor manufacturers may also provide a linearized output from their electronics package. This is very useful for metrology, but a designer should confirm the update rate of this output before using for feedback. Often this additional output is no longer a native analog signal, but it may have been quantized, corrected, and re-synthesized through a digital-to-analog converter. All of these processes add time and phase lag to the measurement making it less suitable for use as a feedback device.

13.3.6 Control Algorithms

The application of control algorithms separates mechatronics systems and mechatronic design from traditional electromechanical systems. Modern controllers do more than simply close a PID loop. They provide additional options on feedback filtering and feedforward control, allow for coordination between multiple axes, and should provide an array of tuning and data collection utilities. An overview of control algorithms used with precision machines is provided here.

13.3.6.1 System Modeling

Dynamic models of the system should be developed for creating the controller and verifying its performance. Each controller design should begin with a simple rigid-body model of the overall structure.

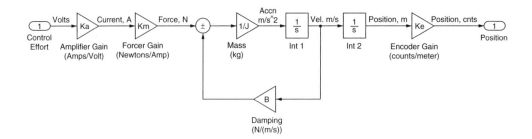

FIGURE 13.9 Simplified plant model used for initial single-axis tuning and motor sizing.

This simplified model is justified when, at a typical closed-loop bandwidth for mechanical systems of 10–100 Hz, friction is low and any structural resonances are located at significantly higher frequencies. While not complete, the simplified model shows the bulk rigid-body motion that can be expected given the mass of the stage, the selection of motors and amplifiers, and the types of commanded moves. It can also be used to generate a more-accurate estimation of motor currents (and power requirements) for typical machine moves. Figure 13.9 shows a block diagram of a simplified linear-motor-driven system. The voltage from the analog output of the controller is converted into current by the amplifiers and into a force by the linear motors. In this first model, we assume no dynamics in the amplifier and that commutation is implemented properly to eliminate force ripple. The motor force accelerates the stage mass, and an encoder measures the resulting stage position. The continuous-time transfer function from control-effort to stage position (in counts) is therefore

$$P(s) = \frac{k}{s(ms+b)} \tag{13.19}$$

and this transfer function can be used to generate controller gains. Scaling factors will always be specific to the supplier of the control electronics, and it may take some research to find these exactly.

The simplified model is adequate for the initial generation of gain parameters, but determining performance at the nanometer level always involves augmenting the model with additional information. This advanced model should include the discrete-time controller, quantized position feedback and control effort, higher order dynamics in the amplifier, higher-frequency mechanical modes (obtained either from a finite-element study or experimentally through a modal analysis), force ripple in the motors, error in the feedback device, and friction and process-force effects. Some can be predicted analytically before building the system, but all should be measured analytically as the parts of the system come together to refine the model.

13.3.6.2 Feedback Compensation

Commercially available controllers generally use some form of classical proportional-integral-derivative (PID) control as their primary control architecture. This PID controller is usually coupled with one or more second-order digital filters to allow the user to design notch and low-pass filters for loop-shaping requirements. In continuous-time notation, the preferred form of the PID compensator is

$$C(s) = K_p \left(1 + T_D s + \frac{1}{T_I s}\right) = K_p \frac{T_D T_I s^2 + T_I s + 1}{T_I s} \tag{13.20}$$

Separating the proportional term by placing it in front of the compensator allows the control engineer to compensate for the most common system changes (payload, actuator, and sensor resolution) by changing a single gain term (K_p). Techniques for designing continuous-time PID loops are well documented in almost any undergraduate-level feedback controls textbook. The preferred method for designing, analyzing, and characterizing servo system performance is by measuring *loop-transmissions*, otherwise known as open-loop transfer functions. This measure quantitatively expresses the bandwidth of the system in terms of the *crossover frequency* and the damping in terms of the *phase margin*.

13.3.6.3 Feedforward Compensation

Feedforward compensation uses a quantitative model of the plant dynamics to prefilter a command to improve tracking of a reference signal. Feedforward control is typically not studied as often as feedback control, but it can cause tremendous differences in the ability of a system to follow a position command. Feedforward control differs from feedback in that feedforward control does not affect system stability. If a perfect model of the plant were available with precisely known initial conditions, then the feedforward filter would conceptually consist of an inverse model of the plant. Preshaping the reference signal with this inverse model before commanding it to the plant will effectively lead to unity gain (prefect tracking) at all frequencies. If properly tuned, the feedback system is required only to compensate for modeling inaccuracies in the feedforward filter and to respond to external disturbances. In practical implementations, the plant cannot be exactly inverted over all frequencies, but the feedforward gain will instead boost tracking ability at low frequencies.

Formal techniques exist for creating feedforward filters for generalized discrete-time plants. Tomizuka [37] developed the zero phase error tracking controller (ZPETC) that inverts the minimum-phase zeroes of a plant model and cancels the phase shift of the nonminimum phase components. Nonminimum phase zeroes are a common occurrence in discrete-time systems. The ZPETC technique begins with a discrete-time model of the closed-loop system (plant and controller),

$$G_p(z^{-1}) = \frac{z^{-d} B_c^-(z^{-1}) B_c^+(z^{-1})}{A_c(z^{-1})} \qquad (13.21)$$

in which the numerator is factored in minimum-phase zeros B_c^+ and non-minimum-phase zeros B_c^-. The z^{-d} term is the d-step delay incurred due to computation time or any phase lag. The ZPETC filter for this system becomes

$$G_{ZPETC}(z^{-1}) = \frac{z^d A_c(z^{-1}) B_c^-(z)}{B_c^+(z^{-1}) B_c^-(1)^2} \qquad (13.22)$$

which cancels the poles and minimum-phase zeros of the system, and cancels the phase of the nonminimum phase zeros. This is a noncausal filter, but it is made causal with a look ahead from the trajectory generator. In multi-axis machine, it is generally allowable to delay the position command by a few servo samples provided the commands to all axes are delayed equally and the motions remain synchronized.

Feedforward control techniques are also available that modify the input command to cancel trajectory-induced vibrations. One such technique, trademarked under the name Input Shaping was developed by Singer and Seering [31] at MIT. Input Shaping,™ and related techniques, convolve the input trajectory with a series of carefully timed and scaled impulses. These impulses are timed to be spaced by a half-period of the unwanted oscillation and are scaled so that the oscillation produced by the first impulse is cancelled by the second impulse. Much of the research work in this area has involved making the filter robust to changes in the oscillation frequency. The downside to this technique, in very general terms, is that the time allowed for the rigid-body move must be lengthened by one-half period of the oscillation to be cancelled.

Specialized feedforward techniques can be applied when the commanded move is cyclic and repetitive. In this case, the move command can be expressed as a Fourier series and decomposed into a summation of sinusoidal components [19]. The steady-state response of a linear system to a sinusoid is well defined as a magnitude and phase shift of the input signal, and so each sinusoidal component of the command can be pre-shifted by the required amount such that the actual movement matches the required trajectory.

13.4 Conclusions

Precision positioning of rotary and linear systems requires a mechatronic approach to the overall system design. The philosophy of determinism in machine design holds that the performance of the system can be predicted using familiar engineering principles. In this chapter, we have presented some of the components that enter into a deterministic precision machine design. Concepts such as alignment errors, force and metrology loops, and kinematic constraint guide the design of repeatable and analytic mechanics. Often-neglected elements in the mechanical design are the cable management system, heat transfer and thermal management, environmental protection (protecting system from the environment it operates in and vice-versa), and inevitable serviceability and maintance. Mechatronic systems require precision mechanics; high-quality sensors, actuators, and amplifiers; and a supervisory controller. This controller is usually implemented with a microprocessor, and the discrete-time effects of sampling, aliasing, and quantization must be accounted for by the system design engineer. Finally, the feedback control algorithms must be defined, implemented, and tested. The elements used to create precision mechatronic systems are widely available, and it is left to the designer to choose among them trading complexity and cost between mechanics, electronics, and software.

References

[1] The American Society for Precision Engineering, P.O. Box 10826, Raleigh, NC 27605-0826 USA.

[2] Auslander, D.M. and Kempf, C.J. *Mechatronics: Mechanical System Interfacing.* Prentice Hall, Upper Saddle River, NJ, 1996.

[3] Blanding, D.L. *Exact Constraint: Machine Design Using Kinematic Principles.* American Society of Mechanical Engineers, New York, NY, 1999.

[4] Bryan, J.B. The deterministic approach in metrology and manufacturing. In: *Proceedings of the ASME 1993 International Forum on Dimensional Tolerancing and Metrology,* Dearborn, MI, Jun 17–19 1993.

[5] Carbone, P. and Petri, D. Effect of additive dither on the resolution of ideal quantizers. *IEEE Trans. Instrum. Meas.,* 43(3):389–396, Jun 1994.

[6] DeBra, D.D. Vibration isolation of precision machine tools and instruments. *Ann. CIRP,* 41(2):711–718, 1992.

[7] Donaldson, R.R. and Patterson, S.R. Design and construction of a large, vertical axis diamond turning machine. *Proc. SPIE – Int. Soc. Opt. Eng.,* 433:62–67, 1983.

[8] Edlen, B. The dispersion of standard air. *J. Opt. Soc. Am.,* 43(5):339–344, 1953.

[9] Evans, C. *Precision Engineering: An Evolutionary View.* Cranfield Press, Bedford, UK, 1989.

[10] Evans, C.J., Hocken, R.J., and Estler, W.T. Self-calibration: reversal, redundancy, error separation, and "absolute testing." *CIRP Ann.,* 45(2):617–634, 1996.

[11] Franklin, G.F., Powell, J.D., and Workman, M.W. *Digital Control of Dynamic Systems,* 3rd ed., Addison-Wesley, Reading, MA, 1998.

[12] Fuller, C.R., Elliot, S.J., and Nelson, P.A. *Active Control of Vibration.* Academic Press, London, 1996.

[13] Gordon, C.G. Generic criteria for vibration-sensitive equipment. In: *Vibration Control in Microelectronics, Optics, and Metrology. Proc. SPIE,* 1619:71–85, 1991.

[14] Hale, L.C. Friction-based design of kinematic couplings. In: *Proceedings of the 13th Annual Meeting of the American Society for Precision Engineering,* St. Louis, MO, 18, 45–48, 1998.

[15] Hale, L.C. *Principles and Techniques for Designing Precision Machines.* Ph.D. Thesis, MIT, Cambridge, 1999.

[16] Jones, D.I.G. *Handbook of Viscoelastic Vibration Damping.* John Wiley & Sons, New York, 2001.

[17] Kiong, T.K., Heng, L.T., Huifang, D., and Sunan, H. *Precision Motion Control: Design and Implementation.* Springer-Verlag, Heidelberg, 2001.

[18] Kurfess, T.R. and Jenkins, H.E. Ultra-high precision control. In: William S. Levine, editor, *The Control Handbook,* 1386–1404. CRC Press and IEEE Press, Boca Raton, FL, 1996.

[19] Ludwick, S.J. *A Rotary Fast Tool Servo for Diamond Turning of Asymmetric Optics.* Ph.D. Thesis, MIT, Cambridge, 1999.

[20] Marsh, E.R. and Slocum, A.H. An integrated approach to structural damping. *Precision Eng.,* 18(2/3):103–109, 1996.

[21] Nayfeh, S.A. *Design and Application of Damped Machine Elements.* Ph.D. Thesis, MIT, Cambridge, 1998.

[22] Oppenheim, A.V. and Shafer, R.W. *Discrete Time Signal Processing.* Prentice Hall, Englewood Cliffs, NJ, 1989.
[23] Oppenheim, A.V., Willsky, A.S., and Young, I.T. *Signals and Systems.* Prentice Hall, Englewood Cliffs, NJ, 1983.
[24] Åstrom K.J. and Wittenmark, B. *Computer Controlled Systems: Theory and Design,* 3rd ed., Prentice Hall, 1996.
[25] Riven, E.I. Vibration isolation of precision equipment. *Precision Engineering,* 17(1):41–56, 1995.
[26] Riven, E.I. *Stiffness and Damping in Mechanical Design.* Marcel Dekker, New York, 1999.
[27] Riven, E.I. *Passive Vibration Isolation.* ASME Press, New York, 2003.
[28] Schmiechen, P. and Slocum, A. Analysis of kinematic systems: a generalized approach. *Precision Engineering,* 19(1):11–18, 1996.
[29] Schouten, C.H., Rosielle, P.C.J.N., and Schellekens, P.H.J. Design of a kinematic coupling for precision applications. *Precision Engineering,* 20(1):46–52, 1997.
[30] Shigley, J.E. and Mischke, C.R. *Standard Handbook of Machine Design.* McGraw-Hill, New York, 1986.
[31] Singer, N.C. and Seering, W.P. Preshaping command inputs to reduce system vibration. *J. Dynamic Syst., Meas., Control,* 112:76–82, March 1990.
[32] Slocum, A.H. Kinematic couplings for precision fixturing — part 1: Formulation of design parameters. *Precision Engineering,* 10(2), Apr 1988.
[33] Slocum, A.H. Kinematic couplings for precision fixturing — part 2: Experimental determination of repeatability and stiffness. *Precision Engineering,* 10(3), Jul 1988.
[34] Slocum, A.H. *Precision Machine Design.* Prentice Hall, Englewood Cliffs, NJ, 1992.
[35] Smith, S.T. and Chetwynd, D.G. *Foundations of Ultraprecision Mechanism Design.* Gordon and Breach Science Publishers, Amsterdam, the Netherlands, 1992.
[36] Stone, J., Phillips, S.D., and Mandolfo, G.A. Corrections for wavelength variations in precision interferometric displacement measurements. *J. Res. Natl. Inst. Stand. Technol.,* 101(5):671–674, 1996.
[37] Tomizuka, M. Zero phase error tracking algorithm for digital control. *ASME J. Dynamic Sys., Meas., Control,* 109:65–68, 1987.
[38] Wirsching, P.H., Paez, T.L., and Ortiz, K. *Random Vibrations: Theory and Practice.* John Wiley & Sons, New York, 1995.

14
Modeling and Identification for Robot Motion Control

14.1	Introduction	**14**-1
14.2	Robot Modeling for Motion Control	**14**-3
	Kinematic Modeling • Modeling of Rigid-Body Dynamics • Friction Modeling	
14.3	Estimation of Model Parameters	**14**-6
	Estimation of Friction Parameters • Estimation of BPS Elements • Design of Exciting Trajectory • Online Reconstruction of Joint Motions, Speeds, and Accelerations	
14.4	Model Validation	**14**-11
14.5	Identification of Dynamics Not Covered with a Rigid-Body Dynamic Model	**14**-12
14.6	Case-Study: Modeling and Identification of a Direct-Drive Robotic Manipulator	**14**-14
	Experimental Set-Up • Kinematic and Dynamic Models in Closed Form • Establishing Correctness of the Models • Friction Modeling and Estimation • Estimation of the BPS • Dynamics Not Covered with the Rigid-Body Dynamic Model	
14.7	Conclusions	**14**-24

Dragan Kostić
Technische Universiteit Eindhoven

Bram de Jager
Technische Universiteit Eindhoven

Maarten Steinbuch
Technische Universiteit Eindhoven

14.1 Introduction

Accurate and fast motions of robot manipulators can be realized via model-based motion control schemes. These schemes employ models of robot kinematics and dynamics. This chapter discusses all the necessary steps a control engineer must take to enable high-performance model-based motion control. These steps are (*i*) kinematic and rigid-body dynamic modeling of the robot, (*ii*) obtaining model parameters via direct measurements and/or identification, (*iii*) establishing the correctness of the models and validating the estimated parameters, and (*iv*) deducing to what extent the rigid-body model covers the real robot dynamics and, if needed for high-performance control, the identification of the dynamics not covered by the derived model. Better quality achieved in each of these steps contributes to higher performance of motion control.

The robotics literature offers various tutorials on kinematic and dynamic modeling [1–4]. A number of modeling methods are available, meeting various requirements. As for robot kinematics, the model is a mapping between the task space and the joint space. The task space is the space of the robot-tip coordinates:

end-effector's Cartesian coordinates and angles defining orientation of the end-effector. The joint space is the space of joint coordinates: angles for revolute joints and linear displacements for prismatic joints. A robot configuration is defined in the joint space. The mapping from the joint to the task space is called the forward kinematics (FK). The opposite mapping is the inverse kinematics (IK). The latter mapping reveals singular configurations that must be avoided during manipulator motions. Both FK and IK can be represented as recursive or closed-form algebraic models. The algebraic closed-form representation facilitates the manipulation of a model, enabling its straightforward mathematical analysis. As high accuracy of computation can be achieved faster with the closed-form models, they are preferable for real-time control.

A dynamic model relates joint motions, speeds, and accelerations with applied control inputs: forces/torques or currents/voltages. Additionally, it gives insight how particular dynamic effects, e.g., nonlinear inertial and Coriolis/centripetal couplings, as well as gravity and friction effects, contribute to the overall robot dynamic behaviour. A variety of methods are available for derivation of a dynamic model, which admits various forms of representation. Algebraic recursive models require less computational effort than the corresponding algebraic closed-form representations. Recursive models have compact structure, requiring low storage resources. Computational efficiency and low storage requirements are traditional arguments recommending the online use of recursive models [3]. However, the power of modern digital computers makes such arguments less relevant, as online use of models in closed-form is also possible. Furthermore, an algebraic closed-form representation simplifies manipulation of a model, enabling an explicit analysis of each dynamic effect. Direct insight into the model structure leads to more straightforward control design, as compensation for each particular effect is facilitated. For example, a closed-form model of gravity can be used in the compensation of gravitational loads.

Unfortunately, the derivation of closed-form models in compact form, in particular IK and dynamic models, is usually not easy, even if software for symbolic computation is used. The derivation demands a series of operations, with permanent combining of intermediate results to enhance compactness of the final model. To simplify derivation, a control designer sometimes approximates robot kinematic and/or inertial properties. For example, by neglecting link offsets according to Denavits-Hartenberg's (DH) notation [1,3], a designer deliberately disregards dynamic terms related with these offsets, thus creating a less involved but incomplete dynamic model. An approach to model manipulator links as slender beams can approximate robot inertial properties, since some link inertia or products of inertias may not be taken into account. Consequently, the resulting model is just an approximation of the real dynamic behavior. It is well-known that with transmissions between joint actuators and links, nonlinear coupling and gravity effects can become small [1–3]. If reduction ratios are high enough, these effects can become strongly reduced and thus ignored in a dynamic model. If robot joints are not equipped with gearboxes, which is characteristic for direct-drive robots [5], then the dynamic couplings and the gravity have critical influence on the quality of motion control, and they must be compensated via control action. More accurate models of these effects contribute to more effective compensation. Additional methods to simplify dynamic modeling are neglecting friction effects or using conservative models for friction, e.g., adopting just viscous and ignoring Coulomb effects. The approximate dynamic modeling is not preferable for high-performance motion control, as simplified models can compensate just for a limited number of actual dynamic effects. The noncompensated effects have to be handled by stabilizing feedback control action. The feedback control design is more involved if more dynamic effects are not compensated with the model.

Obviously, when a model is derived, it is useful to establish its correctness. Comparing it with some recursive representation of the same kinematics or dynamics is a straightforward procedure with available software packages. For example, software routines specialized for robotic problems are presented in [6,7].

With a model available, the next step is to estimate the model parameters. Kinematic parameters are often known with a sufficient accuracy, as they can be obtained by direct measurements and additional kinematic calibration [8]. In this chapter, emphasis will be on estimating the parameters of a robot dynamic model, i.e., the inertial and friction parameters. The estimation of these parameters is facilitated if a model of robot dynamics is represented in so-called regressor form, i.e., linearly in the parameters to be estimated. The estimation itself is a process that requires identification experiments performed directly on a robot. To establish experimental conditions allowing for the simplest and the most time-efficient least-squares

(LS) estimation of the parameters, joint motions, speeds, and accelerations can be reconstructed via an observer [9,10]. Care should be taken when designing robot trajectories to be realized during identification. Guidelines for their design are highlighted in [11,12].

After the estimation is finished, experimental validation of the model has to be done. Its objective is to test how accurately the model represents actual robot dynamics. If the accuracy is satisfactory, the usefulness of the model for model-based control purposes is established. For validation purposes, a manipulator should execute motions similar to those it is supposed to perform in practice. In the task space, these are the sequences of straight-line and curved movements. Hence, a writing task defined in the task space might be used as a representative validation trajectory. The choice of such a task is intuitive because of at least two reasons: writing letters is, indeed, a concatenation of straight-line and curved primitives, similar to common robot movements, and such a task may strongly excite the robot dynamics due to abrupt changes in accelerations and nonuniform speed levels. An example of a writing task applicable for model validation can be found in [13].

A dynamic model can be further used in model-based control algorithms. The model contributes to the performance of the applied control algorithms to the extent that it matches the real robot dynamics. In practice, the rigid-body models may cover the real dynamics only within the low-frequency range, as the real dynamics feature flexible effects at higher frequencies. These effects are caused by elasticity in joints and by distributed link flexibilities. For high-performance motion control, the influence of flexibilities cannot be disregarded. Here we present a technique for the identification and modeling of these effects.

The rest of the chapter is organized as follows. In Section 14.2, kinematic and rigid-body dynamic models will be derived for a general robot manipulator. A regressor form of a dynamic model will be introduced, and a set of often used friction models will be presented. An estimation of robot inertial and friction parameters will be explained in Section 14.3. A way to validate a dynamic model, suitable for motion control purposes, will be presented in Section 14.4. Identification and modeling of the dynamics not covered by a rigid-body model will be discussed in Section 14.5. Section 14.6 is a case-study. It demonstrates modeling and identification strategy for a realistic direct-drive manipulator. Conclusions will come at the end.

14.2 Robot Modeling for Motion Control

14.2.1 Kinematic Modeling

Consider a serial robot manipulator with n-joints, shown in Figure 14.1. A minimal kinematic parameterisation for this manipulator can be established according to the well-known DH (Denavits-Hartenberg) convention [1,3]. This convention systematically assigns coordinate frames to the robot joints and induces the following DH parameters: twist angles α_i, link lengths a_i, joint displacements q_i, and link offsets d_i,

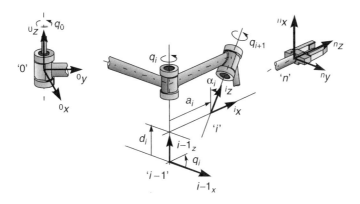

FIGURE 14.1 A general serial manipulator with n joints.

where $i = 1, \ldots, n$. The position and orientation of the ith coordinate frame with respect to the previous one $(i - 1)$ can be specified by the homogenous transformation matrix:

$$^{i-1}\mathbf{T}_i(\mathbf{q}) = \begin{bmatrix} ^{i-1}\mathbf{O}_i(\mathbf{q}) & ^{i-1}\mathbf{x}_i(\mathbf{q}) \\ \mathbf{0}_{1\times 3} & 1 \end{bmatrix} \tag{14.1}$$

Here, $\mathbf{q}$ is the vector of generalized coordinates:

$$\mathbf{q} = [q_1 \; q_2 \; \ldots \; q_n]^T \tag{14.2}$$

The coordinates are linear and angular displacements of prismatic and revolute joints, respectively. The skew-symmetric rotation matrix $^{i-1}\mathbf{O}_i \in \mathbb{R}^{3\times 3}$ defines the orientation of the ith frame with respect to frame $i - 1$:

$$^{i-1}\mathbf{O}_i(\mathbf{q}) = \begin{bmatrix} \cos q_i & -\cos \alpha_i \sin q_i & \sin \alpha_i \sin q_i \\ \sin q_i & \cos \alpha_i \cos q_i & -\sin \alpha_i \cos q_i \\ 0 & \sin \alpha_i & \cos \alpha_i \end{bmatrix} \tag{14.3}$$

The position vector $^{i-1}\mathbf{x}_i \in \mathbb{R}^{3\times 1}$ points from the origin of the frame $i - 1$ to the ith frame:

$$(^{i-1}\mathbf{x}_i(\mathbf{q}))^T = [a_i \cos q_i \;\; a_i \sin q_i \;\; d_i] \tag{14.4}$$

The forward kinematics (FK) can be computed as a product of homogeneous transformations between the circumjacent coordinate frames:

$$^0\mathbf{T}_n(\mathbf{q}) = {}^0\mathbf{T}_1(\mathbf{q})\,^1\mathbf{T}_2(\mathbf{q}) \ldots {}^{n-1}\mathbf{T}_n(\mathbf{q}) = \begin{bmatrix} ^0\mathbf{O}_n(\mathbf{q}) & ^0\mathbf{x}_n(\mathbf{q}) \\ \mathbf{0}_{1\times 3} & 1 \end{bmatrix} \tag{14.5}$$

The orientation and position of the tip coordinate frame n with respect to the base (inertial) frame 0 are determined by $^0\mathbf{O}_n$ and $^0\mathbf{x}_n$, respectively. In general case, both $^0\mathbf{O}_n$ and $^0\mathbf{x}_n$ nonlinearly depend on the generalized coordinates, and thus it is not always possible to explicitly express $\mathbf{q}$ in terms of the tip position and orientation coordinates. Consequently, there is no general closed-form representation of the inverse kinematics (IK), and numerical techniques are often used to solve IK [1].

14.2.2 Modeling of Rigid-Body Dynamics

The rigid-body model of the robot dynamics can be derived using the Euler-Lagrange formulation [1–3]. The standard form of the model is as follows:

$$\mathbf{M}(\mathbf{q}(t))\ddot{\mathbf{q}}(t) + \mathbf{c}(\mathbf{q}(t), \dot{\mathbf{q}}(t)) + \mathbf{g}(\mathbf{q}(t)) + \boldsymbol{\tau}^f(t) = \boldsymbol{\tau}(t) \tag{14.6}$$

where $\mathbf{q}$ is defined by Equation (14.2), $\dot{\mathbf{q}}$ and $\ddot{\mathbf{q}}$ are vectors of the joint speeds and accelerations, respectively, $\mathbf{M}$ is the inertia matrix, $\mathbf{c}$, $\mathbf{g}$, and $\boldsymbol{\tau}^f$ are the vectors of Coriolis/centripetal, gravitational, and friction effects, respectively, and $\boldsymbol{\tau}$ is the vector of joint control inputs (forces/torques). Elements of $\mathbf{M}$, $\mathbf{c}$, and $\mathbf{g}$ can be determined using the homogenous transformation (14.1):

$$m_{i,k} = \sum_{j=\max(i,k)}^{n} \mathrm{Tr}\left[\frac{\partial ^0\mathbf{T}_j}{\partial q_k}\mathbf{J}_j\left(\frac{\partial ^0\mathbf{T}_j}{\partial q_i}\right)^T\right], \quad c_i = \sum_{k=1}^{n}\sum_{l=1}^{n} c_{i,k,l}\dot{q}_k\dot{q}_l$$

$$c_{i,k,l} = \sum_{j=\max(i,k,l)}^{n} \mathrm{Tr}\left[\frac{\partial^2 {}^0\mathbf{T}_j}{\partial q_k \partial q_l}\mathbf{J}_j\left(\frac{\partial ^0\mathbf{T}_j}{\partial q_i}\right)^T\right]$$

$$g_i = \sum_{j=i}^{n}\left(-m_j[0\; 0\; -g\; 0]\frac{\partial ^0\mathbf{T}_j}{\partial q_i}\,^i\mathbf{r}_j\right), \quad i,k,l = 1,\ldots,n \tag{14.7}$$

Here, m_j is the mass of the jth link, g is acceleration due to gravity, ${}^i\mathbf{r}_j$ contains the homogenous coordinates of the center of mass of link j expressed in the ith coordinate frame

$$ {}^i\mathbf{r}_j = [x_j\ y_j\ z_j\ 1]^T \qquad (14.8) $$

while $\mathbf{J}_j$ is the inertia tensor with elements:

$$ j_{j,11} = \frac{-I_{xx,j} + I_{yy,j} + I_{zz,j}}{2};\quad j_{j,22} = \frac{I_{xx,j} - I_{yy,j} + I_{zz,j}}{2};\quad j_{j,33} = \frac{I_{xx,j} + I_{yy,j} - I_{zz,j}}{2} $$

$$ j_{j,12} = j_{j,21} = I_{xy,j};\quad j_{j,13} = j_{j,31} = I_{xz,j};\quad j_{j,23} = j_{j,32} = I_{yz,j};\quad j_{j,14} = j_{j,41} = m_j x_j $$

$$ j_{j,24} = j_{j,42} = m_j y_j;\quad j_{j,34} = j_{j,43} = m_j z_j;\quad j_{j,44} = j_{j,44} = m_j \qquad (14.9) $$

$I_{xx,j}, I_{yy,j}$, and $I_{zz,j}$ denote principal moments of inertia for the link j, and $I_{xy,j}, I_{yz,j}$, and $I_{xz,j}$ are products of inertia of the same link.

The dynamic model (14.6), excluding friction, can be represented linearly in elements of the so-called base parameter set (BPS). These elements are combinations of the inertial parameters used in the equations above, and they constitute the minimum set of parameters whose values can determine the dynamic model uniquely [1,11,12,14]. A linear parameterization of the rigid-body dynamic model (14.6) has the form:

$$ \mathbf{R}(\mathbf{q}(t), \dot{\mathbf{q}}(t), \ddot{\mathbf{q}}(t))\mathbf{p} + \boldsymbol{\tau}^f(t) = \boldsymbol{\tau}(t) \qquad (14.10) $$

where $\mathbf{R} \in \mathbb{R}^{n \times p}$ is the regression matrix and $\mathbf{p} \in \mathbb{R}^{p \times 1}$ is the vector of the BPS elements. The regressor form of the dynamic model is suitable for the estimation of inertial parameters.

14.2.3 Friction Modeling

The dynamic models (14.6) and (14.10) also feature the friction term $\boldsymbol{\tau}^f$. In the real system, friction cannot be disregarded if high-performance motion control is required. Hence, modeling and compensation of friction effects are important issues in robot motion control. A basic friction model covers just Coulomb and viscous effects:

$$ \alpha_{1,i}\operatorname{sgn}\dot{q}_i + \alpha_{2,i}\dot{q}_i = \tau_i^f \quad (i = 1, \ldots, n) \qquad (14.11) $$

where τ_i^f is the ith element of $\boldsymbol{\tau}^f$, while $\alpha_{1,i}$ and $\alpha_{2,i}$ are Coulomb and viscous friction parameters, respectively. This model admits a linear representation in the parameters $\alpha_{1,i}$ and $\alpha_{2,i}$. This enables a joined regressor form of the dynamic model (14.10), where the new regression matrix is formed by vertical stacking $\mathbf{R}$ and the rows containing friction terms, while the parameter vector consists of $\mathbf{p}$ and friction parameters. The joined regressor form allows simultaneous estimation of the BPS and friction parameters [11,12].

However, friction is a very complex interaction between the contact surfaces of the interacting objects and needs a comprehensive model to be described correctly [15–19]. Presliding and sliding are two commonly recognized friction regimes [15,16]. The level of friction prior to the sliding regime is called the static friction, or the stiction. At the beginning of the sliding regime, the decreasing friction force for increasing speed is known as the Stribeck effect. Beyond this stage, the friction force increases as the speed is increasing, which is caused by the viscous effect. A time lag between the speed and friction force can be detected, which implies a dynamic, rather than just static nature of the friction. Several static and dynamic models were proposed to represent the given properties of friction.

A frequently used static friction model has the form:

$$ \gamma(\dot{q}_i)\operatorname{sgn}\dot{q}_i + \alpha_{2,i}\dot{q}_i = \tau_i^f $$
$$ \gamma_i(\dot{q}_i) = \alpha_{0,i}\exp[-(\dot{q}_i/v_{s,i})^2] + \alpha_{1,i} \quad (i = 1, \ldots, n) \qquad (14.12) $$

where $\gamma(\dot{q}_i)$ is the Stribeck curve for steady-state speeds, $v_{s,i}$ is the Stribeck velocity, and $\alpha_{0,i}, \alpha_{1,i}$, and $\alpha_{2,i}$ are static, Coulomb, and viscous friction parameters, respectively. The given model describes the sliding friction regime. Because of the parameter $v_{s,i}$, the model (14.12) cannot be represented linearly in all friction parameters ($v_{s,i}$, $\alpha_{0,i}$, $\alpha_{1,i}$, and $\alpha_{2,i}$). Thus, a joined regressor form of a robot dynamic model is not possible with the model (14.12). This prevents simultaneous estimation of the friction parameters and the BPS elements, using, for example, some time-efficient least-squares technique.

Dynamic friction models aim at covering more versatile friction phenomena. Among others (see surveys in [15,16]), the Dahl and the LuGre models are the most frequently cited, analysed, and implemented dynamic models. They cover a number of static and dynamic friction phenomena and facilitate their representation in a compact way. The LuGre model is more comprehensive than the Dahl one, and it has the following form:

$$\sigma_{0,i} z_i + \sigma_{1,i} \dot{z}_i + \alpha_{2,i} \dot{q}_i = \tau_i^f$$
$$\dot{z}_i = \dot{q}_i - \sigma_{0,i} \frac{|\dot{q}_i|}{\gamma(\dot{q}_i)} z_i \quad (i = 1, \ldots, n) \quad (14.13)$$

where z_i is a presliding deflection of materials in contact, $\gamma(\dot{q}_i)$ is defined by Equation (14.12), while $\sigma_{0,i}$ and $\sigma_{1,i}$ are the stiffness of presliding deflections and viscous damping of the deflections, respectively. Notice that the static model (14.12) corresponds to the sliding behaviour ($\dot{z}_i = 0$) of the LuGre model. A drawback of the LuGre model is that for some choices of its coefficients, the passivity condition from speed $\dot{q}_i$ to friction force τ_i^f is easily violated [18]. This condition reflects the dissipative nature of friction. Without the passivity property, the LuGre model admits periodic solutions, which is not realistic friction behavior. On the other hand, a linear parameterization of the LuGre model is obviously not possible, and consequently, the joined regressor form of a dynamic model cannot be created.

A neural-network friction model is also used [19]

$$\sum_{k=1}^{3} f_{i,k} [1 - 2/(\exp(2 w_{i,k} \dot{q}_i) + 1)] + \alpha_{2,i} \dot{q}_i = \tau_i^f \quad (i = 1, \ldots, n) \quad (14.14)$$

where $f_{i,k}$, $w_{i,k}$, and $\alpha_{2,i}$ are unknown parameters. This static model can represent Stribeck, Coulomb, and viscous effects, and it admits a straightforward online implementation. However, it cannot be linearly parameterized, and consequently, the joined regressor form of a dynamic model is not possible with this model.

When the robot model is completed, the next step is to obtain realistic values for the model parameters. These parameters are acquired via estimation procedures, that are explained in the next section.

14.3 Estimation of Model Parameters

An estimation of friction parameters will be explained first in this section. Friction parameters are estimated separately from the BPS elements if the joined regressor form of a robot dynamic model cannot be created. Then, brief descriptions of two methods for estimation of the BPS elements will be given.

14.3.1 Estimation of Friction Parameters

Unknown parameters of a friction model associated with the joint i can be collected in the vector $\mathbf{p}_i^f$. Elements of this vector can be estimated on the basis of identification experiments in open-loop. In each experiment, the considered joint is driven with a sufficiently exciting control input τ_i, and the corresponding motion q_i is measured. The remaining joints are kept locked by means of appropriate hardware or by feedback control. In the latter case, the feedback controllers must be tuned such that the motions in the remaining joints, possibly caused by the couplings effects and by gravity, are prevented. The dynamics for joint i can be represented as

$$\ddot{q}_i = [\tau_i - \tau_i^f(\mathbf{p}_i^f) - c_i \sin q_i]/J_i \quad (14.15)$$

where J_i is the equivalent moment of inertia of the joint, and c_i is a coefficient of the gravity term. This term vanishes if the axis of joint motion is parallel with the direction of gravity. The parameters J_i and c_i are usually not known exactly, so they need to be estimated together with the friction parameters.

An efficient time-domain identification procedure elaborated in [19] can be carried out to estimate friction parameters of a static friction model together with J_i and c_i from Equation (14.15). This procedure engages a recursive Extended Kalman filtering technique for nonlinear continuous-time processes. The procedure recursively updates estimates, until convergence of all unknown parameters is reached. Friction parameters characteristic to the presliding friction regime of the LuGre model (14.13), i.e., $\sigma_{0,i}$ and $\sigma_{1,i}$, can be estimated using frequency response function (FRF) measurements, as explained in [20]. This procedure requires an experiment where the considered joint is driven with a random excitation (noise) and the FRF from the excitation τ_i to the motion q_i is measured. A low noise level is needed, so as to keep friction within the presliding regime. Joint position sensors of high resolution are thus necessary, because very small displacements have to be detected. The estimation procedure employs a linearization of (14.13) around zero $\dot{q}_i$, z_i, and $\dot{z}_i$, defined in Laplace domain:

$$H_i(s) = \frac{Q_i(s)}{\tau_i(s)} = \frac{1}{J_i s^2 + (\sigma_{1,i} + \alpha_{2,i})s + \sigma_{0,i} + c_i} \tag{14.16}$$

The parameters $\alpha_{2,i}$ and c_i must be known beforehand. The transfer function (14.16) can be fitted into the measured FRF to determine the unknown $\sigma_{0,i}$ and $\sigma_{1,i}$.

14.3.2 Estimation of BPS Elements

Assume that friction models associated with the robot joints reconstruct realistic friction phenomena sufficiently accurately. If this is the case, then the friction component $\boldsymbol{\tau}^f$ is completely determined in a robot rigid-body dynamic model (14.6), as well as in the corresponding regressor representation (14.10). This facilitates estimation of the elements of the BPS set, using identification experiments. During the experiment the robot is controlled with decoupled PD (proportional, derivative) controllers to realize a specific trajectory. This trajectory must be carefully chosen such that the corresponding control input $\boldsymbol{\tau}(t)$ sufficiently excites all system dynamics. If the applied $\boldsymbol{\tau}(t)$ and the resulting $\mathbf{q}(t)$, $\dot{\mathbf{q}}(t)$, and $\ddot{\mathbf{q}}(t)$ are collected during the experiment, and postprocessed to determine elements of the BPS, then we speak about batch estimation. If the estimation of the BPS is performed as the identification experiment is going on, then we speak about online estimation.

14.3.2.1 Batch LS Estimation of BPS

Assume ζ samples of each element of $\boldsymbol{\tau}$, $\mathbf{q}$, $\dot{\mathbf{q}}$, and $\ddot{\mathbf{q}}$ have been collected. These samples correspond to time instants $t_1, t_2, \ldots, t_\zeta$. By virtue of (14.10), we may form the following system of equations:

$$\boldsymbol{\Phi}\, \mathbf{p} = \mathbf{y} \tag{14.17}$$

$$\boldsymbol{\Phi} = \begin{bmatrix} \mathbf{R}(\mathbf{q}(t_1), \dot{\mathbf{q}}(t_1), \ddot{\mathbf{q}}(t_1)) \\ \mathbf{R}(\mathbf{q}(t_2), \dot{\mathbf{q}}(t_2), \ddot{\mathbf{q}}(t_2)) \\ \vdots \\ \mathbf{R}(\mathbf{q}(t_\zeta), \dot{\mathbf{q}}(t_\zeta), \ddot{\mathbf{q}}(t_\zeta)) \end{bmatrix} \tag{14.18}$$

$$\mathbf{y} = [(\boldsymbol{\tau}(t_1) - \boldsymbol{\tau}^f(t_1))^T,\ (\boldsymbol{\tau}(t_2) - \boldsymbol{\tau}^f(t_2))^T, \ldots, (\boldsymbol{\tau}(t_\zeta) - \boldsymbol{\tau}^f(t_\zeta))^T]^T \tag{14.19}$$

If the input $\boldsymbol{\tau}(t)$ sufficiently excites the manipulator dynamics, then the LS (least-squares) estimate of $\mathbf{p}$ is

$$\hat{\mathbf{p}} = \boldsymbol{\Phi}^\# \mathbf{y} \tag{14.20}$$

where # denotes a matrix pseudoinverse [21]. There are several criteria evaluating sufficiency of excitation. The usual one is the condition number of Φ, defined as the ratio between maximal and minimal singular values of Φ. If this number is closer to 1, the excitation is considered better for a reliable estimation of **p**. In [4] several other performance criteria are also listed, such as the largest singular value of Φ, the Frobenius condition number of Φ, and the determinant of the weighted product between Φ and its transpose.

14.3.2.2 Online Gradient Estimator of BPS

Among several solutions for online estimation of the BPS elements [22], the gradient estimator is the simplest choice. It resorts to the concept of prediction error. If an estimate $\hat{\mathbf{p}}$ of the BPS is available and the friction component τ^f is known, then the model (14.10) can be used to predict the difference between the control input and friction:

$$\mathbf{R}(\mathbf{q}(t), \dot{\mathbf{q}}(t), \ddot{\mathbf{q}}(t))\hat{\mathbf{p}} = \hat{\tau}(t) - \tau^f(t) = \hat{\tau}_d(t) \tag{14.21}$$

The prediction error $\tilde{\tau}_d$ is formed as the difference between $\hat{\tau}_d(t)$ and the measured $\tau_d(t) = \tau(t) - \tau^f(t)$:

$$\begin{aligned}\tilde{\tau}_d(t) &= \hat{\tau}_d(t) - \tau_d(t) \\ &= \mathbf{R}(\mathbf{q}, \dot{\mathbf{q}}, \ddot{\mathbf{q}})\hat{\mathbf{p}} - \mathbf{R}(\mathbf{q}, \dot{\mathbf{q}}, \ddot{\mathbf{q}})\mathbf{p} \\ &= \mathbf{R}(\mathbf{q}, \dot{\mathbf{q}}, \ddot{\mathbf{q}})\tilde{\mathbf{p}}\end{aligned} \tag{14.22}$$

where $\tilde{\mathbf{p}} = \hat{\mathbf{p}} - \mathbf{p}$ is the BPS estimation error. The following update law enables online estimation of the BPS:

$$\dot{\hat{\mathbf{p}}} = -\alpha \mathbf{R}^T(\mathbf{q}, \dot{\mathbf{q}}, \ddot{\mathbf{q}})\tilde{\tau}_d \tag{14.23}$$

where α is a positive scalar estimation gain. In the given law, the unknown parameters are updated in the converse direction of the gradient of the squared prediction error with respect to the parameters [22].

As in the batch LS estimation, the quality of the BPS estimates achieved with (14.23) strongly depends on the choice of a trajectory the robot realizes during the identification experiment. Design of such a trajectory deserves special attention.

14.3.3 Design of Exciting Trajectory

The robot trajectory, which is realized during the identification experiment, is called the exciting trajectory. There are several designs of such a trajectory available in literature, and they are briefly summarized in [12]. A design suggested in [11] and [12] postulates the exciting trajectory as finite Fourier series with predefined fundamental frequency and with free coefficients:

$$q_i^{exc}(t) = q_{0,i} + \sum_{j=1}^{N} \frac{1}{j 2\pi \Delta f}[a_{ij}\sin(j 2\pi \Delta f t) - b_{ij}\cos(j 2\pi \Delta f t)] \quad (i = 1, \ldots, n) \tag{14.24}$$

Here, Δf [Hz] is the desired resolution of the frequency content, the number of harmonics N determines the bandwidth of the trajectory, and the free coefficients $q_{0,i}$, a_{ij}, and b_{ij} ($i = 1, \ldots, n; j = 1, \ldots, N$) should be determined such that all manipulator rigid-body dynamics are sufficiently excited during the identification experiment. The given exciting trajectory is periodic, which enables continuous repetition of the identification experiment and averaging of measured data to improve the signal-to-noise ratio. Additional advantages are direct control of the resolution and of the bandwidth of the trajectory harmonic content. The control of the bandwidth is an important precaution for avoiding frequency components that excite the robot flexible modes.

The exciting trajectory can be realized on the robotic systems using a PD feedback controller:

$$\tau = -\mathbf{K}_p(\mathbf{q} - \mathbf{q}^{exc}) - \mathbf{K}_d(\dot{\mathbf{q}} - \dot{\mathbf{q}}^{exc}) \tag{14.25}$$

where $\mathbf{K}_p = diag[k_{p,1}, \ldots, k_{p,n}]$ and $\mathbf{K}_d = diag[k_{d,1}, \ldots, k_{d,n}]$ are matrices of positive position and velocity gains, respectively. The applied τ sufficiently excites the system rigid-body dynamics if the free coefficients in (14.24) are determined by optimising some property of the information matrix Φ (e.g., the condition number of Φ). No matter which property (performance criterion) is adopted, the resulting optimization problem is nonconvex and nonlinear. Different initial conditions may lead to different exciting trajectories, all corresponding to local minima of the performance criterion. Consequently, the optimization determines a suboptimal, rather than the globally optimal exciting trajectory. Still, experience shows that use of suboptimal trajectories gives satisfactory results.

The presence of disturbances in the collected $\mathbf{q}$, $\dot{\mathbf{q}}$, and $\ddot{\mathbf{q}}$ affects the quality of the estimation methods (14.20) and (14.23). The quality is better if disturbances are reduced to a minimum. High-frequency quantization noise and vibrations caused by flexibilities are disturbances commonly found in robot manipulators. By reconstruction of joint motions, speeds, and accelerations with carefully tuned Kalman filters [9,10], a sufficient rejection of the disturbances can be achieved.

14.3.4 Online Reconstruction of Joint Motions, Speeds, and Accelerations

Online reconstruction of $\mathbf{q}$, $\dot{\mathbf{q}}$, and $\ddot{\mathbf{q}}$ using a Kalman filter will be formulated in discrete-time, assuming a digital setup for data-acquisition from a robot joint. It is proven in [10] that at sufficiently high sampling rates, usual in modern robotics, the discrete Kalman filter associated with the motion of a robot joint i can be constructed assuming an all-integrator model for this motion. This model includes a correcting term ν_i, representing the model uncertainty. It accounts for the difference between the adopted model and the real joint dynamics. The correcting term is a realization of the white process noise ξ_i, filtered through a linear, stable, all-pole transfer function. With the correcting term used, the reconstruction of $\mathbf{q}$, $\dot{\mathbf{q}}$, and $\ddot{\mathbf{q}}$ requires at least a third-order model.

As a case-study, we assume that ξ_i is filtered through just a single integrator. A continuous-time model associated to the motion in the joint i has the form:

$$\begin{aligned} \ddot{q}_i &= \ddot{q}_{r,i} + \nu_i \\ \dot{\nu}_i &= \xi_i \\ \tilde{y}_i &= q_i + \eta_i \end{aligned} \tag{14.26}$$

where $\ddot{q}_{r,i}$ is the desired joint acceleration, and η_i is the measurement noise. In the model (14.26), the joint motion q_i is regarded as the only measured coordinate. In the design of the Kalman filter the deviation from the desired joint motion $q_{r,i}$

$$e_i = q_i - q_{r,i} \tag{14.27}$$

and its time derivative can be adopted as the states to be reconstructed, rather than the motion coordinates themselves. In such a way, the state reconstruction process is merely for the deviation from the expected, or modelled, trajectory [10]. Let T_s denote themselves the sampling time used for both data-acquisition and control. We may determine a discrete-time system involving identical solutions with the model that arises after substituting Equation (14.26) into Equation (14.27) at $t = kT_s$:

$$\begin{aligned} \mathbf{x}_i(k+1) &= \mathbf{E}_i(T_s)\mathbf{x}_i(k) + \mathbf{g}_i(T_s)\xi_i(k) \\ \tilde{y}_i(k) &= \mathbf{c}_i\mathbf{x}_i(k) + q_{r,i}(k) + \eta_i(k) \end{aligned} \tag{14.28}$$

where

$$\mathbf{x}_i(k) = [e_i(k), \dot{e}_i(k), v_i(k)]^T \tag{14.29a}$$

$$\mathbf{E}_i = \begin{bmatrix} 1 & T_s & 0.5T_s^2 \\ 0 & 1 & T_s \\ 0 & 0 & 1 \end{bmatrix}, \quad \mathbf{g}_i = \begin{bmatrix} T_s^3/6 \\ 0.5T_s^2 \\ T_s \end{bmatrix} \tag{14.29b}$$

$$\mathbf{c}_i = [1 \ 0 \ 0] \tag{14.29c}$$

Here, k abbreviates kT_s. A Kalman filter that optimally reconstructs the state $\mathbf{x}_i$ in the presence of the model uncertainty v_i and measurement noise η_i has the form:

$$\begin{aligned} \hat{\mathbf{x}}_i(k+1) &= \mathbf{E}_i(T_s)\bar{\mathbf{x}}_i(k) \\ \bar{\mathbf{x}}_i(k) &= \hat{\mathbf{x}}_i(k) + \mathbf{k}_i[\tilde{y}_i(k) - q_{r,i}(k) - \mathbf{c}_i\hat{\mathbf{x}}_i(k)] \end{aligned} \tag{14.30}$$

where $\bar{\mathbf{x}}_i$ denotes updated state estimate and $\mathbf{k}_i$ is a $n \times 1$ vector of constant gains. As the filter reconstructs deviations from the reference motions and speeds, together with the model uncertainty, the motion coordinates can be reconstructed as follows:

$$\begin{aligned} \bar{q}_i &= q_{r,i} + \bar{e}_i \\ \dot{\bar{q}}_i &= \dot{q}_{r,i} + \dot{\bar{e}}_i \\ \ddot{\bar{q}}_i &= \ddot{q}_{r,i} + \bar{v}_i \end{aligned} \tag{14.31}$$

To compute the vector $\mathbf{k}_i$, one needs covariances of the measurement noise η_i and of the process noise ξ_i. Assuming white, zero mean quantization noise, uniformly distributed within a resolution increment $\Delta\theta_m$ of the position sensor, the straightforward choice for η_i is $\Delta\theta_m^2/12$ (see [10] for details). The reference [9] suggests a simple rule to choose ξ_i: it should equal max $\dddot{q}_{r,i}^2$ (maximum value of the square jerk of the motion reference). In practice, these rules are used as initial choices, while the final tuning is most effectively done online. However, use of Bode plots from the measured e_i to the reconstructed $\bar{e}_i$ — an illustrative example is presented in Figure 14.2 — could be instructive for filter tuning. The Bode plots reveal which harmonics of e_i will be amplified after reconstruction, as well as which harmonics will be attenuated. By analyzing the power spectrum of e_i, a control engineer may locate spectral components due to noise and other disturbing effects, that need to be filtered out. The gain vector $\mathbf{k}_i$ can be retuned to reshape the Bode plot according to the needs.

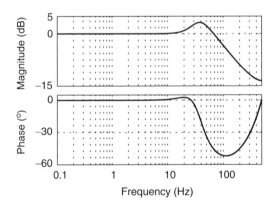

FIGURE 14.2 Bode plots of the Kalman filter, from the measured e_i to the reconstructed $\bar{e}_i$.

14.4 Model Validation

Writing and drawing tasks can be useful for model validation, because they require complex non-smooth motions, and their execution may induce significant dynamic forces/torques. An extensive study of human handwriting is given in [13]. This reference considers one representative writing task: the sequence of letters shown in Figure 14.3a. The given sequence is a concatenation of motion primitives typically executed by robots. Mathematically, the sequence can be piecewise described in closed-form, with the possibility of imposing an arbitrary velocity profile. One may use this freedom to pose very demanding dynamic tasks, that can be used for a rigorous verification of robot kinematic and dynamic models.

For example, the robot-tip should realize the sequence of letters with the velocity profile

$$v(t) = \sqrt{\dot{x}^2(t) + \dot{z}^2(t)} \tag{14.32}$$

presented in Figure 14.3b. The given profile has fast and slow phases. Figure 14.3a and Figure 14.3b define the reference trajectory of the robot-tip. An inverse kinematics model can be used to determine the corresponding reference joint motions $\mathbf{q}_r(t)$. Along these motions a dynamic model can be validated. The organization of the validation experiment is illustrated in Figure 14.4. The objective is to test how accurately the output $\bar{\tau}$ of the dynamic model, which is fed by online reconstructed joint motions, speeds, and accelerations (see formulas (14.30) and (14.31)), reconstructs control signals generated with PD controllers:

$$\tau = -\mathbf{K}_p(\bar{\mathbf{q}} - \mathbf{q}_r) - \mathbf{K}_d(\dot{\bar{\mathbf{q}}} - \dot{\mathbf{q}}_r) \tag{14.33}$$

If $\bar{\tau}$ is close enough to τ, the dynamic model is considered appropriate for control purposes.

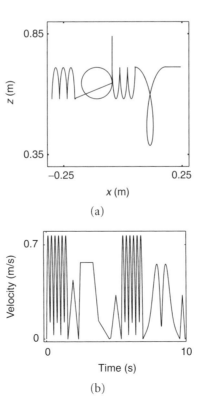

FIGURE 14.3 Reference trajectory in the task space: (a) tip path, (b) adopted tip-velocity profile.

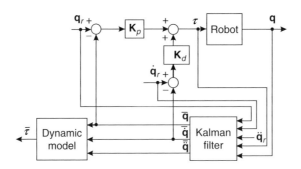

FIGURE 14.4 Experimental validation of a robot dynamic model.

14.5 Identification of Dynamics Not Covered with a Rigid-Body Dynamic Model

In model-based manipulator motion control, a dynamic model is used to compensate for the nonlinear and coupled manipulator dynamics. To illustrate this, we may consider a motion control problem solved, for example, using the feedback linearization [23]:

$$\boldsymbol{\tau}(t) = \mathbf{M}(\mathbf{q}(t))\mathbf{u}(t) + \mathbf{c}(\mathbf{q}(t), \dot{\mathbf{q}}(t)) + \mathbf{g}(\mathbf{q}(t)) + \boldsymbol{\tau}^f(t) \qquad (14.34)$$

Here, $\mathbf{u}(t)$ represents a feedback control action which should stabilize the manipulator motion and enforce the desired accuracy in realizing a reference trajectory $\mathbf{q}_r(t)$. The remaining variables are as in (14.6). When applied, the control (14.34) should decouple the manipulator dynamics. The problem of controlling the nonlinear-coupled system shifts to the problem of controlling the linear plants, defined in the Laplace domain:

$$P_i(s) = q_i(s)/u_i(s) \quad (i = 1, \ldots, n) \qquad (14.35)$$

where u_i is the ith component of $\mathbf{u}$.

Ideally, the plant transfer function P_i would be modelled as a single mass (double integrator). The related frequency response function (FRF) should have a linear amplitude plot with a -2 slope and constant phase at $-180°$. This holds if perfectly rigid-body dynamics are completely linearized by Equation (14.34). In practice this may not happen, as a perfectly rigid mechanical construction is difficult to achieve, and hence, the rigid-body model is hardly an accurate description of the realistic manipulator dynamics. Rather than a double integrator, P_i has a more involved FRF, incorporating resonances and antiresonances due to flexibilities. The frequencies of these resonances and antiresonances may vary as the manipulator changes its configuration. The flexible dynamics not covered with the rigid-body model influence feasible performance and robustness in motion control. Therefore, it is worthwhile spending additional effort on identifying these dynamics.

In the following we sketch two standard techniques capable of, at least, local identification of manipulator dynamics not covered with the rigid-body model. The first technique is based on spectrum analysis, while the second one is established in time-domain. Both of them utilize separate identification experiments for each joint (n experiments for the whole manipulator). Each experiment is performed under closed-loop conditions, because the dynamics not covered with the rigid-body model might not be asymptotically stable. For example, when identifying dynamics in joint i, this joint is feedback controlled to realize a slow motion of a constant speed $\dot{q}_{r,i}$, while the remaining ones take some static posture. A low constant speed is preferable to reduce the influence of friction effects on the quality of identification. The slow motion is performed within the joint range. If, for example, the range in the revolute joint is not limited, a full revolution can be made. The direction of movement is changed after reaching the joint limit, or after the full revolution. Denote with $\mathbf{q}_r(t)$ an $(n \times 1)$ vector of set-points, used in the identification experiment.

Elements of $\mathbf{q}_r(t)$ should satisfy $\dot{q}_{r,i}(t) = \dot{q}_{r,i}$ and $\dot{q}_{r,j}(t) \equiv 0$ if $i \neq j$. The specified set-points can be achieved by applying the control law (14.34) with

$$\mathbf{u} = -\mathbf{K}_p(\mathbf{q} - \mathbf{q}_r) - \mathbf{K}_d(\dot{\mathbf{q}} - \dot{\mathbf{q}}_r) + \mathbf{n} \tag{14.36}$$

where $\mathbf{K}_p = diag[k_{p,1}, \ldots, k_{p,n}]$ and $\mathbf{K}_d = diag[k_{d,1}, \ldots, k_{d,n}]$ are matrices of positive position and velocity gains, respectively, and the $(n \times 1)$ vector $\mathbf{n}$ contains a random excitation (noise) as ith element and zeros elsewhere. Identification is more reliable if the influence of disturbances (e.g., quantization noise) affecting the manipulator motion and speed is low. For that purpose, $\mathbf{M}$, $\mathbf{c}$, $\mathbf{g}$, and τ^f in Equation (14.34) can be computed along $\mathbf{q}_r(t)$ and $\dot{\mathbf{q}}_r(t)$.

The spectrum analysis technique identifies an FRF of the plant dynamics $P_i(j\omega) = q_i(j\omega)/u_i(j\omega)$, where ω denotes angular frequency. A comprehensive description of techniques for FRF identification is available in [24]. If computed directly from the measured $u_i(t)$ and $q_i(t)$, the obtained FRF may appear unreliable. We characterize an FRF as unreliable if the coherence function

$$\Gamma = \frac{\Phi_{nq}^2}{\Phi_{qq}\Phi_{nn}} \tag{14.37}$$

is small. Φ_{nn} and Φ_{qq} denote the autopower spectra of the excitation signal and of the joint response, respectively, while Φ_{nq} is the cross power spectrum between the excitation and the response. Γ becomes small if the excitation and response data are correlated, which is usual if these data are measured in closed-loop. If this is the case, one should carry out indirect, instead of the direct identification of the plant's FRF.

In the indirect identification, an FRF of the transfer from $n_i(t)$ to $u_i(t)$ is estimated first (n_i is the ith element of $\mathbf{n}$ in (14.36)). This estimation represents the sensitivity function $S_i(j\omega) = [1 + (j\omega k_{d,i} + k_{p,i})P_i(j\omega)]^{-1}$ for the feedback loop in the joint i. The coherence of this approach is normally sufficient for reliable estimation in the frequency range beyond the bandwidth of the closed-loop system in the joint i. Hence, a low bandwidth for the closed-loop in the joint i is preferable. On the other hand, sufficiently high bandwidths of the feedback loops in the remaining joints are needed to prevent any motion in these joints. Therefore, separate sets of controller settings are required. With the identified sensitivity, for a known setting of the PD gains, one may find the plant's FRF:

$$P_i(j\omega) = \frac{1/S_i(j\omega) - 1}{j\omega k_{d,i} + k_{p,i}} \tag{14.38}$$

The obtained FRF represents dynamics of the joint i, which are not compensated with the rigid-body dynamic model, as a set of data. For the control design it is more convenient to represent these dynamics with some parametric model in a transfer function or state-space form. There are a number of solutions to fit a parametric model into FRF data [24]. The least-square fit using an output error model structure model is one possibility [25].

The time-domain technique directly determines a parametric model of the plant dynamics. Here, it is only sketched for the case of direct identification, which is based on the measured excitation $u_i(t)$ and response $q_i(t)$ data. Extension to indirect identification is straightforward. If collected at discrete time instants $t_1, t_2, \ldots, t_\zeta$, the measurement pairs $\{u_i(t_1), q_i(t_1)\}, \{u_i(t_2), q_i(t_2)\}, \ldots, \{u_i(t_\zeta), q_i(t_\zeta)\}$ can be used for identification of the plant dynamics assuming a transfer function model of the form

$$\frac{Q_i(z)}{U_i(z)} = P_i(z) = \frac{b_1 z^{-1} + b_2 z^{-2} + \cdots + b_n z^{-n}}{1 - a_1 z^{-1} - \cdots - a_n z^{-n}} \tag{14.39}$$

This model, defined in Z-domain, is often referred to as the ARMA (autoregressive moving-average) model [26]. Provided $n < \zeta$, the least-squares estimates of the model coefficients

$$\vartheta = [a_1 \; a_2 \ldots a_n \; b_1 \; b_2 \ldots b_n]^T \tag{14.40}$$

can be determined using the formula

$$\hat{\vartheta} = \mathbf{F}^{\#} \mathbf{Q}_i \qquad (14.41)$$

where # denotes matrix pseudoinverse and

$$\mathbf{Q}_i = \begin{bmatrix} q_i(t_{n+1}) \\ q_i(t_{n+2}) \\ \vdots \\ q_i(t_\zeta) \end{bmatrix}$$

$$\mathbf{F} = \begin{bmatrix} q_i(t_n) & q_i(t_{n-1}) & \cdots & q_i(t_1) & u_i(t_n) & \cdots & u_i(t_1) \\ q_i(t_{n+1}) & q_i(t_n) & \cdots & q_i(t_2) & u_i(t_{n+1}) & \cdots & u_i(t_2) \\ \vdots & \vdots & \ddots & \vdots & \vdots & \ddots & \vdots \\ q_i(t_{\zeta-1}) & q_i(t_{\zeta-2}) & \cdots & q_i(t_{\zeta-n+1}) & u_i(t_{\zeta-1}) & u_i(t_{\zeta-2}) & u_i(t_{\zeta-n+1}) \end{bmatrix} \qquad (14.42)$$

The determined parametric model of the plant is used for design of a feedback controller. A more realistic model contributes to higher performance of robot motion control and better robustness against uncertainty in the manipulator dynamics. Experience shows that robust model-based motion control of high performance is possible only if all notable peculiarities of the manipulator dynamics are recognized and taken into account during the feedback control design. This certainly increases the time needed for system identification, control design, and feedback tuning. However, a benefit from application of the resulting controller is a robotic system of improved quality.

14.6 Case-Study: Modeling and Identification of a Direct-Drive Robotic Manipulator

Practical application of the theoretical concepts presented so far will be illustrated for the case of a realistic robotic manipulator of spatial kinematics and direct-drive actuation.

14.6.1 Experimental Set-Up

The robotic arm used for experimental identification is shown in Figure 14.5. This experimental facility is used for research on modeling, identification, and control of robots [27–32]. Its three revolute joints ($n = 3$ – RRR kinematics) are actuated by gearless brushless DC direct-drive motors. The joints have infinite range of motions, because power and sensor signals are transferred via sliprings. The coordinate frames indicated in Figure 14.5 are assigned according to the DH convention. Numerical values of the DH parameters are given in Table 14.1. The actuators are Dynaserv DM-series servos with nominal torques of 60, 30, and 15 Nm. These servos are equipped with incremental optical encoders with a resolution of $\sim 10^{-5}$ rad. Each actuator is driven by a power amplifier, which ensures a linear proportionality between the input voltage and the output motor torque. Both encoders and amplifiers are connected to a PC-based

TABLE 14.1 DH Parameters of the RRR Robot

i	α_i [rad]	a_i [m]	q_i	d_i [m]
1	$\pi/2$	0	q_1	$C_0 C_1 = 0.560$
2	0	$P_1 C_2 = 0.200$	q_2	$C_1 P_1 = 0.169$
3	0	$P_2 C_3 = 0.415$	q_3	$C_2 P_2 = 0.090$

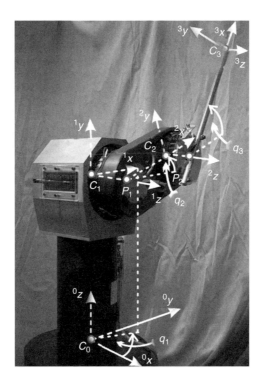

FIGURE 14.5 The RRR robot.

control system. This system consists of a MultiQ I/O board from Quanser Consulting (8 × 13 bits ADC, 8 × 12 bits DAC, eight digital I/O, six encoder inputs, and three hardware timers), combined with the soft real-time system Wincon from Quanser Consulting. The control system facilitates online data acquisition and control directly from Matlab/Simulink.

The rigid-body dynamics has a dominant role in the overall dynamic behavior of the RRR robot. Friction in the robot joints can be observed, too. Moreover, several effects not covered by the rigid-body dynamics with friction can be excited, e.g., vibrations at the robot base and high frequency resonances. These effects cannot be ignored if a high quality feedback control design is desired.

14.6.2 Kinematic and Dynamic Models in Closed Form

Kinematic and rigid-body dynamic models of the RRR robot were derived following the procedure explained in subsections 14.2.1 and 14.2.2. The models are available in closed form, which highly facilitates their analysis and manipulation, as well as a model-based control design.

The elements of the matrix $^0O_3$, which maps orientation of the tip coordinate frame '3' to the base frame '0,' are as follows:

$$o_{1,1} = \cos q_1 \cos(q_2 + q_3),\ o_{1,2} = -\cos q_1 \sin(q_2 + q_3),\ o_{1,3} = \sin q_1, o_{2,1} = \sin q_1 \cos(q_2 + q_3)$$
$$o_{2,2} = -\sin q_1 \sin(q_2 + q_3),\ o_{2,3} = -\cos q_1,\ o_{3,1} = \sin(q_2 + q_3),\ o_{3,2} = \cos(q_2 + q_3),\ o_{3,3} = 0 \quad (14.43)$$

The orthogonal projections of the vector $^0\mathbf{x}_3(\mathbf{q})$ pointing from the origin of '0' to the origin of '3' are

$$x = \cos q_1 (a_3 \cos(q_2 + q_3) + a_2 \cos q_2) + (d_2 + d_3) \sin q_1$$
$$y = \sin q_1 (a_3 \cos(q_2 + q_3) + a_2 \cos q_2) - (d_2 + d_3) \cos q_1$$
$$z = a_3 \sin(q_2 + q_3) + a_2 \sin q_2 + d_1 \quad (14.44)$$

Equation (14.43) and Equation (14.44) define forward kinematics (FK) of the RRR robot. The inverse kinematics (IK) is determined by solving the Equations (14.44) for q_1, q_2, and q_3:

$$q_1 = \mathrm{asin}\frac{x(d_2 + d_3) + y\sqrt{x^2 + y^2 - (d_2 + d_3)^2}}{x^2 + y^2}$$

$$q_3 = \mathrm{atan}\frac{\pm\sqrt{1 - \left(p_{wh}^2 + p_{wv}^2 - a_2^2 - a_3^2\right)^2/(2a_2a_3)^2}}{\left(p_{wh}^2 + p_{wv}^2 - a_2^2 - a_3^2\right)/(2a_2a_3)}$$

$$q_2 = \mathrm{atan}\frac{(a_2 + a_3\cos q_3)p_{wv} - a_3\sin q_3\, p_{wh}}{(a_2 + a_3\cos q_3)p_{wh} + a_3\sin q_3\, p_{wv}}$$

$$p_{wh} = \sqrt{[x - (d_2 + d_3)\sin q_1]^2 + [y + (d_2 + d_3)\cos q_1]^2}, \quad p_{wv} = z - d_1 \quad (14.45)$$

The closed-form representation (14.45) has three advantages: (*i*) an explicit recognition of kinematic singularities, (*ii*) a direct selection of the actual robot posture by specifying a sign in the expression for q_3, as the arm can reach each point in the Cartesian space either with the elbow up ($-$ sign) or with the elbow-down ($+$ sign), and (*iii*) a faster computation of a highly accurate IK solution than using any recursive numerical method.

A rigid-body dynamic model of the RRR robot was initially derived in the standard form (14.6). After that, for the sake of identification, it was transformed into regressor form (14.10). The number of BPS elements is 15, coinciding with the number obtained using the formula derived in [14]. For a given kinematics, this formula computes the minimum number of inertial parameters whose values can determine the dynamic model uniquely. The elements of **p** and **R** are expressed in closed-form:

$$p_1 = a_2^2 m_3 + a_2 m_2(a_2 + 2x_2) - I_{xx,2} + I_{yy,2}$$
$$p_2 = -2(a_2 m_2 y_2 + I_{xy,2})$$
$$p_3 = m_3(a_3 + x_3)$$
$$p_4 = m_3 y_3$$
$$p_5 = a_3 m_3(a_3 + 2x_3) - I_{xx,3} + I_{yy,3}$$
$$p_6 = 2(a_3 m_3 y_3 + I_{xy,3})$$
$$p_7 = 2d_2 m_2 z_2 + d_2^2 m_2 + 2(d_2 + d_3)m_3 z_3 + (d_2 + d_3)^2 m_3 + I_{yy,1} + I_{xx,2} + I_{xx,3}$$
$$p_8 = -(d_2 + d_3)m_3 y_3 - I_{yz,3}$$
$$p_9 = -a_3 m_3(z_3 + d_2 + d_3) - (d_2 + d_3)m_3 x_3 - I_{xz,3}$$
$$p_{10} = -d_2 m_2 y_2 - I_{yz,2}$$
$$p_{11} = -a_2[m_3(z_3 + d_2 + d_3) + m_2(d_2 + z_2)] - d_2 m_2 x_2 - I_{xz,2}$$
$$p_{12} = \left(a_2^2 + a_3^2 + 2a_3 x_3\right)m_3 + a_2 m_2(a_2 + 2x_2) + I_{zz,2} + I_{zz,3}$$
$$p_{13} = a_3 m_3(a_3 + 2x_3) + I_{zz,3}$$
$$p_{14} = m_2(x_2 + a_2) + m_3 a_2$$
$$p_{15} = m_2 y_2 \quad (14.46)$$

$$r_{1,1} = \ddot{q}_1 \cos^2 q_2 - \dot{q}_1 \dot{q}_2 \sin(2q_2)$$
$$r_{1,2} = 0.5\ddot{q}_1 \sin(2q_2) + \dot{q}_1 \dot{q}_2 \cos(2q_2)$$
$$r_{1,3} = 2a_2\{[\ddot{q}_1 \cos(q_2 + q_3) - \dot{q}_1 \dot{q}_3 \sin(q_2 + q_3)]\cos q_2 - \dot{q}_1 \dot{q}_2 \sin(2q_2 + q_3)\}$$
$$r_{1,4} = -2a_2[(\ddot{q}_1 \sin(q_2 + q_3) + \dot{q}_1 \dot{q}_3 \cos(q_2 + q_3))\cos q_2 + \dot{q}_1 \dot{q}_2 \cos(2q_2 + q_3)]$$
$$r_{1,5} = \ddot{q}_1 \cos^2(q_2 + q_3) - \dot{q}_1(\dot{q}_2 + \dot{q}_3)\sin(2q_2 + 2q_3)$$
$$r_{1,6} = -0.5\ddot{q}_1 \sin(2q_2 + 2q_3) - \dot{q}_1(\dot{q}_2 + \dot{q}_3)\cos(2q_2 + 2q_3)$$
$$r_{1,7} = \ddot{q}_1$$

$$r_{1,8} = (\ddot{q}_2 + \ddot{q}_3)\cos(q_2 + q_3) - (\dot{q}_2 + \dot{q}_3)^2 \sin(q_2 + q_3)$$
$$r_{1,9} = (\ddot{q}_2 + \ddot{q}_3)\sin(q_2 + q_3) + (\dot{q}_2 + \dot{q}_3)^2 \cos(q_2 + q_3)$$
$$r_{1,10} = -\dot{q}_2^2 \sin q_2 + \ddot{q}_2 \cos q_2$$
$$r_{1,11} = \dot{q}_2^2 \cos q_2 + \ddot{q}_2 \sin q_2$$
$$r_{2,1} = 0.5\dot{q}_1^2 \sin(2q_2)$$
$$r_{2,2} = -0.5\dot{q}_1^2 \cos(2q_2)$$
$$r_{2,3} = 2a_2\left[(\ddot{q}_2 + 0.5\ddot{q}_3)\cos q_3 - (0.5\dot{q}_3^2 + \dot{q}_2\dot{q}_3)\sin q_3 + 0.5\dot{q}_1^2 \sin(2q_2 + q_3)\right] + g \cdot \cos(q_2 + q_3)$$
$$r_{2,4} = 2a_2\left[-(\ddot{q}_2 + 0.5\ddot{q}_3)\sin q_3 - (0.5\dot{q}_3^2 + \dot{q}_2\dot{q}_3)\cos q_3 + 0.5\dot{q}_1^2 \cos(2q_2 + q_3)\right] - g \cdot \sin(q_2 + q_3)$$
$$r_{2,5} = 0.5\dot{q}_1^2 \sin(2q_2 + 2q_3)$$
$$r_{2,6} = 0.5\dot{q}_1^2 \cos(2q_2 + 2q_3)$$
$$r_{2,8} = \ddot{q}_1 \cos(q_2 + q_3)$$
$$r_{2,9} = \ddot{q}_1 \sin(q_2 + q_3)$$
$$r_{2,10} = \ddot{q}_1 \cos q_2$$
$$r_{2,11} = \ddot{q}_1 \sin q_2$$
$$r_{2,12} = \ddot{q}_2$$
$$r_{2,13} = \ddot{q}_3$$
$$r_{2,14} = g \cdot \cos q_2$$
$$r_{2,15} = -g \cdot \sin q_2$$
$$r_{3,3} = a_2\left[(0.5\dot{q}_1^2 + \dot{q}_2^2)\sin q_3 + \cos q_3 \ddot{q}_2 + 0.5\dot{q}_1^2 \sin(2q_2 + q_3)\right] + g \cdot \cos(q_2 + q_3)$$
$$r_{3,4} = a_2\left[(0.5\dot{q}_1^2 + \dot{q}_2^2)\cos q_3 - \ddot{q}_2 \sin q_3 + 0.5\dot{q}_1^2 \cos(2q_2 + q_3)\right] - g \cdot \sin(q_2 + q_3)$$
$$r_{3,5} = 0.5\dot{q}_1^2 \sin(2q_2 + 2q_3)$$
$$r_{3,6} = 0.5\dot{q}_1^2 \cos(2q_2 + 2q_3)$$
$$r_{3,8} = \ddot{q}_1 \cos(q_2 + q_3)$$
$$r_{3,9} = \ddot{q}_1 \sin(q_2 + q_3)$$
$$r_{3,13} = \ddot{q}_2 + \ddot{q}_3$$
$$r_{1,12} = r_{1,13} = r_{1,14} = r_{1,15} = r_{2,7} = r_{3,1} = r_{3,2} = r_{3,7} = r_{3,10} = r_{3,11} = r_{3,12} = r_{3,14} = r_{3,15} = 0$$
(14.47)

Given expressions reveal the complexity of the nonlinear robot dynamics, enabling analysis of each dynamic effect. With the elements of (14.47) containing the gravitational constant g, one may assemble gravity terms; with the elements containing joint accelerations, the inertial terms can be recovered, while the remaining elements define Coriolis and centripetal terms.

14.6.3 Establishing Correctness of the Models

To establish correctness of the models (14.45)–(14.47), the writing task presented in Figure 14.3a and Figure 14.3b can be considered. With the numerical values of the DH parameters given in Table 14.1, the IK model (14.45) computes the joint motions, shown in Figure 14.6a. The equivalent joint speeds are presented in Figure 14.6b. These plots reveal non-smooth joint motions, with significant speed levels. These motions are compared with the IK solution computed using a recursive numerical routine, implemented in *A Robotics Toolbox for Matlab* [7]. The guaranteed accuracy of the recursive solution was better than 10^{-10} rad. The differences between the IK solutions computed with the closed-form model (14.45) and with the Robotics Toolbox are within the accuracy of the latter ones, see Figure 14.7. This verifies the correctness of (14.45).

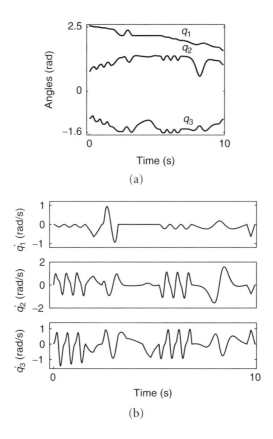

FIGURE 14.6 The joint space trajectory for the writing task: (a) motions, (b) speeds.

To establish the correctness of the rigid-body model (14.46) and (14.47), all inertial parameters are assigned arbitrary nonzero values, positive for masses and principal moments of inertia, without sign constraint for products of inertia and coordinates of centers of masses. The model computes torques corresponding to the joint space trajectory shown in Figure 14.6a and Figure 14.6b. The torques are also calculated using a recursive numerical routine implemented in the Robotics Toolbox. The differences between the two

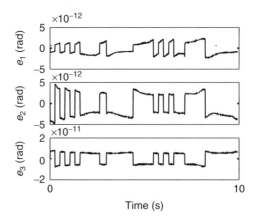

FIGURE 14.7 Differences between the closed-form and the recursive IK solutions.

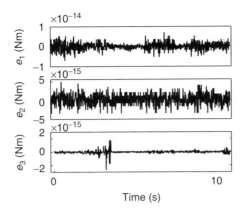

FIGURE 14.8 Differences in torques computed using the closed-form and recursive representations of the rigid-body dynamic model.

solutions are caused by round-off errors only, see Figure 14.8. This verifies correctness of Equation (14.46) and Equation (14.47).

14.6.4 Friction Modeling and Estimation

Two friction models are considered: the dynamic LuGre model (14.13) and the static neural-network model (14.14). The parameters of the LuGre model $v_{s,i}, \alpha_{0,i}, \alpha_{1,i}$, and $\alpha_{2,i}$ ($i = 1, 2, 3$), which correspond to the sliding friction regime, as well as the parameters of the neural-network model $f_{i,k}, w_{i,k}$, and $\alpha_{2,i}$ ($i = 1, 2, 3; k = 1, 2, 3$), were estimated using a recursive Extended Kalman filtering method addressed in subsection 14.3.1. The parameters of the LuGre model corresponding to the presliding regime were estimated using the frequency-domain identification method, also mentioned in subsection 14.3.1. More details about both identification methods can be found in [29] and [30]. Only one way for model validation is presented here, for the neural-network model, because no essential improvement in representing friction effects was observed using the LuGre model.

For illustration, the validation results for the third joint are given. For this validation, angular and speed responses to the input torque τ_3 shown in Figure 14.9a were measured. The identical input was applied to the model (14.14), whose parameters were assigned to the estimates. The model was integrated to compute the corresponding joint motion and speed. The results are shown in Figure 14.9b with thin lines, while thick lines represent the experimentally measured data. By inspection we observe high agreement between experimental and simulated data, which validates the estimated friction model. Similar results hold for the other joints of the RRR robot.

14.6.5 Estimation of the BPS

As explained in subsection 14.3.3, a reliable estimation of the elements of the BPS (14.46) requires custom design of an exciting trajectory. This trajectory is executed in the identification experiment. Motions of the exciting trajectory for the RRR robot were postulated as in (14.24). For frequency resolution we chose $\Delta f = 0.1$ Hz, imposing a cycle period of 10 s. From an experimental study it was known that joint motions within the bandwidth of 10 Hz do not excite flexible dynamics. Consequently, the number of harmonics was selected as $N = 100$. The next step was computing the free coefficients $q_{0,i}, a_{ij}$, and b_{ij} ($i = 1, \ldots, n$; $j = 1, \ldots, N$) in (14.24), that minimize the condition number of the information matrix (14.18). This matrix was created by vertical stacking the regression matrices, that correspond to discrete time instants $0, 0.1, 0.2, \ldots, 10$ s. Elements of each regression matrix are given by (14.47), and they were computed along the exciting trajectory (14.24). The free coefficients were found using a constrained optimization

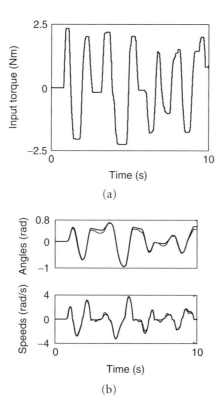

FIGURE 14.9 (a) Torque applied to the third joint for the friction model validation; (b) measured (thick) and simulated (thin) angular motions and speeds in the third joint.

algorithm that takes care about constraints on permissible speed and acceleration ranges in the robot joints: $|\dot{q}_1| \leq 2\pi$ rad/s, $|\dot{q}_2| \leq 3\pi$ rad/s, $|\dot{q}_3| \leq 3\pi$ rad/s, and $|\ddot{q}_i| \leq 60$ rad/s² ($i = 1, 2, 3$). The achieved (suboptimal) condition number was 3.1. The determined coefficients were inserted into (14.24) to give the motions shown in Figure 14.10.

The exciting trajectory was realized with the feedback controller (14.25), which used online reconstructed joint motions and speeds as feedback coordinates. The reconstruction was done according to the Equation (14.30) and Equation (14.31). The PD gains were tuned with "trial and error" to ensure

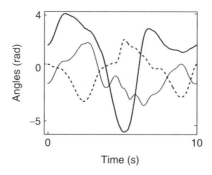

FIGURE 14.10 Motions of the exciting trajectory: q_1^{exc}—thick solid, q_2^{exc}—thin solid, and q_3^{exc}—dotted.

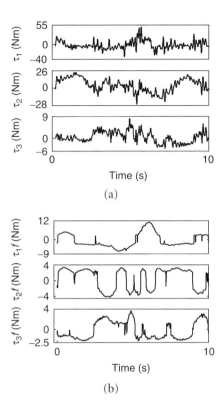

FIGURE 14.11 (a) Control inputs from the identification experiments; (b) reconstructed friction torques.

reasonable tracking of the exciting trajectory, with a low level of noise in the control inputs τ_i ($i = 1, 2, 3$) presented in Figure 14.11a. The sampling time was $T_s = 1$ ms. Because the parameters of the friction models (14.14) in all three joints had been estimated already, the friction torques $\tau_i^f(t)$ ($i = 1, 2, 3$) were reconstructed, too. These are shown in Figure 14.11b.

After collecting all the data used in Equation (14.18) and Equation (14.19), the batch LS estimation (14.20) of the BPS elements could be carried out. To facilitate estimation, one may take advantage of the symmetric mass distribution in the links 2 and 3, along the positive and negative directions of the axes ^2y and ^3y, respectively (see Figure 14.5). These symmetries imply that y_2 and y_3 coordinates of the centres of masses of the links 2 and 3 are identically zero. By virtue of Equation (14.46), it follows that the elements p_4 and p_{15} are also identically zero. This reduces the dimensionality of the problem, as instead of all 15 parameters, only 13 elements of the BPS need estimation. As apparent from the plots shown in Figure 14.12, the LS method results in the convergence of all 13 estimated elements of $\hat{\mathbf{p}}$.

To evaluate the quality of the estimates, a validation experiment depicted in Figure 14.4 was performed. The robot was realizing the writing task shown in Figure 14.3a and Figure 14.3b, with the joint motions and speeds presented in Figure 14.6a and Figure 14.6b. The objective was to check how good the estimated model reconstructed signals generated with the PD controllers. The torques generated with the PD controllers and the torques computed from the model are presented in Figure 14.13a to Figure 14.13c. Apparently, there is a close match between the experimental and the reconstructed signals in all three joints. This implies a satisfactory estimation of the rigid-body dynamics. Moreover, since the model reconstructed the control torques as the experiment was in progress, it follows that the dynamic model itself admits online implementation. Sufficient quality in reconstructing the realistic dynamic behavior and the real-time capability qualify the estimated model for model-based control purposes.

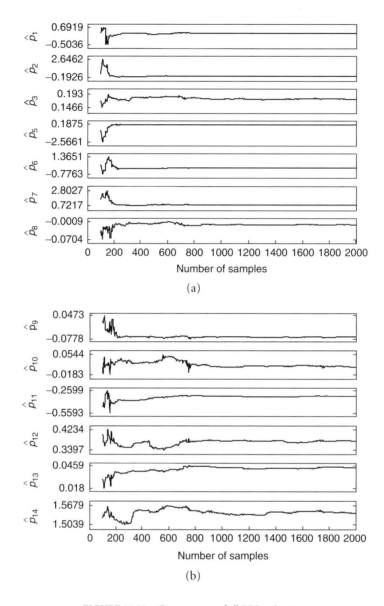

FIGURE 14.12 Convergence of all BPS estimates.

14.6.6 Dynamics Not Covered with the Rigid-Body Dynamic Model

A motivation and procedures to identify and model dynamics not covered with a rigid-body dynamic model are explained in Section 14.5. The frequency domain identification was applied on the RRR robot. For illustration, Figure 14.14 presents frequency response functions (FRF) measured in the first joint, for two different configurations in the remaining joints. The first FRF was obtained for the configuration $[q_2, q_3] = [0, \pi/2]$ [rad], which, according to the kinematic parameterisation shown in Figure 14.5, corresponds to a horizontal orientation of the second link and vertical upwards position of the last link. The second FRF corresponds to a fully horizontally stretched arm, i.e., $[q_2, q_3] = [0, 0]$ [rad].

Both FRFs reveal rigid low frequency dynamics, i.e., the dynamics of a double integrator. Flexible effects can be observed at higher frequencies. A modest resonance is apparent at 28 Hz, caused by insufficient

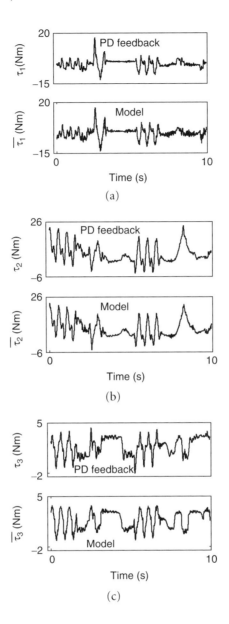

FIGURE 14.13 Control torques produced with decentralized PD controllers and with the rigid-body dynamic model in closed-form (measured in the experimental writing task).

stiffness in mounting the robot base to the floor. Being highly damped and with a temporary phase lead (see the phase plots), it is not a problem for stability. However, it may cause vibrations of the robot mechanism and consequently degrade the robot performance. At higher frequencies (above 80 Hz), more profound resonances can be noticed. Their frequencies and damping factors depend on the robot posture. They can contribute to amplification of vibrations and noise levels in the covered frequency ranges, which would degrade the performance of motion control. If amplified too much by feedback, these resonances may even destabilize the closed-loop system. To achieve motion control of high performance, the observed flexible effects must be taken into account during the feedback control design.

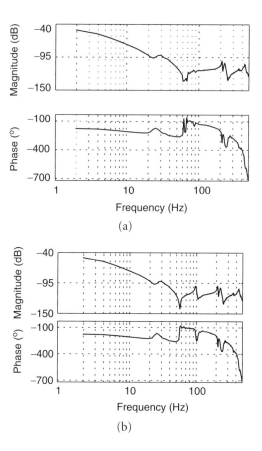

FIGURE 14.14 Flexible dynamics for first joint corresponding to two fixed configurations of the remaining joints: (a) $[q_2, q_3] = [0, \pi/2]$ and (b) $[q_2, q_3] = [0, 0]$.

14.7 Conclusions

This chapter deals with the problems of robot modeling and identification for high-performance model-based motion control. A derivation of robot kinematic and dynamic models was explained. Modeling of friction effects was also discussed. Use of a writing task to establish correctness of the models was suggested. Guidelines for design of an exciting identification trajectory were given. A Kalman filtering technique for online reconstruction of joint motions, speeds, and accelerations was explained. A straightforward but efficient estimation of parameters of the rigid-body dynamic model with friction effects was described. The effectiveness of the procedure was experimentally demonstrated on a direct-drive robot with three revolute joints. For this robot, models of kinematics and rigid-body dynamics were derived in closed-form, and presented in full detail. The correctness of the models was established in simulation. Results of experimental estimation of the dynamic model parameters were presented. The appropriateness of the dynamic model for model-based control purposes was verified. However, it was also indicated that this model is still not sufficient for a perfect match to the real robot dynamics, as these dynamics may contain more effects than covered by the rigid-body model. A procedure to identify the dynamics not covered by the rigid-body model was proposed. With these additional dynamics available, more advanced feedback control designs become possible.

References

[1] Sciavicco, L. and Siciliano, B., *Modeling and Control of Robot Manipulators,* McGraw-Hill, London, 1996.
[2] Vukobratović, M. and Potkonjak, V., *Dynamics of Manipulation Robots: Theory and Application,* Springer-Verlag, Berlin, 1982.
[3] Fu, K.S., Gonzales, R.C., and Lee, C.S.G., *Robotics: Control, Sensing, Vision, and Intelligence,* McGraw-Hill, London, 1987.
[4] Kozlowski, K., *Modeling and Identification in Robotics,* Springer-Verlag, London, 1998.
[5] Asada, H. and Youcef-Toumi, K., *Direct-Drive Robots: Theory and Practice,* MIT Press, London, 1987.
[6] Nethery, J. and Spong, M.W., Robotica: A Mathematica package for robot analysis, *IEEE Rob. Autom. Mag.,* 1, 13, 1994.
[7] Corke, P.I., A robotics toolbox for MATLAB, *IEEE Rob. Autom. Mag.,* 3, 24, 1996.
[8] Bennnett, D.J. and Hollerbach, J.M., Autonomous calibration of single-loop closed kinematic chains formed by manipulators with passive endpoint constraints, *IEEE Trans. Rob. Autom.,* 7, 597, 1991.
[9] Bélanger, P.R., Estimation of angular velocity and acceleration from shaft encoder measurements, in *Proc. IEEE Int. Conf. Rob. Autom.,* Nice, France, 1992, 585.
[10] Bélanger, P.R., Dobrovolny, P., Helmy, A., and Zhang, X., Estimation of angular velocity and acceleration from shaft-encoder measurements, *Int. J. Rob. Res.,* 17, 1225, 1998.
[11] Swevers, J., Ganseman, C., Tukel, D.B., de Schutter, J., and van Brussel, H., Optimal robot excitation and identification, *IEEE Trans. Rob. Autom.,* 13, 730, 1997.
[12] Calafiore, G., Indri, M., and Bona, B., Robot dynamic calibration: Optimal excitation trajectories and experimental parameter estimation, *J. Rob. Syst.,* 18, 55, 2001.
[13] Potkonjak, V., Tzafestas, S., Kostic, D., Djordjevic, G., and Rasic, M., Illustrating man-machine motion analogy in robotics — the handwriting problem, *IEEE Rob. Autom. Mag.,* 10, 35, 2003.
[14] Mayeda, H., Yoshida, K., and Osuka, K., Base parameters of manipulator dynamic models, *IEEE Trans. Rob. Autom.,* 6, 312, 1990.
[15] Armstrong-Hélouvry, B., Dupont, P., and Canudas de Wit, C., A survey of models, analysis tools and compensation methods for the control of machines with friction, *Automatica,* 30, 1083, 1994.
[16] Olsson, H., Åström, K.J., Canudas de Wit, C., Gäfvert, M., and Lischinsky, P., Friction models and friction compensation, *Europ. J. Control,* 4, 176, 1998.
[17] Swevers, J., Al-Bender, F., Ganseman, C.G., and Prajogo, T., An integrated friction model structure with improved presliding behavior for accurate friction compensation, *IEEE Trans. Autom. Control,* 45, 675, 2000.
[18] Barabanov, N. and Ortega, R., Necessary and sufficient conditions for passivity of the LuGre friction model, *IEEE Trans. Aut. Control,* 45, 830, 2000.
[19] Hensen, R.H.A., Angelis, G.Z., Van de Molengraft, M.J.G., De Jager, A.G., and Kok, J.J., Grey-box modeling of friction: an experimental case study, *Europ. J. Contr.,* 6, 258, 2000.
[20] Hensen, R.H.A., Van de Molengraft, M.J.G., and Steinbuch, M., Frequency domain identification of dynamic friction model parameters, *IEEE Trans. on Contr. Syst. Tech.,* 10, 191, 2001.
[21] Golub, G.H. and van Loan, C.F., *Matrix computations,* John Hopkins University Press, London, 1996.
[22] Slotine, J.J.E. and Li, W., *Applied Nonlinear Control,* Prentice Hall, Upper Saddle River, NJ, 1991.
[23] Nijmeijer, H. and Van der Schaft, A., *Nonlinear Dynamical Control Systems,* Springer-Verlag, Berlin, 1991.
[24] Pintelon, R. and Schoukens, J., *System Identification: A Frequency Domain Approach,* IEEE Press, Piscataway, NJ, 2001.
[25] Sanathanan, C.K. and Koerner, J., Transfer function synthesis as a ratio of two complex polynomials, *IEEE Trans. Autom. Control,* 8, 56, 1963.

[26] Phillips, C.L. and Nagle, H.T., *Digital Control System Analysis and Design*, Prentice Hall, Englewood Cliffs, NJ, 1995.
[27] Van Beek, B. and De Jager, B., RRR-robot design: Basic outlines, servo sizing, and control, in *Proc. IEEE Int. Conf. Control Appl.*, Hartford, U.S.A., 1997, 36.
[28] Van Beek, B. and De Jager, B., An experimental facility for nonlinear robot control, in *Proc. IEEE Int. Conf. Control Appl.*, Hawaii, U.S.A., 1999, 668.
[29] Kostić, D., Hensen, R., De Jager, B., and Steinbuch, M., Modeling and identification of an RRR-robot, in *Proc. IEEE Conf. Dec. Control*, Orlando, FL, 2001, 1144.
[30] Kostić, D., Hensen, R., De Jager, B., and Steinbuch, M., Closed-form kinematic and dynamic models of an industrial-like RRR robot, in *Proc. IEEE Conf. Rob. Autom.*, Washington D.C., 2002, 1309.
[31] Kostić, D., De Jager, B., and Steinbuch, M., Experimentally supported control design for a direct drive robot, in *Proc. IEEE Int. Conf. Control Appl.*, Glasgow, Scotland, 2002, 186.
[32] Kostić, D., De Jager, B., and Steinbuch, M., Robust attenuation of direct-drive robot-tip vibrations, in *Proc. IEEE/RSJ Int. Conf. Intellig. Robots Syst.*, Lausanne, Switzerland, 2002, 2206.

15
Motion Control by Linear Feedback Methods

Dragan Kostić
Technische Universiteit Eindhoven

Bram de Jager
Technische Universiteit Eindhoven

Maarten Steinbuch
Technische Universiteit Eindhoven

15.1 Introduction **15**-1
15.2 Decentralized Conventional Feedback Control **15**-3
15.3 Linear Feedback Control Applied with Nonlinear Model-Based Dynamic Compensators **15**-5
 Computed-Torque Control Design • Control Designs with Feedback Linearization
15.4 Case-Study: Motion Control of the RRR Direct-Drive Manipulator **15**-10
 Nominal Plant Model • Conventional Feedback Control Design • μ-Synthesis Feedback Control Design • Experimental Evaluation
15.5 Conclusions **15**-22

15.1 Introduction

This chapter deals with the robot motion control problem, whose objective is accurate execution of a reference robot trajectory. This trajectory is executed by the robot end-effector in the operational space, i.e., in the space where the robot performs its motion tasks. The dimension of this space is not higher than six, as unique positioning and orientation of the end-effector require at most three Cartesian and three angular coordinates. A desired operational trajectory is realized by controlling the motions in the robot joints. The operational reference is mapped into the joint reference trajectory via the inverse kinematics mapping [1,2]. The joint reference is realized by the joint servo control loops. Better quality of motion control in the joint space contributes to higher motion performance in the operational space.

There are a number of criteria to evaluate the performance of motion control. A few will be listed here. Accuracy of tracking a motion reference is the usual performance criterion. Intuitively, it is expected that if the joint reference is realized with a higher accuracy, then the operational motion should be accurate as well. However, the relation between accuracies in the joint and operational spaces is not as straightforward as the previous sentence implies. The mapping from the joint to the operational motion (forward kinematics [1,2]) is intrinsically nonlinear, and hence the accuracies of tracking these two motions are nonlinearly related. Apart from joint configuration, flexibilities and disturbances may influence position of the robot end-effector. In the presence of these effects, even highly accurate execution of the joint reference may give unsatisfactory accuracy in the operational space. Therefore, these effects must be taken into account in the motion control design.

The control bandwidth is another performance criterion. Although often addressed in manipulator motion control, there is no unique definition of it. In [3] it is formulated as a range of frequencies contained in the control input. A higher bandwidth should imply the potential of a manipulator control system to realize faster motions, together with the possibility to compensate for dynamic effects and disturbances of broader frequency content. If dynamics in the manipulator joints are linear and linear feedback control strategies are used, then one may resort to conventional definitions of the control bandwidth [4]. Linear dynamics can be realized by decoupling the robot axes either using high-gear transmission mechanisms between the joint actuators and the robot links or by some nonlinear control compensation. The control bandwidth in a given robot joint can be formulated as the cross-over frequency, i.e., the first zero-crossing of the open-loop gain of the servo-loop in this joint. Because of linearity, a higher bandwidth implies a higher open-loop gain, which contributes to better motion accuracy.

Imposing a high feedback gain for the sake of better tracking accuracy is unacceptable if it also excites parasitic dynamics (flexibilities) and/or amplifies disturbances (e.g., measurement noise). Therefore, a level of rejection of the parasitics and disturbances can be adopted as additional control performance criterion.

Stability and robustness issues are naturally important in robot motion control. Stabilization of robot motions is achieved with a feedback control component. Motion control is robustly stable if despite uncertainty in the robot dynamics and disturbances, the stability is preserved. There are two kinds of uncertainties. The first ones are parametric, and they arise if physical values of robot inertial and/or friction parameters are not known exactly. The second ones are unmodelled dynamic effects, e.g., flexibilities and friction. As examples of disturbances, one may think of cogging forces and quantization noise. The former are typical for direct-drive manipulators [5], while the latter arises if incremental encoders are used as position sensors. Better stability margins enhance robustness of motion control, because they guarantee the safe operation despite uncertainties and disturbances. With linear decoupled joint dynamics and linear feedback, control solutions, standard gain, and phase margins [4] can be considered as control performance criteria.

Increasing demands on the performance and robustness of manipulator motion control has led to the development of versatile motion control approaches. Robust control strategies are introduced to stabilize manipulator motions under uncertainty and disturbance conditions. More advanced strategies should ensure that the desired performance of motion control is preserved despite uncertainties and disturbances. This property is referred to as robust performance.

Decentralized PID (proportional, integral, derivative) and its variants PD and PI are conventional feedback control solutions. The majority of industrial robot manipulators are equipped with high-gear transmission mechanisms, so they can be described by linear and decoupled dynamics. Linear conventional controllers are suitable for such dynamics. They are appealing because of their efficiency in tuning and low computational costs. However, if high-quality motion control performance is needed, conventional controllers may lead to unsatisfactory results. Use of these controllers means making serious tradeoffs among feasible static accuracy, system stability, and damping of high-frequency disturbances. For example, smaller proportional action gives larger gain and phase margins for system stability, sacrificing static accuracy. A higher proportional gain improves static accuracy, but also amplifies quantization noise and other high-frequency disturbances. Inclusion of integral action improves static accuracy, but often reduces stability margins. It is shown in [6] that these tradeoffs become critically conflicting with a digital implementation of the control law, as the sampling rate decreases. The simplicity of conventional controllers can be too restrictive to provide compensation of each dynamic effect encountered in the robotic system: dynamic couplings between the joints, friction, backlash, flexibility, and time-delay. Usually, the conventional controllers can handle only a limited number of these effects, together with other control objectives, e.g., prescribed control bandwidth and reduction of the position error. As non-compensated effects may strongly influence motion performance, more sophisticated control strategies are needed. The situation becomes even more critical if no transmission elements are present between joint actuators and links, which is typical for direct-drive robots. Then, the nonlinear dynamic couplings directly apply upon each joint actuator and their compensation requires use of more advanced control schemes.

Advanced manipulator control methods are capable of simultaneously realizing several control objectives: stability robustness, disturbance rejection, controlled transient behavior, optimal performance, etc.

Of course, it is not possible to achieve arbitrary quality with each control objective independently, because they are mutually dependent and are often conflicting. A detailed survey of advanced control methods is available in [7]. Nonlinear representatives, such as adaptive and sliding-mode methods, can improve system performance in the presence of uncertainty in the robot dynamics and external disturbances (e.g., variable load). The non-linearity of these methods, however, does not facilitate a quantitative prediction of system performance for a given robustness level. This is a limiting factor for their widespread application in practice, where it is often very important to know in advance worst-case motion accuracy for a given bandwidth of reference trajectories.

Stability robustness, disturbance rejection, and controlled transient response can be jointly and directly imposed using feedback schemes based on H_∞ control theory. These schemes enable quantitative prediction of motion performance, given bounds on modeling uncertainty and disturbances. Moreover, with the available knowledge of the system dynamics, parasitic effects, and disturbances, motion performance can be optimized. These are the reasons that make H_∞ controllers appealing solutions for practical problems and motivate their application in robotic systems.

Industrial practice teaches us that ultimate limits of performance and robustness of motion control can be achieved only if all peculiarities of a controlled plant are taken into account during the control design [8]. To illustrate practical effects of such reasoning, in this chapter we will present a detailed motion control design applied to a direct-drive robotic manipulator with revolute joints. The dynamic behavior of such a manipulator is highly nonlinear [5], which particularly impedes robust high-performance control. Undesirable effects, such as flexibility, time-delay, and measurement noise, affect the manipulator operation. As a remedy, we carry out a control design that relies on a comprehensive model of the rigid manipulator dynamics and takes into account experimental characteristics of the manipulator. This design consists of the following steps: (*i*) compensation of the manipulator dynamics via feedback linearization, (*ii*) identification and modeling of remaining (flexible) dynamics, (*iii*) design of feedback controllers for the identified dynamics to meet performance and robustness specifications, and (*iv*) evaluation of the design.

The theory behind the control design will be formulated for the general case of a robotic manipulator with n degrees of freedom. In the case study, the practical demonstration of the design will be given for a direct-drive robotic manipulator. This demonstration should emphasize the feedback control tuning part. The considered manipulator has three rotational joints, implemented as waist, shoulder, and elbow [9,10]. As a similar kinematic structure is often met in industry, results obtained for the case study should be relevant for industrial cases. In the case study, the control design will effectively handle several objectives. Namely, it will robustly preserve a specified accuracy of trajectory tracking, despite vibrations at the robot base, high-frequency flexible effects, time-delay, and measurement noise. The achieved accuracy and bandwidth of motion control will be higher than feasible with conventional feedback design.

The chapter is organized as follows. Decentralized conventional feedback control will be considered in Section 15.2. Use of a dynamic model in the motion control design and feedback control designs applied with nonlinear model-based dynamic compensation will be presented in Section 15.3. Section 15.4 is the case study. Conclusions will come at the end.

15.2 Decentralized Conventional Feedback Control

Let us represent the rigid-body dynamics of a robot manipulator with n joints using Euler-Lagrange's equations of motion [1,2,11]:

$$\mathbf{M}(\mathbf{q}(t))\ddot{\mathbf{q}}(t) + \mathbf{Q}(\mathbf{q}(t),\dot{\mathbf{q}}(t))\dot{\mathbf{q}}(t) + \mathbf{g}(\mathbf{q}(t)) + \boldsymbol{\tau}^f(t) = \boldsymbol{\tau}(t) \qquad (15.1)$$

where $\boldsymbol{\tau}$ is the $(n \times 1)$ vector of generalized forces acting at joints (torques for rotational joints and forces for prismatic joints); $\mathbf{q}$, $\dot{\mathbf{q}}$ and $\ddot{\mathbf{q}}$ are the $(n \times 1)$ vectors of joint generalized motions, speeds, and accelerations, respectively; $\mathbf{M}$ is the $(n \times n)$ inertia matrix; $\mathbf{Q}(\mathbf{q}(t),\dot{\mathbf{q}}(t))\dot{\mathbf{q}}(t)$; $\mathbf{g}(\mathbf{q}(t))$, and $\boldsymbol{\tau}^f(t)$ are the $(n \times 1)$ vector of Coriolis/centripetal, gravity, and friction effects, respectively, and t denotes time.

The dynamics (15.1) are highly nonlinear and coupled. Use of transmission elements between the joint actuators and robot links can have linearizing effect on these dynamics [2]. Let $\boldsymbol{\theta}$ and $\boldsymbol{\tau}_m$ denote the $(n \times 1)$ vectors of actuator displacements and actuator driving torques, respectively; the transmissions then establish the following relationships:

$$\boldsymbol{\theta}(t) = \mathbf{N}\mathbf{q}(t)$$
$$\boldsymbol{\tau}_m(t) = \mathbf{N}^{-1}\boldsymbol{\tau}(t) \tag{15.2}$$

where $\mathbf{N}$ is a diagonal matrix containing the gear reduction ratios n_i. If the moments of inertia $J_{m,i}$ of each actuator shaft are collected in the diagonal matrix $\mathbf{J}_m$, then the inertia matrix $\mathbf{M}$ can be split as follows:

$$\mathbf{M}(\mathbf{q}(t)) = \mathbf{N}\mathbf{J}_m\mathbf{N} + \Delta\mathbf{M}(\mathbf{q}(t)) \tag{15.3}$$

where $\mathbf{N}\mathbf{J}_m\mathbf{N}$ is the constant matrix, and $\Delta\mathbf{M}$ is the configuration dependent matrix. Substituting Equation (15.2) and Equation (15.3) into Equation (15.1) yields:

$$\boldsymbol{\tau}_m(t) = \mathbf{J}_m\ddot{\boldsymbol{\theta}}(t) + \mathbf{d}(t) \tag{15.4}$$

where

$$\mathbf{d}(t) = \mathbf{N}^{-1}\Delta\mathbf{M}(\mathbf{q}(t))\mathbf{N}^{-1}\ddot{\boldsymbol{\theta}}(t) + \mathbf{N}^{-1}\mathbf{Q}(\mathbf{q}(t),\dot{\mathbf{q}}(t))\dot{\boldsymbol{\theta}}(t)\mathbf{N}^{-1} + \mathbf{N}^{-1}(\mathbf{g}(\mathbf{q}(t)) + \boldsymbol{\tau}^f(t)) \tag{15.5}$$

represents the nonlinear dynamic interaction between joint actuators. In case of high reduction ratios ($n_i \gg 1$), the contribution of the interaction is small compared with the linear dynamics $\mathbf{J}_m\ddot{\boldsymbol{\theta}}$. Consequently, the manipulator dynamics (15.4) are practically decoupled and linearized, as each component of $\boldsymbol{\tau}_m$ dominantly influences the corresponding component of $\boldsymbol{\theta}$, while $\mathbf{d}$ just plays the role of a disturbance for each joint servo.

Control of the decoupled dynamics (15.4) can be done using a decentralized feedback method. Denote with $\boldsymbol{\theta}_r(t)$ a reference trajectory of the actuator shaft, which is determined from the joint reference $\mathbf{q}_r(t)$ via Equation (15.2). Define the position error as the difference between the actual and the reference motion:

$$\mathbf{e}_m(t) = \boldsymbol{\theta}(t) - \boldsymbol{\theta}_r(t) \tag{15.6}$$

A decentralized control treats each robot joint as a separate plant to be controlled and applies decoupled linear feedback control actions u_i ($i = 1, \ldots, n_i$) as follows:

$$\boldsymbol{\tau}_m(s) = \mathbf{J}_m\mathbf{G}_a(s)\mathbf{u}(s) \tag{15.7}$$

where $\mathbf{G}_a(s) = diag[G_{a,1}(s), \ldots, G_{a,n}(s)]$ is a diagonal matrix containing the actuator dynamics, $\mathbf{u}$ is the $(n \times 1)$ vector of the feedback control actions, and s is the Laplace operator. Each u_i is decoupled from the remaining ones as it is based just on the position error θ_i corresponding to the actuator i.

For convenience, we may assume no actuator dynamics ($\mathbf{G}_a$ is identity) and a conventional PD feedback control law with acceleration feedforward:

$$\mathbf{u}(t) = \ddot{\boldsymbol{\theta}}_r(t) - \mathbf{K}_p\mathbf{e}_m(t) - \mathbf{K}_d\dot{\mathbf{e}}_m(t) \tag{15.8}$$

where $\mathbf{K}_p = diag[k_{p,1}, \ldots, k_{p,n}]$ and $\mathbf{K}_d = diag[k_{d,1}, \ldots, k_{d,n}]$ are matrices of positive position and velocity gains, respectively. The decentralized control approach of (15.7) and (15.8) is illustrated in Figure 15.1.

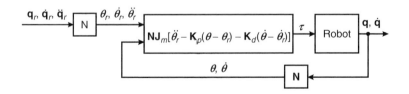

FIGURE 15.1 Decentralized motion control with PD feedback and acceleration feedforward.

The application of the control law (15.7) and (15.8) to the system (15.4) gives the error dynamics:

$$\mathbf{J}_m(\ddot{\mathbf{e}}_m(t) + \mathbf{K}_p \mathbf{e}_m(t) + \mathbf{K}_d \dot{\mathbf{e}}_m(t)) = -\mathbf{d}(t) \qquad (15.9)$$

It follows that the position error falls into a region of size proportional to the magnitude of $\mathbf{d}(t)$. If this region is narrow, which is the case when high reduction ratios are used, then the error can be made small by appropriate selection of the PD gains. If the gear reduction ratios are small, or if $\mathbf{N}$ is the identity matrix (direct-drive actuation), then the nonlinear coupling terms become more competitive or even dominant for fast motions in comparison with the linear part in Equation (15.9). Consequently, the magnitude of $\mathbf{d}(t)$ is not small, and position errors become larger for the same feedback controller gains. As a remedy, one can apply a model-based compensation for $\mathbf{d}(t)$, which is explained in the next section.

15.3 Linear Feedback Control Applied with Nonlinear Model-Based Dynamic Compensators

Nonlinear dynamic couplings between robot joints can be compensated using control algorithms that are based on the knowledge of the dynamic model (15.1). The manipulator dynamics that are not covered with the model are handled by a feedback control action. There are two common approaches to model-based dynamic compensation [2,3,12]:

- computed-torque control

$$\mathbf{M}(\mathbf{q}(t))\ddot{\mathbf{q}}(t) + \mathbf{Q}(\mathbf{q}(t),\dot{\mathbf{q}}(t))\dot{\mathbf{q}}(t) + \mathbf{g}(\mathbf{q}(t)) + \boldsymbol{\tau}^f(t) + \mathbf{u}(t) = \boldsymbol{\tau}(t) \qquad (15.10)$$

- feedback linearization

$$\mathbf{M}(\mathbf{q}(t))\mathbf{u}(t) + \mathbf{Q}(\mathbf{q}(t),\dot{\mathbf{q}}(t))\dot{\mathbf{q}}(t) + \mathbf{g}(\mathbf{q}(t)) + \boldsymbol{\tau}^f(t) = \boldsymbol{\tau}(t) \qquad (15.11)$$

In both approaches $\mathbf{u}(t)$ stands for the linear feedback control action, based on the error between the actual and the reference joint motions:

$$\mathbf{e}(t) = \mathbf{q}(t) - \mathbf{q}_r(t) \qquad (15.12)$$

possibly extended with acceleration feedforward term. The role of $\mathbf{u}$ is to stabilize the manipulator motion and ensure the desired performance when realizing the reference $\mathbf{q}_r(t)$.

The laws (15.10) and (15.11) are often characterised as "centralized control," because each component of the control inputs $\boldsymbol{\tau}$ contains dynamic interaction between the robot joints. The model-based terms given in (15.10) and (15.11) are computed along the actual joint motions, speeds, and accelerations. However, the reference motions and their time derivatives can also be used for their computation. Some feedback designs applicable in the control laws (15.10) and (15.11) are presented next.

15.3.1 Computed-Torque Control Design

Assuming perfect model-based compensation of nonlinear dynamics including friction, a control law that guarantees global uniform asymptotic stability of the origin $(\mathbf{e}, \dot{\mathbf{e}}) = (\mathbf{0}, \mathbf{0})$ is given by [13]

$$\mathbf{M}(\mathbf{q}(t))\ddot{\mathbf{q}}_r(t) + \mathbf{Q}(\mathbf{q}(t),\dot{\mathbf{q}}(t))\dot{\mathbf{q}}_r(t) + \mathbf{g}(\mathbf{q}(t)) + \boldsymbol{\tau}^f(t) - \mathbf{K}_p \mathbf{e}(t) - \mathbf{K}_d \dot{\mathbf{e}}(t) = \boldsymbol{\tau}(t) \qquad (15.13)$$

Here, $\mathbf{K}_p = diag[k_{p,1}, \ldots, k_{p,n}]$ and $\mathbf{K}_d = diag[k_{d,1}, \ldots, k_{d,n}]$ are matrices of positive position and velocity gains, respectively. Obviously, the model-based terms employ both the reference and the actual

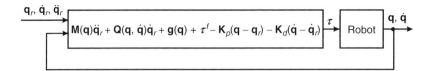

FIGURE 15.2 Computed-torque motion control with PD feedback.

joint trajectory. The control law (15.13), illustrated in Figure 15.2, achieves a passive mapping from the error dynamics

$$\mathbf{M}(\mathbf{q}(t))\ddot{\mathbf{e}}(t) + \mathbf{Q}(\mathbf{q}(t),\dot{\mathbf{q}}(t))\dot{\mathbf{e}}(t) - \mathbf{K}_p \mathbf{e}(t) - \mathbf{K}_d \dot{\mathbf{e}}(t) = \mathbf{\Psi}(t) \quad (15.14)$$

to the time derivative of the position error $\dot{\mathbf{e}}(t)$. This was proven in [13].

Although globally stabilizable and yielding better control performance than a decentralized controller, it is not obvious how to tune the PD gains in (15.13) to realize the desired performance specifications, e.g., prescribed error reduction, bandwidth constraint, and attenuation of undesired high-frequency effects (e.g., quantization noise and flexibilities). The feedback tuning is performed by trial and error on the real system, until a satisfactory trade-off among stability, amount of error reduction, and impact of disturbances is reached.

15.3.2 Control Designs with Feedback Linearization

The feedback linearization approach can establish a situation where common engineering reasoning can be used for tuning the feedback control action. When applied, the control law (15.11) reduces the motion control problem to the linear case, defined in the Laplace domain:

$$\mathbf{q}(s) = \mathbf{P}(s)\mathbf{u}(s), \quad \mathbf{P}(s) = \mathit{diag}\left[\frac{1}{s^2},\ldots,\frac{1}{s^2}\right] \quad (15.15)$$

The dynamics from the feedback action u_i to the displacement q_i describe the plant P_i to be controlled. The plant defined by (15.15) has perfectly rigid dynamics. Its frequency response function (FRF) has a linear amplitude plot with -2 slope and constant phase at $-180°$. Such an FRF can be established only if the law (15.11) perfectly matches the real robot dynamics. In practice, this rarely happens, as no robot is perfect in its realization, and hence, the rigid-body model used in (15.11) is hardly an accurate description of the real robot dynamics. Instead of (15.15), one should rather consider control of a more general transfer matrix $\mathbf{P}$:

$$\mathbf{q}(s) = \mathbf{P}(s)\mathbf{u}(s), \quad \mathbf{P}(s) = \begin{bmatrix} P_{1,1}(s) & \cdots & P_{1,n}(s) \\ \vdots & \ddots & \vdots \\ P_{n,1}(s) & \cdots & P_{n,n}(s) \end{bmatrix} \quad (15.16)$$

The transfer functions on the main diagonal of $\mathbf{P}$ represent dynamics between the feedback control action at each robot joint and the displacement of that joint. The cross-terms represent interactions between the joints that remain after imperfect compensation of the real robot dynamics. If any of these cross-terms is not negligible when compared with the transfer functions on the main diagonal, we can conclude that the dynamic model used in the feedback linearization is not sufficiently accurate. The inclusion of the cross-terms does not principally limit the feedback design procedure, although it makes it more involved. If the cross-terms are quantitatively smaller than the diagonal elements, then the transfer matrix $\mathbf{P}$ can be considered as diagonal. This implies total decoupling of the robot dynamics (15.1) via (15.11), which simplifies the feedback control design to n single-input, single-output cases. When dealing with practical problems, the diagonality of **P** must be analysed before proceeding to the feedback control design.

Each feedback design is focused on a particular transfer function $P_{i,i}(s)$ $(i = 1, \ldots, n)$, representing a plant to be controlled. For convenience, the indices in the subscript will be replaced with a single one: the plant for the ith joint will be denoted with $P_i(s)$. It is not a mere double integrator, as by virtue of (15.15) one would expect, but a higher-order dynamical system with resonance and antiresonance frequencies that emerge because of flexibilities. A time delay can also be experienced in the plant dynamics. These dynamics change for various operating conditions, as resonance frequencies and their relative damping vary with the manipulator configuration.

A technique to identify dynamics of the plant, i.e., FRF $G_i^k(j\omega)$ (ω is angular frequency), is explained in the chapter on modeling and identification. Briefly, this technique is based on analysis of spectra of input and output signals measured under closed-loop conditions. For joint i, several FRFs are identified, each corresponding to a different configuration of the remaining joints. Denote the number of the identified FRFs with N_i, i.e., $k = 1, \ldots, N_i$. For the feedback control design, it is preferable to determine one nominal FRF, for which a parametric model of the plant P_i could be devised. A natural choice for the nominal FRF is some average of all identified FRFs. Such a choice minimizes the difference between the plant model and the identified FRFs, which represents uncertainty in the plant dynamics. For example, one may use the following averages to determine the nominal FRF:

$$G_i^0(j\omega) = \arg \min_{G_i(j\omega)} \max_k \left| \frac{G_i^k(j\omega) - G_i(j\omega)}{G_i(j\omega)} \right| \quad \text{or} \quad (15.17a)$$

$$G_i^0(j\omega) = \frac{1}{N_i} \sum_{k=1}^{N_i} G_i^k(j\omega) \quad (15.17b)$$

For each ω, the solution (15.17a) minimizes the distance in the complex plane between the nominal FRF $G_i^0(j\omega)$ and all $G_i^k(j\omega)$, $k = 1, \ldots, N_i$. The solution (15.17b) is an average of all identified FRFs. A choice between (15.17a) and (15.17b) can be case dependent, although, in principle, the latter one gives smoother magnitude and phase FRF plots.

The nominal $G_i^0(j\omega)$ represents the plant dynamics as a set of data points. For a parametric description of the nominal plant model P_i^0, any technique of fitting the parametric transfer function $P_i^0(j\omega)$ into the nominal data $G_i^0(j\omega)$ can be used. A least-squares fit using an output error model structure [14] is one possibility. Once a parametric nominal plant model is available, it can be used in the feedback control design.

The servosystem for joint i is depicted in Figure 15.3. Here, P_i presents actual plant dynamics that accounts for all variations from the nominal plant model P_i^0. The objective of the feedback control design is to determine the controller C_i which realizes accurate tracking of the reference motion $q_{r,i}(t)$. To enhance the performance of motion control, the known acceleration $\ddot{q}_{r,i}(t)$ is used as the feedforward control input. There are several options for the design of the feedback controller C_i: LQ, LQG, loop-shaping, etc. In this chapter the loop-shaping approach is explained in more detail, because it enables specification of the performance objectives and feedback tuning in a very intuitive way. LQG theory will be used for state reconstruction in the case-study.

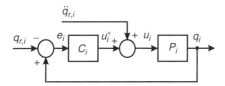

FIGURE 15.3 Servo control system for joint i.

The objective of loop-shaping is to design a feedback controller which enforces the desired shape of the closed-loop transfer functions for the servo-system shown in Figure 15.3. Standard transfer functions are considered:

- Sensitivity function $S_i = 1/(1 + L_i)$
- Input sensitivity function $U_i = C_i S_i$
- Complementary sensitivity function $T_i = 1 - S_i$

where $L_i = P_i C_i$ is the open-loop gain of the servo-loop. Key is to use the open-loop as the primary loop to shape, i.e., closed-loop specifications are transformed in a desired shape of the open-loop gain L_i. From the desired shape of L_i, the required C_i follows directly given the knowledge of the plant P_i. The shaping is performed according to standard rules [8]:

r.1 Stabilize the plant P_i
r.2 Ensure high open-loop gain L_i within the largest possible cross-over frequency
r.3 Allow maximum peaking in the sensitivity function S_i of 6 dB
r.4 Minimize influence of unmodelled dynamics and disturbances on the motion control performance

The simplest case is the use of the loop-shaping technique for tuning the $k_{p,i}$ and $k_{d,i}$ gains of the conventional PD feedback controller:

$$u_i(t) = \ddot{q}_{r,i}(t) + u_i^*(t) \tag{15.18a}$$

$$u_i^*(t) = -k_{p,i} e_i(t) - k_{d,i} \dot{e}_i(t) \tag{15.18b}$$

where e_i is the ith element of the position error defined by (15.12). More advanced use of loop-shaping is the design of a feedback controller by custom selection of the shaping filters, e.g., lead/lag, notch, low-pass, etc. Such a design can be found in [15]. Finally, an optimal loop-shaping design provides a feedback controller minimizing the H_2 or H_∞ norm of a matrix based on the weighted closed-form transfer functions given above. The weightings are frequency-dependent filters, suitably selected to enforce the desired performance specifications. The availability of various software packages that enable calculation of the optimal controller facilitates an optimal loop-shaping design. For illustration, two such designs will be sketched next.

The first control design was presented in [16], and it aims at shaping the sensitivity and input sensitivity functions. It uses two frequency dependent weighting functions W_i^S and W_i^U ($i = 1, \ldots, n$) for performance specification in each servo system. Constraining the sensitivity function enforces a desired low-frequency behavior in the closed loop:

$$|S_i(j\omega)| \leq 1/|W_i^S(j\omega)|, \quad \forall \omega \in \mathbb{R}^+ (i = 1, \ldots, n) \tag{15.19}$$

where ω denotes angular frequency defined on $\mathbb{R}^+ = \{x \in \mathbb{R} | x \geq 0\}$. By shaping the sensitivity, one may impose a minimum bandwidth requirement and integral control. Attenuation of resonances at low frequencies can be enforced, as well. Bounding the input sensitivity enforces the high-frequency roll-off, which is important for robustness against high-frequency resonances and measurement noise:

$$|U_i(j\omega)| \leq 1/|W_i^U(j\omega)|, \quad \forall \omega \in \mathbb{R}^+ (i = 1, \ldots, n) \tag{15.20}$$

The conditions (15.19) and (15.20) jointly hold if the cost function

$$\sqrt{\sup_{\omega \in \mathbb{R}} \left(|W_i^S(j\omega) S_i(j\omega)|^2 + |W_i^U(j\omega) U_i(j\omega)|^2 \right)} \ (i = 1, \ldots, n) \tag{15.21}$$

is less than 1. Thus, the optimal loop-shaping procedure seeks for a controller C_i which stabilizes the servo-loop shown in Figure 15.3 and minimizes the given cost function, ensuring the minimum is below 1. This optimization problem is known as the mixed-sensitivity problem [17]. It can be solved, for example, using software routines presented in [18].

If needed, both W_i^S and W_i^U can be tuned to enforce compensation of parasitic and disturbance effects observed in the controlled system. This is done through disturbance-based control design cycles [8], according to the following algorithm:

1. A conventional PD feedback controller is used to stabilize the manipulator in the initial design cycle; the manipulator is moving along trajectories spanning the whole configuration space; the position error (15.12) is measured and accumulated, and *cumulative power spectra* (CPSs — cumulative sums of all components of the auto-power spectrum) of all error components are calculated; frequencies above the bandwidth of the joint references at which the CPSs have steeper slopes reveal disturbance effects (noise, cogging force, vibrations), which should be tackled with the feedback.
2. The weightings W_i^S and W_i^U are defined to account for the observed disturbances.
3. Optimal controllers C_i minimizing (15.21) for $i = 1, \ldots, n$ are computed and implemented on the manipulator to realize the reference trajectory in the new design cycle.
4. CPSs of newly measured position errors are calculated and evaluated; if the design specifications have not been met yet, weightings W_i^S and W_i^U are adjusted.
5. The steps 3 and 4 are repeated until further improvement of the motion performance cannot be achieved.
6. The latest determined C_i ($i = 1, \ldots, n$) are adopted as the feedback controllers.

The second control design computes the feedback controllers via μ-synthesis [19]. Its objectives are robust stability and robust performance of the servo-loop shown in Figure 15.3. Assume that a nominal parametric model P_i^o of the plant (joint) dynamics is available. The nominal model and actual plant dynamics can be related using a multiplicative plant perturbation model:

$$P_i^o(s) \rightarrow P_i^o(s)(1 + \Delta_i(s)) \quad (15.22)$$

Here, $\Delta_i(s)$ is the multiplicative uncertainty, representing the relative difference between P_i^o and the real dynamics. The uncertainty can be expanded as follows:

$$\Delta_i(s) = W_i^\delta(s) \delta_i(s) \quad (15.23)$$

where the stable parametric weighting function W_i^δ satisfies

$$\left| W_i^\delta(j\omega) \right| \geq |\Delta_i(j\omega)|, \quad \forall \omega \in \mathbb{R}^+ \quad (15.24)$$

to have the normalized scalar uncertainty $|\delta_i(j\omega)| \leq 1$, $\forall \omega \in \mathbb{R}^+$.

A bound for the uncertainty can be determined from the identified FRFs:

$$|\Delta_i(j\omega)| = \max_k \left| \frac{G_i^k(j\omega) - P_i^o(j\omega)}{P_i^o(j\omega)} \right|, \quad \forall \omega \in \mathbb{R}^+ (k = 1, \ldots, N_i) \quad (15.25)$$

Any stable weighting function, preferably of low order, can be adopted for W_i^δ if its magnitude closely bounds the perturbations $|\Delta_i|$ from above. According to the small gain theorem [19], the control of the perturbed plant (15.22) is robustly stable if the following holds

$$|T_i(j\omega)| < 1/\left| W_i^\delta(j\omega) \right|, \quad \forall \omega \in \mathbb{R}^+ \quad (15.26)$$

Simultaneous specification of the closed-loop objectives (15.19), (15.20), and (15.26) (shaping of S_i, U_i, and T_i) can be done using the block-diagram shown in Figure 15.4. In this block-diagram, the channel

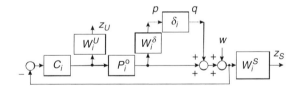

FIGURE 15.4 Setup for servo design using μ-synthesis.

from q to p is the uncertainty channel with the scalar scaled complex uncertainty δ_i. The channel from w to z_S and z_U is the performance channel. If we adopt q and w as the input variables, and p, z_S, and z_U as the output variables, then from Figure 15.4 we may determine the interconnecting transfer function matrix $\mathbf{H}_i$:

$$\begin{bmatrix} p \\ z_S \\ z_U \end{bmatrix} = \mathbf{H}_i \begin{bmatrix} q \\ w \end{bmatrix}, \quad \mathbf{H}_i = \begin{bmatrix} -W_i^\delta T_i & -W_i^\delta T_i \\ W_i^S S_i & W_i^S S_i \\ -W_i^U U_i & -W_i^U U_i \end{bmatrix} \quad (15.27)$$

According to μ-analysis theory [19], to have the objectives (15.19), (15.20), and (15.26) robustly satisifed, it is sufficient if the structured singular value of the transfer function matrix $\mathbf{H}_i$ satisfies

$$\sup_\omega \mu_{\tilde{\Delta}_i}(\mathbf{H}_i) < 1 \quad (15.28)$$

for the extended plant perturbation structure

$$\tilde{\Delta}_i = \begin{bmatrix} \delta_i & \mathbf{0}_{1\times 2} \\ 0 & \delta_{SU} \end{bmatrix} \quad (15.29)$$

where δ_{SU} denotes a complex uncertainty of dimension 1×2.

The objective of μ-synthesis is to construct a controller C_i which stabilizes the feedback loop shown in Figure 15.4 and yields Equation (15.28). If such a controller exists, robust stability and robust performance are realized. The controller can be calculated using the routines provided, for example, in the *μ-Analysis and Synthesis Toolbox for Matlab* [20]. Design of such a controller will be demonstrated in the case study.

15.4 Case-Study: Motion Control of the RRR Direct-Drive Manipulator

The RRR robotic manipulator (Figure 15.5) described in the chapter on modeling and identification is the subject of the case-study. The models of the manipulator kinematics and dynamics are available in closed-form. The kinematic parameters were obtained by direct measurements, while the inertial and friction parameters were estimated with sufficient accuracy. Both models are used for motion control of the RRR robot: the kinematic model computes the reference joint motions given a trajectory of the robot-tip; the rigid-body dynamic model is used in the control laws (15.10) and (15.11) to compensate for nonlinear dynamic couplings between the robot joints.

Because the dynamic model covers only rigid-body dynamics, it cannot counteract flexible effects. To illustrate this, we may inspect the FRF shown in Figure 15.6. It was determined for the first robot joint using the identification procedure explained in the previous chapter. The sampling period in the identification experiment was $T_s = 1$ ms. During identification, joints 2 and 3 were kept locked such that the distal links were fully stretched upwards. As obvious from the magnitude plot, the rigid dynamics hold only at lower frequencies, where the -2 slope can be observed as from 28 Hz, flexible effects become apparent. A modest resonance at 28 Hz is caused by insufficient stiffness in mounting the robot base to the floor. If excited, it may cause vibrations of the robot mechanism and thus degrade robot performance. At higher frequencies, we may observe more profound resonances. Location and damping factors of these resonances are different for other positions of joints 2 and 3.

Apart from flexibilities, an additional peculiarity of the RRR dynamics can be observed by inspection of the phase plot shown in Figure 15.6: a frequency dependent phase lag, superimposed to the phase changes due to flexibilities. The phase lag can be related to the time-delay between the feedback control action and the joint angular response. The time-delay can be easily identified from the FRFs measured in the third joint, as these contain less flexible effects. The phase plot shown in Figure 15.7 illustrates how the delay has

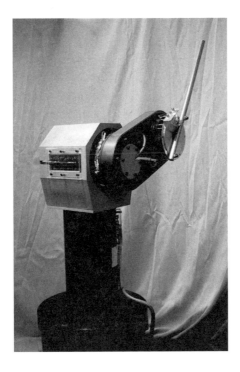

FIGURE 15.5 The RRR robot.

been identified. The phase is plotted linearly against the frequency to reveal a linear phase lag. The linear lag can be described as $-360°\delta f$, where δ is the time-delay and f is the frequency in Hz. As the phase drops for $360°$ until 500 Hz, it follows that $\delta = 1/500 = 2$ ms, i.e., $\delta = 2T_s$. The same time delay is also present in the other two joints. Resonance effects and time-delay are problems often met in practice, and they can be captured by the modelling part of the control design.

Feedback controllers should realize stabilization and desired performance of the manipulator motions. Two practical designs of these controllers are demonstrated in this section. The conventional feedback design will be presented first, just to illustrate its limited effects on motion performance in the presence of parasitic effects and disturbances. Then, design of a robust feedback via μ-synthesis will be demonstrated. It is sufficient to present the designs for the first robot joint only, as similar ones are applied to the other joints. Both feedback designs use the loop-shaping methodology explained in subsection 15.3.2, which means that the compensation of the manipulator nonlinear dynamics is performed via feedback linerization

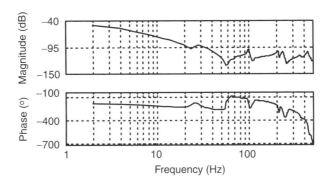

FIGURE 15.6 Flexible dynamics measured in the 1st joint.

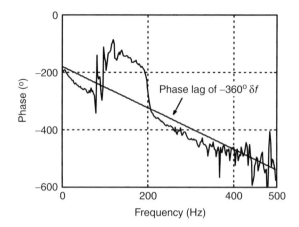

FIGURE 15.7 Phase plot of dynamics measured in the third joint reveals time delay of $2T_s$.

(15.11). Additionally, these designs assume availability of the nominal plant model. Calculation of this model is demonstrated next.

15.4.1 Nominal Plant Model

To determine a nominal model of the dynamics in the first joint, we first identify FRFs $G_1^k(j\omega)$ ($k = 1, \ldots, N_1$) for $N_1 = 16$ static positions in the joints 2 and 3: $[0\ 0], [0\ \pi/2], \ldots, [3\pi/2\ 3\pi/2]$ [rad]. These positions span a complete revolution in both joints. The identification is done under closed-loop conditions, as explained in the previous chapter. The identified FRFs are depicted in Figure 15.8 with the grey lines. By inspecting their magnitude plots, we may notice that they are less steep than -2 in the frequency range up to 4 Hz. This range is within the bandwidth of the closed-loop system established in identification experiments. As pointed out in the chapter on modelling and identification, the spectral components of an FRF that are below the closed-loop bandwidth are not reliable. However, these components can be easily determined from the slope of the FRF components that belong to the frequency range of the rigid dynamics. Once the distinct FRFs $G_1^k(j\omega)$ ($k = 1, \ldots, 16$) are available, the nominal response $G_1^o(j\omega)$ can be computed using (15.17b). The obtained $G_1^o(j\omega)$ is shown in Figure 15.8 with the black line, and it represents the nominal plant as a set of data points. A parametric model (transfer function) of the plant $P_1^o(j\omega)$ is fitted to the data using an output error model structure with a least-square criterion [14]. The

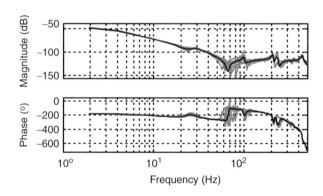

FIGURE 15.8 Identified FRFs $G_1^1, \ldots, G_1^{16}$ (grey) and the nominal FRF G_1^o (black).

Motion Control by Linear Feedback Methods

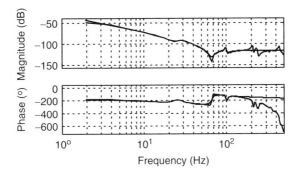

FIGURE 15.9 Bode plots of the nominal data G_1^o (thick) and of the parametric fit P_1^o (thin).

Bode plots of $G_1^o(j\omega)$ and of the corresponding $P_1^o(j\omega)$ are shown in Figure 15.9. The discrepancy between the unreliable data below 4 Hz and the fit is obvious. Above 4 Hz, the fit is good up to 160 Hz. In addition to determining $P_1^o(j\omega)$, FRFs corresponding to the cross couplings between joint 1 and the remaining joints were measured. At each frequency within the range of the rigid dynamics (below 20 Hz), the amplitude of $P_1^o(j\omega)$ is at least two orders of magnitude higher than the amplitude of any of the interaction terms in (15.16). To illustrate this, in Figure 15.10 we present FRFs of the dynamics in joint 1 and of the interaction between joints 1 and 2, identified for $[q_2 \; q_3] = [3\pi/2 \; 0]$. Because the cross-couplings in (15.16) are considerably smaller than the diagonal terms, it is justified to apply single-input, single-output feedback control designs for each manipulator joint separately.

15.4.2 Conventional Feedback Control Design

The conventional PD controller is defined by (15.18). Practically, the position error e_i at time t is formed as the difference between the position measurement (from the incremental encoder) and the reference position. The measurement is usually corrupted with quantization noise. Apart from noise, the time delay of $\delta = 2T_s$ is present in the system. If this delay is associated with the position measurements, we can formulate the following output equation:

$$\tilde{y}_i(t) = q_i(t - \delta) + \eta_i(t) \; (i = 1, 2, 3) \quad (15.30)$$

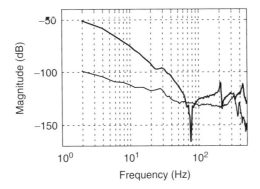

FIGURE 15.10 Magnitude plots of FRFs corresponding to dynamics in joint 1 (thick) and to the cross coupling between joints 1 and 2 (thin), identified for $[q_2 \; q_3] = [3\pi/2 \; 0]$ [rad].

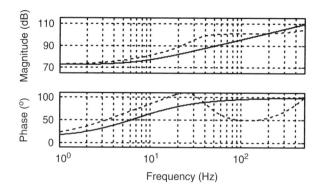

FIGURE 15.11 Bode plots of the conventional controllers defined by Equation (15.32) (solid) and Equation (15.36) (dotted).

where $\tilde{y}_i$ is the position measurement, and η_i is the measurement noise. These may degrade the motion performance if the feedback law (15.18) is based on the position error

$$\tilde{e}_i(t) = \tilde{y}_i(t) - q_{r,i}(t) \tag{15.31}$$

i.e.,

$$u_i(t) = \ddot{q}_{r,i}(t) + u_i^*(t) \tag{15.32a}$$
$$u_i^*(t) = -k_{p,i}\tilde{e}_i(t) - k_{d,i}\dot{\tilde{e}}_i(t) \tag{15.32b}$$

Bode plots of the PD controller designed for the first joint are shown in Figure 15.11 with the solid lines. These plots represent the transfer function from the error (15.31) to the control input $u_i(t)$ (15.32). The PD gains were chosen to meet the design rules r.1–r.4 (subsection 15.3.2) as closely as possible. Fine-tuning of the gains was manually done on the real system, to minimize influences of the quantization noise, vibrations at the base, and flexibilities at higher frequencies.

Negative influence of the time delay and the measurement noise can be optimally reduced with the use of the LQG theory [21]. This theory suggests a Kalman observer for optimal reconstruction of the motion coordinates used in (15.18). The Kalman observer is based on the manipulator dynamics that remain after application of the control law (15.11), with (15.18a) taken as the feedback input:

$$\left. \begin{array}{l} \ddot{q}_i(t) = \ddot{q}_{r,i}(t) + u_i^*(t) + v_i(t) \\ \dot{v}_i(t) = \varsigma_i(t) \\ \tilde{y}_i(t) = q_i(t - \delta) + \eta_i(t) \end{array} \right\} \quad (i = 1, 2, 3) \tag{15.33}$$

Variable v_i represents the modeling error caused by imperfect compensation of the real robot dynamics using the nonlinear part of the control law (15.11). Here, v_i is a realization of the white process noise ς_i, filtered through a single integrator [22]. The variables $\tilde{y}_i$ and η_i have been defined already by (15.30). In the design of the observer, the position error $e_i(t) = q_i(t) - q_{r,i}(t)$ is regarded as a state coordinate, together with its time derivative $\dot{e}_i$ and v_i. We determine a discrete-time system having identical solutions with (15.33) at $t = kT_s$:

$$\left. \begin{array}{l} \mathbf{x}_i(k+1) = \mathbf{E}_i(T_s)\mathbf{x}_i(k) + \mathbf{f}_i(T_s)u_i^*(k) + \mathbf{g}_i(T_s)\varsigma_i(k) \\ \tilde{y}_i(k) = \mathbf{c}_i\mathbf{x}_i(k) + q_{r,i}(k-2) + \eta_i(k) \end{array} \right\} \quad (i = 1, 2, 3) \tag{15.34a}$$

where

$$\mathbf{x}_i(k) = [e_i(k-2),\ e_i(k-1),\ e_i(k),\ \dot{e}_i(k),\ v_i(k)]^T \quad (15.34b)$$

$$\mathbf{E}_i = \begin{bmatrix} 0 & 1 & 0 & 0 & 0 \\ 0 & 0 & 1 & 0 & 0 \\ 0 & 0 & 1 & T_s & T_s^2/2 \\ 0 & 0 & 0 & 1 & T_s \\ 0 & 0 & 0 & 0 & 1 \end{bmatrix}, \quad \mathbf{f}_i = \begin{bmatrix} 0 \\ 0 \\ T_s^2/2 \\ T_s \\ 0 \end{bmatrix}, \quad \mathbf{g}_i = \begin{bmatrix} 0 \\ 0 \\ T_s^3/6 \\ T_s^2/2 \\ T_s \end{bmatrix}, \quad \mathbf{c}_i = [1\ 0\ 0\ 0\ 0]$$

(15.34c)

Here, k abbreviates kT_s. The delayed position reference is present in the output equation of (15.34a), as the following holds: $\tilde{y}_i(k) = q_i(k-2) + \eta_i(k) = e_i(k-2) + q_{r,i}(k-2) + \eta_i(k)$.

The Kalman observer for optimal reconstruction of the motion coordinates in the presence of the modelling uncertainty and the measurement noise has the form:

$$\hat{\mathbf{x}}_i(k+1) = \mathbf{E}_i(T_s)\bar{\mathbf{x}}_i(k) + \mathbf{f}_i(T_s)u_i^*(k)$$
$$\bar{\mathbf{x}}_i(k) = \hat{\mathbf{x}}_i(k) + \mathbf{k}_i(\tilde{y}_i(k) - q_{r,i}(k-2) - \mathbf{c}_i\hat{\mathbf{x}}_i(k)) \quad (15.35)$$

where $\bar{\mathbf{x}}_i$ denotes updated estimate of all states, and $\mathbf{k}_i$ is a 5×1 vector of constant Kalman gains. The observer predicts position and velocity errors at $t = kT_s$, from the difference between the measurement $\tilde{y}_i(k)$ and the delayed position reference $q_{r,i}(k-2)$. The predicted coordinates are used in PD feedback:

$$u_i(t) = \ddot{q}_{r,i}(t) + u_i^*(t) \quad (15.36a)$$
$$u_i^*(k) = -k_{p,i}\bar{e}_i(k) - k_{d,i}\dot{\bar{e}}_i(k) \quad (15.36b)$$

The block-diagram in Figure 15.12 visualizes the considered approach, featuring the following vectors: $\mathbf{h}(\mathbf{q}_r,\dot{\mathbf{q}}_r) = \mathbf{Q}(\mathbf{q}_r,\dot{\mathbf{q}}_r)\dot{\mathbf{q}}_r + \boldsymbol{\tau}^f$, $\mathbf{u}^* = [u_1^*\ u_2^*\ u_3^*]^T$, $\bar{\mathbf{e}} = [\bar{e}_1\ \bar{e}_2\ \bar{e}_3]^T$, $\dot{\bar{\mathbf{e}}} = [\dot{\bar{e}}_1\ \dot{\bar{e}}_2\ \dot{\bar{e}}_3]^T$, and $\tilde{\mathbf{y}} = [\tilde{y}_1\ \tilde{y}_2\ \tilde{y}_3]^T$. The matrices of position and velocity gains are formed as $\mathbf{K}_p = diag[k_{p,1}, k_{p,2}, k_{p,3}]$ and $\mathbf{K}_d = diag[k_{d,1}, k_{d,2}, k_{d,3}]$, respectively. These gains were determined following the rules r.1–r.4 from subsection 15.3.2, including the on-line fine-tuning to reduce influence of base vibrations and high-frequency resonances. Bode plots of the controller for the first joint are shown in Figure 15.11 with the dotted lines. These plots correspond to the transfer function from $\tilde{y}_i(t) - q_{r,i}(t - 2T_s)$ to the control input $u_i(t)$ in (15.36a). By inspecting the Bode plots, we may notice that at low frequencies the PD controller (15.36) introduces

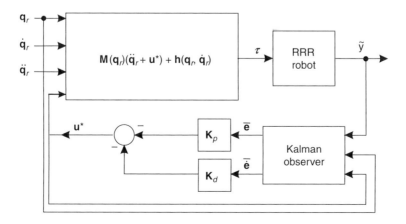

FIGURE 15.12 PD control of the RRR robot with the Kalman observer in the feedback loop.

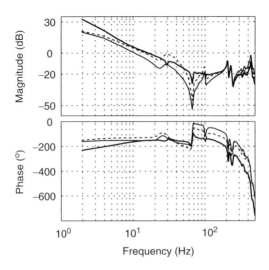

FIGURE 15.13 Open-loop gains for the first joint achieved by three loop-shaping feedback control designs: PD defined by Equation (15.32) (solid thin), PD defined by Equation (15.36) (dotted), and μ-synthesis (solid thick).

a higher gain than the PD controller (15.32). The phase characteristic of the controller (15.36) is also advancing the phase of (15.32) at low frequencies. Higher gain implies better tracking performance, while the phase lead implies a possibility for a higher cross-over frequency in the closed-loop. These can be verified from the Bode plots of the open-loop gains depicted in Figure 15.13. By inspection of the magnitude plots, it is apparent that the cross-over frequency of the controller (15.32) is lower, and its gain is smaller below the cross-over. The higher cross-over frequency of the PD defined by (15.36) indicates a potential of this controller to realize faster robot motion. The magnitudes of the sensitivity functions of both controllers are presented in Figure 15.15. The design rule r.3 on the maximum peaking of 6 dB is obviously realized in both cases. At the lower frequency range, the sensitivity achieved with (15.32) is above the sensitivity achieved with (15.36), implying a better reduction of the tracking error with the latter controller. However, the achieved performance is still below that feasible with the feedback designed by μ-synthesis, which is presented next.

15.4.3 μ-Synthesis Feedback Control Design

The loop-shaping control design via μ-synthesis should provide motion control of robust stability and robust performance. The grey lines in Figure 15.15 present relative differences (perturbations) in magnitudes between $G_1^1, \ldots, G_1^{16}$, and P_1^o. By virtue of (15.25), at each ω the maximum perturbation is represented

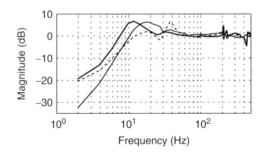

FIGURE 15.14 Sensitivity functions for the first joint achieved by three loop-shaping feedback control designs: PD defined by Equation (15.32) (solid thin), PD defined by Equation (15.36) (dotted), and μ-synthesis (solid thick).

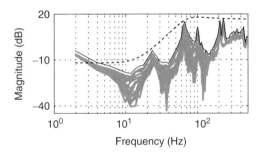

FIGURE 15.15 Magnitude plots of the weighting function W_1^δ (dotted), of relative differences between P_1^o and $G_1^1, \ldots, G_1^{16}$ (grey), and envelope of all differences Δ_1 (solid).

by Δ_1. It is shown in Figure 15.15 with the bold line. The dotted line in the same figure represents the magnitude of the parametric weighting function W_1^δ, which was chosen in accordance with (15.24). As already emphasized, the identified FRFs are not reliable below 4 Hz, and hence W_1^δ does not bound the uncertainty in the lower frequency range. Figure 15.16 shows the performance weightings W_1^S and W_1^U, that were chosen to adequately meet rules r.1–r.4. from subsection 15.3.2. The weighting W_1^S should enforce integral action in the feedback controller, together with as high as possible reduction of the position error at low frequencies. Furthermore, it should enforce attenuation of the vibrations at 28 Hz. The weighting W_1^U should enforce the high-frequency roll-off.

Finally, a feedback controller was designed using μ-synthesis that employs iterative scaling of the 2nd order (D-scaling [19,20]) and H_∞ optimization. Five DK iterations were necessary to determine C_1 which ensures robust performance specified by W_1^S and W_1^U. The Bode plots of the obtained C_1 are shown in Figure 15.17. The controller introduces integral action at low frequencies and deals with various resonance frequencies over a broad frequency range. Its magnitude and phase plots have more involved shapes than the plots of the conventional controllers depicted in Figure 15.11, indicating a potential of the μ-synthesis controller to simultaneously handle more versatile control objectives. It is also worth noticing that the design of the μ-synthesis controller does not require an explicit treatment of the time-delay problem, which was the case with the PD feedback. The μ-synthesis design considers the delay as an inherent property of the controlled plant, and creates an optimal controller that tackles this problem together with other control objectives. The block diagram shown in Figure 15.18 illustrates the control approach with the μ-synthesis controllers $\mathbf{C} = diag[C_1 \; C_2 \; C_3]$ in the feedback loop.

High-quality fulfilment of several control goals with μ-synthesis is obvious from the Bode plots of the open-loop gain and from the magnitude plots of the sensitivity function, shown in Figure 15.13 and Figure 15.14, respectively. The cross-over frequencies achieved with the μ-synthesis controller are higher than with the conventional feedbacks, see Table 15.1, which means that the μ-synthesis controller can

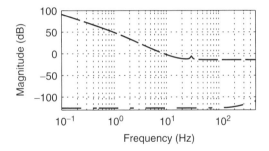

FIGURE 15.16 Magnitude plots of the performance weightings W_1^S (dashed) and W_1^U (dash-dotted).

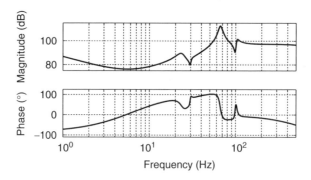

FIGURE 15.17 Bode plots of the feedback controller C_1.

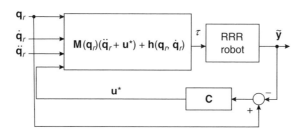

FIGURE 15.18 μ-synthesis feedback control of the RRR robot.

realize faster motions more accurately. Only in the last joint do the PD determined by Equation (15.36) and the μ-synthesis controller yield identical cross-over. As seen from Figure 15.13, the μ-synthesis controller has higher open-loop gain than the PDs below the cross-over. The higher gain is due to the integral action. Consequently, Figure 15.14 indicates smaller sensitivity with the μ-synthesis controller than with the conventional ones below the cross-over. The higher gain and smaller sensitivity imply that the μ-synthesis feedback will reduce the tracking error better. Beyond the cross-over, this controller reduces peaking around 200 Hz and 410 Hz. The μ-synthesis controllers in the other two joints also achieve higher open-loop gains and smaller sensitivities below the cross-over, as well as attenuation of high-frequency resonances. These will be verified by experiments, presented in the next subsection.

The plots shown in Figure 15.13 and Figure 15.14 correspond to the nominal plant model P_1^o. We are interested how robust C_1 is for application within the complete configuration space of the robot. To investigate this, in Figure 15.19 we plot the complementary sensitivity, the sensitivity, and the input sensitivity functions of the closed-loop system with the compensator C_1, calculated for P_1^o and for all $G_1^1, \ldots, G_1^{16}$. The top plots show that all the complementary sensitivities are bounded from above by the magnitude of W_1^δ, implying that the system remains stable for all postures–the condition (15.26) is satisfied. The middle and the bottom plots in Figure 15.19 show the sensitivity functions and the input sensitivity functions, respectively. In the frequency range of reliable measurements (above 4 Hz), it is obvious that magnitudes of all the sensitivity functions are bounded with the magnitude of $1/W_1^S$, and magnitudes of all the input sensitivity functions are bounded with the magnitude of $1/W_1^U$. These imply that the performance

TABLE 15.1 The Cross-Over Frequencies (in hertz)

i	PD Equation (15.32)	PD Equation (15.36)	μ
1	10	14	16
2	10	14	18
3	10	20	20

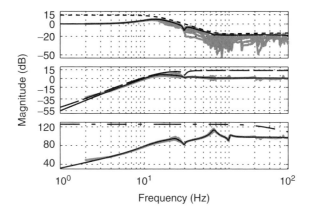

FIGURE 15.19 Magnitude plots of the nominal (bold) and of the perturbed (grey) complementary sensitivity functions (top), of the nominal (bold) and of the perturbed (grey) sensitivity functions (middle), and of the nominal (bold) and of the perturbed (grey) input sensitivity functions (bottom) are all below the prespecified bounds (dotted, dashed, and dash-dotted lines, respectively).

specifications are robustly satisfied, i.e., the conditions (15.19) and (15.20) are satisfied for all postures. Results similar to those presented in Figure 15.19 also hold for the other two joints of the RRR robot.

15.4.4 Experimental Evaluation

The designed controllers were tested in two different experiments. In the first experiment, the reference motion task was defined in the configuration space, as shown in Figure 15.20. It required all the joints to be displaced for π radians in 1 s, with zero initial/terminal speed and accelerations. These demand the full

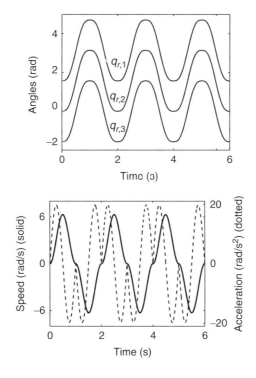

FIGURE 15.20 Reference motion task in the first experiment.

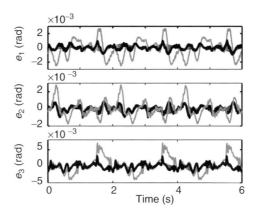

FIGURE 15.21 Position errors in the first experiment achieved with the PD defined by Equation (15.36) (grey) and with the μ-synthesis controller (black).

authority of the drives. The position errors achieved with the PDs (15.36) are presented in Figure 15.21 with the grey lines. Because of the lower open-loop gains within the bandwidth, the errors achieved with the PDs (15.32) were even higher, so there is no need to plot them here. The errors achieved using the μ-synthesis controllers are shown in Figure 15.21 with the black lines. The given plots confirm that the μ-synthesis controllers realize the reference motion more accurately. They better reduce the tracking errors because of the higher gains at low frequencies. Integral action of the μ-synthesis controllers especially contributes to the higher gains. The error ranges, given in Table 15.2, verify the superior tracking performance with the μ-synthesis controllers. In joints 1 and 2, these controllers achieve the errors within the range $[-10^{-3}, 10^{-3}]$ radians. The range of the error in joint 3 is twice that of the first two joints. Such accuracy can be considered as very good for a direct-drive robot.

To evaluate if the reduction of the tracking error was below the prescribed one, we found the ratio between the spectra (determined by fast Fourier transform) of the error e_1 and of the reference $q_{r,1}$. The ratio is plotted on the left in Figure 15.22 (solid line), together with inverse of the weighting function W_1^S (dashed line). Apparently, the curve corresponding to the ratio is always below the weighting. Having in mind the relation $S_1 = e_1/q_{r,1}$, it follows that condition (15.19) is satisfied, i.e., the specified error reduction is realized. It appears that robust performance control of the RRR robot was realized in the experiment.

The contribution of parasitic effects to the tracking performance can be evaluated from the *cumulative power spectra* (CPSs) of the tracking errors. On the right in Figure 15.22 we show the CPSs of the errors in the first robot joint achieved with both PDs and with the μ-synthesis controller. We may observe that most of the signals' energy lies within the bandwidth of the joint references (up to 5 Hz). Since above the bandwidth none of the CPSs abruptly changes its slope, it appears that influence of the parasitic effects on the trajectory tracking is eliminated, i.e., the design rule r.4 from subsection 15.3.2 is met with all feedback controllers. Inspection of the given plots reveals the lowest energy content of the CPS related with the μ-synthesis controller. This can be quantitatively verified by comparing the final CPS values in all joints, presented in Table 15.3. The final CPSs represent the variances (squares of the standard deviations) of

TABLE 15.2 Ranges of the Joint Errors (in $\times 10^{-3}$ radians)

i	PD Equation (15.32)	PD Equation (15.36)	μ
1	$[-5.3, 4.3]$	$[-2.5, 2.7]$	$[-1.0, 1.0]$
2	$[-3.9, 2.9]$	$[-1.7, 2.8]$	$[-1.0, 0.87]$
3	$[-10.2, 17.7]$	$[-4.1, 7.0]$	$[-2.0, 2.2]$

TABLE 15.3 Variances of the Joint Errors (in $\times 10^{-6}$ rad^2/s^2)

i	PD Equation (15.32)	PD Equation (15.36)	μ
1	7.19	1.57	0.11
2	3.08	0.87	0.14
3	46.84	5.59	0.72

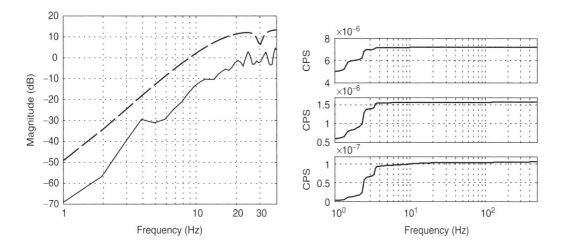

FIGURE 15.22 Results obtained for joint 1 in the first experiment: (left) the ratio between spectra of the position error achieved with the μ-synthesis controller and of the position reference (solid) is below the prescribed weighting $1/W_1^S$ (dashed); (right) the CPSs of the position errors achieved with PD (15.32) (top), PD (15.36) (middle), and μ-synthesis controller (bottom).

the position errors. As obvious from the table, the variances achieved with the μ-synthesis controllers are dramatically lower than the variances obtained with the conventional feedbacks.

In the second experiment the RRR robot performed the writing task [23], also considered in the chapter on modeling and identification. For convenience, the reference tip path and the corresponding joint motions are shown in Figure 15.23. The Cartesian position errors achieved with the PD (15.36) and with the μ-synthesis controller are shown in Figure 15.24 with the grey and black lines, respectively. The latter

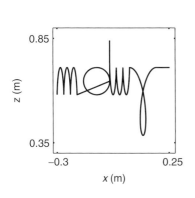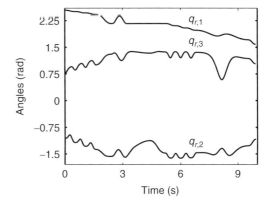

FIGURE 15.23 The writing task as the second experiment: (left) the tip path; (right) the joint motions.

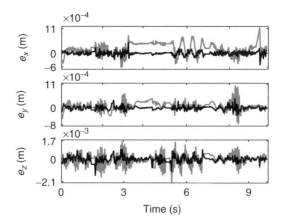

FIGURE 15.24 Cartesian errors in the second experiment achieved with the PD defined by Equation (15.36) (grey) and with the μ-synthesis controller (black).

controller performed better than the former one along each direction. It achieved errors (black lines) less than 0.5 mm, which is very accurate for a direct-drive robot.

15.5 Conclusions

Various aspects of manipulator motion control with linear feedback are considered in this chapter: performance criteria, robustness issues, feedback control designs, etc. It is noted that the decentralized conventional control is generally applicable, but its effects on motion performance can be limited if dynamic couplings between the manipulator joints are emphasized. To enhance the performance, a model-based compensation of the nonlinear dynamics should be applied. Two compensation approaches are considered — computed-torque and feedback linearization. The latter one facilitates an experimental approach to the feedback control design. In this approach, the manipulator dynamics that remains after feedback linearization is characterized from several frequency response functions identified directly on the system. The identified dynamics are then used for feedback control designs. Several feedback control designs are considered, with the emphasis on the tuning parts of these designs. Practical demonstration of the control designs is done for an experimental direct-drive robotic manipulator. The presented experimental results show superior performance of the advanced control design over the conventional ones.

References

[1] Fu, K.S., Gonzales, R.C., and Lee, C.S.G., *Robotics: Control, Sensing, Vision, and Intelligence,* McGraw-Hill, London, 1987.
[2] Sciavicco, L. and Siciliano, B., *Modeling and Control of Robot Manipulators,* McGraw-Hill, London, 1996.
[3] Slotine, J.J.E. and Li, W., *Applied Nonlinear Control,* Prentice Hall, Upper Saddle River, NJ, 1991.
[4] Franklin, G.F., Powell, J.D., and Emami-Naeini, A., *Feedback Control of Dynamic Systems,* Prentice Hall, London, 2001.
[5] Asada, H. and Youcef-Toumi, K., *Direct-Drive Robots: Theory and Practice,* MIT Press, London, 1987.
[6] Chen, Y., Replacing a PID controller by a lag-lead compensator for a robot — a frequency response approach, *IEEE Trans. Rob. Autom.,* 5, 174, 1989.
[7] Sage, H.G., de Mathelin, M.F., and Ostertag, E., Robust control of robot manipulators: a survey, *Int. J. Control,* 72, 1498, 1999.

[8] Steinbuch, M. and Norg, M.L., Advanced motion control, *Europ. J. Control*, 4, 278, 1998.
[9] Van Beek, B. and De Jager, B., RRR-robot design: basic outlines, servo sizing, and control, in *Proc. IEEE Int. Conf. Control Appl.*, Hartford, U.S.A., 1997, 36.
[10] Van Beek, B. and De Jager, B., An experimental facility for nonlinear robot control, in *Proc. IEEE Int. Conf. Control Appl.*, Hawaii, U.S.A., 1999, 668.
[11] Vukobratović, M. and Potkonjak, V., *Dynamics of Manipulation Robots: Theory and Application*, Springer-Verlag, Berlin, 1982.
[12] Nijmeijer, H. and van der Schaft, A., *Nonlinear Dynamical Control Systems*, Springer-Verlag, Berlin, 1991.
[13] Paden, B. and Panja, R., Globally asymptotically stable PD+ controller for robot manipulators, *Int. J. Control*, 47, 1697, 1988.
[14] Sanathanan, C.K. and Koerner, J., Transfer function synthesis as a ratio of two complex polynomials, *IEEE Trans. Autom. Control*, 8, 56, 1963.
[15] Koot, M., Kostić, D., De Jager, B., and Steinbuch, M., A systematic approach to improve robot motion control, in *Proc. Conf. on Mechantronics*, Enschede, The Netherlands, 2002, 1377.
[16] Kostić, D., De Jager, B., and Steinbuch, M., Experimentally supported control design for a direct drive robot, in *Proc. IEEE Int. Conf. Control Appl.*, Glasgow, Scotland, 2002, 186.
[17] Kwakernaak, H., Robust control and H_∞-optimization-tutorial paper, *Automatica*, 29, 255, 1993.
[18] Kwakernaak, H., *MATLAB Marcos for Polynomial H_∞ Control System Optimization*, Memorandum No. 881, Department of Applied Mathematics, University of Twente, Enschede, The Netherlands, 1990.
[19] Zhou, K., Doyle, J.C., and Glover, K., *Robust and Optimal Control*, Prentice Hall, Upper Saddle River, NJ, 1996.
[20] Balas, G.J., Doyle, J.C., Glover, K., Packard, A., and Smith, R., *μ-Analysis and Synthesis Toolbox*, The MathWorks, Natick, MA, 1998.
[21] Anderson, B.D.O. and Moore, J.B., *Optimal Control: Linear Quadratic Methods*, Prentice Hall, Englewood Cliffs, NJ, 1990.
[22] Bélanger, P.R., Dobrovolny, P., Helmy, A., and Zhang, X., Estimation of angular velocity and acceleration from shaft-encoder measurements, *Int. J. Rob. Res.*, 17, 1225, 1998.
[23] Potkonjak, V., Tzafestas, S., Kostic, D., Djordjevic, G., and Rasic, M., Illustrating man-machine motion analogy in robotics — the handwriting problem, *IEEE Rob. Autom. Mag.*, 10, 35, 2003.

16
Force/Impedance Control for Robotic Manipulators

Siddharth P. Nagarkatti
MKS Instruments, Wilmington, MA

Darren M. Dawson
Clemson University

16.1	Introduction	16-1
16.2	Dynamic Model	16-2
	Joint-Space Model • Task-Space Model and Environmental Forces	
16.3	Stiffness Control	16-5
	Controller Design	
16.4	Hybrid Position/Force Control	16-6
	Controller Design	
16.5	Hybrid Impedance Control	16-9
	Types of Impedance • Duality Principle • Controller Design	
16.6	Reduced Order Position/Force Control	16-14
	Holonomic Constraints • Reduced Order Model • Reduced Order Controller Design	
16.7	Background and Further Reading	16-18

16.1 Introduction

The convergence of the fields of robotics and automation in many industrial applications is well established. During the execution of many industrial automation tasks, the robot manipulator is often in contact with its environment (either directly or indirectly via an end-effector payload). For purely positioning tasks such as spray painting that have negligible force interaction with the environment, controlling the manipulator end-effector position results in satisfactory performance. However, in applications such as polishing or grinding, the manipulator end effector will experience interaction forces from the environment. Furthermore, due to contact with the environment, the motion of the end effector in certain directions is restricted. In the field of robotics, this resulting motion is often referred to as constrained motion or compliant motion. This chapter focuses on the control strategies of manipulators for constrained motion.

While there are many techniques that may be used to design controllers for constrained motion, a popular active compliant approach is the use of force control. The fundamental philosophy is to regulate the contact force of the environment. This is often supplemented by a position control objective that regulates the orientation and the location of the end effector to a desired configuration in its environment. For example, in the case of grinding applications, the motion of the manipulator arm is constrained by the grinding surface. It can easily be seen that it is vital to control not only the position of the manipulator end effector to ensure contact with the grinding surface but also the interaction forces to ensure sufficient contact force to enable grinding action.

Another example that illustrates the need for combining force and position controllers is drawing on a blackboard with chalk. If a robot manipulator is required to draw multiple instances of one shape (e.g., a circle) on the blackboard, the end-effector position must follow a specific (in this case, circular) trajectory while ensuring that the chalk applies a certain amount of force on the blackboard. We know that if the force is too small, the drawing may not be visible; however, if the force is too large, the chalk will break. Moreover, as multiple instances of shapes are drawn, the length of the chalk reduces. Thus, one can appreciate complexities and intricacies of simple tasks such as drawing/writing using chalk. This example clearly illustrates the need for desired force trajectory in addition to a desired position trajectory. Another example that motivates the use of a position/force control strategy is the pick-and-place operations using grippers for glass tubes. Clearly, if the gripper closes too tight, the glass may shatter, and if it is too lose, it is likely the glass may slip. The above examples establish the motivation for position and force control.

In many applications, the position/force control objectives are considerably more intertwined than in the examples discussed thus far. Consider the example of a circular diamond-tipped saw cutting a large block of metal, wherein the saw moves from one end to the other. The motion from one end to the other manifests itself as the position control objective, whereas cutting the block of metal without the saw blade binding is the force control objective. The speed of cutting (position control objective) depends on many parameters, such as geometric dimensions and material composition. For example, for a relatively softer metal such as aluminum, the saw can move safely from one end to the other much faster than in the case of a relatively harder material such as steel. It is obvious that the cutting speed would be much faster for thinner blocks of metal.

To achieve control objectives such as those discussed above, the controller must be designed to regulate the dynamic behavior between the force exerted on the environment (force control objective) and the end effector motion (position control objective). This approach forms the basis of impedance control.

In this chapter, we present the following four control strategies that control the force exerted by the end effector on its environment in conjunction with the typical end effector position control for an n-link robot manipulator:

- Stiffness control
- Hybrid position/force control
- Hybrid impedance control
- Reduced order position/force control

Furthermore, for each section, an example for a two-degree-of-freedom (2-DOF) Cartesian manipulator is also presented.

Remark 16.1 It is important to note that for all the aforementioned strategies, the desired velocity and force trajectories should be consistent with the environmental model [1]. If such is not the case, the trajectories may be modified by techniques such as "kinestatic filtering" (see [1] for more details).

16.2 Dynamic Model

This section presents the formulation of the manipulator dynamics that facilitates the design of strategies to control position and force. First, the widely known joint-space model is presented. Next, we define some notation for the contact forces (sometimes referred to as interactive forces) exerted by the manipulator on the environment by employing a coordinate transformation.

16.2.1 Joint-Space Model

The dynamics of motion for an n-link robotic manipulator [2] are constructed in the Euler-Lagrange form as follows:

$$M(q)\ddot{q} + V_m(q,\dot{q})\dot{q} + N(q,\dot{q}) + \tau_e = \tau \quad (16.1)$$

where

$$N(q,\dot{q}) = F(\dot{q}) + G(q) \tag{16.2}$$

$M(q) \in \Re^{n\times n}$ denotes the inertia matrix, $V_m(q,\dot{q}) \in \Re^{n\times n}$ contains the centripetal and Coriolis terms, $F(\dot{q}) \in \Re^n$ contains the static (Coulomb) and dynamic (e.g., viscous) friction terms, $G(q) \in \Re^n$ is the gravity vector, $\tau_e(t) \in \Re^n$ represents the joint-space end-effector forces exerted on the environment by the end-effector, $q(t) \in \Re^n$ represents the joint-space variable vector, and $\tau(t) \in \Re^n$ is the torque input vector.

16.2.2 Task-Space Model and Environmental Forces

In order to facilitate the design of force controllers, the forces are commonly transformed into the task-space via a Jacobian matrix by using a coordinate transformation [2]. Note that this Jacobian matrix is defined in terms of the task-space coordinate system and dependent on the robot application. Specifically, depending on the type of application, the axes (directions) of force control and motion may vary and are captured by the definition of the Jacobian.

Toward formulating the task-space dynamics, we define a task-space vector $x \in \Re^n$ as follows:

$$x = h(q) \tag{16.3}$$

where $h(q)$ is obtained from the manipulator kinematics and the joint and task-space relationships. The derivative of the task-space vector is defined as

$$\dot{x} = J(q)\dot{q} \tag{16.4}$$

where the task-space Jacobian matrix $J(q) \in \Re^{n\times n}$ is defined as [2]

$$J(q) = \begin{bmatrix} I & 0 \\ 0 & T \end{bmatrix} \frac{\partial h(q)}{\partial q} \tag{16.5}$$

with I being the identity matrix, 0 being the zero matrix, and the transformation matrix T is used to convert joint velocities to derivatives of roll, pitch, and yaw angles associated with end-effector orientation. It should be noted that the joint-space representation of the force exerted on the environment (defined first in Equation (16.1)) can be rewritten as

$$M(q)\ddot{q} + V_m(q,\dot{q})\dot{q} + N(q,\dot{q}) + J^T(q)f = \tau \tag{16.6}$$

where $f(t) \in \Re^n$ denotes the vector of forces and torques exerted on the environment in task-space.

Example 16.1 Task-Space Formulation for a Slanted Surface

The following example presents the task-space formulation of the dynamics for a 2-DOF Cartesian manipulator (i.e., both joints are prismatic) moving along a slanted surface as shown in Figure 16.1. To that end, the objectives are to formulate the manipulator dynamics and to decompose the forces exerted on the slanted surface into tangential and normal components. The motion component of the dynamics can easily be formulated from the unconstrained robot model. Thus, after removing the surface, and hence the interaction forces f_1 and f_2, the manipulator dynamics are constructed as follows:

$$M\ddot{q} + F(\dot{q}) + G = \tau \tag{16.7}$$

where

$$q = \begin{bmatrix} q_1 \\ q_2 \end{bmatrix}, \quad M = \begin{bmatrix} m_1 & 0 \\ 0 & m_1 + m_2 \end{bmatrix}, \quad F = \begin{bmatrix} F_1(\dot{q}_1) \\ F_2(\dot{q}_2) \end{bmatrix}, \quad \tau = \begin{bmatrix} \tau_1 \\ \tau_2 \end{bmatrix}, \quad G = \begin{bmatrix} 0 \\ (m_1 + m_2)g \end{bmatrix} \tag{16.8}$$

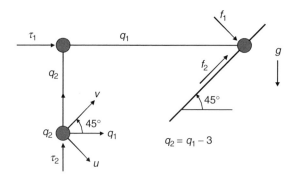

FIGURE 16.1 Manipulator moving along a slanted surface [4].

Note that the explicit structure of the friction vector F is beyond the scope of the material presented in this chapter.

The next step is to incorporate the contact forces. To that end, let $x \in \Re^2$ denote the task-space vector defined as

$$x = \begin{bmatrix} u \\ v \end{bmatrix} \tag{16.9}$$

where u represents the normal distance to the surface and v represents the tangential distance to the surface of a fixed coordinate system. As discussed in Equation (16.3), the representation of the joint-space coordinates into task-space is given by

$$x = h(q) \tag{16.10}$$

where $h(q)$ is constructed based on the system geometry as follows:

$$h(q) = \frac{1}{\sqrt{2}} \begin{bmatrix} q_1 - q_2 \\ q_1 + q_2 \end{bmatrix} \tag{16.11}$$

Based on Equation (16.5) and Equation (16.11), the task-space Jacobian matrix can be constructed as follows:

$$J = \frac{\partial h(q)}{\partial q} = \frac{1}{\sqrt{2}} \begin{bmatrix} 1 & -1 \\ 1 & 1 \end{bmatrix} \tag{16.12}$$

It should be noted that the matrix T in Equation (16.5) is an identity matrix because the end-effector angles are irrelevant to this problem. Following the procedure outlined in the previous section, the robot manipulator equation can be formulated as follows:

$$M\ddot{q} + F(\dot{q}) + G + J^T f = \tau \tag{16.13}$$

where

$$f = \begin{bmatrix} f_1 \\ f_2 \end{bmatrix} \tag{16.14}$$

Remark 16.2 It should be noted that the normal force f_1 and the tangential force f_2 in Figure 16.1 are drawn with respect to the task-space coordinate system.

16.3 Stiffness Control

Most early robot manipulators used in industrial automation were required to perform positional tasks (e.g., spray painting). As a result, the robot manipulators were designed to be very rigid, and the control strategies of choice were simple position controllers that achieved satisfactory positioning accuracy. However, in applications such as sanding and grinding, a rigid (or "stiff") manipulator does not lend itself favorably to the force control objective. If, however, the "stiffness" of the manipulator can be controlled [3], the force control objective can be more easily accomplished. This section presents the stiffness control formulation for an n-DOF robot manipulator and is followed by an example for a 2-DOF Cartesian robot.

16.3.1 Controller Design

The stiffness control objective for an n–link is formulated in this section. First, the force exerted by the manipulator on the environment is defined as

$$f \equiv K_e(x - x_e) \tag{16.15}$$

where $K_e \in \Re^{n \times n}$ is a diagonal, positive semidefinite,[1] constant matrix used to denote the environmental stiffness with $x_e \in \Re^n$ being the task-space vector denoting static location of the environment.

A proportional-derivative type of controller can be designed as follows to achieve the multi-dimensional stiffness control objective:

$$\tau = J^T(q)(-K_v \dot{x} + K_p \tilde{x}) + N(q, \dot{q}) \tag{16.16}$$

where the task-space tracking error is defined as

$$\tilde{x} = x_d - x \tag{16.17}$$

with x_d denoting the desired constant end-effector position. After formulating the closed-loop error system and performing a comprehensive stability analysis [4], it can be shown that for the ith diagonal element of matrices K_e and K_e:

$$\lim_{t \to \infty} f_i \approx K_{pi}(x_{di} - x_{ei}) \quad \text{in the constrained directions} \tag{16.18}$$
$$\lim_{t \to \infty} x_i = x_{di} \quad \text{in the unconstrained directions}$$

The result in Equation (16.18) can be interpreted as follows. In the unconstrained task-space direction, the desired steady-state setpoint is reached and the position control objective is achieved. In the constrained task-space direction (with the assumption that $K_{ei} \gg K_{pi}$ for force control), the force control objective is achieved and K_{pi} may be interpreted as the manipulator stiffness in the task-space direction.

Example 16.2 Stiffness Control of a Cartesian Manipulator

The following example presents the design of a stiffness control strategy for the 2-DOF Cartesian manipulator shown in Figure 16.1. The control objective is to move the end effector to a desired final position v_d while exerting a desired normal force f_{d1}. In this example, the surface friction f_2 and the joint friction are assumed to be negligible, and the normal force f_1 is assumed to satisfy the following relationship:

$$f_1 = k_e(u - u_e) \tag{16.19}$$

---

[1] A diagonal element of the matrix K_e is zero if the manipulator is not constrained in the corresponding task-space direction.

To accomplish the control objective, the stiffness control algorithm is designed as follows:

$$\tau = J^T(q)(-K_v \dot{x} + K_p \tilde{x}) + G(q) \tag{16.20}$$

where

$$\tilde{x} = \begin{bmatrix} u_d - u \\ v_d - v \end{bmatrix} \tag{16.21}$$

u_d denotes the desired normal position, the quantities τ, $J(q)$, and $G(q)$ retain their definitions of Example 1, and the gain matrices can simply be selected as $K_v = k_v I$ and $K_p = k_p I$ with k_v and k_p being positive constants. It is important to select k_p such that $k_p \ll k_e$ as required by the stiffness control formulation. To satisfy the control objective, the desired normal position u_d needs to be specified. To that end, by using the first condition in Equation (16.18), we can solve for u_d from the following equation:

$$f_{d1} = k_p(u_d - u_e) \tag{16.22}$$

16.4 Hybrid Position/Force Control

An important disadvantage of the stiffness controller discussed in the previous section is that the desired manipulator position and the desired force exerted on the environment are constants. That is, the controller is restricted to a constant setpoint regulation objective. In applications such as polishing and deburring, the end effector must track a prescribed path while tracking some desired force trajectory on the environment. In such a scenario, the stiffness controller will not achieve satisfactory results.

To that end, another approach that *simultaneously* achieves position and force tracking objectives is used. The hybrid position/force controller developed in [5, 6] decouples the position and force control problems into subtasks via a task-space formulation. This formulation is critical in determining the directions in which force or position should be controlled. Once these subtasks are identified, separate position and force controllers can be developed. In this section, we first present a hybrid position/force controller for an n-link manipulator and then discuss this control strategy with respect to the 2-DOF Cartesian manipulator shown in Figure 16.1.

16.4.1 Controller Design

The following hybrid position/force controller first uses feedback-linearization to globally linearize the robot manipulator dynamics and then employs linear controllers to track the desired position and force trajectories. To that end, by using the task-space transformation of Equation (16.3) to decompose the normal and tangential surface motions and after some mathematical manipulation, the robot dynamics can be expressed as

$$M(q)J^{-1}(q)(\ddot{x} - \dot{J}(q)\dot{q}) + V_m(q,\dot{q})\dot{q} + N(q,\dot{q}) + J^T(q)f = \tau \tag{16.23}$$

Based on the structure of Equation (16.23) and the control objectives, a feedback-linearizing controller for the above system can be constructed as follows:

$$\tau = M(q)J^{-1}(q)(\bar{y} - \dot{J}(q)\dot{q}) + V_m(q,\dot{q})\dot{q} + N(q,\dot{q}) + J^T(q)f \tag{16.24}$$

where $\bar{y} \in \Re^n$ denotes the linear position and force control strategies. Note that from Equation (16.23) and Equation (16.24), we have

$$\ddot{x} = \bar{y} \tag{16.25}$$

Force/Impedance Control for Robotic Manipulators

Given that the system dynamics in Equation (16.25) have been decoupled in the task-space into tangential (position) and normal (force) components denoted by subscripts T and N, respectively, we can design separate position and force control algorithms.

16.4.1.1 Tangential Position Control Component

The tangential task-space components of $\bar{y}$ can be represented as follows:

$$\ddot{x}_{Ti} = \bar{y}_{Ti} \tag{16.26}$$

where $\bar{y}_{Ti}$ denotes the ith linear tangential task-space position controller. The corresponding tangential task-space tracking error is defined as follows:

$$\tilde{x}_{Ti} = x_{Tdi} - x_{Ti} \tag{16.27}$$

where x_{Tdi} represents the desired position trajectory. Based on the control objective and the structure of Equation (16.27), we can design the following linear algorithm:

$$\bar{y}_{Ti} = \ddot{x}_{Tdi} + k_{Tvi}\dot{\tilde{x}}_{Ti} + k_{Tpi}\tilde{x}_{Ti} \tag{16.28}$$

where k_{Tvi} and k_{Tpi} are the ith positive control gains. After substituting this controller defined in Equation (16.28) into Equation (16.26), we obtain the following closed-loop dynamics:

$$\ddot{\tilde{x}}_{Ti} + k_{Tvi}\dot{\tilde{x}}_{Ti} + k_{Tpi}\tilde{x}_{Ti} = 0 \tag{16.29}$$

Given that k_{Tvi} and k_{Tpi} are positive, an asymptotic position tracking result is obtained:

$$\lim_{t \to \infty} \tilde{x}_{Ti} = 0 \tag{16.30}$$

16.4.1.2 Normal Force Control Component

The normal task-space components of $\bar{y}$ can be represented as follows:

$$\ddot{x}_{Nj} = \bar{y}_{Nj} \tag{16.31}$$

where $\bar{y}_{Nj}$ denotes the jth linear normal task-space force controller. The corresponding normal task-space tracking error is defined as follows:

$$f_{Nj} = k_{ej}(x_{Nj} - x_{ej}) \tag{16.32}$$

where the environment is modeled as a spring, k_{ej} denotes the jth component of the environmental stiffness, and x_{ej} represents the static location of the environment in the normal direction. The force dynamics can be formulated by taking the second time derivative of Equation (16.32) as follows:

$$\ddot{x}_{Nj} = \frac{1}{k_{ej}} \ddot{f}_{Nj} = \bar{y}_{Nj} \tag{16.33}$$

which relates the acceleration in the normal direction to the second time derivative of the force. To facilitate the construction of a linear controller, we define the following force tracking error:

$$\tilde{f}_{Nj} = f_{Ndj} - f_{Nj} \tag{16.34}$$

where f_{Ndj} denotes the jth of the desired force trajectory in the normal direction. A linear controller that achieves the force tracking control objective is defined as

$$\bar{y}_{Nj} = \frac{1}{k_{ej}}(\ddot{f}_{Ndj} + k_{Nvj}\dot{\tilde{f}}_{Nj} + k_{Npj}\tilde{f}_{Nj}) \tag{16.35}$$

where k_{Nvj} and k_{Npj} are jth positive control gains. After substituting the controller into Equation (16.33), we obtain the following closed-loop dynamics

$$\ddot{\tilde{f}}_{Nj} + k_{Nvj}\dot{\tilde{f}}_{Nj} + k_{Npj}\tilde{f}_{Nj} = 0 \tag{16.36}$$

that achieves asymptotic force tracking as shown below:

$$\lim_{t \to \infty} \tilde{f}_{Nj} = 0 \tag{16.37}$$

Example 16.3 Hybrid Position/Force Control along a Slanted Surface

This example discusses the hybrid position/force control strategy for the 2-DOF Cartesian manipulator shown in Figure 16.2. The control objective is to move the end effector with a desired surface position trajectory $v_d(t)$ while exerting a desired normal force trajectory denoted by $f_{d1}(t)$. Note that in the case of the stiffness controller, the desired position and desired force were constant setpoints. With the assumption of negligible joint friction, we also assume that the normal force f_1 satisfies the following relationship:

$$f_1 = k_e(u - u_e) \tag{16.38}$$

To accomplish the control objective, the hybrid position/force controller is defined as follows:

$$\tau = MJ^{-1}\bar{y} + G + J^T f \tag{16.39}$$

where $\bar{y} \in \Re^2$ represents the linear position and force controllers and the quantities τ, $J(q)$, and $G(q)$ have been defined in Example 1. This controller decouples the robot dynamics in the task-space as follows:

$$\ddot{x} = \begin{bmatrix} \ddot{u} \\ \ddot{v} \end{bmatrix} = \begin{bmatrix} \ddot{x}_{N1} \\ \ddot{x}_{T1} \end{bmatrix} = \begin{bmatrix} \bar{y}_{N1} \\ \bar{y}_{T1} \end{bmatrix} = \bar{y} \tag{16.40}$$

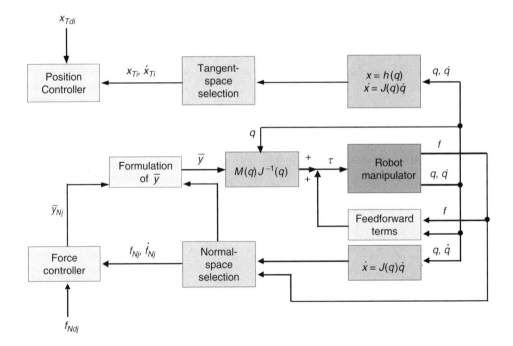

FIGURE 16.2 Hybrid position/force controller [4].

where u represents the normal component of the task-space and v represents the tangential component of the task-space. The corresponding linear position and force controller is designed as follows:

$$\bar{y}_{T1} = \ddot{x}_{Td1} + k_{Tv1}\dot{\tilde{x}}_{T1} + k_{Tp1}\tilde{x}_{T1} \tag{16.41}$$

and

$$\bar{y}_{T1} = \frac{1}{k_{e1}}(\ddot{f}_{Nd1} + k_{Nv1}\dot{\tilde{f}}_{N1} + k_{Np1}\tilde{f}_{N1}) \tag{16.42}$$

16.5 Hybrid Impedance Control

In many applications, such as the circular saw cutting through a block of metal example discussed in the introduction, it is necessary to regulate the dynamic behavior between the force exerted on the environment and the manipulator motion. As opposed to the hybrid position/force control that facilitates the separate design of the position and force controllers, the hybrid impedance control exploits the Ohm's law type of relationship between force and motion; hence, the use of the "impedance" nomenclature.

The model of the environmental interactions is critical to any force control strategy [7]. In previous sections, the environment was modeled as a simple spring. However, in many applications, it can easily be seen that such a simplistic model may not capture many significant environmental interactions. To that end, to classify various types of environments, the following linear transfer function is defined:

$$f(s) = Z_e(s)\dot{x}(s) \tag{16.43}$$

where the variable s denotes the Laplace transform variable, f is the force exerted on the environment, $Z_e(s)$ denotes the environmental impedance, and $\dot{x}$ represents the velocity of the manipulator at the point of contact with the environment.

16.5.1 Types of Impedance

The term $Z_e(s)$ is referred to as the impedance because Equation (16.43) represents an Ohm's law type of relationship between motion and force. Similar to circuit theory, these environmental impedances can be separated into various categories, three of which are defined below.

Definition 16.1 Impedance is inertial if and only if $|Z_e(0)| = 0$.
Figure 16.3 (a) depicts a robot manipulator moving a payload of mass m with velocity $\dot{q}$. The interaction force is defined as

$$f = m\ddot{q}$$

As a result, we can construct the inertial environmental impedance as follows:

$$Z_e(s) = ms$$

Definition 16.2 Impedance is resistive if and only if $|Z_e(0)| = k$ where $0 < k < \infty$.
Figure 16.3 (b) depicts a robot manipulator moving through a liquid medium with velocity $\dot{q}$. In this example of a resistive environment, the liquid medium applies a damping force with a damping coefficient b. The interaction force is defined as

$$f = b\dot{q}$$

which yields a resistive environmental impedance of

$$Z_e(s) = b$$

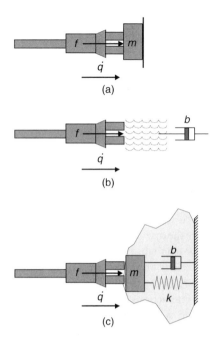

FIGURE 16.3 Types of environmental impedances.

Definition 16.3 Impedance is capacitive if and only if $|Z_e(0)| = \infty$.

A capacitive environment is depicted in Figure 16.3 (c), wherein a robot manipulator pushes against an object of mass m with a velocity $\dot{q}$. A damper-spring component is assumed with a damping coefficient b and a spring constant k. Thus, the interaction force is defined as

$$f = m\ddot{q} + b\dot{q} + kq$$

with a capacitive environmental impedance defined as

$$Z_e(s) = ms + b + \frac{k}{s}$$

From the discussion in the previous section, we have seen that the impedance is defined by a force-velocity relationship. The impedance control formulations discussed hereon will assume that the environmental impedance is inertial, resistive, *or* capacitive. The procedure for designing an impedance controller for any given environmental impedance is as follows. The manipulator impedance $Z_m(s)$ is selected after the environmental impedance has been modeled [8]. The selection criterion for the manipulator impedance is based on the dynamic performance of the manipulator. Specifically, the manipulator impedance is selected so as to result in zero steady-state error for a step input command of force or velocity, and this can be achieved if the manipulator impedance is the dual of the environmental impedance.

16.5.2 Duality Principle

16.5.2.1 Velocity Step-Input

For position control [8], the relationship between the velocity and force is modeled by

$$f(s) = Z_m(s)(\dot{x}_d(s) - \dot{x}(s)) \tag{16.44}$$

where $\dot{x}_d$ denotes the input velocity of the manipulator at the point of contact with the environment and $Z_m(s)$ represents the manipulator impedance, which is selected to achieve zero steady-state error to a step

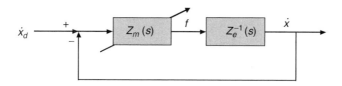

FIGURE 16.4 Position control block diagram [4].

input by utilizing the relationship between $\dot{x}_d$ and $\dot{x}$. The steady-state velocity error can be defined as

$$E_{ss} = \lim_{s \to 0} s(\dot{x}_d(s) - \dot{x}(s)) \quad (16.45)$$

where $\dot{x}_d(s) = 1/s$ for a step velocity input. From Figure 16.4, we can reduce the above expression to

$$E_{ss} = \lim_{s \to 0} \frac{Z_e(s)}{Z_e(s) + Z_m(s)} \quad (16.46)$$

For E_{ss} to be zero in the above equation, $Z_e(s)$ must be noncapacitive (i.e., $|Z_e(0)| < \infty$) and $Z_m(s)$ must be a noninertial impedance (i.e., $|Z_m(0)| > 0$).

The zero steady-state error can be achieved for a velocity step-input if:

- Inertial environments are position controlled with noninertial manipulator impedances
- Resistive environments are position controlled with capacitive manipulator impedances

and the duality principle nomenclature arises from the fact that inertial environmental impedances can be position-controlled with capacitive manipulator impedances.

16.5.2.2 Force Step-Input

To establish the duality principle for force step-input, the dynamic relationship between force and velocity is modeled by

$$\dot{x}(s) = Z_m^{-1}(s)(f_d(s) - f(s)) \quad (16.47)$$

where f_d is used to represent the input force exerted at the contact point with the environment. As outlined previously, the manipulator impedance $Z_m(s)$ is selected to achieve zero stead-state error to a step-input by utilizing the dynamic relationship between f_d and f. To establish the duality concept, we define the steady-state force error as follows:

$$E_{ss} = \lim_{s \to 0} s(f_d(s) - f(s)) \quad (16.48)$$

where $f_d(s) = 1/s$ for a step force input. From Figure 16.5, we can reduce the above expression to

$$E_{ss} = \lim_{s \to 0} \frac{Z_m(s)}{Z_m(s) + Z_e(s)} \quad (16.49)$$

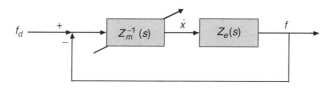

FIGURE 16.5 Force control block diagram [4].

For E_{ss} to be zero in the above equation, $Z_e(s)$ must be noninertial (i.e., $|Z_e(0)| > 0$) and $Z_m(s)$ must be a noncapacitive impedance (i.e., $|Z_m(0)| < \infty$).

The zero steady-state error can be achieved for a force step-input if:

- Capacitive environments are position controlled with noncapacitive manipulator impedances
- Resistive environments are position controlled with inertial manipulator impedances

and the duality principle nomenclature arises from the fact that capacitive environmental impedances can be force-controlled with inertial manipulator impedances.

16.5.3 Controller Design

As in the previous section, we can show that the torque controller of the form

$$\tau = M(q)J^{-1}(q)(\bar{y} - \dot{J}(q)\dot{q}) + V_m(q,\dot{q})\dot{q} + N(q,\dot{q}) + J^T(q)f \tag{16.50}$$

yields the linear set of equations:

$$\ddot{x} = \bar{y} \tag{16.51}$$

where $\bar{y} \in \Re^n$ denotes the impedance position and force control strategies.

To separate the force control design from the position control design, we define the equations to be position controlled in the task-space directions as follows:

$$\ddot{x}_{pi} = \bar{y}_{pi} \tag{16.52}$$

where $\bar{y}_{pi}$ denotes the ith position-controlled task-space direction variable.

Assuming zero initial conditions, the Laplace transform of the previous equation is given by

$$s\dot{x}_{pi}(s) = \bar{y}_{pi}(s) \tag{16.53}$$

From the position control model defined in (16.44),

$$s\dot{x}_{pi}(s) = s\left(\dot{x}_{pdi}(s) - Z_{pmi}^{-1}(s)f_{pi}(s)\right) \tag{16.54}$$

where Z_{pmi} is the ith position-controlled manipulator impedance. After equating Equation (16.53) and Equation (16.54), we obtain the following ith position controller:

$$\bar{y}_{pi} = L^{-1}\left\{s\left(\dot{x}_{pdi}(s) - Z_{pmi}^{-1}(s)f_{pi}(s)\right)\right\} \tag{16.55}$$

where L^{-1} denotes the inverse Laplace transform operation.

We now define the equations to be force controlled in the task-space directions as follows:

$$\ddot{x}_{fj} = \bar{y}_{fj} \tag{16.56}$$

where $\bar{y}_{fj}$ represents the jth force-controlled task-space direction variable.

Assuming zero initial conditions, the Laplace transform of the previous equation is given by

$$s\dot{x}_{fj}(s) = \bar{y}_{fj}(s) \tag{16.57}$$

From the force control model in Equation (16.47), we can construct the following:

$$s\dot{x}_{fj}(s) = s Z_{fnj}^{-1}(s)(f_{fdj}(s) - f_{fj}(s)) \tag{16.58}$$

where Z_{fnj} represents the jth force controlled manipulator impedance. After equating Equation (16.57) and Equation (16.58), we obtain the following jth force controller:

$$\bar{y}_{fj} = L^{-1}\left\{s Z_{fnj}^{-1}(s)(f_{fdj}(s) - f_{fj}(s))\right\} \tag{16.59}$$

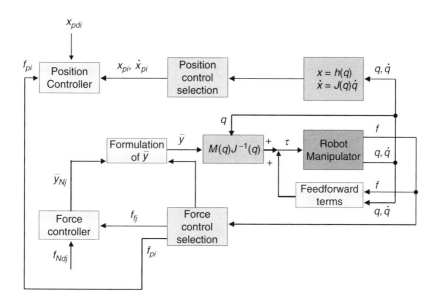

FIGURE 16.6 Hybrid impedance controller [4].

Example 16.4 Hybrid Impedance Control along a Slanted Surface

This example discusses the hybrid impedance control strategy for the 2-DOF Cartesian manipulator shown in Figure 16.6. The joint and surface frictions are neglected, and we assume that the normal force f_1 and the tangential force f_2, respectively, satisfy the following relationships:

$$f_1 = h_e \ddot{u} + b_e \dot{u} + k_e u \tag{16.60}$$

and

$$f_2 = d_e \dot{v} \tag{16.61}$$

where h_e, b_e, k_e, and d_e are positive scalar constants. From Equation (16.50), the hybrid impedance controller is constructed as follows:

$$\tau = M J^{-1} \bar{y} + G + J^T f \tag{16.62}$$

where $\bar{y} \in \Re^2$ represents the separate position and force controllers, and the quantities τ, $J(q)$, and $G(q)$ are defined in Example 1. This torque controller decouples the robot dynamics in the task-space as follows:

$$\ddot{x} = \begin{bmatrix} \ddot{u} \\ \ddot{v} \end{bmatrix} = \begin{bmatrix} \bar{y}_{f1} \\ \bar{y}_{p1} \end{bmatrix} = \bar{y} \tag{16.63}$$

based on which a determination can be made as to which task-space directions should be force or position controlled.

After applying Definition 16.2, we determine that the environmental impedance in the task-space direction of v is resistive in nature. Therefore, after invoking the duality principle, we can select a position controller that utilizes the capacitive manipulator impedance as follows:

$$Z_{pm1}(s) = h_m s + b_m + \frac{k_m}{s} \tag{16.64}$$

where h_m, b_m, and k_m are all positive scalar constants. Given that the task-space variable v is position controlled, we can formulate the following notations:

$$\ddot{x}_{p1} = \ddot{v} = \bar{y}_{p1} \tag{16.65}$$

and

$$f_{p1} = f_2 \tag{16.66}$$

After using Equation (16.64) and the definition for $\bar{y}_{p1}$ in Equation (16.55), it follows that

$$\bar{y}_{p1} = \ddot{x}_{pd1} + \frac{b_m}{h_m}(\dot{x}_{pd1} - \dot{x}_{p1}) + \frac{k_m}{h_m}(x_{pd1} - x_{p1}) - \frac{1}{h_m}f_{p1} \tag{16.67}$$

After applying Definition 3 to the task-space direction given by u, we can state that the environmental impedance is capacitive. Again, by invoking the duality principle, we can select a force controller that utilizes the inertial manipulator impedance as follows:

$$Z_{fn1}(s) = d_m s \tag{16.68}$$

where d_m is a positive scalar constant. Given that the task-space variable u is to be force controlled, we can formulate the following notations:

$$\ddot{x}_{f1} = \ddot{u} = \bar{y}_{f1} \tag{16.69}$$

and

$$f_{f1} = f_1 \tag{16.70}$$

After using Equation (16.68) and the definition for $\bar{y}_{f1}$ in Equation (16.59), it follows that

$$\bar{y}_{f1} = \frac{1}{d_m}(f_{fd1} - f_{f1}) \tag{16.71}$$

The overall impedance strategy is obtained by combining $\bar{y}_{p1}$ and $\bar{y}_{f1}$ defined in Equation (16.67) and Equation (16.71), respectively, which is as follows:

$$\bar{y} = \begin{bmatrix} \bar{y}_{f1} \\ \bar{y}_{p1} \end{bmatrix} \tag{16.72}$$

16.6 Reduced Order Position/Force Control

The end effector of a rigid manipulator that is constrained by a rigid environment cannot move freely through the environment and has its degrees of freedom reduced. As a result, at least one degree of freedom with respect to position is lost. This occurs at the point of contact with the environment (that is, when the environmental constraint is engaged) when contact (interaction) forces develop. Thus, the reduction of positional degrees of freedom while developing interaction forces motivates the design of position/force controllers that incorporate this phenomenon.

16.6.1 Holonomic Constraints

For the reduced order control strategy discussed in this section, an assumption about the environment being holonomic and frictionless is made. Specifically, we assume that a constraint function $\bar{\psi}(q) \in \Re^p$ exists in joint-space coordinates that satisfies the following condition that establishes holonomic environmental constraints:

$$\bar{\psi}(q) = 0 \tag{16.73}$$

where the dimension of the constrained function is assumed to be smaller than the number of joints in the manipulator (i.e., $p < n$). The structure of a constraint function is related to the robot kinematics and the environmental configuration.

The constrained robot dynamics for an n-link robot manipulator with holonomic and frictionless constraints are described by

$$M(q)\ddot{q} + V_m(q,\dot{q})\dot{q} + F(\dot{q}) + G(q) + A^T(q)\lambda = \tau \tag{16.74}$$

where $\lambda \in \Re^p$ represents the generalized force multipliers associated with the constraints and the constrained Jacobian matrix $A(q) \in \Re^{p \times n}$ is defined by

$$A(q) = \frac{\partial \bar{\psi}(q)}{\partial q} \tag{16.75}$$

16.6.2 Reduced Order Model

An n-joint robot manipulator constrained by a rigid environment has $n - p$ degrees of freedom. However, the joint-space model of Equation (16.74) contain n position variables (i.e., $q \in \Re^n$) which when combined with the p force variables (i.e., $\lambda \in \Re^p$) result in $n + p$ controlled states. As a result, the controlled states are greater than the number of inputs. To rectify this situation, a variable transformation can be used to reduce the control variables from $n + p$ to n.

The constrained robot model can be reduced by assuming that a function $g(x) \in \Re^n$ exists that relates the constrained space vector $x \in \Re^{n-p}$ to the joint-space vector $q \in \Re^p$ and is given by

$$q = g(x) \tag{16.76}$$

where $g(x)$ is selected to satisfy

$$\left[\frac{\partial g(x)}{\partial x}\right]^T A^T(q)\bigg|_{q=g(x)} = 0 \tag{16.77}$$

and the Jacobian matrix $\Sigma(x) \in \Re^{n \times (n-p)}$ is defined as

$$\Sigma(x) = \frac{\partial g(x)}{\partial x} \tag{16.78}$$

The reader is referred to the work presented by McClamroch and Wang in [9] for a detailed analysis that establishes the above arguments. The reduced order model is constructed as follows [4, 9]:

$$M(x)\Sigma(x)\ddot{x} + N(x,\dot{x}) + A^T(x)\lambda = \tau \tag{16.79}$$

where

$$N(x,\dot{x}) = (V_m(x,\dot{x})\Sigma(x) + M(x)\dot{\Sigma}(x))\dot{x} + F(x,\dot{x}) + G(x) \tag{16.80}$$

16.6.3 Reduced Order Controller Design

To design a position/force controller based on the reduced order model, we first define the constrained position tracking error and the force multiplier tracking error, respectively, as follows:

$$\tilde{x} = x_d - x \tag{16.81}$$

and

$$\tilde{\lambda} = \lambda_d - \lambda \tag{16.82}$$

It is important to note that the desired constrained space position trajectory x_d and its first two time derivatives are bounded and that the desired force multiplier trajectory λ_d is known and bounded.

Based on the structure of the error system and the control objective, the feedback linearizing reduced order position/force controller [9] is defined as follows:

$$\tau = M(x)\Sigma(x)(\ddot{x}_d + K_v\dot{\tilde{x}} + K_p\tilde{x}) + N(x,\dot{x}) + A^T(x)(\lambda_d + K_f\tilde{\lambda}) \qquad (16.83)$$

where $K_v, K_p \in \Re^{(n-p)\times(n-p)}$, and $K_f \in \Re^{p\times p}$ are diagonal and positive definite matrices. After constructing the closed-loop error system, the following asymptotic stability result for the constrained position tracking error and the constrained force tracking error can be obtained

$$\lim_{t\to\infty} \ddot{\tilde{x}}, \dot{\tilde{x}}, \tilde{x} = 0 \qquad (16.84)$$

and

$$\lim_{t\to\infty} \tilde{\lambda} = 0 \qquad (16.85)$$

Example 16.5 Reduced Order Position/Force Control along a Slanted Surface

This example discusses the reduced order control strategy for the 2-DOF Cartesian manipulator shown in Figure 16.7, where the joint and surface friction is neglected. As given in Example 1, the constrained function for this system is given by

$$\bar{\psi}(q) = q_1 - q_2 - 3 = 0 \qquad (16.86)$$

Based on the structure of Equation (16.79) and Equation (16.80), the manipulator dynamics on the constrained surface can be formulated as follows:

$$M\ddot{q} + G + A^T\lambda = \tau \qquad (16.87)$$

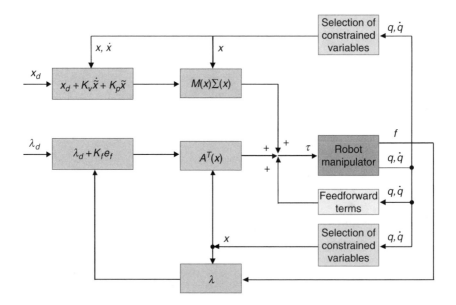

FIGURE 16.7 Reduced order position/force controller [4].

where $M, q, G,$ and τ are defined in Example 1 and

$$A^T = \begin{bmatrix} 1 \\ -1 \end{bmatrix} \quad (16.88)$$

For this problem, we assume that $x = q_1$, and as per Equation (16.76), we must find a function $g(x)$ that satisfies $q = g(x)$. For holonomic constraints, we can verify that following kinematic relationships hold:

$$\begin{bmatrix} q_1 \\ q_2 \end{bmatrix} = \begin{bmatrix} x \\ x-3 \end{bmatrix} \quad (16.89)$$

From the definition of the Jacobian in Equation (16.78), we obtain

$$\Sigma = \begin{bmatrix} 1 \\ 1 \end{bmatrix} \quad (16.90)$$

and the reduced order position/force controller is defined as follows:

$$\tau = M\Sigma(\ddot{x}_d + K_v \dot{\tilde{x}} + K_p \tilde{x}) + G + A^T(\lambda_d + K_f \tilde{\lambda}) \quad (16.91)$$

where x_d represents the desired trajectory of q_1 and λ_d represents the desired force multiplier.

For this problem, we can scrutinize the relationship between λ and the normal force exerted on the surface by equating the expressions for the forces in this example and that in Example 1 as follows:

$$A^T \lambda = J^T f \quad (16.92)$$

where J and f are defined in Example 1. Given that the surface friction has been neglected, from the above equation, it follows that:

$$\lambda = \frac{f_1}{\sqrt{2}} \quad (16.93)$$

where f_1 is the normal force exerted on the surface. Similarly, from kinematic relationships, we observe that

$$q_1 = \frac{\sqrt{2}}{2} v + \frac{3}{2} \quad (16.94)$$

where v denotes the end-effector position measured along the surface. The significance of the expressions in Equation (16.93) and Equation (16.94) lies in the fact that they would be used for trajectory generation because the position and force control objectives are formulated in terms of f_1 and v. Specifically, λ_d is obtained from the desired normal force f_{d1} in Equation (16.93) and q_{d1} is obtained from the desired end-effector surface position v_d in Equation (16.94).

16.7 Background and Further Reading

Over the last three decades, many researchers have developed a variety of force control algorithms for robot manipulators. Raibert and Craig in [6] originally proposed the hybrid position/force control approach. Khatib and Burdik [10] incorporated dynamic coupling effects between the robot joints. In [9], McClamroch and Wang developed a reduced-order model by applying nonlinear transformation to the constrained robot dynamics, which facilitated the separate design of position and force control strategies. For a comprehensive review and comparison of [9] with other position/force control strategies, the reader is referred to Grabbe et al. [11].

More advanced adaptive control strategies (not discussed in this chapter) have also been developed for position and force control, such as in [12], where Carelli and Kelly designed adaptive controllers that ensured asymptotic position tracking but only bounded force tracking error (the reader is referred to [13] and [14] for similar work). Adaptive full-state feedback controllers designed in Arimoto et al. [15], Yao and Tomizuka [16], and Zhen and Goldenberg [17] achieved a stronger stability result of asymptotic position *and* force tracking. Later de Queiroz et al. [18] designed partial-state feedback adaptive position/force controllers that compensated for parametric uncertainty in the manipulator dynamics while eliminating the need for link velocity measurements. In [19], Lanzon and Richards designed robust trajectory/force controllers for non-redundant rigid manipulators using sliding-mode and adaptive control techniques.

Furthermore, position/force control strategies are also supporting other critical technology initiatives. For example, in the field of medical robotics, the UC Berkeley/UCSF Telesurgical Workstation controller [20] is designed to incorporate the motion and force requirements of complex tasks such as tying a knot and suturing (e.g., a sufficiently large force is required to manipulate the suture through tissue). Another interesting application is the use of position/force control strategies in visual servo applications of robot manipulators [21].

In summary, we have presented several position/force control strategies for rigid robot manipulators. It is important to note that while this chapter presents the development of position/force controllers in chronological order, many other sophisticated control strategies have been developed and implemented (some of which have been cited in this section). With the advances in computer processing technology and real-time control, researchers in the robotics and control engineering fields will be able to design strategies that capture and process more information about the environment and compensate for uncertainties in the manipulator model. This would certainly result in significant performance enhancements and as a result, the use of manipulators in force control applications will be more common.

References

[1] Lipkin, H. and Duff, J., Hybrid twist and wrench control for a robot manipulator, *Trans. ASME J. Mech. Transmissions Automation Design*, 110, 138–144, June 1988.

[2] Spong, M. and Vidyasagar, M., *Robot Dynamics and Control,* John Wiely & Sons, New York, 1989.

[3] Salisbury, J. and Craig, J., Active stiffness control of a manipulator in Cartesian coordinates, *Proceedings of the IEEE Conference on Decision and Control*, Dec. 1980.

[4] Lewis, F.L., Abdallah, C.T., and Dawson, D.M., *Control of Robot Manipulators,* Macmillan, New York, 1993.

[5] Chae, A., Atkeson, C., and Hollerbach, J., *Model-Based Control of a Robot Manipulator,* MIT Press, Cambridge, MA, 1988.

[6] Raibert, M. and Craig, J., Hybrid position/force control of manipulators, *ASME J. Dynamic Syst., Meas., Control,* 102, 126–132, June 1981.

[7] Hogan, N., Stable execution of contact tasks using impedance control, *Proceedings of the IEEE Conference on Robotics and Automation,* Raleigh, NC, 595–601, March 1987.

[8] Anderson, R. and Spong, M., Hybrid impedance control of robotic manipulators, *J. Robotics Automation,* 4, 5, 549–556, Oct. 1988.

[9] McClamroch, N. and Wang, D., Feedback stabilization and tracking of constrained robots, *IEEE Transactions on Automatic Control,* 33, 5, 419–426, May 1988.

[10] Khatib, O. and Burdick, J., Motion and force control for robot manipulators, *Proceedings of the IEEE Conference on Robotics and Automation,* 1381–1386, 1986.

[11] Grabbe, M.T., Carroll, J.J., Dawson, D.M., and Qu, Z., Review and unification of reduced order force control methods, *J. Robotic Syst.,* 10, 4, 481–504, June 1993.

[12] Carelli, R. and Kelly, R., Adaptive control of constrained robots modeled by singular system, *Proceedings of the IEEE Conference on Decision and Control,* 2635–2640, Dec. 1989.

[13] Jean, J.H. and Fu, L.-C., Efficient adaptive hybrid control strategies for robots in constrained manipulators, *Proceedings of the IEEE Conference on Robotics and Automation,* 1681–1686, 1991.

[14] Su, C.Y., Leung, T.P., and Zhou, Q.J., Adaptive control of robot manipulators under constrained motion, *Proceedings of the IEEE Conference on Decision and Control,* 2650–2655, Dec. 1990.

[15] Arimoto, S., Naniwa, T., and Liu, Y.-H., Model-based adaptive hybrid control of manipulators with geometric endpoint constraint, *Advanced Robotics,* 9, 1, 67–80, 1995.

[16] Yao, B. and Tomizuka, M., Adaptive control of robot manipulators in constrained motion, *Proceedings of the IEEE Conference on Decision and Control,* 2650–2655, Dec. 1990.

[17] Zhen, R.R.Y. and Goldenberg, A.A., An adaptive approach to constrained robot motion control, *Proceedings of the IEEE Conference on Robotics and Automation,* 2, 1833–1838, 1995.

[18] de Queiroz, M.S., Hu, J., Dawson, D., Burg, T., and Donepudi, S., Adaptive position/force control of robot manipulators without velocity measurements: Theory and experimentation, *IEEE Transactions on Systems, Man, and Cybernetics,* 27-B, 5, 796–809, Oct. 1997.

[19] Lanzon, A. and Richards, R.J., Trajectory/force control for robotic manipulators using sliding-mode and adaptive control, *Proceedings of the American Control Conference,* San Diego, 1940–1944, June 1999.

[20] Cavusoglu, M., Williams, W., Tendick, F., and Shankar, S. Sastry, Robotics for telesurgery: second generation Berkeley/UCSF laparoscopic telesurgical workstation and looking towards the future applications, *Industrial Robot, Special Issue on Medical Robotics,* 30, 1, Jan. 2003.

[21] Xiao, D., Ghosh, B.K., Xi, N., and Tarn, T.J., Sensor-based hybrid position/force control of a robot manipulator in an uncalibrated environment, *IEEE Control Systems Technology,* 8, 4, 635–645, July 2000.

17
Robust and Adaptive Motion Control of Manipulators*

Mark W. Spong
University of Illinois at Urbana-Champaign

17.1 Introduction **17**-1
17.2 Background **17**-2
   Kinematics • Dynamics • Properties of Rigid Robot Dynamics
17.3 Taxonomy of Control Design Methods **17**-6
   Control Architecture • Feedback Linearization Control • Joint Space Inverse Dynamics • Task Space Inverse Dynamics • Robust Feedback Linearization • Adaptive Feedback Linearization • Passivity-Based Approaches • Passivity-Based Robust Control • Passivity-Based Adaptive Control • Hybrid Control
17.4 Conclusions **17**-20

17.1 Introduction

This chapter concerns the robust and adaptive motion control of robotic manipulators. The goals of both robust and adaptive control are to maintain performance in terms of stability, tracking error, or other specifications, despite parametric uncertainty, external disturbances, unmodeled dynamics, or other uncertainties present in the system. In distinguishing between robust control and adaptive control, we follow the commonly accepted notion that a robust controller is usually a fixed controller, static or dynamic, designed to satisfy performance specifications over a given range of uncertainties, whereas an adaptive controller incorporates some sort of online parameter estimation. This distinction is important. For example, in a repetitive motion task the tracking errors produced by a fixed robust controller would tend to be repetitive as well, whereas tracking errors produced by an adaptive controller might be expected to decrease over time as the plant and/or control parameters are updated based on runtime information. At the same time, adaptive controllers that perform well in the face of parametric uncertainty may not perform well in the face of other types of uncertainty such as external disturbances or unmodeled dynamics. An understanding of the trade-offs involved is therefore important in deciding whether to employ robust or adaptive control design methods in a given situation.

---

*Partial support from the National Science Foundation under grant ECS-0122412 is gratefully acknowledged.

Many of the fundamental theoretical problems in motion control of robot manipulators were solved during an intense period of research from about the mid-1980s until the early 1990s during which researchers first began to exploit fundamental structural properties of manipulator dynamics, such as feedback linearizability, passivity, multiple time-scale behavior, and other properties that we discuss below.

Within the scope of the present article, it is impossible to survey the entire range of results in manipulator control. We will focus on analytical design methods, including Lyapunov based design, variable structure control, operator theoretic methods, and passivity-based control for rigid robots. We will briefly touch upon more recent hybrid control methods based on multiple-models but will not discuss control design based on so-called soft computing methods, such as neural networks, fuzzy logic, genetic algorithms, or statistical learning methods. We will also not discuss issues involved in force feedback control, friction compensation, or control of elastic manipulators. The references at the end of this article may be consulted for additional information on these and other topics.

17.2 Background

Robot manipulators are basically multi-degree-of-freedom positioning devices. Robot dynamics are multi-input/multi-output, highly coupled, and nonlinear. The main challenges in the motion control problem are the complexity of the dynamics resulting from nonlinearity and multiple degrees-of-freedom, and uncertainty, both parametric and dynamic. Parametric uncertainty arises from imprecise knowledge of inertia parameters, while dynamic uncertainty arises from joint and link flexibility, actuator dynamics, friction, sensor noise, and unknown environment dynamics.

We consider a robot manipulator with **n-links** interconnected by **joints** into a **kinematic chain**. Figure 17.1 shows a serial link (left) and a parallel link (right) manipulator. A parallel robot, by definition,

FIGURE 17.1 A serial manipulator (left), the ABB IRB1400, and a parallel manipulator (right), the ABB IRB940Tricept. (Photos courtesy of ABB Robotics.)

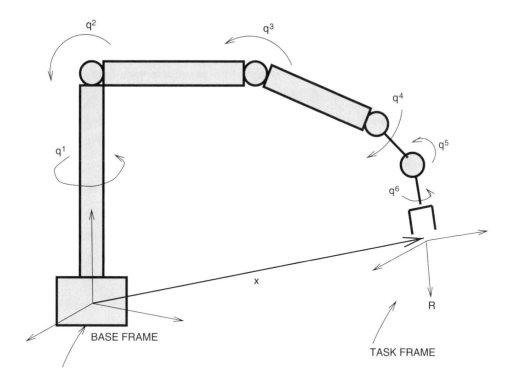

FIGURE 17.2 A serial link manipulator showing the attached base frame, task frame, and configuration variables.

contains two or more independent serial link chains. For simplicity of exposition we shall confine our discussion to serial link manipulators with only rotational or **revolute** joints as shown in Figure 17.2. Most of the discussion in this article remains valid for parallel robots and for robots with sliding or **prismatic** joints.

We define the joint variables, $q_1, \ldots, q_n$, as the relative angles between the links, for example, q_i is the angle between link i and link $i-1$. A vector $q = (q_1, \ldots, q_n)^T$ with each $q_i \in (0, 2\pi)$, is called a **configuration**. The set of all possible configurations is called **configuration space** or **joint space**, which we denote as $\mathcal{C}$. The configuration space for a revolute joint robot is an n-dimensional torus, $\mathcal{T}^n = S^1 \times \cdots \times S^1$, where S^1 is the unit circle.

The **task space** is the space of all positions and orientations (called **poses**) of the end-effector. We attach a coordinate frame, called the **base frame**, or **world frame**, at the base of the robot and a second frame, called the **end-effector frame** or **task frame**, at the end-effector. The end-effector position can then be described by a vector $x \in \Re^3$ specifying the coordinates of the origin of the task frame in the base frame. The end-effector orientation can be described by a 3×3 matrix $R \in SO(3)$. The task space is then isomorphic to the **special Euclidean group**, $SE(3) = \Re^3 \times SO(3)$, elements of which are called **rigid motions** [41].

17.2.1 Kinematics

The **forward kinematics map** is a function

$$X_0 = \begin{pmatrix} x(q) \\ R(q) \end{pmatrix} = f_0(q) : \mathcal{T}^n \to SE(3) \tag{17.1}$$

from configuration space to task space which gives the end-effector pose in terms of the joint configuration. The **inverse kinematics map** gives the joint configuration as a function of the end-effector pose. The forward kinematics map is many-to-one, so that several joint space configurations may give rise to the same end-effector pose. This means that the forward kinematics always has a unique pose for each configuration, while the inverse kinematics has multiple solutions, in general.

The kinematics problem is compounded by the difficulty of parametrizing the rotation group, $SO(3)$. It is well known that there does not exist a minimal set of coordinates to "cover" $SO(3)$, i.e., a single set of three variables to represent all orientations in $SO(3)$ uniquely. The most common representations used are **Euler angles** and **quaternions**. Representational singularities, which are points at which the representation fails to be unique, give rise to a number of computational difficulties in motion planning and control.

Given a minimal representation for $SO(3)$, for example, a set of Euler angles ϕ, θ, ψ, the forward kinematics map may also be defined by a function

$$X_1 = \begin{pmatrix} x(q) \\ o(q) \end{pmatrix} = f_1(\cdot) : \mathcal{T}^n \to \Re^6 \tag{17.2}$$

where $x(q) \in \Re^3$ gives the Cartesian position of the end-effector and $o(q) = (\phi(q), \theta(q), \psi(q))^T$ represents the orientation of the end-effector. The nonuniqueness of the inverse kinematics in this case will include multiplicities due to the particular representation of $SO(3)$ in addition to multiplicities intrinsic to the geometric structure of the manipulator.

Velocity kinematics refers to the relationship between the joint velocities and the end-effector velocities. If the mapping f_0 from Equation (17.1) is used to represent the forward kinematics, then the velocities are given by

$$V = \begin{pmatrix} v \\ \omega \end{pmatrix} = J_0(q)\dot{q} \tag{17.3}$$

where $J_0(q)$ is a $6 \times n$ matrix, called the **manipulator Jacobian**. The vectors v and ω represent the linear and angular velocity, respectively, of the end-effector. The linear velocity $v \in \Re^3$ is just $\frac{d}{dt} x(q)$, where $x(q)$ is the end-effector position vector from Equation (17.1). It is a little more difficult to see how the angular velocity vector ω is computed since the end-effector orientation in Equation (17.1) is specified by a matrix $R \in SO(3)$. If $\omega = (\omega_x, \omega_y, \omega_z)^T$ is a vector in $\Re^3$, we may define a skew-symmetric matrix, $S(\omega)$, according to

$$S(\omega) = \begin{bmatrix} 0 & -\omega_z & \omega_y \\ \omega_z & 0 & -\omega_x \\ -\omega_y & \omega_x & 0 \end{bmatrix} \tag{17.4}$$

The set of all skew-symmetric matrices is denoted by $so(3)$. Now, if $R(t)$ belongs to $SO(3)$ for all t, it can be shown that

$$\dot{R} = S(\omega(t))R \tag{17.5}$$

for a unique vector $\omega(t)$ [41]. The vector $\omega(t)$ thus defined is the angular velocity of the end-effector frame relative to the base frame.

If the mapping f_1 is used to represent the forward kinematics, then the velocity kinematics is written as

$$\dot{X}_1 = J_1(q)\dot{q} \tag{17.6}$$

where $J_1(q) = \partial f_1/\partial q$ is the $6 \times n$ Jacobian of the function f_1. In the sequel we will use J to denote either the matrix J_0 or J_1.

17.2.2 Dynamics

The dynamics of n-link manipulators are conveniently described via the Lagrangian approach. In this approach, the joint variables, $q = (q_1, \ldots, q_n)^T$, serve as **generalized coordinates**. The **manipulator kinetic energy** is given by a symmetric, positive definite quadratic form,

$$\mathcal{K} = \frac{1}{2} \sum_{i,j=1}^{n} d_{ij}(q) \dot{q}_i \dot{q}_j = \frac{1}{2} \dot{q}^T D(q) \dot{q} \tag{17.7}$$

where $D(q)$ is the **inertia matrix**. The **manipulator potential energy** is given by a continuously differentiable function $\mathcal{P}: \mathcal{C} \to \Re$. For a rigid robot, the potential energy is due to gravity alone while for a flexible robot the potential energy will also contain elastic potential energy.

The dynamics of the manipulator are then described by Lagrange's equations [41]

$$\frac{d}{dt} \frac{\partial \mathcal{L}}{\partial \dot{q}_k} - \frac{\partial \mathcal{L}}{\partial q_k} = \tau_k, \quad k = 1, \ldots, n \tag{17.8}$$

where $\mathcal{L} = \mathcal{K} - \mathcal{P}$ is the **Lagrangian** and $\tau_1, \ldots, \tau_n$ represent input generalized forces. In local coordinates Lagrange's equations can be written as

$$\sum_{j=1}^{n} d_{kj}(q) \ddot{q}_j + \sum_{i,j=1}^{n} \Gamma_{ijk}(q) \dot{q}_i \dot{q}_j + \phi_k(q) = \tau_k, \quad k = 1, \ldots, n \tag{17.9}$$

where

$$\Gamma_{ijk} = \frac{1}{2} \left\{ \frac{\partial d_{kj}}{\partial q_i} + \frac{\partial d_{ki}}{\partial q_j} - \frac{\partial d_{ij}}{\partial q_k} \right\} \tag{17.10}$$

are known as **Christoffel symbols of the first kind**, and

$$\phi_k = \frac{\partial \mathcal{P}}{\partial q_k} \tag{17.11}$$

In matrix form we can write Lagrange's Equation (17.9) as

$$D(q)\ddot{q} + C(q, \dot{q})\dot{q} + g(q) = \tau \tag{17.12}$$

In addition to the link inertias represented by the inertia matrix, $D(q)$, the inertias of the actuators are important to include in the dynamic description, especially for manipulators with large gear reduction. The actuator inertias are specified by an $n \times n$ diagonal matrix

$$I = \mathrm{diag}\left(I_1 r_1^2, \ldots, I_n r_n^2\right) \tag{17.13}$$

where I_i and r_i are the actuator inertia and gear ratio, respectively, of the ith joint. Defining $M(q) = D(q) + I$, we may modify the dynamics to include these additional terms as

$$M(q)\ddot{q} + C(q, \dot{q})\dot{q} + g(q) = \tau \tag{17.14}$$

17.2.3 Properties of Rigid Robot Dynamics

The equations of motion (17.14) possess a number of important properties that facilitate analysis and control system design. Among these are

1. The inertia matrix $M(q)$ is symmetric and positive definite, and there exist scalar functions, $\mu_1(q)$ and $\mu_2(q)$, such that

$$\mu_1(q)\|\xi\| \leq \xi^T M(q)\xi \leq \mu_2(q)\|\xi\| \quad (17.15)$$

   for all $\xi \in \Re^n$ where $\|\cdot\|$ denotes the usual Euclidean norm in $\Re^n$. Moreover, if all joints are revolute then μ_1 and μ_2 are constants.

2. For suitable definition of $C(q,\dot{q})$, the matrix $W(q,\dot{q}) = \dot{M}(q) - 2C(q,\dot{q})$ is skew symmetric. (See [41] for the appropriate definition of $C(q,\dot{q})$ and the proof of the skew symmetry property.) Related to the skew symmetry property is the so-called **passivity property**.

3. The mapping $\tau \to \dot{q}$ is **passive**, i.e., there exists $\beta \geq 0$ such that

$$\int_0^T \dot{q}^T(\zeta)\tau(\zeta)d\zeta \geq -\beta, \quad \forall \, T > 0 \quad (17.16)$$

   To show the passivity property, let H be the total energy of the system

$$H = \frac{1}{2}\dot{q}^T M(q)\dot{q} + \mathcal{P}(q) \quad (17.17)$$

   Then it is easily shown using the skew-symmetry property that the change in energy, $\dot{H}$, satisfies

$$\dot{H} = \dot{q}^T \tau \quad (17.18)$$

   Integrating both sides of Equation (17.18) with respect to time gives

$$\int_0^T \dot{q}^T(u)\tau(u)du = H(T) - H(0) \geq -H(0) \quad (17.19)$$

   since the total energy $H(T)$ is nonnegative. Passivity then follows with $\beta = H(0)$.

4. Rigid robot manipulators are fully actuated, i.e., there is an independent control input for each degree-of-freedom. By contrast, robots possessing joint or link flexibility are no longer fully actuated and the control problems are more difficult, in general.

5. The equations of motion (17.12) are linear in the inertia parameters. In other words, there is a constant vector $\theta \in \Re^p$ and a function $Y(q,\dot{q},\ddot{q}) \in \Re^{n\times p}$ such that

$$M(q)\ddot{q} + C(q,\dot{q})\dot{q} + g(q) = Y(q,\dot{q},\ddot{q})\theta = \tau \quad (17.20)$$

The function $Y(q,\dot{q},\ddot{q})$ is called the **regressor**. The parameter vector θ is composed of link masses, moments of inertia, and the like, in various combinations. Historically, the appearance of the passivity and linear parametrization properties in the early 1980s marked watershed events in robotics research. Using these properties researchers have been able to prove elegant global convergence and stability results for robust and adaptive control. We will detail some of these results below.

17.3 Taxonomy of Control Design Methods

The remainder of the article surveys several of the most important control design methods for rigid robots beginning with the notion of computed torque or feedback linearization and ending with hybrid control. For reasons of space, we give only the simplest versions of these ideas applied to rigid robots. We will assume that the full state $q(t), \dot{q}(t)$ of the robot is available whenever needed. Several of the results presented here, particularly the passivity based control results, have been extended to the case of output feedback, i.e., to the case that only the position $q(t)$ is directly measured. In this case, either an observer or some sort of velocity

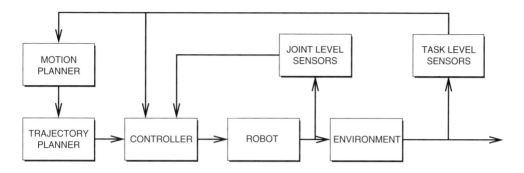

FIGURE 17.3 Block diagram of the robot control problem.

filter must be incorporated into the control design. We will not discuss the inclusion of actuator dynamics or joint or link flexibility into the control design. We will also not discuss the basic PD and PID control methods, which can be considered as limiting cases of robust control when no nominal model is used. PD and PID control is useful for the regulation problem, i.e., for point-to-point motion, especially when used in conjunction with gravity compensation but is less suitable for tracking time varying trajectories. The reader is referred to the references at the end of this article for further reading in these subjects.

17.3.1 Control Architecture

The motion control problem for robots is generally heirarchically decomposed into three stages, **Motion Planning**, **Trajectory Generation**, and **Trajectory Tracking** as shown in Figure 17.3. In the motion planning stage, desired paths are generated in the task space, $SE(3)$, without timing information, i.e., without specifying velocity or acceleration along the paths. Of primary concern is the generation of collision free paths in the workspace. The motion planner may generate a plan offline using background knowledge of the robot and environment, or it may incorporate task level sensing (e.g., vision or force) and modify the motion plan in real-time.

In the trajectory generation stage, the desired position, velocity, and acceleration of the manipulator along the path as a function of time or as a function of arclength along the path are computed. The trajectory planner may parametrize the end-effector path directly in task space, either as a curve in $SE(3)$ or as a curve in $\Re^6$ using a particular minimal representation for $SO(3)$, or it may compute a trajectory for the individual joints of the manipulator as a curve in the configuration space $\mathcal{C}$.

In order to compute a joint space trajectory, the given end-effector path must be transformed into a joint space path via the inverse kinematics mapping. A standard approach is to compute a discrete set of joint vectors along the end-effector path and to perform an interpolation in joint space among these points in order to complete the joint space trajectory. Common approaches to trajectory interpolation include polynomial spline interpolation, using trapezoidal velocity trajectories or cubic polynomial trajectories, as well as trajectories generated by reference models. The trajectory generator should take into account realistic velocity and acceleration constraints of the manipulator.

The computed reference trajectory is then presented to the controller, which is designed to cause the robot to track the given trajectory as closely as possible. The joint level controller will, in general, utilize both joint level sensors (e.g., position encoders, tachometers, torque sensors) and task level sensors (e.g., vision, force). This article is mainly concerned with the design of the tracking controller assuming that the path and trajectory have been precomputed.

17.3.2 Feedback Linearization Control

The goal of feedback linearization is to transform a given nonlinear system into a linear system by use of a nonlinear coordinate transformation and nonlinear feedback. The roots of feedback linearization in robotics predate the general theoretical development by nearly a decade, going back to the early notion of feedforward computed torque [29].

In the robotics context, feedback linearization is also known as **inverse dynamics** or **computed torque** control [41]. The idea is to exactly compensate all of the coupling nonlinearities in the Lagrangian dynamics in a first stage so that a second stage compensator may be designed based on a linear and decoupled plant. Any number of techniques may be used in the second stage. The feedback linearization may be accomplished with respect to the joint space coordinates or with respect to the task space coordinates. Feedback linearization may also be used as a basis for force control, such as hybrid position/force control and impedance control.

17.3.3 Joint Space Inverse Dynamics

Given the plant model

$$M(q)\ddot{q} + C(q,\dot{q})\dot{q} + g(q) = \tau \qquad (17.21)$$

a joint space inverse dynamics control is given by

$$\tau = M(q)a_q + C(q,\dot{q})\dot{q} + g(q) \qquad (17.22)$$

where $a_q \in \Re^n$ is, as yet, undetermined. Because the inertia matrix $M(q)$ is invertible for all q, the closed loop system reduces to the decoupled **double integrator** system

$$\ddot{q} = a_q \qquad (17.23)$$

The term a_q, which has units of acceleration, is now the control input to the double integrator system. We shall refer to a_q as the **outer loop control** and τ in Equation (17.22) as the **inner loop control**. This **inner loop/outer loop** architecture, shown in Figure 17.4, is important for several reasons. The structure of the inner loop control is fixed by Lagrange's equations. What control engineers traditionally think of as **control system design** is contained primarily in the outer loop, and one has complete freedom to modify the outer loop control to achieve various goals without the need to modify the dedicated inner loop control. In particular, additional compensation, linear or nonlinear, may be included in the outer loop to enhance the robustness to parametric uncertainty, unmodeled dynamics, and external disturbances. The outer loop control may also be modified to achieve other goals such as tracking of task space trajectories instead of joint space trajectories, regulating both motion and force. The inner loop/outer loop architecture thus unifies many robot control strategies from the literature.

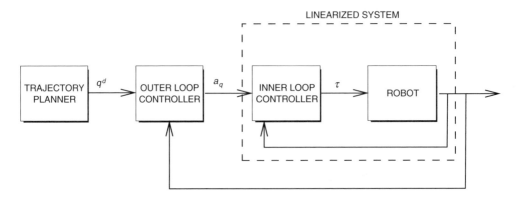

FIGURE 17.4 Inner loop/outer loop architecture.

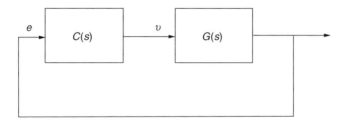

FIGURE 17.5 Block diagram of the error dynamics.

Let us assume that the joint space reference trajectory, $q^d(t)$, is at least twice continuously differentiable and that $q^d(t)$ along with its higher derivatives $\dot{q}^d(t)$ and $\ddot{q}^d(t)$ are bounded. Set $a_q = \ddot{q}^d + v$ and define

$$e(t) = \begin{bmatrix} \tilde{q} \\ \dot{\tilde{q}} \end{bmatrix} = \begin{bmatrix} q(t) - q^d(t) \\ \dot{q}(t) - \dot{q}^d(t) \end{bmatrix}$$

as the joint position and velocity tracking errors. Then we can write the **error equation** in state space as

$$\dot{e} = Ae + Bv \tag{17.24}$$

where

$$A = \begin{bmatrix} 0 & I \\ 0 & 0 \end{bmatrix}; \quad B = \begin{bmatrix} 0 \\ I \end{bmatrix} \tag{17.25}$$

A general linear control for the system (17.24) may take the form

$$\dot{z} = Fz + Ge \tag{17.26}$$
$$v = Hz + Ke \tag{17.27}$$

where F, G, H, K are matrices of appropriate dimensions. The block diagram of the resulting closed loop system is shown in Figure 17.5 with

$$G(s) = (sI - A)^{-1} B$$
$$C(s) = H(sI - F)^{-1} G + K$$

An important special case arises when $F = G = H = 0$. In this case the outer loop control, a_q, is a static state feedback control with gain matrix $K = [K_p, K_d]$ and feedforward acceleration $\ddot{q}^d$,

$$a_q = \ddot{q}^d(t) + v = \ddot{q}^d(t) + K_p(q^d - q) + K_d(\dot{q}^d - \dot{q}) \tag{17.28}$$

Substituting (17.28) into (17.23) yields the linear and decoupled closed loop system

$$\ddot{\tilde{q}} + K_d \dot{\tilde{q}} + K_p \tilde{q} = 0 \tag{17.29}$$

Taking K_p and K_d as positive definite diagonal matrices, we see that the joint tracking errors are decoupled and converge exponentially to zero as $t \to \infty$. Indeed, one can arbitrarily specify the closed loop natural frequencies and damping ratios. We will investigate other choices for the outer loop term a_q below.

17.3.4 Task Space Inverse Dynamics

As an illustration of the importance of the inner loop/outer loop paradigm, we will show that tracking in task space can be achieved by modifying our choice of outer loop control a_q in (17.23) while leaving the inner loop control unchanged. Let $X \in R^6$ represent the end-effector pose using any minimal representation of $SO(3)$. Since X is a function of the joint variables $q \in \mathcal{C}$, we have

$$\dot{X} = J(q)\dot{q} \tag{17.30}$$
$$\ddot{X} = J(q)\ddot{q} + \dot{J}(q)\dot{q} \tag{17.31}$$

where $J = J_1$ is the Jacobian defined in Section 17.2.1. Given the double integrator system, (17.23), in joint space we see that if a_q is chosen as

$$a_q = J^{-1}\{a_X - \dot{J}\dot{q}\} \tag{17.32}$$

the result is a double integrator system in task space coordinates

$$\ddot{X} = a_X \tag{17.33}$$

Given a task space trajectory $X^d(t)$, satisfying the same smoothness and boundedness assumptions as the joint space trajectory $q^d(t)$, we may choose a_X as

$$a_X = \ddot{X}^d + K_p(X^d - X) + K_d(\dot{X}^d - \dot{X}) \tag{17.34}$$

so that the Cartesian space tracking error, $\tilde{X} = X - X^d$, satisfies

$$\ddot{\tilde{X}} + K_d \dot{\tilde{X}} + K_p \tilde{X} = 0 \tag{17.35}$$

Therefore, a modification of the outer loop control achieves a linear and decoupled system directly in the task space coordinates, without the need to compute a joint trajectory and without the need to modify the nonlinear inner loop control.

Note that we have used a minimal representation for the orientation of the end-effector in order to specify a trajectory $X \in \Re^6$. In general, if the end-effector coordinates are given in $SE(3)$, then the Jacobian J in the above formulation will be the Jacobian J_0 defined in Section 17.1. In this case

$$V = \begin{pmatrix} v \\ \omega \end{pmatrix} = \begin{pmatrix} \dot{x} \\ \omega \end{pmatrix} = J(q)\dot{q} \tag{17.36}$$

and the outer loop control

$$a_q = J^{-1}(q)\left\{\begin{pmatrix} a_x \\ a_\omega \end{pmatrix} - \dot{J}(q)\dot{q}\right\} \tag{17.37}$$

applied to (17.23) results in the system

$$\ddot{x} = a_x \in \Re^3 \tag{17.38}$$
$$\dot{\omega} = a_\omega \in \Re^3 \tag{17.39}$$
$$\dot{R} = S(\omega)R, \quad R \in SO(3), \ S \in so(3) \tag{17.40}$$

Although, in this latter case, the dynamics have not been linearized to a double integrator, the outer loop terms a_v and a_ω may still be used to achieve global tracking of end-effector trajectories in $SE(3)$.

In both cases we see that nonsingularity of the Jacobian is necessary to implement the outer loop control. If the robot has more or fewer than six joints, then the Jacobians are not square. In this case, other schemes have been developed using, for example, the pseudoinverse in place of the inverse of the Jacobian. See [10] for details.

The inverse dynamics control approach has been proposed in a number of different guises, such as **resolved acceleration control** [26] and **operational space control** [21]. These seemingly distinct approaches have all been shown to be equivalent and may be incorporated into the general framework shown above [24].

17.3.5 Robust Feedback Linearization

The feedback linearization approach relies on exact cancellation of nonlinearities in the robot equations of motion. Its practical implementation requires consideration of various sources of uncertainties such as modeling errors, unknown loads, and computation errors. Let us return to the Euler-Lagrange equations of motion

$$M(q)\ddot{q} + C(q,\dot{q})\dot{q} + g(q) = \tau \qquad (17.41)$$

and write the control input τ as

$$\tau = \hat{M}(q)(\ddot{q}^d + v) + \hat{C}(q,\dot{q})\dot{q} + \hat{g}(q) \qquad (17.42)$$

where $(\hat{\cdot})$ represents the computed or nominal value of $(\cdot)$ and indicates that the theoretically exact feedback linearization cannot be achieved in practice due to the uncertainties in the system. The error or mismatch $(\tilde{\cdot}) = (\cdot) - (\hat{\cdot})$ is a measure of one's knowledge of the system dynamics.

If we now substitute (17.42) into (17.41) we obtain, after some algebra,

$$\ddot{q} = \ddot{q}^d + v + \eta(q,\dot{q},\ddot{q}^d,v) \qquad (17.43)$$

where

$$\eta = (M^{-1}\hat{M} - I)(\ddot{q}^d + v) + M^{-1}(\tilde{C}\dot{q} + \tilde{g}) \qquad (17.44)$$
$$=: Ev + y \qquad (17.45)$$

where

$$E = (M^{-1}\hat{M} - I) \text{ and}$$
$$y = F(e,t) = E\ddot{q}^d + M^{-1}(\tilde{C}\dot{q} + \tilde{g})$$

The error equations are now written as

$$\dot{e} = Ae + B\{v + \eta\} \qquad (17.46)$$

where e, A, B are defined as before. The block diagram representation of the system must now be modified to include the uncertainty η as shown in Figure 17.6. The robust control design problem is now to design the compensator $C(s)$ in Figure 17.6 to guarantee stability and tracking while rejecting the disturbance η. We will detail three primary approaches that have been used to tackle this problem.

17.3.5.1 Stable Factorizations

The method of stable factorizations was first applied to the robust feedback linearization problem in [39, 40]. In this approach, the so called Youla parametrization [18] is used to generate the entire class, Ω, of

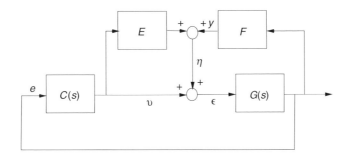

FIGURE 17.6 Uncertain double integrator system.

stabilizing compensators for the unperturbed system, i.e., Equation (17.46) with $\eta = 0$. Given bounds on the uncertainty, the **Small Gain Theorem** [17] is used to generate a sufficient condition for stability of the perturbed system, and the design problem is to determine a particular compensator, $C(s)$, from the class of stabilizing compensators Ω that satisfies this sufficient condition.

The interesting feature of this problem is that the perturbation terms appearing in (17.46) are so-called **persistent disturbances**, i.e., they are bounded but do not vanish at $t \to \infty$. This is chiefly due to the properties of the gravity terms and the reference trajectories. As a result, one may assume that the uncertainty η is finite in the L_∞-norm but not necessarily in the L_2-norm since under some mild assumptions, an L_2 signal converges to zero as $t \to \infty$.

The problem of stabilizing a linear system while minimizing the response to an L_∞-bounded disturbance is equivalent to minimizing the L_1 norm of the impulse response of the transfer function from the disturbance to the output [30]. For this reason, the problem is now referred to as the L_1-optimal control problem [30, 44]. The results in [39, 40] predate the general theory and, in fact, provided early motivation for the more general theoretical development first reported in [44]. We sketch the basic idea of this approach below. See [5, 40] for more details.

We first require some modelling assumptions in order to bound the uncertainty term η. We assume that there exists a positive constant $\alpha < 1$ such that

$$||E|| = ||M^{-1}\hat{M} - I|| \leq \alpha \tag{17.47}$$

We note that there always exists a choice for $\hat{M}$ satisfying this assumption, for example $\hat{M} = \frac{\mu_1 + \mu_2}{2} I$, where μ_i are the bounds on the intertia matrix in Equation (17.15) [40]. From this and the properties of the manipulator dynamics, specifically the quadratic in velocity form of the Coriolis and centrifugal terms, we may assume that there exist positive constants δ_1, δ_2, and b such that

$$||\eta|| \leq \alpha ||v|| + \delta_1 ||e|| + \delta_2 ||e||^2 + b \tag{17.48}$$

Let δ be a positive constant such that

$$\delta_1 ||e|| + \delta_2 ||e||^2 \leq \delta ||e|| \tag{17.49}$$

so that

$$||\eta|| \leq \alpha ||v|| + \delta ||e|| + b \tag{17.50}$$

We note that this assumption restricts the set of allowable initial conditions as shown in Figure 17.7. With this assumption we are restricted to so-called **semiglobal**, rather than global, stabilization. However, the region of attraction can, in theory, be made arbitrarily large. For any region in error space, $|e \leq R|$, we may take $\delta \geq \delta_1 + \delta_2 R$ in order to satisfy Equation (17.49).

Next, from Figure 17.6 it is straightforward to compute

$$e = G(I - CG)^{-1}\eta =: P_1 \eta \tag{17.51}$$
$$v = CG(I - CG)^{-1}\eta =: P_2 \eta \tag{17.52}$$

The above equations employ the common convention of using $P\eta$ to mean $(p * \eta)(t)$ where $*$ denotes the convolution operator and $p(t)$ is the impulse response of $P(s)$. Thus

$$||e|| \leq \beta_1 ||\eta|| \tag{17.53}$$
$$||v|| \leq \beta_2 ||\eta|| \tag{17.54}$$

where β_i denotes the operator norm of the transfer function, i.e.,

$$\beta_i = \sup_{x \in L_\infty^n - \{0\}} \frac{||P_i x||_\infty}{||x||_\infty} \tag{17.55}$$

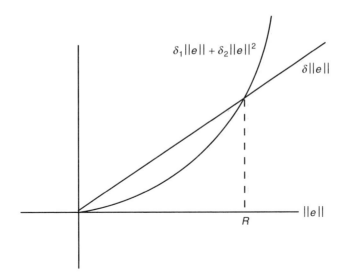

FIGURE 17.7 Linear vs. quadratic error bounds.

As the quantities e, v, η are functions of time, we can calculate a bound on the uncertainty as follows

$$||\eta||_{T\infty} \leq \alpha ||v||_{T\infty} + \delta ||e||_{T\infty} + b \tag{17.56}$$
$$\leq (\alpha\beta_2 + \delta\beta_1)||\eta||_{T\infty} + b \tag{17.57}$$

where the norm denotes the truncated L_∞ norm [17] and we have suppressed the explicit dependence on time of the various signals. Thus, if

$$\alpha\beta_2 + \delta\beta_1 < 1 \tag{17.58}$$

the uncertainty and hence both the control signal and tracking error are bounded as

$$||\eta||_{T\infty} \leq \frac{b}{1 - (\alpha\beta_2 + \delta\beta_1)} \tag{17.59}$$

$$||e||_{T\infty} \leq \frac{\beta_1 b}{1 - (\alpha\beta_2 + \delta\beta_1)} \tag{17.60}$$

$$||v||_{T\infty} \leq \frac{\beta_2 b}{1 - (\alpha\beta_2 + \delta\beta_1)} \tag{17.61}$$

Condition (17.58) is a special case of the small gain theorem [17, 31]. In [40] the stable factorization approach is used to design a compensator $C(s)$ to make the operator norms β_1 and β_2 arbitrarily close to zero and one, respectively. It then follows from Equation (17.60), letting $t \to \infty$, that the tracking error can be made arbitrarily small in norm.

Specifically, the set of all controllers that stabilize $G(s)$ is given by

$$\{-(Y - R\tilde{N})^{-1}(X + R\tilde{D})\} \tag{17.62}$$

where $N, D, \tilde{N}, \tilde{D}$ are stable proper rational transfer functions satisfying

$$G(s) = N(s)[D(s)]^{-1} = [\tilde{D}(s)]^{-1}\tilde{N}(s) \tag{17.63}$$

X, Y are stable rational transfer functions found from the **Bezout Identity**

$$\begin{bmatrix} Y(s) & X(s) \\ -\tilde{N}(s) & \tilde{D}(s) \end{bmatrix} \begin{bmatrix} D(s) & -\tilde{X}(s) \\ N(s) & \tilde{Y}(s) \end{bmatrix} = I \quad (17.64)$$

and $R(s)$ is a free parameter, an arbitrary matrix of appropriate dimensions whose elements are stable rational functions. Defining

$$C_k = \{-(Y - R_k \tilde{N})^{-1}(X + R_k \tilde{D})\} \quad (17.65)$$

a constructive procedure in [40] gives a sequence of matrices, R_k, such that the corresponding gains β_{1k} and β_{2k} converge to zero and one, respectively, as $k \to \infty$.

17.3.5.2 Lyapunov's Second Method

A second approach to the robust control problem is the so-called theory of **guaranteed stability of uncertain systems**, which is based on Lyapunov's second method and sometimes called the **Corless-Leitmann** [13] approach. In this approach the outer loop control, a_q is a static state feedback control rather than a dynamic compensator as in the previous section. We set

$$a_q = \ddot{q}^d(t) + K_p(q^d - q) + K_d(\dot{q}^d - \dot{q}) + \delta a \quad (17.66)$$

which results in the system

$$\dot{e} = Ae + B\{\delta a + \eta\} \quad (17.67)$$

where

$$A = \begin{bmatrix} 0 & I \\ -K_p & -K_d \end{bmatrix}, \quad B = \begin{bmatrix} 0 \\ I \end{bmatrix} \quad (17.68)$$

Thus the double integrator is first stabilized by the linear feedback, $-K_p e - K_d \dot{e}$, and δa is an additional control input that must be designed to overcome the destabilizing effect of the uncertainty η. The basic idea is to compute a time-varying scalar bound, $\rho(e,t) \geq 0$, on the uncertainty η, i.e.,

$$||\eta|| \leq \rho(e,t) \quad (17.69)$$

and design the additional input term δa to guarantee ultimate boundedness of the state trajectory $x(t)$ in Equation (17.67).

Returning to our expression for the uncertainty

$$\eta = E a_q + M^{-1}(\tilde{C}\dot{q} + \tilde{g}) \quad (17.70)$$
$$= E \delta a + E(\ddot{q}^d - K_p \tilde{q} - K_d \dot{\tilde{q}}) + M^{-1}(\tilde{C}\dot{q} + \tilde{g}) \quad (17.71)$$

we may compute

$$||\eta|| \leq \alpha ||\delta a|| + \gamma_1 ||e|| + \gamma_2 ||e||^2 + \gamma_3 \quad (17.72)$$

where $\alpha < 1$ as before and γ_1 are suitable nonnegative constants. Assuming for the moment that $||\delta a|| \leq \rho(e,t)$, which must then be checked a posteriori, we have

$$||\eta|| \leq \alpha \rho(e,t) + \gamma_1 ||e|| + \gamma_2 ||e||^2 + \gamma_3 =: \rho(e,t) \quad (17.73)$$

which defines ρ as

$$\rho(e,t) = \frac{1}{1-\alpha}(\gamma_1 ||e|| + \gamma_2 ||e||^2 + \gamma_3) \quad (17.74)$$

Since K_p and K_d are chosen so that A in Equation (17.67) is a Hurwitz matrix, we choose $Q > 0$ and let $P > 0$ be the unique symmetric matrix satisfying the Lyapunov equation,

$$A^T P + PA = -Q \qquad (17.75)$$

Defining the control δa according to

$$\delta a = \begin{cases} -\rho(e,t) \dfrac{B^T P e}{||B^T P e||}, & \text{if } ||B^T P e|| \neq 0 \\ 0, & \text{if } ||B^T P e|| = 0 \end{cases} \qquad (17.76)$$

it follows that the Lyapunov function $V = e^T P e$ satisfies $\dot{V} \leq 0$ along solution trajectories of the system (17.67). To show this result, we compute

$$\dot{V} = -e^T Q e + 2 e^T P B \{\delta a + \eta\} \qquad (17.77)$$

For simplicity, set $w = B^T P e$ and consider the second term, $w^T \{\delta a + \eta\}$ in the above expression. If $w = 0$ this term vanishes and for $w \neq 0$, we have

$$\delta a = -\rho \frac{w}{||w||} \qquad (17.78)$$

and Equation (17.77) becomes, using the Cauchy-Schwartz inequality,

$$w^T \left(-\rho \frac{w}{||w||} + \eta \right) \leq -\rho ||w|| + ||w|| ||\eta|| \qquad (17.79)$$

$$= ||w||(-\rho + ||\eta||) \leq 0 \qquad (17.80)$$

and hence

$$\dot{V} < -e^T Q e \qquad (17.81)$$

and the result follows. Note that $||\delta a|| \leq \rho$ as required.

Since the above control term δa is discontinuous on the manifold defined by $B^T P e = 0$, solution trajectories on this manifold are only defined in a generalized sense, the so-called **Filippov solutions** [18], and the control signal exhibits chattering. One may implement a continuous approximation to the discontinuous control as

$$\delta a = \begin{cases} -\rho(e,t) \dfrac{B^T P e}{||B^T P e||}, & \text{if } ||B^T P e|| > \epsilon \\ -\dfrac{\rho(e,t)}{\epsilon} B^T P e, & \text{if } ||B^T P e|| \leq \epsilon \end{cases} \qquad (17.82)$$

and show that the tracking error is now uniformly ultimately bounded with the size of the ultimate bound being a function of ϵ [41].

17.3.5.3 Sliding Modes

The sliding mode theory of variable structure systems has been extensively applied to the design of δa in Equation (17.46). This approach is very similar in spirit to the Corless-Leitmann approach above. In the sliding mode approach, we define a **sliding surface** in error space

$$s := \dot{\tilde{q}} + \Lambda \tilde{q} = 0 \qquad (17.83)$$

where Λ is a diagonal matrix of positive elements. Return to the uncertain error equation

$$\ddot{\tilde{q}} + K_d \dot{\tilde{q}} + K_p \tilde{q} = \delta a + \eta \qquad (17.84)$$

the goal is to design the control δa so that the trajectory converges to the sliding surface and remains constrained to the surface. The constraint $s(t) = 0$ then implies that the tracking error satisfies

$$\dot{\tilde{q}} = -\Lambda \tilde{q} \tag{17.85}$$

If we let $K_p = K_d \Lambda$, we may write Equation (17.84) as

$$\ddot{\tilde{q}} = -K_d(\dot{\tilde{q}} + \Lambda \tilde{q}) + \delta a + \eta = -K_d s + \delta a + \eta \tag{17.86}$$

Define

$$V = \tfrac{1}{2} s^T s \tag{17.87}$$

and compute

$$\dot{V} = s^T \dot{s} = s^T(\ddot{\tilde{q}} + \Lambda \dot{\tilde{q}}) \tag{17.88}$$
$$= s^T(-K_d s + \delta a + \eta + \Lambda \dot{\tilde{q}}) \tag{17.89}$$
$$= -s^T K_d s + s^T(\delta a + \eta + \Lambda \dot{\tilde{q}}) \tag{17.90}$$
$$= -s^T K_d s + s^T(v + \eta) \tag{17.91}$$

where δa has been chosen as $\delta a = -\Lambda \dot{\tilde{q}} + v$

Examining the term $s^T(v + \eta)$ in the above, we choose the ith component of v as

$$v_i = \rho_i(e, t)\mathrm{sgn}(s_i), \quad i = 1, \ldots, n \tag{17.92}$$

where ρ_i is a bound on the ith component of η, $s_i = \dot{\tilde{q}}_i + \lambda_i \tilde{q}_i$ is the ith component of s, and $\mathrm{sgn}(\cdot)$ is the signum function

$$\mathrm{sgn}(s_i) = \begin{cases} +1 & \text{if } s_i > 0 \\ -1 & \text{if } s_i < 0 \end{cases} \tag{17.93}$$

Then

$$s^T(v + \eta) = \sum_{i=1}^{n}(s_i \eta_i - \rho_i \mathrm{sgn}(s_i)) \tag{17.94}$$
$$\leq |s_i|(|\eta_i| - \rho_i) \leq 0 \tag{17.95}$$

It therefore follows that

$$\dot{V} = -s^T K_d s < 0 \tag{17.96}$$

and $s \to 0$. The discontinuous control input v_i switches sign on the sliding surface (Figure 17.8). In the ideal case of infinitely fast switching, once the trajectory hits the sliding surface, it is constrained to lie on the sliding surface and the closed loop dynamics are thus given by Equation (17.85), and hence the tracking error is globally exponentially convergent. Similar problems arise in this approach, as in the Corless-Leitmann approach, with respect to existence of solutions of the state equations and chattering of the control signals. Smoothing of the discontinuous control results in removal of chattering at the expense of tracking precision, i.e., the tracking error is then once again ultimately bounded rather than asymptotically convergent to zero [37].

17.3.6 Adaptive Feedback Linearization

Once the linear parametrization property for manipulators became widely known in the mid-1980s, the first globally convergent adaptive control results began to appear. These first results were based on the inverse dynamics or feedback linearization approach discussed above. Consider the plant (17.41) and

FIGURE 17.8 A sliding surface.

control (17.42) as above, but now suppose that the parameters appearing in Equation (17.42) are not fixed as in the robust control approach but are time-varying estimates of the true parameters. Substituting Equation (17.42) into Equation (17.41) and setting

$$a_q = \ddot{q}^d + K_d(\dot{q}^d - \dot{q}) + K_p(q^d - q) \qquad (17.97)$$

it can be shown, using the linear parametrization property, that

$$\ddot{\tilde{q}} + K_d \dot{\tilde{q}} + K_p \tilde{q} = \hat{M}^{-1} Y(q, \dot{q}, \ddot{q}) \tilde{\theta} \qquad (17.98)$$

where Y is the regressor function, and $\tilde{\theta} = \hat{\theta} - \theta$, and $\hat{\theta}$ is the estimate of the parameter vector θ. In state space we write the system (17.98) as

$$\dot{e} = Ae + B\Phi\tilde{\theta} \qquad (17.99)$$

where

$$A = \begin{bmatrix} 0 & I \\ -K_p & -K_d \end{bmatrix}, \quad B = \begin{bmatrix} 0 \\ I \end{bmatrix}, \quad \Phi = \hat{M}^{-1} Y(q, \dot{q}, \ddot{q}) \qquad (17.100)$$

with K_p and K_d chosen so that A is a Hurwitz matrix. Let P be that unique symmetric, positive definite matrix P satisfying

$$A^T P + PA = -Q \qquad (17.101)$$

and choose the parameter update law as

$$\dot{\hat{\theta}} = -\Gamma^{-1} \Phi^T B^T Pe \qquad (17.102)$$

where $\Gamma = \Gamma^T > 0$. Then global convergence to zero of the tracking error with all internal signals remaining bounded can be shown using the Lyapunov function

$$V = e^T Pe + \frac{1}{2} \tilde{\theta}^T \Gamma \tilde{\theta} \qquad (17.103)$$

Calculating $\dot{V}$ yields

$$\dot{V} = -e^T Qe + \tilde{\theta}^T \{\Phi^T B^T Pe + \Gamma \dot{\hat{\theta}}\} \qquad (17.104)$$

the latter term following since θ is constant, i.e., $\dot{\tilde{\theta}} = \dot{\hat{\theta}}$. Using the parameter update law (17.102) gives

$$\dot{V} = -e^T Q e \qquad (17.105)$$

From this it follows that the position tracking errors converge to zero asymptotically and the parameter estimation errors remain bounded. Furthermore, it can be shown that the estimated parameters converge to the true parameters provided the reference trajectory satisfies the condition of **persistency of excitation**,

$$\alpha I \leq \int_{t_0}^{t_0+T} Y^T(q^d, \dot{q}^d, \ddot{q}^d) Y(q^d, \dot{q}^d, \ddot{q}^d) dt \leq \beta I \qquad (17.106)$$

for all t_0, where α, β, and T are positive constants.

In order to implement this adaptive feedback linearization scheme, however, one notes that the acceleration $\ddot{q}$ is needed in the parameter update law and that $\hat{M}$ must be guaranteed to be invertible, possibly by the use of projection in the parameter space. Later work was devoted to overcome these two drawbacks to this scheme by using so-called **indirect** approaches based on a (filtered) prediction error.

17.3.7 Passivity-Based Approaches

One may also exploit the passivity of the rigid robot dynamics to derive robust and adaptive control algorithms for manipulators. These methods are qualitatively different from the previous methods which were based on feedback linearization. In the passivity-based approach, we modify the inner loop control as

$$\tau = \hat{M}(q)a + \hat{C}(q,\dot{q})v + \hat{g}(q) - Kr \qquad (17.107)$$

where v, a, and r are given as

$$v = \dot{q}^d - \Lambda \tilde{q}$$
$$a = \dot{v} = \ddot{q}^d - \Lambda \dot{\tilde{q}}$$
$$r = \dot{q}^d - v = \dot{\tilde{q}} + \Lambda \tilde{q}$$

with K, Λ diagonal matrices of positive gains. In terms of the linear parametrization of the robot dynamics, the control (17.107) becomes

$$\tau = Y(q,\dot{q},a,v)\hat{\theta} - Kr \qquad (17.108)$$

and the combination of Equation (17.107) with Equation (17.41) yields

$$M(q)\dot{r} + C(q,\dot{q})r + Kr = Y\tilde{\theta} \qquad (17.109)$$

Note that, unlike the inverse dynamics control, the modified inner loop control (17.41) does not achieve a linear, decoupled system, even in the known parameter case $\hat{\theta} = \theta$. However, in this formulation the regressor Y in Equation (17.109) does not contain the acceleration $\ddot{q}$ nor is the inverse of the estimated inertia matrix required. These represent distinct advantages over the feedback linearization based schemes.

17.3.8 Passivity-Based Robust Control

In the robust passivity-based approach of [36], the term $\hat{\theta}$ in Equation (17.108) is chosen as

$$\hat{\theta} = \theta_0 + u \qquad (17.110)$$

where θ_0 is a fixed nominal parameter vector and u is an additional control term. The system (17.109) then becomes

$$M(q)\dot{r} + C(q,\dot{q})r + Kr = Y(a,v,q,\dot{q})(\tilde{\theta} + u) \qquad (17.111)$$

where $\tilde{\theta} = \theta_0 - \theta$ is a constant vector and represents the parametric uncertainty in the system. If the uncertainty can be bounded by finding a nonnegative constant, $\rho \geq 0$, such that

$$\|\tilde{\theta}\| = \|\theta_0 - \theta\| \leq \rho \tag{17.112}$$

then the additional term u can be designed according to the expression,

$$u = \begin{cases} -\rho \frac{Y^T r}{\|Y^T r\|}, & \text{if } \|Y^T r\| > \epsilon \\ -\frac{\rho}{\epsilon} Y^T r, & \text{if } \|Y^T r\| \leq \epsilon \end{cases} \tag{17.113}$$

The Lyapunov function

$$V = \frac{1}{2} r^T M(q) r + \tilde{q}^T \Lambda K \tilde{q} \tag{17.114}$$

is used to show uniform ultimate boundedness of the tracking error. Note that $\tilde{\theta}$ is constant and so is not a state vector as in adaptive control. Calculating $\dot{V}$ yields

$$\dot{V} = r^T M \dot{r} + \frac{1}{2} r^T \dot{M} r + 2 \tilde{q}^T \Lambda K \dot{\tilde{q}} \tag{17.115}$$

$$= -r^T K r + 2 \tilde{q}^T \Lambda K \dot{\tilde{q}} + \frac{1}{2} r^T (\dot{M} - 2C) r + r^T Y(\tilde{\theta} + u) \tag{17.116}$$

Using the passivity property and the definition of r, this reduces to

$$\dot{V} = -\tilde{q}^T \Lambda^T K \Lambda \tilde{q} - \dot{\tilde{q}}^T K \dot{\tilde{q}} + r^T Y(\tilde{\theta} + u) \tag{17.117}$$

Uniform ultimate boundedness of the tracking error follows with the control u from (17.113). See [36] for details.

Comparing this approach with the approach in the section (17.3.5), we see that finding a constant bound ρ for the constant vector $\tilde{\theta}$ is much simpler than finding a time-varying bound for η in Equation (17.44). The bound ρ in this case depends only on the inertia parameters of the manipulator, while $\rho(x,t)$ in Equation (17.69) depends on the manipulator state vector and the reference trajectory and, in addition, requires some assumptions on the estimated inertia matrix $\hat{M}(q)$.

17.3.9 Passivity-Based Adaptive Control

In the adaptive approach the vector $\hat{\theta}$ in Equation (17.109) is now taken to be a time-varying estimate of the true parameter vector θ. Combining the control law (17.107) with (17.41) yields

$$M(q)\dot{r} + C(q, \dot{q}) r + K r = Y \tilde{\theta} \tag{17.118}$$

The parameter estimate $\hat{\theta}$ may be computed using standard methods such as gradient or least squares. For example, using the gradient update law

$$\dot{\hat{\theta}} = -\Gamma^{-1} Y^T(q, \dot{q}, a, v) r \tag{17.119}$$

together with the Lyapunov function

$$V = \frac{1}{2} r^T M(q) r + \tilde{q}^T \Lambda K \tilde{q} + \frac{1}{2} \tilde{\theta}^T \Gamma \tilde{\theta} \tag{17.120}$$

results in global convergence of the tracking errors to zero and boundedness of the parameter estimates since

$$\dot{V} = -\tilde{q}^T \Lambda^T K \Lambda \tilde{q} - \dot{\tilde{q}}^T K \dot{\tilde{q}} + \tilde{\theta}^T \{\Gamma \dot{\tilde{\theta}} + Y^T r\} \tag{17.121}$$

See [38] for details.

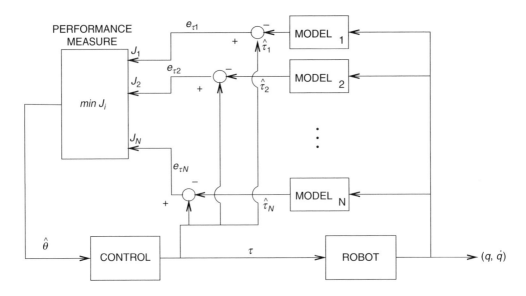

FIGURE 17.9 Multiple-model-based hybrid control architecture.

17.3.10 Hybrid Control

A **Hybrid System** is one that has both continuous-time and discrete-event or logic-based dynamics. **Supervisory Control**, **Logic-Based Switching Control**, and **Multiple-Model Control** are typical control architectures in this context. In the robotics context, hybrid schemes can be combined with robust and adaptive control methods to further improve robustness. In particular, because the preceeding robust and adaptive control methods provide only asymptotic (i.e., as $t \to \infty$) error bounds, the transient performance may not be acceptable. Hybrid control methods have been shown to improve transient performance over fixed robust and adaptive controllers.

The use of the term *Hybrid Control* in this context should not be confused with the notion of Hybrid Position/Force Control [41]. The latter is a familiar approach to force control of manipulators in which the term *hybrid* refers to the combination of pure force control and pure motion control.

Figure 17.9 shows the Multiple-Model approach of [12], which has been applied to the adaptive control of manipulators. In this architecture, the multiple models have the same structure but may have different nominal parameters in case a robust control scheme is used, or different initial parameter estimates if an adaptive control scheme is used. Because all models have the same inputs and desired outputs, the identification errors e_{I_j} are available at each instant for the jth model. The idea is then to define a performance measure, for example,

$$J(e_{I_j}(t)) = \gamma e_{I_j}^2(t) + \beta \int_0^t e_{I_j}^2(\sigma) d\sigma \quad \text{with} \quad \gamma, \beta > 0 \qquad (17.122)$$

and switch into the closed loop the control input that results in the smallest value of J at each instant.

17.4 Conclusions

We have given a brief overview of the basic results in robust and adaptive control of robot manipulators. In most cases, we have given only the simplest forms of the algorithms, both for ease of exposition and for reasons of space. An extensive literature is available that contains numerous extensions of these basic results. The attached list of references is by no means an exhaustive one. The book [10] is an excellent and highly detailed treatment of the subject and a good starting point for further reading. Also, the reprint book [37] contains several of the original sources of material surveyed here. In addition, the two survey papers [1] and [28] provide additional details on the robust and adaptive control outlined here.

Several important areas of interest have been omitted for space reasons including **output feedback control**, **learning control**, **fuzzy control**, **neural networks**, and **visual servoing**, and control of **flexible robots**. The reader should consult the references at the end for background on these and other subjects.

References

[1] Abdallah, C. et al., Survey of robust control for rigid robots, *IEEE Control Systems Magazine*, Vol. 11, No. 2, pp. 24–30, Feb. 1991.

[2] Amestegui, M., Ortega, R., and Ibarra, J.M., Adaptive linearizing-decoupling robot control: A comparative study, *Proc. 5th Yale Workshop on Applications of Adaptive Systems Theory*, New Haven, CT, 1987.

[3] Balestrino, A., De Maria, G., and Sciavicco, L., An adaptive model following control for robotic manipulators, *ASME J. Dynamic Syst., Meas., Contr.*, Vol. 105, pp. 143–151, 1983.

[4] Bayard, D.S. and Wen, J.T., A new class of control laws for robotic manipulators-Part 2. Adaptive case, *Int. J. Contr.*, Vol. 47, No. 5, pp. 1387–1406, 1988.

[5] Becker, N. and Grimm, W.M., On L_2 and L_∞-stability approaches for the robust control of robot manipulators, *IEEE Trans. Automat. Contr.*, Vol. 33, No. 1, pp. 118–122, Jan. 1988.

[6] Berghuis, H. and Nijmeijer, H., Global regulation of robots using only position measurements, *Syst. Contr. Lett.*, Vol. 1, pp. 289–293, 1993.

[7] Berghuis, H., Ortega, R., and Nijmeier, H., A robust adaptive robot controller, *IEEE Trans. Robotics Automat.*, Vol. 9, No. 6, pp.825–830, Dec. 1993.

[8] Brogliato, B., Landau, I.D., and Lozano, R., Adaptive motion control of robot manipulators: A unified approach based on passivity, *Int. J. Robust Nonlinear Contr.*, Vol. 1, pp. 187–202, 1991.

[9] Campion, G. and Bastin, G., Analysis of an adaptive controller for manipulators: Robustness versus flexibility, *Syst. Contr. Lett.*, Vol. 12, pp. 251–258, 1989.

[10] Canudas de Wit, C. et al., *Theory of Robot Control*, Springer-Verlag, London, 1996.

[11] Canudas de Wit, C. and Fixot, N., Adaptive control of robot manipulators via velocity estimated state feedback, *IEEE Trans. Automatic Contr.*, Vol. 37, pp. 1234–1237, 1992.

[12] Ciliz, M.K. and Narendra, K.S., Intelligent control of robotic manipulators: A multiple model based approach, *IEEE Conf. on Decision and Control*, New Orleans, LA, pp. 422–427, December 1995.

[13] Corless, M. and Leitmann, G., Continuous state feedback guaranteeing uniform ultimate boundedness for uncertain dynamic systems, *IEEE Trans. Automatic Contr.*, Vol. 26, pp. 1139–1144, 1981.

[14] Craig, J.J., *Adaptive Control of Mechanical Manipulators*, Addison-Wesley, Reading, MA, 1988.

[15] Craig, J.J., Hsu, P., and Sastry, S., Adaptive control of mechanical manipulators, *Proc. IEEE Int. Conf. Robotics Automation*, San Francisco, CA, March 1986.

[16] De Luca, A., Dynamic control of robots with joint elasticity, *Proc. IEEE Conf. on Robotics and Automation*, Philadelphia, PA, pp. 152–158, 1988.

[17] Desoer, C.A. and Vidyasagar, M., *Feedback Systems: Input-Output Properties*, Academic Press, New York, 1975.

[18] Filippov, A.F., Differential equations with discontinuous right-hand side, *Amer. Math. Soc. Transl.*, Vol. 42, pp. 199–231, 1964.

[19] Hager, G.D., A modular system for robust positioning using feedback from stereo vision, *IEEE Trans. Robotics Automat.*, Vol. 13, No. 4, pp. 582–595, August 1997.

[20] Ioannou, P.A. and Kokotović, P.V., Instability analysis and improvement of robustness of adaptive control, *Automatica*, Vol. 20, No. 5, pp. 583–594, 1984.

[21] Khatib, O., A unified approach for motion and force control of robot manipulators: the operational space formulation, *IEEE J. Robotics Automat.*, Vol. RA–3, No. 1, pp. 43–53, Feb. 1987.

[22] Kim, Y.H. and Lewis, F.L., Optimal design of CMAC neural-network controller for robot manipulators, *IEEE Trans. on Sys. Man and Cybernetics*, Vol. 30, No. 1, pp. 22–31, Feb. 2000.

[23] Koditschek, D., Natural motion of robot arms, *Proc. IEEE Conf. on Decision and Control*, Las Vegas, NV, pp. 733–735, 1984.

[24] Kreutz, K., On manipulator control by exact linearization, *IEEE Trans. Automat. Contr.*, Vol. 34, No. 7, pp. 763–767, July 1989.

[25] Latombe, J.C., *Robot Motion Planning*, Kluwer, Boston, MA, 1990.

[26] Luh, J., Walker, M., and Paul, R., Resolved-acceleration control of mechanical manipulators, *IEEE Trans. Automat. Contr.*, Vol. AC–25, pp. 468–474, 1980.

[27] Middleton, R.H. and Goodwin, G.C., Adaptive computed torque control for rigid link manipulators, *Syst. Contr. Lett.*, Vol. 10, pp. 9–16, 1988.

[28] Ortega, R. and Spong, M.W., Adaptive control of rigid robots: a tutorial, *Proc. IEEE Conf. on Decision and Control*, Austin, TX, pp. 1575–1584, 1988.

[29] Paul, R.C., Modeling, trajectory calculation, and servoing of a computer controlled arm, Stanford A.I. Lab, A.I. Memo 177, Stanford, CA, Nov. 1972.

[30] Dahleh, M.A. and Pearson, J.B., L^1-optimal compensators for continuous-time systems, *IEEE Trans. Automat. Contr.*, Vol. AC-32, No. 10, pp. 889–895, Oct. 1987.

[31] Porter, D.W. and Michel, A.N., Input-output stability of time varying nonlinear multiloop feedback systems, *IEEE Trans. Automat. Contr.*, Vol. AC-19, No. 4, pp. 422–427, Aug. 1974.

[32] Sadegh, N. and Horowitz, R., An exponentially stable adaptive control law for robotic manipulators, *Proc. American Control Conf.*, San Diego, pp. 2771–2777, May 1990.

[33] Schwartz, H.M. and Warshaw, G., On the richness condition for robot adaptive control, *ASME Winter Annual Meeting*, DSC-Vol. 14, pp. 43–49, Dec. 1989.

[34] Slotine, J.-J.E. and Li, W., On the adaptive control of robot manipulators, *Int. J. Robotics Res.*, Vol. 6, No. 3, pp. 49–59, 1987.

[35] Spong, M.W., Modeling and control of elastic joint manipulators, *J. Dyn. Sys., Meas. Contr.*, Vol. 109, pp. 310–319, 1987.

[36] Spong, M.W., On the robust control of robot manipulators, *IEEE Trans. Automat. Contr.*, Vol. 37, pp. 1782–1786, Nov. 1992.

[37] Spong, M.W., Lewis, F., and Abdallah, C., *Robot Control: Dynamics, Motion Planning, and Analysis*, IEEE Press, 1992.

[38] Spong, M.W., Ortega, R., and Kelly, R., Comments on 'Adaptive manipulator control', *IEEE Trans. Automat. Contr.*, Vol. AC–35, No. 6, pp. 761–762, 1990.

[39] Spong, M.W. and Vidyasagar, M., Robust nonlinear control of robot manipulators, *Proc. 24th IEEE Conf. Decision and Contr.*, Fort Lauderdale, FL, pp. 1767–1772, Dec. 1985.

[40] Spong, M.W. and Vidyasagar, M., Robust linear compensator design for nonlinear robotic control, *IEEE J. Robotics Automation*, Vol. RA-3, No. 4, pp. 345–350, Aug. 1987.

[41] Spong, M.W. and Vidyasagar, M., *Robot Dynamics and Control*, John Wiley & Sons, New York, 1989.

[42] Su, C.-Y., Leung, T.P., and Zhou, Q.-J., A novel variable structure control scheme for robot trajectory control, *IFAC World Congress*, Vol. 9, pp. 121–124, Tallin, Estonia, August 1990.

[43] Vidyasagar, M., *Control Systems Synthesis: A Factorization Approach*, MIT Press, Cambridge, MA, 1985.

[44] Vidyasagar, M., Optimal rejection of persistent bounded disturbances, *IEEE Trans. Automat. Contr.*, Vol. AC-31, No. 6, pp. 527–534, June 1986.

[45] Yaz, E., Comments on On the robust control of robot manipulators by M.W. Spong, *IEEE Trans. Automatic Control*, Vol. 38, No. 3, pp. 511–512, Mar. 1993.

[46] Yoo, B.K. and Ham, W.C., Adaptive control of robot manipulator using fuzzy compensator, *IEEE Trans. on Fuzzy Systems*, Vol. 8, No. 2, pp. 186–199, Apr. 2000.

[47] Yoshikawa, T., *Foundations of Robotics: Analysis and Control*, MIT Press, Cambridge, MA, 1990.

[48] Youla, D.C., Jabr, H.A., and Bongiorno, J.J., Modern Wiener-Hopf design of optimal controllers–Part 2: the multivariable case, *IEEE Trans. Automatic Control*, Vol. AC-21, pp. 319–338, June 1976.

[49] Zhang, F., Dawson, D.M., deQueiroz, M.S., and Dixon, W.E., Global adaptive output feedback tracking control of robot manipulators, *IEEE Trans. Automat. Contr.*, Vol. AC–45, No. 6, pp. 1203–1208, June 2000.

18
Sliding Mode Control of Robotic Manipulators

Hector M. Gutierrez
Florida Institute of Technology

18.1 Sliding Mode Controller Design–An Overview **18**-1
18.2 The Sliding Mode Formulation of the Robot Manipulator Motion Control Problem **18**-2
18.3 Equivalent Control and Chatter Free Sliding Control **18**-4
18.4 Control of Robotic Manipulators by Continuous Sliding Mode Laws **18**-6
  Sliding Mode Formulation of the Robotic Manipulator Motion Control Problem • Sliding Mode Controller Design
18.5 Conclusions **18**-8

18.1 Sliding Mode Controller Design – An Overview

Sliding mode design [1–6] has several features that make it an attractive technique to solve tracking problems in motion control of robotic manipulators, the most important being its robustness to parametric uncertainties and unmodelled dynamics, and the computational simplicity of the algorithm. One way of looking at sliding mode controller design is to think of the design process as a two-step procedure. First, a region of the state space where the system behaves as desired (sliding surface) is defined. Then, a control action that takes the system into such surface and keeps it there is to be determined. Robustness is usually achieved based on a switching control law. The design of the control action can be based on different strategies, a straightforward one being to define a condition that makes the sliding surface an attractive region for the state vector trajectories. Consider a nonlinear affine system of the form:

$$x_i^{(n_i)} = f_i(\vec{x}) + \sum_{j=1}^{m} b_{ij}(\vec{x}) u_j, \quad i = 1, \ldots, m, \quad j = 1, \ldots, m \tag{18.1}$$

where $\vec{u} = [u_1, \ldots, u_m]^T$ is the vector of m control inputs, and the state $\vec{x}$ is composed of the x_i coordinates to be tracked and their first $(n_i - 1)$ derivatives. Such systems are called square systems because they have as many control inputs u_j as outputs to be controlled x_i [3]. The motion control problem to be addressed is the one of making the state vector $\vec{x}$ track a desired trajectory $\vec{r}$. Consider first the time-varying manifold σ given by the intersection of the surfaces $s_i(\vec{x}, t) = 0, i = 1, \ldots, m$, specified by the components of

$S(\vec{x},t) = [s_1(\vec{x},t),\ldots,s_m(\vec{x},t)]^T$, where

$$s(\vec{x},t) = \left(\frac{d}{dt} + \lambda_i\right)^{n_i-1}(x_i - r_i) = 0 \qquad (18.2)$$

which can be computed from $\vec{x}$ and $\vec{r}$, λ_i being some positive constant. Such manifold is usually called *sliding surface*; any state trajectory lying on it tracks the reference $\vec{r}$ since Equation (18.2) defines a differential equation on the error vector that stably converges to zero. An integral term can be incorporated to the sliding surface to further penalize tracking error. For instance, Equation (18.2) can be rewritten for $n_i = 3$ as

$$s(\vec{x},t) = \left(\frac{d}{dt} + \lambda_i\right)^2 \left(\int_0^t (x_i - r_i)dt\right) = (\dot{x}_i - \dot{r}_i) + 2\lambda_i(x_i - r_i) + \lambda_i^2 \int_0^t (x_i - r_i)dt = 0 \qquad (18.3)$$

There are also several possible strategies to design the control action u_j that takes the system to the sliding surface. One such strategy is to find u_j in a way that each component of the sliding surface s_i is an attractive region of the state space by forcing the control action to satisfy a geometric condition such as

$$\frac{1}{2}\frac{d}{dt}(s_i^2) \leq -\eta_i|s_i| \Leftrightarrow s_i\frac{ds_i}{dt} \leq -\eta_i|s_i| \leq 0 \qquad (18.4)$$

where η_i is a strictly positive constant. This condition forces the squared distance to the surface (as measured by s_i^2) to decrease along all state trajectories [3]. In other words, all the state trajectories are constrained to point toward σ. A simple solution that satisfies the sliding condition (18.4) in spite of the uncertainty on the dynamics (18.1) is the switching control law:

$$u_{sw} = k(\vec{x},t)\mathrm{sgn}(s), \qquad \mathrm{sgn}(s) = \begin{cases} 1, & s > 0 \\ -1, & s < 0 \end{cases} \qquad (18.5)$$

The gain $k(\vec{x},t)$ of this infinitely fast switching control law increases with the extent of parametric uncertainty, introducing a discontinuity in the control signal called chatter. This is an obviously undesirable practical problem that will be discussed in Section 18.3.

18.2 The Sliding Mode Formulation of the Robot Manipulator Motion Control Problem

The sliding mode formulation of the robot manipulator motion control problem starts from the basic dynamic equations that link the input torques to the vector of robot joint coordinates:

$$H(q)\ddot{q} + g(q,t) = \tau \qquad (18.6)$$

where τ is the $n \times 1$ vector of input torques, q is the $n \times 1$ vector of joint coordinates, $H(q)$ is the $n \times n$ diagonal positive definite inertia matrix, and $g(q,t)$ is the $n \times 1$ load vector that describes all other mechanical effects acting on the robot. This basic formulation can be expanded to describe more specific motion control problems, e.g., for an ideal rigid-link rigid-joint robot with n links interacting with the environment [7]:

$$H(q)\ddot{q} + c(q,\dot{q}) + g(q) - J^T(q)F = \tau \qquad (18.7)$$

where $c(q,\dot{q})$ is an $n \times 1$ vector of centrifugal/Coriolis terms, $g(q)$ is the $n \times 1$ vector of gravitational forces, $J^T(q)$ is the $n \times m$ transpose of the robot's Jacobian matrix, and F is the $m \times 1$ vector of forces that the (m-dimensional) environment exerts in the robot end-effector. The motion control problem is

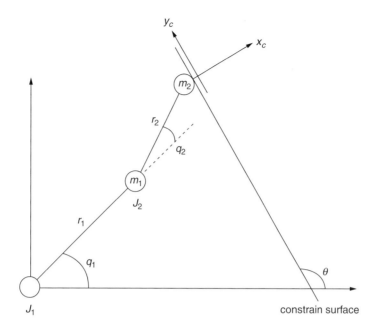

FIGURE 18.1 Rigid-link rigid-joint robot interacting with a constrain surface.

therefore either a position tracking problem in the q-state space or a force control problem or a hybrid of both.

Example 18.1

Consider the force control problem of the two-link robot depicted in Figure 18.1 [7]. A control scheme to track a desired contact force F_d is shown in Figure 18.2, where J_c is the robot's Jacobian matrix relative to the coordinate system fixed at the point of contact $\langle x_c, y_c \rangle$, T is the $n \times n$ diagonal selection matrix with elements equal to zero in the force controlled directions, F is the measured contact force, τ_{ff} is the output of the feed-forward controller, and τ_{sm} the output of the sliding mode controller (Figure 18.2).

The matrices used to estimate the torque vector are

$$H(q) = \begin{bmatrix} (m_1 + m_2)r_1^2 + m_2 r_2^2 + 2m_2 r_1 r_2 C_2 + J_1 & m_2 r_2^2 + m_2 r_1 r_2 C_2 \\ m_2 r_2^2 + m_2 r_1 r_2 C_2 & m_2 r_2^2 + J_2 \end{bmatrix} \quad (18.8)$$

$$c(q, \dot{q}) = \begin{bmatrix} -m_2 r_1 r_2 S_2 \dot{q}_2^2 - 2m_2 r_1 r_2 S_2 \dot{q}_1 \dot{q}_2 \\ m_2 r_1 r_2 S_2 \dot{q}_1^2 \end{bmatrix} \quad (18.9)$$

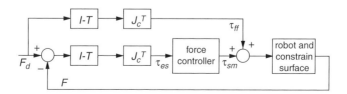

FIGURE 18.2 Force controller with feed-forward compensation.

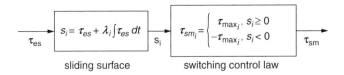

FIGURE 18.3 Sliding mode force controller.

$$g(q) = \begin{bmatrix} g(m_1 r_1 C_1 + m_2 r_1 C_1 + m_2 r_2 C_{12}) \\ g m_2 r_2 C_{12} \end{bmatrix} \quad (18.10)$$

$$J^T(q) = \begin{bmatrix} -r_1 S_1 - r_2 S_{12} & r_1 C_1 + r_2 C_{12} \\ -r_2 S_{12} & r_2 C_{12} \end{bmatrix} \quad (18.11)$$

where $C_i = \cos(q_i)$, $C_{ij} = \cos(q_i + q_j)$, $S_i = \sin(q_i)$, $S_{ij} = \sin(q_i + q_j)$.

A similar compensation scheme can be used to solve the position tracking problem.

18.3 Equivalent Control and Chatter Free Sliding Control

Sliding mode control comprises two synergistic actions: reaching the sliding surface and letting the state vector ride on it. These are represented as two different components of the control action: a switching term (u_{sw}) and a so-called equivalent control (u_{eq}). The robustness of sliding control is given by the switching term, u_{sw}, derived from the sliding condition (18.4). This term takes the system from its arbitrary set of initial conditions to the sliding surface and accommodates parametric uncertainties and unmodelled dynamics by virtue of the switching action [3, 4, 5]. A purely switching controller can only provide high performance at the expense of very fast switching. While in theory infinitely fast switching can provide asymptotically perfect tracking, in practice it implies a trade-off between tracking performance and control signal chatter, which is either undesirable due to noise considerations or impossible to achieve with real actuators.

After sliding has been achieved, the system dynamics match that of the sliding surface. Equivalent control (u_{eq}) is the control action that drives the system once sliding has been achieved and is calculated from solving $dS/dt = 0$. It is a nonrobust control law because it assumes the plant model to be exact, but does not generate control chatter because it is a continuous function [5]. In systems that are affine in the input, such as the one described by Equation (18.1), both actions can be combined into a single control law ($u = u_{eq} + u_{sw}$), and several possible switching terms can be derived by forcing the compound control, u, to satisfy the sliding condition [5]. An equivalent control term improves tracking performance while reducing chatter, alleviating the performance trade-off implied by the switching gain $k(\vec{x}, t)$ in (18.5).

Several different methods have been proposed to deal with control chatter. Two basic approaches can be distinguished.

1. The boundary layer method, which essentially consists of switching the control action in finite time as the system's trajectory reaches an ε-vicinity of the sliding surface.
2. The equivalent control method, which consists of finding a continuous function that satisfies the sliding condition (18.4). This is particularly important in systems where the control action is continuous in nature, such as motion control systems where torque, force, or current are treated as the control input.

A boundary layer [3, 5] is defined as

$$b_L(t) = \{\vec{x}, \; s(\vec{x}, t)| \leq \Phi\} \quad (18.12)$$

where $\Phi > 0$ is the boundary layer thickness. $\varepsilon = \Phi/\lambda^{n-1}$ is the corresponding boundary layer width, where λ is the design parameter of the sliding surface (18.2) for the case $\lambda_1 = \lambda_2 = \cdots = \lambda_n$. The switching control law (18.5) remains the same outside the boundary layer, and inside $b_L(t)$ the control action is interpolated by using, e.g., $u_{sw} = s/\Phi$, providing an overall piece-wise continuous control law. This creates an upper bound in tracking error given by λ (as opposed to perfect tracking): $\forall t \geq 0, |\vec{x}^{(i)}(t) - \vec{r}^{(i)}(t)| \leq (2\lambda)^i \varepsilon; i = 0, \ldots, n-1$, for all trajectories starting inside $b_L(t)$. The boundary layer thickness Φ can be made time-varying to tune up the control law to exploit the control bandwidth available, and in that case the sliding condition (18.4) becomes

$$|s(\vec{x},t)| \geq \Phi \Rightarrow \frac{1}{2}\frac{d}{dt}s^2 \leq (\dot{\Phi} - \eta)|s(\vec{x},t)| \tag{18.13}$$

A simple switching term that satisfies (18.13) is

$$u_{sw} = (k(\vec{x},t) - \dot{\Phi})\,sat\left(\frac{s}{\Phi}\right), \quad sat(z) = \begin{cases} z, & |z| \leq 1 \\ sgn(z), & \text{otherwise} \end{cases} \tag{18.14}$$

which replaces Equation (18.5). This method eliminates control chatter provided that high-frequency unmodelled dynamics are not excited [3] and that the corresponding trade-off in tracking accuracy is acceptable.

The equivalent control (u_{eq}) [3, 5, 9] is the continuous control law that would maintain $dS/dt = 0$ if the dynamics were exactly known. Consider the nonlinear affine system (18.15) and (18.16) with the associated sliding surface (18.17):

$$\dot{\vec{x}}_1 = \vec{f}_1(\vec{x}_1, \vec{x}_2) \tag{18.15}$$

$$\dot{\vec{x}}_2 = \vec{f}_2(\vec{x}_1, \vec{x}_2) + B_2(\vec{x})\vec{u} + B_2(\vec{x})\vec{d}(t) \tag{18.16}$$

$$\vec{S} = \{\vec{x} : \vec{\varphi}(t) - \vec{s}_a(\vec{x}) = \vec{s}(\vec{x},t) = 0\} \tag{18.17}$$

where $\vec{u}$ is the $m \times 1$ vector of control inputs, $\vec{d}$ is the $m \times 1$ vector of input disturbances, $\vec{x}_2$ is the vector of m states to be controlled, $\vec{x}_1$ is the $(n-m)$ vector of autonomous states, $\vec{f}$ is the $n \times 1$ vector of system's equations, $B_2(\vec{x})$ is a $m \times m$ input gain matrix, $\vec{s}_a(\vec{x})$ is a $m \times 1$ continuous function of the states to be tracked, and $\vec{\varphi}(t)$ is the $m \times 1$ vector of desired trajectories. The equivalent control is obtained from

$$\frac{d}{dt}(\vec{\varphi}(t) - \vec{s}_a(\vec{x}, \vec{u} = \vec{u}_{eq})) = 0 \tag{18.18}$$

by calculating $dS/dt = 0$ from Equation (18.18), replacing the system's Equation (18.15) and Equation (18.16) in the resulting expression, and finally solving for u_{eq}. This yields

$$\vec{u}_{eq} = -\vec{d} + (G_2 B_2)^{-1}\left(\frac{d\vec{\varphi}}{dt} - G_2 \vec{f}_2 - G_1 \vec{f}_1\right) \tag{18.19}$$

where $d\vec{s}_a/dt = G_1 \dot{\vec{x}}_1 + G_2 \dot{\vec{x}}_2$ and G_1, G_2 are defined as $[\partial \vec{s}_a/\partial \vec{x}_1] = G_1, [\partial \vec{s}_a/\partial \vec{x}_2] = G_2$.

The continuous control (18.19) is a nonrobust control law because it assumes the plant model to be exact. Several different techniques have been proposed to achieve robustness against parametric variations and unmodelled dynamics (disturbances) [6,9] by a *continuous* control law (as opposed to a switching control action such as Equation (18.5)), which is obviously essential in systems where the control inputs are continuous functions and hence Equation (18.5) cannot be realized. One such technique [9] is based on

- Given the Lyapunov function candidate $v = S^T S/2$, if the Lyapunov stability criteria are satisfied, the solution $\vec{\varphi}(t) - s_a(\vec{x}) = 0$ is stable for all possible trajectories of systems (18.15) and (18.16).
- The problem then becomes that of finding the control u that satisfies the Lyapunov condition $\frac{dv}{dt} = -S^T DS \leq 0$ for some positive definite matrix D.

From the sliding surface (18.17) where $\vec{s}_a(\vec{x})$ is defined as $\vec{s}_a(\vec{x}) = \vec{x}_2 + G\vec{x}_1$ and replacing $dS/dt = \dot{\varphi} - \vec{f}_2 - B_2\vec{u} - B_2\vec{d} - G\vec{f}_1$ in the Lyapunov condition $dv/dt = -S^T DS \leq 0$, the following control law is obtained

$$\vec{u} = sat\left[-\vec{d} + B_2^{-1}\left(\frac{d\vec{\varphi}}{dt} - \vec{f}_2 - G\vec{f}_1\right) + B_2^{-1}D\vec{s}\right] = sat\left[\vec{u}_{eq} + B_2^{-1}D\vec{s}\right] \quad (18.20)$$

which is a continuous control law and chatter is therefore eliminated. Because Equation (18.20) requires knowledge of the equivalent control law (18.19), this solution is not directly useful. After some algebra, and considering that $B_2\vec{u}_{eq} = B_2\vec{u} + d\vec{s}/dt$, it can be shown that Equation (18.19) and Equation (18.20) imply

$$\vec{u}(t) = sat\left[\vec{u}(t - \Delta) + B_2^{-1}\left(D\vec{s} + \frac{d\vec{s}}{dt}\right)\right], \quad \Delta \to 0 \quad (18.21)$$

i.e., the control u at instant t is calculated from the value at time $(t - \Delta)$. The control law (18.21) is continuous everywhere, except at the discontinuities of $\vec{s}(\vec{x}, t)$.

18.4 Control of Robotic Manipulators by Continuous Sliding Mode Laws

The technique described in Section 18.3 to achieve robust sliding control of a nonlinear plant using a continuous control law will be used here to propose solutions to the most common motion control problems in robotic manipulators, namely, the trajectory tracking, force control, and impedance control problems.

A robotic manipulator actuated by electrical machines can be described by the nonlinear Equation (18.6), where the input torque is given by

$$\vec{\tau} = K_t(\vec{i}_d)\vec{i}_q \quad (18.22)$$

where $\vec{i}_q$ is the $n \times 1$ vector of input currents (control input), $\vec{i}_d$ the $n \times 1$ vector of current disturbances, and $K_t(\vec{i}_d)$ a $n \times n$ diagonal positive matrix. The load vector in Equation (18.6) is

$$g(q, t) = -K_t(\vec{i}_d)\vec{i}_L \quad (18.23)$$

where $\vec{i}_L$ is the $n \times 1$ vector of currents associated to the mechanical load acting on the manipulator. The joint space and work space descriptions can be written as follows:

$$\frac{d^2\vec{z}}{dt^2} = \vec{f}(\vec{z}, t) + B(\vec{i}_d)\vec{i}_q - B(\vec{i}_d)\vec{i}_L \quad (18.24)$$

where the state vector $\vec{z}$ is defined as $\vec{z}^T = [q_1, \ldots, q_n]$, $B = J^{-1}K_T$ with inertia matrix J (for the joint space formulation), and as $\vec{z}^T = [x_1, \ldots, x_n]$, $B = J_0 J^{-1} K_T$, with J_0 the Jacobian defined by the direct kinematics $\vec{x} = \vec{F}(\vec{q})$; $J_0 = [\frac{\partial \vec{F}(\vec{q})}{\partial \vec{q}}]$ (for the work space formulation). The model (18.24) is referred to as the "reduced order dynamics" if the control input is a vector of motor currents, and as "full order dynamics" if the control input is motor voltages (i.e., motor currents are considered internal states of the system) [9].

18.4.1 Sliding Mode Formulation of the Robotic Manipulator Motion Control Problem

The control goal is to find the vector of input currents that will force the nonlinear function of the states $\vec{s}_a(\vec{z}, \dot{\vec{z}})$ to track the reference function $\vec{\varphi}(t)$ or, in other words, that will force the system states to remain

Sliding Mode Control of Robotic Manipulators

TABLE 18.1 Sliding Mode Manifolds for Robotic Manipulator Motion Control

	Joint Space	Work Space
Position tracking	$\vec{s}_a = C_1 \vec{q} + \dot{\vec{q}}$	$s_a = C_1 \vec{x} + \dot{\vec{x}}$
	$\vec{\varphi} = C_1 \vec{q}^{ref} + \dot{\vec{q}}^{ref}$	$\vec{\varphi} = C_1 \vec{r} + \dot{\vec{r}}$
	$C_1 > 0$	$C_1 > 0$
Impedance control	$\vec{s}_a = D\dot{\vec{q}} + K\vec{q} + M\ddot{\vec{q}}$	$\vec{s}_a = D\dot{\vec{x}} + K\vec{x} + M\ddot{\vec{x}}$
	$\vec{\varphi} = \vec{F}^{ext}$	$\vec{\varphi} = \vec{F}^{ext}$
	$M, K, D > 0$	$M, K, D > 0$
Force control	$\vec{s}_a = \vec{F}^{ext}$	$\vec{s}_a = \vec{F}^{ext}$
	$\vec{\varphi} = \vec{F}^{ref}$	$\vec{\varphi} = \vec{F}^{ref}$

on the smooth manifold:

$$S = \left\{ \vec{z}, \frac{d\vec{z}}{dt} : \quad \vec{\varphi}(t) - \vec{s}_a(\vec{z}, \dot{\vec{z}}) = \vec{s}(\vec{z}, t) = \vec{0} \right\} \tag{18.25}$$

Simple examples of sliding mode manifolds for the trajectory tracking, force control, and impedance control problems are shown in Table 18.1 [9].

18.4.2 Sliding Mode Controller Design

Controller design is based on the dynamic equations of the manipulator (18.24), the sliding mode manifold (as those presented in Table 18.1), and the sliding mode control algorithm (18.21). The output of the algorithm is $\vec{i}_q^{ref}$, the vector of reference currents to be used by the motor current controller:

$$\vec{i}_q^{ref} = sat\left[\vec{i}_L + (GB)^{-1}\left(D\vec{s} + \frac{d\vec{\varphi}}{dt} - G\vec{f}\right)\right], \quad G = \left[\frac{\partial \vec{s}}{\partial \vec{z}}\right] \tag{18.26}$$

which in terms of the practical form (18.21) becomes

$$\vec{i}_q^{ref} = sat\left[\vec{i}_q + (GB)^{-1}\left(D\vec{s} + \frac{d\vec{s}}{dt}\right)\right] \tag{18.27}$$

For the trajectory tracking problem in the joint space, the sliding surface becomes

$$\vec{S} = \vec{\varphi} - \vec{s}_a = C_1(\vec{q}^{ref} - \vec{q}) + (\dot{\vec{q}}^{ref} - \dot{\vec{q}}), \quad C_1 > 0 \tag{18.28}$$

and the corresponding control current

$$\vec{i}_q^{ref} = sat\left[\vec{i}_q + K_T^{-1} J\left(D\vec{S} + \frac{d\vec{S}}{dt}\right)\right] \tag{18.29}$$

Similarly, for trajectory tracking in the work space,

$$\vec{S} = \vec{\varphi} - \vec{s}_a = C_1(\vec{r} - \vec{x}) + (\dot{\vec{r}} - \dot{\vec{x}}), \quad C_1 > 0 \tag{18.30}$$

$$\vec{i}_q^{ref} = sat\left[\vec{i}_q + K_T^{-1} J J_0^{-1}\left(D\vec{S} + \frac{d\vec{S}}{dt}\right)\right] \tag{18.31}$$

Notice that the proposed algorithms (18.29) and (18.31) are rather simple and do not require estimation of either acceleration or disturbance for their implementation.

The mechanical impedance of a robotic manipulator is defined as a linear combination of its state coordinates. A corresponding sliding mode manifold can be defined as

$$\vec{S} = \{\vec{z}, \dot{\vec{z}}, \ddot{\vec{z}};\quad F^{ext} - (D\vec{x} + K\dot{\vec{x}} + M\ddot{\vec{x}}) = \vec{s}(\vec{x}, t) = \vec{0}\} \tag{18.32}$$

The corresponding control current can be obtained by solving $\vec{s}(\vec{x}, t) = \vec{0}$:

$$\vec{i}_q^{ref} = sat[\vec{i}_q + (MJ_0^{-1}K_T)^{-1}(D\vec{x} + K\dot{\vec{x}} + M\ddot{\vec{x}} - F^{ext})] \tag{18.33}$$

The same concept can be easily extended to the force control problem (whether in full order or reduced order formulation). A mathematical model of the contact force such as the one described in [13] will be necessary.

18.5 Conclusions

This chapter has introduced the basic concepts and formulations involved in the use of sliding mode control techniques for robotic manipulators. Sliding mode controller design has been extensively used in robotics due to its remarkable robustness to parametric uncertainties and unmodelled dynamics. Several other aspects of the use of sliding mode for motion control of robotic manipulators have been extensively discussed in the literature, such as the use of nonlinear observers to estimate unmeasurable states [10], model reference adaptive control in variable structure systems [12], perturbation estimation [8], and path following [11].

References

[1] Slotine, J.J., Sliding controller design for nonlinear systems, *Int. J. Control*, Vol. 40, No. 2, pp. 421–434, 1984.
[2] Slotine, J.J. and Coetsee, J.A., Adaptive sliding controller synthesis for nonlinear systems, *Int. J. Control*, Vol. 43, No. 6, 1986.
[3] Slotine, J.J. and Li, W., *Applied Nonlinear Control*, Prentice Hall, Englewood Cliffs, NJ, 1987.
[4] Utkin, V.I., Variable structure systems with sliding modes, *IEEE Trans. Autom. Control*, Vol. 22, No. 2, pp. 212–222, 1977.
[5] De Carlo, R.A., Zak, S.H., and Drakunov, S.V., Variable structure, sliding mode controller design, *IEEE Control Engineering Handbook*, pp. 941–951, CRC Press, Boca Raton, FL, 1995.
[6] Olgac, N. and Chang, J., Constrained sliding mode control for nonlinear dynamic systems without chatter, *Proc. 32nd IEEE Conference on Decision and Control*, San Antonio, TX, pp. 414–415, 1993.
[7] Azenha, A. and Machado, J.A.T., Variable structure position/force hybrid control of manipulators, *Proc. 4th IFAC Workshop on Algorithms and Architectures for Real-Time Control*, Vilamoura, Portugal, pp. 347–352, 1997.
[8] Curk, B. and Jezernik, K., Sliding mode control with perturbation estimation: application on DD robot mechanism, *Robotica*, Vol. 19, No. 6, pp. 641–648, 2000.
[9] Sabanovic, A., Jezernik, K., and Wada, K., Chattering-free sliding modes in robotic manipulator control, *Robotica*, Vol. 14, pp. 17–29, 1996.
[10] Jezernik, K., Curk, B., and Harnik, J., Observer-based sliding mode control of a robotic manipulator, *Robotica*, Vol. 12, pp. 443–448, 1994.
[11] Dagci, O.H., Ogras, U.Y., and Ozguner, U., Path following controller design using sliding mode control theory, *Proc. 2003 American Control Conference*, Denver, CO, Vol. 1, pp. 903–908, June 2003.

[12] Tso, S.K., Xu, Y., and Shum, H.Y., Variable structure model reference adaptive control of robot manipulators, *Proc. 1991 IEEE International Conference on Robotics and Automation,* Sacramento, CA, Vol. 3, pp. 2148–2153, April 1991.

[13] Arai, F., Fukuda,T., Shoi, W., and Wada, H., Models of mechanical systems for controller design, *Proc. IFAC Workshop on Motion Control for Intelligent Automation,* Perugia, Italy, pp. 13–19, 1992.

19

Impedance and Interaction Control

19.1 Introduction: Controlling Mechanical Interaction **19**-1
   The Effect of Interaction on Performance and Stability •
   Interaction as Disturbance Rejection • Interaction as Modeling
   Uncertainty

19.2 Regulating Dynamic Behavior **19**-5
   Mechanical Impedance and Admittance • Port Behavior and
   Transfer Functions

19.3 Analyzing Coupled Systems **19**-8
   Causal Analysis of Interaction Port Connection • Impedance
   vs. Admittance Regulation • Coupled System Stability Analysis
   • Passivity and Coupled Stability

19.4 Implementing Interaction Control **19**-14
   Virtual Trajectory and Nodic Impedance • "Simple"
   Impedance Control • Direct Impedance Modulation

19.5 Improving Low-Impedance Performance **19**-18
   Force Feedback • Natural Admittance Control • Series
   Dynamics

19.6 More Advanced Methods **19**-22

19.7 Conclusion: Interaction Control, Force Control,
   and Motion Control **19**-23

Neville Hogan
Massachusetts Institute of Technology

Stephen P. Buerger
Massachusetts Institute of Technology

19.1 Introduction: Controlling Mechanical Interaction

Mechanical interaction with objects is arguably one of the fundamentally important robot behaviors. Many current robot applications require it; for example, mechanical interaction is essential for manipulation, the core task of assembly systems. Future robot applications, such as versatile use of tools or close cooperation with humans, may be enabled by improved control of mechanical interaction.

Interaction with the environment may serve sensory or motor functions (or both), and the most appropriate mechanical interaction is different for sensory or motor tasks. Mechanical interaction dynamics may be characterized by mechanical impedance, which may loosely be considered a dynamic extension of stiffness.[1] Lower mechanical impedance reduces interaction forces due to encountering an unpredicted object, thereby protecting both the robot and any object it manipulates (interaction forces on each being

---

[1]Conversely, mechanical admittance may be considered a dynamic generalization of compliance. A more rigorous definition is provided below.

opposite but equal). Using a human analogy, by this reasoning, tactile exploration and manipulation of fragile objects should evoke the use of our lowest-impedance limb segments, and while we can (and routinely do) interact with objects using other body parts (the elbow, the knee, the foot, etc.), we naturally tend to use our fingers for gentle, delicate tasks.

Conversely, wielding an object as a tool often requires it to be stabilized and that requires higher mechanical impedance. This is especially important if the interaction between the manipulator and the object is destabilizing, as is the case for many common tools. Again using a human analogy, consider, for example, the simple task of pushing on a surface with a rigid stick. If force is exerted on the stick normal to the surface, then the stick is statically unstable; small displacements from the configuration in which stick axis and force vector co-align result in torques that act to drive the stick further from that configuration. Success at this task requires a stabilizing mechanical impedance, and because pushing harder exacerbates the problem (the magnitude of the destabilizing torque is proportional to the applied force), the minimum required impedance grows with the force applied; see Rancourt and Hogan (2001) for a detailed analysis. Simple though this task may be, it is an essential element of the function of many tools (e.g., screwdrivers, power drills) and any manipulator — human or robotic — must provide a stabilizing mechanical impedance to operate them.

In other applications the robot's interactive behavior may be the main objective of control. For example, to use a robot to serve as a force-reflecting haptic display (Miller et al., 2000) or to deliver physiotherapy (Volpe et al., 2000) requires intimate physical interaction with humans. In these applications the "feel" of the robot becomes an important performance measure, and "feel" is determined by mechanical interaction dynamics. Versatile interaction with objects (whether tools or humans or other robots), therefore, requires an ability to modulate and control the dynamics of interaction.

19.1.1 The Effect of Interaction on Performance and Stability

When interaction occurs, the dynamic properties of the environment are important. If we attempt to control motion or force, interaction affects the controlled variable, introducing error upon which the controller must act. Errors clearly mar performance but even more importantly, though perhaps less obvious, stability may also be compromised. That is, a system that is stable in isolation can become unstable when coupled to an environment that is itself stable. Such instabilities, known as coupled or contact instabilities, appear even in simple systems with basic controllers contacting simple environments. To illustrate this key point consider the following examples.

Example 19.1

The first example is based on a common design for a robot motion controller. One of the simplest models of a robot is depicted in Figure 19.1. A single mass m represents the robot's inertial properties. It is subject

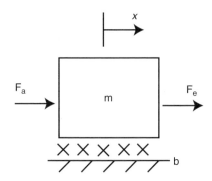

FIGURE 19.1 Model with inertial, frictional, actuator, and environmental forces.

to actuator forces F_a and environmental forces F_e. A single damper b connected to ground represents frictional losses. The Laplace-transformed equation of motion for this simple model is as follows:

$$(ms^2 + bs)X = F_a + F_e \tag{19.1}$$

where X is the Laplace transform of the mass position x. A proportional-integral motion controller is applied

$$F_a = K_p(R - X) + \frac{K_I}{s}(R - X) \tag{19.2}$$

where R is the Laplace transform of the reference position and K_p and K_I are proportional and integral gains, respectively. In isolation, $F_e = 0$ and the closed-loop transfer function is

$$\frac{X}{R} = \frac{K_p s + K_I}{ms^3 + bs^2 + K_p s + K_I} \tag{19.3}$$

From the Routh-Hurwitz stability criterion, a condition for *isolated* stability is the following upper bound on the integral gain:

$$K_I < \frac{bK_p}{m} \tag{19.4}$$

However, this condition is not sufficient to ensure that the robot will remain stable when it interacts with objects in its environment. Even the simple act of grasping an object may be destabilizing. If the system is not isolated, and instead is coupled to a mass m_e, this is equivalent to increasing the mass from m to $(m + m_e)$. Hence a condition for stability of the coupled system is

$$K_I < \frac{bK_p}{(m + m_e)} \tag{19.5}$$

Any controller that satisfies Equation (19.4) results in a stable isolated system. Prudent controller design would use a lower integral gain than the marginal value, providing robustness to uncertainty about system parameters and perhaps improved performance. However, for any fixed controller gains, coupling the robot to a sufficiently large m_e will violate the condition for stability. Interaction with a large enough mass can always destabilize a system that includes integral-action motion control, even if that system is stable in isolation.

This difficulty may be obscured by the facts that, at present, most robots have inertia far exceeding that of the payloads they may carry, the total mass $m + m_e$ is not much greater than the mass m of the robot alone, and the bounds of Equation (19.4) and Equation (19.5) are similar. However, in some applications, e.g., in space or under water, a robot need support little or none of a payload's weight, and objects of extremely large mass may be manipulated. In these situations the vulnerability of integral-action motion control to coupled instability may become an important consideration.

Example 19.2

The second example illustrates one of the common difficulties of force control. A robot with a single structural vibration mode is represented by the model shown in Figure 19.2. Two masses m_1 and m_2 are connected by a spring of stiffness k. Frictional losses are represented by dampers b_1 and b_2 connected from the masses to ground and damper b_3 in parallel with the spring. One mass is driven by the actuator force F_a and the other is subject to F_e, an interaction force with the environment.

A proportional-derivative (PD) controller acting on the error between the position of the mass at the actuator x_1 and a reference r is applied to control motion. A proportional controller acting on force fed back from the environment is applied to control force and improve interactive behavior. The control law

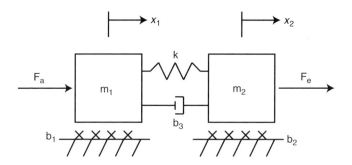

FIGURE 19.2 Model including a single structural resonance.

is as follows:

$$F_a = K(r - x_1) + B(\dot{r} - \dot{x}_1) + K_f(F_e + K(r - x_1) + B(\dot{r} - \dot{x}_1)) \qquad (19.6)$$

where K and B are related to the proportional and derivative motion feedback gains, respectively, and K_f is the proportional force feedback gain.[2] When the system is isolated from any environment, $F_e = 0$ and Equation (19.6) reduces to PD motion control. Using the Routh-Hurwitz criterion, the isolated system's closed-loop characteristic polynomial can be shown to have all its poles in the open left half plane, providing at worst marginal stability for arbitrary nonnegative real parameters and controller gains.

If the system is not isolated but instead is connected to a spring to ground, such that $F_e = -k_e x_2$, the closed loop characteristic polynomial is changed. It is now easy to find parameters such that this polynomial has right-half-plane roots. For example, if $m_1 = m_2 = 10$, $b_1 = b_2 = b_3 = 1$, $k = 100$, $K = 10$, $B = 1$, $K_f = 10$, and $k_e = 100$, the closed loop poles are at $-2.78 \pm 5.51i$ and $2.03 \pm 5.53i$, where the latter pair are unstable. Once again, a system that is stable when isolated (corresponding to $k_e = 0$) is driven to instability when coupled to an object in its environment, in this case, a sufficiently stiff spring ($k_e = 100$).

Structural vibrations are a common feature of robot dynamics and present a substantial challenge to controller design. Though simple, the PD motion controller in this example has the merit that it is not destabilized by structural vibration because the actuator and assumed motion sensor are co-located.[3] However, the addition of a force feedback loop renders the robot control system vulnerable to coupled instability. In part this is because the actuator and force sensor are not co-located. Given the difficulty of designing a robot with no significant dynamics interposed between its actuators and the points at which it contacts its environment (see Tilley and Cannon, 1986, Sharon et al., 1988), the simple model in this example represents a common situation in robot interaction control, and we will return to this model several times in this chapter.

These two examples show that to ensure stability when interacting even with simple environments, it is not sufficient to design for stability of the manipulator in isolation. The environment's dynamics must also be considered — dynamics that are generally not known exactly. Furthermore, in many applications a robot must be capable of stable interaction with a variety of environments. For example, an assembly robot might pick up a component, move it across free space, bring it into contact with a kinematic constraint used to guide placement of the component, move it along the constraint, release the component, and return to get another. In this case, at different times in the process, the robot must remain stable when moving unloaded in free space, moving to transport the component, and moving to comply with the kinematic constraint. Each of these three contexts poses a different stability challenge which illustrates one

---

[2] This controller is discussed further below in the section on force feedback.
[3] An alternative interpretation is presented below in the section on simple impedance control.

key reason why, even if the environment can be closely modeled, there are strong benefits to designing a controller that is insensitive to ignorance of its properties. If, for example, a different controller is required for each of the three contexts, the system must decide which controller to use at each instant and manage the transition between them. Usually the easiest way to identify the environment is to interact with it; thus, interaction might be required before the appropriate controller is in place. Stable interaction is needed to identify the environment, but stability cannot be guaranteed without a well-characterized environment. A *single* controller that can perform satisfactorily within all expected contexts without compromising stability would have significant advantages.

19.1.2 Interaction as Disturbance Rejection

Control theory offers several tools to endow controllers with the ability to deal with unknown or poorly characterized interference (Levine, 1996). Using a disturbance rejection approach, the environment's dynamics could be included as disturbance forces. The success of this approach depends on bounding the disturbance forces, but for many interactive applications the environmental forces may equal or exceed the robot's nominal capacity; for example, a kinematic constraint can introduce arbitrarily large forces depending on the robot's own behavior. Furthermore, environmental forces generally depend on the robot's state, and traditionally disturbances are assumed to be state-independent. Thus, treating interaction as a disturbance rejection problem does not seem promising.

19.1.3 Interaction as Modeling Uncertainty

Modeling the environment as an uncertain part of the robot and using robust control tools (Levine, 1996) to guarantee stability is another reasonable approach. Interacting with an environment effectively changes the robot plant by adding some combination of elastic, dissipative, and inertial properties, perhaps including kinematic constraints. If the effect of interaction is only to alter the parameters of the robot model (e.g., by adding to the endpoint mass) a robust control approach may succeed, though robustifying the system to a large range of environment parameters might require an unacceptable sacrifice of performance. However, interaction may change the *structure* of the model. For example, it may reduce the order of the system (e.g., moving from free motion to contact with a kinematic constraint reduces the degrees of freedom of the parts contacting the constraint) or increase it (e.g., contact with an elastic object adds a new mode of behavior due to interaction between the robot inertia and the object elasticity). If interaction changes the model structure, the applicability of a robust control approach is unclear. For example, the environment forces and motions may be of the same magnitude and in the same frequency range as the known robot dynamics.[4] As robust control methods commonly assume that uncertain dynamics lie outside the frequency range of interest, treating interaction as a robustness problem does not seem promising.

As mentioned above, in some applications the robot's dynamic behavior (colloquially termed its "feel") may be a key performance measure and hence a legitimate objective of control system design. Rather than attempt to overwhelm the consequences of interaction by building in robustness, the approach described herein aims to regulate the interaction itself by controlling the robot's dynamic behavior at the places where it interacts with its environment.

19.2 Regulating Dynamic Behavior

The general approach here termed "interaction control" refers to regulation of the robot's dynamic behavior at its *ports of interaction* with the environment. An interaction port is a place at which energy may be exchanged with the environment. It is therefore defined by a set of motion and force variables such that

---

[4]To use a human example, consider two people shaking hands: each has a similar response speed or bandwidth and produces comparable force and motion; each individual control system is coupled to a system with approximately double the large number of degrees of freedom that it controls when isolated.

TABLE 19.1 Laplace-Transformed Impedance and Admittance Functions for Common Mechanical Elements

Element	Impedance $F/\dot{x}$	Admittance $\dot{x}/F$
Spring k	$\dfrac{k}{s}$	$\dfrac{s}{k}$
Mass m	ms	$\dfrac{1}{ms}$
Damper b	b	$\dfrac{1}{b}$

$P = F^t v$, where v is a vector of velocities (or angular velocities) along different degrees of freedom at the contact point, F is a corresponding vector of forces (or torques), and P is the power that flows between robot and environment.[5] Interaction control involves specifying a dynamic relationship between motion and force at the port, and implementing a control law that attempts to minimize deviation from this relationship.

19.2.1 Mechanical Impedance and Admittance

The most common forms of interaction control regulate the manipulator's impedance or admittance. Assuming force is analogous to voltage and velocity is analogous to current, mechanical impedance is analogous to electrical impedance, characterized by conjugate variables that define power flow, and defined as follows.

Definition 19.1 Mechanical impedance at a port (denoted Z) is a dynamic operator that determines an output force (torque) time function from an input velocity (angular velocity) time function at the same port. Mechanical admittance at a port (denoted Y) is a dynamic operator that determines an output velocity (angular velocity) time function from a force (torque) time function at the same port.

The terms "driving point impedance" or "driving point admittance" are also commonly used. Because they are defined with reference to an interaction port, impedance and admittance are generically referred to as "port functions."

If the system is linear, admittance is the inverse of impedance, and both can be represented in the Laplace domain by transfer functions, $Z(s)$ or $Y(s)$. Table 19.1 gives example admittances and impedances for common mechanical elements. For a linear system impedance can be derived from the Laplace-transformed dynamic equations by solving for the appropriate variables.

While most often introduced for linear systems, impedance and admittance generalize to nonlinear systems. In the nonlinear case, using a state-determined representation of system dynamics, mechanical impedance may be described by state and output equations relating input velocity (angular velocity) to output force (torque) as follows.

---

[5]Equivalently, an interaction port may be defined in terms of "virtual work," such that $dW = F^t dx$ where dx is a vector of infinitesimal displacements (or angular displacements) along different degrees of freedom at the contact point, F is a corresponding vector of forces (or torques), and dW is the infinitesimal virtual work done on or by the robot.

Impedance and Interaction Control

$$\dot{z} = Z_s(z, v)$$
$$F = Z_o(z, v) \quad (19.7)$$
$$P = F^t v$$

where z is a finite-dimensional vector of state variables and $Z_s(\ )$ and $Z_o(\ )$ are algebraic (or memoryless) functions. The only difference from a general state-determined system model is that the input velocity and output force must have the same dimension and define by their inner product the power flow into (or out of) the system.

In the nonlinear case, mechanical admittance is the causal[6] dual of impedance in that the roles of input and output are exchanged; mechanical admittance may be described by state and output equations relating input force (torque) to output velocity (angular velocity). However, mechanical admittance may not be the inverse of mechanical impedance (and *vice versa*) as the required inverse may not be definable. Some examples of nonlinear impedances and admittances with no defined inverse forms are presented by Hogan (1985).

A key point to note is that, unlike motion or force, the dynamic port behavior is *exclusively a property of the robot system, independent of the environment* at that port. This is what gives regulating impedance or admittance its appeal. Motion and force depend on both robot and environment as they meet at an interaction port and cannot be described or predicted in the absence of a complete characterization of both systems. Indeed, this is the principal difficulty, as illustrated above, in regulating either of these quantities. Impedance, on the other hand, can (in theory) be held constant regardless of the environment; impedance defines the relationship between the power variables and does not by itself determine either.

Impedance serves to completely describe how the manipulator will interact with a variety of environments. In principle, if arbitrary impedance can be achieved, arbitrary behavior can be achieved; it remains only to sculpt the impedance to yield the desired behavior. Of course, as with all controller designs, the goal of achieving a desired impedance is an ideal; in practice, it can be difficult. The problem of achieving a desired impedance (or admittance) is central to the study of interaction control and is discussed in Section 19.4 and Section 19.5.

19.2.2 Port Behavior and Transfer Functions

Port impedance and admittance are just two of the possible representations of a linear system's response, and it is important to highlight the distinction between these expressions and general input/output transfer functions. By way of example, consider again the system in Figure 19.2. This system is an example of a "2-port" because it has two power interfaces, one characterized by F_a and $\dot{x}_1$, the other by F_e and $\dot{x}_2$. If such an element is part of a robot, one side is generally connected to an actuator and the other to a downstream portion of the robot or directly to the environment. Any mechanical 2-port has four transfer functions relating motion to force. Two of them, $\frac{F_a}{\dot{X}_2}(s)$ and $\frac{F_e}{\dot{X}_1}(s)$ (or their inverses), are input/output transfer functions and define the force produced by motion at the opposite port, assuming certain boundary conditions to determine the other power variables. This type of transfer function is used in traditional block-diagram analysis; if the left port is driven by a controllable force and the right uncoupled, then $\frac{\dot{X}_2}{F_a}(s)$ describes the motion of the right port as a result of actuator force, as shown in Figure 19.3. The other two transfer functions, $\frac{F_a}{\dot{X}_1}(s)$ and $\frac{F_e}{\dot{X}_2}(s)$, each represent the impedance at a port (and their inverses the corresponding admittance), depending on the boundary conditions at the other port.

The principal distinction between these two types of transfer functions is in their connection. If we have two systems like that shown in Figure 19.2, their input/output transfer functions cannot be properly

---

[6]The causal form (or *causality*) of a system's interaction behavior refers to the choice of input and output variables. By convention, input variables are referred to as "causing" output variables, though the physical system equations need not describe a causal relation in the usual sense of a temporal order; changes of input and output variables may occur simultaneously.

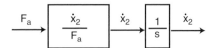

FIGURE 19.3 Block diagram of a forward-path or transmission transfer function.

connected in cascade because the dynamics of the second system affect the input-output relationship of the first. At best, cascade combination of input/output transfer functions may be used as an approximation if the two systems satisfy certain conditions, specifically that the impedance (or admittance) at the common interaction port are maximally mismatched to ensure no significant loading of one system by the other.

19.3 Analyzing Coupled Systems

To analyze systems that are coupled by interaction ports, it is helpful to consider the causality of their interaction behavior, that is, the choice of input and output for each of the systems. The appropriate causality for each system is constrained by the nature of the connection between systems, and in turn affects the mathematical representation of the function that describes the system. Power-based network modeling approaches (such as bond graphs [Brown, 2001]) are useful, though not essential, to understand this important topic.

19.3.1 Causal Analysis of Interaction Port Connection

Figure 19.4 uses bond graph notation to depict direct connection of two mechanical systems, S_1 and S_2, such that their interaction ports have the same velocity (the most common connection in robotic systems), denoted by the **1** in the figure. If we choose to describe the interaction dynamics of one system as an impedance, causal analysis dictates that the other system must have admittance causality. The causal stroke indicates that S_1 is in impedance causality; in other words, S_1 takes motion as an input and produces force output, and S_2 takes force as an input and produces motion output. If we use the common sign convention that power flow is positive into each system (denoted by the half arrows in the figure) this is equivalent to a negative feedback connection of the two power ports, as shown in Figure 19.5.

More complicated connections are also possible. Consider three mechanical systems connected such that their common interaction ports have the same velocity (e.g., consider two robots S_3 and S_4 pushing

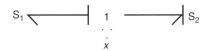

FIGURE 19.4 Bond graph representation of two systems connected with common velocity.

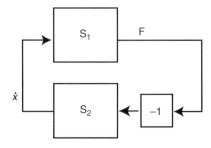

FIGURE 19.5 Feedback representation of two systems connected with common velocity.

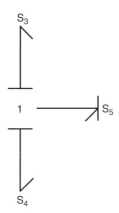

FIGURE 19.6 Bond graph representation of three systems connected with common velocity.

on the same workpiece S_5). Figure 19.6 shows a bond graph representation. If we choose to describe the workpiece dynamics as an admittance (and this may be the only option if the workpiece includes a kinematic constraint) then causal analysis dictates that the interaction dynamics of the two robots must be represented as impedances. Once again, this connection may be represented by feedback network. S_3 and S_4 are connected in parallel with each other and with the workpiece S_5 in a negative feedback loop, as shown in Figure 19.7. Note that S_3 and S_4 can be represented by a single equivalent impedance, and the system looks like that in Figure 19.4 and Figure 19.5. S_3 and S_4 need not represent distinct pieces of hardware and, in fact, might represent superimposed control algorithms acting on a single robot (Andrews and Hogan, 1983; Hogan, 1985).

19.3.2 Impedance vs. Admittance Regulation

Causal analysis provides insight into the important question whether it is better to regulate impedance or admittance. For most robotics applications, the environment consists primarily of movable objects, most simply represented by their inertia and surfaces or other mechanical structures that kinematically constrain their motion. The interaction dynamics in both cases may be represented as an admittance. An unrestrained inertia determines acceleration in response to applied force, yielding a proper admittance transfer function (see Table 19.1). A kinematic constraint imposes zero motion in one or more directions regardless of applied force; it does not admit representation as an impedance. Inertias prefer admittance causality; kinematic constraints require it. Because the environment has the properties best represented

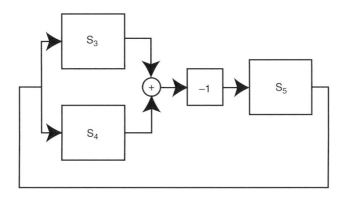

FIGURE 19.7 Feedback representation of three systems connected with common velocity.

as an admittance, the ideal robot behavior is an impedance, which can be thought of as a dynamic generalization of a spring, returning force in response to applied displacement. Initially assuming that arbitrary port impedance could be achieved, Hogan argued for this approach (Hogan, 1985).

However, most robots consist of links of relatively large mass driven by actuators. If the actuators are geared to amplify motor torque (a common design), the total inertia apparent at the end effector is increased by the reflected inertia of the motor; indeed for high gear ratios this can dwarf the mass of the links. These inertial properties are difficult to overcome (see Section 19.5.2), and tend to dominate the robot's response. Thus it is difficult to make a robot behave as a spring (or impedance), and it is usually more feasible to make it behave as primarily a mass (or admittance), an argument for admittance control (Newman, 1992). In other words, impedance behavior is usually ideal, but admittance behavior is often more easily implemented in real hardware, which itself prefers admittance causality because of its inertial nature. The choice of particular controller structure for an application must be based on the anticipated structure of environment and manipulator, as well as the way that they are coupled.

19.3.3 Coupled System Stability Analysis

If S_1 and S_2 are a linear time-invariant impedance and admittance, respectively, Figure 19.4 and Figure 19.5 depict the junction of the two systems via a power-continuous coupling. In other words the coupling is lossless and does not require power input to be implemented; all power out of one system goes into the other. Colgate has demonstrated the use of classical feedback analysis tools to evaluate the stability of this coupled system, using only the impedance and admittance transfer functions (Colgate, 1988; Colgate and Hogan, 1988). By interpreting the interaction as a unity negative feedback as shown in Figure 19.5, Bode and/or Nyquist frequency response analysis can be applied. The "open-loop" transfer function in this case is the product of admittance and impedance $S_1 S_2$. Note that unlike the typical cascade of transfer functions, this product is exact, independent of whether or how the two systems exchange power or "load" each other. The Nyquist stability criterion ensures stability of the coupled (closed-loop) system if the net number of clockwise encirclements of the -1 point (where magnitude $= 1$ and phase angle $= \pm 180°$) by the Nyquist contour of $S_1(j\omega)S_2(j\omega)$ plus the number of poles of $S_1(j\omega)S_2(j\omega)$ in the right half-plane is zero. The slightly weaker Bode condition ensures stability if the magnitude $|S_1(j\omega)S_2(j\omega)| < 1$ at any point where $\angle(S_1(j\omega)S_2(j\omega)) = \angle S_1(j\omega) + \angle S_2(j\omega) = \pm 180°$.

19.3.4 Passivity and Coupled Stability

This stability analysis requires an accurate representation of the environment as well as the manipulator, something that we have argued is difficult to obtain and undesirable to rely upon. The principles of this analysis, however, suggest an approach to guarantee stability if the environment has certain properties — particularly if the environment is passive.

There are a number of definitions of a system with passive port impedance, all of which quantify the notion that the system cannot, for any time period, output more energy at its port of interaction than has in total been put into the same port for all time. Linear and nonlinear systems can have passive interaction ports; here we restrict ourselves to the linear time-invariant case and use the same definition as Colgate (1988). For nonlinear extensions, see Wyatt et al. (1981) and Willems (1972).

Definition 19.2 A system defined by the linear 1-port[7] impedance function $Z(s)$ is passive iff:

1. $Z(s)$ has no poles in the right half-plane.
2. Any imaginary poles of $Z(s)$ are simple, and have positive real residues.
3. $\mathrm{Re}\{Z(j\omega)\} \geq 0$.

---

[7] An extension to multi-port impedances and admittances is presented in Colgate (1988).

These requirements ensure that $Z(s)$ is a positive real function and lead to the following interesting and useful extensions:

1. If $Z(s)$ is positive real, so is its inverse, the admittance function $Y(s) = Z^{-1}(s)$, and $Y(s)$ has the same properties.
2. If equality is restricted from condition 3, the system is dissipative and is called *strictly passive*.
   a. The Nyquist contours of $Z(s)$ and $Y(s)$ each lie wholly within the closed right half-plane.
   b. $Z(s)$ and $Y(s)$ each have phase in the closed interval between $-90°$ and $+90°$.
3. If condition 3 is met in equality, the system is passive (but not strictly passive).
   a. The Nyquist contours of $Z(s)$ and $Y(s)$ each lie wholly within the open right half-plane.
   b. $Z(s)$ and $Y(s)$ each have phase in the open interval between $-90°$ and $+90°$.

Note that pure masses and springs, as shown in Table 19.1, are passive but not strictly passive. Pure dampers are strictly passive, as they have zero phase. Furthermore, any collection of passive elements (springs, masses, dampers, constraints) assembled with any combination of power-continuous connections is also passive. Passive systems comprise a broad and useful set, including all combinations of mechanical elements that either store or dissipate energy without generating any — even when they are nonlinear. Passivity and related concepts have proven useful for other control system applications, including robust and adaptive control (Levine, 1996).

Colgate has shown that the requirement for a manipulator to interact stably with any passive environment is that the manipulator itself be passive (Colgate, 1988; Colgate and Hogan, 1988). The proof is described here intuitively and informally. If two passive systems are coupled in the power-continuous way described above and illustrated in Figure 19.4, the phase property of passive systems constrains the total phase of the "open-loop" transfer function to between $-180°$ and $+180°$ (the phase of each of the two port functions is between $-90°$ and $+90°$, and the two are summed). Because the phase never crosses these bounds, the coupled system is never unstable and is at worst marginally stable. This holds true *regardless of the magnitudes* of the port functions of both systems. In the Nyquist plane, as shown in Figure 19.8, since the total phase never exceeds $180°$, the contour cannot cross the negative real axis, and therefore can never encircle -1, regardless of its magnitude. This result shows that if a manipulator can be made to behave with passive driving point impedance, coupled stability is ensured with *all* passive environments, provided the coupling obeys the constraints applied above. The magnitude of the port functions is irrelevant; if passivity (and therefore the phase constraint) is satisfied, the coupled system is stable. The indifference to magnitude also means that requiring a robot to exhibit passive interaction behavior need not compromise performance; in principle the system may be infinitely stiff or infinitely compliant.

It is worthy of mention that if either of the two systems is *strictly* passive, the total phase is strictly less than $\pm 180°$, and the coupled system is asymptotically stable. Different types of coupling can produce slightly different results. If the act of coupling requires energy to be stored or if contact is through a mechanism such as sliding friction, local asymptotic stability may not be guaranteed. However, the coupled system energy remains bounded. This result follows from the fact that neither system can generate energy or supply it continuously; the two can only pass it back and forth and the total energy can never grow.

Example 19.3

Because both examples presented at the start of this paper were unstable when interacting with passive elements, both systems must be nonpassive. This is in fact true; for the second case, for example, Figure 19.9 shows the real part of the port admittance, evaluated as a function of frequency, and it is clearly negative between 2 and 10 rad/sec, hence violates the third condition for passivity.

Colgate has proven another useful result via this argument, particularly helpful in testing for coupled stability of systems. As can easily be determined from Table 19.1, an ideal spring in admittance causality produces $+90°$ of phase, and an ideal mass produces $-90°$, both for all frequencies, making each passive

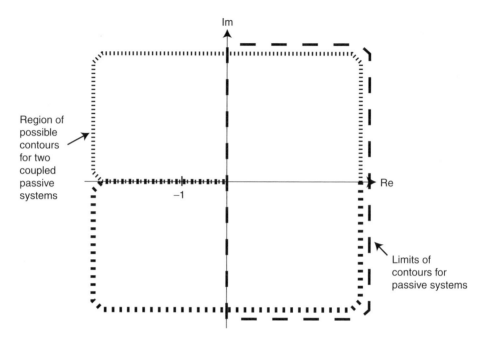

FIGURE 19.8 Region of the Nyquist plane that contains contours for passive systems (dashed), and region that can contain coupled passive systems (dotted). The coupled Nyquist contour may lie on the negative real axis but cannot encircle the -1 point. If a system is *strictly* passive, its Nyquist plot cannot lie on the imaginary axis. If at least one of two coupled passive systems is *strictly* passive, then its shared Nyquist contour cannot intersect or lie on the negative real axis.

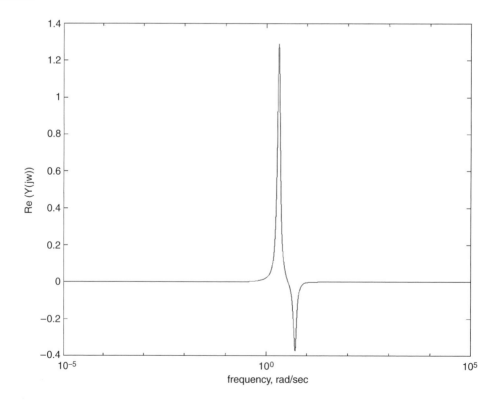

FIGURE 19.9 Real part of the interaction port admittance of the system from Example 2.

Impedance and Interaction Control

but not strictly passive. If the manipulator is nonpassive, its phase must necessarily exceed either −90° or +90° at some frequency, so that coupling it to a pure spring or a pure mass results in phase that exceeds −180° or +180° at that frequency. The value of the environmental stiffness or mass acts as a gain and can be selected such that the magnitude exceeds 1 at a frequency where the phase crosses the boundary, proving instability by the Bode criterion. Alternatively by the Nyquist criterion, the gain expands or contracts the contour, and can be selected so as to produce an encirclement of the −1 point. Conversely, it is impossible for the manipulator to be unstable when in contact with any passive environment if it does not become unstable when coupled to any possible spring or mass. Thus the passivity of a manipulator can theoretically be evaluated by testing stability when coupled to all possible springs and masses. If no spring or mass destabilizes the manipulator, it is passive. Much can be learned about a system by understanding which environments destabilize it (Colgate, 1988; Colgate and Hogan, 1988).

Example 19.4

The destabilizing effect of springs or masses can be observed by examining the locus of a coupled system's poles as an environmental spring or mass is changed. Figure 19.10 shows the poles of a single-resonance system (as in Example 19.2) coupled to a mass at the interaction port, as the magnitude of the mass is varied from 0.1 to 10 kg. The system remains stable. Figure 19.11 shows the coupled system poles as a stiffness connected to the interaction port is varied from 1 to 150 N/m. When the stiffness is large, the system is destabilized.

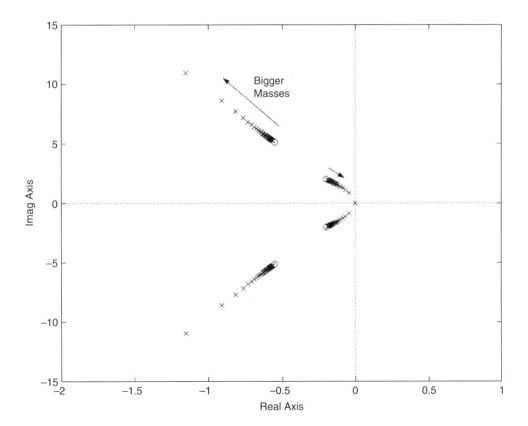

FIGURE 19.10 Locus of coupled system poles for the system from Example 19.2 as environment mass increases.

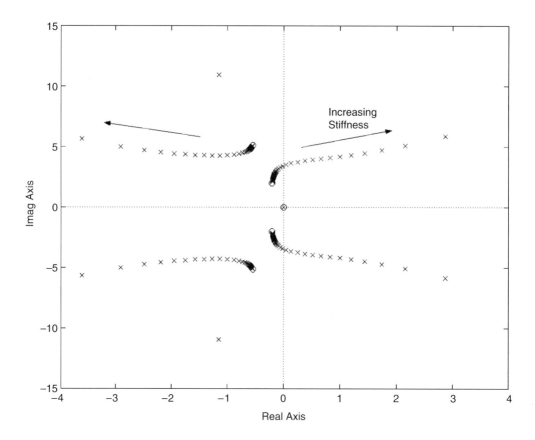

FIGURE 19.11 Locus of coupled system poles for the system from Example 19.2 as environment stiffness increases.

19.4 Implementing Interaction Control

Like any control system, a successful interactive system must satisfy the twin goals of stability and performance. The preceding sections have shown that stability analysis must necessarily include a consideration of the environments with which the system will interact. It has further been shown that stability can in principle be assured by sculpting the port behavior of the system. Performance for interactive systems is also measured by dynamic port behavior, so both objectives can be satisfied in tandem by a controller that minimizes error in the implementation of some target interactive behavior. For this approach to work for manipulation, any controller must also include a way to provide a motion trajectory. The objectives of dynamic behavior and motion can, to some degree, be considered separately.

19.4.1 Virtual Trajectory and Nodic Impedance

Motion is typically accomplished using a virtual trajectory, a reference trajectory that specifies the desired movement of the manipulator. A virtual trajectory is much like a motion controller's nominal trajectory, except there is no assumption that the controlled machine's dynamics are fast in comparison with the motion, which is usually required to ensure good tracking performance. The virtual trajectory specifies which way the manipulator will "push" or "pull" to go, but actual motion depends on its impedance properties and those of the environment. It is called "virtual" because it does not have to be a physically realizable trajectory.

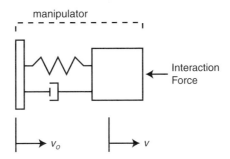

FIGURE 19.12 Virtual trajectory v_o and nodic impedance. v is the interaction port velocity.

Deviation in the manipulator's trajectory from its virtual trajectory produces a force response that depends on the interactive properties of the manipulator, its *nodic*[8] impedance/admittance. If the robot is represented as a mass subjected to actuator and environmental forces (similar to the simple model of Example 19.1 above) and the nodic impedance of the controller is a parallel spring and damper, the resulting behavior is as pictured in Figure 19.12 in which v_o represents the virtual trajectory and v the actual manipulator motion. In many applications, the controller will be required to specify manipulator position x and to be consistent, the virtual trajectory must also be specified as a position x_o. In that case, the noninertial (nodic) behavior is strictly described as a dynamic operator that produces output force in response to input displacement (the difference between virtual and actual positions, $\Delta x = x_o - x$), which might be termed a "dynamic stiffness." However, the term "impedance" is loosely applied to describe either velocity or displacement input. To the interaction port, the system behaves as a mass with a spring and damper, connecting the port to some potentially moving virtual trajectory. The dynamics of the spring, mass, and dashpot are as much a part of the prescribed behavior as is the virtual trajectory, distinguishing this strategy from motion control.

19.4.2 "Simple" Impedance Control

One primitive approach to implementing impedance control, proposed by Hogan (1985), has been used with considerable success. Termed "simple" impedance control, it consists of driving an intrinsically low-friction mechanism with force- or torque-controlled actuators, and using motion feedback to increase output impedance. This approach makes no attempt to compensate for any physical impedance (mass, friction) in the mechanism, so the actual output impedance consists of that due to the controller plus that due to the mechanism.

If a robot is modeled as a multi-degree-of-freedom inertia retarded by damping and subject to actuator and environmental torques (a multi-degree-of-freedom version of the model used in example 1 above), the robot with the simple impedance controller is as follows:

$$I(\Theta)\ddot{\Theta} + C(\Theta, \dot{\Theta}) + D(\dot{\Theta}) = T_a + T_e \tag{19.8}$$

where Θ is a vector of robot joint variables (here assumed to be angles, though that is not essential), I is the robot inertia matrix (which depends on the robot's pose), C denotes nonlinear inertial coupling torques (due to Coriolis and/or centrifugal accelerations), D is a vector of dissipative velocity-dependent torques (e.g., due to friction), T_a is a vector of actuator torques, and T_e a vector of environmental torques. The target behavior is impedance; if the behavior of a spring with stiffness matrix K_j and a damper with

---

[8]The term *nodic* refers to "transportable" behavior that may be defined relative to a nonstationary (or even accelerating) reference frame. Nodicity is discussed by Hogan (1985) and Won and Hogan (1996).

damping matrix B_j is chosen, the control law is simply

$$T_a(\Theta, \dot{\Theta}) = K_j(\Theta_o - \Theta) + B_j(\dot{\Theta}_o - \dot{\Theta}) \qquad (19.9)$$

where Θ_o is a virtual trajectory in robot joint space. Combining the controller impedance (Equation 19.9) with the robot dynamics (Equation 19.8), the result is as follows:

$$I(\Theta)\ddot{\Theta} + C(\Theta, \dot{\Theta}) + D(\dot{\Theta}) + B_J \dot{\Theta} + K_j \Theta = B_J \dot{\Theta}_o + K_j \Theta_o + T_e \qquad (19.10)$$

This controller implements in robot joint space a dynamic behavior analogous to that depicted in Figure 19.12; the controller spring and damper serve to push or pull the robot pose toward that specified by the virtual trajectory.

Due to the nonlinear kinematics of a typical robot, constant stiffness and damping matrices in the control law of Equation (19.9) result in stiffness and damping apparent at the robot's end-effector (the usual interaction port) that vary with the robot's pose. However, a straightforward extension of Equation (19.9) may be defined to control end-effector impedance, using the robot's kinematic transformations to express a desired end-effector impedance in robot joint space. The required transformations are the forward kinematic equations, denoted $X = L(\Theta)$, relating robot pose to end-effector position (and orientation), X; the pose-dependent Jacobian of this transformation, denoted $\dot{X} = J(\Theta)\dot{\Theta}$, relating robot velocity (e.g., joint rates) to end-effector velocity (and angular velocity); and the transformation relating end-effector forces (and torques), F_e, to actuator torques. For the common arrangement in which each robot actuator acts independently on a robot joint variable, the required transformation is the transpose of the Jacobian: $T_a = J^t(\Theta)F_e$. To achieve a target impedance with constant end-effector stiffness K and damping B, the control law is as follows:

$$T_a(x_o, \dot{x}_o, \Theta, \dot{\Theta}) = J^t(K(x_o - L(\Theta)) + B(\dot{x}_o - J(\Theta)\dot{\Theta})) \qquad (19.11)$$

This is a nonlinear counterpart of the control law in Equation (19.9). An important point is that all of the required nonlinear transformations are guaranteed to be well-defined throughout the robot workspace — even where the Jacobian loses rank.[9] Thus the simple impedance controller can operate near or at the robot's kinematic singularities.

Note that the simple impedance controller does not rely on force feedback and does not require a force sensor. If the forces due to the robot's inertia and friction are sufficiently small, the robot's interaction port behavior will be close to the desired impedance. Inertial forces decline for slower movements and vanish if the robot is at a fixed pose. As a result a simple impedance controller can be quite effective in some applications. Frictional forces may also decline for slower movements but, especially due to dry friction, need not vanish when the robot stops moving. This is one reason why, for applications requiring low impedance, low inertia and friction are desirable in the design of interactive robots.

From a controller design viewpoint, the simple impedance control law of Equation (19.9) closely resembles a PD motion controller acting on the error between actual and virtual trajectory in joint space. Given the assumptions outlined above (a robot modeled as a mass driven by controllable forces) the impedance controller stiffness K and damping B correspond, respectively, to proportional and derivative gains. The nonlinear simple impedance controller of Equation (19.11) resembles a gain-scheduled PD motion controller in which the nonlinear transformations adjust the proportional and derivative gains to achieve constant end-effector behavior.

---

[9]The term *joint space* refers to a set of generalized coordinates of the robot mechanism, which uniquely define the configuration of the robot and hence the position and orientation of any interaction port. Similarly any force and torque applied to the interaction port can be projected to determine the corresponding forces and torques on the actuators.

The simple impedance controller has another important feature. Insofar as the controller exactly mimics the behavior of a spring and damper, the robot behaves exactly as it would with a real spring and damper connecting it to the virtual trajectory. If the virtual trajectory specifies a constant pose, the entire system (robot plus controller) is passive and, therefore, guarantees stable interaction with all passive environments. Regardless of the actual robot mass, damping, and controller gains, coupled instability is completely eliminated. In fact, this is true even if the interaction occurs at points other than the end-effector, whether inadvertently or deliberately (e.g., consider using an elbow or other body part to manipulate) (Hogan, 1988). In addition, because the simple impedance controller does not rely on force feedback control to shape impedance, it is not vulnerable to the loss of passivity that can occur when structural modes interact with a force feedback loop (as illustrated in Example 19.2).

Thus, although it is primitive, a simple impedance controller goes a long way toward solving the stability problem, and its performance gets better as the inherent robot impedance is reduced. In practice, though this implementation performs well in some applications, it has limitations. Several factors make the controller impedance nonpassive, including discrete-time controller implementation and unmodeled dynamics between actuator and sensor. Under these conditions stability cannot be guaranteed with all environments. At the same time, the creation of low-impedance hardware can be difficult, particularly for complex geometries. Still, the approach has been quite successful, particularly when used in conjunction with highly back-drivable designs (for example, using robots for physiotherapy (Volpe et al., 2000)). Feedback methods to reduce intrinsic robot impedance are discussed below.

19.4.3 Direct Impedance Modulation

A complementary approach to implementing interaction control without relying on feedback is to modulate impedance directly. For example, as discussed below, feedback control of interaction-port inertia can be especially challenging. However, apparent inertia is readily modulated by taking advantage of kinematic redundancies in the robot mechanism. Using the human musculo-skeletal system as an example, all aspects of mechanical impedance, including inertia, may be varied by using different body segments as the interaction port (e.g., the hand vs. the foot) and/or by varying the pose of the rest of the musculo-skeletal system. For example, with the phalanges partially flexed the fingertips present extremely low mechanical impedance to their environment, while with the elbow fully extended, the arm presents a high impedance along the line joining hand to shoulder.

An alternative way to modulate mechanical impedance is by taking advantage of actuator redundancies. Because muscles pull and don't push they are generally arranged in opposition about the joints so that both positive and negative torques may be generated. Consequently, the number of muscles substantially exceeds the number of skeletal degrees of freedom. Interestingly, the force developed by mammalian muscle varies with muscle length and its rate of change, giving it mechanical impedance similar to a damped spring, and this mechanical impedance increases with muscle activation. While the torque contributions of different muscles may counteract each other, impedance contributions of muscles always add to the joint impedance (Hogan, 1990). As a result, mechanical impedance about a joint may be directly modulated without adjusting pose by simultaneous contraction of opposing muscles. For example, a "firm grip" on a tool handle may be achieved by squeezing it. That requires simultaneous activation of opposing muscles (at least in the hand and forearm but likely throughout the upper arm) which can be done without altering the net force and/or torque on the tool, changing only the mechanical impedance of the hand where it interacts with the tool and, in turn, stabilizing the pose of the tool relative to the hand.

The human ability to modulate neuromuscular mechanical impedance at will by co-contracting muscles is readily observed, and human control of mechanical impedance appears to be quite sophisticated and selective. For example, recent experimental results (Burdet et al., 2001) have shown that human subjects making point-to-point movements holding the handle of a planar robot that presented a stabilizing impedance along the direction of motion and a destabilizing impedance in the normal direction increased their arm impedance preferentially along the normal but not the tangent.

A similar approach to direct modulation of actuator impedance may be implemented in robotic systems. There are many ways this may be accomplished. For example, a closed volume of air in a pneumatic actuator acts as a spring; because of the nonlinear relation between air pressure and volume or due to a change of shape, increasing pressure increases the apparent stiffness. Fluid flowing through a passage (e.g., in a hydraulic actuator) dissipates energy, and varying the geometry of the passage changes the apparent damping (Kleidon, 1983). Alternatively, field-responsive fluids (e.g., electro-rheologic or magneto-rheologic) permit variation of viscous damping without changing geometry and have been used in automotive applications (Carlson et al., 1999). Changing excitation current and hence magnetic field strength in an electromagnetic actuator (e.g., a rotary or linear solenoid) may be used to change both apparent stiffness and damping, and electromechanical actuator designs that provide for direct modulation of impedance have been proposed (Fasse et al., 1994).

To the best of our knowledge, actuators with directly-modulated impedance have yet to see widespread applications in robotics. However, the development of new actuator technologies to serve as "artificial muscles" is presently a topic of active research, and new applications may emerge from this work.

19.5 Improving Low-Impedance Performance

A large class of applications, including robots that interact with humans, demands interactive robots with low mechanical impedance. The most direct approach is to design low-impedance hardware and use a simple impedance control algorithm; in fact, this is the recommended approach. However, intrinsically low-impedance hardware can be difficult to create, particularly with complex geometries and large force or power outputs. Most robotic devices have intrinsically high friction and/or inertia, and the simple impedance control technique described above does nothing to compensate for intrinsic robot impedance. Considerable effort has been devoted to designing controllers to reduce the apparent endpoint impedance of interactive robots.

19.5.1 Force Feedback

Force feedback is probably the most appealing approach for reducing apparent impedance. If a simple impedance controller is applied to a one-dimensional inertial mass as in Figure 19.1, that is also subject to some nonlinear friction force $F_n(x,\dot{x})$, and is augmented with a proportional force feedback controller, the control law is given by Equation (19.6), repeated here for convenience.

$$F_a = K(r - x) + B(\dot{r} - \dot{x}) + K_f[F_e + K(r - x) + B(\dot{r} - \dot{x})] \tag{19.12}$$

K and B are the scalar desired stiffness and damping, respectively, and r now represents the virtual position.[10] The force feedback term serves to minimize the deviation of the actual endpoint force from the desired endpoint force, which looks like a damped spring characterized by K and B connected to the virtual trajectory. The equation of motion for the uncontrolled system with nonlinear friction is

$$m\ddot{x} + b\dot{x} + F_n(x,\dot{x}) = F_a + F_e \tag{19.13}$$

Combining Equation (19.12) and Equation (19.13), the controlled equation of motion is as follows:

$$\frac{m}{1 + K_f}\ddot{x} + \frac{b}{1 + K_f}\dot{x} + \frac{F_n(x,\dot{x})}{1 + K_f} + K(x - r) + B(\dot{x} - \dot{r}) = F_e \tag{19.14}$$

---

[10] As outlined above, given the assumed robot model, a simple impedance controller closely resembles a PD motion controller. In Equation (19.6) and Equation (19.14) the corresponding proportional and derivative gains are $K(1 + K_f)$ and $B(1 + K_f)$, respectively. This form is chosen so that force feedback does not change the stiffness and damping introduced by the controller.

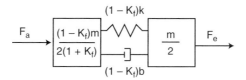

FIGURE 19.13 Equivalent physical system for a single-resonance model under force feedback control. (Modified from Colgate, J.E., The control of dynamically interacting systems, Ph.D. thesis, Massachusetts Institute of Technology, 1988, with permission.)

Equation (19.14) shows that the introduction of proportional force feedback reduces the apparent mass and friction by a factor of the force feedback gain plus one, so that (in principle) arbitrarily large force gains drive the actual impedance to the desired impedance specified by the controller. Friction, inertia, or any other behavior, whether linear or nonlinear, is minimized.

This one-degree-of-freedom combination of force and motion feedback is readily extended to multiple degrees of freedom. Assuming the robot may be modeled as a multi-degree-of-freedom inertial mechanism, retarded by friction and subject to actuator and environmental forces (as in Equation (19.8) above), a controller based on feedback of endpoint position, velocity, and force can be formulated to replace the manipulator's inherent dynamics with arbitrary inertia, stiffness, and damping (Hogan, 1985).

Despite its appeal, this approach has fundamental limitations. Clearly, the largest possible force feedback gain is desirable to minimize unwanted components of intrinsic robot behavior, e.g., due to nonlinear friction. However, experienced control system designers intuit that increasing gains toward infinity almost always leads to instability, and the following observation shows why this holds true in the case of force feedback. Colgate has derived a *physical equivalent representation* of force feedback. A physical equivalent representation is a model of a mechanical system that exactly replicates the behavior of a system under feedback control. Figure 19.13 shows a physical equivalent representation for a model of a robot with a single structural vibration mode (as shown in Figure 19.2 but with $b_1 = b_2 = 0$, $m_1 = m_2 = m/2$) under force feedback control (Equation (19.12) with $K = B = 0$). It also exhibits a single structural mode but with parameters that depend on the force feedback gain. The result has *negative* stiffness, mass, and damping parameters for any force feedback gain greater than 1 (Colgate, 1988). Because endpoint force feedback is not state feedback, *isolated* stability is unaffected; this system remains neutrally stable as long as the interaction force F_e is zero. However, for $K_f > 1$, it is not passive; it can exhibit negative visco-elastic behavior upon contact, resulting in coupled instability. Note that this gain limit is independent of the value of the structural stiffness and damping. As any real robot almost inevitably exhibits resonance, this analysis shows that any system under force feedback becomes nonpassive for a force feedback loop gain greater than unity and is, therefore, vulnerable to coupled instability.

19.5.2 Natural Admittance Control

One effective solution to the loss of passivity due to force feedback is natural admittance control (NAC) developed by Newman (1992). In essence, this approach is based on the observation that a system under force feedback becomes nonpassive when the controller reduces the apparent inertia (to less than about half its physical value), but passivity is not compromised by the elimination of friction. Natural admittance control specifies a desired inertia that is close to that of the physical system (the "natural admittance") and focuses on reducing friction as much as possible, preserving passivity.

A method for the design of natural admittance controllers is detailed in Newman (1992). Here the procedure is sketched for the simple robot model shown in Figure 19.1 consisting of a single mass retarded by friction and driven by actuator and environmental forces. The system might also have nonlinear friction that the NAC seeks to eliminate, that need not be modeled at all in the design of the compensator; the controller treats such friction as a disturbance from desired port behavior and rejects it. To ease notation, the substitutions $v = v_a = \dot{x} = v_e$ are made. The equation of motion for the system velocity neglecting nonlinear friction is

$$m\dot{v} + bv = F_a + F_e \qquad (19.15)$$

In the Laplace domain:

$$v = \frac{1}{ms + b}(F_a + F_e) \qquad (19.16)$$

A generic form for the controller is assumed, that incorporates some velocity feedback with compensator $G_v(s)$ and endpoint force feedback with some compensator $G_f(s)$:

$$F_a = G_v v + G_f F_e \qquad (19.17)$$

Using Equation (19.16) and Equation (19.17) the actual endpoint admittance $Y(s) = v/F_e$ is determined:

$$Y(s) = \frac{G_f + 1}{ms + b - G_v} \qquad (19.18)$$

This is equated to the target port admittance. The target stiffness K and damping B are chosen at will but the target mass is equal to the physical mass m of the system:

$$Y(s) = Y_{des}(s) = \frac{1}{ms + B + K/s} \qquad (19.19)$$

Equation (19.18) and Equation (19.19) can be solved for G_f in terms of G_v:

$$G_f = \frac{(b - G_v - B)s - K}{ms^2 + Bs + K} \qquad (19.20)$$

A simple form for G_v can be assumed, such as a constant, and the compensator design is complete. Equation (19.20) may be thought of as a force feedback "filter" that yields the desired admittance of Equation (19.19). Although the mass is not reduced, the "disturbance" to the desired behavior due to friction, b, is rejected with the feedback gain of the compensator, and its effect is smaller as the velocity loop gain, G_v, is increased. The approach serves equally well to minimize disturbances due to nonlinear frictional forces, e.g., dry friction.[11]

In principle, the controlled system is passive even if the velocity gain is increased to arbitrary size, minimizing the unwanted frictional effects. In practice unmodeled dynamic effects limit the gain that may be used without compromising passivity, but the technique affords significant practical improvement over simple proportional force feedback. A more general formulation and discussion of the method can be found in Newman (1992) and Newman and Zhang (1994).

19.5.3 Series Dynamics

An alternative approach to stabilizing interaction under force feedback is to place compliant and/or viscous elements in series between a manipulator and its environment. Variants of this approach have been used since the earliest attempts at robot force control (Whitney, 1977), e.g., in the form of compliant pads on the robot end-effector. A drawback is that this may compromise fine motion control. More recently a series spring has been incorporated within force control actuators to facilitate force feedback and absorb impacts (Pratt and Williamson, 1995). The following example illustrates the beneficial effect of series dynamic elements.

---

[11] The Laplace domain representation above was used to simplify the presentation and is not essential.

Impedance and Interaction Control

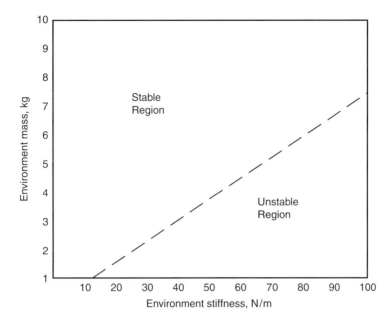

FIGURE 19.14 Stable and unstable parameter values for spring-and-mass environment coupled to the system of Example 19.2. The coupled system is stable when parameters fall in the region above the line and unstable when they fall in the region below the line.

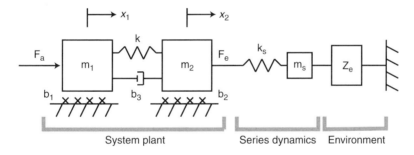

FIGURE 19.15 The single-resonance model of Figure 19.2 with a spring and mass interposed in series between interaction port and environment.

Example 19.5

A series spring's stabilizing effect on a force-controlled system can be illustrated by considering again the simple model of a robot with a single structural vibration mode shown in Figure 19.2. Under the control law of Equation (19.6) or Equation (19.12), this system may have a nonpassive impedance and be destabilized by interaction with some environments. However, it remains stable when interacting with a broad range of spring and mass environments,[12] as depicted by the stability boundary in Figure 19.14. In this particular example, high-stiffness springs are destabilizing while masses are stabilizing. If a spring and mass are placed in series between the interaction port and the environment, as depicted by k_s and m_s in Figure 19.15, the apparent stiffness that the controlled system experiences cannot exceed k_s. By appropriate

---

[12] Recall that the set of all positive springs and masses includes the most destabilizing passive environments a robot may encounter.

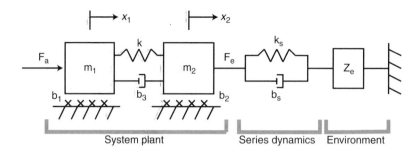

FIGURE 19.16 The single-resonance model of Figure 19.2 with a parallel spring and damper interposed in series between interaction port and environment.

choice of k_s the resulting interaction port impedance is made passive, stabilizing the interaction. In effect, a piece of the controller's "environment" is carried around with the robot, acting to "filter" the environment dynamics to ensure that the controller cannot encounter any environment capable of destabilizing it.

The implementation of "series elastic actuators" (Pratt and Williamson, 1995) also includes an inertia m_s between k_s and the environment, endowing the controller's environment with not only a maximum stiffness but also a minimum inertia, further stabilizing the system. Considering Figure 19.15, the stiffness k_s and inertia m_s may be chosen so that the uncoupled robot behavior corresponds to a point in the stable region. Assuming interaction with passive environments, the interaction port is then guaranteed to encounter only environments that reside above and to the left of this point, thereby ensuring coupled stability.

Although the method described above can guarantee stability for certain types of nonpassive systems, it lacks the generality to be applicable to all nonpassive systems. Dohring and Newman have provided a formalization that includes damping to drain the energy introduced by nonpassive behavior of the controller or the environment (Dohring and Newman, 2002). The recommended approach includes a parallel spring and damper in series with the environment, as shown in Figure 19.16. Analysis of passivity of the admittance function at the interaction port dictates the selection of the spring and damper values. By modeling the parallel spring and damper as a two-port, the admittance at the outboard side of the spring and damper, $Y(s)$, is computed in terms of the original robot admittance, $Y_o(s)$, and the series elements:

$$Y(s) = Y_o(s) + \frac{s}{b_s s + k_s} \qquad (19.21)$$

A small series damper b_s contributes a large positive-real term to the net admittance at high frequencies, compensating for negative-real, nonpassive behavior in $Y_o(s)$. The spring determines the frequency range at which the damping dominates. Together they can be selected to make $Y(s)$ passive.

Intentional series dynamics can be thought of as "mechanical filters," perhaps loosely analogous to anti-aliasing filters used with analog-to-digital converters, that attenuate undesirable effects in peripheral frequency ranges that make the system vulnerable to coupled instability, while preserving desired dynamic behavior in the frequency range of interest. As with signal filters, something is lost; here, it might be bandwidth or positioning accuracy, a tradeoff that may be acceptable for interactive robots.

19.6 More Advanced Methods

The foregoing considered some of the basic approaches to interaction control. Alternative and more advanced techniques and applications have been developed. The following is a small sample of this extensive work.

Nonlinear approaches to impedance control have been pursued with success, including adaptive (Singh and Popa, 1995), learning (Cheah and Wang, 1998), robust sliding-mode (Ha et al., 2000), and decentralized control (Kosuge et al., 1997) methods. Stability properties of different implementations of impedance control have been investigated by Lawrence (1988). An alternative approach to design for coupled stability based on the small-gain theorem has been pursued by Kazerooni et al. (1990). More powerful methods to guarantee stable interaction have been studied in depth by Miller et al. (2000).

For spatial rotations and translations, the specification of task-relevant impedance (and especially stiffness) is particularly challenging. By characterizing impedance using a "spatially affine" family of compliance and damping, Fasse and Broenink (1997) develop a method that substantially simplifies specification and implementation of spatial impedance. A quaternion representation of rotational impedance is presented by Caccavale et al. (1999).

19.7 Conclusion: Interaction Control, Force Control, and Motion Control

The fundamental goal of the interaction controllers described in this chapter is to minimize deviations from some specified ideal of a system's dynamic response to inputs at its ports of interaction. In a sense, both motion and force controllers attempt the same task, though each focuses on only one port variable in particular. A system with zero admittance, or infinite impedance, gives zero motion in response to any applied force. In other words, it imposes motion and its motion is unaffected by any applied force. This ideal is the goal of a purely motion-controlled manipulator. Conversely, the force applied by a system with zero impedance, or infinite admittance, has zero dependence on motion at its interaction port. In other words, it is the ideal of a pure force controller. Those designing motion or force controllers in effect decide to forego a guarantee of interaction stability in favor of maximizing or minimizing port impedance. This may be appropriate for applications in which the work exchanged between the robot and its environment is negligible. For applications where power exchange cannot be ignored, interaction control provides techniques to accommodate its effects.

References

Andrews, J.R. and Hogan, N. (1983). Impedance control as a framework for implementing obstacle avoidance in a manipulator. In: D. Hardt and W.J. Book (eds.), *Control of Manufacturing Processes and Robotic Systems*, ASME, New York, pp. 243–251.

Brown, F.T. (2001). *Engineering System Dynamics*. Marcel Dekker, New York.

Burdet, E., Osu, R., Franklin, D.W., Milner, T.E., and Kawato, M. (2001). The central nervous system stabilizes unstable dynamics by learning optimal impedance. *Nature*, 414:446–449.

Caccavale, F., Natale, C., Siciliano, B., and Villani, L. (1999). Six-DOF impedance control based on angle/axis representations. *IEEE Trans. Robotics and Automation*, 15(2):289–300.

Carlson, J.D., St. Clair, K.A., Chrzan, M.J., and Prindle, D.R. (1999). Controllable vibration apparatus. U.S. Patent No. 5,878,851.

Cheah, C.C. and Wang, D. (1998). Learning impedance control for robotic manipulators. *IEEE Trans. Robotics and Automation*, 14(3):452–465.

Colgate, J. E. (1988). The control of dynamically interacting systems. Ph.D. thesis, Deptartment of Mechanical Engineering, Massachusetts Institute of Technology.

Colgate, J.E. and Hogan, N. (1988). Robust control of dynamically interacting systems, *Int. J. Control*, 48(1):65–88.

Dohring, M. and Newman, W.S. (2002). Admittance enhancement in force feedback of dynamic systems. *Proc. IEEE Int. Conf. Robotics and Automation*, pp. 638–643.

Fasse, E.D., Hogan, N., Gomez, S.R., and Mehta, N.R. (1994). A novel variable mechanical-impedance, electromechanical actuator. In: Radcliffe, C. (ed.), *Dynamic Systems and Control 1994*, vol. 1, ASME, New York, pp. 311–318.

Fasse, E.D. and Broenink, J.F. (1997). A spatial impedance controller for robotic manipulation. *IEEE Trans. Robotics and Automation*, 13(4):546–556.

Ha, Q.P., Nguyen, Q.H., Rye, D.C., and Durrant-Whyte, H.F. (2000). Impedance control of a hydraulically actuated robotic excavator. *Automation in Construction,* 9:421–435.

Hogan, N. (1985). Impedance control: an approach to manipulation. *ASME J. Dynamic Syst. Meas. Control,* 107:1–24.

Hogan, N. (1988). On the stability of manipulators performing contact tasks. *IEEE J. Robotics and Automation,* 4:677–686.

Hogan, N. (1990). Mechanical impedance of single- and multi-articular systems. In: J. Winters and S. Woo (eds.), *Multiple Muscle Systems: Biomechanics and Movement Organization,* Springer-Verlag, New York, pp. 149–164.

Kazerooni, H., Waibel, B., and Kim, S. (1990). On the stability of compliant motion control: theory and experiments. *ASME J. Dynamic Syst. Meas. Control,* 112:417–426.

Kleidon, M. (1983). Modeling and performance of a pneumatic/hydraulic hybrid actuator with tunable mechanical impedance. S.M. Thesis, Department of Mechanical Engineering, Massachusetts Institute of Technology, September.

Kosuge, K., Oosumi, T., and Seki, H. (1997). Decentralized control of multiple manipulators handling an object in coordination based on impedance control of each arm. *Proc. IEEE/RSJ Int. Conf. Intelligent Robots and Systems,* pp. 17–22.

Lawrence, D.A. (1988). Impedance control stability properties in common implementations. *Proc. IEEE Int. Conf. Robotics and Automation,* 2:1185–1190.

Levine, W.S. (1996). *The Control Handbook.* CRC Press and IEEE Press.

Miller, B.E., Colgate, J.E., and Freeman, R.A. (2000). Guaranteed stability of haptic systems with nonlinear virtual environments. *IEEE Trans. Robotics and Automation,* 16(6):712–719.

Newman, W.S. (1992). Stability and performance limits of interaction controllers. *ASME J. Dynamic Syst. Meas. Control,* 114(4):563–570.

Newman, W.S. and Zhang, Y. (1994). Stable interaction control and coulomb friction compensation using natural admittance control. *J. Robotic Systems,* 11(1):3–11.

Pratt, G.A. and Williamson, M.M. (1995). Series elastic actuators. *Intelligent Robots and Systems,* Proc. Human Robot Interaction and Cooperative Robots, pp. 399–406.

Rancourt, D. and Hogan, N. (2001). Stability in force-production tasks. *J. Motor Behavior,* 33(2):193–204.

Sharon, A., Hogan, N., and Hardt, D.E. (1988). High-bandwidth force regulation and inertia reduction using a macro/micro manipulator system. *Proc. IEEE Int. Conf. Robotics and Automation,* 1:126–132.

Singh, S.K. and Popa, D.O. (1995). An analysis of some fundamental problems in adaptive control of force and impedance behavior: theory and experiments. *IEEE Trans. Robotics and Automation,* 11(6):912–921.

Tilley, S.W. and Cannon, R.H., Jr. (1986). End point position and force control of a flexible manipulator with a quick wrist. *Proc. AIAA Guidance, Navigation and Control Conference,* pp. 41–49.

Volpe, B.T., Krebs, H.I., Hogan, N., Edelstein, O.L., Diels, C., and Aisen, M. (2000). A novel approach to stroke rehabilitation: robot-aided sensorimotor stimulation, *Neurology,* 54:1938–1944.

Whitney, D. (1977). Force feedback control of manipulator fine motions. *ASME J. Dynamic Syst. Meas. Control,* 99:91–97.

Willems, J.C. (1972). Dissipative dynamical systems part I: general theory. *Arch. Ration. Mech. Anal.,* 45(5):321–351.

Won J. and Hogan N. (1996). Nodicity and nonlinear interacting dynamic systems. *Proc. ASME Int. Mech. Eng. Conf. Exposition,* DSC-58:615–621.

Wyatt, J.L., Chua, L.O., Gannett, J.W., Göknar, I.C., and Green, D.N. (1981). Energy concepts in the state-space theory of nonlinear n-ports: part I — passivity. *IEEE Trans. on Circuits and Systems,* CAS-28(1):48–61.

20
Coordinated Motion Control of Multiple Manipulators

20.1 Introduction 20-1
20.2 Manipulation of an Object 20-2
20.3 Coordinated Motion Control of Multiple
     Manipulators for Handling an Object 20-5
     Motion of Object and Control of Internal Force/Moment
     • Load Sharing Problem
20.4 Coordinated Motion Control Based on Impedance
     Control Law 20-7
     Design of Impedance for Handling an Object • Design
     of Impedance for RCC Dynamics
20.5 Decentralized Motion Control of Multiple Mobile
     Manipulators 20-10
     Multiple Mobile Manipulators • Leader-Follower Type Control
     Algorithm

Kazuhiro Kosuge
Tohoku University

Yasuhisa Hirata
Tohoku University

20.1 Introduction

With the development of robot technologies, many control algorithms for multiple manipulators have been proposed so far to realize several tasks using manipulators in coordination as humans do using their arms. Especially the control algorithms proposed by Nakano et al. [1] and by Kurono [2] are known as the pioneering research or handling of a single object by multiple manipulators in coordination. By using multiple manipulators in a cooperative way, we could execute several tasks which could not be done by a single manipulator in addition to the handling of a single object. Figure 20.1 shows typical examples of tasks executed by multiple manipulators.

Figure 20.1 (a) shows the handling of an object by multiple manipulators in coordination. With an appropriately designed control system, multiple manipulators in coordination could handle a large and/or heavy object, which could not be handled by a single manipulator because the load to each manipulator is reduced by distributing it among the manipulators. Figure 20.1 (b) shows an example of tasks using a tool by two manipulators in coordination. In this case, the use of the tool becomes possible with the improved rigidity of the system. The kinematic closed loop is formed by the manipulators, and the tool is grasped by them in coordination. Figure 20.1 (c) shows an assembly task of two parts, as an example of dexterous tasks which could not be done by a single manipulator. Two parts could be assembled without using any jig and additional devices by the dual manipulators.

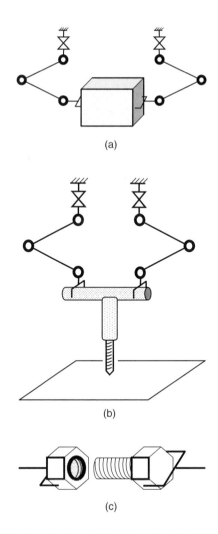

FIGURE 20.1 Tasks of manipulators: (a) load capacity, (b) rigidity, and (c) dexterity.

In this chapter, we first consider problems of coordinated motion control of multiple manipulators having physical interaction among the manipulators including an object. Then, we outline several types of control algorithms for multiple manipulators for handling an object in coordination. The impedance-based motion control algorithm is introduced for the manipulation of an object in coordination and the assembly of two parts. Finally, we introduce decentralized control algorithms of multiple mobile manipulators in coordination.

20.2 Manipulation of an Object

Consider motion of an object supported by multiple manipulators. Suppose that n manipulators grasp an object rigidly and each manipulator applies force/moment to the object as shown in Figure 20.2. We define coordinate systems as shown in Figure 20.3 and several parameters as follows:

$O_u - x_u y_u z_u$ absolute coordinate system
$O_0 - x_0 y_0 z_0$ object coordinate system
$O_{hi} - x_{hi} y_{hi} z_{hi}$ ith arm coordinate system
$r_0 \in R^3$ position vector from the origin of the absolute coordinate system to the origin of the object coordinate system

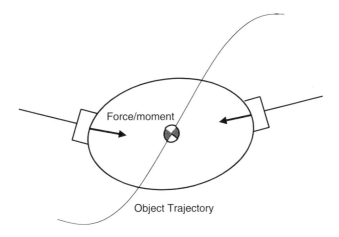

FIGURE 20.2 Manipulation of an object.

$v_0 \in R^3$ linear velocity of the object
$w_0 \in R^3$ angular velocity of the object
m mass of the object
$M \in R^3$ inertia matrix of the object
$g \in R^3$ acceleration of gravity
$I_N \in R^{N \times N}$ $N \times N$ identity matrix

As is well known, the motion equation of an object supported by multiple manipulators is expressed as follows:

$$m\ddot{r}_0 = F_0 + mg \tag{20.1}$$

$$M\dot{w}_0 + w_0 \times (Mw_0) = N_0 \tag{20.2}$$

where F_0 and N_0 are the resultant force and the resultant moment applied to the object by all manipulators and are expressed as follows:

$$F_0 = \sum_{i=1}^{n} F_i \tag{20.3}$$

$$N_0 = \sum_{i=1}^{n} N_i \tag{20.4}$$

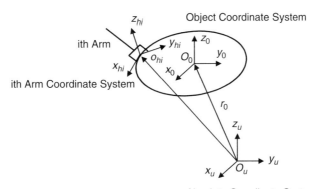

FIGURE 20.3 Coordinate systems.

F_i and N_i are the force and the moment applied to the object by the ith manipulator in the object coordinate system with respect to the absolute coordinate system. Note that the motion of the object is generated based on the force F_0 and the moment N_0, which are the resultant force and the resultant moment applied to the object by the manipulators.

Putting Equation (20.1) and Equation (20.2) together, we have the following equation:

$$L = \begin{bmatrix} mI_3 & 0 \\ 0 & M \end{bmatrix} \begin{bmatrix} \ddot{r}_0 l \\ \dot{w}_0 \end{bmatrix} + \begin{bmatrix} -mg \\ w_0 \times (Mw_0) \end{bmatrix} \qquad (20.5)$$

where

$$L \equiv \begin{bmatrix} F_0 \\ N_0 \end{bmatrix} \qquad (20.6)$$

F_0 and N_0 are also expressed as

$$\begin{bmatrix} F_0 \\ N_0 \end{bmatrix} = KF \qquad (20.7)$$

where

$$K = [I_6 \, I_6 \cdots I_6] \in R^{6 \times 6n} \qquad (20.8)$$
$$F = \begin{bmatrix} F_1^T N_1^T F_2^T N_2^T \cdots F_n^T N_n^T \end{bmatrix}^T \in R^{6n} \qquad (20.9)$$

When each manipulator applies the force F_i and the moment N_i to the object, the resultant force F_0 and the resultant moment N_0 applied to the object by all of the manipulators are shown by Equation (20.3) and Equation (20.4). Thus, the motion of the object is defined uniquely by Equation (20.5) based on the resultant force and the resultant moment applied to the object coordinate system with respect to the absolute coordinate system.

On the other hand, when the motion of the object is generated by multiple manipulators in coordination, we cannot determine uniquely the force F_i and the moment N_i, which are applied by each manipulator to the object from Equation (20.7), although we can derive the resultant force and the resultant moment applied to the object by all of the manipulators. The force/moment in the null space of K represented by Equation (20.8) is referred to as the internal force/moment [3]. The internal force/moment is the force and the moment that do not contribute to the motion of the object.

The solution of Equation (20.7) depends on how to distribute the load required for the manipulation of the object among the manipulators and how to apply the internal force/moment to the object. The following shows an example of the solutions of Equation (20.7) for the dual manipulators case ($n = 2$) proposed by Uchiyama [3].

Solving Equation (20.7) using the generalized inverse matrix, we have the following equation:

$$F = W^- L + [I_6 - I_6]^T \xi \qquad (20.10)$$

where $\xi (\in R^{12})$ is any vector and W^- is the generalized inverse matrix defined as

$$W^- = \begin{bmatrix} R \\ I_6 - R \end{bmatrix} \qquad (20.11)$$

From $W^{-1}L$ in Equation (20.10) and Equation (20.11), we can see that the load of the manipulator L is distributed to two manipulators based on ratio of $R : I_6 - R$.

When we distribute the load of the manipulator L equally to two manipulators, we can solve Equation (20.7) with $R = \frac{1}{2} I_6$ as

$$F = \frac{1}{2} \begin{bmatrix} I_6 \\ I_6 \end{bmatrix} L + \begin{bmatrix} I_6 \\ -I_6 \end{bmatrix} \xi \tag{20.12}$$

In the general case ($n \geq 3$), we can solve Equation (20.7) using the pseudo inverse matrix of K, K^+ as

$$F = K^+ L + (I_{6n} - K^+ K)\xi \tag{20.13}$$

The second terms on the right hand sides of Equation (20.10), Equation (20.12), and Equation (20.13) do not influence the force/moment applied to the object; that is, these terms do not affect the motion of the object. By using these terms, we can specify the internal force/moment applied to the object.

20.3 Coordinated Motion Control of Multiple Manipulators for Handling an Object

When multiple robots manipulate a single object in coordination as shown in Figure 20.2, we have to consider the following problems:

(1) How to grasp the object
(2) How to generate the motion of the object
(3) How to apply the internal force/moment to the object
(4) How to distribute the load of the object among robots

In this section, we consider the problems of (2), (3), and (4), under the assumption that the manipulators grasp the object firmly and can apply the force/moment to the object arbitrarily.

20.3.1 Motion of Object and Control of Internal Force/Moment

Many researchers have discussed the coordinated motion control problems of multiple manipulators. Roughly speaking, the coordinated motion control algorithms for manipulators proposed so far could be classified into five types: the master-slave type of control algorithms, the hybrid type of control algorithms, the compliance-based control algorithms, the object dynamics–based control algorithms, and the augmented dynamics–based control algorithms. In this section, we briefly introduce the outlines of these control algorithms.

1. Master-slave type of control algorithms [1,4]: Master-slave type of control algorithm has been proposed by Nakano [1] as the pioneering research of multiple robots coordination. In this method, a manipulator referred to as a master controls the pose of the object based on the position control law, and the other manipulators referred to as slave control the force/moment applied to the object based on the force control law as shown in Figure 20.4. This method controls the position of the object and the force/moment applied to the object precisely. In this method, however, the distribution of the load among robots could not be realized appropriately. The servo stiffness of

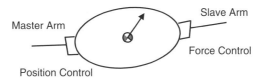

FIGURE 20.4 Master-slave type of control algorithm.

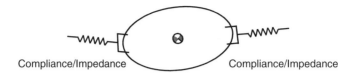

FIGURE 20.5 Compliance based control algorithm.

the position-controlled master manipulator is very high, the servo stiffness of the force-controlled slave manipulators is almost zero, and the object is supported through the stiffness. Eventually, the position-controlled master manipulator has to support the entire load required for the object motion.

2. Hybrid type of control algorithms [5–7]: To control the motion of the object in 3D space, the robot has to have six degrees of freedom (6-DOF). In addition, to control the internal force/moment applied to the object, the robot has to have 6-DOF. In the hybrid type of control algorithms, multiple robots control the 6-DOF with respect to the motion of the object and the 6-DOF with respect to the internal force/moment applied to the object by using the $6n$-DOF of n manipulators.

    This type is similar to the hybrid position/force control algorithm and is regarded as a generalization of the master-slave type of control algorithm. Takase [5] has proposed this type of control algorithm, which is derived based on how to constrain the motion of the object by manipulators. In the practical use of this method, however, we have to consider several problems such as position and orientation errors among the coordinate systems of manipulators or kinematic modeling errors of each robot and the manipulated object. Unless we could control the manipulators exactly, the manipulators might apply excessive internal force/moment to the object.

3. Compliance-based control algorithms [2,8,9]: In the compliance-based control algorithm, the object is compliantly grasped through manipulator compliances or impedances realized by the hardware or the software as shown in Figure 20.5. This type of control algorithm is robust against the kinematic modeling errors of the manipulators and the object. Even if the modeling errors exist, the robots would not apply the excessive internal force/moment to the object. The effect of the modeling errors is reduced by the compliances or the impedances, through which the robots control the position/orientation of the object and the internal force/moment applied to the object.

4. Object dynamics–based control algorithms [10]: In this algorithm, the motion of the object is controlled dynamically based on Equation (20.5), under the assumption that the robots could be regarded as actuators which generate the force/moment at grasping points of the object as shown in Figure 20.6. The control algorithm has been proposed by Nakamura et al. [10]. When the mass of the manipulated object is small, the precise manipulation of the object is not easy. To manipulate the

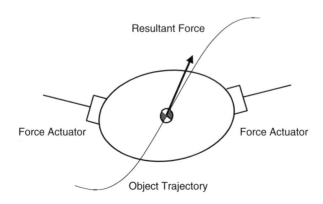

FIGURE 20.6 Object dynamics–based control algorithm.

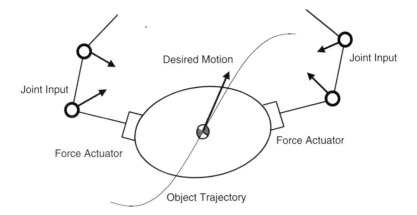

FIGURE 20.7 Augmented dynamics–based control algorithm.

    object with small mass, the manipulators have to apply small force/moment to the object precisely. In the practical robots; however, it is difficult to apply the small force/moment precisely to the manipulated object at their grasping points.

5. Augmented dynamics–based control algorithm [11]: This control algorithm has been proposed by Khatib [11]. This is an extension of their hybrid position/force control algorithm based on the manipulator dynamics at its endeffector. They derived a resultant dynamics including the manipulator dynamics and the object dynamics at the representative point of the object as shown in Figure 20.7. Then, they designed a hybrid position/force control algorithm so that both the pose of the object and the force/moment applied to the object are controlled. Note that they do not consider the internal force/moment applied to the object, but they do consider force/moment applied among grasping points.

20.3.2 Load Sharing Problem

When two manipulators manipulate an object, the load sharing between them is realized based on Equation (20.10). Several researches for the load sharing have been proposed so far. Zheng and Luh [12] have proposed an optimal algorithm for load distribution with minimum exerted forces on the object. These algorithms require less computational time, which makes them attractive for real-time applications. The adaptive load-sharing controller developed by Pittelkau [13] optimizes the distribution of joint torques over two manipulators to increase the efficiency and combined load-carrying capacity of the two manipulators. Uchiyama and Yamashita [14] have considered the adaptive load sharing of two cooperative manipulators holding a common object from the viewpoint of robustness of the holding.

20.4 Coordinated Motion Control Based on Impedance Control Law

In this section, we introduce the control algorithm based on the impedance control to realize the handling of a single object and the assembly of two parts.

20.4.1 Design of Impedance for Handling an Object

Kosuge et al. have proposed a coordinated motion control algorithm of manipulators handling an object based on impedance control of each manipulator [9,15]. In the algorithm based on the impedance control law, the object is supported by multiple manipulators that are controlled so as to have impedance dynamics

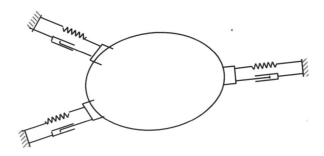

FIGURE 20.8 Compliant support of object.

around desired trajectories of their endeffectors as shown in Figure 20.8. The compliant support of the object using mechanical impedance is used for decreasing the modeling errors of the robot system.

In this section, we are going to reconsider the algorithm and introduce an alternative coordinated motion control algorithm of manipulators handling an object in coordination, so that we can control a mechanical impedance of a manipulated object as well as the external force/moment sharing of the object among the manipulators.

First, let us consider a control problem of an apparent mechanical impedance of the manipulated object by multiple manipulators. We consider a problem to realize a mechanical impedance of the object around its desired compliance center as shown in Figure 20.9. We assume that each manipulator supports the object rigidity with its mechanical impedance around offset force/moment required for the manipulation of the object which consists of the desired internal force/moment and the shared load.

Let the mechanical impedance of the object be expressed by the following equation:

$$M\Delta\ddot{x} + D\Delta\dot{x} + K\Delta x = F_{ext} \tag{20.14}$$

where Δx is the deviation of the manipulated object from the desired trajectory of the compliance center of the object, F_{ext} is the external force/moment applied to the object around its desired compliance center, and M, D, K are 6×6 positive definite matrices.

We assume that each manipulator grasps the object firmly and that no relative motion between the object and each manipulator exists. To realize the impedance of the manipulated object, we consider the case in which each manipulator is controlled, based on the impedance control law around the desired compliance center of the object as shown in Figure 20.9. Let the mechanical impedance of the ith manipulator around the desired compliance center of the object be expressed by

$$M_i\Delta\ddot{x}_i + D_i\Delta\dot{x}_i + K_i\Delta x_i = F_{ei} \tag{20.15}$$

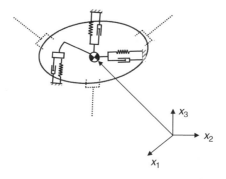

FIGURE 20.9 Desired object impedance.

where F_{ei} is the external force/moment applied to the ith manipulator around the desired compliance center of the object and Δx_i is the deviation of the compliance center of the ith manipulator. From the assumption that there is no relative motion between each manipulator and the object, we have the following relation:

$$\Delta x = \Delta x_i \quad (i = 1, 2, \ldots, n) \tag{20.16}$$

The external force/moment F_{ext} is shared by each manipulator which is as follows:

$$F_{ei} = \delta_i F_{ext} \quad (i = 1, 2, \ldots, n) \tag{20.17}$$

where $\delta_i > 0$. Concerned with the external force/moment applied to the object, the following relation holds:

$$\sum_{i=1}^{n} F_{ei} = F_{ext} \tag{20.18}$$

and we have

$$\sum_{i=1}^{n} \delta_i = 1 \quad (\delta_i > 0) \tag{20.19}$$

From these equations, we obtain the mechanical impedance of each manipulator, which is necessary to realize the desired mechanical impedance of the manipulated object expressed by Equation (20.14), as follows:

$$M_i = \delta_i M \quad (i = 1, 2, \ldots, n) \tag{20.20}$$
$$D_i = \delta_i D \quad (i = 1, 2, \ldots, n) \tag{20.21}$$
$$K_i = \delta_i K \quad (i = 1, 2, \ldots, n) \tag{20.22}$$

20.4.2 Design of Impedance for RCC Dynamics

As shown in Figure 20.10, the remote compliance center (RCC) is a well-known device for realizing the assembly tasks of two parts. We are going to consider realizing a dynamic motion of the RCC between two parts by using dual manipulators. Let the dynamics of the RCC device be expressed by the following equation:

$$M_{RCC} \Delta \ddot{x} + D_{RCC} \Delta \dot{x} + K_{RCC} \Delta x = F \tag{20.23}$$

where $M_{RCC}, D_{RCC}, K_{RCC}$ are 6×6 positive definite matrices.

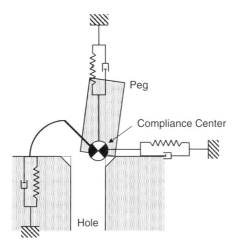

FIGURE 20.10 Remote compliance center.

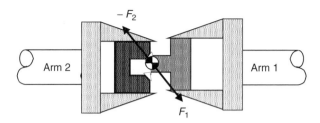

FIGURE 20.11 Assembly task of two parts by two arms.

Suppose that each manipulator has the same impedance structure as the one of the RCC device and let the impedance of each manipulator be expressed by the following equations:

$$M_1 \Delta \ddot{x}_1 + D_1 \Delta \dot{x}_1 + K_1 \Delta x_1 = F_1 \tag{20.24}$$

$$M_2 \Delta \ddot{x}_2 + D_2 \Delta \dot{x}_2 + K_2 \Delta x_2 = F_2 \tag{20.25}$$

The force/moment applied to each manipulator has the following relation as shown in Figure 20.11.

$$F_1 = -F_2 (= F) \tag{20.26}$$

From Equation (20.23), Equation (20.24), and Equation (20.25), we have

$$\Delta x = \Delta x_1 - \Delta x_2 \tag{20.27}$$

$$M_1 = M_2 = 2M_{RCC} \tag{20.28}$$

$$D_1 = D_2 = 2D_{RCC} \tag{20.29}$$

$$K_1 = K_2 = 2K_{RCC} \tag{20.30}$$

By specifying the impedance parameters expressed in Equation (20.28), Equation (20.29), and Equation (20.30), the impedance dynamics expressed in Equation (20.23) is realized between two parts supported by two manipulators. Under the assumption that we could design the impedance dynamics appropriately for realizing the assembly tasks of two parts, two manipulators could realize the assembly tasks successfully.

20.5 Decentralized Motion Control of Multiple Mobile Manipulators

In this section, we consider the case of multiple mobile manipulators handling an object in coordination. As explained below, the coordination of multiple mobile manipulators has several control problems different from the coordination of multiple manipulators. We briefly introduce the concept of a leader-follower type of control algorithm designed for multiple mobile manipulators for handling a single object in coordination as an example of the solution.

20.5.1 Multiple Mobile Manipulators

The mobile manipulators are expected to do tasks in an ordinary environment. Mobility is an important function to cover the working space in the ordinary environment. In the ordinary environment, multiple small robots are more appropriate than a large and heavy robot because the small robot has less kinetic energy than the large and heavy one. When they are moving with the same speed, the small one is less harmful to a human if its collisions occur.

As mentioned above, much research has been done for the motion control of multiple robots. However, most of the control algorithms proposed so far are designed based on the centralized control system; that is, a single controller is supposed to control all of the robots in a centralized way based on the global information. The centralized control system may be effective in case of the coordinated motion control of manipulators since the number of the manipulators in coordination is usually limited to two or three.

The use of mobile manipulators in the real world is one of the challenging topics in recent years [16–18]. Considering the case where such mobile manipulators transport a single object in coordination, a single controller could not control a large number of robots because of the real-time communication problem around the robots and the computational burden of the single controller. A centralized controller is no more realistic to control a large number of mobile manipulators.

In addition, deferent from the case of the manipulator, the dead reckoning system of a mobile manipulator is not so reliable, because the mobile manipulator has a slippage between its wheels and the ground and we could not position the mobile manipulator precisely. Therefore, we could not apply the same control principle of manipulators for controlling the multiple mobile manipulators, and the control system of the mobile manipulators has to be redesigned robust against the inevitable positioning error of each mobile manipulator.

20.5.2 Leader-Follower Type Control Algorithm

Many control algorithms have been proposed for the handling of a single object by multiple robots in coordination. Most of these control algorithms proposed so far have been designed under the assumption that the geometric relations among the robots are known precisely. However, it is not easy to know the geometric relations among them precisely, especially when the robots handle an unknown object in coordination in a real environment.

Errors in position/orientation of each mobile robot detected by a dead reckoning system are inevitable because of a slippage between a mobile mechanism and the ground. In addition to that, even if we knew geometric relations among the robots, the geometric relations could not be kept precisely any more because of the errors included in position/orientation information of each robot. To overcome these problems, we have to design a coordinated motion control algorithm robust against the inevitable positioning errors.

Kume et al. [18] have proposed the leader-follower type control algorithm of multiple mobile manipulators for handling a single object in coordination as shown in Figure 20.12. In this algorithm, the desired trajectory of the object is given to one of the mobile manipulators, referred to as a leader, and the other robots, referred to as followers, are controlled so as to have a virtual caster-like dynamics. The followers

FIGURE 20.12 Coordination of multiple mobile manipulators.

estimate the desired trajectory of the leader along the heading direction of the virtual caster to handle the object in coordination with the leader. By using the virtual caster-like dynamics, which generates motion of each follower passively based on the force/moment applied to it similar to the real caster, multiple mobile manipulators could handle a single object in coordination without using the geometric relations among robots. That is, this type of control algorithm is robust against the inevitable positioning errors of mobile manipulators.

References

[1] Nakano, E., Ozaki, S., Ishida, T., and Kato, I., Cooperational control of the anthropomorphous manipulator "MELARM," *Proc. 4th Int. Symp. Industrial Robots,* pp. 251–260, 1974.

[2] Kurono, S., Coordinated computer control of a pair of artificial arms, *Biomechanism,* Vol. 3, pp. 182–193, Tokyo University Press, Tokyo, 1975.

[3] Uchiyama, M., A unified approach to load sharing, motion decomposing, and force sensing of dual arm robots, robotic research, *5th Int. Symp.,* pp. 225–232, MIT Press, Cambridge, MA, 1990.

[4] Luh, J.Y.S. and Zheng, Y.F., Constrained relations between two coordinated industrial robots for motion control, *Int. J. Robotics Res.,* Vol. 6, No. 3, pp. 60–70, 1987.

[5] Takase, K., Representation of constrained motion and dynamic control of manipulators under constrained, *Trans. SICE,* Vol. 21, No. 5, pp. 508–513, 1985.

[6] Hayati, S., Hybrid position/force control of multi-arm cooperating robots, *Proc. IEEE Int. Conf. Robotics and Automation,* pp. 82–89, 1986.

[7] Tarn, T.J., Bejczy, A.K., and Yun, X., Design of dynamic control of two coordinated robots in motion, *Proc. 24th IEEE Conf. Decision and Control,* pp. 1761–1765, 1985.

[8] Hanafusa, H. and Asada, H., A robot hand with elastic fingers and its application to assembly process, *Proc. IFAC Symp. Inf. Control Probl. Manuf. Technol.,* pp. 127–138, 1977.

[9] Koga, M., Kosuge, K., Furuta, K., and Nosaki, K., Coordinated motion control of robot arms based on the virtual internal model, *IEEE Trans. Robotics and Automation,* Vol. 8, No. 1, pp. 77–85, 1992.

[10] Nakamura, Y., Nagai, K., and Yoshikawa, T., Mechanics of coordinative manipulation by multiple robotic mechanisms, *Proc. IEEE Int. Conf. Robotics and Automation,* pp. 991–998, 1987.

[11] Khatib, O., Object manipulation in a multi-effector robot system, *Int. Symp. Robotics Res., Santa Cruz,* CA, August 1987.

[12] Zheng, Y.F. and Luh, J.Y.S., Optimal load distribution for two industrial robots handling a single object, *Proc. 1988 IEEE Int. Conf. Robotics and Automation,* pp. 344–349, 1988.

[13] Pittelkau, M.E., Adaptive load sharing force control for two-arm manipulators, *Proc. Int. Conf. Robotics and Automation,* pp. 498–503, 1988.

[14] Uchiyama, M. and Yamashita, T., Adaptive load sharing for hybrid control two cooperative manipulators, *Proc. IEEE Int. Conf. Robotics and Automation,* pp. 986–991, 1991.

[15] Kosuge, K., Ishikawa, J., Furuta, K., and Sakai, M., Control of single-master multi-slave manipulator system using VIM, *Proc. 1990 IEEE Int. Conf. Robotics and Automation,* pp. 1172–1177, 1991.

[16] Khatib, O., Yokoi, K., Chang, K., Ruspini, D., Holmberg, R., and Casal, A., Vehicle/arm coordination and multiple mobile manipulator decentralized cooperation, *Proc. IEEE/RSJ Int. Conf. Intelligent Robots Syst.,* pp. 546–553, 1996.

[17] Yamamoto, Y. and Xiaoping, Y., Effect of the dynamic interaction on coordinated control of mobile manipulators, *IEEE Trans. Robotics and Automation,* Vol. 12, No. 5, pp. 816–824, 1996.

[18] Kume, Y., Hirata, Y., Wang, Z., and Kosuge, K., Decentralized control of multiple mobile manipulators handling a single object in coordination, *Proc. IEEE/RSJ Int. Conf. Intelligent Robots Syst.,* pp. 2758–2763, 2002.

21
Robot Simulation

21.1	Introduction	**21**-1
21.2	Example Applications	**21**-2
	Factory Floor Simulation • Nuclear Waste Remediation Simulation • Surgical Simulation	
21.3	Robot Simulation Options	**21**-5
	Where Do You Start • Robot Design Packages • Animation of Robot Motion • Simulation of Robot Dynamics	
21.4	Build Your Own	**21**-11
	Graphical Animation • Numerical Simulation	
21.5	Conclusion	**21**-22

Lonnie J. Love
Oak Ridge National Laboratory*

21.1 Introduction

Mention the topic robot simulation, and, like many phrases, it means many different things to different people. To some practitioners, the simulation of robotic systems covers the visualization of how a robot moves through its environment. These simulators are heavily based on CAD and graphical visualization tools. In some cases, there is a man-machine interface so that the human can interact with the simulation. Likewise, to some, robot simulation covers the numerical simulation of the dynamics, sensing, and control of robots. Instead of having an interface that consists of three-dimensional (3D) graphical objects, these simulators consist of software commands and/or graphical tools used to construct block diagrams of the system. No matter the definition, simulation allows the practitioner to design and test robots in a variety of environments for a fraction of the time and cost of building real systems.

In this chapter, we provide an overview of robot simulation. We begin with a short historical review and examples of practical applications where simulation aided in solving challenging problems. While in no way complete, the survey is intended to give the reader a flavor for the problems that many practitioners address with simulation. In most cases, there are two levels of robot simulation. The first form of simulation is based primarily on kinematics. Kinematic simulations neglect the dynamics and low level control of the manipulator. The user specifies a target trajectory for the robot to follow and the simulation translates Cartesian motion of the robot to the resulting joint motion. This form of robot simulation aids in addressing a variety of issues in both the manipulator design and integration into the environment. The ability to visualize the motion of the manipulator can quickly answer questions regarding path planning, kinematic/workspace constraints, and coordination issues with other systems (scheduling, workspace layout, and collision detection). A few simulation packages provide off-line programming capabilities. The traditional approach to programming a robot consists of using an interface, such as a teach pendant,

---

*Oak Ridge National Laboratory is managed by UT-Battelle, LLC, for the U.S. Department of Energy under Contract DE-AC05-00OR22725.

to define the path and actions a robot is to execute during a task. This approach is time consuming and requires taking the robot out of production during the programming phase. Off-line programming integrates simulation with the robot's programming language. Advantages of off-line programming include a single programming system for multiple robots, verification of programs, and the ability to rapidly change the program and predict performance without taking the system off-line (Yong and Bonney, 1999).

In most cases, it is sufficient to neglect the robot's dynamics when simulating robot motion. This is perfectly valid when simulating commercial robots operating within the manufacturer's specifications (payloads, accelerations, speeds). The primary motivation for including dynamics in the simulation is to aid in the robot's mechanical and/or control design. When this is the case, there are numerical simulation tools based on physical principles that attempt to provide as high fidelity a prediction of the actual behavior of the robot as possible. A good numerical simulation tool can not only aid in mechanical and control design, but also establish actuator power requirements, selection of sensors, and performance metrics such as maximum bandwidth, accuracy, resolution, and speed. One additional utility of robot simulation is to aid in visualizing a concept to a colleague or sponsor. A good simulation can sometimes easily convey a concept that is not clear with pictures or verbal explanations.

In reality, it is very difficult to list all of the possible advantages that can be obtained with a good robotic simulation tool. The best gage for impact is simply experience: develop a high fidelity simulation and experiment with the system before any hardware is designed and fabricated or components are purchased. Only then can you fully appreciate the impact simulation can have on success. Our objective in this chapter is to provide the reader with enough information to make an informed decision on the selection of a suitable simulation platform and how best to get started.

21.2 Example Applications

21.2.1 Factory Floor Simulation

Robots are traditionally used for two types of tasks: boring or dangerous. In terms of boring, 90% of robots operate in factories, of which 50% are devoted toward the automotive industry. In the early 1980s, many automotive and aerospace companies began to integrate robotic systems into their production lines. Simulation became a primary focus in terms of both arranging resources and off-line programming. The designer of a robot system must determine an appropriate choice of robots and associated workspace components (feeders, tools, grippers, fixtures, and sensors). As a recent example, Nissan used the robot simulation program CimStation® Robotics to reduce production lead times by as much as a third for its new Primera Sedan car. Robotic simulation software (see Figure 21.1 and Figure 21.2) was used to program robots ahead of production, identifying potential problem areas such as spot welding access and workspace configurations. According to Craig Grafton, BIW planning engineer at Nissan, robot simulation enabled "offline programming to reduce downtime incurred by online programming." Without simulation, it is likely that a proper solution would require many iterations on the factory floor before a satisfactory solution is obtained. This is where powerful simulation tools are invaluable.

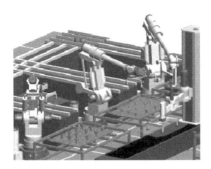

FIGURE 21.1 CimStation simulated floor (courtesy AC&E, Ltd.).

FIGURE 21.2 Factory floor (courtesy AC&E, Ltd.).

21.2.2 Nuclear Waste Remediation Simulation

In terms of dangerous applications of robotics, traditional areas focused on space based applications, deep sea exploration, and handling radioactive equipment. The primary mission of Oak Ridge National Laboratory (ORNL) during World War II was the processing of pure plutonium metal in support of the Manhattan Project. Byproducts of this process include radioactive cesium-137 and strontium-90. Between 1943 and 1951, the Gunite and Associated Tanks (GAAT) at ORNL were built to collect, neutralize, and store these byproducts. These tanks held approximately 75,000 gal of radioactive sludge and solids and over 350,000 gal of liquid. Characterization studies of these tanks in 1994 indicated that the structural integrity of some of the tanks was questionable. Consequently, there was a potential threat to human health through possible contamination of soil and groundwater. These risks provided the motivation for remediation and relocation of waste stored in the ORNL tanks. There were a number of factors that complicated the remediation process. The material stored in these tanks ranged from liquid to sludge and solid and was composed of organic, heavy metals, and radionuclides. Furthermore, the tanks, which ranged from 12 to 50 ft in diameter, were located below ground and in the middle of the ORNL complex. The only access to these tanks was through one to three access ports that were either 12 or 24 in. in diameter (Van Hoesen et al., 1995). These characteristics provide a daunting challenge: How can material be safely removed from such a confined structure?

An integrated system model of the hardware, illustrated in Figure 21.3, was developed under DELMIA's IGRIP® simulation program to aide in designing the system and investigating any potential problems that might arise during the remediation process. The system model included a detailed model of the spar aerospace modified light duty utility arm (MLDUA), a dynamic model of the hose management arm (HMA), and a flexible exhaust hose connecting the two arms. The goal of the simulation was threefold.

1. The model would provide a tool for operator training. The system could simulate teleoperation input commands through the same interface available on the hardware aiding in operator training.
2. The model provided a benchmark for identifying potential mining strategies. As the operator moved the system through the waste, the texture of the waste surface varied with the amount of material remaining.
3. The model provided an interface capable of investigating alternative control strategies. Alternative control strategies were easily imbedded in the simulation and executed during robotic and/or teleoperation tasks.

21.2.3 Surgical Simulation

As illustrated in the previous example, robot simulation can aid in training. One emerging area that is growing in terms of its dependence on robotics is surgery. Since its introduction in 1987, minimally

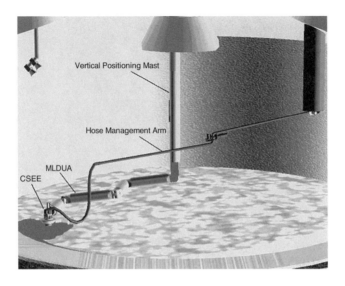

FIGURE 21.3 Gunite and associated tank hardware (courtesy Oak Ridge National Laboratory).

invasive surgery (MIS) has revolutionized the surgical field in a very short period of time (Cushieri, 1995). However, laparoscopy requires a unique subset of surgical skills that, for the inexperienced, can significantly delay the development of basic operative skills. Unfortunately financial constraints limit the time a physician can spend in the operating room for training. The cost of using operating room time for training surgical residents in the United States is estimated at $53 million per year (Bridges and Diamond, 1999). Subsequently, there has been growing interest in the development of robotic surgical simulators that combine virtual reality and haptics for realistic laparoscopic surgical training. The Mentice Corporation sells advanced medical simulation software and systems for a wide range of applications. As an example, Procedicus® MIST, shown in Figure 21.4, is a computer-based system designed to teach and assess minimally invasive surgical skills. The system comprises a frame holding two standard laparoscopic instruments electronically linked to a low-cost personal computer. The screen displays the movement of the surgical instruments in real-time 3D graphics. Advancements in computing and animation permit visual displays that provide smooth and realistic views of the internal organs. Figure 21.5 shows an example of the animation capabilities of Simbionix LapMentor™ software.

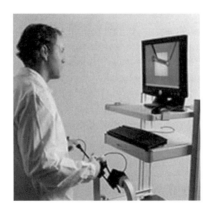

FIGURE 21.4 Procedicus MIST surgical trainer (courtesy Mentice, Inc.).

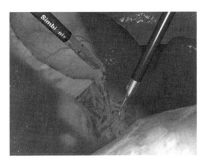

FIGURE 21.5 Simbionix virtual patient (courtesy Simbionix, Inc.).

21.3 Robot Simulation Options

21.3.1 Where Do You Start

Figuring out where to start is always the most important step. Making the wrong choice can end up costing valuable time and money. In this section, we provide a summary of some of the systems available today along with a list of distinguishing features. In addition, we will provide the reader with a list of questions to consider. Sometimes, knowing the right question is as important as finding the right answer.

In robotic simulation, there are potentially three types of software packages the practioner may use. First is a design (CAD) package such as Pro/Engineer,® SolidWorks,® or AutoCAD®. Each of these packages provides the ability to construct 3D solid models, machine drawings, and assemblies. Some of these packages include the ability to simulate motion. In many cases, this may be sufficient in terms of the design of the system. However, it is possible that the practitioner is interested in the programming of the robot and its behavior in its environment. In this case, we need to consider using a more traditional robot simulation package that is devoted towards animation for visualization of robot motion such as AC&E's CimStation Robotics and DELMIA's IGRIP. These packages focus more on off-line programming and animation than design. Both packages include a modeling component that enables the construction of the robot kinematics. In some cases it is possible to import the robot kinematics from many of the previously mentioned CAD packages. As mentioned earlier, some of these simulation packages have the ability to program and test the robot in its factory setting prior to implementation of any hardware. This is extremely attractive when purchasing off-the-shelf components and integrating into a factory setting.

However, if the practitioner is designing his own robot and wants to explore the low level control and behavior of the robot, he may be in need of something a little more sophisticated (and complex). In this case, the practitioner will be working with a numerical simulation tool that provides insight into the dynamics and control of the system. In the past, this process required a great deal of effort to formulate the dynamic equations of motion for a multi-body system. However, computation tools have greatly simplified this process. First, some CAD packages, such as SolidWorks, have developed interface software that automates the computation of the dynamic equations of motion and translates the equations into numerical simulation tools such as Matlab® and Simulink®. As an example, the engineer designs the mechanical system in SolidWorks then uses Simulink to design and test the controller. If the CAD package does not enable this translation, it is possible to use symbolic tools to rapidly and accurately compute the dynamic equations of motion and import them into simulation packages. In the last part of this chapter, we provide a simple example of how to symbolically compute the dynamic equations of motion of multi-degree of freedom serial link manipulators.

21.3.2 Robot Design Packages

If the practitioner is designing robots, the latest CAD tools provide a great deal of flexibility in terms of simulating robot motion. First, in some cases, the design software has direct animation capabilities.

Second, many CAD packages provide the ability to export the assembly (for example in a VRML format) and animate the motion with another software package. Finally, there are a few instances in which it is possible to export the system dynamics to the simulation directly from the CAD package. In each of these cases, it is possible to provide some insight into the behavior of the robot during the design process. There are three popular design software packages that have similar capabilities and costs: SolidWorks,[1] Pro/Engineer,[2] and AutoCAD.[3] For brevity, we provide a brief description of SolidWorks to provide insight into the flexibility and capabilities of present-day computer aided design (CAD) packages. The primary intent is to provide the reader with both a starting point, if interested in design, and an overview of how each of these packages can be exploited for both the animation and dynamic simulation of robotic systems.

The latest CAD packages provide a wide range of capabilities. As an example, SolidWorks is a commercial, windows-based, 3D mechanical design software package. The designer can use this package to develop 3D parts, draft machine drawings, and seamlessly integrate the parts into an assembly. SolidWorks provides the ability to define constraints between parts. The constraints include joint motion: revolute, cylindrical, spherical, universal, translational, screw, planar, fixed, and isotropic bushings. After defining the constraints, it is possible to reconfigure the assembly in SolidWorks to identify any possible conflicts in terms of the system's kinematics. In addition, it is possible to export the geometry for animation in a different package. For example, SolidWorks, ProE®, and AutoCAD provide the ability to export their respective assemblies in a VRML format. It is then possible to import these models directly into a VRML-compatible emulator to animation motion. The advantage of this approach is that the same package that is used for the mechanical design can be exploited for animation. Many of these packages also include the ability to calculate mass properties of each component. As with the animation, it is possible to import these properties and models into dynamic simulation packages (such as Adams® and Cosmos) to provide realistic dynamic simulation of the system. In some cases (as will be discussed shortly), it is possible to embed the dynamic simulation capabilities directly into the design software, eliminating any file compatability issues.

So, some possible questions the reader may wish to consider for the CAD package:

- Does the package include the ability to simulate motion?
  If designing a robot, the ability to rapidly simulate motion can identify possible flaws in the design.
- What types of file formats are compatible?
  If you want to quickly iterate on a kinematic design, you want to make sure that there is a seamless translation between the CAD and simulation package.
- Does the package include the ability to estimate and export dynamic properties?
  Many packages provide an interface to estimate mass properties of the parts. In addition, some packages, like SolidWorks, include interfaces to Adams and Simulink to enable estimation of the multi-body dynamics. This is important only if you are designing a robot or experimenting with the low-level controls.
- Can you get a trial version of their software?
  Some CAD packages can take a long time to learn. No company is going to tell you that their software is not user friendly. So, obtaining a trial version and spending a little time up front can save you a lot of time and money in the long run.
- Does the package have backwards compatibility to current/previous CAD applications?
  You will want the ability to use legacy work.

---

[1] http://www.solidworks.com
[2] http://www.ptc.com/
[3] http://www.autodesk.com/autocad

21.3.3 Animation of Robot Motion

The advancement of computing and graphical animation technologies has provided tremendous advancements in the area of robot animation. Early versions of robot simulation packages were based on simple wireframe models of the robot's kinematics. Today's simulation packages provide shading, textures, levels of sophistication, and fidelity that rival computer animation. In this section, we attempt to provide a sampling of a few commercial packages that range in both capabilities as well as costs. Our objective is to provide you, the reader, with an idea of what you can achieve given your budgetary constraints.

21.3.3.1 High End Robot Simulation Packages

DELMIA's IGRIP[4] is an example of the high end simulation packages. In the introductory section, one of the examples (nuclear waste cleanup) exploited IGRIP to answer basic questions prior to implementation of complex robotic systems. Typical issues that IGRIP was designed to address include process planning, ergonomics, factory floor layout, and process simulation. IGRIP includes the ability to include commercial robotic and peripheral equipment, motion attributes, I/O logic, and kinematic motion limitations. The primary utility of IGRIP is to develop simulations and programs that aid in optimization of robot locations, motions, and cycle times and identify possible collisions between robots, parts, tools, and fixtury (see Figure 21.6 and Figure 21.7).

Packages such as IGRIP and AC&E's CimStation Robotics (formerly Silma) are designed more for the simulation and modeling of robotic systems in workcells. In both cases, the software has a CAD interface to aid in the modeling of the robots and workcells. A robot is defined as a collection of parts. Each part is made up of a number of primitives (blocks, cylinders, cones, wedges, spheres...), enabling a great deal of details. Each of these primitives can be translated and rotated relative to each other in the CAD world. When constructing the model, joints (prismatic or rotary) are defined to specify the relationship between adjoining parts on the robot. Models can be translated from other CAD systems using formats such as IGES, VDA, and DES. Once the model of the robot is defined, there are a number of ways to command motion. The easiest approach is to move the robot joint by joint, incrementally changing each joint displacement and observing the resulting motion. However, in many cases we are concerned with how the robot behaves when commanding a specific motion of the tip of the robot. Most robots are defined as a combination of rotary and prismatic joints and links. In IGRIP, the robot kinematic parameters are represented using the Denavit-Hartenberg (D-H) notation. This notation provides a specific procedure for specifying the coordinate system of each robot link at appropriate locations. The motivation for using this approach is its universal appeal. All robots can be defined as a combination of rotary and prismatic joints. The D-H parameters provide a well-defined parametric approach to specify the kinematics of the robot.

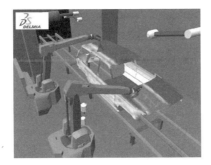

FIGURE 21.6 Simulated workcell (courtesy DELMIA Corp.).

---

[4]http://www.delmia.com/index.html

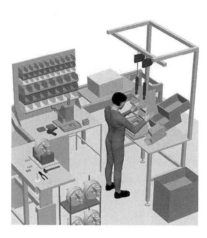

FIGURE 21.7 Eronomic simulation (courtesy DELMIA Corp.).

IGRIP provides, as of the late 1990s, over 32 kinematic routines (e.g., solutions to both the forward and inverse kinematics for a host of robot configurations). If the robot can adopt one of these configurations, the solution for the forward and inverse kinematics can be directly incorporated into the simulation. However, the package also enables the inclusion of your own kinematic routine if desired. Basically, the user writes C code that contains the solution for the forward and inverse kinematics. The IGRIP interface provides the ability to use this code instead of the canned kinematic solutions. If using a commercial robot, IGRIP has over 55 models that can be directly imported into your simulated workcell. Once the robots and workcell are defined, the user defines tag points. The tag points represent the target points of the path the robot tip is to follow. The simulation uses simple path planning routines to generate desired positions and velocities for the robot as a function of time. The inverse kinematic solution transforms these desired tip motions into joint motion. The simulation then animates the resulting motion of the robot. During the simulation, the software checks for problems such as collisions, singularities, and joint limits. Furthermore, some packages include the ability, if using a commercial robot, to export the resulting path planning and program to the actual robot controller. These capabilities can be extremely valuable, especially when modifying an assembly line; it is possible to test a workcell layout without stopping the assembly line. Furthermore, you can test a wide variety of programs before finally exporting the code to the actual hardware. All of these capabilities come at a cost. A commercial license from AC&E's CimStation Robotics costs approximately $35,000.

21.3.3.2 Low Cost Robot Simulation Packages

An example of a lower cost solution is RoboWorks® by Newtonium.[5] This package clearly lacks the sophistication and flexibility of the more mature packages such as IGRIP and AC&E's CimStation. However, with this lack of sophistication is a lower cost ($249) and shorter learning curve. Unlike IGRIP and CimStation, it is not possible at this time to import the geometries of your target robotic system from CAD software into the RoboWorks simulation package. RoboWorks provides two windows to define the robot, shown in Figure 21.8. One window shows the model and the other is a tree structure that you use to define the robot's geometry. A robot is defined by two basic elements: shapes and transformations. Shapes are the geometry associated with the links of the robot. Transformations provide the joints (both translation and orientation), links, and shapes on the links. These transformations can be either static or dynamic. Static transformations are used to define the transformation from one shape on a link to the next. Dynamic transformations are used to define the motion of a joint. When building the model, the tree structure is used to define the robot (geometry and transformations).

---

[5] http://www.newtonium.com/

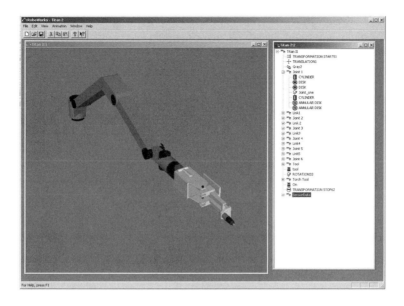

FIGURE 21.8 ORNL's RoboWorks model of a schilling Titan II.

There are three basic methods of commanding joint motion on the simulation: keyboard commands, reading a text file of time vs. displacements for each joint, or through a standard TCP/IP interface.

So, some possible questions the reader may wish to consider for the animation package:

- How do you construct the model of the robot?
  Most simulation packages have an internal CAD system for constructing the model. Try to understand its limitations (portability, degrees of freedom, joint types, ease of use).
- Can you import other file formats?
  Many CAD packages have translators that export the robot assembly into a VRML (or similar) file format that is used in many animations packages.
- Does the package support any off-the-shelf robots?
  Some robot manufacturers provide very accurate CAD models to some simulation packages. For example, Adept has a close association with CimStation.
- Is it possible to provide some level of off-line programming?
  One of the key benefits of robot simulation for manufacturing is the ability to program the robot prior to installation in the factory.
- What methods are available for planning motion? How do you actually plan motion?
  If you can only provide joint level motion commands, you will need to know how to solve for the inverse kinematics and have the ability to import these data into the simulation. Some packages have inverse kinematic solutions that enable the user to command Cartesian motion.
- Can you get a free trial version of the software?
  This is the best way to understand the capabilities and limitations of the software prior to committing resources (time and money).

21.3.4 Simulation of Robot Dynamics

The emphasis of robot simulation up to this point has been kinematic visulation. Basically, we assume the dynamics of the robot have no role in the behavior of the robot. When planning a workcell in a factory, using off-the-shelf components operating within their specified limits, most questions (workcell layout, collision detection, kinematic constraints) can be directly answered using kinematic simulation. Inclusion of the dynamics becomes important when the engineer is interested in more detailed information such as reaction forces, sizing motors/actuators, sensor selection, control design, estimating power consumption and determining how contacting parts behave. We consider two approaches to simulation of the robot dynamics. First is the canned approach that handles the computation of the dynamic equations of motion based on a CAD model of the system. The second approach is based on using a time domain simulation package such as Matlab's Simulink or MatrixX's SystemBuild®. The first approach is generally much easier and provides sufficient information to answer fundamental questions. In addition, in many cases, it is possible to integrate the dynamic simulation package with a control package such as Simulink or SystemBuild. The second approach requires a deeper understanding of the system's dynamics but enables greater flexibility in terms of modeling the system (including control algorithms, sensory feedback, arbitrary inputs).

21.3.4.1 Multibody Dynamic Packages

Two examples of multibody dynamic simulation packages are LMS International's Dynamic Motion Simulation (DADS) and MSC Software's Adams. With both packages, you begin by either building a parameterized model of the robot or importing the model from a CAD system, much like the simulation packages described earlier. The software automatically formulates and solves the dynamic equations of motion. Both MSC Adams and DADS® have integration capabilities with control simulation software such as Simulink and SystemBuild. The engineer can take the geometrically defined models and easily incorporate them within block diagrams created with the control system software. It is then possible to evaluate a wide class of mechanical and control designs and estimate the system's performance prior to ever cutting metal.

SolidWorks provides an option for integrating the Adams simulation technology directly into the SolidWorks environment. This interface, CosmosMotion®, is a physics-based motion simulation package that is fully embedded into the SolidWorks environment. The part's geometry and mass properties are defined inside SolidWorks. CosmosMotion reads these characteristics, along with all of the defined constraints. The engineer specifies joint motion or forces, and the software displays the resulting forces and/or motion of the system based on fundamental physical principles. The advantage of this approach is twofold. First, the design and simulation is embedded in the same interface. The user is not concerned with file transfers or compatability issues. Second, the same software package that the engineer uses to design the robot is used to define the geometry and mass properties for the simulation. These advantages come at a cost. A professional version of SolidWorks, at the time of this writing, is $4000. Likewise, the CosmosMotion package is $6000.

Another option is to import the CAD model into a time domain simulation package such as Matlab and Simulink. SimMechanics® provides an effortless importation of SolidWorks CAD assemblies into the Simulink time domain simulation environment. This approach enables the ability to simulate the behavior of mechanical, control, and other dynamic systems. As an example, we constructed a simple three-degrees-of-freedom manipulator in the SolidWorks environment. This model, shown in Figure 21.9, consists of four parts: a base and three links. These four parts are combined in an assembly. The assembly includes the ability to define constraints (what SolidWorks calls Mates) between parts. In the case of the robot which has three revolute joints, we require six constraints: three constraints forcing the joints on adjoining parts to be collinear and three constraints forcing each face on adjoining parts to be as close as possible. Now, when moving a part, the whole assembly moves, illustrating the constrained motion due to the robot's kinematics. It is also possible to include collision detection and restrict motion when constraints are reached. At this point, SolidWorks is restricted to constraint detection and ranges of motion. However, when defining parts, you can include not only the geometry but the types of materials as well (specifically the material's density). SolidWorks, like many CAD packages, includes the ability to compute, based upon the geometry and materials, the resulting mass properties for each part (mass, location of center of mass, principal axes of inertia,

FIGURE 21.9 SolidWorks robot model (courtesy Oak Ridge National Laboratory).

moments of inertia...). SimMechanics takes this information, along with the constraints, derives the dynamic equations of motion, and exports the results into a Simulink model. It is then straightforward to include actuation forces, sensors, and control to enable a high fidelity numerical simulation of the robot.

The minimum requirements for this approach is the cost associated with SolidWorks in addition to Matlab ($1900), Simulink ($2800), and SimMechanics ($4000). If these prices are beyond budgetary constraints, there are other options. First, there are software packages that are lower cost. In terms of dynamic simulation, a professional license for ProgramCC®,[6] is $750. The lower cost is offset by reduced flexibility. There is no graphical interface for developing simulation block diagrams as there is in Simulink or SystemBuild. The basic package is predominately script based. It is possible to construct sophisticated dynamic models and simulate the response of the system to various inputs, but it is not as straightforward as with Simulink.

21.4 Build Your Own

One last possibility is to build your own simulation capabilities. The clear disadvantage is time and know-how. Our motivation in this last section is to provide some insight into the know-how. There are a few obvious, and not so obvious, advantages to building your own simulation package. Clearly, cost is one consideration. If you already have a good understanding of programming, there are quite a few free or inexpensive utilities available that can ease the process of writing a robot simulation package. Another reason is flexibility. It is quite possible that there are no robot simulation packages available that exactly answer your needs. If this is the case, you may need to develop your own program.

---

[6]http://www.programcc.com/

21.4.1 Graphical Animation

There are a number of different ways to visually display robot motion. In this section, we will focus on where to begin if we wish to develop a package similar to the systems covered in the previous section. Many of these packages exploit the OpenGL interface. OpenGL was originally developed by Silicon Graphics, Inc. (SGI) as a multi-purpose, platform-independent graphical API. OpenGL is a collection of approximately 150 distinct commands that you can use to specify objects and operations needed to produce interactive three-dimensional applications. To provide a starting point, note that OpenGL is based on object oriented programming, C++. To efficiently develop a robot simulation package, you develop a set of classes that can construct simple objects such as spheres, cylinders, rectangles, and whatever basic objects you need to define a robot. OpenGL uses a number of vector and matrix operations to define objects in three-dimensional space. As a simple example, the following function would draw a cube.

```
Void DrawCube(float x, float y, float z)
{
    glPushMatrix();
            glTranslatef(x,y,z);
            glBegin(GL\_POLYGON);
            glVertex3f(0.0f, 0.0f, 0.0f); // top face
            glVertex3f(0.0f, 0.0f, -1.0f);
            glVertex3f(-1.0f, 0.0f, -1.0f);
            glVertex3f(-1.0f, 0.0f, 0.0f);
            glVertex3f(0.0f, 0.0f, 0.0f); // front face
            glVertex3f(-1.0f, 0.0f, 0.0f);
            glVertex3f(-1.0f, -1.0f, 0.0f);
            glVertex3f(0.0f, -1.0f, 0.0f);
            glVertex3f(0.0f, 0.0f, 0.0f); // right face
            glVertex3f(0.0f, -1.0f, 0.0f);
            glVertex3f(0.0f, -1.0f, -1.0f);
            glVertex3f(0.0f, 0.0f, -1.0f);
            glVertex3f(-1.0f, 0.0f, 0.0f); // left face
            glVertex3f(-1.0f, 0.0f, -1.0f);
            glVertex3f(-1.0f, -1.0f, -1.0f);
            glVertex3f(-1.0f, -1.0f, 0.0f);
            glVertex3f(0.0f, 0.0f, 0.0f); // bottom face
            glVertex3f(0.0f, -1.0f, -1.0f);
            glVertex3f(-1.0f, -1.0f, -1.0f);
            glVertex3f(-1.0f, -1.0f, 0.0f);
            glVertex3f(0.0f, 0.0f, 0.0f); // back face
            glVertex3f(-1.0f, 0.0f, -1.0f);
            glVertex3f(-1.0f, -1.0f, -1.0f);
            glVertex3f(0.0f, -1.0f, -1.0f);
    glEnd();
    glPopMatrix();

}
```

OpenGL uses what is called a matrix stack to construct complicated models of many simple objects. In the case of the cube, the function first gets the most recent coordinate frame, possibly pushed onto the stack prior to this function call. Then the function glTranslate() performs a translation on the coordinate frame. Likewise there are functions such as glRotate() and glScale() that provide the ability to rotate and scale the object. The glBegin() and glEnd() functions provide bounds on the definition of the verticies, defined by the glVertex3f() function, of the object.

Your robot will consist of a number of these primitives, thus the motivation for object-oriented programming. Animation of motion will require the coordination of these primitives. It is now clear that a visual robot simulation program requires a combination of programming for animation (using utilities such as OpenGL) and an understanding of robot kinematics. There are a number of references, including this handbook, that provide excellent guidelines for writing the algorithms necessary for providing

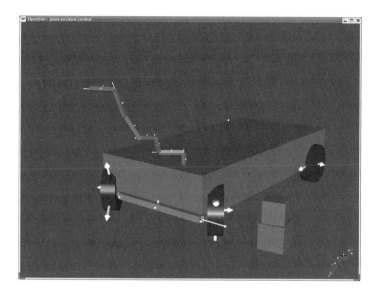

FIGURE 21.10 OpenSim simulation of vehicle and arm (courtesy of Oak Ridge National Laboratory).

coordinated motion of a robot (Asada and Slotine, 1989; Craig, 1989; Spong and Vidyasagar, 1989; and Yoshikawa, 1990). Basically, you will have some form of interface that provides a desired motion for your robot. This can be from a script file, mouse motion, slider bars on a dialog box, or any basic form of input into your application. The kinematic routines will transform the desired motion of the robot into the required transformations for each primitive or link on your robot. The results of these transformations are passed to functions such as glTranslate() and glRotate() to provide the animation of the coordinated motion.

Clearly, the above description is not sufficient to enable you to immediately start writing software for robot animation. However, it does provide a good starting point. There are many examples where research groups have used OpenGL to develop their own robot simulation package.[7,8] In addition, there are a number of good references on programming in OpenGL (Hawkings and Astle, 2001; Woo et al., 1999). These references, combined with the above robotics references, provide guidance on exactly how to write software that animates objects, for example, the robot shown in Figure 21.10.

21.4.2 Numerical Simulation

In Section 21.3, we briefly described the utility of programs such as Simulink and SystemBuild for simulating the motion of a robotic system. However, this approach required the use of a secondary software package for computing the dynamic equations of motion. In this final section, we will describe a simple procedure for computing the equations of motion for a serial link manipulator and embedding the dynamics in a Simulink simulation. The basic methodology requires the use of a symbolic software package and a deeper understanding of dynamics. The approach is based on using homogenous transforms to describe key points on the robot: joint locations and centers of mass. Using the transformations, we construct linear and angular velocity vectors about each center of mass. The displacement terms are used to construct a potential energy term while the velocity terms are used in the computation of the kinetic energy. We use the LaGrange approach to compute the resulting equations of motion. We then describe how to include joint

---

[7]ORNL's simulation package: http://marvin.ornl.gov/opensim/
[8]Georgia Tech's simulation package: http://www.imdl.gatech.edu/mbsim/index.html

and tip forces as well as partitioning the equations so that they are suitable for any numerical integration routine. While this approach is presently limited to serial link manipulators, it should give the reader a better understanding of the mechanics associated with simulation of robot dynamics.

Two primary elements in any time domain simulation consist of the state equations and an integration routine. The state equations are derived using basic principles in dynamics. Our objective is to formulate the dynamics into a general form, given in Equation (21.1), where $\bar{x}$ represents a vector of states for the system and $\bar{u}$ is a vector of control inputs. In the case of a robot, the vector $\bar{x}$ could represent the position and velocities of each of the joints. The vector $\bar{u}$ would contain the joint forces due to the actuators driving the robot. The equation is derived from the system's dynamic equations of motion.

$$\dot{\bar{x}} = F(\bar{x}, \bar{u}, t) \qquad (21.1)$$

During each time step of the simulation, we compute the forces driving the robot, possibly due to a control law, and compute the derivative of the state using Equation (21.1). We then use an integration routine to compute the state of the system, $\bar{x}$, for the next time step. The basic elements are shown in Figure 21.11.

Clearly, the computation of the state equations requires quite a bit of effort. The following section provides a basic procedure for computing the dynamic equation of motion, in the state equation format described above, for a general serial link manipulator. The approach is based on computing, symbolically, the position and orientation of each joint and center of mass of the robot. While not necessary, we use the Denavit-Hartenburg approach. First, as a review, the homogeneous transform is expressed using the traditional Denavit-Hartenburg (D-H) representation found in most robotics texts where the four quantities θ_i (angle), α_i (twist), d_i (offset), l_i (length) are parameters of link and joint i.

$$H_i = \begin{bmatrix} c_{\theta i} & -s_{\theta i} c_{\alpha i} & s_{\theta i} s_{\alpha i} & a_i c_{\theta i} \\ s_{\theta i} & c_{\theta i} c_{\alpha i} & -c_{\theta i} s_{\alpha i} & a_i s_{\theta i} \\ 0 & s_{\alpha i} & c_{\alpha i} & d_i \\ 0 & 0 & 0 & 1 \end{bmatrix} \qquad (21.2)$$

The conventional use of the homogeneous transform treats each subsequent transformation as a body fixed rotation and translation.

We also assume that we can define an additional set of homogeneous transforms from each joint to a point on each link where the associated mass properties (mass and inertia matrix) are known. So, our basic methodology consists of using two sets of homogeneous transformations; one to describe the pose of the arm and one to identify the displacements and velocities, both translation and rotation, of the center of mass of each link and payload with respect to the manipulators state (joint position and velocity). We extract from the transforms the vertical displacement of each center of mass of each link for an expression of the total potential energy of the system. Likewise, computation of the system's kinetic energy is based on computing the linear and angular and velocity of each links center of gravity with respect to the inertial frame. Once the kinetic and potential energy terms are derived, we simply use the Matlab `jacobian()` command to symbolically calculate the mass matrix and nonlinear dynamic

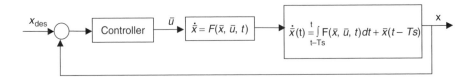

FIGURE 21.11 General simulation block diagram.

terms following the Lagrange formulation. Symbolic packages such as Maple and Mathmatica do all the rest.

First, the position of the center of mass for each link, with respect to the system's inertial coordinate system, is computed by iteratively postmultiplying each of the homogeneous transforms, starting from the base and working to the tip of the robot.

$$\begin{aligned} H^i_{\text{base}} &= H^{i-1}_{\text{base}} H^i_{i-1} \\ &= \begin{bmatrix} R^{i-1}_{\text{base}} & \bar{x}^{i-1}_{\text{base}} \\ \bar{0} & 1 \end{bmatrix} \begin{bmatrix} R^i_{i-1} & \bar{x}^i_{i-1} \\ \bar{0} & 1 \end{bmatrix} \\ &= \begin{bmatrix} R^i_{\text{base}} & \bar{x}^i_{\text{base}} \\ \bar{0} & 1 \end{bmatrix}, \quad \bar{x}^i_{\text{base}} = \begin{Bmatrix} x^i_{\text{base}} \\ y^i_{\text{base}} \\ z^i_{\text{base}} \end{Bmatrix} \end{aligned} \quad (21.3)$$

The potential energy due to gravity for link i is the vertical component (z^i_{base} in direction of gravity) of $\bar{x}^i_{\text{base}}$ times the mass of the link.

$$V^i = m_i g z^i_{\text{base}} \quad (21.4)$$

To compute the kinetic energy, we must first derive expressions for the linear and angular velocity of the center of gravity for each link as a function of the states of the manipulator. We have an expression, $\bar{x}^i_{\text{base}}$ in Equation (21.3), for the position of the center of gravity of link i with respect to the base inertial frame. The velocity vector $\bar{v}_i$ is computed by multiplying the Jacobian (with respect to the combined states of the manipulator $\bar{q}^i_{\text{base}}$) of $\bar{x}^i_{\text{base}}$, $J(\bar{x}^i_{\text{base}}, \bar{q})$, by the state velocity vector, $\dot{\bar{q}}$. Note from Equation (21.5), based on the chain rule, that $\bar{q}^i_{\text{base}}$ is a vector of robot joint displacements from the base of the robot (joint q_0) to the ith joint (q_i).

$$\begin{aligned} \bar{v}_i &= \frac{\partial \bar{x}^i}{\partial t} \\ &= \sum_{j=1}^{i} \frac{\partial \bar{x}^i}{\partial q_j} \dot{q}_j \\ &= J(\bar{x}^i, \bar{q}) \dot{\bar{q}} \quad \text{where } \bar{q} = [q_0 q_1 \ldots q_i] \end{aligned} \quad (21.5)$$

The rotational velocity is a little more involved but can be simplified by again using the homogeneous transform. Starting at the base of the robot, project the net rotational velocity vector forward to the center of mass of each link using the rotational matrix R^i_{base}. Each subsequent joint consists of projecting the total angular velocity vector of the previous joint to the current joint's coordinate system, using the rotational component of that joint's homogenous transform, and adding the joint angular velocity.

$$\bar{\omega}_i = \dot{\bar{q}}_i + R^i_{i-1} \bar{\omega}_{i-1} \quad (21.6)$$

We now have expressions for the linear and angular velocity of the center of mass for each link. The total kinetic energy of the system is

$$T = \frac{1}{2} \sum_{i=1}^{N} m_i v_i^t v_i + R^i_{\text{base}} \bar{\omega}_i^t [I_i] \bar{\omega}_i \quad (21.7)$$

where m_i is the mass of link i and I_i is the inertia matrix of link i about the center of gravity. As a final step, we add external forces applied to the system. For now, we assume forces are applied only to the joints and tip of the robot. We use the principle of virtual work to lump these terms together.

$$\partial W = \sum_{i=1}^{N} \tau_i \partial q_i + \bar{F}_{\text{tip}} \partial \bar{x}_{\text{tip}} + \bar{M}_{\text{tip}} \partial \theta_{\text{tip}}$$

$$= \left[\bar{\tau} + J_{\text{tip}}^t(\bar{q}) \left\{ \begin{array}{c} \bar{F}_{\text{tip}} \\ \bar{M}_{\text{tip}} \end{array} \right\} \right] \partial \bar{q}$$

$$= \bar{Q} \partial \bar{q} \qquad (21.8)$$

$$J_{\text{tip}}(\bar{q}) = \frac{\partial \bar{x}_{\text{base}}^{\text{tip}}}{\partial \bar{q}}$$

Equation (21.4) and Equation (21.7) provide expressions for the kinetic and potential energy of the system. We start with the classic definition of the Lagrange equations of motion.

$$\frac{\partial}{\partial t}\left(\frac{\partial T}{\partial \dot{\bar{q}}}\right) - \frac{\partial T}{\partial \bar{q}} + \frac{\partial V}{\partial \bar{q}} = \bar{Q} \qquad (21.9)$$

The first term in Equation (21.9) can be expanded using the chain rule.

$$\frac{\partial}{\partial t} = \frac{\partial}{\partial \bar{q}}\frac{\partial \bar{q}}{\partial t} + \frac{\partial}{\partial \dot{\bar{q}}}\frac{\partial \dot{\bar{q}}}{\partial t} \qquad (21.10)$$

Substituting Equation (21.10) into Equation (21.9),

$$\frac{\partial}{\partial \dot{\bar{q}}}\left(\frac{\partial T}{\partial \dot{\bar{q}}}\right)\ddot{\bar{q}} + \frac{\partial}{\partial \bar{q}}\left(\frac{\partial T}{\partial \dot{\bar{q}}}\right)\dot{\bar{q}} - \frac{\partial T}{\partial \bar{q}} + \frac{\partial V}{\partial \bar{q}} = \bar{Q} \qquad (21.11)$$

As with the velocity computation in Equation (21.5), we can exploit the jacobian() function in Matlab for the evaluation of many of the terms in Equation (21.11). First, the term $\partial T/\partial \dot{\bar{q}}$ is the differential of the scalar kinetic energy term with respect to the full state velocity vector defined in Equation (21.5). This results in the vector, L_v, in the script files shown below. The first term in Equation (21.11) represents the mass matrix. This expression is easily computed by taking the Jacobian on L_v with respect to the full state velocity vector. The remaining elements in Equation (21.11) represent the nonlinear dynamics (coriolis, centripetal, gravitational) of the system. Thus, it should be clear that once the kinetic and potential energy terms are defined, it is straightforward to symbolically evaluate the dynamic equations. The Jacobian for projecting external forces to the generalized coordinates can similarly be computed using the tip position of the robot and the Matlab jacobian() function.

This basic methodology was developed to compute the dynamics of a class of robots operating on the deck of a ship (a moving platform). We provide two examples to verify this methodology: a simple one-degree-of-freedom (dof) system operating on a three-dof moving platform and a three-dof system experiencing all six degrees of motion from the sea state. The first example is simple enough to verify through hand calculations. The second example is more complex, yet practical. Figure 21.12 shows the basic kinematic model of the one degree of freedom system experiencing three sea states in the X-Y plane. We are assuming a one-dof system with mass M and rotary inertia I_z located at the tip of a link of length L. The system is experiencing only three of the six sea states: surge (x_s), heave (y_s), and pitch (θ_s). The only external force applied to the system is a joint torque τ, applied at joint 1. Appendix A shows the listing of Matlab code used to generate the dynamic equations of motion. The two output variables of interest are the MassMatrix and NLT (nonlinear terms). The resulting output is listed in Equation (21.12) and can be

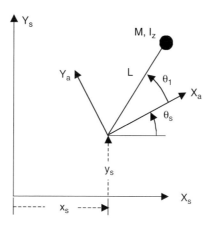

FIGURE 21.12 One DOF model.

easily verified by the reader. Clearly, for this simple case, there is not a great advantage to using a symbolic package over hand calculations.

MassMatrix $= ML1^2 + I_{1z}$

$$\text{NLT} = ML1^2\ddot{\theta}_p + I_{1z}\ddot{\theta}_p + ML1\cos(\theta_p + \theta_1)\ddot{y}_s - ML1\sin(\theta_p + \theta_1)\ddot{x}_s \\ + MgL1\cos(\theta_p + \theta_1)$$

combining (21.12)

$$(ML1^2 + I_{1z})(\ddot{\theta}_1 + \ddot{\theta}_p) + ML1\cos(\theta_p + \theta_1)\ddot{y}_s - ML1\sin(\theta_p + \theta_1)\ddot{x}_s \\ + MgL1\cos(\theta_p + \theta_1) = \tau_1$$

The power of this approach is more evident as we progress to more complex systems. First, all that is required is the definition of joint transformations. All of the differentiation for velocity and acceleration terms is handled by the symbolic computational package (such as Maple or Mathematica). As a second example, we derive the dynamic equations of motion for the three-dof system, shown in Figure 21.12, with the full six-dof from the sea state. A listing of the Matlab code used for computing the dynamics of this system on the deck of a ship is shown in Appendix B. It should be clear comparing the listings in Appendix A and Appendix B that there is only a slight difference in the formulation of the transformations, but the methodology for deriving the dynamics is the same. The resulting equations of motion can be partitioned into a compact form, Equation (21.13),

$$\begin{bmatrix} M_{rr} & M_{rs} \\ M_{rs}^t & M_{ss} \end{bmatrix} \begin{Bmatrix} \ddot{q}_r \\ \ddot{q}_s \end{Bmatrix} + \begin{Bmatrix} \text{NLT}_r \\ \text{NLT}_s \end{Bmatrix} = \begin{Bmatrix} \bar{Q}_r \\ \bar{0} \end{Bmatrix} + \begin{Bmatrix} J^t(q_r)F_{ext} \\ \bar{0} \end{Bmatrix} \quad (21.13)$$

where M_{rr} is the 3×3 mass matrix for the robot with respect to the robot's state acceleration, M_{ss} is the 6×6 mass matrix of the robot with respect to the sea state acceleration, M_{rs} is the inertial coupling of the sea state to the robot state, NLT_r is a 3×1 vector of the nonlinear terms (gravitational, Coriolis, centripetal) as a function of both the robot's state and the sea state, Q_r is the joint force input vector to the system, F_{ext} is an external force vector applied to the end effector, and $J^t(q_r)$ is the Jacobian from the end effector to the joint space. In order to include the dynamic equations of motion in Simulink, we use Equation (21.14) to solve for the acceleration of the robot's state vector as a function of all of the inputs (external forces and joint torques), system state (position and velocity), and external disturbances (sea state position, velocity,

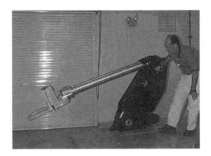

FIGURE 21.13 Force controlled hydraulic manipulator (courtesy Oak Ridge National Laboratory).

and acceleration). This expression is in the same form as Equation (21.1).

$$\ddot{\bar{q}}_r = M_{rr}^{-1}\{\bar{Q}_r + J^t(\bar{q}_r)\bar{F}_{ext} - \overline{\text{NLT}}_r - M_{rs}\ddot{\bar{q}}_s\} \tag{21.14}$$

While the output of the single degree of freedom case can be listed in Equation (21.12), the results of the dynamic equations of motion for the second system generates 84 pages of C code and would require considerable effort to derive by hand. Fortunately, the code can be directly imported into Simulink through the S-Function builder. The motivation for computing the dynamics equations of motion are threefold. First, by having the dynamics in a symbolic form, it is possible to aid in the design process, changing parameters to optimize the system. Second, a model of the system dynamics can aid in increasing the fidelity of simulation for control design and analysis. Finally, there is no doubt about how the simulation derived the system's equations of motion.

The system modeled in this investigation is displayed in Figure 21.13. This system has a 500-lb payload capacity and has three active degrees of freedom. The actuator models include nonlinear dynamic modeling of the hydraulic system (servo-valve orifice equations, asymetric cylinders, fluid compliance . . .), controls, and the dynamic equations of motion computed above.

The hydraulic actuator models generate force as a function of the servovalve current, actuator position, and velocity. The Simulink model of the full system, including the sea state inputs, hydraulic models, controls, and dynamic equations of motion, is shown in Figure 21.14.

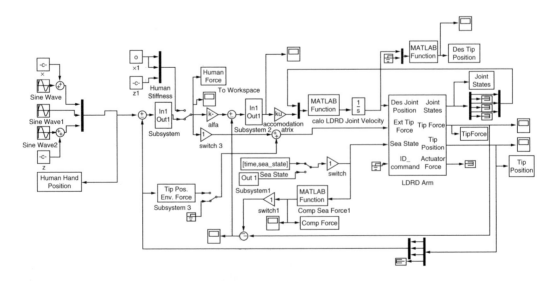

FIGURE 21.14 HAT simulation model (courtesy Oak Ridge National Laboratory).

Robot Simulation 21-19

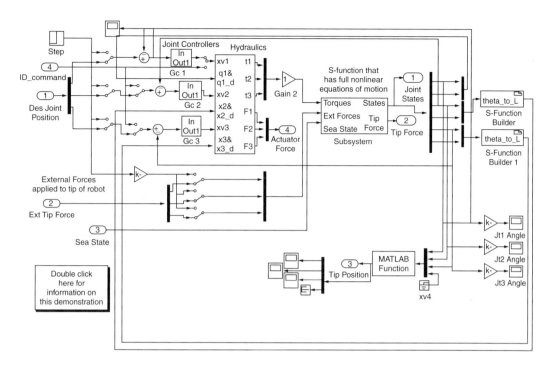

FIGURE 21.15 Details of HAT controller and manipulator model.

The nonlinear model of the hydraulic manipulator, shown as the LDRD arm block in Figure 21.14, is expanded in Figure 21.15. The inputs to the model include the desired joint positions, the eighteen elements of the sea state (displacement, velocity, and acceleration of roll, pitch, yaw, heave, surge, and sway), and the external force applied to the robot. In addition, this model includes auxiliary inputs for system identification.

There are two primary blocks to note in Figure 21.15. The first is the system dynamics block. This contains the C code generated previously that represents the forward dynamics of the LDRD arm. The second is the hydraulics block. From the expanded hydraulics block, each of the three joints has two primary elements: the servo valve and the actuator. In the case of the second and third joints, there is also a transmission associated with the coupling of the linear actuator with the joint (the two "theta_to_L" blocks in the upper right). The servo valve models are based on the full nonlinear orifice equations and regulate the fluid flow to the actuators as a function of excitation signal (command to the moving coil on the servo vavle), the supply, and return pressure, as well as the pressure on both sides of the actuator. The actuator model generates a force based on the position, velocity, and bulk modulus (stiffness) of the fluid. The position and velocity of the actuator come from the dynamic model of the manipulator. The force from the actuator is the excitation to the dynamic model of the manipulator. It should now be clear that the full nonlinear behavior of the manipulator is embodied in the simulation of the manipulator.

To complete our example, we now consider the impact sea state has on the performance of a force controlled system. Inputs to the system consist of a human command regulating about a single point and a sea state with conditions commensurate with a destroyer moving at 20 knots in sea state 5. Figure 21.16 shows the force provided by the human while attempting to regulate the tip position. There is the expected DC bias on the Z-direction force required to offset the gravitational load. The system is used for strength amplification. The human only senses a fraction of the total payload weight. For this example, the payload weight is 2224 N which projects to a human force of 22.4 N when in static equilibrium with a 100:1 force amplification ratio. However, during the 120 sec simulation run, the maximum perturbation felt by

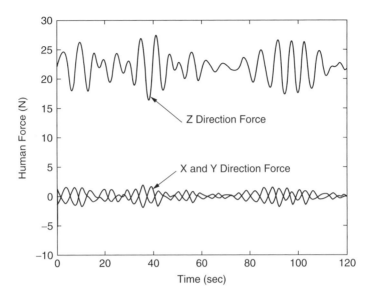

FIGURE 21.16 Human force without compensation.

the operator, after the initial transient, was 11.38 N, approximately 51% of the actual load. In addition, there are orthogonal disturbances in the X and Y directions due to the rolling, swaying and surging motion of the ship. The resulting tip motion is displayed in Figure 21.17, Figure 21.18, Figure 21.19, and Figure 21.20. It is clear that the vertical direction is the most sensitive to the sea state. The variation in the vertical tip position is 22.8 mm. The tracking error above is a function of two inputs: the force (due to the sea state) applied to the arm and the commanded motion from the force applied by the operator.

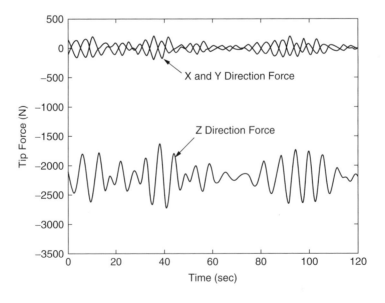

FIGURE 21.17 Tip force without compensation.

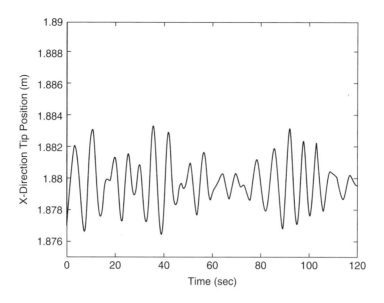

FIGURE 21.18 X tip direction.

The primary motivation for this exercise is to show the potential for robot simulation. In the case above, studying the control of a robot on the deck of a moving ship could prove quite costly in terms of either deploying hardware on a ship or designing and constructing a testbed capable of emulating ship motion. Clearly, the final step in the process is validation of the simulation through experimentation. However, a high fidelity simulation can provide insight into many problems at a fraction of the cost. In addition, simulation reduces the hazards associated with testing immature control concepts. No one gets hurt if the simulation fails.

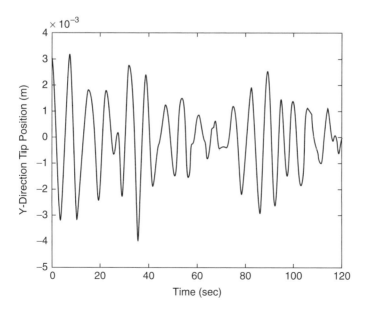

FIGURE 21.19 Y tip direction.

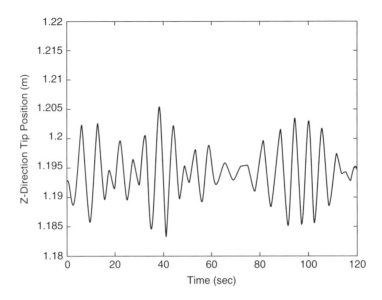

FIGURE 21.20 Z tip direction.

21.5 Conclusion

This chapter focused on recent developments in robot simulation. The traditional role of robot simulation in the automotive and aerospace industries still exists but has expanded in terms of its utility and flexibility. However, more advanced engineering tools are providing the ability for rapid prototyping, design optimization, and hardware in the loop-simulation. With the expanding use of robotics in areas such as medicine and the battlefield, as well as advancements in haptics and man-machine interfaces, it is likely that there will be greater interaction between humans and simulation in the near future.

References

Asada, H. and Slotine, J. (1989). *Robot Analysis and Control,* Wiley Interscience, New York.
Bridges, M. and Diamond, D. (1999). The financial impact of teaching surgical residents in the operating room, *Am. J. Surg.,* Vol. 177, pp. 28–32.
Craig, J. (1989). *Introduction to Robotics: Mechanics and Control,* 2nd ed., Addison-Wesley, Reading, MA.
Cushieri, A. (1995). Whither minimal access surgery: tribulations and expectations, *Am. J. Surg.,* Vol. 69, pp. 9–19.
Hawkings, K. and Astle, D. (2001). *OpenGL Game Programming,* Prima Publishing, Roseville, CA.
Spong, M. and Vidyasagar, M. (1989). *Robot Dynamics and Control,* John Wiley & Sons, New York.
Van Hoesen, D., Bzorgi, F., Kelsey, A., Wiles, C., and Beeson, K. (1995). Underground radioactive waste tank remote inspection and sampling, DOE EM Fact Sheet, Gunite and Associated Tanks Operable Unit at Waste Area Grouping 1 at Oak Ridge National Laboratory, The Department of Energy Environmental Program.
Woo, M., Neider, J., Davis, T., and Shreiner, D. (1999). *OpenGL Programming Guide: The Official Guide to Learning OpenGL,* 3rd ed., Addison-Wesley, Reading, MA.
Yong, Y. and Bonney, M. (1999). Offline programming, In: *Handbook of Industrial Robotics,* S. Nof, ed., John Wiley & Sons, New York, Chapter 19.
Yoshikawa, T. (1990). *Foundations of Robotics: Analysis and Control,* MIT Press, Cambridge.

Appendix A: Matlab Code for The Single DOF Example

```
syms ai alfai di thi mass g
syms q1 q1_d q1_dd
syms roll pitch yaw heave surge sway
syms roll_d pitch_d yaw_d heave_d surge_d sway_d
syms roll_dd pitch_dd yaw_dd heave_dd surge_dd sway_dd
syms th1 th1d th1dd
syms th2 th2d th2dd
syms th3 th3d th3dd
syms L1 I1x I1y I1z M

pi = sym('pi');

% Symbolically derive motion of base of robot on deck of ship experiencing
% 6 dof of sea motion (roll, pitch, yaw, heave, surge, sway).
R1s=[1  0          0;
     0  cos(roll)  sin(roll);
     0 -sin(roll)  cos(roll)];

R2s=[cos(pitch)  0  -sin(pitch);
     0           1   0;
     sin(pitch)  0   cos(pitch)];

R3s=[cos(yaw)   sin(yaw)  0;
    -sin(yaw)   cos(yaw)  0;
     0          0         1];

Hsea=[[simple(R1s*R2s*R3s) [surge;sway;heave;]];[0 0 0 1]];

% mask off 3 of 6 sea states (keep sea states in x-z plane)
Hsea=subs(Hsea,'roll',0);
Hsea=subs(Hsea,'yaw',0);
Hsea=subs(Hsea,'sway',0);

% compute kinematics of arm, start off by rotating z from vertical to horizontal
ai=0;alfai=pi/2;di=0;thi=0;
H1=[cos(thi)   -sin(thi)*cos(alfai)  sin(thi)*sin(alfai)   ai*cos(thi);
    sin(thi)    cos(thi)*cos(alfai) -cos(thi)*sin(alfai)   ai*sin(thi);
    0           sin(alfai)           cos(alfai)            di;
    0           0                    0                     1];

ai=L1;alfai=0;di=0;thi=th1;
H2=[cos(thi)   -sin(thi)*cos(alfai)  sin(thi)*sin(alfai)   ai*cos(thi);
    sin(thi)    cos(thi)*cos(alfai) -cos(thi)*sin(alfai)   ai*sin(thi);
    0           sin(alfai)           cos(alfai)            di;
    0           0                    0                     1];

% Homogeneous transform for the robot
H=simple(H1*H2);
% Full homogeneous transform of robot include sea state
H_full=simple(Hsea*H);

% state vector of robot
q=[th1];
% derivitive of state vector
qd=[th1d];

% sea state vector
qs=[pitch;heave;surge];
% sea state velocity
```

```
qsd=[pitch_d;heave_d;surge_d];
% sea state acceleration
qsdd=[pitch_dd;heave_dd;surge_dd];

% velocity computation. Each velocity is velocity of the cg of the line wrt base
coordinate system
% velocity of cg of 2nd link
Ht=simple(Hsea*H1*H2); % homogeneous transform from base to cg of link 2
R2c=Ht((1:3),4); % pull out x, y, z (vector from base to link 2 cg)
V2c=Jacobian(R2c,[q(1);qs])*[qd(1);qsd]; % calculate velocity of cg
of link 2 wrt inertial frame (V = dR/dt = dR/dq * dq/dt)

% rotation matricies from base to each associated coordinate system
R1=transpose(H1(1:3,1:3));
R2=transpose(H2(1:3,1:3));

% angular velocity of each link about cg wrt local coordinate frame
Q1=[0;0;th1d]+[0;0;qsd(1)];

% inertia matrix for each link about center of gravity wrt coordinate frame of
line (same as homogeneous transform, just translated to cg)
I1=[I1x 0 0;0 I1y 0;0 0 I1z];

% Payload information (position/velocity)
Rtip=H_full(1:3,4);
Vtip=Jacobian(Rtip,[q;qs])*[qd;qsd];

% total kinetic energy: T = 1/2 qdot' * I * qdot + 1/2 V' M V
T=1/2*M*(transpose(Vtip)*Vtip)+1/2*transpose(Q1)*I1*Q1;

% potential energy due to gravity
V=M*g*Rtip(3);

%=====================================================================
% Energy approach: d/dt(dT/dqd)-dT/dq+dV/dq=Q
% Note: d/dt() = d()/dq * qd + d()/dqd * qdd.
% thus, energy expansion: d(dT/dqd)/dqd * qdd + d(dT/dqd)/dq * qd
-dT/dq + dV/dq = Q.
% first term d(dT/dqd)/dqd is mass matrix, nonlinear terms (coriolis,
centripetal, gravity...)
%=====================================================================

% calculate dT/dqdot
dT_qdot=Jacobian(T,qd);

% extract out mass matrix
MassMatrix= simple(Jacobian(dT_qdot,qd));

% now finish off with remaining terms
NLT1=simple(Jacobian(dT_qdot,[q])*[qd]);
NLT2=simple(Jacobian(dT_qdot,transpose(qs))*qsd);
NLT3=simple(Jacobian(dT_qdot,transpose(qsd))*qsdd);
NLT4=simple(transpose(-1*(Jacobian(T,q))));
NLT5=simple(((Jacobian(V,q))));
NLT=simple(NLT1+NLT2+NLT3+NLT4+NLT5);
```

Appendix B: Matlab Code for The 3-DOF System, Full Sea State

```
syms ai alfai di thi mass g
syms q1 q1_d q1_dd
syms roll pitch yaw heave surge sway
```

```
syms roll_d pitch_d yaw_d heave_d surge_d sway_d
syms roll_dd pitch_dd yaw_dd heave_dd surge_dd sway_dd
syms th1 th1d th1dd
syms th2 th2d th2dd
syms th3 th3d th3dd
syms L1 L2 L3 L4 L1c L2c L3c L3x L3h L3y L4c
syms I1x I2x I3x I1y I2y I3y I1z I2z I3z
syms m1 m2 m3

pi = sym('pi');

% Symbolically derive motion of base of robot on deck of ship experiencing
% 6 dof of sea motion (roll, pitch, yaw, heave, surge, sway). Use DH
  parameters
% to describe this motion in terms of homogeneous transforms.
R1s=[1   0           0;
     0   cos(roll)   sin(roll);
     0   -sin(roll)  cos(roll)];

R2s=[cos(pitch)  0   -sin(pitch);
     0           1   0;
     sin(pitch)  0   cos(pitch)];

R3s=[cos(yaw)   sin(yaw)  0;
     -sin(yaw)  cos(yaw)  0;
     0          0         1];

Hsea=[[simple(R1s*R2s*R3s) [surge;sway;heave;]];[0 0 0 1]];

ai=-L1;alfai=pi/2;di=0;thi=th1;
H1=[cos(thi)   -sin(thi)*cos(alfai)  sin(thi)*sin(alfai)   ai*cos(thi);
    sin(thi)   cos(thi)*cos(alfai)   -cos(thi)*sin(alfai)  ai*sin(thi);
    0          sin(alfai)            cos(alfai)            di;
    0          0                     0                     1];

ai=L2;alfai=0;di=0;thi=th2+pi/2;
H2=[cos(thi)   -sin(thi)*cos(alfai)  sin(thi)*sin(alfai)   ai*cos(thi);
    sin(thi)   cos(thi)*cos(alfai)   -cos(thi)*sin(alfai)  ai*sin(thi);
    0          sin(alfai)            cos(alfai)            di;
    0          0                     0                     1];

ai=L2c;alfai=0;di=0;thi=th2+pi/2;
H2c=[cos(thi)   -sin(thi)*cos(alfai)  sin(thi)*sin(alfai)   ai*cos(thi);
     sin(thi)   cos(thi)*cos(alfai)   -cos(thi)*sin(alfai)  ai*sin(thi);
     0          sin(alfai)            cos(alfai)            di;
     0          0                     0                     1];

ai=L3x;alfai=0;di=0;thi=th3-pi/2;
H3=[cos(thi)   -sin(thi)*cos(alfai)  sin(thi)*sin(alfai)   ai*cos(thi);
    sin(thi)   cos(thi)*cos(alfai)   -cos(thi)*sin(alfai)  ai*sin(thi);
    0          sin(alfai)            cos(alfai)            di;
    0          0                     0                     1];

ai=L3h;alfai=0;di=0;thi=th3-pi/2;
H3h=[cos(thi)   -sin(thi)*cos(alfai)  sin(thi)*sin(alfai)   ai*cos(thi);
     sin(thi)   cos(thi)*cos(alfai)   -cos(thi)*sin(alfai)  ai*sin(thi);
     0          sin(alfai)            cos(alfai)            di;
     0          0                     0                     1];

ai=L3c;alfai=0;di=0;thi=th3-pi/2;
H3c=[cos(thi)   -sin(thi)*cos(alfai)  sin(thi)*sin(alfai)   ai*cos(thi);
     sin(thi)   cos(thi)*cos(alfai)   -cos(thi)*sin(alfai)  ai*sin(thi);
```

```
                0         sin(alfai)           cos(alfai)         di;
                0         0                    0                  1];

ai=L3y;alfai=0;di=0;thi=pi/2;
H4=[cos(thi)    -sin(thi)*cos(alfai) sin(thi)*sin(alfai)  ai*cos(thi);
    sin(thi)    cos(thi)*cos(alfai)  -cos(thi)*sin(alfai) ai*sin(thi);
    0           sin(alfai)           cos(alfai)           di;
    0           0                    0                    1];

ai=L4;alfai=-pi/2;di=0;thi=th4-pi/2;
H5=[cos(thi)    -sin(thi)*cos(alfai) sin(thi)*sin(alfai)  ai*cos(thi);
    sin(thi)    cos(thi)*cos(alfai)  -cos(thi)*sin(alfai) ai*sin(thi);
    0           sin(alfai)           cos(alfai)           di;
    0           0                    0                    1];

H=simple(H1*H2*H3*H4*H5);    % Homogeneous transform for the robot
H_full=simple(Hsea*H);       % Full homogeneous transform of robot include sea state

q=[th1;th2;th3]; % state vector of robot
qd=[th1d;th2d;th3d]; % derivitive of state vector

qs=[roll;pitch;yaw;heave;surge;sway]; % sea state vector
qsd=[roll_d;pitch_d;yaw_d;heave_d;surge_d;sway_d]; % sea state velocity
qsdd=[roll_dd;pitch_dd;yaw_dd;heave_dd;surge_dd;sway_dd]; % sea
state acceleration

% velocity computation. Each velocity is the velocity of the cg of the line wrt
base coordinate system
% velocity of cg of 2nd link
Ht=simple(Hsea*H1*H2c); % homogeneous transform from base to cg of link 2
R2c=Ht((1:3),4); % pull out x, y, z (vector from base to link 2 cg)
V2c=Jacobian(R2c,[q(1:2);qs])*[qd(1:2);qsd]; % calculate velocity of cg of link 2
wrt inertial frame (V = dR/dt = dR/dq * dq/dt)

% velocity of cg of 3rd link}
Ht=simple(Hsea*H1*H2*H3c);
R3c=Ht((1:3),4);
V3c=simple(Jacobian(R3c,[q;qs])*[qd;qsd]);

% rotation matricies from base to each associated coordinate system
R1=transpose(H1(1:3,1:3));
R2=transpose(H2(1:3,1:3));
R3=transpose(H3(1:3,1:3));

% angular velocity of each link about cg wrt local coordinate frame
Q1=[0;0;th1d]+qsd(1:3);
Q2=simple(R2*R1*Q1+[0;0;th2d]);
Q3=simple([0;0;th3d]+R3*Q2);

% inertia matrix for each link about center of gravity wrt coordinate frame of
line (same as homogeneous transform, translated to cg)
I1=[I1x 0 0;0 I1y 0;0 0 I1z];
I2=[I2x 0 0;0 I2y 0;0 0 I2z];
I3=[I3x 0 0;0 I3y 0;0 0 I3z];

% Payload information (position/velocity)
syms M_payload;
Rtip=H_full(1:3,4);
Vtip=Jacobian(Rtip,[th1;th2;th3;roll;pitch;yaw;heave;surge;sway])*[th1d;th2d;th3d;
roll_d;pitch_d;yaw_d;heave_d;surge_d;sway_d];
```

```
% total kinetic energy: T = 1/2 qdot' * I * qdot + 1/2 V' M V
T=(1/2*transpose(Q1)*I1*Q1 + 1/2*transpose(Q2)*I2*Q2 + 1/2*transpose(Q3)*I3*Q3
 + ... 1/2*m2*transpose(V2c)*V2c + 1/2*m3*transpose(V3c)*V3c) +
1/2*M_payload*(transpose(Vtip)*Vtip);;

% potential energy due to gravity
V=m2*g*R2c(3)+m3*g*R3c(3)+M_payload*g*Rtip(3);

% calculate dT/dqdot
dT_qdot=Jacobian(T,qd);

% extract out mass matrix
MassMatrix= simple(Jacobian(dT_qdot,qd));

% now finish off with remaining terms
NLT1=simple(Jacobian(dT_qdot,[q])*[qd]);
NLT2=simple(Jacobian(dT_qdot,transpose(qs))*qsd);
NLT3=simple(Jacobian(dT_qdot,transpose(qsd))*qsdd);
NLT4=simple(-1*transpose((Jacobian(T,q))));
NLT5=simple(((Jacobian(V,q))));

% translate to C-code
MassMatrix_cc=ccode(MassMatrix);
NLT1_cc=ccode(NLT1);
NLT2_cc=ccode(NLT2);
NLT3_cc=ccode(NLT3);
NLT4_cc=ccode(NLT4);
NLT5_cc=ccode(NLT5);

% calculation of jacobian from tip frame to joint space
LDRDJacobian=simple(Jacobian(H(1:3,4),[th1;th2;th3]));
LDRDJacobian_cc=ccode(LDRDJacobian);
```

22
A Survey of Geometric Vision

22.1	Introduction .... 22-1 Camera Model and Image Formation • 3-D Reconstruction Pipeline • Further Readings
22.2	Two-View Geometry .... 22-4 Epipolar Constraint and Essential Matrix • Eight-Point Linear Algorithm • Further Readings
22.3	Multiple-View Geometry .... 22-8 Rank Condition on Multiple Views of Point Feature • Linear Reconstruction Algorithm • Further Readings
22.4	Utilizing Prior Knowledge of the Scene — Symmetry 22-13 Symmetric Multiple-View Rank Condition • Reconstruction from Symmetry • Further Readings
22.5	Comprehensive Examples and Experiments .... 22-17 Automatic Landing of Unmanned Aerial Vehicles • Automatic Symmetry Cell Detection, Matching and Reconstruction • Semiautomatic Building Mapping and Reconstruction • Summary

Kun Huang
Ohio State University

Yi Ma
University of Illinois

22.1 Introduction

Vision is one of the most powerful sensing modalities. In robotics, machine vision techniques have been extensively used in applications such as manufacturing, visual servoing [7, 30], navigation [9, 26, 50, 51], and robotic mapping [58]. Here the main problem is how to reconstruct both the pose of the camera and the three-dimensional (3-D) structure of the scene. This reconstruction inevitably requires a good understanding of the geometry of image formation and 3-D reconstruction. In this chapter, we provide a survey of the basic theory and some recent advances in the geometric aspects of the reconstruction problem. Specifically, we introduce the theory and algorithms for reconstruction from two views (e.g., see [29, 31, 33, 40, 67]), multiple views (e.g., see [10, 12, 23, 37, 38, 40]), and a single view (e.g., see [1, 3, 19, 25, 28, 70, 73, 74]). Since this chapter can only provide a brief introduction to these topics, the reader is referred to the book [40] for a more comprehensive treatment. Without any knowledge of the environment, reconstruction of a scene requires multiple images. This is because a single image is merely a 2-D projection of the 3-D world, for which the depth information is lost. When multiple images are available from different known viewpoints, the 3-D location of every point in the scene can be determined uniquely by triangulation (or stereopsis). However, in many applications (especially those for robot vision), the viewpoints are also unknown. Therefore, we need to recover both the scene structure and the camera poses. In computer vision literature, this is referred to as the "structure from motion" (SFM) problem. To solve this problem, the theory of *multiple-view geometry* has been developed (e.g., see [10, 12, 23, 33, 37, 38, 40, 67]). In this chapter, we introduce the basic theory of multiple-view geometry and show how it can be used to develop algorithms for reconstruction purposes. Specifically, for the two-view case, we introduce in Section 22.2 the *epipolar constraint* and the *eight-point structure from motion algorithm* [29, 33, 40]. For the multiple-view case, we introduce in Section 22.3 the *rank conditions* on *multiple-view matrix* [27, 37, 38, 40] and a multiple-view factorization algorithm [37, 40].

Since many robotic applications are performed in a man-made environment such as inside a building and around an urban area, much of prior knowledge can be exploited for a more efficient and accurate reconstruction. One kind of prior knowledge that can be utilized is the existence of "regularity" in the man-made environment. For example, there exist many parallel lines, orthogonal corners, and regular shapes such as rectangles. In fact, much of the regularity can be captured by the notion of *symmetry*. It can be shown that with sufficient symmetry, reconstruction from a single image is feasible and accurate, and many algorithms have been developed (e.g., see [1, 3, 25, 28, 40, 70, 73]). Interestingly, these symmetry-based algorithms, in fact, rely on the theory of multiple-view geometry [3, 25, 70]. Therefore, after the multiple-view case is studied, we introduce in Section 22.4 basic geometry and reconstruction algorithms associated with imaging and symmetry.

In the remainder of this section, we introduce in Section 22.1.1 basic notation and concepts associated with image formation that help the development of the theory and algorithms. It is not our intention to give in this chapter all the details about how the algorithms surveyed can be implemented in real vision systems. While we will discuss briefly in Section 22.1.1 a pipeline for such a system, we refer the reader to [40] for all the details.

22.1.1 Camera Model and Image Formation

The camera model we adopt in this chapter is the commonly used pinhole camera model. As shown in Figure 22.1A, the camera comprises a camera center o and an image plane. The distance from o to the image plane is the *focal length* f. For any 3-D point p in the opposite side of the image plane with respect to o, its image x is obtained by intersecting the line connecting o and p with the image plane. In practice, it is more convenient to use a mathematically equivalent model by moving the image plane to the "front" side of the camera center as shown in Figure 22.1B.

There are usually three coordinate frames in our calculation. The first one is the *world frame*, also called *reference frame*. The description of any other coordinate frame is the motion between that frame and the reference frame. The second is the *camera frame*. The origin of the camera frame is the camera center, and the z-axis is along the perpendicular line from o to the image plane as shown in Figure 22.1B. The last one

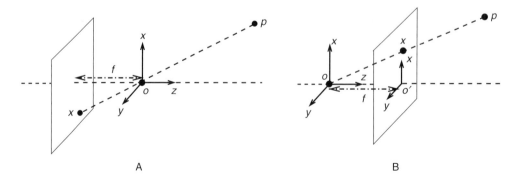

FIGURE 22.1 A: Pinhole imaging model. The image of a point p is the intersecting point x between the image plane and the ray passing through camera center o. The distance between o and the image plane is f. B: Frontal pinhole imaging model. The image plane is in front of the camera center o. An image coordinate frame is attached to the image plane.

is the 2-D *image frame*. For convenience we choose its origin o' as the projection of o on the image plane and set its x-axis and y-axis to be parallel to the x-axis and y-axis of the camera frame. As a convention, the focal length f is set to be of unit 1.

For points in 3-D space and 2-D image, we use *homogeneous coordinates*; i.e., a 3-D point p with coordinate $[x, y, z]^T \in \mathbb{R}^3$ is denoted as $X = [x, y, z, 1]^T \in \mathbb{R}^4$, and an image point with coordinate $[x, y]^T \in \mathbb{R}^2$ is denoted as $x = [x, y, 1]^T \in \mathbb{R}^3$. The motion of the camera frame with respect to the reference frame is sometimes referred to as the *camera pose* and is denoted as $[R, T] \in SE(3)$ with $R \in SO(3)$ being the rotation ($RR^T = I$) and $T \in \mathbb{R}^3$ being the translation.[1] Therefore, for a point with coordinate X in the reference frame, its image is obtained by

$$\lambda x = [R, T] X \in \mathbb{R}^3 \tag{22.1}$$

where $\lambda > 0$ is the depth of the 3-D point with respect to the camera center. This is the *perspective projection* model of image formation.

22.1.1.1 The Hat Operator

One notation that will be extensively used in this chapter is the hat operator "$\hat{\ }$" that denotes the skew-symmetric matrix associated to a vector in $\mathbb{R}^3$. More specifically, for a vector $u = [u_1, u_2, u_3]^T \in \mathbb{R}^3$, we define

$$\hat{u} \doteq \begin{bmatrix} 0 & -u_3 & u_2 \\ u_3 & 0 & -u_1 \\ -u_2 & u_1 & 0 \end{bmatrix} \in \mathbb{R}^{3 \times 3}$$

such that $\hat{u}v = u \times v, \forall v \in \mathbb{R}^3$. In particular, $\hat{u}u = 0 \in \mathbb{R}^3$.

22.1.1.2 Similarity

We use "$\sim$" to denote similarity. For any pair of vectors or matrices x and y of the same dimension, $x \sim y$ means $x = \alpha y$ for some (nonzero) scalar $\alpha \in \mathbb{R}$.

22.1.2 3-D Reconstruction Pipeline

Before we delve into the geometry, we first need to know how the algorithms to be developed can be used. Reconstruction from multiple images often consists of three steps: *feature extraction*, *feature matching*, and *reconstruction using multiple-view geometry*.

Features, or image primitives, are the conspicuous image entities such as corner points, line segments, or structures. The most commonly used image features are points and line segments. Algorithms for extracting these features can be found in most image processing papers and handbooks [4, 18, 22, 40]. At the end of this chapter we also give an example of using (symmetric) structures as image features for reconstruction.

Feature matching is to establish correspondence of features across different views, which is usually a difficult task. Many techniques have been developed to match features. For instance, when the motion (baseline) between adjacent views is small, feature matching is often called *feature tracking* and typically involves finding an affine transformation between the image patches around the feature points to be matched [34, 54]. Matching across large motion is a much more challenging task and is still an active research area. If a large number of image points is available for matching, some statistical technique such as the RANSAC type of algorithms [13] can be applied. Readers can refer to [23, 40] for details.

Given image features and their correspondences, the camera pose and the 3-D structure of these features can then be recovered using the methods that will be introduced in the rest of this chapter.

---

[1] By setting the motion in $SE(3)$, we consider only rotations and translations. Reflections are not included since it is in $E(3)$, but not in $SE(3)$.

22.1.3 Further Readings

22.1.3.1 Camera Calibration

In reality, there are at least two major differences between our camera model and the real camera. First, the focal length f of the real camera is not 1. Second, the origin of the image coordinate frame is usually chosen at the top-left corner of the image instead of at the center of the image. Therefore, we need to map the real image coordinate to our homogeneous representation. This process is called *camera calibration*. In practice, camera calibration is more complicated due to the pixel aspect ratio and nonlinear radial distortion of the image. The simplest calibration scheme is to consider only the focal length and location of the image center. So the actual image coordinates of a point are given by

$$\lambda \boldsymbol{x} = K[R, T]\boldsymbol{X}, \quad K = \begin{bmatrix} f & 0 & x_0 \\ 0 & f & y_0 \\ 0 & 0 & 1 \end{bmatrix} \in \mathbb{R}^{3\times 3} \quad (22.2)$$

where K is the *calibration matrix* with f being the real focal length and $[x_0, y_0]^T$ being the location of the image center in the image frame. The related theory and algorithms for camera calibration have been studied extensively in the computer vision literature, readers can refer to [5, 20, 41, 60, 71]. In the rest of this chapter, unless otherwise stated, we assume the camera is always calibrated, and we will use Equation (22.1) as the camera model.

22.1.3.2 Different Image Surfaces

In the pinhole camera model (Figure 22.1), we assume that the image plane is a planar surface. However, there are other types of image surfaces such as spheres in omni-directional cameras. For different image surfaces, the theory that we are going to introduce in this chapter still holds with only slight modification. Interested readers please refer to [16, 40] for details.

22.1.3.3 Other Types of Projections

Besides the perspective projection model in Equation (22.1), other types of projections have also been adopted in the literature for various practical or analytical purposes. For example, there are *affine projection* for an uncalibrated camera and *orthographic projection* and *weak perspective projection* for far away objects. For a detailed treatment of these projection models, please refer to [11, 23, 59].

22.2 Two-View Geometry

Let us first study the two-view case. German mathematician Erwin Kruppa [32] is among the first who studied this problem. He showed that given five pairs of corresponding points, the camera motion and structure of the scene can be solved up to a finite number of solutions [24, 32, 46]. In practice, however, we usually can obtain more than five points, which may significantly reduce the complexity of the solution and increase the accuracy. In 1980, Longuet-Higgins [33], based on the epipolar constraint — an algebraic constraint governing two images of a point, developed an efficient linear algorithm that requires eight pairs of corresponding points. The algorithm has since been refined several times to reach the current standard *eight-point linear algorithm* [29, 40]. Several variations to this algorithm for coplanar point features and for continuous motions have also been developed [40, 67]. In the remainder of this section, we introduce the epipolar constraint and the eight-point linear algorithm.

22.2.1 Epipolar Constraint and Essential Matrix

The key issue in solving a two-view SFM problem is to identify the algebraic relationship between corresponding image points and the camera motion. Figure 22.2 shows the relationship between the two

A Survey of Geometric Vision

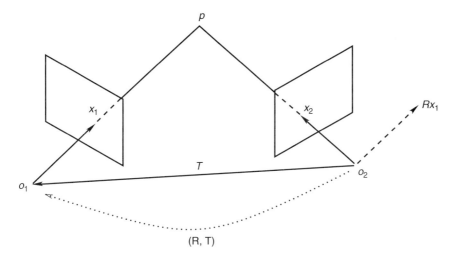

FIGURE 22.2 Two views of a point p. The vectors T, x_2, and Rx_1 are the three vectors all expressed in the second camera frame and are coplanar.

camera centers o_1, o_2, the 3-D point p with coordinate $X \in \mathbb{R}^4$, and its two images x_1 and x_2. Obviously, the three points o_1, o_2, and p form a plane, which implies that the vectors T, Rx_1, and x_2 are coplanar. Mathematically, it is equivalent to the triple product of T, Rx_1, and x_2 being zero, i.e.,

$$x_2^T \widehat{T} R x_1 = 0 \tag{22.3}$$

This relationship is called *epipolar constraint* on the pair of images and the camera motion. We denote $E = \widehat{T}R$ and call E the *essential matrix*.

22.2.2 Eight-Point Linear Algorithm

Given image correspondences for $n(\geq 8)$ 3-D points in general positions, the camera motion can be solved linearly. Conceptually it comprises two steps: first, recover the matrix E using n epipolar constraints; then decompose E to obtain motion R and T. However, due to the presence of noise, the recovered matrix E may not be an essential matrix. An additional step of projecting E into the space of essential matrices is necessary prior to the decomposition.

First, let us see how to recover E using the epipolar constraint. Denote

$$E = \begin{bmatrix} e_1 & e_2 & e_3 \\ e_4 & e_5 & e_6 \\ e_7 & e_8 & e_9 \end{bmatrix}$$

and let $E^s = [e_1, e_4, e_7, e_2, e_5, e_8, e_3, e_6, e_9]^T$ be a "stacked" version of E. The epipolar constraint in Equation (22.3) can be written as

$$(x_1 \otimes x_2)^T E^s = 0 \tag{22.4}$$

where $\otimes$ denotes the *Kronecker product* of two vectors such that

$$x_1 \otimes x_2 = [x_1 x_2, x_1 y_2, x_1 z_2, y_1 x_2, y_1 y_2, y_1 z_2, z_1 x_2, z_1 y_2, z_1 z_2]^T$$

given $x_i = [x_i, y_i, z_i]^T$ ($i = 1, 2$). Therefore, in the absence of noise, given $n(\geq 8)$ pairs of image correspondences x_1^j and x_2^j ($j = 1, 2, \ldots, n$), we can linearly solve E^s up to a scale using the following equation:

$$L E^s \doteq \begin{bmatrix} (x_1^1 \otimes x_2^1)^T \\ (x_1^2 \otimes x_2^2)^T \\ \vdots \\ (x_1^n \otimes x_2^n)^T \end{bmatrix} E^s = 0, \quad L \in \mathbb{R}^{n \times 9} \quad (22.5)$$

and choosing E^s as the eigenvector of $L^T L$ associated with the eigenvalue 0.[2] E can then be obtained by "unstacking" E^s.

After obtaining E, we can project it to the space of essential matrix and decompose it to extract the motion, which is summarized in the following algorithm [40].

Algorithm 22.1 (Eight-point structure from motion algorithm.) Given $n(\geq 8)$ pairs of image correspondence of points x_1^j and x_2^j ($j = 1, 2, \ldots, n$), this algorithm recovers the motion $[R, T]$ of the camera (with the first camera frame being the reference frame) in three steps.

1. **Compute a first approximation of the essential matrix.** Construct the matrix $L \in \mathbb{R}^{n \times 9}$ as in (22.5). Choose E^s to be the eigenvector of $L^T L$ associated to its smallest eigenvalue: compute the SVD of $L = U_L S_L V_L^T$ and choose E^s to be the 9th column of V_L. Unstack E^s to obtain the 3×3 matrix E.

2. **Project E onto the space of essential matrices.** Perform SVD on E such that

$$E = U \operatorname{diag}\{\sigma_1, \sigma_2, \sigma_3\} V^T$$

where $\sigma_1 \geq \sigma_2 \geq \sigma_3 \geq 0$ and $U, V \in SO(3)$. The projection onto the space of essential matrices is $U \Sigma V^T$ with $\Sigma = \operatorname{diag}\{1, 1, 0\}$.

3. **Recover motion from by decomposing the essential matrix.** The motion R and T of the camera can be extracted from the essential matrix using U and V such that

$$R = U R_z^T \left(\pm \frac{\pi}{2} \right) V^T, \quad \widehat{T} = U R_z \left(\pm \frac{\pi}{2} \right) \Sigma U^T$$

where $R_z(\alpha)$ means rotation around z-axis by α counterclockwise.

The mathematical derivation and justification for the above projection and decomposition steps can be found in [40].

The above eight-point algorithm in general gives rise to four solutions of motion $[R, T]$. However, only one of them guarantees that the depths of all the 3-D points reconstructed are positive with respect to both camera frames [40]. Therefore, by checking the depths of all the points, the unique physically possible solution can be obtained. Also notice that T is recovered up to a scale. Without any additional scene knowledge, this scale cannot be determined and is often fixed by setting $\|T\| = 1$.

Given the motion $[R, T]$, the next thing is to recover the structure. For point features, that means to recover its depth with respect to the camera frame. For the jth point with depth λ_i^j in the ith ($i = 1, 2$) camera frame, from the fact $\lambda_2^j x_2^j = \lambda_1^j R x_1^j + T$, we have

$$M^j \vec{\lambda}^j \doteq \begin{bmatrix} \widehat{x}_2^j R x_1^j & \widehat{x}_2^j T \end{bmatrix} \begin{bmatrix} \lambda_1^j \\ 1 \end{bmatrix} = 0 \quad (22.6)$$

So λ_1^j can be solved by finding the eigenvector of $M^{j^T} M^j$ associated to its smallest eigenvalue.

---

[2]It can be shown that for $n(\leq 8)$ points in general positions, $L^T L$ has only one zero eigenvalue.

A Survey of Geometric Vision

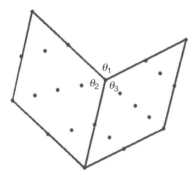

FIGURE 22.3 The two images of a calibration cube and two views of the reconstructed structure. The three angles are $\theta_1 = 89.7°$, $\theta_2 = 92.3°$, and $\theta_3 = 91.9°$.

Example 22.1

As shown in Figure 22.3, two images of a calibration cube are present. Twenty-three pairs of feature points are extracted using Harris corner detector, and the correspondences are established manually. The reconstruction is performed using the eight-point algorithm and depth calculation. The three angles for an orthogonal corner are close to right angles. The coplanarity of points in each plane are almost preserved. Overall, the structure is reconstructed fairly accurately.

22.2.2.1 Coplanar Features and Homography

Up to now, we assume that all the 3-D points are in general positions. In practice, it is not unusual that all the points reside on the same 3-D plane, i.e., they are *coplanar*. The eight-point algorithm will fail due to the fact that the matrix $L^T L$ (see (22.5)) will have more than one eigenvalue being zero. Fortunately, besides epipolar constraint, there exists another constraint for coplanar points. This is the *homography* between the two images, which can be described using a matrix $H \in \mathbb{R}^{3 \times 3}$:

$$H = R + \frac{1}{d} T N^T \tag{22.7}$$

with $d > 0$ denoting the distance from the plane to the first camera center and N being the unit normal vector of the plane expressed in the first camera frame with $\lambda_1 N^T x_1 + d = 0$.[3] It can be shown that the

---

[3]The homography is not limited to two image points in two camera frames; it is for the coordinates of the 3-D point on the plane expressed in any two frames (with one frame being the reference frame). In particular, if the second frame is chosen with its origin lying on the plane, then we have a homography between the camera and the plane.

two images x_1 and x_2 of the same point are related by

$$\widehat{x}_2 H x_1 = 0 \tag{22.8}$$

Using the homography relationship, we can recover the motion from two views with a similar procedure to the epipolar constraints: First H can also be calculated linearly using $n(\geq 4)$ pairs of corresponding image points. The reason that the minimum number of point correspondences is four instead of eight is that each pair of image points provides two independent equations on H through (22.8). Then the motion $[R, T]$ as well as the plane normal vector N can be obtained by decomposing H. However, the solution for the decomposition is more complicated. This algorithm is called *four-point algorithm* for coplanar features. Interested readers please refer to [40, 67].

22.2.3 Further Readings

The eight-point algorithm introduced in this section is for general situations. In practice, however, there are several caveats.

22.2.3.1 Small Baseline Motion and Continuous Motion

If $\| T \|$ is small and data are noisy, the reconstruction algorithm often would fail. This is the *small baseline* case. Readers can refer to [40, 65] for special algorithms dealing with this situation. When the baseline become infinitesimally small, we have the case of *continuous motion*, for which the algebra becomes somewhat different from the discrete case. For a detailed analysis and algorithm, please refer to [39, 40].

22.2.3.2 Multiple-Body Motions

For the case in which there are multiple moving objects in the scene, there exists a more complicated *multiple-body epipolar constraint*. The reconstruction algorithm can be found in [40, 66].

22.2.3.3 Uncalibrated Camera

If the camera is uncalibrated, the essential matrix E in the epipolar constraint should be substituted by the *fundamental matrix* F with $F = K^{-T} E K^{-1}$, where $K \in \mathbb{R}^{3\times 3}$ is the calibration matrix of the camera, defined in Equation (22.2). The analysis and affine reconstruction for the uncalibrated camera can be found in [23, 40].

22.2.3.4 Critical Surface

There are certain degenerate positions of the points for which the reconstruction algorithm would fail. These configurations are called *critical surfaces* for the points. Detailed analysis is available in [40, 44].

22.2.3.5 Numerical Problems and Optimization

To obtain accurate reconstruction, some numerical issues such as data normalization need to be addressed before applying the algorithm. These are discussed in [40]. Notice that the eight-point algorithm is only a suboptimal algorithm; various nonlinear "optimal" algorithms have been designed, which can be found in [40, 64].

22.3 Multiple-View Geometry

In this section, we study the case for reconstruction from more than two views. Specifically, we present a set of *rank conditions* on a *multiple-view matrix* [27, 37, 38]. The epipolar constraint for two views is just a special case implied by the rank condition. As we will see, the multiple-view matrix associated to a geometric entity (point, line, or plane) is exactly the 3-D information that is missing in a single 2-D image but encoded in multiple ones. This approach is compact in representation, intuitive in geometry, and simple in computation. Moreover, it provides a unified framework for describing multiple views of all types of features and incidence relations in 3-D space [40].

A Survey of Geometric Vision

22.3.1 Rank Condition on Multiple Views of Point Feature

First, let us look at the case of multiple images of a point. As shown in Figure 22.4, multiple images x_i of a 3-D point X with camera motions $[R_i, T_i]$ $(i = 1, 2, \ldots, m)$ satisfy

$$\lambda_i x_i = [R_i, T_i] X \tag{22.9}$$

where λ_i is the point depth in the ith camera frame. Multiplying $\widehat{x}_i$ on both sides of Equation (22.9), we have

$$[\widehat{x}_i R_i \quad \widehat{x}_i T_i] X = 0 \tag{22.10}$$

Without loss of generality, we choose the first camera frame as the reference frame such that $R_1 = I$, $T_1 = 0$, and $X = \begin{bmatrix} \lambda_1 x_1 \\ 1 \end{bmatrix}$. Therefore, for $i = 2, 3, \ldots, m$, Equation (22.10) can be transformed into

$$[\widehat{x}_i R_i x_1 \quad \widehat{x}_i T_i] \begin{bmatrix} \lambda_1 \\ 1 \end{bmatrix} = 0 \tag{22.11}$$

Stacking the left side of the above equations for all $i = 2, \ldots, m$, we have

$$M \begin{bmatrix} \lambda_1 \\ 1 \end{bmatrix} \doteq \begin{bmatrix} \widehat{x}_2 R_2 x_1 & \widehat{x}_2 T_2 \\ \widehat{x}_3 R_3 x_1 & \widehat{x}_3 T_3 \\ \vdots & \vdots \\ \widehat{x}_m R_m x_1 & \widehat{x}_m T_m \end{bmatrix} \begin{bmatrix} \lambda_1 \\ 1 \end{bmatrix} = 0 \quad \in \mathbb{R}^{3(m-1)} \tag{22.12}$$

The matrix $M \in \mathbb{R}^{3(m-1) \times 2}$ is called the *multiple-view matrix* for point features. The above relationship is summarized in the following theorem.

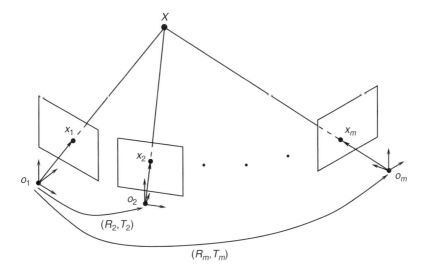

FIGURE 22.4 Multiple images of a 3-D point X in m camera frames.

Theorem 22.1 (Rank condition on multiple-view matrix for point features.) *The multiple-view matrix M satisfies the following rank conditions:*

$$\text{rank}(M) \leq 1 \tag{22.13}$$

Furthermore, $M\begin{bmatrix}\lambda_1\\1\end{bmatrix} = 0$ with λ_1 being the depth of the point in the first camera frame.

The detailed proof of the theorem can be found in [37, 40].

The implication of the above theorem is multifold. First, the rank of M can be used to detect the configuration of the cameras and the 3-D point. If $\text{rank}(M) = 1$, the cameras and the 3-D point are in the general positions, and the location of the point can be determined (up to a scale) by triangulation. If $\text{rank}(M) = 0$, then all the camera centers and the point are collinear; and the point can only be determined up to a line. Second, all the algebraic constraints implied by the rank condition involve no more than three views, which means that for point features a fourth image no longer imposes any new algebraic constraint. Last but not the least, the rank condition on M uses all the data simultaneously, which significantly simplifies the calculation. Also note that the rank condition on the multiple-view matrix implies the epipolar constraint. For the ith ($i > 1$) view and the first view, the fact that $\widehat{x}_i R_i x_1$ and $\widehat{x}_i T_i$ are linearly dependent is equivalent to the epipolar constraint between the two views.

22.3.2 Linear Reconstruction Algorithm

Now we demonstrate how to apply the multiple-view rank condition in reconstruction. Specifically, we show the linear reconstruction algorithm for point features. Given $m(\geq 3)$ images of $n(\geq 8)$ points in 3-D space with x_i^j ($i = 1, 2, \ldots, m$ and $j = 1, 2, \ldots, n$), the structure (depths λ_1^j with respect to the first camera frame) and the camera motions $[R_i, T_i]$ can be recovered in two major steps.

First for the jth point, a multiple-view matrix M^j associated with its images satisfies

$$M^j \begin{bmatrix} 1 \\ \frac{1}{\lambda_1^j} \end{bmatrix} = \begin{bmatrix} \widehat{x}_2^j R_2 x_1^j & \widehat{x}_2^j T_2 \\ \widehat{x}_3^j R_3 x_1^j & \widehat{x}_3^j T_3 \\ \vdots & \vdots \\ \widehat{x}_m^j R_m x_1^j & \widehat{x}_m^j T_m \end{bmatrix} \begin{bmatrix} 1 \\ \frac{1}{\lambda_1^j} \end{bmatrix} = 0 \tag{22.14}$$

The above equation implies that if the set of motions $g_i = [R_i, T_i]$'s are known for all $i = 2, \ldots, m$, the depth λ_1^j of the jth point with respect to the first camera frame can be recovered by computing the kernel of M^j. We denote $\alpha^j = \frac{1}{\lambda_1^j}$.

Similarly, for the ith image ($i = 2, \ldots, m$), if α^j's are known for all $j = 1, 2, \ldots, n$, the estimation of

$$R_i = \begin{bmatrix} r_{11} & r_{12} & r_{13} \\ r_{21} & r_{22} & r_{23} \\ r_{31} & r_{32} & r_{33} \end{bmatrix}$$

and T_i is equivalent to solving the stacked vectors

$$R_i^s = [r_{11}, r_{21}, r_{31}, r_{12}, r_{22}, r_{33}, r_{13}, r_{23}, r_{33},]^T \in \mathbb{R}^9$$

and $T_i \in \mathbb{R}^3$ using the equation

$$P_i \begin{bmatrix} R_i^s \\ T_i \end{bmatrix} \doteq \begin{bmatrix} \widehat{x}_i^1 \otimes x_1^1 & \alpha^1 \widehat{x}_i^1 \\ \widehat{x}_i^2 \otimes x_1^2 & \alpha^2 \widehat{x}_i^2 \\ \vdots & \vdots \\ \widehat{x}_i^n \otimes x_1^n & \alpha^n \widehat{x}_i^n \end{bmatrix} \begin{bmatrix} R_i^s \\ T_i \end{bmatrix} = 0 \in \mathbb{R}^{3n}, \quad P_i \in \mathbb{R}^{3n \times 12} \tag{22.15}$$

A Survey of Geometric Vision

where $\otimes$ is the *Kronecker product* between two matrices. It can be shown that P_i has rank 11 if $n \geq 6$ points in general positions are present. Therefore the solution of $[R_i, T_i]$ will be unique up to a scale for $n \geq 6$.

Obviously, if no noise is present on the images, the recovered motion and structure will be the same with what can be recovered from the two-view eight-point algorithm. However, in the presence of noise, it is desired to use the data from all the images. In order to use the data simultaneously, a reasonable approach is to iterate between reconstructions of motion and structure, i.e., initializing the structure or motions, then alternating between Equation (22.14) and Equation (22.15) until the structure and motion converge.

Motions $[R_i, T_i]$ can then be estimated up to a scale by performing SVD on P_i as in (22.15).[4] Denote $\tilde{R}_i$ and $\tilde{T}_i$ to be the estimates from the eigenvector of $P_i^T P_i$ associated to the smallest eigenvalue. Let $\tilde{R}_i = U_i S_i V_i^T$ be the SVD of $\tilde{R}_i$. Then $R_i \in SO(3)$ and T_i are given by

$$R_i = \text{sign}\left(\det\left(U_i V_i^T\right)\right) U_i V_i^T \in SO(3) \quad \text{and} \quad T_i = \frac{\text{sign}\left(\det\left(U_i V_i^T\right)\right)}{(\det(S_i))^{\frac{1}{3}}} \in \mathbb{R}^3 \quad (22.16)$$

In this algorithm, the initialization can be done using the eight-point algorithm for two views. The initial estimate on the motion of second frame $[R_2, T_2]$ can be obtained using the standard two-view eight-point algorithm. Initial estimate of the point depth is then

$$\alpha^j = -\frac{\left(\widehat{\boldsymbol{x}}_2^j T_2\right)^T \widehat{\boldsymbol{x}}_2^j R_2 \boldsymbol{x}_1^j}{\left\|\widehat{\boldsymbol{x}}_2^j T_2\right\|^2}, \quad j = 1, \ldots, n \quad (22.17)$$

In the multiple-view case, the least square estimates of point depths $\alpha^j = 1/\lambda_1^j, j = 1, \ldots, n$ can be obtained from Equation (22.14) as

$$\alpha^j = -\frac{\sum_{i=2}^m \left(\widehat{\boldsymbol{x}}_i^j T_i\right)^T \widehat{\boldsymbol{x}}_i^j R_i \boldsymbol{x}_1^j}{\sum_{i=2}^m \left\|\widehat{\boldsymbol{x}}_i^j T_i\right\|^2}, \quad j = 1, \ldots, n \quad (22.18)$$

By iterating between the motion estimation and structure estimation, we expect that the estimates on structure and motion converge. The convergence criteria may vary for different situations. In practice we choose the reprojected images error as the convergence criteria. For the jth 3-D point, the estimate of its 3-D location can be obtained as $\lambda_1^j \boldsymbol{x}_1^j$, and the reprojection on the ith image is obtained $\tilde{\boldsymbol{x}}_i^j \sim \lambda_1^j R_i \boldsymbol{x}_1^j + T_i$. Its reprojection error is then $\sum_{i=2}^m \| \boldsymbol{x}_i^j - \tilde{\boldsymbol{x}}_i^j \|^2$. So the algorithm keep iterating until the summation of reprojection errors over all points are below some threshold ϵ.

The algorithm is summarized below:

Algorithm 22.2 (A factorization algorithm for multiple-view reconstruction.) Given $m(\geq 3)$ images $\boldsymbol{x}_1^j, \boldsymbol{x}_2^j, \ldots, \boldsymbol{x}_m^j, j = 1, 2, \ldots, n$ of $n(\geq 8)$ points, the motions $[R_i, T_i], i = 2, \ldots, m$ and the structure of the points with respect to the first camera frame $\alpha^j, j = 1, 2, \ldots, n$ can be recovered as follows:

1. Set the counter $k = 0$. Compute $[R_2, T_2]$ using the eight-point algorithm, then get an initial estimate of α^j from Equation (22.17) for each $j = 1, 2, \ldots, n$. Normalize $\alpha^j \leftarrow \alpha^j/\alpha^1$ for $j = 1, 2, \ldots, n$.
2. Compute $[\tilde{R}_i, \tilde{T}_i]$ from the eigenvector of $P_i^T P_i$ corresponding to its smallest eigenvalue for $i = 2, \ldots, m$.
3. Compute $[R_i, T_i]$ from Equation (22.16) for $i = 2, 3, \ldots, m$.
4. Compute α_{k+1}^j using Equation (22.18) for each $j = 1, 2, \ldots, n$. Normalize so that $\alpha^j \leftarrow \alpha^j/\alpha^1$ and $T_i \leftarrow \alpha_{k+1}^1 T_i$. Use the newly recovered α^j's and motion $[R_i, T_i]$'s to compute the reprojected image $\tilde{\boldsymbol{x}}$ for each point in all views.
5. If the reprojection error $\sum \| \boldsymbol{x} - \tilde{\boldsymbol{x}} \|^2 > \epsilon$ for some threshold $\epsilon > 0$, then $k \leftarrow k + 1$ and go to step 2, otherwise stop.

---

[4]Now we assume that the cameras are all calibrated, which is the case of Euclidean reconstruction. This algorithm also works for uncalibrated case.

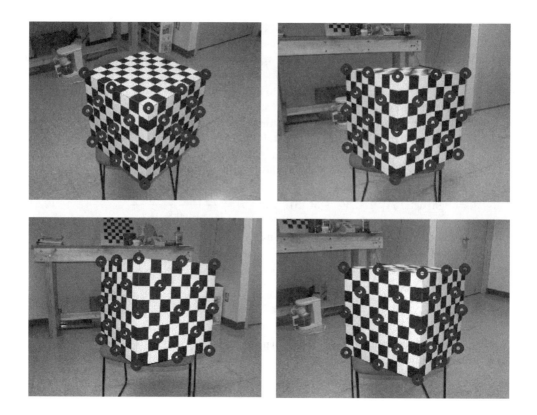

FIGURE 22.5 The four images used to reconstruct the calibration cube.

The above algorithm is a direct derivation from the rank condition. There are techniques to improve its numerical stability and statistical robustness for specific situations [40].

Example 22.2

The algorithm proposed in previous section has been tested extensively in both simulation [37] and experiments [40]. Figure 22.5 shows the four images used to reconstruct the calibration cube. The points are marked in circles. Two views of the reconstruction results are shown in Figure 22.6. The recovered angles are more accurate than the results from Figure 22.3. Visually, the coplanarity of the points is preserved well.

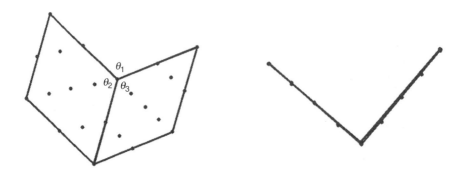

FIGURE 22.6 The two views of the reconstructed structure. The three angles are $\theta_1 = 89.9°$, $\theta_2 = 91.0°$, and $\theta_3 = 90.6°$.

22.3.2.1 Coplanar Features

The treatment for coplanar point features is similar to the general case. Assume the equation of the plane in the first camera frame is $[\pi_1^T, \pi_2]X = 0$ with $\pi_1 \in \mathbb{R}^3$ and $\pi_2 \in \mathbb{R}$. By simply appending the 1×2 block $[\pi_1^T x_1, \pi_2]$ to the end of M in Equation (22.12), the rank condition in Theorem 22.1 still holds. Since $\pi_1 = N$ is the unit normal vector of the plane and $\pi_2 = d$ is the distance from the first camera center to the plane, the rank condition implies $d(\widehat{x}_i R_i x_1) - (N^T x_1)(\widehat{x}_i T_i) = 0$, which is obviously equivalent to the homography between the ith and the first views (see Equation (22.8)). As for reconstruction, we can use the four-point algorithm to initialize the estimation of the homography and then perform a similar iteration scheme to obtain motion and structure. The algorithm can be found in [36, 51].

22.3.3 Further Readings

22.3.3.1 Multilinear Constraints and Factorization Algorithm

There are two other approaches dealing with multiple-view reconstruction. The first approach is to use the so-called *multilinear constraints* on multiple images of a 3-D point or line. For small number of views, these constraints can be described in terms of tensorial notations [23, 52, 53]. For example, the constraints for $m = 3$ can be described using *trifocal tensors*. For large number of views ($m \geq 5$), the tensor is difficult to describe. The reconstruction is then to calculate the trifocal tensors first and factorize the tensors for camera motions [2]. An apparent disadvantage is that it is hard to choose the right "three-view sets" and also difficult to combine the results. Another approach is to apply some factorization scheme to iteratively estimate the structure and motion [23, 40, 57], which is in the same spirit as Algorithm 22.2.

22.3.3.2 Universal Multiple-View Matrix and Rank Conditions

The reconstruction algorithm in this section was only for point features. Algorithms have also been designed for lines features [35, 56]. In fact the multiple-view rank condition approach can be extended to all different types of features such as line, plane, mixed line, point and even curves. This leads to a set of rank conditions on a *universal multiple-view matrix*. For details please refer to [38, 40].

22.3.3.3 Dynamical Scenes

The constraint we developed in this section is for static scene. If the scene is *dynamic*, i.e., there are moving objects in the scene, a similar type of rank condition can be obtained. This rank condition is obtained by incorporating the dynamics of the objects in 3-D space into their own descriptions and lifting the 3-D moving points into a higher dimensional space in which they are static. For details please refer to [27, 40].

22.3.3.4 Orthographic Projection

Finally, note that the linear algorithm and the rank condition are for the perspective projection model. If the scene is far from the camera, then the image can be modeled using orthographic projection, and the Tomasi-Kanade factorization method can be applied [59]. Similar factorization algorithm for other types of projections and dynamics have also been developed [8, 21, 48, 49].

22.4 Utilizing Prior Knowledge of the Scene — Symmetry

In this section we study how to incorporate scene knowledge into the reconstruction process. In our daily life, especially in a man-made environment, there exist all types of "regularity." For objects, regular shapes such as rectangle, square, diamond, and circle always attract our attention. For spatial relationship between objects, orthoganality, parallelism, and similarity are the conspicuous ones. Interestingly, all the above regularities can be described using the notion of *symmetry*. For instance, a rectangular window has one rotational symmetry and two reflective symmetry; the same windows on the same wall have translational symmetry; the corner of a cube displays rotational symmetry.

22.4.1 Symmetric Multiple-View Rank Condition

There are many studies using instances of symmetry in the scene for reconstruction purposes [1, 3, 6, 19, 25, 28, 70, 73, 74]. Recently, a set of algorithms using symmetry for reconstruction from a single image has been developed [25]. The main idea is to use the so-called *equivalent images* encoded in a single image of a symmetric object. Figure 22.7 illustrates this notion for the case of a reflective symmetry. We attach an *object coordinate frame* to the symmetric object and set it as the reference frame. 3-D points X and X' are related by some symmetric transformation g (in the object frame) with $g(X) = X'$. If the image is obtained from viewpoint O, then the image of X' can be interpreted as the image of X viewed from the virtual viewpoint O' that is the correspondence of O under the same symmetry transformation g. The image of X' is called an *equivalent image* of X viewed from O'. Therefore, given a single image of a symmetric object, we have multiple equivalent images of this object. The number of all equivalent images is the number of all symmetry of the object.

In modern mathematics, symmetry of an object are characterized by a *symmetry group* with each element in the group representing a transformation under which the object is invariant [25, 68]. For example, a rectangle possesses two reflective symmetry and one rotational symmetry. We can use the group theoretic notation to define a 3-D symmetric object as in [25].

Definition 22.1 Let S be a set of 3-D points. It is called a *symmetric structure* if there exists a non-trivial subgroup G of the Euclidean group $E(3)$ acting on S such that for any $g \in G$, g defines an isomorphism from S to itself. G is called the *symmetry group* of S.

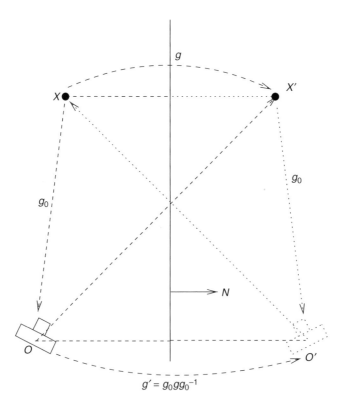

FIGURE 22.7 X and X' are corresponding points under reflective symmetry transformation g (expressed in the object frame) such that $X' = g(X)$. N is the normal vector of the mirror plane. The motion between the camera frame and the object frame is g_0. Hence, the image of X' in real camera can be considered as the image of X viewed by a virtual camera with pose $g_0 g$ with respect to the object frame or $g_0 g g_0^{-1}$ with respect to the real camera frame.

Under this definition, the possible symmetry on a 3-D object are reflective symmetry, translational symmetry, rotational symmetry, and any combination of them. For any point $p \in S$ on the object, its symmetric correspondence for $g \in G$ is $g(p) \in S$. The images of p and $g(p)$ are denoted as $\boldsymbol{x}$ and $g(\boldsymbol{x})$.

Let the symmetry transformation in the object frame be $g = [R, T] \in G$ ($R \in O(3)$ and $T \in \mathbb{R}^3$). As illustrated in Figure 22.7, if the transformation from the object (reference) frame to the real camera frame is $g_0 = [R_0, T_0]$, the transformation from the reference frame to the virtual camera frame is $g_0 g$. Furthermore, the transformation from the real camera frame to the virtual camera frame is

$$g' = g_0 g g_0^{-1} = [R', T'] = \left[R_0 R R_0^T, (I - R_0 R R_0^T) T_0 + R_0 T \right] \tag{22.19}$$

For the symmetric object S, assume its symmetry group G has m elements $g_i = [R_i, T_i], i = 1, 2, \ldots, m$. Then the transformation between the ith virtual camera and the real camera is $g'_i = [R'_i, T'_i]$ ($i = 1, 2, \ldots, m$) as can be calculated from Equation (22.19). Given any point $X \in S$ with image $\boldsymbol{x}$ and its equivalent images $g_i(\boldsymbol{x})$'s, we can define the *symmetric multiple-view matrix*

$$M(\boldsymbol{x}_i) = \begin{bmatrix} \widehat{g_1(\boldsymbol{x})} R'_1 \boldsymbol{x} & \widehat{g_1(\boldsymbol{x})} T'_1 \\ \widehat{g_2(\boldsymbol{x})} R'_2 \boldsymbol{x} & \widehat{g_2(\boldsymbol{x})} T'_2 \\ \vdots & \vdots \\ \widehat{g_m(\boldsymbol{x})} R'_m \boldsymbol{x} & \widehat{g_m(\boldsymbol{x})} T'_m \end{bmatrix} \tag{22.20}$$

According to Theorem 22.1, it satisfies the *symmetric multiple-view rank condition*:

$$\mathrm{rank}(M(\boldsymbol{x})) \leq 1 \tag{22.21}$$

22.4.2 Reconstruction from Symmetry

Using the symmetric multiple-view rank condition and a set of n symmetric correspondences (points), we can solve for $g'_i = [R'_i, T'_i]$ and the structure of the points in S in the object frame using an algorithm similar to Algorithm 22.2. However, a further step is still necessary to recover $g_0 = [R_0, T_0]$ for the camera pose with respect to the object frame. This can be done by solving the following Lyapunov type of equations:

$$g'_i g_0 - g_0 g_i = 0, \quad \text{or} \quad R'_i R_0 - R_0 R = 0 \quad \text{and} \quad T_0 = (I - R')^\dagger (T' - R_0 T) \tag{22.22}$$

Since g'_i and g_i are known, g_0 can be solved. The detailed treatment can be found in [25, 40].

As can be seen in the example at the end of this section, symmetry-based reconstruction is very accurate and requires a minimum amount of data. A major reason is that the baseline between the real camera and the virtual one is often large due to the symmetry transformation. Therefore, degenerate cases will occur if the camera center is invariant under the symmetry transformation. For example, if the camera lies on the mirror plane for a reflective symmetry, the structure can only be recovered up to some ambiguities [25].

The reconstruction for some special cases can be simplified without explicitly using the symmetric multiple-view rank condition (e.g., see [3]). For the reflective symmetry, the structure with respect to the camera frame can be calculated using only two pairs of symmetric points. Assume the image of two pairs of reflective points are $\boldsymbol{x}, \boldsymbol{x}'$ and $\boldsymbol{y}, \boldsymbol{y}'$. Then the image line connecting $\boldsymbol{x}$ and $\boldsymbol{x}'$ is obtained by $\boldsymbol{l}_x \sim \widehat{\boldsymbol{x}} \boldsymbol{x}'$. Similarly, $\boldsymbol{l}_y$ connecting $\boldsymbol{y}$ and $\boldsymbol{y}'$ satisfies $\boldsymbol{l}_y \sim \widehat{\boldsymbol{y}} \boldsymbol{y}'$. It can be shown that the unit normal vector N (see Figure 22.7) of the reflection plane satisfies

$$\begin{bmatrix} \boldsymbol{l}_x^T \\ \boldsymbol{l}_y^T \end{bmatrix} N = 0 \tag{22.23}$$

from which N can be solved. Assume the depth of $\boldsymbol{x}$ and $\boldsymbol{x}'$ are λ and λ', then they can be calculated using the following relationship

$$\begin{bmatrix} \widehat{N}x & -\widehat{N}x' \\ N^T x & N^T x' \end{bmatrix} \begin{bmatrix} \lambda \\ \lambda' \end{bmatrix} = \begin{bmatrix} 0 \\ 2d \end{bmatrix} \qquad (22.24)$$

where d is the distance from the camera center to the reflection plane.

22.4.2.1 Planar Symmetry

Planar symmetric objects are widely present in man-made environment. Regular shapes such as square and rectangle are often good landmarks for mapping and recognition tasks. For a planar symmetric object, its symmetry forms a subgroup G of $E(22.2)$ instead of $E(22.3)$. Let $g \in E(2)$ be one symmetry of a planar symmetric object. Recall that there exists a homography H_0 between the object frame and camera frame. Also between the original image and the equivalent image generated by g, there exists another homography H. It can be shown that $H = H_0 g H_0^{-1}$. Therefore all the homographies generated from the equivalent images form a *homography group* $H_0 G H_0^{-1}$, which is conjugate to G [3]. This fact can be used in testing if an image is the image of an object with desired symmetry. Given the image, by calculating the homographies from all equivalent images based on the hypothesis, we can check the group relationship of the homography group and decide if the desired symmetry exists [3]. In case the object is a rectangle, the calculation can be further simplified using the notion of *vanishing point* as illustrated in the following example.

Example 22.3 (Symmetry-based reconstruction for a rectangular object.)

For a rectangle in 3-D space, the two pairs of parallel edges generate two vanishing points v_1 and v_2 in the image. As a 3-D vector, v_i $(i = 1, 2)$ can also be interpreted as the vector from the camera center to the vanishing point and hence must be parallel to the pair of parallel edges. Therefore, we must have $v_1 \perp v_2$. So by checking the angle between the two vanishing points, we can decide if a region can be the image of a rectangle. Figure 22.8 demonstrates the reconstruction based on the assumption of a rectangle. The two sides of the cube are two rectangles (in fact squares). The angles between the vanishing points are 89.1° and 92.5°, respectively. The reconstruction is performed using only one image and six points. The angle between the two planes is 89.2°.

22.4.3 Further Readings

22.4.3.1 Symmetry and Vision

Symmetry is a strong vision cue in human vision perception and has been extensively discussed in psychology and cognition research [43, 45, 47]. It has been noticed that symmetry is useful for face recognition [61, 62, 63].

22.4.3.2 Symmetry in Statistical Context

Besides geometric symmetry that has been discussed in this section, symmetry in the sense of statistics have also been studied and utilized. Actually, the computational advantages of symmetry were first explored in the statistical context, such as the study of isotropic texture [17, 42, 69]. It was the work of [14, 15, 42] that provided a wide range of efficient algorithms for recovering the orientation of a textured plane based on the assumption of isotropy or weak isotropy.

22.4.3.3 Symmetry of Surfaces and Curves

While we only utilized symmetric points in this chapter, the symmetry of surfaces has also been exploited. References [55, 72] used the surface symmetry for human face reconstruction. References [19, 25] studied reconstruction of symmetric curves.

A Survey of Geometric Vision   **22**-17

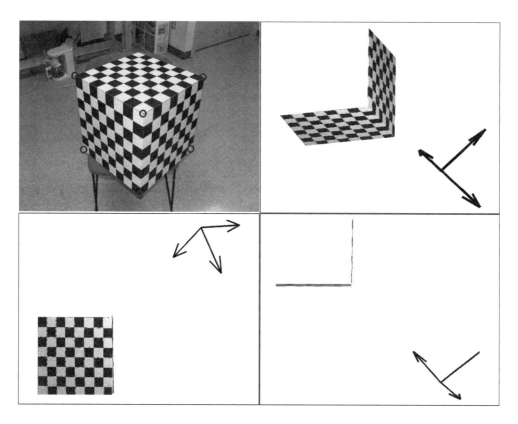

FIGURE 22.8 Reconstruction from a single view of the cube with six corner points marked. An arbitrary view, a side view, and a bird view of the reconstructed and rendered rectangles are displayed. The coordinate frame shows the recovered camera pose.

22.5 Comprehensive Examples and Experiments

Finally, we discuss several applications of using the techniques introduced in this chapter. These applications all involve reconstruction from a single or multiple images for the purpose of robotic navigation and mapping. While not all of them are fully automatic, they demonstrate the potential of the techniques discussed in this chapter and some future directions for improvement.

22.5.1 Automatic Landing of Unmanned Aerial Vehicles

Figure 22.9 displays the experiment setup for applying the multiple view rank condition based algorithm in automatic landing of UAV in University of California at Berkeley [51]. In this experiment, the UAV is a model helicopter (Figure 22.9, top) with an on-board video camera facing downwards searching for the landing pad (Figure 22.9, bottom left) on the ground. When the landing pad is found, corner points are extracted from each image in the video sequence (Figure 22.9, bottom right). The pose and the motion of the camera (and the UAV) are estimated using the feature points in the images. Previously the four-point two-view reconstruction algorithm for planar features was tested. However, its results were noisy (Figure 22.10, top). Later a nonlinear two-view algorithm was developed, but it was time consuming. Finally, by adopting the rank condition based multiple-view algorithm, the landing problem was solved.

The multiple-view reconstruction is performed at 10 Hz for every four images. The algorithm is a modification of the Algorithm 22.2 introduced in this chapter, which is specifically designed for coplanar features. Figure 22.10 shows the comparison of this algorithm with other algorithms and sensors. The multiple-view algorithm is more accurate than either two-view algorithm (Figure 22.10, top) and is close

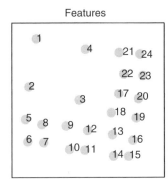

FIGURE 22.9 Top: A picture of the UAV in landing process. Bottom left: An image of the landing pad viewed from an on-board camera. Bottom right: Extracted corner features from the image of the landing pad. (Photo courtesy of O. Shankeria.)

to the results obtained from differential GPS and INS sensors (Figure 22.10, bottom). The overall error for this algorithm is less than 5 cm in distance and $4°$ for rotation [51].

22.5.2 Automatic Symmetry Cell Detection, Matching and Reconstruction

In Section 22.4, we discussed symmetry-based reconstruction techniques from a single image. Here we present a comprehensive example that performs symmetry-based reconstruction from multiple views [3, 28]. In this example, the image primitives are no longer point or line. Instead we use the *symmetry cells* as features. The example includes three steps.

22.5.2.1 Feature Extraction

By *symmetry cell* we mean a region in the image that is an image of a desired symmetric object in 3-D. In this example, the symmetric object we choose is rectangle. So the symmetry cells are images of rectangles. Detecting symmetry cells (rectangles) includes two steps. First, we perform color-based segmentation on the image. Then, for all the detected four-sided regions, we test if its two vanishing points are perpendicular to each other and decide if it can be an image of a rectangle in 3-D space. Figure 22.11 demonstrates this for the picture of an indoor scene. In this picture, after color segmentation, all four-sided regions with reasonable sizes are detected and marked with darkened boundaries. Then each four-sided polygon is passed through the vanishing point test. For those polygons passing the test, we denote them as symmetry cells and recover the object frames with the symmetry-based algorithm. Each individual symmetry cell is recovered

A Survey of Geometric Vision

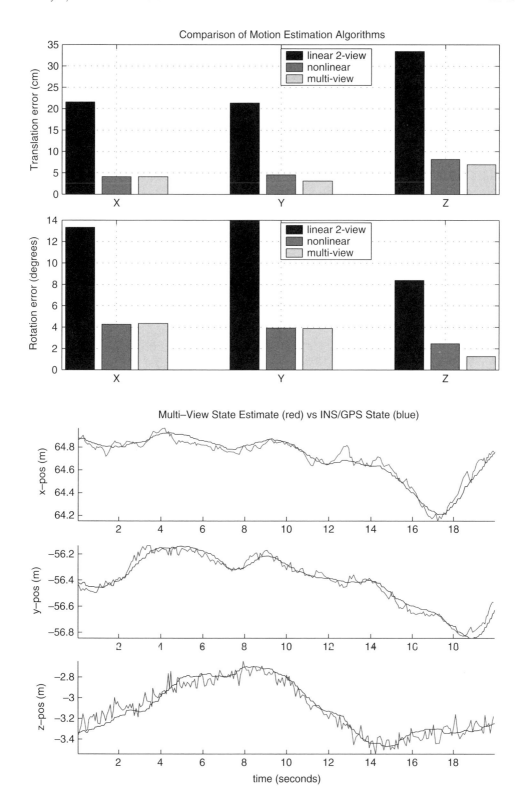

FIGURE 22.10 Top: Comparison of the multiple-view rank condition based algorithm with two-view linear algorithm and nonlinear algorithm. Bottom: Comparison of the results from multiple-view rank condition based algorithm (red line) with GPS and INS results (blue line). (Image courtesy of O. Shankeria.)

FIGURE 22.11 Left: original image. Middle: image segmentation and polygon fitting. Right: symmetry cells detected and extracted — an object coordinate frame is attached to each symmetry cell.

with a different scale. Please note that this test is a hypothesis testing step, it cannot verify if the object in 3-D is actually a desired symmetric object, instead it can confirm that its image satisfies the condition of the test.

22.5.2.2 Feature Matching

For each cell we can calculate its 3-D shape. Then for two cells in two images, by comparing their 3-D shapes and colors, we can decide if they are matching candidates. In the case of existence of many similar symmetry cells, we need to use a matching graph [28]; that is, set the nodes of the graph to be all pairs of possible matched cells across the two images. For each pair of matched pair, we can calculate a camera motion between the two views; then correct matching should generate the same (up to scale) camera motion. We draw an edge between two nodes if these two pairs of matched cells generate similar camera motions. Therefore, the problem of finding the correctly matched cells becomes the problem of finding the (maximum) cliques in the matching graph. Figure 22.12 shows the matching results for two cells in three images. The cells that have no match in other images are discarded. The scales for all pairs of matched cells are unified.

22.5.2.3 Reconstruction

Given the set of correctly matched cells, only one more step is needed for reconstruction. Note that the camera motions from different pairs of matched cells have different scales. To unify the scale, we pick one pair of matched cells as the reference to calculate the translation and use this translation to scale the remaining matched pairs. The new scales can further be passed to resize the cells. Figure 22.13 shows the reconstructed 3-D cells and the camera poses for the three views. The physical dimension of the cells are calculated with good accuracy. The aspect ratios for the white board and the table top are calculated as 1.51 and 1.01, respectively, with the ground truth being 1.50 and 1.00.

In this example, all the features are matched correctly without any manual intervention. Please note that establishing feature matching among the three views using other approaches would otherwise be very difficult due to the large baseline between the first and second images and the pure rotational motion

FIGURE 22.12 Two symmetry cells are matched in three images. From the raw images, symmetry cell extraction, to cell matching, the process needs no manual intervention.

A Survey of Geometric Vision

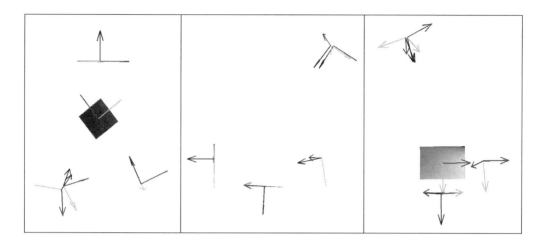

FIGURE 22.13 Camera poses and cell structure recovered. From left to right: top, side, and frontal views of the cells and camera poses.

between the second and third images. For applications such as robotic mapping, the symmetric cells can serve as "landmarks." The pose and motion of the robot can be easily derived using a similar scheme.

22.5.3 Semiautomatic Building Mapping and Reconstruction

If a large number of similar objects is present, it is usually hard for the detection and matching scheme in the above example to work properly. For instance, for the symmetry of window complexes on the side of a building shown in Figure 22.14, many ambiguous matches may occur. In such cases, we need to take manual intervention to obtain a realistic 3-D reconstruction (e.g., see [28]). The techniques discussed so far, however, help to minimize the amount of manual intervention. For images in Figure 22.14, the user only needs to point out cells and provide the cell correspondence information. The system will then automatically generate a consistent set of camera poses from the matched cells, as displayed in Figure 22.15

FIGURE 22.14 Five images used for reconstruction of a building. For the first four image, we mark a few cells manually. The last image is only used for extracting roof information.

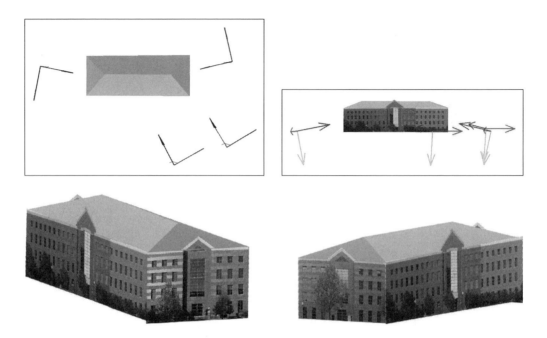

FIGURE 22.15 Top: The recovered camera poses and viewpoints as well as the cells. The four coordinate frames are the recovered camera poses from the four images in Figure 22.14. The roof was substituted by a "virtual" one based on corners extracted from the fifth image in Figure 22.14 using the symmetry-based algorithm. Blue arrows are the camera optical axes. Bottom: 3-D model of the building reconstructed from the original set of four images.

top. For each cell, a camera pose is recovered using the symmetry-based algorithm. Similar to the previous example, motion between cameras can be recovered using the corresponding cell. When the camera poses are recovered, the 3-D structure of the building can be recovered as shown in Figure 22.15 bottom. The 3-D model is rendered as piecewise planar parts. The angles between the normal vectors of any two orthogonal walls differ from 90° by an average of 1° error without any nonlinear optimization. The whole process (from taking the images to 3-D rendering) takes less than 20 min.

22.5.4 Summary

The above examples demonstrate the potential of the theory and algorithms introduced in this chapter. Besides these examples, we can expect that these techniques will be very useful in many other applications such as surveillance, manipulation, navigation, and vision-based control. Finally, we like to point out that we are merely at the beginning stage of understanding the geometry and dynamics associated with visual perception. Much of the topological, geometric, metric, and dynamical relations between 3-D space and 2-D images is still largely unknown, which, to some extent, explains why the capability of existing machine vision systems is still far inferior to that of human vision.

References

[1] Francois, A.R.J., Medioni, G.G., and Waupotitsch, R., Reconstructing mirror symmetric scenes from a single view using 2-view stereo geometry, in *Proc. Int. Conf. Pattern Recognition*, pp. 40012–40015, 2002.

[2] Avidan, S. and Shashua, A., Novel view synthesis by cascading trilinear tensors, *IEEE Trans. Visualization and Computer Graphics*, 4(4):293–306, 1998.

[3] Yang, A.Y., Rao, S., Huang, K., Hong, W., and Ma, Y., Geometric segmentation of perspective images based on symmetry groups, in *Proc. IEEE Int. Conf. Computer Vision*, Nice, France, Vol. 2, pp. 1251–1258, 2003.

[4] Canny, J.F., A computational approach to edge detection, *IEEE Trans. Pattern Anal. & Machine Intelligence,* 8(6):679–698, 1986.
[5] Caprile, B. and Torre, V., Using vanishing points for camera calibration, *Int. J. Computer Vision,* 4(2):127–140, 1990.
[6] Carlsson, S., Symmetry in perspective, in *Proc. Eur. Conf. Computer Vision,* pp. 249–263, 1998.
[7] Corke, P.I., *Visual Control of Robots: High-Performance Visual Servoing,* Robotics and Mechatronics Series, Research Studies Press, Somerset, England, 1996.
[8] Costeira, J. and Kanade, T., A multi-body factorization method for motion analysis, in *Proc. IEEE Int. Conf. Computer Vision,* pp. 1071–1076, 1995.
[9] Das, A.K., Fierro, R., Kumar, V., Southball, B., Spletzer, J., and Taylor, C.J., Real-time vision-based control of a nonholonomic mobile robot, in *Proc. IEEE Int. Conf. Robotics and Automation,* pp. 1714–1719, 2002.
[10] Faugeras, O., *Three-Dimensional Computer Vision,* MIT Press, Cambridge, MA, 1993.
[11] Faugeras, O., Stratification of three-dimensional vision: projective, affine, and metric representations, *J. Opt. Soc. Am.,* 12(3):465–84, 1995.
[12] Faugeras, O. and Luong, Q.-T., *Geometry of Multiple Images,* MIT Press, Cambridge, MA, 2001.
[13] Fischler, M.A. and Bolles, R.C., Random sample consensus: a paradigm for model fitting with application to image analysis and automated cartography, *Commn. ACM,* 24(6):381–395, 1981.
[14] Gärding, J., Shape from texture for smooth curved surfaces in perspective projection, *J. Math. Imaging and Vision,* 2(4):327–350, 1992.
[15] Gärding, J., Shape from texture and contour by weak isotropy, *J. Artificial Intelligence,* 64(2):243–297, 1993.
[16] Geyer, C. and Daniilidis, K., Properties of the catadioptric fundamental matrix, in *Proc. Eur. Conf. Computer Vision,* Copenhagen, Denmark, Vol. 2, pp. 140–154, 2002.
[17] Gibson, J., *The Perception of the Visual World,* Houghton Mifflin, Boston, 1950.
[18] Gonzalez, R. and Woods, R., *Digital Image Processing,* Addison-Wesley, Reading, MA, 1992.
[19] Gool, L.V., Moons, T., and Proesmans, M., Mirror and point symmetry under perspective skewing, in *Proc. IEEE Int. Conf. Computer Vision & Pattern Recognition,* San Francisco, pp. 285–292, 1996.
[20] Gruen, A. and Huang, T., *Calibration and Orientation of Cameras in Computer Vision.,* Information Sciences, Springer-Verlag, Heidelberg, 2001.
[21] Han, M. and Kanade, T., Reconstruction of a scene with multiple linearly moving objects, in *Int. Conf. Computer Vision & Pattern Recognition,* Vol. 2, pp. 542–549, 2000.
[22] Harris, C. and Stephens, M., A combined corner and edge detector, in *Proc. Alvey Conf.,* pp. 189–192, 1988.
[23] Hartley, R. and Zisserman, A., *Multiple View Geometry in Computer Vision.,* Cambridge, 2000.
[24] Heyden, A. and Sparr, G., Reconstruction from calibrated cameras — a new proof of the Kruppa Demazure theorem, *J. Math. Imaging and Vision,* pp. 1–20, 1999.
[25] Hong, W., Yang, Y., Huang, K., and Ma, Y., On symmetry and multiple view geometry: Structure, pose and calibration from a single image, *Int. J. Computer Vision,* in press.
[26] Horn, B., *Robot Vision,* MIT Press, Cambridge, MA, 1986.
[27] Huang, K., Fossum, R., and Ma, Y., Generalized rank conditions in multiple view geometry with applications to dynamical scene, in *Proc. 6th Eur. Conf. Computer Vision,* Copenhagen, Denmark, Vol. 2, pp. 201–216, 2002.
[28] Huang, K., Yang, A.Y., Hong, W., and Ma, Y., Large-baseline matching and reconstruction using symmetry cells, in *Proc. of IEEE Int. Conf. Robotics and Automation,* pp. 1418–1423, 2004.
[29] Huang, T. and Faugeras, O., Some properties of the E matrix in two-view motion estimation, *IEEE Trans. Pattern Anal. Mach. Intelligence,* 11(12):1310–12, 1989.
[30] Hutchinson, S., Hager, G.D., and Corke, P.I., A tutorial on visual servo control, *IEEE Trans. Robotics and Automation,* pp. 651–670, 1996.
[31] Kanatani, K., *Geometric Computation for Machine Vision,* Oxford Science Publications, 1993.

[32] Kruppa, E., Zur ermittlung eines objecktes aus zwei perspektiven mit innerer orientierung, *Sitz.-Ber. Akad. Wiss., Math. Naturw., Kl. Abt. IIa*, 122:1939–1948, 1913.

[33] Longuet-Higgins, H.C., A computer algorithm for reconstructing a scene from two projections, *Nature*, 293:133–135, 1981.

[34] Lucas, B.D. and Kanade, T., An iterative image registration technique with an application to stereo vision, in *Proc. Seventh Int. J. Conf. Artificial Intelligence*, pp. 674–679, 1981.

[35] Ma, Y., Huang, K., and Kosecka, J., New rank deficiency condition for multiple view geometry of line features, *UIUC Technical Report, UILU-ENG* 01-2209 (DC-201), May 8, 2001.

[36] Ma, Y., Huang, K., and Vidal, R., Rank deficiency of the multiple view matrix for planar features, *UIUC Technical Report, UILU-ENG* 01-2209 (DC-201), May 18, 2001.

[37] Ma, Y., Huang, K., Vidal, R., Kosecka, J., and Sastry, S., Rank conditions of multiple view matrix in multiple view geometry, *Int. J. Computer Vision*, 59:115–137, 2004.

[38] Ma, Y., Kosecka, J., and Huang, K., Rank deficiency condition of the multiple view matrix for mixed point and line features, in *Proc. Asian Conf. Computer Vision*, Sydney, Australia, 2002.

[39] Ma, Y., Kosecka, J., and Sastry, S., Motion recovery from image sequences: discrete viewpoint vs. differential viewpoint, in *Proc. Eur. Conf. Computer Vision*, Vol. 2, pp. 337–353, 1998.

[40] Ma, Y., Soatto, S., Kosecka, J., and Sastry, S., *An Invitation to 3D Vision: From Images to Geometric Models*, Springer-Verlag, Heidelberg, 2003.

[41] Ma, Y., Vidal, R., Kosecká, J., and Sastry, S., Kruppa's equations revisited: its degeneracy, renormalization and relations to cherality, in *Proc. Eur. Conf. Computer Vision*, Dublin, Ireland, 2000.

[42] Malik, J., and Rosenholtz, R., Computing local surface orientation and shape from texture for curved surfaces, *Int. J. Computer Vision*, 23:149–168, 1997.

[43] Marr, D., *Vision: A Computational Investigation into the Human Representation and Processing of Visual Information*, W.H. Freeman, San Fancisco, 1982.

[44] Maybank, S., *Theory of Reconstruction from Image Motion*, Springer Series in Information Sciences. Springer-Verlag, Heidelberg, 1993.

[45] Morales, D. and Pashler, H., No role for colour in symmetry perception, *Nature*, 399:115–116, May 1999.

[46] Nister, D., An efficient solution to the five-point relative pose problem, in *CVPR*, Madison, U.S.A., 2003.

[47] Plamer, S.E., *Vision Science: Photons to Phenomenology*, MIT Press, Cambridge, MA, 1999.

[48] Poelman, C.J. and Kanade, T., A paraperspective factorization method for shape and motion recovery, *IEEE Trans. Pattern Anal. Mach. Intelligence*, 19(3):206–18, 1997.

[49] Quan, L. and Kanade, T., A factorization method for affine structure from line correspondences, in *Proc. Int. Conf. Computer Vision & Pattern Recognition*, pp. 803–808, 1996.

[50] Robert, L., Zeller, C., Faugeras, O., and Hebert, M., Applications of nonmetric vision to some visually guided tasks, in Aloimonos, I., Ed., *Visual Navigation*, pp. 89–135, 1996.

[51] Shakernia, O., Vidal, R., Sharp, C., Ma, Y., and Sastry, S., Multiple view motion estimation and control for landing an unmanned aerial vehicle, in *Proc. Int. Conf. Robotics and Automation*, 2002.

[52] Shashua, A., Trilinearity in visual recognition by alignment, in *Proc. Eur. Conf. Computer Vision*, Springer-Verlag, Heidelberg, pp. 479–484, 1994.

[53] Shashua, A. and Wolf, L. On the structure and properties of the quadrifocal tensor, in *Proc. Eur. Conf. Computer Vision, vol. I*, Springer-Verlag, Heidelberg, pp. 711–724, 2000.

[54] Shi, J. and Tomasi, C., Good features to track, in *Proc. IEEE Conf. Computer Vision and Pattern Recognition*, pp. 593–600, 1994.

[55] Shimshoni, I., Moses, Y., and Lindenbaum, M., Shape reconstruction of 3D bilaterally symmetric surfaces, *Int. J. Computer Vision*, 39:97–112, 2000.

[56] Spetsakis, M. and Aloimonos, Y., Structure from motion using line correspondences, *Int. J. Computer Vision*, 4(3):171–184, 1990.

[57] Sturm, P. and Triggs, B., A factorizaton based algorithm for multi-image projective structure and motion, in *Proc. Eur. Conf. Computer Vision*, IEEE Comput. Soc. Press, Washington, pp. 709–720, 1996.

[58] Thrun, S., Robotic mapping: a survey, *CMU Technical Report, CMU-CS-02-111,* February 2002.
[59] Tomasi, C. and Kanade, T., Shape and motion from image streams under orthography, *Int. J. Computer Vision,* 9(2):137–154, 1992.
[60] Triggs, B., Autocalibration from planar scenes, in *Proc. IEEE Conf. Computer Vision and Pattern Recognition,* 1998.
[61] Troje, N.F. and Bulthoff, H.H., How is bilateral symmetry of human faces used for recognition of novel views, *Vision Research,* 38(1):79–89, 1998.
[62] Vetter, T. and Poggio, T., Symmetric 3d objects are an easy case for 2d object recognition, *Spatial Vision,* 8:443–453, 1994.
[63] Vetter, T., Poggio, T., and Bulthoff, H.H., The importance of symmetry and virtual views in threedimensional object recognition, *Current Biology,* 4:18–23, 1994.
[64] Vidal, R., Ma, Y., Hsu, S., and Sastry, S., Optimal motion estimation from multiview normalized epipolar constraint, in *Proc. IEEE Int. Conf. Computer Vision,* Vancouver, Canada, 2001.
[65] Vidal, R. and Oliensis, J., Structure from planar motion with small baselines, in *Proc. Eur. Conf. Computer Vision,* Copenhagen, Denmark, pp. 383–398, 2002.
[66] Vidal, R., Soatto, S., Ma, Y., and Sastry, S., Segmentation of dynamic scenes from the multibody fundamental matrix, in *Proc. ECCV workshop on Vision and Modeling of Dynamic Scenes,* 2002.
[67] Weng, J., Huang, T.S., and Ahuja, N., *Motion and Structure from Image Sequences,* Springer-Verlag, Heidelberg, 1993.
[68] Weyl, H., *Symmetry,* Princeton University Press, 1952.
[69] Witkin, A.P., Recovering surface shape and orientation from texture, *J. Artif. Intelligence,* 17:17–45, 1988.
[70] Yang, A.Y., Hong, W., and Ma, Y., Structure and pose from single images of symmetric objects with applications in robot navigationi, in *Proc. Int. Conf. Robotics and Automation,* Taipei, 2003.
[71] Zhang, Z., A flexible new technique for camera calibration, *Microsoft Technical Report MSR-TR-98-71,* 1998.
[72] Zhao, W.Y. and Chellappa, R., Symmetric shape-from-shading using self-ratio image, *Int. J. Computer Vision,* 45(1):55–75, 2001.
[73] Zisserman, A., Mukherjee, D.P., and Brady, J.M., Shape from symmetry-detecting and exploiting symmetry in affine images, *Phil. Trans. Royal Soc. London A,* 351:77–106, 1995.
[74] Zisserman, A., Rothwell, C.A., Forsyth, D.A., and Mundy, J.L., Extracting projective structure from single perspective views of 3d point sets, in *Proc. IEEE Int. Conf. Computer Vision,* 1993.

23
Haptic Interface to Virtual Environments

	23.1	Introduction **23**-1
		Related Technologies • Some Classifications • Applications
	23.2	An Overview: System Performance Metrics and Specifications **23**-4
		Characterizing the Human User • Specification and Design of the Haptic Interface • Design of the Virtual Environment
	23.3	Human Haptics **23**-9
		Some Observations • Some History in Haptics • Anticipatory Control
	23.4	Haptic Interface **23**-13
		Compensation Based on System Models • Passivity Applied to Haptic Interface
	23.5	Virtual Environment **23**-17
		Collision Detector • Interaction Calculator • Forward Dynamics Solver
R. Brent Gillespie		
University of Michigan	23.6	Concluding Remarks **23**-21

23.1 Introduction

A haptic interface is a motorized and instrumented device that allows a human user to touch and manipulate objects within a virtual environment. As shown in Figure 23.1, the haptic interface intervenes between the user and virtual environment, making a mechanical contact with the user and an electrical connection with the virtual environment. At the mechanical contact, force and motion are either measured by sensors or driven by motors. At the electrical connection, signals are transmitted that represent the force and motion occurring at a *simulated* mechanical contact with a virtual object. A controller within the haptic interface processes the various signals and attempts to ensure that the force and motion signals describing the mechanical contact track or in some sense follow the force and motion signals at the simulated contact. The motors on the haptic interface device provide the authority by which the controller ensures tracking. A well-designed haptic device and controller will cause the behaviors at the mechanical and simulated contacts to be "close" to one another. In so doing it will extend the mechanical sensing and manipulation capabilities of a human user into the virtual environment.

23.1.1 Related Technologies

Naturally, the field of haptic interface owes much to the significantly older field of telerobotics. A telerobot intervenes between a human user and a remote *physical* environment rather than between a user and a computationally mediated *virtual* environment. In addition to a *master* manipulator (to which the haptic

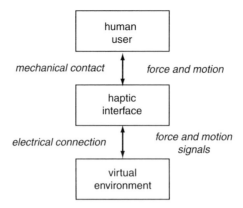

FIGURE 23.1 The haptic interface makes a mechanical contact with the user and an electrical connection with the virtual environment.

interface device is analogous), a telerobot includes a *slave* manipulator that makes mechanical contact with the remote environment. Telerobots were first developed in the 1940s for radioactive materials handling [1] and in their early realizations, were purely mechanical linkages designed to extend the touch and manipulation capabilities of a human operator past a protective wall. When electromechanical control first made its way into telerobots, only the slave was motorized so that it could follow the motions picked up by sensors in the master. But without a motorized master, the touch and manipulation capabilities of an operator were diminished when compared with the mechanical linkage, because interaction forces at the slave could not be felt. Later, with the introduction of motors on the master, forces (or motion errors) picked up by sensors on the slave could be reflected back to the operator. Thus the notion of a *bilateral* or *force-reflecting* teleoperator came into being, and the full manipulation *and feel* capacities of the mechanical linkage were restored. In analogy to the development of the bilateral telerobot, virtual reality gloves and other pointing devices support the positioning of virtual objects, but not until such devices are also outfitted with motors can the user feel interaction forces reflected back from the virtual environment. Once outfitted with motors, these devices graduate from mere pointing tools to become haptic interfaces.

Another prior technology related to the field of haptic interface is flight simulation. Flight simulators intervene between a human pilot trainee and a virtual aircraft and airfield using a yoke or joystick input device, a visual display, and often a motion display. See [2] for a full review of this technology. Haptic interface and flight simulation share much in terms of the virtual environment and real-time simulation technology, but an important difference exists with regard to the kind of interface made up by an input and display device. A haptic interface is simultaneously an input and display device, with two-way information flow across a single mechanical contact. In contrast, the motion display that moves the seat or cockpit and the visual display in a flight simulator do not share the same *port* with the yoke or joystick input device. If the yoke or joystick is motorized (or hydraulically driven) so that the pilot-trainee can feel the loads on the simulated aircraft control surfaces, then that yoke or joystick is actually a haptic interface. Indeed, such technology, both in flight simulation and fly-by-wire aircraft [3], pre-dates the field of haptic interface by many years.

23.1.2 Some Classifications

The term *haptic* derives from the Greek word *haptesthai*, meaning "to touch"[1] and is used broadly in engineering to describe mechanical interaction, especially between a human and machine. Naturally,

---

[1]The verb *haptesthai* is an archaic version of the modern Greek verb $\alpha\pi\tau o\mu\alpha\iota$ ("to touch") that in turn is derived from the modern noun $\alpha\phi\eta$ ("touch" or "feeling"). The phonetic spellings of the verb *haptesthai* and noun *haptikos* propagated in English circa 1890, with an "h" in front of the initial Greek "α" to denote the accent grave.

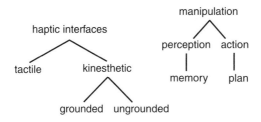

FIGURE 23.2 A taxonomy of haptic terms.

mechanical interaction encompasses a very rich set of behaviors, including actions that may be planned and perceived quantities that may be remembered.

As depicted in Figure 23.2, mechanical interactions may be roughly categorized by the percepts involved, whether *tactile* (served by sensors located in the skin, called cutaneous mechanoceptors) or *kinesthetic* (served by sensors in the muscles and joints: muscle spindles and Golgi organs). However, haptic interaction also includes manipulation, a process that involves both perception and action. Therefore, a strict classification in terms of tactile and kinesthetic sensing is difficult to achieve. Nevertheless, haptic interface devices may be roughly categorized as tactile or kinesthetic interfaces.

Tactile devices generally involve multiple, individually motor-driven localized contacts that describe an area contact with the skin, invoking texture and shape percepts and supporting haptic explorations such as stroking and contour-following. Kinesthetic interfaces, though they may also involve area contact with the skin, use a single motor-driven rigid body such as a knob, stylus, or thimble that is touched or grasped in the user's hand. The paradigm for interaction with virtual objects is then through that rigid body, an image of which is projected into the virtual environment. Thus, kinesthetic interfaces are used to invoke relationships between force and motion that pertain to interactions between a single rigid body and the virtual environment, with percepts such as stiffness, damping, and inertia. Kinesthetic interfaces support haptic explorations such as pushing, squeezing, indenting, and throwing. Kinesthetic interfaces may be further classified as to whether they are *grounded* (include a mechanical linkage to a fixed table or floor) or *ungrounded* (lacking such a linkage). Ungrounded interfaces are capable of producing only inertia forces and torques, whose frequency response lacks a DC component. For example, an ungrounded motor with an eccentric on its shaft such as one might find in a cell phone vibration ringer can be used to produce certain haptic or so-called vibrotactile impressions. In the present chapter, we will concentrate on kinesthetic haptic interfaces. For a full overview of haptic device technology, covering both tactile and kinesthetic interfaces, see [4].

23.1.3 Applications

Applications of haptic interface include training and assessment of cognitive and motor skills, rehabilitation, scientific visualization, and entertainment. Industries in which haptic devices are beginning to appear include medical training (e.g., surgical skill training simulators), automotive (so-called infotronic devices or haptic interfaces to climate and other controls that keep the eyes on the road), and gaming industries (immersive and shared virtual environments). Entertainment is one of the fastest growing markets for commercial development of haptic interface, where the ability to engage more of the user's senses leads to new dimensions of interactivity. A number of industrial concerns are quickly growing to serve these markets. Also, because human haptic psychophysics is a relatively new and very rich field of academic and industrial research, a small but significant commercial activity has grown to provide devices for basic and applied haptics research.

Additional uses of haptic interface are on the horizon. For example, haptic interface can be used in an *augmented reality* approach to overlay mechanical features or *virtual fixtures* on one's interaction with physical objects. The user can use these virtual fixtures to develop more efficient, more ergonomic, or more robust strategies for manipulating objects. For example, a virtual fixture might be used to eliminate tremor when performing retinal surgery through a teleoperator [5].

Closely related to the idea of virtual fixtures is the use of haptic interface to realize shared control between a human and an automatic controller. A human and an automatic control system can collaborate to complete a task if they both apply their control inputs as forces on a single manual interface (e.g., joystick or steering wheel). The user can monitor the actions of the automatic controller by feel using his hand resting on the joystick or he can augment or override the actions of the automatic controller by applying his own forces. Shared control through a haptic interface is being explored in the context of driver assist for agricultural and road vehicles [6] and surgical assist [5, 7].

Thinking more generally about the training of motor skills, a haptic interface can present the force-motion relationship (including history-dependent or dynamical relationship) that is deemed appropriate or representative of the physical world, yet carries no risk when errors are committed. Alternatively, a force-motion relationship that is adapted to the user, or contains specially designed responses to errors, may be created. The particulars of the mechanical interaction are all programmable, as is the relationship to the accompanying visual and audio responses. Further, a haptic interface can be used to test various hypotheses in human motor control, the answers to which may be used to develop therapies for rehabilitation or motor skill training [8].

Haptic interface has also been used to teach cognitive skills or concepts in certain disciplines. For example, concepts in system dynamics for engineering undergraduates can be demonstrated in a hands-on manner with the hopes that physical intuition might kick-in to boost understanding and the development of good modelling skills [9, 10]. By programming their own virtual environments using the differential equations studied in class and then interacting with them through a haptic interface, students should be able to more quickly establish appropriate relationships between mathematical models and observed mechanical behavior.

23.2 An Overview: System Performance Metrics and Specifications

A haptic interface, precisely because its role is to intervene between a user and virtual environment, is difficult to evaluate by itself. Its performance in a given application depends on certain characteristics and behaviors of the human user (e.g., size of hand, grip, experience) and on certain properties of the virtual environment (e.g., update rate, model fidelity, integration with visual display). What really determines the success of a given design is, of course, the performance of the coupled system comprising all three elements: user, haptic interface, and virtual environment. Yet, performance measures that are tied to the haptic interface alone are much to be preferred over system measures, for purposes of modularity. With measures defined only in terms of the haptic interface, one could design for any user and then interchange various haptic devices and virtual environment rendering algorithms. Also, such metrics are necessary to compare devices with one another. To define such haptic interface performance standards and to relate them to the system performance that really counts is one of the chief challenges of this field.

In the body of this chapter, performance measures to be attached to the haptic interface alone will be presented and discussed in detail. However, it is necessary to keep in mind that the most relevant performance measures in a given application depend on a user performing a particular task through a haptic interface on objects within a virtual environment. The contributions of the specification of the task, the selection of a particular user, and the use of a particular virtual environment rendering algorithm to the determination of the system performance cannot be underestimated. The haptic interface is only one of several determinants. Examples of *system* performance measures include: discriminability of two rendered textures or impedances, pursuit tracking performance, and peg-in-hole task completion times. Production of these performance measures requires some form of experiment with human subjects and a particular virtual environment.

Haptic Interface to Virtual Environments

To emphasize the roles of the user and virtual environment in system performance, the central section of this chapter (whose topic is haptic interface metrics) is flanked by a section on human haptics and a section on virtual environment design. Thus, the organization of this chapter reflects the three-element system comprising human, haptic interface, and virtual environment. But before launching into these three sections, a quick overview with additional unifying commentary is given here.

23.2.1 Characterizing the Human User

Truly it is the human user whose role is central and ties together the various applications of haptic interface. A very useful viewpoint is that the ground for the success of virtual reality (and by extension haptic interface) was laid by the failure of robotics. So far, the robot has not been able to enter the world of the human. Although the assembly line and paint shop have been opened to robots, by comparison the significantly less structured environments of the kitchen and laundry room have remained closed. Coping with the complexities of the physical world has proven quite difficult in practice. On the other hand, the human can easily suspend disbelief and adapt to synthetic worlds and, thus, in a very real sense, can enter the world of the robot. In most successful applications or demonstrations of haptic interface to date, the adaptability and forgiving nature of the human were harnessed, notwithstanding the various technical challenges and technical achievements that were also brought to bear. Thus, a key to design for any given application is an understanding of human haptic perception and its cooperative role with other perceptual modalities.

Likewise, the significant ability of a human when acting as a controller to adapt to new circumstances and respond to unforeseen events has given rise to the use of robots that extend human capabilities rather than replace them. Thus, the field of teleoperation is ever active despite advances in autonomous robots. Even when restricted through limited kinematics, masking dynamics, and destabilizing time-delays, human operators acting through a telerobot are preferred over autonomous robots for many applications. Haptic interface draws motivation for its existence from the capabilities (even needs) of human users to suspend their disbelief, adapt their manipulation strategies, and perform within an otherwise inaccessible or unsafe training environment.

In Section 23.3 below, a small review of human haptic psychophysics will be undertaken. Further results from this branch of cognitive psychology will be discussed, especially those that impact the engineering of haptic interface devices and virtual environments.

23.2.2 Specification and Design of the Haptic Interface

Figure 23.3 shows the three elements: human user, haptic interface, and virtual environment presented in both a network diagram and a block diagram. Associated with both diagrams are the user/interface

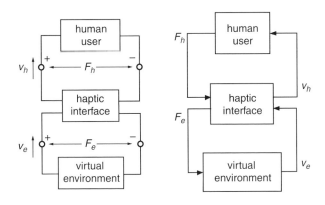

FIGURE 23.3 System network diagram and block diagram.

interaction force F_h and velocity v_h and interface/virtual-environment interaction force F_e and velocity v_e. Note that in the network diagram, a sign convention is specified for each variable that is not explicitly available in the block diagram. In the block diagram, a *causality* or signal flow direction is assumed that is not explicitly available in the network diagram. The sign convention and causality assumption are both choices made by the analyst or modeler.

Certainly the sign convention is arbitrary and has no implication on the eventual system equations. The causality assumption, however, by specifying what variables are inputs and outputs for a particular component, determines the *form*, or what quantities are on the left hand sides of the system equations. Although causality is chosen by the analyst, the particular causality assumption shown in the block diagram of Figure 23.3 is most appropriate for a haptic device designed according to the *impedance display* paradigm and a virtual environment in a *forward dynamics* form. The opposite causality is available for both the haptic interface and virtual environment, with corresponding modifications to the hardware. If the haptic interface uses motors in current-control mode to apply forces and moments, it is generally considered to be an impedance display. A haptic interface that uses force or torque sensors to close a control loop around the interface/human interaction force is called an admittance display. Most haptic interface devices to date are controlled using impedance display, which may be implemented without force or torque sensors using low inertia motors and encoders connected to the mechanism through low friction, zero-backlash transmissions with near-unity mechanical advantages.

A natural way to analyze the haptic interface independent of the user and virtual environment is to look at quantities defined at the boundaries: the two mechanical contacts, the physical contact with the user and the simulated contact with the virtual environment. One can define and analyze the notion of "closeness" between the force-motion relationship observed at the physical contact and the force-motion relationship observed at the virtual contact. One useful viewpoint is that the haptic interface is a kind of *filter* on top of the virtual environment, transmitting certain aspects of mechanical behavior of the virtual environment, but also adding some of its own behavior to what the user feels.

Given linear models of virtual environment impedance, definitions of closeness are not difficult to pin down. For example, *transparency* has been defined for teleoperators as the ratio of the filtered mechanical impedance presented by the master to the impedance presented by the remote environment [11]. Transparency for a haptic interface can be defined as the ratio of the impedance presented by the device to that presented by the virtual environment. Thus transparency is the ability of the haptic interface to faithfully reproduce what is rendered by the virtual environment. If the user feels he is touching the virtual environment directly, then the force and motion measured at the physical contact should have the same relationship as that measured in the simulated contact with the virtual environment. After having defined "closeness" and determined some of the underlying factors, the objective is to optimize a given haptic device and controller design while satisfying certain implementation or cost limits.

When a user contacts a haptic interface, not only is one control loop closed as might be suggested in Figure 23.3, but multiple loops are closed. The human is after all a complex system containing its own hierarchically organized control system. One of the loops closed by the human around the haptic interface is quite tight, or of relatively high bandwidth. It does not involve any perception or voluntary action on the part of the user. That is the loop closed by the mechanics of the human body, taken without its supervisory sensory-motor controller. In other words, it is the loop closed by the passive dynamics of the user, or the nominal mechanical impedance of his hand or hand and arm. This loop determines certain aspects of the force-motion relationship at the user/interface contact, especially the high-frequency aspects. Oscillations that appear in the system with characteristic frequencies beyond the capabilities of human motor control, especially those that are otherwise absent when the user does not contact the interface, will depend only on this inner loop (the passive dynamics) and not the outer loop involving sensory-motor processes.

More will be said about such oscillations in Section 23.4 below. Here it suffices to say that these oscillations are associated with a stability limit and determine the limits of performance through a performance/stability tradeoff. An associated transparency/stability tradeoff has been established for teleoperators [11]. As in the field of teleoperation, the quantification of this tradeoff is one of the chief challenges in the field of haptic interface. Unlike transparency, which can be defined without specifying the human user and virtual environment, stability is a system property. It is difficult to determine the limits of stability in terms independent of the dynamics of the human user and virtual environment. The work in this area will be reviewed in Section 23.4.

23.2.3 Design of the Virtual Environment

The virtual environment is an algorithm that produces commands for the haptic device in response to sensor signals so as to cause the haptic device to behave as if it had dynamics other than its own. As such, the virtual environment is a controller, while the haptic interface is the plant. Naturally, the algorithm is implemented in discrete time on a digital computer and together with the haptic interface and user, the interconnection forms a sampled-data system.

Figure 23.4 presents a schematic view of a haptic interface, a virtual environment, and the link between the two. In the top portion of the Figure, mechanical interaction takes place between the user's fingers and the haptic device end-effector. In the lower portion, an image of the device end-effector E is connected to an end-effector proxy P through what is called the *virtual coupler*. The proxy P in turn interacts with

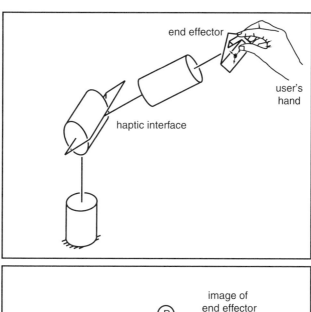

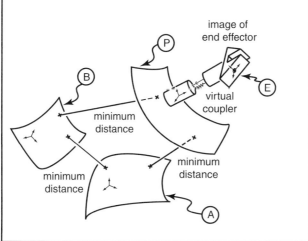

FIGURE 23.4 Schematic representation of haptic rendering.

objects such as A and B in the virtual environment. Proxy P might take on the shape of the fingertip or a tool in the user's grasp.

The virtual coupler is depicted as a spring and damper in parallel, which is a model of its most common computational implementation, though generalizations to 3D involve additional linear and rotary spring-damper pairs not shown. The purpose of the virtual coupler is two-fold. First, it links a forward-dynamics model of the virtual environment with a haptic interface designed for impedance-display. Motion (displacement and velocity) of the virtual coupler determines, through the applicable spring and damper constants, the forces and moments to be applied to the forward dynamics model and the equal and opposite forces and moments to be displayed by the haptic interface. Note that the motion of P is determined by the forward dynamics solution, while the motion of E is specified by sensors on the haptic interface. The second role of the virtual coupler is to filter the dynamics of the virtual environment so as to guarantee stability when display takes place through a particular haptic device. The parameters of the virtual coupler can be set to guarantee stability when parameters of the haptic device hardware are known and certain input-output properties of the virtual environment are met. Thus, the virtual coupler is most appropriately considered part of the haptic interface rather than part of the virtual environment [12, 13]. If an admittance-display architecture is used, an alternate interpretation of the virtual coupler exists, though it plays the same two basic roles. For further discussion of the role played by the virtual coupler in performance/stability tradeoffs in either the impedance or admittance-display cases, see [12–14].

Another note can be made with reference to Figure 23.4: rigid bodies in the virtual environment, including P, have both configuration and shape — they interact with one another according to their dynamic and geometric models. Configuration (including orientation and position) is indicated in Figure 23.4 using reference frames (three mutually orthogonal unit vectors) and reference points fixed in each rigid body, while shape is indicated by a surface patch. Note that the end-effector image E has configuration but no shape. Its interaction with P takes place through the virtual coupler and requires only the configuration of E and P.

Figure 23.5 shows the user, haptic device, virtual coupler, and virtual environment interconnected in a block diagram much like the block diagram in Figure 23.3. Intervening between the human and haptic device, that "live" in the continuous, physical world, and the virtual coupler and virtual environment, that live in the discrete, computed world, are a sampling operator T and zero-order hold. The virtual coupler is shown as a two-port that operates on velocities v_m and v_e to produce the motor command force F_m and force F_e imposed on the virtual environment. Forces F_m and F_e are usually equal and opposite. Finally,

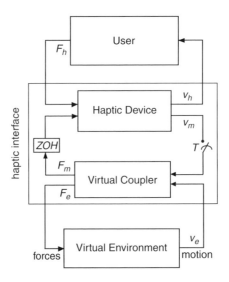

FIGURE 23.5 Block diagram of haptic rendering.

the virtual environment is shown in its forward dynamics form, operating on applied forces F_e to produce response motion v_e. Naturally, the haptic device may use motors on its joints, so the task-space command forces F_m must first be mapped through the manipulator Jacobian before being applied to the motors.

Not apparent in the block diagram is the detail inside the virtual environment block which must be brought to bear to simulate systems with changing contact conditions, including a forward dynamics solver, collision detector, and interaction response algorithm. A collision detector and interaction response algorithm are needed to treat bodies that may collide, rest, slide, or roll on one another.

The virtual environment can easily include memory or the storage of information about past inputs, perhaps in the form of state variables. Thus, a virtual environment can emulate the behavior of a dynamical system or a physical system with energy storage elements. One very convenient way to produce such a virtual environment is to use numerical integration of the differential equations of motion of the modelled dynamical system. Since the integration must run in real-time, explicit numerical routines must be used.

As in the choice of any modelling formalism, numerous choices must be made in the design of a virtual environment, many of which have implications on the set of objects or kinds of interactions that may be simulated. The consequences of each choice are often quite complex and sometimes surprising. Unexpected or nonphysical behaviors sometimes appear, due to inherent limits of a formalism or certain assumptions that are challenged by the range of interaction behaviors that arise under human exploration. Tight computational resources imposed by the real-time requirement mean that the production of efficient and extensible algorithms is a very active field of research. Some of the problems and methods of this field will be explored in Section 23.5 below.

This concludes an overview. Another three sections appear below with the same headings: human haptics, haptic interface, and virtual environment. But this time each appearance treats its topic in a bit more technical detail and in a manner more independent of the others.

23.3 Human Haptics

Haptics research is concerned with understanding the way in which humans gather information from their environment through mechanical interaction. The field of human haptics lies primarily in the domain of cognitive psychology and has many proponents, including Lederman and Klatzky [15–17]. Contributions to the field of haptics follow and are often parallel to earlier contributions in the areas of vision and audition. Understanding haptic perception, however, is significantly more complicated (and interesting) than visual or aural perception, because the haptic sensory apparatus is integrated into the same organ (the hand, or more generally, the body) as the motor apparatus that is used to generate the mechanical stimuli. Haptic perception cannot be divorced from the process of manipulation. Mechanical excitation of objects in the environment is generally accomplished through muscle action so as to generate the sensory stimuli that carry the information. Katz gave this process of coupled motor control and haptic perception the name *active touch* in 1925 [18]. Gibson further developed the idea of active touch in his treatise on "ecological perception" in 1966 [19].

23.3.1 Some Observations

An important theme throughout the development of virtual environments has been the resolution of the cost/benefit tradeoff of including another display technology. This is an especially important question when it comes to investing in haptic display technology, which is typically expensive and difficult to maintain. This motivates a thorough investigation into the value of haptic feedback. How, exactly, is haptic feedback used by a human user to form images in his mind, or to confirm and make more compelling the images first inspired by the other senses? And further, how is haptic feedback used by a human user to create and maintain manipulation strategies as he attempts to carry out an action on an object or set of objects in his environment?

Although very simple, some experiments by Schaal and Sternad [20, 21] shed light on the value of haptic feedback while juggling a ball. In these experiments, human subjects attempted to maintain a stable juggle of a ball on a paddle in their control while being provided selectively with visual feedback, haptic feedback, or both. It was shown that a stable juggle was best maintained when haptic feedback was available, and the case of haptic feedback alone was much better than visual feedback alone and almost as good as combined visual and haptic feedback. It seems to be not just the continuous interaction dynamics, but also the intermittent impulses and changes in dynamics which inspire the most compelling images and inform as to the best strategies for manipulation.

Recent research in haptics has led to an understanding of certain aspects of the relationship between haptic perception and motion production. For example, Klatzky and Lederman [22, 23] showed that subjects will employ characteristic motor patterns when asked to ascertain certain object properties. Klatzky and Lederman called these motor patterns *exploratory procedures*. For example, when asked about an object's texture, subjects will glide or rub their fingers over the object, whereas when asked about shape, they will follow contours with fingertips or enclose the object between their hands. It seems that certain patterns of movement maximize the availability of certain information. Klatzky and Lederman have also conducted experiments on the recognition of object representations that have demonstrated poor apprehension of form in two dimensions but good apprehension in three dimensions [24].

Modern researchers in haptics are fortunate to have haptic interface devices available to them, much like researchers in audition have speakers and researchers in vision have computer displays. Previous researchers in the field, though restricted to experiments and observations with physical objects, nevertheless have produced much of our understanding. Perspective on the history of haptics research and thinking can lead to very valuable insights for haptics researchers, especially engineers working in the field. For this reason, a brief historical overview is given here.

23.3.2 Some History in Haptics

Haptics as an academic discipline dates to the time of Aristotle. His treatise *De Anima* [25], which dates from 350 B.C., still provides very interesting reading to a modern haptics researcher. To Aristotle, touch was the most essential of the five senses and is the one feature that can be used to distinguish an animal from a plant or an inanimate object. While some animals cannot see or hear, all respond to touch. Interestingly (in light of modern thought on haptics), there exists a thread tying together the sense of touch and the capacity for movement in Aristotle's work. To Aristotle, features are closely tied to function so that one may suitably classify an object by describing either its features or its function. Having identified the sense of touch as the distinguishing feature of animals, Aristotle associated touch with the accepted functional definition of animals: objects that move of their own volition. Today the close link between motility and haptics is readily acknowledged, both because the mechanical senses are indispensable in the production of movement and because movement is indispensable in the gathering of haptic information. Another question in haptics that interested Aristotle persists even today: Is skin the organ of touch or is the touch organ situated somewhere else, possibly deeper? Even Aristotle acknowledged that "We are unable clearly to detect in the case of touch what the single subject is which underlies the contrasted qualities and corresponds to sound in the case of hearing."

In 1749, Diderot (of *Diderot's Encyclopedia* fame) published his "Letter on the Blind," a fascinating account of tactile perception in the congenitally blind. He laid the foundation for our understanding of sensory substitution, that one sense gains in power with use or loss of another. Modern neurological evidence also points to the plasticity of the brain: changes in cortical organization occur with changes in use or type of sensory stimulation. Diderot also wrote on the role of memory and the process of learning in touch, noting that an impression of form relies on retention of component sensations.

Ernst H. Weber introduced systematic experimental procedures to the study of haptics and the other senses and is, thus, considered the founder of the field of psychophysics. His famous law, formulated while investigating cutaneous sensation, was reported in *The Sense of Touch* (1834). Weber's law states that one's ability to discriminate differences between a standard and a comparison is a function of the magnitude of the standard. For example, a larger difference is needed to discriminate between two weights when the standard weighs 100 g than when the standard weighs 20 g. Anticipating later work in haptics,

Weber recognized the role of intentional movement in the perception of hardness and distance between objects.

In 1925 David Katz published his influential book *Der Aufbau der Tastwelt* (The World of Touch) [26]. He was interested in bringing the sense of touch back into prominence, because psychological research in vision and audition had already outstripped haptics research. Although Katz was certainly influenced by the work of his contemporaries in Gestalt psychology, he was more concerned with texture and ground than form and figure. Rather than simplicity of the internal response, he was interested in the correspondence of the internal response with the external stimulus. But, consistent with Gestalt thinking, he held that sensations themselves are irrelevant. Rather, the invariants of the object are obtained over time, and an internal impression is formed that is quite isolated from the sensory input.

Katz was particularly interested in the role of movement in haptic perception. Resting your hand against a surface, you may feel that it is flat, but until there is relative movement between your fingertips and the surface, you will not be able to discern its texture. Only with movement do objects "come into view" to the haptic senses. With movement, touch becomes more effective than vision at discerning certain types of texture.

Katz noted that the pressure sense can be excluded by holding a stick or stylus between the teeth and moving it across some material: vibrations are still produced and accurate judgments can be made as to the material "touched." Katz's experiments with styli further suggest that touch is a far sense, like vision and hearing, contrary to our tendency to assume that it requires direct impression on the skin by an object. Vibration of the earth (felt in our feet) may signal the imminent approach of a train or a herd of wild buffalo. In a real sense, a tool becomes an extension of one's body; the sensory site moves out to the tool tip. These comments further underline the claim that understanding haptics has important implications for the effective use and design of tools.

Arguably, Katz's most important contribution to haptics research was on the subject of active and passive touch. When a subject is allowed to independently direct the movements of her hand, she is able to make a much more detailed report of surface texture than when the object is moved under her passive fingertips. Rather boldly, and with much foresight, Katz proposed an altogether different kind of organ for the sense of touch: the hand. By identifying the hand as the seat of haptic perception, he emphasized the role of intentional movement. He essentially coupled the performatory function of the hand to its perceptual function. By naming an organ that includes muscles, joints, and skin, he coupled the kinesthetic sense to the tactile. In certain instances, he claimed, two hands may be considered the organ of touch just as two eyes may be considered the organ of vision.

Geza Revesz (discussed in [27]) was particularly interested in the development of haptic perception in the blind and especially the coding of spatial information. According to Revesz, haptic recognition of objects is not immediate, as it is in vision, but requires constructive processing of sequentially acquired information. In haptics, the construction of the whole is a cognitive process that follows perception of parts. Revesz emphasized the spatial nature of haptics and its possibilities for apprehending an object from all sides. His theories and experiments with blind persons have had important implications for the development of aids for the blind, such as tactile maps. Perspective cues, occlusion, and background fading, each of which work so well in drawings presented to the eyes, do not work well in raised-line drawings presented to the hands. Recognition of three-dimensional models of objects with the hands, in contrast, is very good.

Gibson [19] contributed in subtle but important ways to the field of psychophysics and to haptics in particular. Gibson was interested in fostering a more ecological approach to research in sensory processes and perception, an approach that takes into account all properties of an environment that may have relevance to a person with particular intentions within that environment. He argued that perceptual psychologists should study recognition of objects rather than such "intellectual" processes as memory or imagination, or such low-level phenomena as stimulus response. Gibson proposed that perception is not simply a process of information gathering by the senses and subsequent processing by perceptual centers, but the result of a hierarchical perceptual system whose function depends on active participation by the perceiver. For example, the visual system includes not only the eyes and visual cortex but also the active eye muscles, the actively positioned head, and even the mobile body. The haptic system, in addition to the tactile and kinesthetic sensors and somatosensory cortex, includes the active muscles of the arms, hands, and fingers.

Gibson, like Katz, stressed the importance of intentional movement in haptic perception. He preferred to think of active touch as a separate sense. Even when a subject has no intention of manipulating an object, she will choose to run her fingers over the object when left to her own devices. Certainly the fingertips are to the haptic sense as the fovea is to the visual sense: an area with a high concentration of sensors, and thus particular acuity. The fingers may be moved to place the highest concentration of sensors on the area of interest. Movement may be used to produce vibration and transient stimuli, which we know to be important from the experiments of Katz.

Gibson pointed to yet another reason for exploratory movement of the hand: to "isolate invariants" in the flux of incoming sensory information. Just as the image of an object maintains identity as it moves across the retina, or the sound of an instrument maintains identity as its changing pitch moves the stimulus across the basilar membrane, so an object maintains its identity as its depression moves across the skin. The identity even persists as the object is moved to less sensitive areas of the arm, and it is felt to maintain a fixed position in space as the arm glides by it. These facts, central to Gestalt theory, were underlined by Gibson and used as a further basis for understanding active touch. The exploratory movements are used to produce known changes in the stimulus flux while monitoring patterns that remain self-consistent. Thus, active touch is used to test object identity hypotheses, in Gibson's words, to "isolate the invariants."

Gibson also demonstrated that a subject passively presented with a haptic stimulus will describe an object in subjective terms, noting the sensations on the hand. By contrast, a subject who is allowed to explore actively will tend to report object properties and object identity. Under active exploration, she will tend to externalize the object or ascribe percepts to the object in the external world. For example, when a violin bow is placed on the palm of a subject's passive hand, she will report the sensations of contact on the skin, whereas a subject who is allowed to actively explore will readily identify the object and report object properties rather than describe sensations. Furthermore, when a string is bowed, the contact is experienced at the bow hairs and not in the hand.

Today the field of haptics has many proponents in academe and industry. From our present vantage point in history, we can identify reasons for the earlier lack of research interest in haptics. Certainly the haptic senses are more complex than the auditory or the visual, in that their function is coupled to movement and active participation by the subject. And further, the availability of an experimental apparatus for psychophysical study in haptics, the haptic interface, has been lacking until now.

Many open questions remain in haptics. We are still not sure if we have an answer to the question that Aristotle raised: What is to haptics as sound is to hearing and color is to seeing? As is apparent from experiments with active and passive touch, the notion of haptic sensation cannot be divorced from the notion of manipulation. Furthermore, the spatial and temporal sensitivity of the haptic sensors is not fully understood. Much research, especially using haptic interfaces, will likely lead to new results. As never before, psychologists and mechanical engineers are collaborating to understand human haptic perception. Results in the field have important implications for virtual reality: the effective design of virtual objects that can be touched through a haptic interface requires a thorough understanding of what is salient to the haptic senses.

23.3.3 Anticipatory Control

In the manipulation of objects that are familiar, one can predict the response to a given input. Thus, there is no need to use feedback control. One can use control without feedback, called open loop control or anticipatory control, wherein one anticipates an object's response to a given manipulation. From among many possible manipulations, one is chosen that is expected to give the desired response, according to one's best knowledge or guess. Anticipatory control is closely related to ballistic control. However, in, ballistic control, not only the feedback, but also the feedforward path, is cut before the manipulation task is complete. In the case of a ball aimed at a target and thrown, for instance, one has no control over the ball's trajectory after it leaves the hand. Both anticipatory control and ballistic control require that the human be familiar with the behavior of an environmental object or system, under a variety of input or manipulation strategies.

An object that is unfamiliar might be classifiable into a group of similar objects. If it is similar to another object, say in terms of size, simple scaling of the candidate manipulation/response interaction set (or model) would probably be appropriate. Scaling of an already available candidate input/output pair can be considered a case of using an internal model. Evidence for the existence of internal models is found in a manipulation experience with which most people are familiar firsthand. When grasping and lifting a suitcase, one may be surprised to have the suitcase rise off the table faster and higher than expected, only to realize that the suitcase is empty. The grasp and lift strategy was chosen and planned for a heavy suitcase.

The most important reason that humans and animals use anticipatory control rather than feedback control is that adequate time is not available for feedback control. There are significant delays in the arrival of sensory information, partly because of the slow (compared with electric wires) conduction of impulses along nerve fibers. Likewise, the execution of commands by muscles is accompanied by delays. Human response times, which include both sensory delays and command delays (the subject registers perception by pushing a button) are on the order of 180 msec. Yet many musical events, for example, are faster. A fast trill is 8 to 10 Hz. With less than 100 msec per cycle, the musician must issue commands almost a full cycle ahead of execution. Actuation based on a comparison between desired and feedback information is not possible.

23.4 Haptic Interface

As any person working with haptic interface technology soon finds out, sustained oscillations often hinder the display of virtual environments, especially hard virtual surfaces. Even the simplest virtual wall, whose physical counterpart is passive and well behaved, may nonetheless exhibit chatter when rendered through a haptic interface that is grasped by the user in a certain way. Suppose the following algorithm is used to render a virtual wall:

$$F_e = \begin{cases} k(x_e - x_o), & x_e \geq x_o \\ 0, & x_e < x_o \end{cases} \quad (23.1)$$

where k is a spring constant and x_o is a boundary locating the wall. Using this simple algorithm, and even when the haptic interface is equipped with the most ideal collocated sensors and actuators and lacks significant structural dynamics, one will notice that chatter appears. Chatter, or sustained oscillations often involving repeated contact with the wall, reflects a limit cycle in the coupled dynamics of human, haptic interface, and virtual environment. As the spring stiffness k is increased, the chatter will intensify. Somehow, energy not accounted for in the models of the virtual environment and interface and not provided by the user is leaking into the system to produce the oscillatory behavior. Typically, the frequency of oscillation is higher than the capabilities of human voluntary movement or even involuntary tremor, indicating that the energy required to sustain the oscillations is supplied by the interface/virtual environment system.

Origins of the energy leak include a half-sample delay associated with the zero-order-hold that intervenes between the discrete virtual environment and the haptic interface hardware. Other possible culprits include computational delay, signal quantization imposed by sensors, and delays associated with filtering or numerical differentiation to produce velocity estimates.

A simple graphical explanation of the energy-instilling effects of sampling on the virtual wall is available in Figure 23.6. For simplicity, the wall has been located at the origin: $x_o = 0$. Overlayed on the spring law $F_e = kx_e$ is a trace describing the reaction force of a wall generated using Equation (23.1) that was penetrated a single time to depth x_{max}. The trace has a staircase shape because x_e is subject to sampling and the force F_e is zero-order held. As x_e increases, the held reaction force is less than that given by the continuous spring law, except at the sampling times, where $F_e = kx_e$. Then as x_e decreases again, the held force is greater than the spring law, except at the sampling times. As a result, the work done on the wall

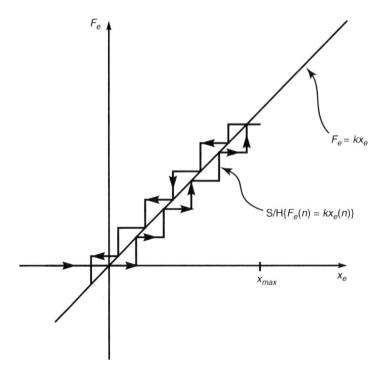

FIGURE 23.6 The sampled and held force vs. displacement curve for a virtual wall.

during compression of the virtual spring is less than the work returned during decompression, and energy may be extracted from the wall. The area inscribed by the pair of staircases is a measure of the energy generated by the wall.

Figure 23.6 leads naturally to the idea of predicting the displacement x_e a half-sample ahead to compensate for the effects of the zero-order-hold. The idea is to use a staircase approximation to the spring law that preserves area under the curve or equivalently preserves the force-displacement integral. The average of $x_e(n)$ and a value $\hat{x}_e(n+1)$ predicted a full sample ahead may be used as a half-sample prediction of x_e, yielding the algorithm

$$F_e(n) = \frac{k}{2}(x_e(n) + \hat{x}_e(n+1)) \tag{23.2}$$

If the velocity $\dot{x}_e$ is available, the prediction $\hat{x}_e(n+1)$ may be obtained using $\hat{x}_e(n+1) \approx x_e(n) + \dot{x}_e(n)T$, which yields the algorithm

$$F_e(n) = kx_e(n) + \frac{kT}{2}\dot{x}_e(n) \tag{23.3}$$

The second term can be interpreted as a virtual damper that is placed in parallel with the virtual spring to dissipate the energy generated by the zero-order-hold.

Very often the velocity $\dot{x}_e$ is not available from a sensor and must instead be estimated based on sampled x_e measurements. If a backward difference is used, $\dot{x}_e(n) \approx \frac{1}{T}(x_e(n) - x_e(n-1))$ which produces

$$\hat{x}_e(n+1) = 2x(n) - x(n-1) \tag{23.4}$$

When substituted into Equation (23.3), this rule produces the following spring law:

$$F_e(n) = k(1.5x(n) - 0.5x(n-1)) \tag{23.5}$$

A virtual wall rendered using this spring law will work significantly better than the wall rendered using a law without prediction, Equation (23.1). But it may be even further improved. Ellis [28] borrowed the idea of prediction-correction from numerical methods to further improve Equation (23.5). At sample n the displacement $x_e(n)$ is available from a sensor and may be compared with the previously estimated $\hat{x}_e(n)$. The error incurred as a result of using $\hat{x}_e$ in place of x_e may be corrected. The difference between the actual $x_e(n)$ and the previously predicted value $\hat{x}_e(n)$ is added as a corrector term, yielding the predictor-corrector algorithm

$$F_e(n) = \frac{k}{2}(2x_e(n) + \hat{x}_e(n+1) - \hat{x}_e(n)) \qquad (23.6)$$

Substituting for the predicted values using Equation (23.4), the following algorithm is produced:

$$F_e(n) = k(2x(n) - 1.5x(n-1) + 0.5x(n-2)) \qquad (23.7)$$

This algorithm can be extended beyond the virtual wall and applied more generally as a filter to the force to be displayed by the haptic interface [28]. Taking $\tilde{F}_e$ as the filtered force and F_e has the name of the constitutive law of the virtual environment,

$$\tilde{F}_e(n) = 2F_e(n) - 1.5F_e(n-1) + 0.5F_e(n-2) \qquad (23.8)$$

23.4.1 Compensation Based on System Models

The prediction Equation (23.4) used in the algorithms above is based on the forward-Euler rule and a first-difference estimate of velocity. It uses a polynomial fit to sensor data from the present and recent past to predict sensor readings slightly into the future and, thus, account for delays. An alternative approach is to model the coupled dynamical system, including the virtual environment, haptic interface, and the human user, and use simulation or analytical solutions to the model to predict future states. This is the approach adopted in [29]. If the action of the human is modeled as a constant force and the human biomechanics as a second order system, the entire coupled system, including virtual wall and inertia and damping in the haptic interface, can be modeled as a lumped second order system with switching dynamics. Justification for the simple models of the human user are drawn from the disparate time-scales of sampling and volitional control, and the numerous demonstrations of good fit between second order models and experimental characterizations of human biomechanics [30]. Analytical solutions of the second order dynamics may be used to predict states between sample times. A half-sample prediction may be used to compensate for the effects of the zero-order hold.

Because the instants at which the wall threshold is crossed generally fall between sampling times, the effect of the zero-order-hold cannot be compensated with a half-sample prediction for those sampling periods that span the threshold crossings. However, the analytic solution may be used to account precisely for the change in dynamics (wall on/wall off) that occurs at the threshold crossings. The solution may be used as a goal state for a dead-beat controller that then drives the sampled data dynamics to the goal in two steps. The dead-beat controller is invoked at each threshold crossing and complements the half-sample prediction to produce sampled data system simulation which fully emulates the continuous system dynamics in [29].

23.4.2 Passivity Applied to Haptic Interface

An alternative approach to the analysis of stability in the coupled dynamics of the human/interface/virtual environment is based on the passivity theorem. When a virtual environment is modelled after a passive physical system, yet its rendering through a haptic interface exhibits sustained oscillations, analytical treatments like passivity, involving the production and dissipation of energy naturally come to mind. Another attractive feature of the passivity approach is that to characterize a system as passive, only an input/output description is necessary. A full state-space model is not needed. Rather, system components can be handled as members of classes with certain properties, and results are significantly more modular in nature. Note that passivity is a property of the virtual environment alone; it is not a system property like stability. Components whose input-output properties are the same but internal realizations differ may be substituted for one another. Also, time delays can be accommodated, whether arising from sampling, computational, or communication delay, or having to do with explicit integration schemes.

By the passivity theorem, if the human and virtual environment are assumed passive and the haptic interface assumed strictly passive (since it certainly has some energy dissipation due to friction and damping), then the coupled system will be stable. The contra-positive statement says that instability of the interconnected system implies that one of the three system components is not passive. Because the haptic interface is constructed of physical components, it cannot be the culprit. At first it would not seem so reasonable to assume that a human operator is a passive system, since, of course, humans metabolize food to produce mechanical energy. But because the unstable, chatter-like behavior so often observed in haptic rendering is typically characterized by frequencies that lie outside the range of human motor capabilities, one typically assumes that the human is in fact passive. Thus, the virtual environment is implicated, and measures to ensure its passivity should be taken to remedy the instability.

A system with input $u(t) \in \Re^n$ and output $y(t) \in \Re^n$ is passive if there exists a nonnegative function $W(x)$, called a storage function, such that

$$W(x(t)) \leq W(x(0)) + \int_0^t y(\tau)^T u(\tau) d\tau \qquad (23.9)$$

for all inputs $u(t)$ and all $t \geq 0$. The term $W(x(0))$ is the initial stored energy in the system and the product $y^T u$ is called the supply rate for the system. If y and u are force and velocity, then the integral of the supply rate is the power supplied to the system up to time t. The stipulation that the inequality must hold for *any* input $u(t)$ means that no matter how the system is driven, its net absorbed energy will always exceed its initial stored energy. Parseval's theorem can be used to transform the passivity definitions above into expressions in the frequency domain. For linear systems, Parseval's theorem shows that passivity corresponds to strict positive realness.

Colgate and Schenkel [31] determined parameter values that guarantee passivity of the virtual wall of Equation (23.1) when it is rendered through a zero-order-hold. Colgate's analysis is based on an application of the small gain theorem (as is the passivity theorem itself). But because the small gain theorem takes only magnitude information into account and completely disregards phase, a linear fractional transformation (LFT) (which has equivalent interpretations as a loop transformation and a coordinate change) must be used to reduce conservativeness. Using only the constraint that the human operator be passive, Colgate first finds the area in the Nyquist plane within which a passive human operator in feedback connection with the haptic interface and linked with a zero-order hold and integrator must lie. This area (a disk) can be mapped to the unit disk (uncertain phase; unity magnitude) by an LFT. A corresponding LFT is found for the discrete controller (the virtual wall) in [32]. The unit disk then becomes a bound in the Nyquist plane for the mapped discrete controller that, if satisfied, guarantees coupled stability by the small gain theorem. The result may be stated simply as

$$b > \frac{KT}{2} + |B| \qquad (23.10)$$

where b is viscous damping in the physical haptic device, K is the virtual wall stiffness, T is the sampling time, and B characterizes a virtual viscous damper in parallel with the virtual spring. The result is both necessary and sufficient to guarantee passivity for the constitutive law without the unilateral nonlinearity, and sufficient for the virtual wall. This equation says that excess damping b in the haptic interface device can be used to account for a deficiency of passivity in the virtual environment (due to half-sample delay in the zero-order-hold). Note the term $KT/2$ which appeared above in Equation (23.3) as the damping coefficient of a virtual damper that, when added to the spring law, compensated for the effects of the zero-order-hold. Here, physical damping in the haptic interface, rather than virtual damping, is used to account for the effects of the zero-order hold.

Equation (23.10) can be interpreted either as a design directive or as an insight. As a directive, Equation (23.10) says that an excess of dissipativity should be built into the haptic device and then its effect compensated by negative virtual damping B (note the absolute value around B). In effect, a physical damper b extracts energy at all frequencies (including the high frequencies above the Nyquist frequency) whereas a negative virtual damper only adds energy at low frequencies (where its effect is felt). Thus, in the end, stability is guaranteed and transparency (wallness) is not compromised. However, viscous dampers (at least reliable, linear ones) are not readily available commercially and are rather difficult to build and maintain.

Adams and Hannaford [12] and also [33] applied a passivity argument to a broader class of virtual environments than the Colgate papers, including general differential ODE models. In addition, the Adams and Hannaford framework brought in the virtual coupler and extended to impedance and admittance implementations. However, their work does not address the excess of passivity of the device that may be used to account for nonpassive behavior of the virtual environment. The virtual environment is, by necessity, nonpassive since explicit integrators must be used in real-time simulation.

Miller et al. [34] extend the passivity arguments of [12] and [31] to include concepts of input-strict and output-strict passivity. Using these tools, an excess of passivity in the haptic device can be used explicitly to account for a lack of passivity in the virtual environment. The results apply to virtual environments with nonlinearities and even time delays.

Hannaford and Ryu in [35] present a general passivity controller and a passivity observer that work together to guarantee that the haptic rendering is passive. The strength of this compensator is its generality; it does not depend on any knowledge of the system driving the interface nor on the behavior of the user. The generality of the solution is also a weakness though. If the human operator is active for a long period of time, then the system driving the interface can be active for a long period of time and exhibit undesirable active interactions with the operator.

23.5 Virtual Environment

A set of surface patches and their interconnection can be used to describe the geometry of each body in the virtual environment. The whole collection of surface patches along with their connected graphs is called the *geometric model*. Computations involving the geometric model, in a process called *collision detection*, identify points in time at which bodies within the virtual environment make contact with one another. At or during times of contact, the collision detector triggers another process called the *interaction calculator*, as shown in the flowchart of Figure 23.7. The interaction calculator uses an interaction model to compute the appropriate impulse response or interaction force between bodies. A third process is the solution of the forward dynamic model for the collection of bodies, producing motion in response to the applied forces and moments. The applied forces and moments include both those applied by the virtual coupler and the results of the interaction calculator.

The interaction calculator passes either new initial conditions (the result of an impulse response computation) or interaction forces (the result of continuous contact) to the forward dynamics solver. Equal and opposite forces and moments are applied to pairs of interacting bodies. The forward dynamics solver operates on the forward dynamics model or equations of motion, which is a set of differential equations in the configuration and motion variables and inertia parameters. The dynamic model might also contain embedded holonomic or nonholonomic constraint equations, written in terms of certain geometric parameters that are not necessarily part of the geometric model.

The use of a collision detector and interaction calculator as in Figure 23.7 ensures that interacting bodies respond to each other's presence. Note that the interaction calculator is called upon only intermittently whereas the collision detector and forward dynamics solver run continually, either alongside or subsequent to each other in computational time.

Let us now consider in turn each of the elements of the flow chart in greater detail.

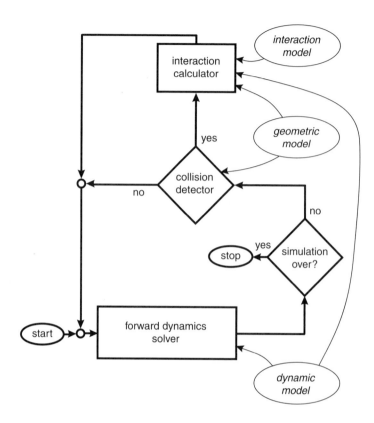

FIGURE 23.7 Flowchart showing interconnection of the dynamics solver, collision detector, and interaction calculator.

23.5.1 Collision Detector

To calculate a global solution in a computationally efficient manner, it is very common to handle the collision detection problem in two parts: a *broad phase*, which involves a coarse global search for potentially interacting surfaces, and a *narrow phase*, which is usually based on a fast local optimization scheme on convex surface patches. To handle nonconvex shapes, preprocessing is performed to decompose them into sets of convex surface patches. However, there exist shapes for which such a decomposition is not possible, for example, a torus.

23.5.1.1 Broad Phase

The broad phase is composed of two major steps. First, a global proximity test is performed using hierarchies of bounding volumes or spatial decompositions for each surface patch. Among the most widely used bounding volumes and space decompositions are the octrees [36], *k-d* trees [37], BSP-trees [38], axis-aligned bounding boxes (AABBs) [39], and oriented bounding boxes (OBBs) [40].

During the global proximity test, the distances between bounding boxes for each pair of surface patches drawn from all pairs of bodies are compared with a threshold distance and surfaces that are too distant to be contacting are pruned away. Remaining surfaces are set to be active.

In the second step of the broad phase, approximate interaction points on active surfaces are calculated. For example, if the geometric models are represented by non-uniform rational B-splines (NURBS), control polygons of the active surface models can be used to calculate a first order approximation to the closest points on these surfaces. Specifically, bounding box centroids of each interacting surface patch can be projected onto the polygonal control mesh of the other patch. Using the strong convex hull property of

NURBS surfaces, one can determine the candidate span and interpolate the surface parameters from the node values. These approximate projections serve as good initialization points for the narrow phase of the collision detector.

23.5.1.2 Narrow Phase

After initialization with the approximations calculated in the broad phase, the narrow phase employs an algorithm to iteratively update the minimum distance between each active pair of surfaces.

Previous work in narrow phase collision detection has concentrated on computing the minimum distance between convex polyhedral objects. State-of-the-art algorithms for computing the distance between convex polyhedra are based on the algorithm by Gilbert, Johnson, and Keerthi (GJK) [41, 42], the algorithm of Lin and Canny [43], and the algorithm by Mirtich [44].

The GJK algorithm makes use of Minkowski difference and (simplex-based) convex optimization techniques to calculate the minimum distance. The iterative algorithm generates a sequence of ever improving intermediate steps within the polyhedra to converge to the true solution. The algorithm of Lin and Canny makes use of Voronoi regions and temporal/spatial coherence between successive queries to navigate along the boundaries of the polyhedra in the direction of decreasing distance. The V-Clip algorithm by Mirtich is reminiscent of the Lin and Canny closest features algorithm but makes several improvements.

Less literature exists on direct collision detection for nonpolygonal models. Gilbert et al. extended their algorithm to general convex objects in [45]. In a related paper [46], Turnbull and Cameron modify the widely used GJK algorithm to handle convex shapes defined using NURBS. Similarly, in [47, 48] Lin and Manocha present an algorithm for curved models composed of spline or algebraic surfaces by extending their earlier algorithm for polyhedra.

Finally, a new class of algorithms, namely, minimum distance tracking algorithms for parametric surfaces, is presented in [49–52]. These algorithms are designed to maintain the minimum distance between two parametric surfaces by integrating the differential kinematics of the surfaces as the surfaces undergo rigid body motion.

23.5.2 Interaction Calculator

The interaction calculator, triggered into action by the collision detector, computes inputs (forces and moments) and new initial conditions (in the case of impulse response) for the forward dynamics solver. In so doing, it realizes interaction between bodies, whether such interaction occurs during a finite or an infinitesimal period of time. As indicated in Figure 23.7, the interaction calculator calls upon an interaction model and the geometric model to compute its response. If impulse response is also used, the interaction calculator additionally calls upon portions of the dynamic model.

The most widely used interaction model in both haptic rendering and multibody dynamics is called the penalty method: it assumes compliant contact characterized by a spring-damper pair. The force response is then proportional to the amount and rate of interpenetration. The interaction calculator need only query the geometric model using these surface parameters to determine the interpenetration vector. To compute a replacement (a resultant applied at the center of mass and a couple applied to the body) for the set of all contact forces and moments acting on a body for use as input to the dynamic model, the interaction calculator queries the geometric model using the surface parameters and then carries out the appropriate dot and cross products and vector sums. Using replacements for the set of interaction forces allows the dynamic model to be fully divorced from the geometric model.

The penalty contact model allows interpenetration between the objects within the virtual environment, which in a certain sense is nonphysical. However, the penalty method also eliminates the need for impulse response models, since all interactions take finite time. Important alternatives to the penalty method include the direct computation of constraint forces using the solution of a linear complementarity problem espoused by Baraff [53, 54] and widely adopted within the computer graphics community. The formulation by Mirtich and Canny [55] uses only impulses to account for interaction, with many tiny repeated impulses for the case of sustained contact. Note that computation of impulses requires that the interaction calculator query the dynamic model with the current extremal point parameters and system configuration to determine the effective mass and inertia at each contact point.

It is important to carefully distinguish between the role of the penalty method and the virtual coupler. The virtual coupler serves as a filter between the virtual environment and haptic device whose design mitigates instability and depends on the energetic properties of the closed-loop system components. The penalty method, on the other hand, though it may also be modelled by a spring-damper coupler, is an empirical law chosen to produce reasonable behavior in multibody system simulation. Parameter values used in the penalty method also have implications for stability, though in this case it is not haptic rendering system but rather numerical stability that is at issue. Differential equation stiffness is, of course, also determined by penalty method parameter values.

Associated with the penalty method is another subtle issue, that of requiring a unique solution to the maximum distance problem when the interpenetrated portions of two bodies are not both strictly convex or there exists a medial axis [56] within the intersection. This is the issue which the god-object or proxy methods address [57, 59].

23.5.3 Forward Dynamics Solver

The final component comprising the haptics-equipped simulator is a differential equation solver that advances the solution of the equations of motion in real time. The equations of motion are a set of differential equations constructed by applying a principal of mechanics to kinematic, inertial, and/or energy expressions derived from a description of the virtual environment in terms of configuration and motion variables (generalized coordinates and their derivatives) and mass distribution properties. Typically the virtual environment is described as a set of rigid bodies and multibody systems interconnected by springs, dampers, and joints. In such case the equations of motion become ordinary differential equations (ODEs), expressing time-derivatives of generalized coordinates as functions of contact or distance forces and moments acting between bodies of the virtual environment. In Figure 23.4, representative bodies A, B, and P comprise the virtual environment, while the forces and moments in the springs and dampers that make up the virtual coupler between bodies P and E are inputs or arguments to the equations of motion. The configuration (expressed by the generalized coordinates) and motion (expressed by the generalized coordinate derivatives) of bodies A, B, and P are then computed by solving the equations of motion (see also Figure 23.5). One may also say that the state (configuration and motion) of the virtual environment is advanced in time by the ODE solver.

Meanwhile, the collision detector runs in parallel with the solution of the equations of motion, monitoring the motion of bodies A, B, and P and occasionally triggering the interaction calculator. The interaction calculator runs between time-steps and passes its results (impulses and forces) to the equations of motion for continued simulation.

Quite often the virtual environment is modeled as a constrained multibody system, and expressed as a set of differential equations accompanied by algebraic constraint equations. For example, constraint appending using the method of Lagrange Multipliers produces differential-algebraic equations (DAEs). In such case, a DAE solver is needed for simulation. Note that DAE solvers are not generally engineered for use in real-time or with constant step-size and usually require stabilization. Alternatively, constrained multibody systems may be formulated as ODEs and then simulated using standard ODE solvers using constraint-embedding techniques. Constraint embedding can take place symbolically (usually undertaken prior to simulation time) or numerically (possibly undertaken during simulation and in response to run-time events).

Alternatives to the forward dynamics/virtual coupler formulation described above have been developed for tying together the dynamic model and the haptic interface. For example, the configuration and motion of the haptic device image (body E in Figure 23.5) might be driven by sensors on the haptic interface. A constraint equation can then be used to impose that configuration and motion on the dynamic model of the virtual environment.

23.6 Concluding Remarks

In this chapter, we have presented a framework for rendering virtual objects for the sense of touch. Using haptic interface technology, virtual objects can not only be seen, but felt and manipulated. The haptic interface is a robotic device that intervenes between a human user and a simulation engine, to create, under control of the simulation engine, an appropriate mechanical response to the mechanical excitation imposed by the human user. 'Appropriate' here is judged according to the closeness of the haptic interface mechanical response to that of the target virtual environment. If the excitation from the user is considered to be motion, the response is force and vice-versa. Fundamentally, the haptic interface is a multi-input multi-output system. From a control engineering perspective, the haptic interface is a plant simultaneously actuated and sensed by two controllers: the human user and the simulation engine. Thus the behavior of the haptic interface is subject to the influence of the human user, the virtual environment simulator, and finally its own mechanics. As such, its ultimate performance requires careful consideration of the capabilities and requirements of all three entities.

While robotics technology strives continually to instill intelligence into robots so that they may act autonomously, haptic interface technology strives to strike all intelligence out of the haptic interface—to get out of the way of the human user. After all, the human user is interested not in interacting with the haptic interface, but with the virtual environment. The user wants to retain all intelligence and authority. As it turns out, to get out of the way presents some of the most challenging design and analysis problems yet posed in the greater field of robotics.

References

[1] Goertz, R.C., Fundamentals of general-purpose remote manipulators, *Nucleonics,* vol. 10, no. 11, pp. 36–45, 1952.

[2] Baarspul, M., Review of flight simulation techniques, *Progress in Aerospace Sciences,* vol. 27, no. 1, pp. 1–20, 1990.

[3] Collinson, R., Fly-by-wire flight control, *Computing and Control Engineering Journal,* vol. 10, pp. 141–152, August 1999.

[4] Burdea, G.C., *Force and Touch Feedback for Virtual Reality,* Wiley Interscience, New York, NY, 1996.

[5] Marayong, P., Li, M., Okamura, A.M., and Hager, G.D. Spatial motion constraints: theory and demonstrations for robot guidance using virtual fixtures, in *IEEE International Conference on Robotics and Automation,* pp. 1954–1959, 2003.

[6] Griffiths, P. and Gillespie, R.B., Shared control between human and machine: Haptic display of automation during manual control of vehicle heading, in *12th International Symposium of Haptic Interfaces for Virtual Environment and Teleoperator Systems (HAPTICS'04),* Chicago, IL, pp. 358–366, March 27–28, 2004.

[7] Bettini, A., Marayong, P., Lang, S., Okamura, A.M., and Hager, G. Vision assisted control for manipulation using virtual fixtures, *IEEE Transactions on Robotics and Automation,* in press.

[8] Reinkensmeyer, D., *Standard Handbook of Biomedical Engineering & Design,* ch. Rehabilitators, pp. 35.1–35.17, McGraw-Hill, New York, 2003.

[9] Gillespie, R.B., Hoffman, M., and Freudenberg, J., Haptic interface for hands-on instruction in system dynamics and embedded control, in *IEEE Virtual Reality Conference,* (Los Angeles, CA), pp. 410–415, March 22–23, 2003.

[10] Okamura, A.M., Richard, C., and Cutkosky, M.R., Feeling is believing: using a force-feedback joystick to teach dynamic systems, *ASEE J. Engineering Education,* vol. 92, no. 3, pp. 345–349, 2002.

[11] Lawrence, D.A., Stability and transparency in bilateral teleoperation, *IEEE Transactions on Robotics & Automation,* vol. 9, pp. 624–637, October 1993.

[12] Seguna, C.M., The design, construction and testing of a dexterous robotic end effector, *Proc. IEEE R8 Eurocon 2001 Int. Conf. Trends Commun.,* vol. 1, pp. XXXVI–XLI, July 2001.

[13] Adams, R.J., Klowden, D., and Hannaford, B., Stable haptic interaction using the excalibur force display, in *Proceedings of International Conference on Robotics and Automation,* vol. 1, pp. 770–775, 2000.

[14] Miller, B.E., Colgate, J.E., and Freeman, R.A., Guaranteed stability of haptic systems with nonlinear virtual environments, *IEEE Transactions on Robotics and Automation*, vol. 16, no. 6, pp. 712–719, 2000.

[15] Klatzky, R.L., Lederman, S.J., and Reed, C., Haptic integration of object properties: texture, hardness, and planar contour, *Journal of Experimental Psychology-Human Perception and Performance*, vol. 15, no. 1, pp. 45–57, 1989.

[16] Klatzky, R.L., Lederman, S.J., Pellegrino, J., Doherty, S., and McCloskey, B., Procedures for haptic object exploration vs. manipulation, *Vision and Action: The Control of Grasping*, pp. 110–127, 1990.

[17] Lederman, S.J. and Klatzky, R.L., An introduction to human haptic exploration and recognition of objects for neuroscience and AI, *Neuroscience: From Neural Networks to Artificial Intelligence*, vol. 4, pp. 171–188, 1993.

[18] Katz, D., *The World of Touch, 1925*, edited and translated by Lester E. Krueger, Lawrence Erlbaum, Hillsdale, NJ, 1989.

[19] Gibson, J.J., *The Senses Considered as Perceptual Systems*, Boston: Houghton Mifflin, vol. 2, pp. 1453–1458, 1966.

[20] Schaal, S. and Atkeson, C.G., Robot juggling: an implementation of memory-based learning, *Control Systems Magazine*, vol. 14, no. 1, pp. 57–71, 1994.

[21] Schaal, S., Sternad, D., and Atkeson, C.G., One-handed juggling: a dynamical approach to a rhythmic movement task, *Journal of Motor Behavior*, vol. 28, no. 2, pp. 165–183, 1996.

[22] Klatzky, R.L. and Lederman, S.J., Identifying objects from a haptic glance, *Perception and Psychophysics*, vol. 57, no. 8, pp. 1111–1123, 1995.

[23] Lederman, S.J. and Klatzky, R.L., Extracting object properties by haptic exploration, *Acta Psychologica*, vol. 84, pp. 29–40, 1993.

[24] Klatzky, J.M., Loomis, R.L., Lederman, S.J., Wake, H., and Fujita, N., Haptic identification of objects and their depictions, *Perception and Psychophysics*, vol. 54, no. 2, pp. 170–178, 1993.

[25] Aristotle, *De Anima*, translated by Hugh Lawson-Tancred, Penguin Books, New York, NY, 1986.

[26] Katz, D., *The World of Touch, 1925*, edited and translated by Lester E. Krueger, Lawrence Erlbaum, Hillsdale, NJ, 1989.

[27] Heller, M.A. and Schiff, W., eds., *The Psychology of Touch*, Lawrence Erlbaum, Hillsdale, NJ, 1991.

[28] Ellis, R.E., Sarkar, N., and Jenkins, M.A., Numerical methods for the force reflection of contact, *Journal of Dynamic Systems, Measurement, and Control*, vol. 119, pp. 768–774, Decemeber 1997.

[29] Gillespie, R.B. and Cutkosky, M.R., Stable user-specific haptic rendering of the virtual wall, in *Proceedings of the ASME Dynamic Systems and Control Division*, vol. 58, pp. 397–406, November 1996.

[30] Hajian, A. and Howe, R.D., Identification of the mechanical impedance at the human finger tip, *Journal of Biomechanical Engineering, Transactions of the ASME*, vol. 119, pp. 109–114, February 1997.

[31] Colgate, J.E. and Schenkel, G.G., Passivity of a class of sampled-data systems: application to haptic interfaces, *Journal of Robotic Systems*, vol. 14, pp. 37–47, January 1997.

[32] Colgate, J.E., Coordinate transformations and logical operations for minimizing conservativeness in coupled stability criteria, *Journal of Dynamic Systems, Measurement, and Control*, vol. 116, pp. 643–649, December 1994.

[33] Colgate, J.E., Stanley, M.C., and Brown, J.M., Issues in the haptic display of tool use, in *Proceedings of IEEE/RSJ International Conference on Intelligent Robots and Control*, pp. 140–145, 1995.

[34] Miller, B.E., Colgate, J.E., and Freeman, R.A., Guaranteed stability of haptic systems with nonlinear virtual environments, *IEEE Transactions on Robotics & Automation*, vol. 16, pp. 712–719, December 2000.

[35] Hannaford, B. and Ryu, J.-H., Time-domain passivity control of haptic interfaces, *IEEE Transactions on Robotics & Automation*, vol. 18, pp. 1–10, February 2002.

[36] Moore, M. and Wilhelms, J. Collision detection and response for computer animation, *Computer Graphics SIGGRAPH 1988*, vol. 22, pp. 289–298, 1988.

[37] Held, M., Klosowski, J.T., and Mitchell, J.S.B., Evaluation of collision detection methods for virtual reality fly-throughs, *Proceedings of Seventh Canadian Conference Computer Geometry*, pp. 205–210, 1995.

[38] Naylor, B., Amatodes, J.A., and Thibault, W., Merging BSP trees yields polyhedral set operations, *Computer Graphics SIGGRAPH 1990*, vol. 24, pp. 115–124, 1990.

[39] Beckmann, N., Kriegel, H.P., Schneider, R., and Seeger, B., The r*-tree: an efficient and robust access method for points and rectangles, *Proceedings of ACM SIGMOD International Conference on Management of Data*, pp. 322–331, 1990.

[40] Barequet, G., Chazelle, B., Guibas, L.J., Mitchell, J.S.B., and Tal, A., Boxtree: a hierarchical representation for surfaces in 3D, *EuroGraphics' 96*, vol. 15, no. 3, pp. 387–484, 1996.

[41] Gilbert, E.G., Johnson, D.W., and Keerthi, S.S., Fast procedure for computing the distance between convex objects in three-dimensional space, *IEEE Journal of Robotics and Automation*, vol. 4, no. 2, pp. 193–203, 1988.

[42] Ong, C.J. and Gilbert, E.G., Fast versions of the Gilbert-Johnson-Keerthi distance algorithm: Additional results and comparisons, *IEEE Transactions on Robotics and Automation*, vol. 17, no. 4, pp. 531–539, 2001.

[43] Lin, M.C. and Canny, J.F. A fast algorithm for incremental distance calculation, in *IEEE International Conference on Robotics and Automation*, vol. 2, pp. 1008–1014, 1991.

[44] Mirtich, B., V-Clip: fast and robust polyhedral collision detection, *ACM Transactions on Graphics*, vol. 17, no. 3, pp. 177–208, 1998.

[45] Gilbert, E.G. and Foo, C.P., Computing the distance between general convex objects in three-dimensional space, *IEEE Transactions on Robotics and Automation*, vol. 6, no. 1, pp. 53–61, 1990.

[46] Turnbull, C. and Cameron, S., Computing distances between NURBS-defined convex objects, in *Proceedings IEEE International Conference on Robotics and Automation*, vol. 4, pp. 3685–3690, 1998.

[47] Lin, M.C. and Manocha, D., Interference detection between curved objects for computer animation, in *Models and Techniques in Computer Animation*, Thalmann, N.M. and Thalmann, D. (eds.), Springer-Verlag, Tokyo, pp. 43–57, 1993.

[48] Lin, M.C. and Manocha, D., Fast interference detection between geometric models, *Visual Computer*, vol. 11, no. 10, pp. 542–561, 1995.

[49] Thompson II, T.V., Johnson, D.E., and Cohen, E., Direct haptic rendering of sculptured models, in *Proceedings Symposium on Interactive 3D Graphics*, pp. 167–176, 1997.

[50] Johnson, D.E. and Cohen, E. An improved method for haptic tracing of a sculptured surface, in *Proceedings of ASME International Mechanical Engineering Congress and Exposition*, vol. 64, pp. 243–248, 1998.

[51] Nelson, D.D., Johnson, D.E., and Cohen, E., Haptic rendering of surface-to-surface sculpted model interaction, in *Proceedings ASME Dynamic Systems and Control Division*, vol. 67, pp. 101–108, 1999.

[52] Patoglu, V. and Gillespie, R.B., Extremal distance maintenance for parametric curves and surfaces, *Internationl Conference on Robotics & Automation 2002*, pp. 2817–2823, 2002.

[53] Baraff, D., Analytical methods for dynamic simulation of nonpenetrating rigid bodies, *Computer Graphics*, vol. 23, no. 3, pp. 223–232, 1989.

[54] Baraff, D., Fast contact force computation for nonpenetrating rigid bodies, in *Computer Graphics Proceedings. Annual Conference Series SIGGRAPH 94*, vol. 1, pp. 23–34, 1994.

[55] Mirtich, B. and Canny, J., Impulse-based simulation of rigid bodies, in *Proceedings of the Symposium on Interactive 3D Graphics*, pp. 181–188, 1995.

[56] Ramamurthy, R. and Farouki, R.T., Voronoi diagram and medial axis algorithm for planar domains with curved boundaries-II. Detailed algorithm description, *Journal of Computational and Applied Mathematics*, vol. 102, pp. 253–277, 1999.

[57] Zilles, C. and Salisbury, J., A constraint based god-object method for haptic display, in *IEE/RSJ International Conference on Intelligent Robots and Systems, Human Robot Interaction, and Cooperative Robots*, vol. 3, pp. 146–151, 1995.

[58] Ruspini, D.C., Kolarov, K., and Khatib, O., The haptic display of complex graphical environments, in *Computer Graphics Proceedings. Annual Conference Series*, vol. 1, pp. 345–352, 1997.

24
Flexible Robot Arms

Wayne J. Book
Georgia Institute of Technology

24.1 Introduction 24-1
  Motivation Based on Higher Performance • Nature of Impacted Systems • Description of Contents
24.2 Modeling 24-2
  Statics • Kinematics • Dynamics • Distributed Models • Frequency Domain Solutions • Discretization of the Spatial Domain • Simulation Form of the Equations • Inverse Dynamics Form of the Equations • System Characteristic Behavior • Reduction of Computational Complexity
24.3 Control 24-27
  Independent Proportional Plus Derivative Joint Control • Advanced Feedback Control Schemes • Open Loop and Feedforward Control • Design and Operational Strategies
24.4 Summary 24-42

24.1 Introduction

Robots, as mechanical devices, are subject to mechanical strain as a result of loads experienced in their application. These loads may result from gravity, acceleration, applied contact forces, or interaction with process dynamics. An effect of strain is deflection and consequent flexibility. No robot escapes this consequence, although the effects may be justifiably ignored in some instances. In the following pages, methods for predicting and minimizing the adverse consequences of flexibility will be presented. Primarily, arm-like, serial link devices will be examined, although lessons may often be extended to parallel link (nonserial) manipulators and mobile robots.

24.1.1 Motivation Based on Higher Performance

Flexibility can be reduced by reducing strain in critical locations. This should first be attempted by sound design practices for structural and drive components. Wise choices for cross sections and materials ultimately reach their limits and some flexibility will remain. If the resulting design is inadequate because of more demanding requirements, further steps must be taken. It is these further steps that are the focus of this section. Higher performance may be required in terms of speed, accuracy, payload mass, arm weight, or arm bulk. Typically, several or all of these measures have an impact on the utility of the design.

24.1.2 Nature of Impacted Systems

There are early indications that a robot design will be challenged in terms of flexibility. Static deflection and vibration natural frequency, for the dynamic case, are numerical indicators of this problem. Long arms are very susceptible to flexibility problems as arm mass and compliance increase with length. The weight of a device that must be carried on a mobile platform or launched into space is directly penalized, leading to severe tradeoffs between flexibility and mobility. When access through narrow openings is desired, as for servicing a nuclear facility, the cross section of an arm is limited with consequences for flexibility. In order to reduce cycle time, high accelerations produce loads that excite flexible behavior. The time for vibrations to settle to the required precision can extend the very cycle time that required the offending accelerations. Static deflections as well as vibrations may impact the arm's credentials in a given task if tip positioning is based on joint measurement. In summary, long, lightweight, thin, quick, and accurate arms are likely to have problems arising from flexibility. For a more complete discussion of the nature of flexible devices, the reader can refer to Book [1].

24.1.3 Description of Contents

Discussed in the following pages is the modeling of elastic behavior of material and of the structures composed of that material. A brief introduction to statics is useful for modeling the modes of failure that can occur, including excessive flexibility, fracture, and buckling. We focus on flexibility, but that may not be the active constraint. Kinematics of a flexible arm, discussed next, describes material deformation as well as joint motion. The dynamic equations are first derived for lumped parameter models consisting of springs and masses, and then controlled joints are added between the springs. Distributed effects are examined in two types of models, frequency domain linear models and discretized time domain models. The discussion of control begins with the simplest feedback controls, and considers their limitations. It progresses to more complex schemes intended to overcome those limitations. Open loop and feed forward techniques are equally important and provide the trajectory that the feedback control will try to follow. Improving the performance may also require rethinking the operational strategies used to deploy the robot arm.

24.2 Modeling

When modeling flexible behavior, we are acknowledging the deformable properties of engineering materials that introduce several practical considerations in addition to flexibility. What are the constraints on the structural geometry and selection of the arm materials? Flexibility must be limited to achieve acceptable behavior (accuracy, natural frequency, etc.) but other constraints on geometry should be examined as well. One or more of these constraints will be the active constraint that limits one in creating a light, long, or fast arm. The arm may yield, it may fail in fatigue, or it may buckle before the flexibility constraint comes into effect. It is not within the scope of this article to comprehensively treat all of these issues, but it is essential that they be identified to the reader.

24.2.1 Statics

24.2.1.1 Mechanics of Deformable Bodies

The classical treatment of deformable bodies is appropriate to the analysis of most manipulator structures. Exceptions are components made of anisotropic materials, that is, materials having a pronounced directionality in their properties. Truly, these are some of the most exciting materials for combating the limitations of flexibility. The material itself, typically a composite of two or more elementary materials, is engineered to contend with the loading conditions expected. The general nature of some arms, however, prevents a highly specialized design. The quantitative aspects of our analysis will only apply to isotropic materials.

24.2.1.1.1 Stress vs. Strain

Hooke's law describes a material where stress (force per unit area) is proportional to strain (change in dimension per unit dimension). The constant of proportionality between normal stress and extension for a material in plane strain is the elastic or Young's modulus, given the symbol E. Another hypothetical deformation is pure shear, in which a rectangular specimen distorts into a parallelogram with a change in the corner right angles by a shear angle γ. The proportionality between shear stress and shear angle is the shear modulus commonly denoted G. Mechanics of materials instructs us that the faces of a cubic differential element in a specimen under stress will have different predictable levels of normal and shear stress depending on the orientation of the face. Mohr's circle is often used to display and compute the values at an arbitrary angle. Failure of the material correlates with these stresses and the manner in which they are applied as well as environmental conditions. The purposes of this article are served by special cases of loading that can be used to compute a reasonable prediction of the stress state at the critical location of the geometry.

The nature of the material (e.g., brittle or ductile) and the local geometry (which leads to stress concentration factors) are also required to produce a reasonable prediction of structural failure. For repeated loading of structural or drive components, one must be concerned about fatigue failure as a result of crack development and propagation. Although various criteria for failure prediction have been developed, the Von Mises effective stress [2] is appropriate and convenient for multi-axial stress. It can be calculated from the principal normal stresses (found on planes with no shear stress) $\sigma_1, \sigma_2, \sigma_3$, or the normal and shear stresses at any surface. The Von Mises stress is found as

$$\sigma' = \sqrt{\sigma_1^2 + \sigma_2^2 + \sigma_3^2 - \sigma_1\sigma_2 - \sigma_2\sigma_3 - \sigma_1\sigma_3} \qquad (24.1)$$

Static failure in ductile materials is predicted based on failure in shear when the Von Mises stress σ' exceeds the shear strength, normally taken to be one-half the tensile yield strength. Fatigue failure of ductile materials is predicted when σ' exceeds the stress value that varies with the number of cycles of loading, and with the mean and alternating stress levels. The Modified Goodman diagram is one accepted way to determine this value [2]. For ferrous materials, a fatigue limit exists. This is a stress at which there is no limit to the number of cycles, i.e., the material will never fail. Aluminum and other nonferrous materials do not exhibit such a limit and will eventually fail (according to this theory) for low levels of stress if the number of cycles of loading is large enough. The purpose here is not to elaborate on these methods of machine design, but to provide the engineer with an indication of when these stress-based failure modes dominate the deflection based failure modes.

Techniques accounting for local factors of stress concentration due to sharp notches and corners will be found in machine design texts also. Multiplying factors may be applied to the stress or the strength to account for these factors. Coefficients are based on a combination of empirical and theoretical results. Again, the purpose here is not served by diversion into these topics. The reader should be aware that they can be applied and do not change the fundamental results of the following discussion.

Brittle materials (e.g., gray cast iron or extremely cold materials) or materials with uneven maximum stress in tension and compression do not generally fail in shear, but in tension. Alternative methods of predicting failure will not be covered here because the components that we analyze for flexibility are seldom constructed from these materials; however, the extreme conditions of temperature might occur for special applications.

24.2.1.1.2 Material Properties

The discussion of stress and strain above incorporates a discussion of material properties without elaboration. It serves us well to examine these properties in more detail as they relate to arm flexibility. In particular we will examine the elastic and shear moduli, the density, damping, and the strength of these materials. Recognize that bulk materials such as steel or aluminum are not directly comparable in all respects to glass or Kevlar® fibers; however, the numbers give one a useful comparison of the potential of the materials for use in robot construction.

24.2.1.1.2.1 Elastic and Shear Modulus

The elastic modulus E and the shear modulus G are related through Poisson's ratio μ as $G = E/(2 + 2\mu)$. For a discussion of flexibility these are the two most important material characteristics, closely

TABLE 24.1 Representative Properties (Various Sources)

Material	Elastic Modulus E (GPa)	Shear Modulus G (GPa)	Density ρ (Mg/m^3)	E/ρ	Yield (2%) Strength (MPa)
Steel-carbon 1040 cold rolled	206.8	80.8	7.8	26.5	490
Stainless steel 301 cold rolled	206.8	80.8	7.8	26.5	1138
Aluminum 6061 heat treated	71.7	26.8	2.8	25.6	276
Kevlar 49®	120	X	1.45	82.8	2300[a]
S Glass fibers	85.5	X	2.49	34.3	X
Magnesium AZ91A die cast	44.8	16.8	1.8	24.9	152
Titanium T1-13 V-11 Cr-3 Al Alloy heat treated	113.8	42.4	4.4	25.9	1172

X = not available or not relevant. [a] Fails at 1.9% strain.

followed by density. For bulk materials the elastic modulus is independent of orientation of the stress and equal in compression and tension. It is almost completely independent of alloying modifications of metal composition that may have a dramatic effect on strength.

24.2.1.1.2.2 Density

Density of the structural material is critical to natural frequency and to the torque requirement of actuators to lift or accelerate an arm. This is especially true when the arm itself is the main load to be carried, as in many cases. If geometry is fixed, the natural frequency is proportional to $\sqrt{E/\rho}$ in typical bending modes. Thus, the ratio of modulus to density is a strong indicator of the material's resistance to dynamic flexibility limitations. We, therefore, see that the lighter weight, lower stiffness material like aluminum is comparable to the heavier but stiffer steel. On the other hand, if a large payload is accommodated, steel gains an advantage because the payload adds less mass on a percentage basis to the steel arm. Materials like Kevlar® are dramatically better in this figure of merit (five times better than most metals). Unfortunately, construction using this fiber to advantage is a serious difficulty and warranted only in exceptional situations. The cost is also high.

24.2.1.1.2.3 Strength

Strength refers in general to the load bearing capacity of the material before failure, and is usually cited in terms of the stress occurring at the point of failure. Since failure may occur in many modes, the reference to strength by itself is insufficient. At low values of stress, structural materials deform elastically and hence reversibly. As stress rises above the elastic range a permanent deformation occurs. Yield strength is the top stress of the elastic range of the material, often defined at a point when 0.2% of strain will not reversibly disappear on removal of the load. Ultimate strength (tensile, compressive, or shear) refers to stress at the point of fracture, or separation of two regions of the material. These failure modes could be the result of a single loading event, whereas fatigue failure is the result of repetitive loading and is more complicated to describe or predict. For steels and some titanium materials a predictable fatigue limit seems to exist beyond which an indefinite number of cycles of loading can be sustained. The fatigue limit is the corresponding stress and is typically 40–50% of the ultimate tensile strength.

24.2.1.1.2.4 Damping

Increased damping reduces the settling time for any vibrations that occur in the arm. Damping inherent in an arm is due to many things, particularly the material and the manner of construction. A welded construction looses less energy, thus has less damping than a bolted construction. Any relative motion between two arm components can remove vibrational energy, and when the damping is very low a small improvement can result in substantial reductions in settling time. Thus, arms that are back drivable tend to absorb vibrational energy better than arms which are not back drivable due to high ratio gearboxes, hydraulic actuators blocked by valves with positive overlap, or even a high position feedback gain. This can create a dilemma for the designer who wants high accuracy in joint positioning and, thus, chooses high feedback control gains to control the joint. In terms of material damping, a wide range of damping values is observed and composites tend to have more damping than metals.

24.2.1.1.3 Idealized Structures and Loading

Exact evaluation of the stress in a robot component may require finite element analysis. For present purposes it will be more valuable to compose the robot from idealized structures tractable to analytical determination of stress and deflection. In particular we will use bars in compression, shafts in torsion, and beams in bending. For static deflection the load and the elastic properties are needed. For dynamic deflection the mass properties are also needed. Geometry will contribute to both mass and elastic properties.

24.2.1.1.3.1 Bars and Compression

A bar element is considered for axial loading. Almost pure compression or tension will result for a strut or similar element pinned at each end. In this case the static deflection δ is also axial and predicted simply and accurately by the assumption of plane strain. If the cross-sectional area A is constant over the length L, deflection δ under applied axial force F_a is

$$\delta = \frac{F_a L}{AE} = \alpha_{XF} F_a \tag{24.2}$$

If A is not constant, the deflection is obtained by integrating a differential deflection (axial strain) over the length with the area $A(x)$ variable. Normal stress is approximately

$$\sigma = F_a/A$$

The dynamic behavior of a bar alone is rarely of concern with robots of conventional size if the bar only experiences compression. This is because there are usually other components exhibiting more severe deflection and because buckling constrains the ratio L/A to high values. The bar element can be used to represent "tension only" components including cables and bands. In these cases buckling does not limit L/A and the deflection can be substantial. Dynamic behavior incorporates mass as well. The mass of the bar in tension is reduced as the cross-sectional area is reduced with the resulting effect that the natural frequency is related to the speed of sound in the bar, which is invariant with the geometry and high relative to other dynamic phenomena that impact robot performance. Thus, the bar plays the role of a spring for most flexible robot analysis; its mass can be considered lumped, if considered at all, and does not warrant consideration as a distributed parameter effect.

24.2.1.1.3.2 Shafts and Torsion

Pure torsion is a fair representation of drive shaft loading. Arm structural elements also may experience torsion in conjunction with bending or compression, and to a reasonable approximation the effects can be added or superimposed. The simple case we will build on is a circular cross section, either solid or hollow. The torque T acting on a circular tube of length l with cross-section polar moment of inertia J produces a rotation of one end with respect to the other of θ, where

$$\theta = \frac{Tl}{JG} = \alpha_{\theta T} T \tag{24.3}$$

The pure shear τ resulting on the plane perpendicular to the axis of the torsion is approximately linear with the radial distance r, giving the maximum at the outer radius R of

$$\tau_{max} = \frac{TR}{J} \tag{24.4}$$

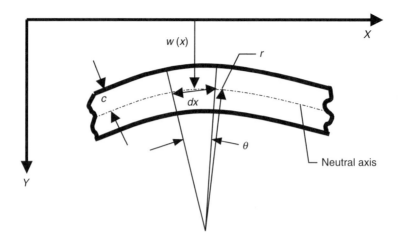

FIGURE 24.1 Geometry of bending deformation.

When the cross section is not circular an approximate analysis is often used that is accurate as long as the cross section is not prone to extreme warping with torsion. In this analysis

$$\theta = \frac{Tl}{KG} = \alpha_{\theta T} T \tag{24.5}$$

$$\tau_{max} = \frac{T}{Q} \tag{24.6}$$

where K and Q are geometry dependent factors found in various handbooks and references such as Norton [2].

24.2.1.1.3.3 Beams and Bending

The simple theory of beams is remarkably accurate for slender beams with cross sections that are not extreme in their departure from a rectangle. If the beam has a deflection w such that at the position x, the second derivative $\partial^2 w/\partial x^2$ is approximately equal to $1/r$, where r is the radius of curvature of the neutral axis. As illustrated in Figure 24.1, the further assumption that planes perpendicular to the neutral axis remain planes dictates the amount of elongation δ that takes place on the outer edge of the element of length dx. With any finite curvature, an angle θ subtends the distance $dx = r\theta$ at a radius r from the center of the local circular arc, that is, along the neutral axis. The strain along the neutral axis is zero by definition. Moving to the outermost fiber of the beam, a distance c from the neutral axis, means the angle θ subtends a greater distance of $dx + \delta = (c + r)\theta$. The combination of these relationships means the strain at the outermost fiber is

$$\varepsilon = c\frac{\partial^2 w}{\partial x^2}\frac{dx}{dx} = c\frac{\partial^2 w}{\partial x^2} \tag{24.7}$$

By the simple beam theory the normal stress σ in the outermost fiber would be $E\varepsilon$ or

$$\sigma = Ec\frac{\partial^2 w}{\partial x^2}dx \tag{24.8}$$

and would vary linearly along y, the distance from the neutral axis. The moment $M(x)$ developed at the cross section by normal stress is found by integrating over the area of the cross section the stress times the

Flexible Robot Arms

distance from the neutral axis times the differential area.

$$M(x) = \int_{Area} \frac{\sigma(x,y)y}{c} y\, dA = EI(x)\frac{\partial^2 w}{\partial x^2} \qquad (24.9)$$

where $I(x)$ is the cross section's area moment of inertia about the neutral axis at x.

$$I = \int_{y\,min}^{y\,max} y^2\, dA \qquad (24.10)$$

The deformation, displacement, and rotation, at the end of a loaded beam is readily computed from the moment equation above when knowing the loading and deflection constraints on the beam. Loading could occur in any location and as a concentrated or distributed force (perpendicular to the neutral axis) or moment (about an axis perpendicular to the plane of bending). Tabulations of these loading effects are found in more specialized handbooks [3] and it is productive here to cite only the response of a beam with one end constrained not to translate or rotate with the other end completely free. Apply to the free end of this beam a force F_y and a moment M_z, and the resulting deflection $w(l)$ and rotation $\theta(l)$ at the end will be

$$w(x)|_{x=l} = \frac{l^3}{3EI}F_y + \frac{l^2}{2EI}M_z = \alpha_{XF}F_y + \alpha_{XM}M_z$$

$$\theta_z(x)|_{x=l} = \frac{l^2}{2EI}F_y + \frac{l}{EI}M_z = \alpha_{\theta F}F_y + \alpha_{\theta M}M_z \qquad (24.11)$$

The second form of these two equations identifies coefficients referred to later in this section and tells you how to compute them for the case of uniform beams. Tapered, shaped, or stepped beams have coefficients that can be found from the deflection equation or in handbooks.

24.2.1.1.3.4 Combinations of Loading

In many situations it is acceptable to superimpose torsion, bending, and compression from the above simple cases to calculate deformation and stress. The worst case stress will be where maximum axial normal stress from bending is increased by the normal stress from axial forces. At this same point the shear from torsion occurs and is maximum. The plane stress approximation is valid here and the von Mises stress is found (assuming a circular cross section of outer radius R) to be

$$\sigma' = \sqrt{\left(\frac{MR}{I} + \frac{F_a}{A}\right)^2 + \left(\frac{TR}{J}\right)^2}$$

Extreme cases require a more complex analysis beyond the scope of this section. The boundary between acceptable and unacceptable is fuzzy and case dependent, but if any component of loading is modified by a significant fraction of the factor of safety, a more detailed analysis should be undertaken. For example, if axial torsion T on a beam is realigned by $10°$, the bending moment on the beam is increased by about 0.18 T. An extremely slender beam could undergo $10°$ of deformation without failure but the addition of an additional bending moment of 0.18 T should be further considered. An exception to this proportional effect is buckling behavior which is treated next.

24.2.1.1.4 Buckling

Bending becomes the most pronounced mode of stress and deflection in most beam type elements, including arm links. The naïve designer is, therefore, tempted to place the maximum material at the maximum distance from the neutral axis. Because bending may take place about any axis perpendicular to the neutral axis, circular cross sections are attractive as they also resist torsion about the beam axis well. Why not carry this to the extreme of very thin walls located at extreme distances from the neutral axis so that the arm cross section is similar to an aluminum beverage can? The limitation may come from one or more of several sources: arm bulk (limiting access), manufacturing difficulty, or the tendency of a thin wall cylinder to buckle. Buckling is a structural instability in which the imperfections of construction and loading result in a sudden drastic change in structure geometry. For the empty aluminum can, shell buckling alters the geometry of the load bearing material, allowing localized stresses to rise and the geometry of the arm to be destroyed. (Note that a pressurized drink can does not buckle so easily, so a thin, pressurized arm might actually have merit.) Another form of buckling, column buckling, is typical of beams with small cross sections. When such beams have a compressive axial load not perfectly aligned with a perfectly straight neutral axis, the beam buckles in "column buckling" that was the subject of a classical analysis by Euler, an analysis which still bears his name. Finally, "torsional buckling" arises from small cross sections subject to imperfections in the straightness of the neutral axis or other loads which deflect the neutral axis a small amount. The critical loads predicted by the simple analyses below are typically too optimistic for arm design and must be generously amplified in the case of robot arms, which are subjected to many unexpected and complex loads. Nevertheless, it is important to realize there are constraints on arm geometry other than flexibility that interplay with our main focus of flexible arms.

24.2.1.1.4.1 Column Buckling

Euler's classical analysis of slender columns determines a bifurcation point at which an alternative shape of the column is possible. The shape for simple uniform columns is a sinusoid, the periods of which depend on the boundary conditions. More restrictive boundary conditions result in a shorter period and in turn a higher critical load. The critical compressive load is

$$P_{critical} = \frac{\pi^2 EI}{l_{eff}^2} = \frac{\pi^2 EAk^2}{l_{eff}^2}, \quad \frac{P_{critical}}{A} = \frac{\pi^2 Ek^2}{l_{eff}^2} \quad (24.12)$$

where
  k = radius of gyration corresponding to the cross section shape
  l_{eff} = the effective column length, dependent on end-constraint conditions
    = l for pinned-pinned ends
    = $2l$ for fixed (clamped)-free ends
    = $0.707l$ for fixed-pinned ends
    = $0.5l$ for fixed-fixed ends

While arm structures will not be sized on buckling alone, this analysis shows a sensitivity to the cross section shape and the link length.

24.2.1.1.4.2 Shell Buckling

The thin drink can is prone to shell buckling at locations of compressive stress. Here the classical analysis yields a critical stress

$$\sigma_{critical} = \frac{E}{\sqrt{3(1-\nu^2)}} \left(\frac{t}{R}\right) \quad (24.13)$$

where
  t = wall thickness
  R = radius of cylindrical shell
  ν = Poisson's ratio for the material
  E = Young's modulus

Note that this is expressed in terms of a critical stress locally in the shell. Where that stress distributed over a thin circular annulus with area $2\pi Rt$, the critical load becomes independent of R and dependent on t^2. In terms of stress, the ratio of thickness to radius is the key geometric feature that controls the susceptibility to shell buckling for a more general geometry.

24.2.1.1.4.3 Torsional Buckling
Torsional buckling is a third possibility for unstable behavior. If a slender shaft is simply supported (free to rotate in flexure but not to translate), the critical longitudinal torsion on the shaft is found to be

$$T_{critical} = \frac{2EI}{l} \quad (24.14)$$

The nomenclature is the same as the preceding. The resulting shape upon buckling is a spiral with orthogonal deflections described by sinusoids. Shaft buckling would not be typically found in a free link but may occur with the transmission of torque from motor to joint.

Other cases where torsion could be involved in link buckling are lateral-torsional buckling and flexural-torsional buckling. These failures are because of a reduction of flexural stiffness resulting from a longitudinal rotation of the cross section [4]. A cross section with a large discrepancy between the area moments about orthogonal directions is most susceptible to this type of failure.

24.2.2 Kinematics

Kinematics will be developed using 4×4 transformation matrices as frequently found in robotics. The relation between positions and velocities is critical to developing dynamic equations and commanding and controlling the joints to achieve desired end effector position. Two classes of relative motion will be considered: discrete joints and distributed deflection.

24.2.2.1 Transformation Matrices

The commonly used 4×4 transformation matrices relate the position of a point as described in two coordinate frames as detailed elsewhere in this book. For the immediate purposes, the nomenclature relating the x, y, z values of two position vectors p_0 and p_1 is clarified by

$$\begin{aligned} p^0 &= T_1^0 p^1 \\ p^i &= [x^i \quad y^i \quad z^i \quad 1]^T \end{aligned} \quad (24.15)$$

Note that the superscripts on position vectors and scalars represent the frame of reference in which the position is described.

The transformation matrix T_i^j can represent a variety of static or dynamic features of the mechanism to which the coordinate frames are attached. For rigid links the key features are the relative location and orientation of joint axes due to construction or motion of the joints. The strategy for efficiently locating and describing the coordinate frames is described elsewhere in this handbook or in Sciavicco and Siciliano [5], a slight variation of Craig [6], and is based on the original scheme proposed by Denavit and Hartenberg [7]. For flexible links, the features will include deflection under static or dynamic loads which we will focus on in this section as employed by Book [8].

The core concept is the composition of the 4×4 transformation matrix as

$$T_i^j = \begin{bmatrix} & & & x_i^j \\ & R_i^j & & y_i^j \\ & & & z_i^j \\ 0 & 0 & 0 & 1 \end{bmatrix} \quad (24.16)$$

R_i^j is a 3×3 rotation matrix describing rotations of the coordinates.

$$\begin{bmatrix} x_i^j \\ y_i^j \\ z_i^j \end{bmatrix}$$

is the translation of the frame origin from i to j in j coordinates. The subscripts of the matrices, vectors, and scalars identify the point of interest.

24.2.2.2 Simple Kinematic Pairs

For the intentional joints of a robot we can use the tools described for rigid robots. Rotational joints introduce terms in the rotation matrix R_i^j that vary as the joint angle changes. Translational joints introduce terms in the translation vector of the matrix T. The orientation of the joint axes and their distance of separation (in terms of a common normal) appear as constant terms. The reader is referred to other chapters for the development of these terms.

24.2.2.3 Closed Kinematic Chains

The serial connection of linkages is readily described by a chain of transformation matrices multiplied together. If the series reattaches to itself, a closed chain is formed. The chain can also branch into multiple series of linkages, which is often the case when a closed chain is introduced so that a distant joint can be actuated with a stationary actuator. The additional constraints that are created by these topologies can be readily described by consistently following a chain to the point of rejoining itself, describing each successive transformation with a transformation matrix, and ultimately relating the initial coordinate frame to itself. This constraint system may enhance the behavior of a compliant manipulator arm. Details of these procedures are not included here.

24.2.2.4 Kinematics of Deformation

When a transformation describes deflection, the terms of the T transformation matrix are dependent on loading and on the geometry of the component which is under deformation. The deformation may be large or small, but the most common and useful case is the small deformation case. With small rotations the rotation matrix of direction cosines can be simplified by considering only first order terms composed of the angles Ψ, Θ, Φ about X, Y, and Z, respectively, equivalent to the roll, pitch, and yaw rotations. Note that for small angles the order of rotation is irrelevant. Hence

$$T_i^j = \begin{bmatrix} \cos(x_i x_j) & \cos(y_i x_j) & \cos(z_i x_j) & \delta x_i^j \\ \cos(x_i y_j) & \cos(y_i y_j) & \cos(z_i y_j) & \delta y_i^j \\ \cos(x_i z_j) & \cos(y_i z_j) & \cos(z_i z_j) & \delta z_i^j \\ 0 & 0 & 0 & 1 \end{bmatrix}, \quad T_i^j \cong \begin{bmatrix} 1 & -\Phi & \Theta & \delta x_i^j \\ \Phi & 1 & -\Psi & \delta y_i^j \\ -\Theta & \Psi & 1 & \delta z_i^j \\ 0 & 0 & 0 & 1 \end{bmatrix} \quad (24.17)$$

The deformation of a beam with bending and torsion, for example, would use the deformations for the bending for $\delta y, \delta z, \Theta$, and Φ and the equation for torsion for Ψ. δx would be zero unless compression of the beam is considered, although this deflection is typically small.

When moving from one point on an arm to another, it is convenient to break the transformations into a rigid transformation A_i and an elastic transformation E_i. The elastic transformation is small, as described above, and dependent on loading. The combined effect is the multiplication of the two transforms. If the typical slender linkage of an arm is envisioned, the elastic transformation due to bending, torsion, and compression of relatively small amounts will be well described by

$$E_i = \begin{bmatrix} 1 & -\alpha_{\theta Fi} F_{Yi}^i - \alpha_{\theta Mi} M_{Zi}^i & -\alpha_{\theta Fi} F_{Zi}^i + \alpha_{\theta Mi} M_{Yi}^i & \alpha_{Ci} F_{Xi}^i \\ \alpha_{\theta Fi} F_{Yi}^i + \alpha_{\theta Mi} M_{Zi}^i & 1 & -\alpha_{Ti} M_{Xii} & \alpha_{XFi} F_{Yi}^i + \alpha_{XMi} M_{Zi}^i \\ \alpha_{\theta Fi} F_{Zi}^i - \alpha_{\theta Mi} M_{Yi}^i & \alpha_{Ti} M_{Xi}^i & 1 & \alpha_{XFi} F_{Zi}^i + \alpha_{XMi} M_{Yi}^i \\ 0 & 0 & 0 & 1 \end{bmatrix}$$

(24.18)

where

α_{Ci} = coefficient in compression, displacement/force
α_{Ti} = coefficient in torsion, angle/moment
$\alpha_{\theta Fi}$ = coefficient in bending, angle/force
$\alpha_{\theta Mi}$ = coefficient in bending, angle/moment
α_{XFi} = coefficient in bending, displacement/force
α_{XMi} = coefficient in bending, displacement/moment
F_{Xi}^{j} = force at the end of link i in the X direction of coordinate frame j
F_{Yi}^{j} = force at the end of link i in the Y direction of coordinate frame j
F_{Zi}^{j} = force at the end of link i in the Z direction of coordinate frame j
M_{Xi}^{j} = moment at the end of link i in the X direction of coordinate frame j
M_{Yi}^{j} = moment at the end of link i in the Y direction of coordinate frame j
M_{Zi}^{j} = moment at the end of link i in the Z direction of coordinate frame j

Note that for readability we have assumed that the beam has a symmetrical cross section and the coefficient of F_{Yi}^{i} is the same as the coefficient of F_{Zi}^{i}. In general these coefficients could be distinct. The numerical values for the simple and useful case of a uniform beam are found in the discussion of bending, torsion, and compression above in Equation (24.2), Equation (24.3), and Equation (24.11).

After deformation the position of the arm's tip, located at the origin of the nth coordinate frame, is found as

$$p^0 = [A_1 E_1 A_2 E_2 \cdots A_i E_i A_{i+1} E_{i+1} \cdots A_N E_N] \begin{bmatrix} 0 \\ 0 \\ 0 \\ 1 \end{bmatrix} \quad (24.19)$$

24.2.3 Dynamics

In moving from kinematics to dynamics, we consider the relationship between forces and motion, that is, the change in position over time. For rigid robots this primarily involves actuator forces and torques, inertia acceleration or kinetic energy storage, frictional dissipation, and possibly gravity or other body forces. For flexible robots, the storage of potential energy in elastic deformation becomes important. One of the early decisions on modeling the dynamics is how the kinetic and elastic potential energy storage are to be spatially distributed. If predominantly separated in lumped components, the separate modeling of springs and inertias will lead directly to ordinary differential equations. If these storage modes are combined throughout the links, the lumped model is inefficient and unsystematic, and the partial differential equations or at least the distributed nature of the bodies should be considered.

24.2.3.1 Lumped Models

Lumped arm models will consider the arm to be made up of inertias and springs. The focus in this section will be the linear behavior resulting from small motions consistent with the modeling of a structure as an elastic member. The reader may be concerned with the choice between lumped and distributed models. Lumped models are approximations that offer convenience, a more intuitive understanding and simpler solution procedures at the cost of accuracy and/or computational efficiency. Components that predominantly store potential energy as a spring with minimum storage of kinetic energy (due to low mass or small velocities) are candidates for representation as a lumped spring. In contrast, components storing kinetic energy with minimum compliance are candidates for lumped inertias. Because structural materials intrinsically have both properties, the lumped approximation is never perfect.

Dissipation effects also warrant modeling. It is typically possible to include friction or viscous damping into either the actuator forcing or to incorporate these losses into the equations for either lumped or distributed models using the variables that are dictated by the mass and compliance representations.

24.2.3.1.1 Lumped Inertia

The definition of inertia could be approached from the distribution of mass and consequently a triple integral over the volume of the body. For our purposes it will be more direct to define the inertia as the matrix coefficient that relates accelerations (translational and rotational) to forces and moments:

$$J \begin{bmatrix} \ddot{p} \\ \ddot{\Theta} \end{bmatrix} = \begin{bmatrix} F \\ M \end{bmatrix} \tag{24.20}$$

While not specific in the equation above, the usual reference point for writing these equations will be the center of mass of the body and the motion will be relative to an inertial reference frame.

24.2.3.1.2 Lumped Springs

Springs are often visualized as a machine element intended to deform under load. More generally, any real machine element will deform to a limited extent, and we should be much more inclusive in our consideration of springs. For robots and other mechanisms there are machine elements that consistently behave as unintentional springs: Shafts for transmission of power by rotation, bearings and bearing housings, gear and other speed reducers, belts for transmission of power, and couplings. These and other elements may be constrained in their cross section and hence are allowed to deform because the space constraints do not permit an adequate stiffness.

It is tempting to enlarge the components that are not constrained in cross section. This mistake is often observed in robot arm design and in machine design in general. In arms, an external beam member will be increased in size and inertia making it very nearly rigid, while aggravating the lack of rigidity of the constrained components that move it: shafts, bearings, joint structures, etc., as well as the links and drive motors inboard of the oversized link. Structural natural frequency is a good indicator of the flexibility problem that ensues, and that natural frequency is predicted by the square root of the ratio of spring constant over inertia. Increasing the cross section of an outboard component will continue to increase inertia after its value in increasing stiffness of the chain of components has diminished. Because these elements are more difficult to analyze, the tendency is to over design them to approximate rigidity. For this reason, considerable attention will be given to analysis of the compliance and distributed flexibility of the beam type elements of an arm's construction.

24.2.3.1.3 Stiffness of a Series of Links

To evaluate the compliance of a series of N links to forces on the end of the chain, the partial derivative of the position of the end can be taken, yielding, in the example case of F_{XN}:

$$\frac{\partial p^0}{\partial F_{XN}^0} = \left[\sum_{i=1}^{N} A_1 A_2 \cdots A_i \frac{\partial E_i}{\partial F_{XN}^0} A_{i+1} \cdots A_N \right] \begin{bmatrix} 0 \\ 0 \\ 0 \\ 1 \end{bmatrix} \tag{24.21}$$

Similar expressions result for all six forces and moments on the arm's end.

To complete this calculation numerically, note that

$$\frac{\partial E_i}{\partial F_{XN}^0} = \frac{\partial E_i}{\partial F_{Xi}^i} \frac{\partial F_{Xi}^i}{\partial F_{XN}^0} \tag{24.22}$$

Flexible Robot Arms

which can be obtained from the vector form

$$\begin{bmatrix} \frac{\partial}{\partial F_N^0} \\ \frac{\partial}{\partial M_N^0} \end{bmatrix} \left[\left(F_i^i\right)^T \left(M_i^i\right)^T \right] = \begin{bmatrix} R_0^i & r_i^i \times R_0^i \\ 0 & R_0^i \end{bmatrix} \quad (24.23)$$

and from substitution into the equation for E_i the value $F_{Xi}^i = 1$ and the value zero for all other forces and moments. That is,

$$\frac{\partial E_i}{\partial F_{Xi}^i} = \begin{bmatrix} 1 & -\alpha_{\theta Fi}F_{Yi}^i - \alpha_{\theta Mi}M_{Zi}^i & -\alpha_{\theta Fi}F_{Zi}^i + \alpha_{\theta Mi}M_{Yi}^i & \alpha_{Ci}F_{Xi}^i \\ \alpha_{\theta Fi}F_{Yi}^i + \alpha_{\theta Mi}M_{Zi}^i & 1 & -\alpha_{Ti}M_{Xii}^i & \alpha_{XFi}F_{Yi}^i + \alpha_{XMi}M_{Zi}^i \\ \alpha_{\theta Fi}F_{Zi}^i - \alpha_{\theta Mi}M_{Yi}^i & \alpha_{Ti}M_{Xi}^i & 1 & \alpha_{XFi}F_{Zi}^i + \alpha_{XMi}M_{Yi}^i \\ 0 & 0 & 0 & 1 \end{bmatrix} \quad (24.24)$$

where

$$F_{Xi}^i = 1, \quad F_{Yi}^i = F_{Zi}^i = 0, \quad M_{Xi}^i = M_{Yi}^i = M_{Zi}^i = 0$$

The vector and scalar forms of forces F, moments M, positions X, and angles Θ are distinguished by having only the single subscript indicating the link index. The expressions for all other forces and moments are found equivalently and the lot assembled into a compliance matrix C, the inverse of the spring constant matrix $K_S = C^{-1}$ according to

$$C = \begin{bmatrix} \frac{\partial}{\partial F_N^0} \\ \frac{\partial}{\partial M_N^0} \end{bmatrix} \left[\left(X_N^0\right)^T \left(\Theta_N^0\right)^T \right] \quad (24.25)$$

Knowledge of the spring constant and the rigid body inertias is all that is needed to specify the equations of motions of the lumped parameter flexible link system if the joints are fixed. As will be seen in the following section, the approach above is also useful if the joints are controlled to move the arm.

24.2.3.2 Dynamics of Lumped Masses and Massless Elastic Links with Servo Controlled Joints

For some applications of lightweight manipulators, particularly in space where a gravity load need not be supported, very light links can be used to position large inertias. The Space Shuttle RMS and the Space Station Arm are excellent examples of this situation. In these cases, the joints function to deform the links (springs) which then apply forces and moments to the essentially rigid masses comprising the vehicle and the payload. The following approach applies directly to that case and can be modified to apply to other cases. For a more complete development see Book [8].

The kinematic relationships will be analyzed using transformation matrices consisting of

R_i^j = rotation matrix from system i to system j
r_i^j = distance vector from the origin of system i to the end of the chain (origin of system N) in terms of coordinates j

We will consider small translations and rotations of the two masses on either end, which will be described by the vectors

$$Z_0 = \begin{bmatrix} d_0^T & \phi_0^T \end{bmatrix}^T, \quad Z_n = \begin{bmatrix} d_n^T & \phi_n^T \end{bmatrix}^T \quad (24.26)$$

where d_0, d_n are the position vectors from the origin of -1 (inertial frame) to the origins of 0 and n, respectively, and Φ_0, Φ_n are the vectors of small rotations of mass 0 and n, respectively, from an initial undeformed orientation.

In the case of joint angle rotations, the rotation will be permitted only about either the Z or the X axes. The angles of rotations are:

γ_i = angle from positive Z_{i-1} to positive Z_i, measured counterclockwise about positive X_i
β_i = angle from positive X_{i-1} to positive X_i measured counterclockwise about positive X_i

Joint position and rate control will be employed so that

$$\text{Torque by actuators} = K_c(\theta_{dJ} - \theta_J) + D_c(\dot{\theta}_{dJ} - \dot{\theta}_J)$$

where

θ_J = vector of m controlled joint angles measured from an arbitrary reference point
θ_{dJ} = vector of m desired joint angles
$\dot{\theta}_J$ = vector of m controlled joint angular velocities
$\dot{\theta}_{dJ}$ = vector of m desired joint angular velocities
K_c = $m \times m$ matrix of position control gains
D_c = $m \times m$ matrix of velocity control gains

The displacement of the end mass gives forces on that mass of

$$\begin{bmatrix} \text{Forces} \\ \text{Moments} \end{bmatrix}_{\substack{\text{on mass at } n \\ n \text{ displaced}}} = -K_S(Z_n - Z_{ns}) \tag{24.27}$$

while that same force gives torques on the actuators of

$$\begin{bmatrix} \text{Torques on actuators} \\ \text{with } n \text{ displaced} \end{bmatrix} = HK_S(Z_n - Z_{ns}) \tag{24.28}$$

We will construct the matrix H element by element depending on the actuated joint, first defining the cross product term

$$XCR_h = r_{hh} \times R_{0h}$$

And the element notation $H(i,j)$ = element in ith row, jth column of H, $1 \leq i \leq m$, $1 \leq j \leq 6$.

$$H(i,j) = XCR_h(k,j), \quad 1 \leq j \leq 3, \quad 1 \leq i \leq m$$
$$H(i,j) = R_{0h}(k, j-3), \quad 3 \leq j \leq 6, \quad 1 \leq i \leq m$$

where

$k = 1$ if γ_h is the ith element of θ_J
$k = 3$ if β_h is the ith element of θ_J
m = total number of controlled joints

The following transformation will also be needed since rotations will also cause translations at a distant point.

$$\begin{bmatrix} I & XC^T \\ 0 & I \end{bmatrix}(Z_0 - Z_{0e}) = T^T(Z_0 - Z_{0e}) \tag{24.29}$$

$$XC = \begin{bmatrix} 0 & -r_{Z0n} & r_{Y0n} \\ r_{Z0n} & 0 & -r_{X0n} \\ -r_{Y0n} & r_{X0n} & 0 \end{bmatrix} \tag{24.30}$$

Flexible Robot Arms

The final equations then incorporate first order equations for the joints (one each) and second order equations for each mass (six each).

$$\dot{\theta}_J = D_c^{-1}[(-HK_sH^T - K_c)\theta_J + HK_s(Z_n - T^TZ_0) + K_c\theta_{dJ}]\dot{\theta}_{dJ} \tag{24.31}$$

$$\frac{d}{dt}(\dot{Z}_n) = -J_n^{-1}K_s(Z_n - T^TZ_0 - G\theta_J) \tag{24.32}$$

$$\frac{d}{dt}(\dot{Z}_0) = J_0^{-1}TK_s(Z_n - T^TZ_0 - G\theta_J) + J_0^{-1}B_TF_0 \tag{24.33}$$

24.2.4 Distributed Models

When a component stores kinetic and elastic potential energy over the same spatially distributed region, the modeler has two choices. A larger number of lumped springs and masses can be interspersed to approximate this distribution, or the distribution can be more fully modeled with partial differential equations or finite elements. The distribution of lumped elements can be kept to a minimum, but the lumped approximation becomes increasingly hard to fit to the physical reality. Alternatively, a finite difference approximation can be made, but then the number of lumped elements increases and the equations become inefficient to analyze or simulate. In this section we shall examine the alternative of recognizing the distributed mass and elasticity in the basic modeling of the structural and drive components. As will be seen, approximations are still necessary to enable analysis and simulation, and the nature and limitations of the approximations must be understood.

24.2.4.1 Distributed Bar Elements

The simplest element to explain is the uniform bar element, justified when a cross section of material with a modest aspect ratio is forced along its longitudinal (X) axis, which is the axis along which the cross section is extruded to form the bar. The assumptions of plane strain and uniform stress over the cross section permit the simple partial differential equation for the displacement ξ of the mass element of length dx at location x along the X axis:

$$\rho\frac{\partial^2\xi}{\partial t^2} = E\frac{\partial^2\xi}{\partial x^2} \tag{24.34}$$

24.2.4.2 Distributed Shaft Elements

Another simple behavior to model is distributed torsional stiffness due to the shear modulus G and rotational inertia J. Again, the spatial variation of the twist angle ϕ is along the X axis. The polar moment of inertia J is typically small for drive shafts relative to driven components such as gears, flywheels, or beams. For this reason the distributed shaft elements can often be replaced with a lumped rotational spring. However, the partial differential equation for the distributed model is

$$\mu\frac{\partial^2\phi}{\partial t^2} - GJ\frac{\partial^2\phi}{\partial x^2} \tag{24.35}$$

24.2.4.3 Distributed Beam Elements in Bending

Beams in bending are a major source of flexible behavior. Also, the kinetic energy and elastic potential energy is often of equal consequence, especially when higher modes of vibration are considered. The static model treated previously yielded the stiffness properties of a slender beam. In the distributed case, a differential mass element is introduced into the partial differential equation. When the slender beam approximation is augmented in this way with negligible shear deformation and negligible mass moment of inertia of the differential cross section (not the entire beam), the first and most frequently suitable of model results, the Bernoulli-Euler beam model. When these two assumptions are not justified, the complexity of the Timoshenko beam equation (still manageable) must be introduced. The Timoshenko model will not be examined here as it is typically not necessary for flexible robotic arms.

For the Bernoulli-Euler beam, deflection $w(x)$ perpendicular to the X axis varies with distance along X, x and time t. We can consider it possible for the cross section to vary along X also, resulting in the mass density per unit length $m(x)$ and the area moment of inertia about the neutral axis, $I(x)$, to also vary. The material properties are assumed constant including the elastic modulus E. The resulting partial differential equation is

$$m(x)\frac{\partial^2 w(x,t)}{\partial t^2} + \frac{\partial^2}{\partial x^2}\left[EI(x)\frac{\partial^2 w(x,t)}{\partial x^2}\right] = 0 \tag{24.36}$$

24.2.4.4 Combined Distributed Effects and Components

By combining distributed effects we mean to combine in a single component the effects of various deformation processes: torsion, compression, and bending, for example. When components are attached to each other, an element of the arm system must interact at its boundary with another element. One beam may be attached to another or to a payload mass, for example. Both problems will be discussed below.

For small displacements a viable assumption is sometimes to treat each effect on a component as if it were independent but occurring at the same value of the independent spatial variables. Hence, bending in two orthogonal planes, axial compression and axial torsion could occur independently in the same slender arm link. This is clearly an imperfect model, since simultaneous bending in the two directions can produce torsion which would not be otherwise accounted for. Nevertheless, the combination allows representative displacements to be considered, and when an element of this type is adjoined to another element, it transfers energy in all the appropriate modes.

Combining elements or components involves matching boundary conditions on the partial differential equations with a possible reorientation of the variables due to angle of attachment. The simultaneous solution of the resulting systems of partial differential equations is extremely difficult in the general case. Two methods around this obstacle are discussed below: (24.2.5) assume small displacement and linear behavior and Laplace transform the partial differential equations (Frequency Domain Solutions), or (24.2.6) convert the partial differential equations into simultaneous ordinary differential equations (assumed mode solutions).

24.2.5 Frequency Domain Solutions

The treatment here will cover only bending of a planar arm. Bending is the most significant distributed flexible effect, and the planar arm permits its illustration in a meaningful way. Extensions to more complex situations of two axis bending, torsion, and compression are also useful. The transform of the differential equation allows a convenient representation of the dynamics between two points on a structure with a single independent spatial variable. This representation is known as the transfer matrix representation. An extensive treatment of this representation is found in Pestel and Leckie [9] and has been applied to arms as documented in Book et al. [10, 11]. Results that can be obtained from this implementation include determination of the frequency response, eigenvalues and eigenfunctions (mode shapes), and impulse response, and through the convolution of the impulse response and response to arbitrary inputs.

The transfer matrix relates variables at two stations along the arm as shown in Figure 24.2. The variables are referred to as state variables of the partial differential equation but are not to be confused with the

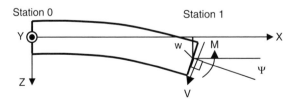

FIGURE 24.2 State vector for bending transfer matrix.

states of the overall system. Hence for Figure 24.2 we have

$$\underline{z}_1 = \begin{bmatrix} -W \\ \Psi \\ M \\ V \end{bmatrix}_1 = \begin{bmatrix} -\text{displacement} \\ \text{angle} \\ \text{moment} \\ \text{shear force} \end{bmatrix}_{\text{at station} 1} \quad (24.37)$$

$$\underline{z}_0 = B\underline{z}_1$$

B is beam transfer matrix.

$$B = \begin{bmatrix} c_0 & -lc_1 & ac_2 & -alc_3 \\ -\beta^4 c_3/l & c_0 & ac_1/l & ac_2 \\ \beta^4 c_2/a & -\beta^4 lc_3/a & c_0 & -lc_1 \\ -\beta^4 c_1/al & \beta^4 c_2/a & -\beta^4 c_3/l & c_0 \end{bmatrix} \quad (24.38)$$

where
$\beta^4 = \omega^2 l^4 \mu/(EI), \quad a = l^2/(EI)$
$c_0 = (\cosh \beta + \cos \beta)/2$
$c_1 = (\sinh \beta + \sin \beta)/(2\beta)$
$c_2 = (\cosh \beta - \cos \beta)/(2\beta^2)$
$c_3 = (\sinh \beta - \sin \beta)/(2\beta^3)$
$\mu = $ density/unit length
$\omega = $ circular frequency of vibration
$E = $ elastic (Young's) modulus
$I = $ cross sectional area moment of inertia

The spatial variable has been transformed to the spatial Laplace variable and replaced with its value at the points of interest. The time Laplace variable remains and is designated as s or as the frequency variable $\omega = -js$ where $j = \sqrt{-1}$. A brief tabulation of important transfer matrices for other planar elements appears below.

For an angle in the plane of angle φ

$$A = \begin{bmatrix} 1/\cos\varphi & 0 & 0 & 0 \\ 0 & 1 & 0 & 0 \\ 0 & 0 & 0 & 0 \\ m_s \omega^2 \sin\varphi \tan\varphi & 0 & 0 & \cos\varphi \end{bmatrix} \quad (24.39)$$

where m_s is the sum of all outboard masses from the angle to the end of the arm.

For a rigid mass,

$$R = \begin{bmatrix} 1 & -l & 0 & 0 \\ 0 & 1 & 0 & 0 \\ m\omega^2(l-h) & (-hl+h^2)\omega^2 m + I\omega^2 & 1 & -l \\ -m\omega^2 & hm\omega^2 & 0 & 1 \end{bmatrix} \quad (24.40)$$

where m is the mass of the body, I is the mass moment of inertia about an axis through the center of mass and perpendicular to the plane of the arm, l is the length of the mass (distance between points of attachment at stations i and $i+1$), and h is the distance to the center of mass from station i.

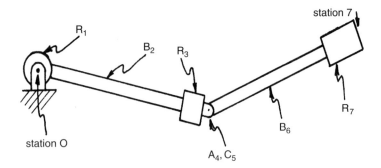

FIGURE 24.3 Transfer matrix representation of an arm.

For a controlled joint with the controller transfer function (joint torque/joint angle) $= k(j\omega)$

$$C = \begin{bmatrix} 1 & 0 & 0 & 0 \\ 0 & 1 & 1/k(j\omega) & 0 \\ 0 & 0 & 1 & 0 \\ 0 & 0 & 0 & 1 \end{bmatrix} \qquad (24.41)$$

When combined together to represent the planar arm with two joints shown in Figure 24.3, the result is a representation of the entire dynamics wrapped into a matrix product which relates the four state variables at the two extremes of the arm. The known boundary conditions at these points can be imposed and the resulting two equations employed to great benefit.

$$\begin{bmatrix} -W \\ \Psi \\ M \\ V \end{bmatrix}_0 = R_1 B_2 R_3 A_4 C_5 B_6 R_7 \begin{bmatrix} -W \\ \Psi \\ M \\ V \end{bmatrix}_7 \qquad (24.42)$$

In the example of Figure 24.3, the left end is pinned and the right end is free, yielding

$$z_0 = \begin{bmatrix} -W \\ \Psi \\ M \\ V \end{bmatrix}_0 = \begin{bmatrix} u_{11} & u_{12} & u_{13} & u_{14} \\ u_{21} & u_{22} & u_{23} & u_{24} \\ u_{31} & u_{32} & u_{33} & u_{34} \\ u_{41} & u_{42} & u_{43} & u_{44} \end{bmatrix} \begin{bmatrix} -W \\ \Psi \\ M \\ V \end{bmatrix}_6$$

$$\begin{bmatrix} u_{11} & u_{12} \\ u_{31} & u_{32} \end{bmatrix} \begin{bmatrix} -W \\ \Psi \end{bmatrix}_6 = \begin{bmatrix} 0 \\ 0 \end{bmatrix} \qquad (24.43)$$

24.2.5.1 Eigenvalues and Corresponding Eigenfunctions

After examination of the four equations a set of homogeneous equations can be selected. Such a set has two solutions, the trivial solution when the system is at rest and the nontrivial solution when the determinant of the coefficient matrix is zero. Changing the boundary conditions on the arm does not affect the Equation (24.42) above but does change the selection of the homogeneous equations and hence

Flexible Robot Arms

the frequency determinant.

$$d(\omega) = \begin{vmatrix} u_{ik} & u_{il} \\ u_{jk} & u_{jl} \end{vmatrix}_{\omega=\omega_i} = 0 \qquad (24.44)$$

$$z_{i0} = z_{j0} = 0 \text{ at station } 0$$
$$z_{kn} \neq 0, \quad z_{ln} \neq 0 \text{ at station } n$$

Note that the frequency determinant is a function of the time Laplace variable. It is zero only for $\omega = \omega_i$, the system eigenvalues. Since the transfer function is transcendental, there will be an infinite number of solutions which may be real, imaginary, or complex. The solution process is a numerical search over the complex plane. Considerable care must be exercised to find all the most important eigenvalues.

If the eigenvalues ω_i are substituted into the transfer matrix for ω, a relationship between the unknown boundary conditions results. An arbitrary amplification factor remains that can be assigned at will, so, for example, z_{jn} could be given the value 1 and the relative value of the other boundary condition found from the equation of the form:

$$\begin{bmatrix} u_{ik} & u_{il} \\ u_{jk} & u_{jl} \end{bmatrix}_{\omega=\omega_i} \begin{bmatrix} z_{in} \\ z_{jn} \end{bmatrix} = \begin{bmatrix} 0 \\ 0 \end{bmatrix} \qquad (24.45)$$

Because valid boundary conditions are known, the product transfer matrix equation can be used to find the states at any station along the arm, and new stations can be inserted along the flexible beams to determine the mode shapes (eigenfunctions) corresponding to any of the eigenvalues.

24.2.5.2 Frequency Response and Impulse Response

If the boundary conditions are not zero, but are instead sinusoidal forcing functions of a frequency ω, the analysis proceeds as follows. First, realize that the state variables occur in complementary pairs, and that if a moment is forced sinusoidally, for example, the angle must also respond sinusoidally and cannot be independently specified. The shear force for this example or the displacement could be specified to have nonzero amplitude, although it is seldom helpful to do so. If the equations are rearranged into the forced boundary variables f_i and unforced, response boundary variables r_i, the equations become

$$z_0 = \begin{bmatrix} f_0 \\ r_0 \end{bmatrix} = \begin{bmatrix} \tilde{U}_{11} & \tilde{U}_{12} \\ \tilde{U}_{21} & \tilde{U}_{22} \end{bmatrix} \begin{bmatrix} f_n \\ r_n \end{bmatrix} = \tilde{U} z_n \qquad (24.46)$$

and can be solved as

$$\begin{bmatrix} r_0 \\ r_n \end{bmatrix} = \begin{bmatrix} \tilde{U}_{22}\tilde{U}_{12}^{-1} & \tilde{U}_{21} - \tilde{U}_{22}\tilde{U}_{12}^{-1}\tilde{U}_{11} \\ \tilde{U}_{12}^{-1} & -\tilde{U}_{12}^{-1}\tilde{U}_{11} \end{bmatrix} \begin{bmatrix} f_0 \\ f_n \end{bmatrix} \qquad (24.47)$$

assuming the inverse exists.

The frequency of excitation is substituted for ω, and the amplitude of the forcing sinusoid(s) is substituted for the forcing vector. A solution for the response variables can thus be evaluated, and as in the case of the mode shapes, additional stations can be arbitrarily generated according to the position along the arm.

Since the frequency response can be generated, it is also quite feasible to produce the impulse response via numerically calculating the inverse transform for any frequency response.

24.2.6 Discretization of the Spatial Domain

It is possible to represent the spatially distributed nature of mass and compliance while retaining the discrete representation of the system state (as opposed to representing both space and time as continuous variables). This is done by describing the continuously variable spatial shapes $w(x,t)$ of arm deflection by an infinite series of fixed basis functions (shapes) $\phi_i(x)$ with time variable amplitudes $q_i(t)$.

$$w(x,t) = \sum_{i=1}^{\infty} q_i(t)\phi_i(x) \qquad (24.48)$$

A wise choice of the shapes achieves the necessary accuracy without a large number of basis functions, which means the infinite series is truncated after a few terms. It is necessary that the basis functions span the space and that they meet the geometric or essential boundary conditions. Convenient basis functions include polynomials and unforced solutions to the partial differential equations with assumed boundary conditions. These later shapes are known as assumed modes and can be particularly efficient choices for the basis functions. The q_i become state variables, along with their time derivatives.

The arm joint variables also can be used as state variables. The selection of joint variables and deflection variables and their corresponding shape functions must be compatible. For example, if the joint angle is measured from tangent of the preceding beam link to the tangent of the following beam link, called link j, a clamped boundary condition is appropriate for the first end of link j. In order to permit relative deflection and rotation of the second end of link j, a free boundary condition should be employed. Solutions of the partial differential equation for a beam with clamped-free boundary conditions are thus a good choice for basis shapes. This combination is convenient because the joint variables of the model are the same as a joint angle sensor would measure.

For other applications of the arm model, it may be preferable to use as joint variables the angle between lines that connect successive joint axes of the arm. In this case the joint angle accounts for the net deflection between joints and the deflection variable accounts for the undulations of the shape between joints. In this case pinned-pinned boundary conditions provide the necessary constraints and are good choices. This is convenient if one is solving for the inverse dynamics of the flexible arm when moving from point to point, since the joint angles and deflection variables can be calculated from simple geometry and static deflections.

Due to space constraints the development of the equations of motion for flexible link manipulators is abbreviated in this volume. The essence of the method will now be described, and the results will be given in sufficient detail to allow them to be used. Lagrange's method will be employed, although other techniques are preferred by some. More detail on Lagrange's method applied to structural systems in general may be found in Meirovitch [12] or as applied specifically to arms with the more complex joint motions included in Book [13]. An outline of the equation development procedure is now presented.

The kinematics of the flexible arm must first be described. This uses the joint variables and the deflection variables to locate every differential element of mass, as well as to express its velocity in terms of the derivative of its position and to construct expressions for describing the elastic and gravitational potential energy of the system as required by Lagrange's method. Knowing the velocity of each particle of mass in a given link, integration over the spatial domain of that link allows one to express the kinetic energy of that link. Similarly, the potential energy of the link can be determined by integrating over the link. We then sum over all the links to obtain the two scalars, kinetic energy (T) and potential energy (V) of the system as functions of the joint and deflection variables. The simple conservative form of Lagrange's equations then states that

$$\frac{d}{dt}\frac{\partial T}{\partial \dot{q}_i} - \frac{\partial T}{\partial q_i} + \frac{\partial V}{\partial q_i} = F_i \qquad (24.49)$$

F_i is the force that does work as q_i is varied.

These equations hold for every state q_i, both the joint and the flexible variables. Note that this simplified form results when T is a function of the variable and its derivative, while V is a function of only the variable and does not include specifically its derivative. Also note that if clamped boundary conditions are employed, there is no work performed by the joint actuators when the deflection variables change. The same is not true when pinned boundary conditions are used and a term for F_i appears in each equation.

Other hints are appropriate for the selection of good basis function shapes. Because we are often interested in the eigenvalues of the overall system and the subsequent system mode shapes (as opposed to the assumed mode shapes), one should choose the shapes that most naturally conform to the differential equation and all interior conditions that form effectively the boundary conditions on the distributed beams. In mechanics terms, we must choose shapes that meet the geometric boundary conditions and that are complete. Assumed mode shapes have the advantage of also satisfying the partial differential equation. If these shapes satisfied the natural boundary conditions (essentially the forces and torques on the beam), the assumed mode shape would be the true system mode shape. Closely matching the natural boundary conditions may be worth extra trouble if the most efficient representation is of importance.

To describe the kinematics, we again resort to coordinate transformations as described by 4×4 matrices.

$$W_i^j = \begin{bmatrix} & & & x_j \text{ component of } O_i \\ & R_i^j & & y_j \text{ component of } O_i \\ & & & z_j \text{ component of } O_i \\ 0 & 0 & 0 & 1 \end{bmatrix} \tag{24.50}$$

$$h_i = W_i^0 h_i^i = W_i h_i^i \tag{24.51}$$

We will consider transforms to be composed of separate transforms for the joints and undeformed links and for the link (or joint) deformation

$$W_j = W_{j-1} E_{j-1} A_j = \widehat{W}_{j-1} A_j \tag{24.52}$$

where

A_j = joint transformation matrix for joint j
E_{j-1} = the link transformation matrix for link $j - 1$ between joints $j - 1$ and j
$\widehat{W}_{j-1}$ = the cumulative transformation from base coordinates to $\widehat{O}_{j-1}$ at the distal end of link $j - 1$.

Because the axes of our links run in the X direction, we can describe the nominal position to consist of an X component η and the deflection terms to be described by a series of shapes in X, Y, and Z with amplitudes δ_{ij}

$$h_i^i(\eta) = \begin{bmatrix} \eta \\ 0 \\ 0 \\ 1 \end{bmatrix} + \sum_{j=i}^{m_i} \delta_{ij} \begin{bmatrix} x_{ij}(\eta) \\ y_{ij}(\eta) \\ z_{ij}(\eta) \\ 0 \end{bmatrix} \tag{24.53}$$

where

x_{ij}, y_{ij}, z_{ij} = the x_i, y_i, and z_i displacement components of mode j of link i's deflection, respectively
δ_{ij} = the time-varying amplitude of mode j of link i (24.54)
m_i = the number of modes used to describe the deflection of link i

$$E_i = \left[H_i + \sum_{j=i}^{m_i} \delta_{ij} M_{ij} \right] \tag{24.55}$$

where

$$H_i = \begin{bmatrix} 1 & 0 & 0 & l_i \\ 0 & 1 & 0 & 0 \\ 0 & 0 & 1 & 0 \\ 0 & 0 & 0 & 1 \end{bmatrix} \tag{24.56}$$

and

$$M_{ij} = \begin{bmatrix} 0 & -\theta_{zij} & \theta_{yij} & x_{ij} \\ \theta_{zij} & 0 & -\theta_{xij} & y_{ij} \\ -\theta_{yij} & \theta_{xij} & 0 & z_{ij} \\ 0 & 0 & 0 & 0 \end{bmatrix} \quad (24.57)$$

$$\frac{dh_i}{dt} = \dot{h}_i = \dot{W}_i^j h_i + W_i^j \dot{h}_j \quad (24.58)$$

$$\dot{W}_j = \dot{\widehat{W}}_{j-1} A_j + \widehat{W}_{j-1} \dot{A}_j \quad (24.59)$$

and

$$\ddot{W}_j = \ddot{\widehat{W}}_{j-1} A_j + 2\dot{\widehat{W}}_{j-1} \dot{A}_j + \widehat{W}_{j-1} \ddot{A}_j \quad (24.60)$$

where

$$\dot{A}_j = U_j \dot{q}_j \quad (24.61)$$
$$\ddot{A}_j = U_{2j} \dot{q}_j^2 + U_j \ddot{q}_j \quad (24.62)$$
$$U_j = \partial A_j / \partial q_j \quad (24.63)$$
$$U_{2j} = \partial^2 A_j / \partial q_j^2 \quad (24.64)$$

q_j is the joint variable of joint j.

$$\begin{aligned} \widehat{W}_j &= W_j E_j \\ \dot{\widehat{W}}_j &= \dot{W}_j E_j + W_j \dot{E}_j \\ \ddot{\widehat{W}}_j &= \ddot{W}_j E_j + 2\dot{W}_j \dot{E}_j + W_j \ddot{E}_j \\ \dot{E}_j &= \sum_{k=1}^{m_j} \dot{\delta}_{jk} M_{jk} \end{aligned} \quad (24.65)$$

and

$$\ddot{E}_j = \sum_{k=1}^{m_j} \ddot{\delta}_{jk} M_{jk}$$

With expressions for the velocity of any point on the slender beam, one can form the kinetic energy of a differential mass element at that point. Instead of the simpler expression $v^T v$ the trace operation $\text{Tr}\{v v^T\}$ is used to enable later simplifications by exchanging the order of summation and integration.

$$dk_i = \frac{1}{2} dm \text{Tr}\{\dot{W}_i h_i^i h_i^{iT} \dot{W}_i^T + 2\dot{W}_i h_i^i \dot{h}_i^{iT} W_i^T + W_i^i \dot{h}_i^i \dot{h}_i^{iT} W_i^T\} \quad (24.66)$$

where

$$\dot{h}_i^i = \sum_{j=1}^{m_j} \dot{\delta}_{ij} [x_{ij} \quad y_{ij} \quad z_{ij} \quad 0]^T \quad (24.67)$$

The system kinetic energy is found by integrating over each link and summing over all links.

$$K = \sum_{i=1}^{n} \int_0^{l_i} dk_i \quad (24.68)$$

Careful examination of the expanded form of this equation shows that the following terms appear.

$$B_{li} = \sum_{j=1}^{m_1} \sum_{k=1}^{m_1} \dot{\delta}_{ij} \dot{\delta}_{ik} C_{ikj} \qquad (24.69)$$

where

$$C_{ikj} = \frac{1}{2} \int_0^{l_1} \mu [x_{ik} \quad y_{ik} \quad z_{ik} 0]^T [x_{ij} \quad y_{ij} \quad z_{ij} \quad 0] d\eta \qquad (24.70)$$

is a 4×4 matrix, it is nonzero only in the 3×3 upper-left-hand corner. It can also be shown that $C_{ikj} = C_{ijk}^T$. Other terms include

$$B_{2i} = \sum_{j=1}^{m_i} \dot{\delta}_{ij} C_{ij} + \sum_{k=1}^{m_i} \sum_{j=1}^{m_i} \delta_{ik} \dot{\delta}_{ij} C_{ikj} \qquad (24.71)$$

where

$$C_{ij} = \frac{1}{2} \int_0^{l_i} \mu [\eta \quad 0 \quad 0 \quad 1]^T [x_{ij} \quad y_{ij} \quad z_{ij} \quad 0] d\eta \qquad (24.72)$$

$$B_{3i} = C_i + \sum_{j=1}^{m_i} \delta_{ij} [C_{ik} + C_{ik}^T] + \sum_{k=1}^{m_i} \sum_{j=1}^{m_i} \delta_{ik} \delta_{ij} C_{ikl} \qquad (24.73)$$

$$C_i = \frac{1}{2} \int_0^{l_i} \mu [\eta \quad 0 \quad 0 \quad 1]^T [\eta \quad 0 \quad 0 \quad 1] d\eta \qquad (24.74)$$

$$D_{ik} = C_{ik} + \sum_{l=1}^{m_1} \delta_{il} C_{ilk} \qquad (24.75)$$

$$G_i = C_i + \sum_{k=1}^{m_i} \delta_{ik} (C_{ik} + C_{ik}^T) \qquad (24.76)$$

Lagrange's equations will require the calculation of derivatives of the kinetic energy

$$\frac{d}{dt}(\partial K/\partial \dot{q}_j) - \partial K/\partial q_j = 2 \sum_{i=j}^{n} \mathrm{Tr} \left\{ \frac{\partial W_i}{\partial q_j} \left[G_i \ddot{W}_i^T + \sum_{k=1}^{m_i} \ddot{\delta}_{ik} D_{ik} W_i^T + 2 \sum_{k=1}^{m_i} \dot{\delta}_{ik} D_{ik} \dot{W}_i^T \right] \right\} \qquad (24.77)$$

$$\frac{d}{dt}(\partial K/\partial \dot{\delta}_{jf}) - \partial K/\partial \delta_{jf} = 2 \sum_{i=j+1}^{n} \mathrm{Tr} \left\{ \frac{\partial W_i}{\partial \delta_{jf}} \left[G_i \ddot{W}_i^T + \sum_{k=1}^{m_i} \ddot{\delta}_{ik} D_{ik} W_i^T + 2 \sum_{k=1}^{m_i} \dot{\delta}_{ik} D_{ik} \dot{W}_i^T \right] \right\}$$

$$+ \mathrm{Tr} \left\{ 2 \left[\dot{W}_j D_{jk} + 2 \dot{W}_j \sum_{k=1}^{m_i} \delta_{jk} C_{jkf} + W_j \sum_{k=1}^{m_i} \ddot{\delta}_{jk} C_{jkf} \right] W_j^T \right\} \qquad (24.78)$$

$E =$ Young's modulus of elasticity of the material.
$G =$ the shear modulus of the material.
$I_x =$ the polar area moment of inertia of the link's cross section about the neutral axis.
$I_y, I_z =$ the area moment of inertia of the link's cross section about the y_i and z_i axes, respectively.

Here θ_{xik} is the angle about the x_i axis corresponding to the kth mode of link i at the point η.

The potential energy terms in Lagrange's equations also require integration over each link and summation over all the links. Integration of the terms results in a "modal stiffness" matrix that is not dependent on amplitude.

$$K_{ikl} = K_{xikl} + K_{yikl} + K_{zikl} \tag{24.79}$$

where

$$K_{xikl} = \int_0^{1_i} GI_x(\eta) \frac{\partial \theta_{xil}}{\partial \eta} \frac{\partial \theta_{xik}}{\partial \eta} d\eta \tag{24.80}$$

$$K_{yikl} = \int_0^{1_i} EI_y(\eta) \frac{\partial \theta_{yil}}{\partial \eta} \frac{\partial \theta_{yik}}{\partial \eta} d\eta \tag{24.81}$$

$$K_{zikl} = \int_0^{1_i} EI_z(\eta) \frac{\partial \theta_{zil}}{\partial \eta} \frac{\partial \theta_{zik}}{\partial \eta} d\eta \tag{24.82}$$

Lagrange's equations then require the calculation of partials of the potential energy

$$\frac{\partial V_e}{\partial q_j} = 0 \tag{24.83}$$

$$\frac{\partial V_e}{\partial \delta_{jf}} = \sum_{k=1}^{m_i} \delta_{jk} K_{jkf} \tag{24.84}$$

The values of the coefficients K_{jkf} can be determined analytically or numerically, for example, by finite-element methods.

The potential energy due to gravity requires that a gravity vector be defined

$$g^T = [g_x \quad g_y \quad g_z \quad 0] \tag{24.85}$$

This is then used in the partials of gravitational potential energy

$$\frac{\partial V_g}{\partial q_j} = -g^T \sum_{i=j}^n \frac{\partial W_i}{\partial q_j} r_i \tag{24.86}$$

$$\frac{\partial V_g}{\partial \delta_{jf}} = -g^T \sum_{i=j+1}^n \left(\frac{\partial W_i}{\partial \delta_{jf}} r_i \right) - g^T W_j \varepsilon_{jf} \tag{24.87}$$

For $j = n$,

$$\frac{\partial V_g}{\partial \delta_{nf}} = -g^T W_n \varepsilon_{nf} \tag{24.88}$$

where

$$r_i = M_i r_{ri} + \sum_{k=1}^{m_i} \delta_{ik} \varepsilon_{ik}$$

M_i is the total mass of link i

$r_{ri} = [r_{xi} \quad 0 \quad 0 \quad 1]$

a vector to the center of gravity from joint i (undeformed)

$$\varepsilon_{ik} = \int_0^{l_i} \mu [x_{ik} \quad y_{ik} \quad z_{ik} \quad 0]^T d\eta$$

Flexible Robot Arms

The generalized force corresponding to joint variable q_i is the joint torque F_i in those cases where the boundary conditions are clamped, as will be assumed here.

$$\frac{d}{dt}(\partial K/\partial \dot{q}_j) - \partial K/\partial q_j + \frac{\partial V_e}{\partial q_j} + \frac{\partial V_g}{\partial q_j} = F_j \tag{24.89}$$

$$\frac{d}{dt}(\partial K/\partial \dot{\delta}_{jf}) - \partial K/\partial \delta_{jf} + \frac{\partial V_e}{\partial \delta_{jf}} + \frac{\partial V_g}{\partial \delta_{jf}} = 0 \tag{24.90}$$

24.2.6.1 Finite Element Representations

Additional detail can be added to a model by using more exact geometry for the structural elements; however, a detailed finite element model will contain at least an order of magnitude more variables than a well-derived assumed modes model. Fortunately, it is possible to combine the best of both techniques. Examination of the above shows that the result of the specification of the geometry (mass and stiffness distribution) and all the analysis that ensues are embodied in a modal mass matrix, a modal stiffness matrix, and similar parameter terms that account for the Coriolis and centrifugal terms. These matrices can also be constructed from basis shapes resulting from a finite element representation of the components followed by a specification of approximate boundary conditions on each element separately and determination of the resulting mode shape. This process is fairly direct for the finite element expert but beyond the scope of this volume.

24.2.6.2 Approximations of This Method

The reader should be aware that the above technique, while satisfactory for most robotic applications, is an approximation that will be inaccurate in some instances. Some of these instances have already been alluded to, such as the case of large elastic deformation. More subtle inaccuracies exist that bear names associated with the conditions of their occurrence. If the base of the arm is rotating rapidly about the Z axis, vibrations in the $X - Y$ plane would not account for the position dependent body forces effectively stiffening the beam and leading to the phenomena of centrifugal stiffening. Hence, these equations are not recommended for modeling helicopter rotors or similar rapidly rotating structures.

24.2.7 Simulation Form of the Equations

Two forms of the equations of motion are commonly found useful. For purposes of simulation the following structure may be used:

$$\begin{bmatrix} M_{rr}(q) & M_{rf}(q) \\ M_{fr}(q) & M_{ff} \end{bmatrix} \begin{bmatrix} \ddot{q}_r \\ \ddot{q}_f \end{bmatrix} = -\begin{bmatrix} 0 & 0 \\ 0 & K_s \end{bmatrix} \begin{bmatrix} q_r \\ q_f \end{bmatrix} + N(q,\dot{q}) + G(q) = R(q,\dot{q},Q) \tag{24.91}$$

where
- q_r = rigid coordinates or joint variables
- q_f = flexible coordinates
- M_{ij} = the "mass" matrix for rigid and flexible coordinates corresponding to the rigid ($i, j = r$) or flexible ($i, j = f$) coordinates and equations
- $N(q,\dot{q})$ = nonlinear Coriolis and centrifugal terms
- $G(q)$ = gravity effects
- Q = externally applied forces
- R = effect of external and restoring forces and all other nonconservative forces including friction

For serial arms it is generally more difficult to produce this form of the equations, because all second time derivatives must be collected. For simulation the second derivatives must be solved for, given the input forces Q, the coordinates, and their first derivatives. An integration scheme, often one capable of handling stiff systems of equations, is then used to integrate these equations, forward in time twice, given initial conditions and external forcing $Q(t)$. The solution for $\ddot{q}$ requires either the inversion of its coefficient matrix or, more efficiently another means such as singular decomposition to solve these equations for known constant values (R is constant) of all other terms. This solution must be repeated at each time step, leading to the major computational cost of the process.

24.2.8 Inverse Dynamics Form of the Equations

The inverse dynamics solution seeks the inputs Q that would yield a known time history of the coordinates q, $\dot{q}$, and $\ddot{q}$. The inverse dynamics equations are less readily applied for flexible arms than for rigid arms, not due to the solution of the equation as given below, but because the values of q_f are not readily obtained, especially for some models. That situation is improved slightly if the rigid coordinates are chosen, not as joint angles, but as the angles connecting the preceding and following axes. In such a way the tip of the arm can be prescribed only in terms of q_r. We know all velocities should come to zero after the arm reaches its final position. If gravity is to be considered, the static deflections at the initial and final times will also be needed. These can be solved for without knowing the dynamics. In the common case that $Q = R$

$$Q = \begin{bmatrix} M_{rr}(q) & M_{rf}(q) \\ M_{fr}(q) & M_{ff} \end{bmatrix} \begin{bmatrix} \ddot{q}_r \\ \ddot{q}_f \end{bmatrix} + \begin{bmatrix} 0 & 0 \\ 0 & K_s \end{bmatrix} \begin{bmatrix} q_r \\ q_f \end{bmatrix} - N(q,\dot{q}) - G(q) \qquad (24.92)$$

In general Q is dependent on the vector T of motor torques or forces at the joints, and the distribution of these effects is such that

$$Q = \begin{bmatrix} B_r \\ B_f \end{bmatrix} T \qquad (24.93)$$

where T is the same dimension as q_r. By rearranging the inverse dynamics equation we obtain

$$\begin{bmatrix} B_r & -M_{rf} \\ B_f & -M_{ff} \end{bmatrix} \begin{bmatrix} T \\ \ddot{q}_f \end{bmatrix} = \begin{bmatrix} M_{rr} \\ M_{rf} \end{bmatrix} \ddot{q}_r + N(q,\dot{q}) + G(q) \qquad (24.94)$$

which is arranged with a vector of unknowns on the left side of the equation. Unknowns may also appear in the vector of nonlinear and gravitational terms which contain q_f and $\dot{q}_f$, and strictly speaking, the solution must proceed simultaneously. In practice these are typically weak functions of the flexible variables and the solution can proceed as a differential-algebraic equation.

It is worth noting what happens if we choose other rigid coordinates, such that the tip position is not described by q_r alone. Now B_f is zero, $B_r = I$, and the tip position is unknown. We may be satisfied with specifying the joint coordinates, however, during the motion of the robot and accept the resulting deflection of the tip. Further note that the equations are likely to be nonminimum phase. Resorting to linear thinking for simplicity, the transfer functions have zeros in the right half plane. The inverse dynamics effectively inverts the transfer function giving poles in the right half plane. Normal solution techniques produce an unstable solution which is causal. Another solution, the acausal, stable solution exists and will be found to be useful when we discuss control via inverse dynamics.

24.2.9 System Characteristic Behavior

Nonminimum phase behavior will be observed at the tip of a uniform beam subject to a torque input at its root. In a flexible arm this manifests itself as an initial reverse action to a step torque input. In other words, the initial movement of the tip is in the opposite direction of its final displacement. The consequences of this behavior are pervasive if the tip position is measured and used for feedback control. To see this effect, a linear analysis is sufficient, and the transfer matrix frequency domain analysis is well suited for this purpose. The nonminimum phase in this case is described by zeros in the right half plane for the transfer function from torque to position. When a feedback loop is closed, the root locus analysis shows that the positive zero must attract one branch of the root locus into the right, unstable region of the complex plane. The usual light damping of the flexible vibration modes allows this to happen with low values of feedback gain. A uniform beam shows these characteristics, and it has been shown (Girvin [14]) that tapering a beam reduces the severity of the nonminimum phase problem but does not eliminate it.

24.2.10 Reduction of Computational Complexity

The equations presented above can result in very complex relationships that need to be solved in order to be used either for simulation or for inverse dynamics. It is worth enumerating several ways of reducing this complexity. The number of basis shapes can be reduced if assumed modes close to the true modes are used for basis shapes. Several terms in the equation involve integration of the product of two basis shapes. If orthogonal assumed modes are used the cross product of two different mode shapes is zero when integrated over the length of the beam.

24.3 Control

The flexibility of a motion system becomes an issue when the controlled natural frequency becomes comparable to the vibrational frequencies of the arm in the direction of movement. One of the first things to know is when this might occur. If it is expected to be an issue, the next thing to consider is ways to ameliorate the problem. This involves a great number of possibilities, many of which involve control, but not exclusively control. Control should not be interpreted to exclusively mean feedback control, because open loop approaches are also very valuable.

24.3.1 Independent Proportional Plus Derivative Joint Control

The vast majority of feedback controlled motion devices, including mechanical arms, use some of proportional plus integral plus derivative (PID) control. Velocity feedback may be an alternative to derivative feedback. The simplest approach here is to apply the PID algorithm to each joint independently. In this case the actuation of other joints can create a disturbance on each joint which could have been anticipated with a centralized approach. This problem was analyzed in the earliest work considering the flexibility of robot arms [11]. Because the parameters involved in a PD controller (no integral action because it has minimal effect) are only two per joint, the problem can be effectively studied by simply sweeping through the design space. The pole positions of the controlled flexible system provide a way to judge the controller effectiveness. PD control of a rigid degree-of-freedom enables arbitrary placement of the system poles. The dominant poles of the flexible arm behave as a rigid system when the lowest natural frequency of the system is about one-fifth the flexible natural frequency with the joints clamped. Critical damping on the dominant modes cannot be obtained by adding additional derivative gain when the servo frequency is above about one third of the structural natural frequency. Damping reaches a maximum and then decreases with additional velocity or derivative feedback. The effect of derivative feedback for three values of proportional feedback is sketched in Figure 24.4. While more joints and links distort this conclusion to some extent, servo bandwidths above one third the system's natural frequency are very hard to achieve.

To the extent that true PD action can be obtained, the independent joint controlled arm will not go unstable but will become highly oscillatory with high gains. The root locus converges to the position of the clamped joint natural frequencies which make up the zeros of this system since this is the condition of zero joint motion. Higher modes that start at the pinned-free natural frequencies are seen to converge to the higher clamped-free natural frequencies. Zero joint motion means the joints are unable to move in response to vibration and thereby unable to remove the vibrational energy. Note that heavily geared joints or joints actuated by hydraulic actuators may not be back drivable in any event, and hence, the shaping of the commanded motion becomes the most effective way to damp vibration. Low gains still have the effect of smoothing out abrupt commands that would excite vibration although at the expense of rapid and precise response.

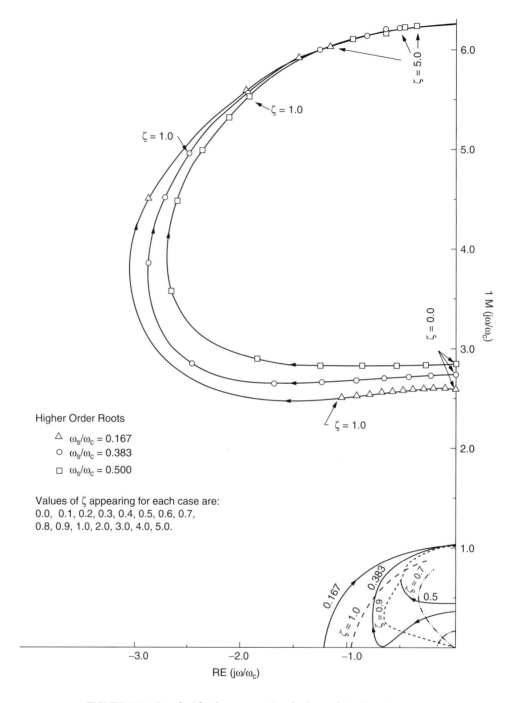

FIGURE 24.4 Root loci for three proportional gains as derivative gain varies.

When multiple axes are controlled with PD control, the details become more complicated but the overall behavior remains the same. If two joints contribute to the same vibrational mode, they both offer points at which energy can be put into or taken out of that mode. The limiting case is that energy can be taken out in a distributed fashion, giving a bandwidth which is a larger percentage of the natural frequency but not exceeding it.

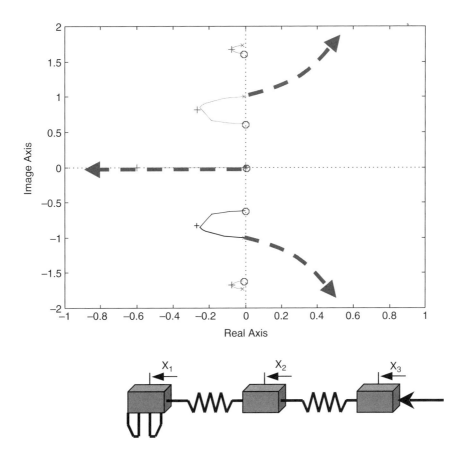

FIGURE 24.5 Noncollocated feedback of dx_1/dt to apply force at x_3 produces unstable (dashed) root locus. (Collocated feedback shown in solid lines.)

To avert the limitations of flexibility a reasonable but naïve approach to improve control performance, PID control based on measuring the end effector position fails because of the nonminimum phase behavior (resulting from noncollocated sensors and actuators) of the system. Even simple systems of this type involving two masses connected by a spring illustrate the effect as shown in Figure 24.5 in a root locus plot.

24.3.2 Advanced Feedback Control Schemes

Because the number of control algorithms that have been applied to flexible arms is extremely large, it is not possible to cover all of them in this treatment. Other texts and handbooks exist to present the details of these algorithms [15]. It is also impossible to verify that all of these approaches have practical merit, although there may be merit in specific cases to all of them. In this subsection the discussion will begin with the objectives and obstacles to practical advantages of these designs.

24.3.2.1 Obstacles and Objectives for Advanced Schemes

The modeling of the flexible arm involves many assumptions and presumptions. We assume that joint friction is linear or at least known. We presume to know the mass, stiffness properties, and geometries of the links and payloads. We have assumed a serial structure or at least one that could be approximated as such. Some of the controls proposed use measurements of the system states or related variables. Sensors to measure these variables over the workspace may not exist, and actuators that apply desired (collocated) inputs are unavailable or inappropriate. Some traditional actuators are not back drivable and hence don't serve as force or torque sources as the algorithms require. Computation to implement some algorithms is enormous, and hardware to perform these computations (typically done digitally) does not always exist at this time although the trend in computational speed is encouraging. Related to the computational complexity is the model order. Because a perfect model of the structure would contain an infinite number of modes, practical control design often violates perfection by working with a limited number of modes. Hence, the unmodeled modes may be excited by our attempts to control the modeled modes (control spillover), or our attempt to measure the modeled modes may be corrupted by the unmodeled modes (measurement spillover) that do affect the measurement. One of the objectives of a practical controller is to overcome these obstacles and thereby move the flexible arm more rapidly, precisely, and reliably through its prescribed motions.

24.3.2.2 State Feedback with Observers

The flexibility is modeled as a collection of second order equations that interact. In addition, the actuators will add dynamics that may be important. By converting this model to a state space model the control task is reduced to moving the arm states to a desired point in state space. That desired point is based on the desired position and velocity of the end effector, thus specifying the output of the system, plus the fact that all vibrations should quickly converge to zero. Very powerful techniques exist for doing this if the model of the system is linear. While the model is never perfectly linear, linearity may be a good approximation in the region around an operating point of larger or smaller size.

To implement a regulator bringing the state to the origin of the state space the following model and feedback law can be used. The output is the vector y. The plant matrix A, the input matrix B, the output matrix C, and the feed through matrix D are all required to be of appropriate dimension.

$$\begin{aligned} \dot{x} &= Ax + Bu \\ u &= Kx \\ y &= Cx + Du \end{aligned} \quad (24.95)$$

An extension of these basic equations can produce a controller that tracks a time varying reference trajectory as well. The linear equations can be written so that the origin of the state space produces the desired value of y. There is no unique choice for states x. This allows one to transform to different representations for different purposes. In terms of the models derived above the state, variables may be joint variables and their derivatives and deflection variables and their derivatives.

The first obstacle to this algorithm is measurement of all the needed states. In particular the deflection variables and their derivatives are not directly available. Strain gages have been used to effectively obtain this information and state feedback is then approximately achievable [16, 17, 18]. For each basis shape used to model a beam, one additional measurement will enable calculation of its amplitude. The derivative of that amplitude can then be approximated with a filtered difference equation.

State observers are dynamic systems that use present and past measurements of some outputs to converge on estimates of the complete system states. Full order observers have the same number of states as the system they observe, while reduced order observers allow the measured variables to be used directly with a proportional reduction of the observer order. Observers, that have been optimized according to a quadratic performance index assuming Gaussian noise is corrupting the measurements and the control input, are often referred to as Kalman filters after the pioneer in the use of this technique.

Evaluation of the feedback matrix K and the gains of the observer system can be done in various ways. The optimization scheme referred to above is popular but requires some knowledge or assumption of the strengths of the noises corrupting inputs and measurements. Pole placement is another approach that allows the engineer more direct influence over the resulting dynamics. The choice of pole positions that are "natural" for the system and yet desirable for the application can be challenging. These approaches are discussed in appropriate references [15].

Flexible Robot Arms

Knowledge of the plant model is required to construct the observer and to design the feedback matrix. Sensitivity to error in this model is one of the major issues when seeking high performance from state feedback. Nonlinear models complicate this issue immensely.

24.3.2.3 Strain and Strain Rate Feedback

Joint position is commonly differentiated numerically to approximate velocity and similarly link deflection can be differentiated numerically to complete the vector of state variables to be used in the state feedback equation. Thus, to control two modes, at least two strain gage readings are needed for effective flexible arm control which is not extremely sensitive to model parameters [16]. This approach has been implemented with good results as shown in Yuan [19]. Those experiments incorporated hydraulic actuation of a 6 m long, two-link arm and roughly cut the settling time of PD control in half. Hence, the complexity of an observer was not required, although filtering of the noisy differentiated signals may be appropriate.

One of the obstacles to the general application of strain feedback, or state feedback in general, is the lack of a credible trajectory for the flexible state variables. The end user knows where the tip should be, but the associated strain will not be zero during the move. See the section below on inverse dynamics for a solution to this dilemma.

24.3.2.4 Passive Controller Design with Tip Position Feedback

While it is natural to want to feed back the position of the point of interest, the nonminimum phase nature of the flexible link arm results in instability. Nonminimum phase in linear cases results from zeros of the transfer function in the right half plane. The output can be modified to be passive and of minimum phase by closing an inner loop or by redefining the output as a modified function of the measured states. For example, if tip position is computed from a measured joint angle θ and a measured deflection variable δ for a link of length l, the form of the equation will be (tip position) $= l\theta + \delta$ and the system is nonminimum phase. If, on the other hand, the (reflected tip position) $= l\theta - \delta$ is used for feedback control, the system is passive and can be readily controlled for moderate flexibility [20]. Obergfell [21] measured deflection of each link of a two link flexible arm and closed each inner loop using this measurement. Then a vision measurement of the tip position was used to control the passified system with a simple controller. He used classical root locus techniques to complete compensator design. The device and the nature of the results are shown in Figure 24.6 and Figure 24.7.

24.3.2.5 Sliding Mode Control

Sliding mode control operates in two phases. First, the system is driven in state space toward a sliding surface. On reaching the sliding surface, the control is switched in a manner to move along the surface toward the origin, set up to be the desired equilibrium point. The switching operation permits an extremely robust behavior for many systems, but if there are unmodeled dynamics, the abrupt changes may excite these modes more strongly than other controllers. Given the concern for robustness with model imperfections and changes, the sliding mode has a natural appeal. The implementation of sliding mode control is subject to some of the same problems as state feedback controllers; that is, knowledge of the state is difficult to obtain by measurement. Frame [22] has used observers based on a combination of joint position, tip acceleration, and tip position (via camera) measurements. The results are recent and, while promising, have not produced a clear advantage over state feedback.

24.3.3 Open Loop and Feedforward Control

Three matters for discussion in this section are the initial generation of the motion trajectory for a flexible link system, either the end point or the joints and the modification of an existing trajectory to create a more compatible input. Finally, by observing the errors in tracking a desired trajectory, the learning control approach can improve the trajectory on successive iterations, reducing the demands on the feedback control.

FIGURE 24.6 The experimental arm RALF (Robotic Arm Large and Flexible). (Total length about 6 m, lowest natural frequency about 3 Hz.)

24.3.3.1 Trajectory Specification

Among the most popular trajectory forms is the trapezoidal velocity profile that results from the desire to move in minimum time from point to point, with constraints on acceleration and velocity. This bang-bang solution creates extreme values of jerk (the third derivative of position) and consequently excites the flexible modes of an arm. A modified approach consists of also limiting the value of jerk and minimizing move time under this new constraint.

24.3.3.2 Filtering of Commands for Improved Trajectories

Small changes in the speed along a given path of motion can have major changes in the resulting vibration of a flexible arm. This is because the excitation is largely the result of jerk, the second derivative of speed, and is not readily apparent to someone planning a velocity profile, much less a position profile.

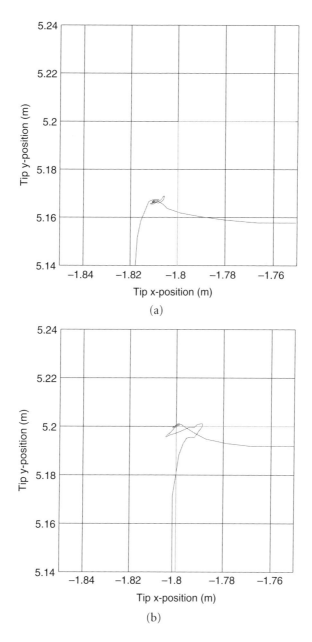

FIGURE 24.7 (a) Path followed without tip position feedback. (Commanded trajectory is a square, partially shown in green and actual trajectory in red.) (b) Path followed with tip position feedback. (Commanded trajectory is a square, partially shown in green and actual trajectory in red.)

Filtering the nominal input trajectory can make small adjustments to all trajectories entered and relieve the end user from this task. Filters commonly used are the low pass filter, notch filter, and the time delay filter.

Low pass filtering can be performed by either digital or analog means. Typically a cutoff frequency and a rate of roll off are determined. The rate of roll off is determined by the order of the filter n (number of integrators required), and the cutoff frequency is determined by the selection of parameters. Standard techniques for parameter selection are available [23], and filter poles are placed according to standard patterns. The analog Butterworth filter of order n, a popular low pass filter, is based on $2n$ poles evenly spaced around a circle, symmetric about the real axis, and no zeros. The stable poles are maintained in the filter design. The filter can then be converted to a digital filter by various mapping techniques. Chebyshev filters have poles similarly placed about an ellipse, and superior frequency domain specifications are achieved in the pass band but with some fluctuations in the magnitude of the response. For filtering the robot trajectory, we are also concerned about the time response of the filter itself and do not want the filter to insert oscillations of its own. Implementation of a low pass filter such as a Butterworth filter results in an infinite impulse response (IIR) filter as a result of the system poles. Theoretically, the response to an input never dies out. Artifacts of the filter include phase shift and an extension of the period of time over which the commanded motion continues. In the design of the low pass filter the engineer must base the cutoff frequency on the vibration frequencies of the arm. All components with frequencies above the cutoff are reduced in magnitude.

Time delay filtering is readily implemented with a digital controller but not with an analog controller. Effectively, every finite impulse response (FIR) filter is a time delay filter. The input is received by the filter, and the output based on that input is given for only a finite time period thereafter. When implemented by a digital delay line with n delays of time T and a zero-order hold, a single input enduring for a period T would produce n outputs that were each piecewise constant for each period of time T. While low pass and other filters can be implemented in FIR form, our interest is with a specific type of filter that places zeros at appropriate places to cancel arm vibrations. The placement of these zeros requires knowledge of the arm's vibrational frequency. When contrasted with the IIR filter, the FIR filter stretches the input by only a known time, at most nT. The operation of this filter is explained by a combination of time response and frequency response.

The early version of a command shaping filter [24, 25] suffered from excessive sensitivity to variations in the plant. Since then a revised form was proposed [26], and a number of researchers have explored variations on this theme [27, 28, 29, 30]. The problem of sensitivity has been addressed in several ways that can be grouped into two approaches: increase robustness or adapt the filter [31]. Increasing robustness allows the resonant frequencies to drift somewhat during operation while remaining in a broadened notch of low response. The notch is broadened by placement of multiple zeros. Unfortunately this requires additional delay in completing the filtered command. The time-delay filter in various forms has become a widely used approach to canceling vibration with minimal demand on controller complexity and design sophistication. Its effectiveness can be understood in terms of the superposition that linear systems obey. An impulse input is separated into several terms, which when combined by superposition cancel the vibration as shown in Figure 24.8.

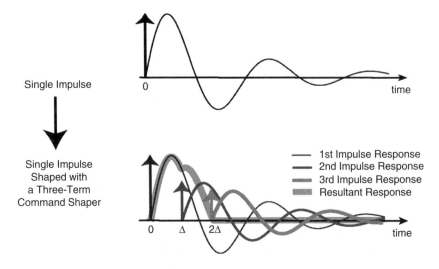

FIGURE 24.8 Effect of three term OAT command shaping filter.

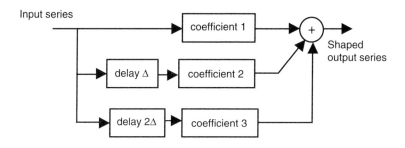

FIGURE 24.9 Time delay filter implementation.

It can also be understood by examining the frequency response of the two- and three-term filters. The two-term filter is effectively the original deadbeat filter with a narrow notch indicating low robustness (high sensitivity). Singer et al. [26] termed this the zero vibration or ZV shaper. A three-term filter with the delay between terms equal to half the vibration period was called the ZVD shaper because the derivative of the magnitude is also zero. The optimal arbitrary time-delay or OAT filter of Magee [27] also shown allows the delay to be selected at will according to the equations for the coefficients given in Equations (24.96). This is highly desirable if short time delay is needed, if delay needs to be matched to sampling time multiples (usual for digital implementation), or if one desires to adapt the gains and thereby track known or measured changes in the system behavior. Note, however, that robustness for the OAT filter is greatest when the time delay is half the vibration period. When this is true, the OAT filter is equivalent to the ZVD filter.

$$\begin{aligned}\text{coefficient } 1 &= 1/M \\ \text{coefficient } 2 &= -(2\cos\omega_d T_d e^{-\zeta\omega_n T_d})/M \\ \text{coefficient } 3 &= (e^{-2\zeta\omega_n T_d})/M\end{aligned} \quad (24.96)$$

where
$M = 1 - 2\cos\omega_d T_d e^{-\zeta\omega_n T_d} + e^{-2\zeta\omega_n T_d}$
$\omega_d = \omega_n\sqrt{1-\zeta^2} =$ damped natural frequency
$\omega_n =$ undamped natural frequency
$\zeta =$ damping ratio
$T_d =$ time delay selected, an integer number of samples $= \Delta$

The OAT filter and other time delay filters are simple to implement as shown schematically in Figure 24.9. A delay line (memory) sufficient to hold $2\pi/\omega_d T$ samples of the input is needed. After the appropriate delay the sample is taken from memory and multiplied by an appropriate coefficient as specified in Equations (24.96), and the terms are added together to compose the output. If more than one mode of vibration needs to be cancelled, two such filters are placed in series, resulting in a delay which is the sum of the two periods. Note that multiple zeros result from a single filter. It may be possible (although maybe not desirable) to cancel multiple modes with a single filter.

Adaptation of time delay filters based on measuring the system response has been proposed in many ways. The obvious way of identifying vibrational frequencies from an FFT and adjusting the gains according to the design equation does not work as well as a direct adaptation of the OAT filter parameters based on the measured residual vibration during periods when the arm should be at rest [32]. This does not work if the arm does not have periods of rest or if the rest periods do not occur frequently relative to the variation in parameters. Adaptation of the filter coefficients occurs during rest and after computation the updated parameters are transferred to the filter for the next motion as shown in Figure 24.10. Adaptation of this form is shown in Figure 24.11 to be very effective after only one repetition of the path.

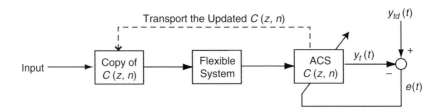

FIGURE 24.10 Adaptive command shaping (ACS) modifies the filter $C(z, n)$.

24.3.3.3 Learning Control

Learning control, in particular repetitive learning control, has been shown to dramatically improve the accuracy of rigid robot tracking [33]. Of special interest here is the use of learning control in conjunction with OAT filtering as described in the previous section [34]. Learning control remembers the error signal that occurred on the last repetition of the motion and thereby supplements the feed forward signal based solely on the input with a learned function appropriate for that corresponding time in the motion cycle. As the motion is repeated the learned function progressively eliminates the need for an error correction. Improvements of the accuracy of joint motion of more than an order of magnitude can result, effectively eliminating all but the noise or unsystematic variation in the signal. While the joint movement becomes more and more accurate, the tip motion may experience more vibration since the input trajectory is followed more precisely, including all sudden changes in velocity. Also, vibration can no longer back drive the joints and thereby dissipate the vibrational energy. If the input to the command has been prefiltered with an OAT or other shaping filter, however, vibration cancellation is shown to improve. This is due to the ability of learning control to overcome the systematic component of such imperfections as friction and cogging of the motor. Figure 24.12 shows a comparison of the tip acceleration of a gantry robot for four cases: PID only, PID plus OAT prefilter, PD and learning, and PD, OAT, and learning. This combination has been shown effective in commercial situations for the often found situation of repeated motions. The referenced paper also shows how to reduce the amount of memory required for storing the learned feed forward command by using multiple rates of sampling for different parts of the control loop. Adding vision or other tip position measurement can further eliminate errors due to static deflection [35].

24.3.3.4 Inverse Dynamics for Trajectory Design

The prior discussion of modeling included the development and discussion of the inverse dynamics equations. The specification of tip motion is preferred or required since the tip carries the tool for the task. Equations capable of doing this are presented above, but with the caveat about working with nonminimum phase systems. For this important subset of flexible arms, which includes arms with flexible links, special solution procedures must be considered. While complete discussion is beyond the scope of this article and is a continuing research topic, an introduction is important to understand the limitations of current techniques and the potential of future developments.

Consider for present purposes the linear form of the dynamic equations for a single joint moving a single link which can be reduced to a transfer function. If the transfer function of the forward dynamics has zeros in the right half plane, we do not want to apply the usual procedure to solve the inverse dynamics for the causal form in which the response always follows the input. The inverse Laplace transform permits another solution that is usually ignored but is appropriate here, the acausal solution. We effectively are asking what the input should have been to get the response we want at time t. The input must precede the output in the physical case, but for the computation, the input is the desired response and the output (of the computation) is the necessary response. A practical solution procedure exists for separating the causal and acausal solutions [36]. The rigid portion is specified to define the tip motion over time (using the pinned-pinned mode shapes). The causal portion of the solution is based on initial conditions and solved by moving forward in time. The anticausal portion is solved backward in time starting with final conditions. The acausal solution for the driving torque is the sum of these three terms. By appropriate shaping of the

Flexible Robot Arms **24**-37

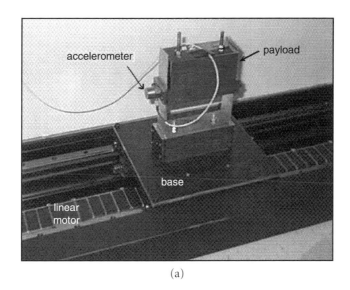

FIGURE 24.11 (a) Test bed for adaptive command shaping (dominant elastic mode at 16 Hz), (b) results of adaptive command shaping. (Note reduced oscillation in acceleration after initial update at approximately 2 seconds as indicated by arrow in acceleration plot. Delay time $T_d = 25$ ms.)

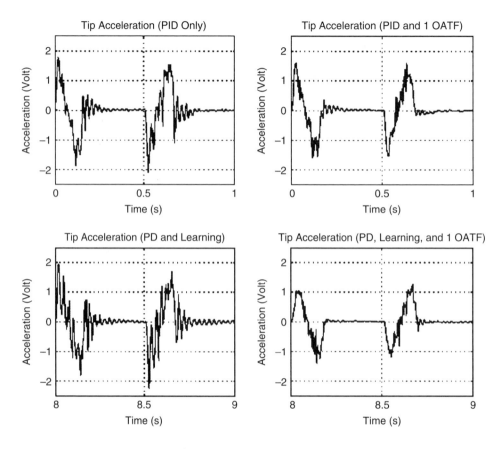

FIGURE 24.12 Comparison of joint PID, OAT filtering, and repetitive learning combinations.

commanded joint motion of the nonminimum phase arm, the tip remains stationary after the joint moves, then carries out the prescribed move and stops with no overshoot or vibration as shown in Figure 24.13.

As mentioned above under the feedback strategy of strain feedback, the desired history of strain or other flexible variables is needed to effectively create a tracking controller. The inverse dynamics solution gives a desired strain profile for the motion that can be applied to this end [37]. Neural networks have

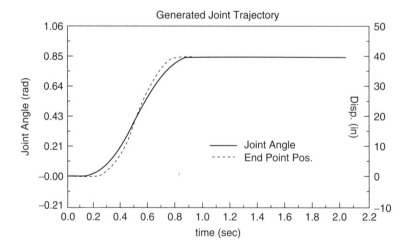

FIGURE 24.13 Inverse dynamics trajectory moves joint before tip and leaves no vibration.

Flexible Robot Arms

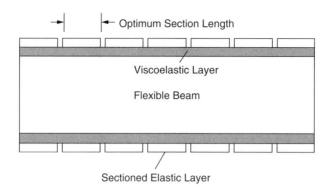

FIGURE 24.14 Sectioned constraining layer for passive damping.

also been applied to learn the necessary inverse for relatively simple single link arms [38]. Extension to multiple links has also been shown to be feasible [39, 40].

24.3.4 Design and Operational Strategies

24.3.4.1 Passive Damping Treatments

In the preceding discussion it has been apparent that we rely heavily on the truncation of the arm's flexible response to a finite number (one or two) of degrees of freedom when a perfect representation of distributed flexibility would take an infinite number. This assumption should not be taken for granted. It can be made more credible if the damping on the flexible modes is increased. Passive damping is an effective way to do this. Constrained layer damping treatments have been shown to be effective in adding damping to all modes of flexible beams used in laboratory robots [41]. These treatments sandwich a viscoelastic damping material between the structural member, typically a beam, and a sectioned elastic constraining layer. The Figure 24.14 illustrates this construction and Figure 24.15 illustrates the improvement in the damping that results. Optimization of the length of the constraining layer sections depends on the frequency to be damped, but damping effect is not extremely sensitive to this choice so that a wide range of frequencies will be treated [42]. These works also show that the treatment can eliminate instability in the higher modes not treated by the active controller.

24.3.4.2 Augmentation of the Arm Degrees of Freedom

How an arm is designed must be based on how it is to be used. In the case of flexible arms, this is also true, with three examples presented below. By incorporating additional actuators on the arm with additional degrees of freedom, a net gain in the performance of the arm can be achieved in spite of the physical inevitability of elasticity.

24.3.4.2.1 Bracing Strategies

An anthropomorphic justification may be the most effective way to introduce bracing. Fine motor skills of the human are concentrated in the fingers and gross motion capabilities in the arms, body, and even the legs. We use them in a modular way on many occasions, bracing our wrists when typing, writing, or threading needles. This permits a more stable base for fine manipulation during these precise motions. Sizing an arm's structure without bracing becomes a tradeoff between gross and fine motion. A short, heavy, stiff, precise structure is best for fine motion and a long, light, and consequently flexible structure is best for gross motion. Bracing enables one to have the second case when it is needed and then transition to the use of the first case for fine motion. This concept was first explored [43] without consideration of the overall implementation technology. The complexity of the maneuvers to move, achieve bracing contact, and then manipulate have been enabled by relevant research [44, 45].

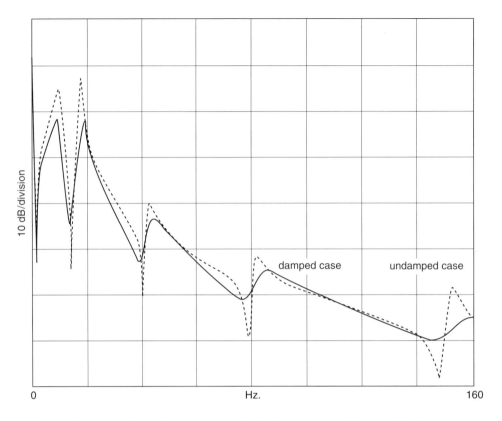

FIGURE 24.15 Resonance reduction from passive damping in the magnitude of the hub velocity/hub torque transfer function.

24.3.4.2.2 Inertial Damping

Bracing requires a stable surface to brace against, but if additional degrees of freedom exist, one can use them to advantage even without a bracing surface. The reaction to the movement of a "micro manipulator" at the end of the "macro manipulator" can be coordinated to absorb vibrational energy as shown by several researchers [46, 47, 48]. Because the micro manipulator is typically short and effectively rigid, it can have a high bandwidth that is needed to damp relatively high frequency vibrations that are above the bandwidth of the actuators of the macro manipulator. Furthermore, the micro manipulator is collocated with the tip of the macro manipulator and avoids the nonminimum phase problems of the macro actuators. Six actuators in the micromanipulator can provide three forces and three torques where the micromanipulator is mounted to the macro manipulator that can couple to any bending mode and remove energy. The complexity results from the fact that complex motions may be needed to generate the forces needed. The inverse dynamics equations to solve for the joint motions and torques to generate the prescribed forces have been generated, but the full equations are complex and must in general be solved in real time to create the reaction to vibration as it occurs. Simplified equations to generate a portion (say, only the forces) of the six components of force and torque can be implemented, but stability cannot be guaranteed in this case because the act of damping one mode with a force can excite another mode with a torque that was not desired. Very effective damping has been generated for vibration of a single beam in two dimensions [49] and two beams in series in three dimensions [50]. Figure 24.16 shows the micromanipulator on a flexible mount that is representative of a macromanipulator. Figure 24.17 shows the effectiveness of the vibration quenching on one of two axes. The other axis was similarly damped.

Flexible Robot Arms   24-41

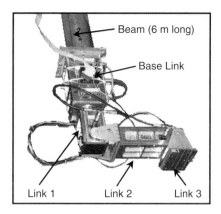

FIGURE 24.16   Three axis arm used as micromanipulator for inertial damping.

24.3.4.2.3 *Piezoelectric Actuation for Damping*

Thin layers of piezoelectric material (polymer, crystal, and ceramic forms exist) will change length upon the application of a voltage. If the material is bonded to a beam the strain created by this action can be used to deform the beam. By the same principle, the deformation of the beam creates a voltage across the piezoelectric material. Vibrational energy can be dissipated by the current that results. In either a passive or an active manner this transduction can be used for vibration control. The forces involved are small but significant in some cases and may provide a solution for some vibration problems as has been known for many years [51]. The ability to place patches of the material in optimum location for critical modes is one of the key advantages of this approach, but instability can also result from incorrect placement.

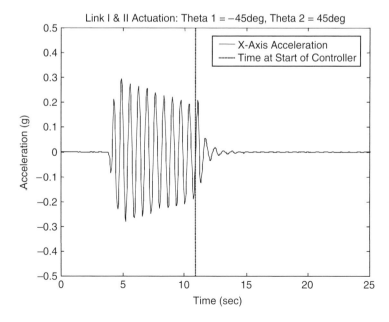

FIGURE 24.17   Inertial damping controller quenches oscillations of flexible base. (Controller action started at vertical line.)

24.4 Summary

Elasticity of structural material is a pervasive and simple phenomenon that leads to unavoidable and complex undesirable behaviors of mechanical arms and other motion devices. It can be treated simply in some cases by adding structural material and consequent mass. Better ways to treat this phenomenon do exist, and some of those techniques are discussed above. The motivation for these alternative approaches includes examination of when the simple approach is inadequate or undesirable. This implies modeling of both static and dynamic phenomena and how they lead to failure of the arm device. The failure of a structure may be due to its strength, stability, or flexibility. If strength or stability (buckling) is the active constraint, then a solution to flexibility is not needed. For fast, precise; and long reach arms, flexibility is often the active constraint. Control, both feedback and feed forward, design strategies; and operational strategies; should be considered to achieve a systematic solution.

References

[1] Book, W.J., Controlled motion in an elastic world, *J. Dynamic Syst., Meas. Control*, 50th Anniversary Issue, pp. 252–261, March 1993.
[2] Norton, R.L., *Machine Design, An Integrated Approach*, Prentice Hall, New York, 1998.
[3] Young, W.C. and Budynas, R.G., *Roark's Formulas for Stress and Strain*, McGraw-Hill, New York, 2002.
[4] Trahair, N.S., *Flexural-Torsional Buckling of Structures*, CRC Press, Boca Raton, FL, 1994.
[5] Sciavicco, L. and Siciliano, B., *Modeling and Control of Robot Manipulators*, 2nd ed., Springer-Verlag, London, 2000.
[6] Craig, J.J., *Introduction to Robotics: Mechanics and Control*, Addison-Wesley, Reading, MA, 1986.
[7] Denavit, J. and Hartenberg, R.S., A kinematic notation for lower-pair mechanisms based on matrices, *ASME J. Applied Mechanics*, vol. 22, pp. 215–221, 1955.
[8] Book, W., Analysis of massless elastic chains with servo controlled joints, *ASME J. Dynamics Syst., Meas. Control*, vol. 101, no. 3, pp. 187–192, Sept. 1979.
[9] Pestel, E.C. and Leckie, F.A., *Matrix Methods in Elastomechanics*, McGraw-Hill, New York, 1963.
[10] Book, W., Mark, M., and Kong, M., Distributed systems analysis package (DSAP) and its application to modeling flexible manipulators, Final Report, Subcontract No. 551 to Charles Stark Draper Laboratory, Contract NAS9-13809 to NASA, July 1979.
[11] Book, W.J., Modeling, design and control of flexible manipulator arms, Ph.D. thesis, Department of Mechanical Engineering, Massachusetts Institute of Technology, April 1974.
[12] Meirovitch, L., *Dynamics and Control of Structures*, John Wiley & Sons, New York, 1990.
[13] Book, W., Recursive lagrangian dynamics of flexible manipulators, *Int. J. Robotics Res.*, vol. 3, no. 3, pp. 87–106, 1984.
[14] Girvin, D.L. and Book, W.J. Analysis of poles and zeros for tapered link designs, in *Advanced Studies of Flexible Robotic Manipulators*, Chapter 8, World Scientific Publishing Company, NJ, 2003.
[15] Levine, W.S., Ed. *The Control Handbook*, CRC Press, Boca Raton, FL, 1996.
[16] Hastings, G.G. and Book, W.J., Reconstruction and robust reduced-order observation of flexible variables, *ASME Winter Annual Meeting*, Anaheim, CA, December, 1986.
[17] Cannon, R.H. and Schmitz, E., Initial experiments on the end-point control of a flexible one-link robot, *Int. J. Robotics Res.*, vol. 3, no. 3, pp. 62–75, 1988.
[18] Truckenbrot, A., Modeling and control of flexible manipulator structures, *Proc 4th CISM-IFToMM Ro.Man.Sy*. Zaborow, Poland, pp. 90–101, Sept. 8–12, 1981.
[19] Yuan, B.S., Huggins, J.D., and Book, W.J., Small motion experiments with a large flexible arm with strain feedback, *Proceedings of the 1989 American Control Conference*, Pittsburgh, PA, pp. 2091–2095, June 21–23, 1989.
[20] Wang, D. and Vidyasagar, M., Passive control of a stiff flexible link, *Int. J. Robotics Res.*, vol. 11, pp. 572–578, 1992.

[21] Obergfell, K. and Book, W.J., Control of flexible manipulators using vision and modal feedback, *Proceedings of the ICRAM,* Istanbul, Turkey, 1995.
[22] Frame, A. and Book, W.J., Sliding mode control of a non-collocated flexible system, *2003 ASME International Congress and Exposition,* Washington D.C., Paper IMECE2003-41386, November 16–22, 2003.
[23] Oppenheim, A.V., Schaefer, R.W., and Buck, J.R., *Discrete-Time Signal Processing,* 2nd ed., Prentice Hall, New York, 1998.
[24] Calvert, J.F. and Gimpel, D.J., Method and apparatus for control of system output in response to system input, Patent 2,801,351, July 30, 1957.
[25] Smith, O.J.M., *Feedback Control Systems,* McGraw-Hill, New York, 1958.
[26] Singer, N. and Seering, W.P., Preshaping command inputs to reduce system vibration, *ASME J. Dynamic Syst., Meas. Control,* vol. 112, no. 1, pp. 76–82, 1990.
[27] Magee, D.P. and Book, W.J., The application of input shaping to a system with varying parameters, *Proceedings of the 1992 Japan-U.S.A. Symposium on Flexible Automation,* San Francisco, CA, pp. 519–526, July 1992.
[28] Rhim, S. and Book, W.J., Noise effect on time-domain adaptive command shaping methods for flexible manipulator control, *IEEE Trans. Control Syst. Technol.,* vol. 9, no. 1 pp. 84–92, Jan. 2001.
[29] Singer, N., Singhose, W., and Seering, W., Comparison of filtering methods for reducing residual vibrations, *Eur. J. Control.,* vol. 5, pp. 208–218, 1999.
[30] U.S. Patent 6,078,844, Optimal arbitrary time-delay (OAT) filter and method to minimize unwanted system dynamics, issued June 2000.
[31] Book, W.J., Magee, D.M., and Rhim, S., Time-delay command shaping filters: robust and/or adaptive, *J. Robotics Soc. Japan,* vol. 17, no. 6, pp. 7–15, Sept. 1999.
[32] Rhim, S. and Book, W.J., Noise effect on time-domain adaptive command shaping methods for flexible manipulator control, *IEEE Trans. Control Syst. Technology,* vol. 9, no. 1, pp. 84–92, Jan. 2001.
[33] Sadegh, N., Synthesis of a stable discrete-time repetitive controller for MIMO systems, *ASME J. Dynamic Syst. Meas. Control,* vol. 117, no. 1, pp. 92–98, 1995.
[34] Rhim, S., Hu, A., Sadegh, N., and Book, W.J., Combining a multirate repetitive learning controller with command shaping for improved flexible manipulator control, *ASME J. Dynamic Syst., Meas. Control,* vol. 123, no. 3, pp. 385–390, Sept. 2001.
[35] Book, W.J., Sadegh, N., and Dickerson, S.L., Robotic assembly machine of low machine precision and weight but high assembly precision and speed, U.S. Patent 5,946,449, issued August 31, 1999.
[36] Book, W.J. and Kwon, D.S., Contact control for advanced applications of light weight arms, *J. Intelligent and Robotic Systems,* vol. 6, no. 1, pp. 121–137, 1992.
[37] Kwon, D.-S. and Book, W.J., A time-domain inverse dynamic tracking control of a single-link flexible manipulator, *J. Dynamic Syst., Meas. Control,* vol. 116, pp. 193–200, June 1994.
[38] Register, A., Book, W.J., and Alford, C.O., Neural network control of nonminimum phase systems based on a noncausal inverse, *Proceedings of the ASME Dynamic Systems and Control Division,* vol. 58, Atlanta, GA, pp 781–788, November 17–22, 1996.
[39] Bayo, E. and Paden B., On trajectory generation for flexible robots, *J. Robotic Syst.,* vol. 4, no. 2, pp. 229–235, 1987.
[40] Bayo, E. and Moulin, H., An efficient computation of the inverse dynamics of flexible manipulators in the time domain, *Proceedings of the IEEE Conference on Robotics and Automation,* pp. 710–715, 1989.
[41] Alberts, T.E., Book, W.J., and Dickerson, S., Experiments in augmenting active control of a flexible structure with passive damping, *AIAA 24th Aerospace Sciences Meeting,* Reno, NV, January 6–9, 1986.
[42] Alberts, T., Augmenting the control of a flexible manipulator with passive mechanical damping, Ph.D. Thesis, School of Mechanical Engineering, Georgia Institute of Technology, September 1986.
[43] Book, W.J., Sangveraphunsiri, V., and Le, S., The bracing strategy for robot operation, *Joint IFToMM-CISM Symposium on the Theory of Robots and Manipulators (RoManSy),* Udine, Italy, June 1984.

[44] West, H. and Haruhiko, A., A method for the desing of hybrid position/force controllers for manipulators constrained by contact with the environment, *Proceedings of the IEEE International Conference on Robotics and Automation,* St. Louis, MO, 1985.

[45] Lew, J. and Book, W.J., Bracing micro/macro manipulator control, *Proceedings of the 1994 IEEE International Conference on Robotics and Automation,* San Diego, CA, pp. 2362–2368, May 8–13, 1994.

[46] Book, W.J. and Lee, S.H., Vibration control of a large flexible manipulator by a small robotic arm, *Proceedings of the American Control Conference,* Pittsburgh, PA, pp. 1377–1380, 1989.

[47] Sharf, I., Active damping of a large flexible manipulator with a short-reach robot, *Proceedings of the American Control Conference,* Seattle WA, pp. 3329–3333, 1995.

[48] Lew, J. and Moon, S.-M., A simple active damping control for compliant base manipulators, *IEEE/ASME Trans. Mechatronics,* vol. 2, pp. 707–714, 1995.

[49] Book, W.J. and Loper, J.C., Inverse dynamics for commanding micromanipulator inertial forces to damp macromanipulator vibration, *1999 IEEE, Robot Society of Japan International Conference on Intelligent Robots and Systems,* Kyongju, Korea, October 17–21, 1999.

[50] George, L. and Book, W.J., Inertial vibration damping control of a flexible base manipulator *IEEE/ASME Transactions on Mechatronics,* pp. 268–271, 2003.

[51] Bailey, T. and Hubbard Jr., J.E., Distributed piezoelectric-polymer active vibration control of a cantilever beam, *J. Guidance, Control and Dynamics,* vol. 8, no. 5, pp. 605–611, 1985.

25
Robotics in Medical Applications

25.1	Introduction 25-1
25.2	Advantages of Robots in Medical Applications 25-1
25.3	Design Issues for Robots in Medical Applications 25-2
25.4	Research and Development Process 25-3
25.5	Hazard Analysis 25-4
	Hazard Identification • Verification and Validation • Initial and Final Risk Legend • Likelihood Determination • Severity Determination • Risk Acceptability
25.6	Medical Applications 25-6
	Noninvasive Robotic Surgery — CyberKnife® Stereotactic Radiosurgery System • Minimally Invasive Robotic Surgery — da Vinci® Surgery System • Invasive Robotic Surgery — ROBODOC® Surgical Assistant • Upcoming Products

Chris A. Raanes
Accuray Incorporated

Mohan Bodduluri
Restoration Robotics, Inc.

25.1 Introduction

The use of robots in medical applications has increased considerably in the last decade. Today, there are robots being used in complex surgeries such as those of the brain, eye, heart, and hip. By one survey, 2285 medical robots were estimated to be in use at the end of 2002, and that number is expected to rise to over 8000 medical robots by 2006. It is also estimated that medical robots may, in the end, have the largest market value among all types of robots.

Complex surgeries have complex requirements, such as high precision, reliability over multiple and long procedures, ease of use for physicians and other personnel, and a demonstrated advantage, to the patient, of using a robot. Furthermore, all new technologies in the medical area have to undergo strict regulatory clearance procedures, which may include clinical trials, as outlined by various government regulatory agencies. In the U.S., the Food and Drug Administration (FDA) has jurisdiction over medical devices. As more and more devices get through the regulatory procedures, there will be more and more robots in the medical world.

25.2 Advantages of Robots in Medical Applications

Before one can consider the usage of robots in medical applications, it is important to understand that medical applications have unique requirements, different from general, or "traditional," robot design. Some of the design issues and associated advantages are described below.

1. **High precision**: Modern robots are demonstrated to be highly precise. The precision range depends on the robot and the application, of course, but it is generally accepted that for a given application,

a robot can be designed to meet or exceed the precision requirements of the application. A typical industrial robot has repeatability specifications measured in tenths of a millimeter. A representative ratio of motion in robotic assisted surgery is that a 1 cm movement of a doctor's hand translates to a 0.1 cm movement of the robotic tool.

2. **Heavy payloads**: Modern robots can carry heavy payloads over large workspaces, at high speeds, with high precision. Industrial robots are available with payload capacity of a few ounces to over 1000 lb.
3. **Workspace**: Medical robot workspace requirements tend to be significantly larger than industrial needs because of patient related factors, such as uncertainty in patient location during the procedure and safety requirements. There is an obvious overriding need to avoid any hazard to the patient, physician, and other medical personnel; this drives an exclusionary zone around the patient, doctor, and other equipment that may be attached to the patient. Thanks to the advances driven by industrial applications, the workspace of most available robots is significant and can be utilized for medical applications.
4. **High Speed**: Most new robots have been designed and optimized for industrial automation, enabling them to move at high speeds with high precision. The majority of medical applications do not require robots to move at high speeds as these robots are working on a patient. Reassurance, comfort, and safety dictate the robot's speed in medical applications.
5. **Reliability**: Industrial robots are designed to work round the clock without stopping; their medical counterparts work only a few hours a day. The nature of medical applications is that most of the time is taken up by other parts of the surgery, such as operating room preparation, patient preparation, and postoperative procedures. The robots actually perform surgeries for only a limited time, around 10% of surgery time. The resulting reliability numbers for medical work are excellent, leading to very limited downtime.
6. **Tedium**: Most of the medical applications where robots are sought involve repetitive tasks over a very long period of time. Some surgeries last for many hours, during which the operators are required to repeat tasks hundreds or thousands of times. Obviously, robots do not have any problems with tedium.
7. **High Quality**: Robotic assisted surgery can help a wide variety of doctors perform complex surgeries with the same high quality previously achieved only by some accomplished surgeons. Additionally, most medical procedures cannot tolerate any degradation in quality due to trembling or unsteadiness of hands. Robotic systems in the operating room can compensate for imperfections in the user due to age, fatigue, or other factors, without degrading the quality of care administered to the patient.
8. **Computer control**: Robotic surgery is able to capitalize on available diagnostic data to calculate an optimized approach to treatment. Most modern systems use fusion of multiple imaging modalities such as CT, PET, and MRI.
9. **Remote operation**: Finally, because robots are typically controlled by computers and/or remote electrical signals, the option exists to remotely operate the units over large distances through direct data links, or even over the internet (telerobotics).

People have recognized many of these obvious advantages; therefore, we have seen a considerable increase in usage of robots in medical applications in recent times. As these advantages are general and apply to many medical procedures, the authors believe that it is just a matter of time before more robots are employed in automating a variety of procedures, ultimately increasing the quality while reducing the cost of medical care in the future.

25.3 Design Issues for Robots in Medical Applications

Using robots in medical applications presents a unique set of challenges. This section briefly discusses the design issues that should be considered in many medical applications.

1. **Safety**: Safety of patients and users is the ultimate concern when using robotics in medical applications. In the industrial world, safety is addressed, most typically, by ensuring that humans are not present in the robot's workspace. Considerable precautions are taken so that no one inadvertently enters the workspace of a working robot, and if someone does enter the area, the robot is automatically stopped. In the case of a medical application, by definition, a human being (the patient) needs to be in the workspace. Moreover, the physician and other medical personnel generally need to be in the workspace as well to attend to other needs of the patient and the surgery. A robotic system, therefore, has to be designed so that it is safe for the patient, physician, and other personnel in the room while it is effectively operating on the patient.
2. **Uncertainty of position**: Most medical applications have a higher level of uncertainty in the position of the target (patient) than their industrial counterparts do. In a typical industrial application, one can expect the workpiece to be aligned and mounted precisely in the same location and orientation. In a typical surgery, the patient and the specific organ that needs to undergo surgery cannot practically be located in the same location and orientation. Furthermore, various steps in the procedure will likely have to be modified and adapted based on the patient's condition.
3. **Fail-safe**: It is required that the medical robotic system operates in a fail-safe mode. By fail-safe, one means that if and when any component fails, the system reaches a safe state, thus minimizing chance for injury or death to the patient or other personnel.
4. **Power/System Failure**: The system needs to be designed in such a way that in case of a power/system failure, the physician can move the robot away to keep the patient safe and be able to attend to the patient.
5. **Record Keeping**: All records related to an operation need to be kept and protected for future use. This issue has become more acute in recent times with current U.S. and international regulations regarding patient data privacy. The opposing needs of the system are to maintain confidentiality, while ensuring that the patient on the table matches the program in the robotic system.
6. **Regulatory Issues**: All design, development, and production activity needs to be done in a controlled fashion following the appropriate regulatory guidelines.

The issues outlined above are in addition to the technical issues one has to deal with for any product design. It is obvious that these additional requirements add a significant cost in terms of time and resources to successfully design, develop, and deploy a product in the field.

25.4 Research and Development Process

It is essential, as well as required (by FDA regulations in the U.S.), that an organization strictly adheres to a process in the design, development, and manufacturing stages of a product. The specificity of the process varies from organization to organization, but all the different processes follow good manufacturing practices (GMP) and good laboratory practices (GLP). See FDA's document Title 21, Part 820 for the guidance on quality systems regulation. Figure 25.1 shows an example of a research and development process.

The research and development process shown in Figure 25.1 shows four stages of the development:

1. Requirements Definition
2. Concept Development
3. Feasibility Studies
4. Product Development

While it is important to maintain good documentation throughout each of these stages of development, design control procedures are especially necessary during the product development phase. Please refer to Title 21, Part 820.30 for guidance from the FDA on design control procedures.

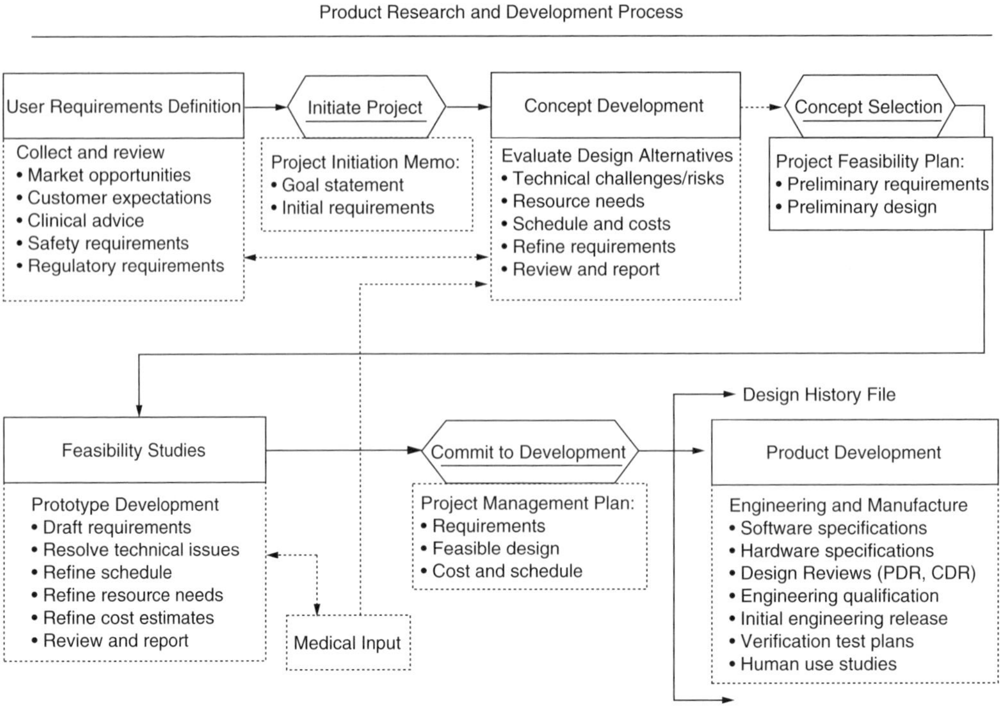

FIGURE 25.1 An example of research and development process.

25.5 Hazard Analysis

As mentioned in the previous sections, safety is the first and foremost concern of using robots in medical applications. To address this issue, a formal hazard analysis is done during the design process. This section describes a typical hazard analysis (based on EN60601-1-4).

25.5.1 Hazard Identification

Preliminary hazards are identified from

- Requirements specifications
- Architecture and design descriptions
- Previous hazard analysis results

The preliminary hazard list is updated into the final hazard list by the addition of hazards identified during fault tree analysis. The final hazard list is revised from time to time as the product undergoes changes.

25.5.2 Verification and Validation

Fault tree analysis (FTA) will be done on the hazards identified. This will identify states within the system that could cause a hazard. Recommendations will be made to test these states. Code or design change recommendations will also be made if necessary.

25.5.3 Initial and Final Risk Legend

The risk codes used in the hazards list initial and final risk columns are composed of a letter indicating *likelihood* and a number indicating *severity*.

25.5.4 Likelihood Determination

Likelihoods of hazards are defined according to the table below. Entries here follow the EN 60601-1-4 suggested rankings for a programmable electrical medical system (PEMS).

ID	Likelihood	Description
A	Frequent	Likely to occur frequently or continuously during the product lifetime
B	Probable	Likely to occur several times during the product lifetime
C	Occasional	Likely to occur sometime during the product lifetime
D	Remote	Unlikely to occur during the product lifetime
E	Improbable	Possible but extremely unlikely to occur during the product lifetime
F	Impossible	Cannot occur given the current product makeup

25.5.5 Severity Determination

Severities of hazards are defined according to the table below. Entries here follow the EN 60601-1-4 suggested rankings for a PEMS.

ID	Severity	Description
1	Catastrophic	Potential for multiple deaths or serious injuries
2	Critical	Potential for death or serious injury
3	Marginal	Potential for minor injury
4	Negligible	Little or no potential for injury

Hazard severities are assigned according to the anticipated severity of a *single incident or accident* resulting from that hazard. In particular, the severity rating does not reflect the possibility that the hazard could cause multiple incidents or accidents over the life of the product or at multiple product installations. These aspects of the hazard are reflected in the likelihood rating.

25.5.6 Risk Acceptability

Acceptability of risk is assessed according the following table:

Likelihood	4 - Negligible	3 - Marginal	2 - Critical	1 - Catastrophic
A - Frequent	ALARP	Unacceptable	Unacceptable	Unacceptable
B - Probable	ALARP	ALARP	Unacceptable	Unacceptable
C - Occasional	ALARP	ALARP	ALARP	Unacceptable
D - Remote	Acceptable	ALARP	ALARP	ALARP
E - Improbable	Acceptable	Acceptable	ALARP	ALARP
F - Impossible	Acceptable	Acceptable	ALARP	ALARP
	\multicolumn{4}{c}{Severity Level}

ALARP: As low as reasonably practical.

25.6 Medical Applications

As pointed out earlier in the chapter, Robots in medicine offer several potential advantages:

- They do not tire.
- They do not become inattentive.
- They do not suffer from human "frailties" such as shaking or jittering.
- They are more accurate.

Studies have shown that a surgeon's skill is one of the most significant factors in predicting patient outcomes. One of the goals of medical robotics is to minimize this difference between surgeons.

In the remainder of this chapter, we review three different products as examples of robotics in medicine as well as provide an insight into new and upcoming medical products. The three products highlighted belong to the categories of noninvasive surgery, minimally invasive surgery, and invasive surgery in terms of the procedure being automated. We study noninvasive applications in more detail than others, as the authors believe that noninvasive surgery will bring unique opportunities in advancing the state of medicine in the future.

25.6.1 Noninvasive Robotic Surgery — CyberKnife® Stereotactic Radiosurgery System

25.6.1.1 Introduction to Stereotactic Radiosurgery

Radiation has been widely used over many decades to treat cancerous tumors and other malformations in various parts of the body in a noninvasive fashion. There are two forms of radiation treatments, *radiotherapy* and *radiosurgery*. Radiotherapy treats the tumors with low amounts of radiation per session over many sessions, while radiosurgery treats the tumors with high amounts of radiation per session over one or a few sessions. Radiotherapy relies on the differential response of the tumor and the surrounding normal tissue, whereas the radiosurgery relies on very high, ablative doses of radiation delivered accurately to the tumor while minimizing the radiation to the surrounding normal tissue.

Delivering high amounts of radiation into a tumor while minimizing the radiation to the surrounding tissue is achieved by radiosurgery systems by delivering narrow beams from numerous (hundreds) angles all pointed into the tumor. The cumulative dose in the tumor where all the beams cross will thus be very high to eradicate the tumor, while the surrounding tissue where the beams are not overlapping receives low amounts of radiation. This technique was pioneered by Leksell (1951) and Larsson et al. (1958).

Because delivering high amounts of radiation in order to knock out the tumor is the primary focus of the radiosurgery, it is important to understand that accuracy immediately becomes a big issue. The error in the delivery system of this high amount of radiation directly increases the amount of surrounding tissue damaged in order for the entire tumor to receive the prescription dose.

25.6.1.2 Project Background

In the early 1990s, a project was launched to develop frameless radiosurgery, which, in principle, can then be applied anywhere in the body (including brain treatments), as the frame limitation would not exist anymore. As described above, radiosurgery is characterized by the following attributes:

1. Narrow beams
2. Numerous (hundreds) beams from various approach angles
3. High accuracy
4. Single or few treatments

Robotics in Medical Applications

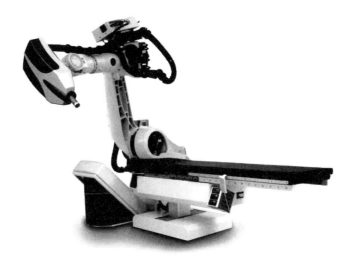

FIGURE 25.2 CyberKnife® stereotactic radiosurgery system. (Source: Accuray, Inc.)

25.6.1.3 CyberKnife® System

The CyberKnife stereotactic radiosurgery system, produced by Accuray, Incorporated, of Sunnyvale, California, achieves frameless stereotactic radiosurgery using robotics and image guidance (see Figure 25.2). The CyberKnife system has the following major components:

1. Linear accelerator (radiation source)
2. Six axis robotic manipulator
3. Stereo x-ray imaging system
4. Patient positioning system
5. Computer software

The linear accelerator produces a 6 MeV radiation beam which can be collimated to various sizes (5 to 60 mm diameter) based on the size of the tumor to be treated. The linear accelerator is mounted as the tool of the robot manipulator; therefore, it can be positioned and aimed at the patient arbitrarily as per the treatment needs. Two x-ray sources and cameras are used to monitor the patient, and patient movements are calculated and communicated to the robot which then compensates for the patient movement. The treatment delivered is accurate, frameless, effective, and elegant.

We briefly describe the major components in the following sections.

25.6.1.3.1 Radiation Source

The radiation source consists of a pulsed electron gun combined with a waveguide that uses microwave energy to accelerate the electrons into a tungsten target. The result is a 6 MeV x-ray beam that can be collimated into a diameter as small as 5 mm. This radiation source is known as a linear accelerator and is commonly called a "linac." The weight of the linac is approximately 165 kg and must be manipulated by the robot to accommodate tumor movement.

25.6.1.3.2 Robot Manipulator

An industrial robot (KUKA KR210), typically used in the automotive industry, is adapted for this medical device. The robot can carry a tool of weight up to 210 kg, with a maximum speed of 2 m/sec and a repeatability specification of 0.2 mm. This robot has a reach of 2.5 m or more which makes it more than adequate for positioning the linear accelerator to precisely aim at the patient from various different approach angles.

25.6.1.3.3 Stereo X-ray Imaging System

The imaging system consists of two x-ray sources mounted to the ceiling of the treatment room and a pair of x-ray cameras mounted in a "V"-shaped frame and fixed to the floor. Amorphous silicon detectors are used to provide 20×20 cm x-ray images. The distance between the camera and patient is approximately 65 cm, which eliminates most of the scatter radiation. The distance between the patient and the x-ray source is approximately 260 cm, which provides the robot ample access to the patient during the patient treatment.

25.6.1.3.4 Patient Positioning System

The patient positioning system provides precise location of the patient relative to the imaging system. Appropriate location of the tumor and landmarks in the field of view of the cameras, with appropriate x-ray illumination, enhances tracking performance. However, once in position the system remains fixed throughout treatment. Patient movement during treatment is accommodated by the robot manipulator.

25.6.1.3.5 Computer Software

A high-end computer provides software integration to the system. It supports both the treatment planning and the treatment delivery software. The system is designed to have a fixed set of nodes on a sphere, from which treatment beams can be delivered. A treatment plan consists of a dose rate, duration, and direction at each node. The robot moves sequentially through each of the nodes and delivers the required dose.

The treatment planning system (see Figure 25.3) allows the physician to specify radiation dose distributions in the tumor and surrounding tissue. The planning algorithm translates this into dose rate, duration, and direction at each of the treatment nodes.

On the day of the treatment, the patient lies on the couch of the patient positioning system. The treatment plan is selected and the system assists the operator in aligning the patient in the center of the imaging system. This is done by acquiring x-ray images using the diagnostic x-ray sources and cameras. Once the patient is aligned the treatment begins.

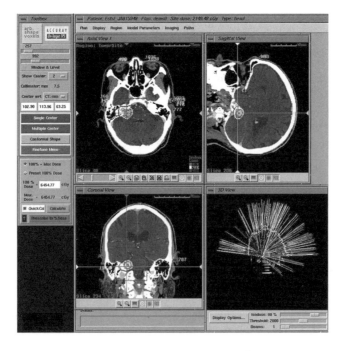

FIGURE 25.3 Treatment planning system for CyberKnife®. (Source: Accuray, Inc.)

During the patient treatment, the robot moves the linear accelerator through each beam position and orientation. Before each irradiation, a pair of orthogonal x-ray images of the patient is obtained and the movement of the patient is calculated. This information is transmitted to the robot, which compensates for the patient movement before the therapeutic beam is turned on. This process is repeated through all the nodes to complete an entire treatment.

25.6.1.4 Patient Safety

The system is designed with patient safety in mind, from treatment planning all the way through the treatment delivery. The treatment planning system limits the use of the robot to its known safe positions (nodes), so the treatment planned by a physician is always guaranteed to be safe during delivery. Furthermore, the treatment planning system provides a computer simulation of treatment and automatically flags any potentially dangerous movements of the equipment close to the patient.

During treatment delivery, the system monitors the robot movement in real time, and if any impending collisions are detected, the machine is stopped immediately. Furthermore, an operator is required to monitor the treatment through multiple video monitors and can issue an emergency stop to the system when any out-of-the-ordinary movements of the system or the patient are observed. Any stopped treatment can be resumed and completed with no adverse effects to the patient.

Fail-safe operation of the CyberKnife has also been designed in. For example, any fault condition, either detected by the system automatically or issued by the operator, immediately stops all the radiation and any robotic movement, thus causing the system to reach a safe condition during a failure. Furthermore, the operator can use the pendant to move the robot away in order to attend to the patient, if necessary.

25.6.1.5 System Accuracy and Calibration

A system centerpoint called the *isocenter* is identified in order to calibrate the installation of the imaging system and the robot in the treatment facility. The alignment of the cameras and location of the sources places the image of the isocenter within 5 mm of the center of both cameras. A calibration process identifies variations in location over the face of the camera to account for small variations. The imaging system's accuracy is better than 1 mm.

The installation of the robot determines the orientation of the user frame which has its origin at the isocenter. The tool frame is identified by measuring the position and orientation of the line from the laser-mounted coaxial to the therapeutic beam in the linear accelerator. Further calibration involves an automated process in which the robot illuminates an isocrystal and searches for the position and orientation that provides the maximum signal. A correction look-up table is maintained by the system. The residual error of the robot positioning system is less than 0.5 mm.

The total clinically relevant system accuracy, from treatment planning through delivery, is measured by exposing a film cube in a phantom and is found to be within 0.95 mm.

25.6.1.6 Robotic Advantages

Using a robot in the CyberKnife brings distinct advantages to the system. The ability to deliver a beam that can originate at an arbitrary point and can be aimed arbitrarily, brings an enormous level of flexibility in designing the treatments. Complex treatments can be generated that conform to the oddly shaped tumors (see Figure 25.4).

In addition, robotics makes the CyberKnife the only technology that is able to treat a moving tumor. Thanks to the advancement of robotics and computer vision, the CyberKnife is able to track a lung tumor accurately through the entire breathing cycle to deliver treatments.

25.6.2 Minimally Invasive Robotic Surgery — da Vinci® Surgery System

As opposed to the CyberKnife system, which is a noninvasive robotic medical device, the da Vinci system is a minimally invasive robotic system. The term "minimally invasive" in the medical world means that the surgery is performed with small incisions (also called laparoscopic surgery) in place of traditional surgery.

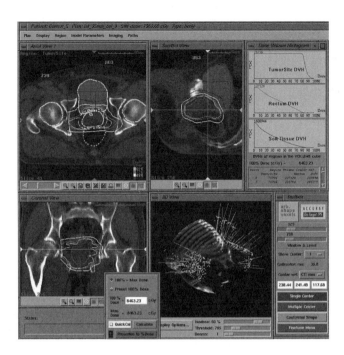

FIGURE 25.4 The dose radiation and dose distributions are planned on the CT slices directly. Simulation of the treatment allows verification of the dose distributions. (Source: Accuray, Inc.)

The da Vinci, a product of Intuitive Surgical, Inc., of Sunnyvale, California, brings telerobotics into the world of medicine. Telerobotics was originally developed to perform remote tasks by the robots. It enables a human to remain in a safe environment while directing the robots to perform complex and dangerous tasks in a remote and/or dangerous environment.

The da Vinci, being a telerobotic device, is operated by the surgeon remotely (see Figure 25.5). The robot system is composed of multiple arms manipulating various instruments inserted into the patient through tiny ports. One of the instruments is a camera with a light source to see the anatomy as well as the surgical instruments. The surgeon operates while seated comfortably (away from the patient, yet still in the operating room) at a console monitoring a 3D image of the surgical field. The surgeon's fingers grasp the master controls with hands and wrists, and the surgeon's movements are translated into the actual movements of the instruments by the robots, thus performing the surgery telerobotically.

FIGURE 25.5 da Vinci® surgery system. (Source: Intuitive Surgical, Inc.)

25.6.3 Invasive Robotic Surgery — ROBODOC® Surgical Assistant

As a final example in this chapter, we will look at the ROBODOC Surgical Assistant offered by Integrated Surgical Systems of Davis, California. The ROBODOC system is used currently for procedures that typically tend to be fully invasive type of surgical procedures—total hip replacement and total knee replacement. The system is designed to aid doctors with hip implants and other bone implants, through more accurate fitting and positioning. The advantage currently offered by ROBODOC system is accuracy, which should translate into better patient outcomes. According to Integrated Surgical Systems' own literature, a typical surgical procedure without robotic assistance will routinely leave a gap of 1 mm or greater between the bone and the implant. ROBODOC aids the surgeon in shaping the patient's bone to match the implant to within 0.5 mm.

The ROBODOC system incorporates a computer planning system combined with a five-axis robot (see Figure 25.6). The robot carries a high-speed end-milling device to do the shaping. One should note the theme of preplanning, which is pervasive in robotic surgery—given adequate information prior to the procedure (CT scans, MR scans, PET scans); a good planning component exploits the precision and degrees of freedom of a robot to offer a better technical option for the procedure.

Follow-up studies on ROBODOC cases support the fundamental thesis of robots in medicine of enhanced outcomes: better fit and positioning of the implant to the bone (based on x-ray evaluations) with fewer fractures, as one might expect based on better fit and more accurate positioning. With development of newer technology, the ROBODOC system offers the potential for performing the surgery through a very small incision [Sahay et al. 2004] of about 3 cm compared to standard incision sizes of about 15 cm. Thus, even in the area of joint surgery that is typically an invasive procedure, robotic systems offer the potential for reducing invasiveness while maintaining the advantage of precision and accuracy.

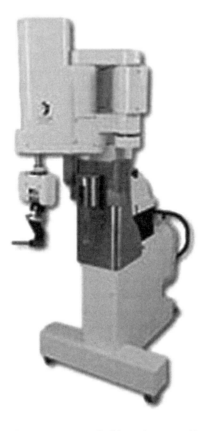

FIGURE 25.6 ROBODOC Surgical Assistant System for hip replacement. (Source: Integrated Surgical Systems)

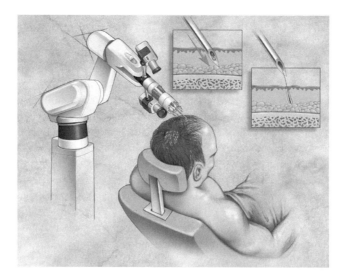

FIGURE 25.7 Artist's rendering of robotic hair transplantation system. (Source: Restoration Robotics, Inc.)

25.6.4 Upcoming Products

Robotics in medicine has been on the rise. There will be newer products that employ robots in various different practices of medicine. Two such new products that are in development are described here.

1. **Hair Transplantation Robot**: A robotic system using image guidance is being developed to perform hair transplants. Hair transplantation is a successful procedure that is performed routinely across the world. The procedure involves transplanting 1000 to 2000 individual follicular units from a donor area of the patient (back of the head) to the target area of the patient (bald spot or thinning area on the head). The procedure is highly tedious, repetitive, and prone to errors due to fatigue in the surgeon as well as the technicians. A robotic system that automates this process is being developed by Restoration Robotics, Inc., Sunnyvale, California, which will eliminate the tedium, thus enhancing the quality of the transplants (Figure 25.7).
2. **Robotic Catheter System**: A telerobotic device is being developed to guide catheters in patients. Cardiac surgery has undergone drastic changes in the past decade. There are fewer and fewer open heart surgeries being performed and most of the problems related to the heart are being addressed by delivering the appropriate treatment using catheters. These procedures have become routine in most of the hospitals. However, guiding the catheter through the patient involves tedious work for the surgeon. Furthermore, in order for the physician to observe the position of the catheter, the patient needs to be monitored using x-rays, which also exposes the surgeon while he or she is guiding the catheter. Hansen Medical, Palo Alto, California, is developing a robotic catheter system with broad capabilities as a standalone instrument or highly-controllable guide catheter to manipulate other minimally invasive instruments via a working lumen formed by the device. The system has very sophisticated control and visualization aspects to enable an operator to navigate and conduct procedures remotely with high degrees of precision. This system removes the tedium in the procedure as well as enables the surgeon to stay out of the radiation field of the x-ray machine.

Bibliography

Adler, J.R., Frameless radiosurgery, in: Goetsch, S.J. and DeSalles, A.A.F. (eds.), *Sterotactic Surgery and Radiosurgery*, Medical Physics Publishing, Wisconsin, vol. 17, pp. 237–248, 1993.

Adler, J.R., Murphy, M.J., Chang, S.D., and Hancock, S.L., Image-guided robotic radiosurgery, *Neurosurgery*, 44(6):1299–1307, June 1999.

Bodduluri, M. and McCarthy, J.M. X-ray guided robotic radiosurgery for solid tumors, *Indus. Robot J.*, 29:3, March 2002.

Carts-Powell, Y., Robotics transforming the operating room, OE Reports (SPIE), 201, September 2000.

Chenery, S.G., Chehabi, H.H., Davis, D.M., and Adler, J.R., The CyberKnife: beta system description and initial clinical results, *J. Radiosurg.*, 1(4):241–249, 1998.

Larsson, B., Leksell, L., and Rexed, B., The high energy proton beam as a neurosurgical tool, *Nature*, 182:1222–1223, 1958.

Leksell, L., The stereotaxic method and radiosurgery of the brain, *Acta Chir. Scand.*, 102:316–319, 1951.

Murphy, M.J. and Cox, R.S., The accuracy of dose localization for an image-guided frameless radiosugery system, *Med. Phys.*, 23(12):2043–2049, 1996.

Murphy, M.J., Adler, J.R., Bodduluri, M., Dooley, J., Forster, K., Hai, J., Le, Q., Luxton, G., Martin, D., and Poen, J., Image-guided radiosurgery for the spine and pancreas, *Comput. Aided Surg.*, 5:278–288, 2000.

Sahay, A., Witherspoon, L., and Bargar, W.L., Computer model-based study for minimally invasive THR femoral cavity preparation using the ROBODOC system, *Proceedings of the Computer-Aided Orthopedic Surgery Meeting*, Chicago, IL, June 2004.

Schweikard, A., Adler, J.R., and Latombe, J.C., Motion planning in stereotaxic radiosurgery, *Proceedings of the International Conference on Robotics and Automation*, vol. 9, pp. 1909–1916, IEEE Press, 1993.

Schweikard, A., Tombropoulos, R.Z., Adler, J.R., and Latombe, J.C., Treatment planning for a radiosurgical system with general kinematics, *Proceedings of the International Conference on Robotics and Automation*, vol. 10, pp. 1720–1727, IEEE Press, 1994.

Sugano, N. and Ochi, T., Medical robotics and computer-assisted surgery in the surgical treatment of patients with rheumatic diseases, www.rheuma21st.com, published April 27, 2000.

Tatter, S.B., History of stereotactic radiosurgery, http://neurosurgery.mgh.harvard.edu/hist-pb.htm, MGH Neurological Service, 1998.

World Robotics 2003, United Nations Economic Commission for Europe, October 2003.

26
Manufacturing Automation

26.1	Introduction	26-1
26.2	Process Questions for Control	26-1
26.3	Terminology	26-2
26.4	Hierarchy of Control and Automation	26-2
	History	
26.5	Controllers	26-4
	PLC: Programmable Logic Controller • DCS: Distributed Control System • Hybrid Controller • Motion Controller • PC-Based Open Controller	
26.6	Control Elements	26-6
	HMI: Human-Machine Interface • I/O: Inputs and Outputs	
26.7	Networking and Interfacing	26-9
	Sensor-Level I/O Protocol • Device-Level Networks • Advanced Process Control Fieldbuses • Controller Networks • Information Networks and Ethernet • Selection of Controllers and Networks	
26.8	Programming	26-13
	Ladder Logic Diagrams • Structured Text • Function Block Diagram • Sequential Flow Chart • IL: Instruction List • Selection of Languages	
26.9	Industrial Case Study	26-17
26.10	Conclusion	26-18

Hodge Jenkins
Mercer University

26.1 Introduction

As the global marketplace demands higher quality goods and lower costs, factory floor automation has been changing from separate machines with simple hardware-based controls, if any, to an integrated manufacturing enterprise with linked and sophisticated control and data systems. For many organizations the transformation has been gradual, starting with the introduction of programmable logic controllers and personal computers to machines and processes. However, for others the change has been rapid and is still accelerating. This chapter discusses the current state of control and data systems that make up manufacturing automation.

26.2 Process Questions for Control

The appropriate level of control and automation depends on the process to be automated. Before this can be accomplished, questions about the physical process and product requirements must be answered.

1. What types of process and product feedback are required to control the process (e.g., line speed, force, pressure, temperature, length, thickness, moisture, color)?
2. How is the process run (continuous, batch, sequential operations)?
3. What is the current level of automation (none, relay logic, programmable controller, etc.)?
4. What is the process operation schedule (single shift or 24-h operation)?
5. What cost opportunities are available from reduction of waste, improvement of quality, reduction of downtime?

The last question, which is financial, is typically the most important. When applied correctly process control and automation will rapidly pay for itself.

26.3 Terminology

AS-i	Actuator sensor interface
CAN	Controller area network
DCS	Distributed control system
DDE	Dynamic data exchange
FBD	Function block diagram
GUI	Graphical user interface
HMI	Human machine interface
IL	Instruction list
I/O	Inputs and outputs (analog or digital)
IEC 61158	International standards for controller fieldbuses
IEC 61131	International standards for programmable controllers
LLD	Ladder logic diagram programming language
OLE	Object linking and embedding
OPC	OLE for process control
P & I D	Piping and instrumentation diagrams
PID	Proportional, integral, and derivative
PLC	Programmable logic controller
SCADA	Supervisory control and data acquisition
SFC	Sequential flow chart
ST	Structured text
TCP/IP	Transmission control protocol/internet protocol
UDP/IP	User datagram protocol/internet protocol

26.4 Hierarchy of Control and Automation

Figure 26.1 depicts a hierarchy of automation and control. While the desired top level of automation depends on the goals of the project, control typically starts with the first two elements of motor control and event control. The top level of manufacturing automation is enterprise information feedback, commonly referred to as a manufacturing management information system, Figure 26.2. For some manufacturers the desired stage of automation is integrated statistical process control, for others it is feedback for product design. In either case, the design of the industrial control system is critical to overall process success.

26.4.1 History

The basic elements of motor control and event (I/O) control have long been thought of as separate actions. Early process control consisted of speed regulation of asynchronous or DC servo motors via analog or

Manufacturing Automation

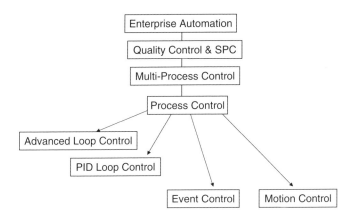

FIGURE 26.1 Hierarchy of automation and control.

relatively simple control methods. Event control was often accomplished with relay logic. Automatic control was all hardware-based, and as such it was not easily changed or improved.

As microprocessors became more prevalent and accepted in the later part of the 20th century, programmable logic controllers (PLC) were introduced and vastly improved process event control and provided the ability to easily modify a process. A separate and parallel action was programmable motion controllers. With the increasing computational power of successive versions of microprocessors, proportional, integral, and derivative (PID) control was easily implemented in these controllers. This allowed relatively easy tuning of servomechanisms. Communication between the two controller types was initially analog signals, then serial data, and most recently one of several data networks. While the first motion and process controllers were great milestones, integrated process and motion control with real-time process data availability did not appear until the late 1990s. Critical processes, such as high speed drawing of optical fiber, required tightly couple motion and process control to manufacture competitively.

Thus, modern manufacturing automation systems joined motion control and process control together for greater flexibility and control potential. Along with this improvement came newer and faster data buses,

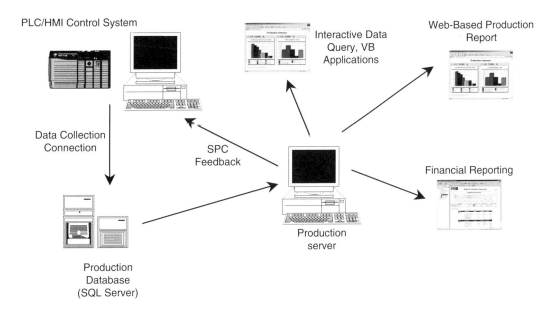

FIGURE 26.2 Manufacturing management information flow.

such as 100 Mbit Ethernet, for sharing real-time process information across the factory. This provided the opportunity to arrive at the next level of automation that uses real-time process and product measurement data for statistical process control using product quality metrics. Many controllers integrate web-based features such as TCP/IP or UDP/IP, detailed data structures, allowing easy flow of information between controllers. Often this type of networking is used in supervisory control and data acquisition (SCADA) applications. With this configuration real-time process data and product quality feedback are available to any computer or person requiring the information. Many processes can be programmed and run remotely via intranets or using secure internet logins (although remote programming and operation of processes is not necessarily a safe way of performing these tasks).

The elements of these modern manufacturing automation systems and their usage are given in the following sections. Programming and appropriate system designs for applications are also discussed.

26.5 Controllers

There are many different distinctions in the area of industrial automation controllers. The most widely used controllers are motion controllers, programmable logic controllers (PLC), distributed control systems (DCS), and PC-based control. Each controller type has special features that make it the controller of choice for different automation projects. Many of the distinctions of the various controller types are beginning to be less noticeable with each successive product generation, as options expand and pitfalls are addressed.

26.5.1 PLC: Programmable Logic Controller

The programmable logic controller (PLC) has been part of manufacturing automation for over two decades, replacing the hard-wired relay logic controllers (Figure 26.3). For smaller-scale, event-driven processes and machines with limited I/O points, stand alone PLCs are the controller of choice. PLCs are rugged, relatively fast, and low cost with excellent sequential control performance.

In the last 10 years the functionality of the PLC and systems using PLCs has been growing rapidly, with integration of networking, peripherals, and expanded programming options. The distinction between PLC

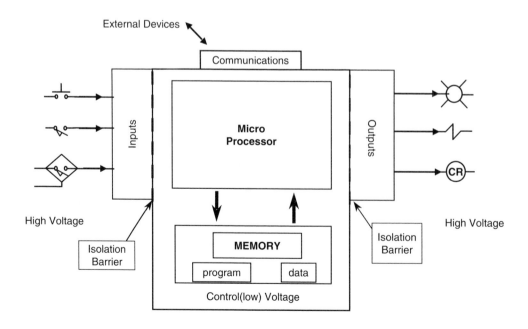

FIGURE 26.3   PLC-I/O structure.

systems and other more complex controllers (e.g., DCS) is diminishing, as PLCs are moving up in function and connectivity. Advanced PLCs and DCSs overlap each other's controller areas. Indeed, the networked PLC-SCADA systems are virtually equivalent to the larger DCS systems [1]. Programming standards and networks have also removed the limitations of PLCs. In addition to the traditional ladder logic, four other standardized programming languages are available (discussed in software).

26.5.2 DCS: Distributed Control System

Traditionally a DCS is a system of controllers linked by a data network, as a single system. Functionality, physical location, or both separate these controllers. The DCS is typically used in complex process applications where large amounts of I/O and data are required, such as chemical plants or oil refineries. They are well suited to batch processes and have an ability to handle complex interlocks and timing between operations. DCS are multitasking systems able to handle large common databases. A DCS allows for various control loops, using graphical representation of function blocks, and is easier to program than ladder logic-based PLCs. Scan rates can be more predictable than the PLCs. PLCs have scan rates dependent on the amount of I/O [2]. A DCS has safety features, redundancy, and even diagnostics built into its control philosophy to be more robust with less down time [1].

Other advantages of a traditional DCS include

Integrated system of hardware and software reduces engineering time.
Database applications are easily integrated with DCS.
Process simulation and advanced control applications are readily available.
System is designed for ease of future expansions.

26.5.3 Hybrid Controller

Some controllers define themselves as hybrid controllers, able to perform control of continuous processes (DCS) and discrete events (PLC) as well as batch and logic. The main difference is in the algorithms that are used by each controller. Hybrids can provide a combination of DCS-like infrastructure as well as SCADA functions, all while being used in process/discrete event automation. Hybrid systems are very flexible. They provide HMI/SCADA functionality, along with complex continuous and discrete control, or demanding chemical processing functions in applications with limited I/O to those with thousands of I/O points.

26.5.4 Motion Controller

Motion controllers are implemented as machine controllers when position, velocity, or other servo-control loop critical functions must occur in a process with limited I/O. Machine tools and robotics are the primary uses of these controllers. The motion controllers have built-in servo control algorithms, typically PID. Motion and process I/O data are easily integrated into control loops, as they reside on the same CPU. Some manufacturers allow user-defined interrupt service routines (ISR) for advanced control algorithms.

The motion controllers offer faster update rates for high bandwidth, servo-control loops. Motion controllers can provide servo update rates greater than 1 kHz. However, as integration of motion control into PLCs advances with new control networks, the need of using a motion controller as a process controller is diminishing for processes requiring high speed servo control with large amounts of process I/O. Motion controllers generally provide significantly fewer I/O points than PLCs or DCSs. PLC manufacturers have specific PLC/motor driver/high speed bus configuration for accessing and controlling motion within the PLC [3].

A major limitation of motion controllers as process controllers is their use of unique and proprietary programming languages. Most of these are textural languages, sometimes based on the C programming language. This limits the portability and maintainability of the controller programs.

26.5.5 PC-Based Open Controller

As personal computer (PC) operating systems have become more stable, the open architecture PC systems have made their way into factories. The drive for data automation and electronic commerce along with lower initial and operating costs, flexibility, and built-in networking ability are some of the reasons for pursuing this approach. PCs have an open systems advantage over both DCS and PLC approaches.

PC deployment in control systems occurs in several different ways. In some applications PCs are used as HMIs linked to process controllers. For other applications the PC is the controller and HMI. A PC-based controller is an integrated package of software (operating system, programs, data structures, communication protocols) and hardware (CPUs, communication/network cards, I/O device networks, I/O modules, bus resident controllers, etc). Another driving reason for PC-based control is the open architecture, allowing the linking and integration of different hardware, software, and networks. While individually excellent components can be purchased and assembled, the success of the system depends on the compatibility of both the hardware and software components [4].

Typically PC-based controls are built using fairly stable operating systems (OS), such as Windows NT/2000/XP/CE or Linux. There are three levels of control/operating systems interactions. First is relying completely on the OS for all levels of control. This is only acceptable in noncritical, low usage systems that do not require real-time control at regular update intervals. Second is the use of a real-time operating system or real-time kernel for Windows/DOS. This provides real-time control with easy access to all the standard PC components. These are sometimes referred to as software PLC applications. Both of these schemes should be avoided in critical applications including life safety systems, to avoid possible shutdowns from suspensions or system reboots. The third PC-based control approach is the use of a separate controller card on the PC bus. Here the real-time control resides in the controller card, which takes its power from the bus. All control I/O are connected through the card. The advantage of this scheme is that the PC may be re-booted without affecting the controller operation. All three PC approaches have cost savings associated with using the PC as a database, communication host, and HMI [5].

26.6 Control Elements

26.6.1 HMI: Human-Machine Interface

An integral part of the modern industrial control system is the human-machine interface (HMI). Before the widespread use of microprocessor-based graphical displays, an HMI consisted of hardwired dial gauges, strip chart and servo recorders, light indicators, and various push buttons. Modern controllers and automation systems interact with people mainly through graphical user interfaces (GUI)[6]. These GUI interfaces most often appear as touch screens with a menu driven display. Minimally HMIs display process information and controls in real-time. Examples of a main menu and subsystem menu are given in Figure 26.4 and Figure 26.5. The graphics also provide some physical system insight, as in the gas flow control menu of Figure 26.5. A low-feature HMI (some are integrated with low cost PLCs) may simply display a current rung of the ladder logic diagram.

With a PC-based HMI, powerful graphics can be used. Historical and real-time data can be displayed, using the databases, file structures, and graphics capabilities of a PC. The use of dynamic data exchange (DDE), multiple open windows, and other standard PC platform software components, such as OPC, augments its functionality [7]. Control parameters and functions are integrated into the HMI screens for easy use by operators and control engineers. Some displays mimic the analog devices they replaced (dial gauges, LED displays, indicator lights, etc.). Other features include charting and trending of data for quality control. Animations are also available for better display process status. System status and other diagnostic tools are also presentable. Using data monitoring with alarms, an HMI may also serve as a supervisory control and data acquisition system (SCADA).

Manufacturing Automation

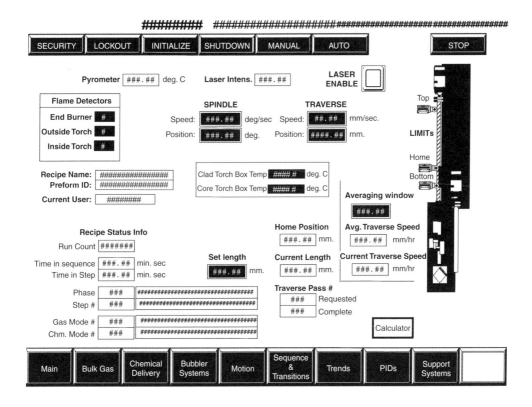

FIGURE 26.4 HMI main menu example.

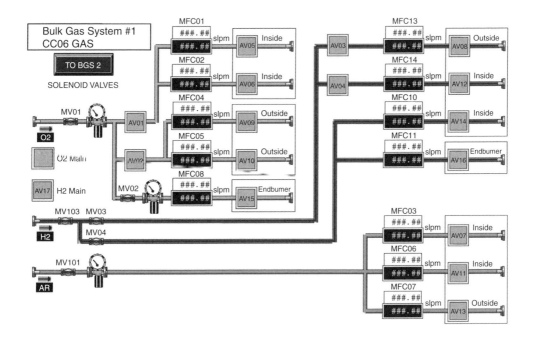

FIGURE 26.5 HMI gas delivery sub-system menu example.

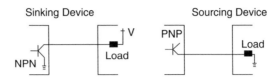

FIGURE 26.6 Sinking and sourcing of I/O devices.

HMIs can be devices connected directly to the controllers or to database servers that pass information back and forth to the process controller via several network protocols. In fact may HMIs are run in a Web-browser windows, allowing remote viewing and control.

26.6.2 I/O: Inputs and Outputs

To interface with a physical process a PLC, DCS, or PC-based controller must have inputs and outputs. Inputs and outputs (I/O) were initially contact relays and instrument circuits. Today nearly all I/Os are optically isolated, solid-state devices, with many specializations. The optical isolation prevents any device voltage or current from damaging the controller. To verify an event or the presence of an object at a particular location, a digital input is used; this is simply the completion of circuit, typically a threshold voltage. Digital outputs provide a voltage to turn on off devices, such as a solenoid.

The digital I/O are designated by voltage thresholds, AC or DC current (typically 24 VDC or 120 VAC), and whether the current flow to the I/O unit is sinking or sourcing. Sinking and sourcing refer to the current flow into the input or output switch (transistor). Note that this is only relevant for DC voltage and is not associated with AC currents. All inputs require a voltage source and a load to operate. A sinking input (NPN transistor: Figure 26.6) requires the voltage and load to be present before connecting it to the circuit. This means that it is "sinking" the current to *ground* for the circuit. A sourcing input (PNP transistor: Figure 26.6) must be before the load in the circuit. This means that it is "sourcing" the current to the circuit, from a positive reference voltage. Voltage and a load must be present in either situation to detect a voltage change at the input. A similar logic follows for sinking or sourcing outputs. A sourcing I/O device is generally considered safer than sinking, as supply voltage is not always present at the device. Although, using proper maintenance safety, both types of I/O are safe. Typical digital inputs are push buttons, selector switches, limit switches, level switches, photoelectric sensors, proximity sensors, motor starter contacts, relay contacts, thumbwheel switches. Typical digital outputs are valves, motor starters, solenoids, control relays, alarms, lights, fans, horns, etc.

Analog inputs and outputs are more diverse than digital inputs and outputs. Many additional parameters are used as these devices are used to measure an output voltage and current. For long distances, the current loop devices (generally 4–20 mA) are preferred over the voltage I/O (most commonly 0–5 V) for accurate measurements without line losses. Special purpose analog inputs are available for ease in connecting various types of thermocouples and resistance temperature transducers (RTD).

Apart from the type of analog signal, the resolution of the signal is the most important consideration. Most applications are 8, 16, or 32 bit resolution for scaling the range of the particular I/O unit. So for maximum resolution chose an input or output range that matches the output of the sensor. In 16-bit applications, one bit is used to denote positive or negative leaving 2^{15} bits for data. For example, a -10 to $+10$ V analog input would result in a resolution of 0.3 mV regardless of the magnitude of the signal.

I/O are linked to the controllers via many routes. All controllers have module connections that are either integral to the controller or connected to the main data bus, for example, the backplane of PLCs. Different I/O modules can be connected for the specific applications (analog, digital, inputs, outputs). There are a limited number of module configuration slots and, thus, limited I/O points with these direct connections, as specified by the manufacturer. Devices are hardwired to the I/O point on the modules and must be configured on the controller. Many controllers allow remote I/O racks with hardwiring back to the controller. Recently device-level networks and field buses have emerged to address problems associated with a central controller with detached I/O racks and other devices. This allows the reduction of control wiring with remote I/O and improved diagnostics with intelligent devices. These networks are discussed in the following section.

26.7 Networking and Interfacing

As networking information technology has evolved and become more robust, it was a logical step for process automation to take advantage of these technologies. Communication networks abound at many levels in modern industrial automation. Figure 26.7 depicts several levels of networking in a typical manufacturing control system. Several somewhat distinct levels of connections and control can be visualized and coexist: sensor/device-level, process control, controller-level, and information-level networks. (Control networks below the information-only level can be referred to as fieldbuses.) As protocols and fieldbus technologies advance, the differences between the sensor, device, process, and controller levels is becoming fuzzy. This subsection on control networks and communication highlights some of the technology in manufacturing today and is not intended as a comprehensive discussion.

26.7.1 Sensor-Level I/O Protocol

The transition from hardwired I/O to direct network connection is rapidly occurring. The first level of networking in an industrial control system is between the controller and its sensors. Sensor-level networks are generally used in smaller systems with limited I/O.

The highway addressable remote transducer (HART) sensor-level communications protocol is an open protocol developed in the 1980s for communication at the I/O level. It was one of the first protocols for

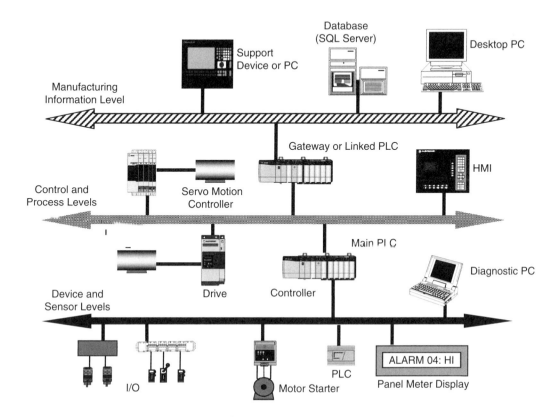

FIGURE 26.7 Multi-bus system architecture.

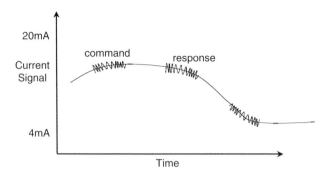

FIGURE 26.8 HART protocol.

device/sensor networking. It supports two-way digital communications between a HART I/O module and a HART capable device. The protocol is designed to complement a standard 4–20 mA analog signal, requiring no additional wires. HART uses a frequency modulation (e.g., 1200 Hz = "1", 2200 Hz = "0") with small amplitude (0.5 mA) on top of a 4–20 mA analog signal for device communication (Figure 26.8). This provides exceptional opportunities for sending device data to the controller, allowing for device identification, status, and diagnostics. While this protocol allows for data communication between devices, it has no more or less wiring than standard 4–20 mA devices. The current HART protocol provides for 15 nodes (devices) per loop [8, 9].

Another current sensor bus protocol present in industry is the actuator sensor interface (AS-i) protocol, [10]. AS-i was developed in Germany in 1994 by a group of industrial automation equipment suppliers to be a low-cost method for addressing discrete sensors in factory automation applications. The physical network layer is a balanced differential voltage (30–24 VDC). Each AS-i 2.0, network can support up to 31 devices or nodes (248 I/O points). The AS-i, version 2.1, specification doubles the maximum allowable nodes to 62 devices per segment for a total network capacity of 434 I/O points.

These sensor networks are cost effective for small systems, using master/slave communication techniques. However, modern manufacturing automation often requires more vertical integration of controllers, sensors, devices, and instruments, using peer-to-peer and publisher-subscriber techniques for higher communication bandwidth. More extensive industrial applications with larger I/O potential and more diverse devices require device networks in lieu of the node-limited sensor level networks.

26.7.2 Device-Level Networks

Device networks are more general and expandable than the lower level sensor networks. They connect a wider range of devices, including all I/O, motor drives, and display panels. Devices may be connected into the controller I/O rack or remotely via device-level networks using one or more of the variety of the available data buses. There are two driving forces behind device-level networking of the I/O units. One is the reduction of the wiring limitations of previously hard-wired remote I/O chassises. The second is the ability to have more device-level information to ease configuration and diagnostics. While complex in function, these network compatible devices are more easily set up than the hard-wired devices.

There are numerous device-level networks and fieldbuses already in use. For example, DeviceNet (Allen-Bradley/Rockwell Automation) and Profibus DP (Siemens) are in wide use. Each has its own rationale for use, much of which depends on the selected controller. Many recent efforts have focused on using Ethernet at the device level. While this is not the dominant approach for lower level networks, it is gaining interest. Ethernet does have a large presence at the controller-level and information networks.

The DeviceNet network is an open, low-level network that provides connections between simple industrial devices (sensors and actuators) and higher-level devices (PLCs and PCs). It currently has the largest installed base of all device networks [11]. DeviceNet is an open device network using controller area network (CAN) protocol to provide the control, configuration, and data collection capabilities for industrial devices. DeviceNet can currently support up to 253 individual nodes at data rates of 125 kb at a maximum distance of 500 m, or 500 kb at shorter distances (without repeaters). The physical layer of CAN protocol uses a differential voltage.

Another device network with a significant installation base is Profibus-DP. Profibus-DP is a device-level bus that supports both analog and discrete signals. Profibus-DP has widespread usage for such items as remote I/O and motor drivers. Profibus-DP communicates at speeds from 94 kbps over distances of 1200 m to 12 Mbps up to 100 m, without repeaters. The physical layer of Profibus-DP is based on RS-485 communication [9,12]. Neither Profibus-DP nor DeviceNet is designed for intrinsically safe applications.

26.7.3 Advanced Process Control Fieldbuses

Process control networks are the most advanced fieldbus networks in use today. They provide connectivity of sophisticated process measuring and control equipment. Modern process control networks can be easily deployed for new or existing process equipment and provide more complex functionality than device networks. The advanced characteristics of the host interfaces and connectivity aid in the ease of configuration and start up.

While the MODBUS (Modicon PLCs) was one of the first process control field buses, Foundation Fieldbus and Profibus-PA are two dominant advanced process control fieldbuses/protocols. (Note: Profibus-PA is distinct from Profibus-DP.) Foundation Fieldbus and Profibus-PA are full-function fieldbuses for process-level instrumentation that has a physical layer based on IEC-61158, low-speed voltage mode [13]. Both provide connectivity of sophisticated process measuring and control equipment. Each provides communication at distances up to 1900 m and is intrinsically safe, with low current on the control wiring. Foundation Fieldbus has data transmission rates of 31.25 kbps, while Profibus-PA communicates at 93.75 kbps. Foundation Fieldbus and Profibus-PA (generally linked with Siemens) are market leaders for process fieldbuses, with significant installation bases in the U.S. and Europe, respectively.

When using a device-level network or process control fieldbus, sensors or other I/O devices can be added anywhere by tapping into the trunk line with a multi-wire connection. Devices can be configured over network. Many diagnostics are available along with large amounts of I/O data. Simple sensors to sophisticated devices, all work on the same network.

The key benefits of device and process fieldbuses are

- Less wiring and installation time
- Few wiring mistakes
- Remote and quick setup of devices
- Diagnostics for predictive failure warnings and rapid troubleshooting
- Interoperability of devices from multiple vendors
- Control I/O and information data transmitted together
- Configuration of devices over the network

26.7.4 Controller Networks

Interlinking of multiple controllers in a manufacturing process has grown rapidly with the demand for tighter real-time process control. Often several controllers are linked to PLCs, PCs, and motion controllers across several subprocesses. The controller level network requirements include deterministic data flow, repeatable (time) transfers of all operation-critical control data, in addition to supporting transfers of non-time-critical data. I/O updates and controller-to-controller interconnects must always take priority over program file transfers and other communication. Transmission times must be constant and unaffected by devices connecting to, or leaving, the network. In the controller-level arena, several popular fieldbuses are viable, including ControlNet, Foundation Fieldbus, Profibus-FMS, and Ethernet. Independent organizations support and maintain these network standards [12,14,15]. Ethernet using TCP/IP or UDP/IP is a strong contender and is rapidly gaining popularity.

Rockwell Automation conceived ControlNet as a high-level fieldbus network. It uses the control and information protocol (CIP) to provide dual functionality of an I/O network and a peer-to-peer network. Up to 99 nodes can be networked with data transfer rate of 5 Mbps at distances up to 5 km. Multiple PLCs can control unique I/O devices (multimaster) or share input devices on the same wire interlocking of PLCs, and this is accomplished using a producer/consumer network model (a.k.a. publisher/subscriber). ControlNet was designed with a time-slice algorithm that manages these functions on one network without affecting performance or determinism.

Profibus-FMS is a control bus generally used for communications between DCS and PLC systems most widely used with Siemens PLCs. It was meant to be a peer-to-peer connection between Profibus masters, exchanging structured data. The Profibus-FMS network is positioned in the architecture similarly to ControlNet. Profibus-FMS has various data rates available, from 9.6 kbps at a maximum distance of 19.2 km to 500 kbps for distances less than 200 m.

26.7.5 Information Networks and Ethernet

With several integrated IP protocols, Ethernet is the most popular network at multiple system levels. Control manufacturers select Ethernet to avoid many of the proprietary or custom networks. Ethernet provides interoperability among products from various vendors. 100 Mbps Ethernet is becoming a logical choice for controller networks as well as for information networks. Ethernet provides high standard data transfer rates, efficient error correction, and industrial Ethernet switches for better determinism. A typical architecture is seen in Figure 26.9. Ethernet also brings with it a large available IT expertise. Many control manufacturer have selected Ethernet for connecting remote I/O rack locations to controllers, effectively making Ethernet a process control fieldbus. There is even a push downward for device-level Ethernet.

There are many competitors to Ethernet including Modbus/TCP, ProfiNet, HSE Fieldbus, and many other proprietary protocols. Arguments against using Ethernet in factory floor automation often stress that Ethernet lacks the level of robustness and determinism needed in control applications. However, recent developments of intelligent switches have largely discounted this argument [16].

Along with the high data transfer rates, Ethernet also provides superior technology compared with others networks, in terms of openness, ease of access to and from the Internet, ease of wiring, setup, and maintenance. A new open application layer protocol, Ethernet/IP, has been created for the industrial environment. It is built on the standard TCP/IP protocols and takes advantage of commercial off-the-shelf Ethernet integrated circuits and physical media. IP stands for "industrial protocol." Ethernet/IP is supported by several networking organizations: ControlNet International (CI), Industrial Ethernet Association (IEA), Open DeviceNet Vendor Association (ODVA), and Industrial Open Ethernet Association (IOANA) [11].

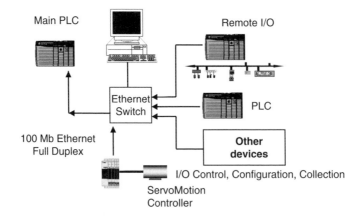

FIGURE 26.9 Control and interconnecting over Ethernet.

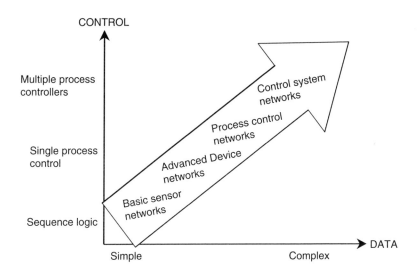

FIGURE 26.10 Fieldbus capabilities.

26.7.6 Selection of Controllers and Networks

The selection of controllers and networks is usually a function of the process specifications, existing controller base, and familiarity of personnel with specific controllers. When starting from scratch, the answers to the questions of Section 26.2 will provide guidance. Process familiarity in terms of the physical functions and requirements will greatly augment the piping and instrumentation diagrams (P&I D) and written specifications. Data and information flow is important in determining the appropriate network configuration and hardware (Figure 26.10). Data transmission speed, information complexity, and distance limits are critical design limits for selection of controllers and associated fieldbuses. There is no unique solution. Often there are several viable alternative configurations.

26.8 Programming

Software is the center of great debate and change. Ladder logic, rung ladder logic, or ladder logic diagrams (LLD) have long been the main mode of PLC programming as the diagrams graphically resemble the relay logic they originally replaced. Other styles of programming have evolved to augment computational abilities and ease program development. In addition to LLD, four other major programming paradigms are in use in PLCS and DCS systems. These are structured text (ST), function block diagrams (FBD), sequential flow charts (SFC), and instruction lists (IL). Special purpose motion controllers with augmented processing logic use proprietary text-based languages. While powerful, these unique control languages are not transportable to other controllers and make replacement difficult.

As particular controller vendors often customize their programming functions, it has become more important to have a programming standard for the five prevalent process control languages. One major reason is so one does not need to learn different instruction sets for different PLC manufacturers. The International Standard IEC 61131 is a complete collection of standards on programmable controllers and their associated peripherals. Section 26.4 of the standard defines a minimum instruction set, the basic programming elements, and syntactic and semantic rules for the most commonly used programming languages [17].

The standard was originally published in 1993, with a recent revision published in 2003. Guidelines for application and implementation of programming languages are presented in IEC 61131-8, a related document published in 2000 [18]. Many products already support three of the five languages. Vendors are not required to support all five languages in order to be IEC 1131-3 compliant. Note that the International

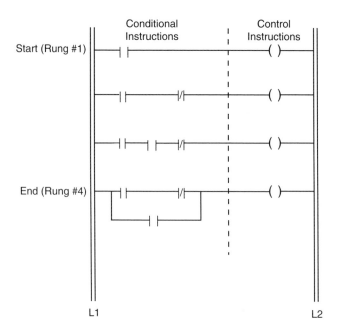

FIGURE 26.11 Ladder diagram.

Electrotechnical Commission (IEC) is a worldwide standardization body. Nearly all countries over the world have their own national standardization bodies, that have agreed to accept the IEC approved and published standards.

26.8.1 Ladder Logic Diagrams

The ladder logic diagram (LLD) is the most common programming language used in PLC applications. LLD is a graphical language resembling wiring diagrams, as seen in Figure 26.11. Each instruction set is a rung on the ladder. Each rung is executed in sequential order for every control cycle. A rung is a logic statement, reading from left to right (Figure 26.12). Rungs can have more than one branch, as seen in the 4th rung of Figure 26.11.

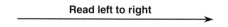

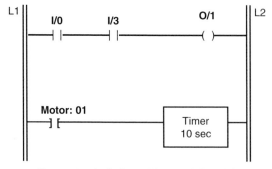

FIGURE 26.12 Reading ladder logic diagrams.

```
;************* force control loop ****************
WHILE (M362>Q5 AND M362<Q4)      ;begin loop
        P6 = (M156*Q2)            ;DESIRED FORCE (N * force bits/N)
        P15 = M1001               ;GET WEIGHT
        P8 = P15–P14              ;TARED FORCE
        P7 = P8–P6                ;FORCE ERROR
        P9 = P9+P7                ;INTEGRAL ERROR

        IF (P9>Q8)
                P9 = Q8           ;LIMIT INTEGRATION
        ENDIF
        IF (P9 < (–Q8))
                P9 = –Q8
        ENDIF
                                  ; UPDATE GAINS
        Q6 = M153*P20             ;40 nom PROPORTIONAL GAIN
        Q7 = M154*P20             ;80 nom INTEGRAL GAIN
        X0Y0Z ((P7*Q6+P9*Q7))     ;COMMAND MOVE

ENDWHILE                          ;end of loop
```

FIGURE 26.13 Structured text example.

As ladder logic was the first replacement for relay logic, it is most easily suited to Boolean I/O variables. However, over many years of use and refinement, the current LLD instruction set is quite large, reducing the need for other languages. Many special instructions are available for motion control, array handling, diagnostics, serial port I/O, and character variable manipulation. LLD is used in continuous and batch processes. Both off-line and on-line editing of individual rungs is possible, allowing changes to a running system (on the next cycle). User interfaces allow off-line programming and reviewing of LLD code at either PLC display panels or HMIs, or at remotely via PC software.

26.8.2 Structured Text

Structured text (ST) is a high-level programming language similar to BASIC, Fortran, Pascal, or C. Most engineers are familiar with structured programming languages. Structured text has standard program flow control command statements such as If/Then, Case, Do/While, Do/Until, and For/Next constructs. An example of a structured text is given in Figure 26.13. Most LLD, SFC, and FBD instructions are supported in ST.

While ST provides great flexibility in programming and editing, it is somewhat more difficult to read and follow logic flow, as compared with the more graphical languages. Thus, it is less maintainable over time.

26.8.3 Function Block Diagram

Function block diagram (FBD) is a highly visual language and is easy to understand because it resembles circuit diagrams. The graphical free-form programming environment is easy for manipulating process variables and control. A typical function block diagram is given in Figure 26.14. The foundation of the FBD program is a set of instruction blocks with predefined structures of inputs and outputs from each block. Different blocks are placed and connections are drawn to pass parameters or variables between blocks. Blocks are positioned and organized, based on the specific application, to improve readability.

Although simple instructions can take as much program space as complex instructions, the architecture simplifies program creation and modification. It is ideal for analog control algorithms, as it graphically represents control loops and other signal conditioning devices. This language is commonly found in distributed control systems (DCS) with continuous or batch process control.

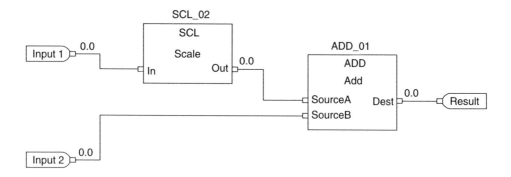

FIGURE 26.14 Function block diagram example.

26.8.4 Sequential Flow Chart

Sequential flow chart (SFC) is another highly visual programming language yielding applications that are easy to create and read. SFC is a graphical flowchart-based programming environment. Logical steps (blocks) are placed on the visual layout in an organized manner. Connections are drawn to determine execution flow of the program. An example is provided in Figure 26.15. This figure shows two branches from the start block of the program. Note: Blocks are descriptively labeled for case in following program logic. I/O and other fuctions are embedded within each block. Multiple branches allow for transitional and simultaneous execution flow. Position and organization of blocks are used to increase readability. Floating or linked text boxes provide application documentation (comments). Embedded structured text in action blocks directly improves readability and maintenance, while reducing the number of subroutine calls.

The flow chart structure is ideal for sequencing of machine states (e.g., Idle, Run, Normal, Stop). High-level program/subroutine flow management provides a more flexible approach to developing process sequences. SFC is ideal for machines with repetitive operations and batch processing.

26.8.5 IL: Instruction List

Instruction list (IL) is the most basic-level programming language for PLCs. All languages can be converted to IL, although it is most often used with LLD. IL resembles assembly language, using logical operator codes, operands, and an instruction stack. This language is difficult to follow and is not typically selected. The major application of IL is found in hand-held program for nonnetworked PLCs.

26.8.6 Selection of Languages

The language of choice is typically based on an organizational standard, existing code base, or personal preference. Typically a graphical language is preferred. However, different languages are preferred in specific instances. Ladder logic and function block diagrams are best suited for continuous operations, while sequential flow charts are more suited to operations using states and events. Structure text is suitable for either. When the majority of process variables are Boolean (on/off), ladder logic diagrams, sequential flow charts, and structure text are favored over function block diagrams

Generally, one language is selected for the entire process control project. However, this is no longer required. Controllers that follow the IEC 61131 standard allow multiple languages to be used within the same programmable controller. This allows the programmer or control engineer to select the language best suited to each particular task.

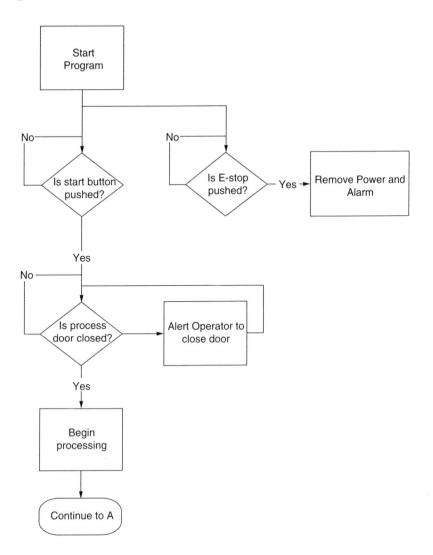

FIGURE 26.15 Sequential flow chart example.

26.9 Industrial Case Study

While it is impossible to completely detail an industrial control system, this section reviews a system architecture and layout for a specific industry case. The facility discussed is a chemical processing facility for creation of high purity materials. The process line is specified for development of process and product improvement. This requires interfacing of many new sensors and control algorithms on a continual basis. The process uses hazardous (reactive and combustible) chemicals. Thus, chemical leak monitoring must occur 24 hours a day. The control system is separated into two major subsystems: life safety and process control. The architecture is depicted in Figure 26.16.

The life safety monitoring system (LSS) must have a high degree of availability and redundancy as this is classified as a SIL3 (Safety Integrity Level 3) process [18, 19]. This is based on the potential consequence of injury to one or more people with a moderate risk for chemical leaks. A specifically designed life safety PLC (SIL1-3) with redundant processors and networking connections is selected. In the event of a fault in the control system, redundant control components take over and chemical leak monitoring continues.

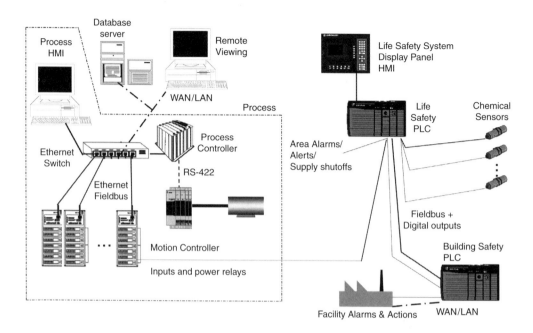

FIGURE 26.16 Chemical process control and life safety systems.

A touch screen panel HMI is provided for status of alarms (no networking is provided through the HMI). Outputs of the LSS go to several places. In the event of a detected leak, outputs trigger alarms in the process area, starting safety measures (e.g., exhaust fans, bells, lights), removing power from chemical supply valves, signaling the process controller to shutdown, as well as alerting building life safety main controller for additional actions.

The separate controls for process and safety allow flexibility in the design and operation of the process. The process has several hundred I/O points, both analog and digital. Since this process is state-driven, a controller and software based on flow charts (SFC) is used [*]. Ethernet is selected as the process control/device bus for remote I/O controllers and HMI interfaces using a switch for collision control. While a direct parallel bus for I/O can be faster, the use of Ethernet allows for greater future system expansion and longer distances between I/O controllers. Motion control is accomplished via a separate (legacy) controller. Because process motions are not time critical (time constants for the process are in terms of minutes), RS-422 serial communication between the process controller and the motion controller is more than sufficient.

Data are recorded for batch processing locally and via data servers on the WAN/LAN. Local and remote PC-based HMIs allow viewing and operation of the process via the Ethernet connections to the process controller. HMI and process control software are provided from the hardware vendor in this case, yielding an integrated approach. The local HMIs also serve a SCADA function by monitoring and recording events and alarms.

26.10 Conclusion

This is only a brief overview of the ever-changing modern manufacturing automation. Other automation aspects include dynamic linking of I/O point data from tag servers and database servers. These tasks require interfacing organizations in a manufacturing enterprise. Manufacturing automation is rapidly integrating control, manufacturing, and business functions as manufacturers pursue improvements of products, processes, and profits in real-time.

References

[1] Bob Waterbury, DCS, PLC, PC, or PAS?, *Control Eng.*, p. 12, July 2001.
[2] Geller, D.A., *Programmable Controllers using the Allen-Bradley SLC-500 Family*, Prentice Hall, Upper Saddle River, NJ, 2000.
[3] Piyevsky, S., Open network and automation products, *Allen-Bradley Automation Fair*, Anaheim, CA, 21 November 2002.
[4] Fielder, P.J. and Schlib, C.J., Open architecture systems for robotic workcell integration, *IWACT 1997 Conference Proceedings*, Columbia, OH, 1997.
[5] Soft PLC Overview, URL: http://www.softplc.com/splcdata.htm.
[6] Mintchell, G.A., HMI/SCADA software-more than pretty pictures, *Control Eng.*, 49, 18, December 2002.
[7] OPTO22 Factory Floor Software, v 3.1,D, *OPTODisplay User Guide*, Form 723-010216, OPTO22, 2001.
[8] Meldrum, N., ControlLogix® and HART protocol an integrated solution, *Spectrum Controls*, 2002.
[9] Fieldbuses, look before you leap, *EDN*, p. 197, 1998.
[10] URL: http://www.as-interface.com, 2003.
[11] Open DeviceNet Vendor Association (ODVA), URL: http://www.odva.org, 2003.
[12] Profibus International, URL: http://www.profibus.org, 2003.
[13] IEC 61158, Digital data communications for measurement and control — Fieldbus for use in industrial control systems — Part 1: Overview and guidance, IEC, Geneva, 2003.
[14] ControlNet International, URL: http://www.controlnet.org, 2003.
[15] Foundation fieldbus, http://www.fieldbus.org, 2003.
[16] Lee, K.C. and Lee, S., "Performance evaluation of switched Ethernet for real-time industrial communications," *Computer Standards Interfaces*, vol. 24, no. 5, pp. 411–423, November 2002.
[17] IEC 61131-3, Programmable controllers — Part 3: Programming languages, IEC, Geneva, 2003.
[18] IEC 61508-1, Functional safety of electrical/electronic/programmable electronic safety-related systems — Part 1, IEC, Geneva, 1998.
[19] ANSI/ISA-S84.01-1996, Application of safety instrumented systems for the process industries, Instrument Society of America S84.01 Standard, Research Triangle Park, NC 27709, February 1996.

Index

A

A465, **11**-4
AABB, **23**-18
ABB, **1**-8
Abbé error (sine error), **13**-5f
Abbé principle, **13**-4–5
Absolute coordinates
    of vector x, **2**-3
Absolute coordinate system, **20**-3f
Absolute encoders, **12**-3
    example, **12**-3f
Acceleration control for payload limits, **11**-18
Accelerations, **4**-9, **12**-9–10
    of center of mass, **4**-6
    online reconstruction of, **14**-9–10
Acceptance procedures, **10**-2
Accuracy, **13**-3f
    definition of, **13**-2–3
AC&E's CimStation Robotics, **21**-7, **21**-8
ACS, **24**-36f, **24**-37f
Active touch, **23**-9, **23**-11
Activity of force F, **6**-4
Activity principle, **6**-4
Actuator forces, **19**-2f
Actuators, **12**-12–18, **13**-17
ADAMS
    Kane's method, **6**-27
Adaptive command shaping (ACS), **24**-36f, **24**-37f
Adaptive feedback linearization, **17**-16–18
Adjoint
    Jacobian matrices, **2**-12
Adjoint transformation, **5**-3
Admittance regulation
    vs. impedance, **19**-9–10
Advanced feedback control schemes, **24**-29–31
    with observers, **24**-30–31
    obstacles and objectives, **24**-29–30
    passive controller design with tip position feedback, **24**-31
    sliding mode control, **24**-31
    strain and strain rate feedback, **24**-31
Advanced process control fieldbuses, **26**-11
Affine connection, **5**-10

Affine projection, **22**-4
AI, **1**-5
AIBO, **1**-11
AIC, **1**-5
Aliasing, **13**-9–10
    frequency-domain view of, **13**-10f
Alignment errors, **13**-4–5
Al Qaeda, **1**-10
Ambient temperature, **10**-2
American Machine and Foundry, **1**-7
AMF Corporation, **1**-7
Analog displacement sensors, **12**-4–5
Analog photoelectric, **12**-7
Analog sensors, **12**-4–10, **13**-18–19
    analog filtering, **13**-19f
Analog-to-digital conversion, **13**-11
Analyzing coupled systems, **19**-8–9
Angular error motions, **10**-6t, **10**-9f
Angular velocity
    and Jacobians associated with parametrized rotations, **2**-8–10
ANSI Y14.5M, **10**-3
Anticipatory control, **23**-12–13
Approximations, **24**-25
ARB IRB1400, **17**-2f
Aristotle, **23**-10
ARMA, **14**-13
Arm controller
    robot end effector integrated into, **11**-4f
Arm degrees of freedom augmentation, **24**-39–41
    bracing strategies, **24**-39
    Inertial damping, **24**-40
    piezoelectric actuation for damping, **24**-41
Articulating fingers, **11**-11
Artificial intelligence (AI), **1**-5
Artificial Intelligence Center (AIC), **1**-5
ASEA, Brown and Boveri (ABB), **1**-8
ASEA Group, **1**-8
Asimov, Isaac, **1**-3–4, **1**-4, **1**-6
Asimov, Janet Jeppson, **1**-4
Asimov, Stanley, **1**-4
Assembly task
    two parts by two arms, **20**-10
Augmented dynamics-based control algorithm, **20**-7, **20**-7f

I-1

Augmented reality, **23**-3
AUTOLEEV
    Kane's method, **6**-27
Automated system
    forming leads on electronic packages, **10**-13f
    leads location, **10**-14f
Automatic calculator invention, **1**-2
Automatic rifle, **1**-2
Automatic symmetry cell
    detection, matching and reconstruction, **22**-18–21
Automaton, **1**-3
Autoregressive moving-average (ARMA), **14**-13
Axis, **5**-3
Axis-aligned bounding boxes (AABB), **23**-18
6-axis robot manipulator with five revolute joints, **8**-13

B

Babbage, Charles, **1**-2
Backward recursion, **4**-2
Ball races, **12**-13
Bar elements
    distributed, **24**-15
Bares, John, **1**-7
Bargar, William, **1**-10
Bars and compression, **24**-5
Base frame, **2**-3, **17**-3
Base parameter set (BPS), **14**-5
    batch LS estimation, **14**-7–8
    element estimation, **14**-7–8
    estimation, **14**-19–21
    online gradient estimator, **14**-8
Batch LS estimation
    of BPS, **14**-7–8
BBN criteria, **13**-15
Beam elements in bending
    distributed, **24**-15–16
Beams and bending, **24**-6–7
Bending deformation
    geometry of, **24**-6f
Bending transfer matrix, **24**-16f
Bernoulli-Euler beam model, **6**-21
Bernoulli-Euler beam theory, **6**-16
Bezout identity, **17**-14
Bilateral or force-reflecting teleoperator, **23**-2
Body, **5**-3–4
Body-fixed coordinate frame, **5**-1
Body manipulator Jacobian matrix, **5**-5
Bolt Beranek & Newman (BBN) criteria, **13**-15
Bond graph modeling, **4**-2
BPS. *See* Base parameter set (BPS)
Bracing strategies
    arm degrees of freedom augmentation, **24**-39
Bridge crane example, **9**-4–6
Broad phase, **23**-18–19
Brooks, Rodney, **1**-10
Brown Boveri LTD, **1**-8
Buckling, **24**-7–9
Building
    reconstruction, **22**-21f

C

Cable-driven Hexaglide, **9**-1
Cable management, **13**-7
CAD and graphical visualization tools, **21**-1
Cadmus, **1**-1
Calibration cube
    four images used to reconstruct, **22**-12f
    two images, **22**-7f
    two views, **22**-7f
Camera calibration, **22**-4
Camera model, **22**-2–3
Camera poses
    cell structure recovered, **22**-21f
CAN, **26**-10
Capacitive displacement sensors, **12**-5–6
    distance and area variation in, **12**-6f
Capek, Jose, **1**-3
Capek, Karel, **1**-3
Carl Sagan Memorial Station, **1**-9
Carnegie Mellon University, **1**-7
Cartesian error, **15**-22f
Cartesian manipulator
    stiffness control of, **16**-5–6
Cell structure recovered
    camera poses, **22**-21f
Centrifugal forces, **4**-8
Centrifugal stiffening, **6**-14
Characterizing human user
    haptic interface to virtual environments, **23**-5
Chasles' Theorem, **2**-5, **2**-6, **5**-3
Chatter free sliding control, **18**-4–6
Chemical process control, **26**-18f
Christoffel symbols, **5**-8, **5**-10
    of first kind, **17**-5
CimStation Robotics, **21**-2
CimStation simulated floor, **21**-2f
Cincinnati Milacron Corporation, **1**-8
Closed-form equations, **4**-7–8
Closed-form solutions
    vs. recursive IK solutions, **14**-18f
Closed kinematic chains, **24**-10
Collision detection, **23**-17, **23**-18–19
Collision detector, **23**-17
    flowchart, **23**-18f
Collision sensors, **11**-17
Column buckling, **24**-8
Combinations of loading, **24**-7–9
Combined distributed effects and components, **24**-16
Command generation, **9**-4
Command shaping filter, **24**-34
Common velocity
    bond graph, **19**-8f, **19**-9f
    feedback representation, **19**-8f, **19**-9f
Compensation based on system models, **23**-15
Compliance based control algorithm, **20**-6, **20**-6f
Compliant support of object, **20**-8f
Composition of motions, **2**-5
Compressed air, **11**-8
Compression
    and bars, **24**-5

Computational complexity reduction, **24**-27
Computed torque, **17**-8
Computed-torque control design, **15**-5–6
Computejacobian.c, **3**-18, **3**-23–24
Conductive brushes, **12**-15
Configuration, **5**-2
   infinite numbers
      with none, **3**-3f
      with one, **3**-3f
Configuration space, **17**-3
Consolidated Controls Corporation, **1**-5
Constrained Euler-Lagrange equation
   geometric interpretation, **5**-12
Constrained layer dampers, **13**-15
Constrained systems, **5**-11–13
Constraint(s), **13**-6
   Kane's method, **6**-14
Constraint connection, **5**-12
Constraint distribution, **5**-12
Constraint forces and torques
   between interacting bodies, **7**-15–16, **7**-15f
Contents description, **24**-2
Continuously elastic translating link, **6**-17f
Continuous motion, **22**-8
Continuous system
   Kane's method, **6**-16
Control, **24**-27
Control algorithms, **13**-19–21
Control architecture, **17**-7
Control bandwidth, **15**-2
Control design, **16**-5–6, **16**-6-8, **16**-12-14
   with feedback linearization, **15**-6–10
   method taxonomy, **17**-6-8
   μ-synthesis feedback, **15**-16–19
Control effort
   tracking of various frequencies
      with feedforward compensation, **9**-20f
      without feedforward compensation, **9**-17
Controller(s)
   experimental evaluation, **15**-19–21
   implementation, **13**-16–17
   networks, **26**-11–12
   selection of, **26**-13
Controller area network (CAN), **26**-10
ControlNet, **26**-11, **26**-12
Control system design, **17**-0
Conventional controllers
   bode plots of, **15**-14f
Coordinated motion control
   algorithm, **20**-7–9
   based on impedance control law, **20**-7–10
   of multiple manipulators
      for handling an object, **20**-5–7
      problems of multiple manipulators, **20**-5–7
Coordinate frames, **8**-3, **8**-13
   schematic, **8**-3
Coordinate measuring machine
   deflection of, **9**-3f
Coordinate systems, **20**-3f
   associated with link n, **4**-3f
Coriolis centrifugal forces, **5**-8

Coriolis effect, **4**-7
Coriolis force, **4**-8
Coriolis matrix, **5**-8
Corless-Leitmann approach, **17**-14
Correlation among multiple criteria, **10**-13–14
Cosine error
   example of, **13**-4f
CosmosMotion, **21**-10
   cost, **21**-10
Coupled stability, **19**-10–13
Coupled system stability analysis, **19**-10
Couples systems poles
   locus of, **19**-13f
Covariant derivative, **5**-10
CPS
   of tracking errors, **15**-20
Craig notation and nomenclature, **3**-3
Crane response to pressing move button, **9**-5f
Crane response to pressing move button twice, **9**-5f
Critical curve, **10**-16
   calculating points on, **10**-18f
Critical surface, **22**-8
Cross-over frequencies, **15**-18t
Ctesibus of Alexandria, **1**-2
Cube
   reconstruction from single view, **22**-17f
Cube drawing
   example, **21**-12
Cumulative power spectra (CPS)
   of tracking errors, **15**-20
Cutting tool, **10**-16f
   envelope surface, **10**-16f
   as surface of revolution, **10**-17f
   swept volume, **10**-16f
CyberKnife stereotactic radiosurgery system, **25**-6–9, **25**-7f
   accuracy and calibration, **25**-9
   computer software, **25**-8–9
   patient positioning, **25**-8
   patient safety, **25**-9
   radiation source, **25**-7
   robotic advantage, **25**-9
   robot manipulator, **25**-7
   stereo x-ray imaging system, **25**-8
   treatment planning system for, **25**-8, **25**-8f

D

DADS, **21**-10
Damping, **24**-4–5
   inertial
      arm degrees of freedom augmentation, **24**-40
      three axis arm as micromanipulator for, **24**-41f
   inertial controller
      quenching flexible base oscillations, **24**-41f
   passive, **24**-39, **24**-40f
      sectioned constraining layer, **24**-39f
   piezoelectric actuation for
      arm degrees of freedom augmentation, **24**-41
Dante, **1**-7
Dante II, **1**-7
DARPA, **1**-6

Dartmouth Summer Research Project on Artificial
    Intelligence, **1**-6
Da Vinci Surgical System, **1**-11, **25**-9–10, **25**-10f
DC brushless motor, **12**-16
DC brush motor, **12**-15–16, **12**-15f
Decentralized conventional feedback control, **15**-3–5
Decentralized motion control
    with PD feedback and acceleration feedforward, **15**-4f
Decentralized PD, **15**-2
    controllers
        control torques produced with, **14**-23f
Defense Advanced Research Projects Agency (DARPA), **1**-6
Deformable bodies mechanics, **24**-2–3
DEMLIA's IGRIP, **21**-7
Denavit-Hartenberg (D-H), **8**-1
    approach, **3**-4
    convention, **8**-1–21
        examples, **8**-8–21
    frame assignment, **3**-8
    framework, **2**-7
    notation, **21**-7
    parameters, **3**-11–13, **8**-1–5
        C-code, **3**-18, **3**-29–30
        determining for Stanford arm, **8**-13
        example PUMA 560, **3**-11t
        flow chart, **8**-5f–6f
        schematic, **8**-4f
        systematic derivation, **8**-4
    pathology, **2**-7
    procedure, **3**-4
    representation, **21**-14
    transformation, **4**-1
Density, **24**-4
Desired object impedance, **20**-8f
Detent torque, **12**-14
Determinism, **13**-4
Device-level networks, **26**-10–11
DeviceNet, **26**-10
Devol, George C., **1**-4–5
Dexterity, **20**-2f
D-H. *See* Denavit-Hartenberg (D-H)
Dh.dat, **3**-18, **3**-28
Different image surfaces, **22**-4
Digital sensors, **12**-10–12
    common uses for, **12**-11–12
    with NPN open collector output, **12**-11f
Digital-to-analog conversion, **13**-13–14
Direct collision detection, **23**-19
Direct-drive robotic manipulator modeling and
        identification, **14**-14–15
    experimental setup, **14**-14–15
Direct impedance modulation, **19**-17–18
Discrete-time samples
    multiple continuous time-frequencies, **13**-10f
Discrete-time system
    sampling and aliasing, **13**-9–10
Discrete-time system fundamentals, **13**-9–14
Discretization of spatial domain, **24**-19–25
Disk and link interaction, **7**-19–21, **7**-20f
Dispensers, **11**-16
Displacement vector, **8**-3

Distributed bar elements, **24**-15
Distributed beam elements in bending, **24**-15–16
Distributed control system (DCS), **26**-5
Distributed models, **24**-15
Distributed shaft elements, **24**-15
Disturbances
    feedforward compensation of, **9**-15f
DOF model, **21**-17f
    single
        Matlab code, **21**-23–24
DOF planar robot
    grasping object, **6**-15f
    with one revolute joint an one prismatic joint, **6**-8–13
    with two revolute joints, **6**-4–8
3-DOF system
    full sea state
        Matlab code, **21**-24–27
Double integrator system, **17**-8
Double pendulum in the plane, **7**-16–18
    associated interaction forces, **7**-16f
Double pole single throw (DPST) switch, **12**-10, **12**-10f
Doubles two matrices
    C-code, **3**-28–29
DPST switch, **12**-10, **12**-10f
Drive related errors, **10**-6t
Drone, **1**-10
Duality principle, **16**-10–12
Ductile materials static failure, **24**-3
Dynamical scenes, **22**-13
Dynamic data exchange (DDE), **26**-6
Dynamic effects, **10**-6t
Dynamic equation, **5**-1, **5**-6–11
    of motion, **21**-17
Dynamic models, **16**-2–4
    in closed form
        and kinematics, **14**-15–17
Dynamic Motion Simulation (DADS), **21**-10
DYNAMICS, **6**-3
Dynamics, **17**-5, **24**-11–15
    error
        block diagram, **17**-9f
Dynamics solver
    flowchart, **23**-18f

E

Eddy current sensors, **12**-5
Edinburgh Modular Arm System (EMAS), **1**-11
Eigenfunctions, **24**-18–19
Eigenvalues and corresponding eigenfunctions, **24**-18–19
Eight-point linear algorithm, **22**-4, **22**-5
    coplanar features, **22**-7–8
    homography, **22**-7–8
Eight-point structure from motion algorithm, **22**-6
Elastic averaging, **13**-6
Elastic modulus, **24**-3–4
Elbow manipulator, **3**-5, **3**-5f
    link frame attachments, **3**-5f
Electrical power, **11**-9
Electromagnetic actuators, **12**-12–17
Electromagnets, **11**-16

Electronic leads
  foot side overhang
    specification, **10**-4f
Electronic numerical integrator and computer (ENIAC), **1**-5
EMAS, **1**-11
Embedding of constraints
  dynamic equations, **5**-12
Encoders, **12**-1, **13**-11–12
  typical design, **12**-2f
Endeffector(s), **5**-4
  attachment precision, **11**-4–5
  design of, **11**-1–19
  grasping modes, forces, and stability, **11**-11–13
  gripper kinematics, **11**-9–11
  grippers and jaw design guidelines, **11**-13–16
  interchangeable, **11**-16
  multi-tool, **11**-17f
  power sources, **11**-7–9
  robot attachment and payload capacity, **11**-3–7
  sensors and control considerations, **11**-17–19
  special environments, **11**-3
  special locations, **11**-5
Endeffector frame, **17**-3
  transformation to base frame, **8**-8f
Endoscopic surgery, **1**-10
Engelberger, Joseph F., **1**-4–5, **1**-10
Engelberger Robotics Awards, **1**-5
ENIAC, **1**-5
Environmental forces, **19**-2f
Environmental impedances
  types of, **16**-10f
Environmental issues, **1**-3
Environmental stiffness
  locus of coupled system poles, **19**-14f
Epipolar constraint, **22**-4–5
Equations of motion
  of rigid body, **7**-13–14
Equivalent control, **18**-4–6
Ergonomic simulation, **21**-8f
Ernst, Heinrich A., **1**-6
Error bounds
  linear *vs.* quadratic, **17**-13f
Error budgeting, **10**-1–20
  accuracy and process capability assessment, **10**-12–15
  error sources, **10**-5–7, **10**-6t
  probability, **10**-2–3
  tolerances, **10**-3–5
Error dynamics
  block diagram, **17**-9f
Error equation, **17**-9
Error sources, **10**-1
  effects on roundness, **10**-15f
  superposition of, **10**-15f
Essential matrix, **22**-4–5, **22**-6
Ethernet, **26**-11, **26**-12, **26**-12f
Euclidean distance, **2**-1
Euler angles, **2**-4, **17**-4
Euler-Lagrange equations, **5**-6
Euler's equation of motion, **4**-3f

Euler's equations
  covariant derivative, **7**-8–11
  disadvantages of, **7**-8
  in group coordinates, **7**-12
  rigid body, **7**-11–13
Exact-constraint, **13**-6
Exciting trajectory
  motions of, **14**-20f
Exciting trajectory design, **14**-8–9
Exploratory procedures, **23**-10
Exponential coordinates, **5**-3
Exponential map, **5**-2
  action on group, **7**-9f
Extended forward kinematics map, **5**-4

F

Factorization algorithm
  multilinear constraints, **22**-13
Factory floor, **21**-3f
Fault tree analysis (FTA), **25**-4
FBD, **26**-15
Feasibility, **10**-1
Feature extraction, **22**-3
Feature matching, **22**-3
Feature tracking, **22**-3
Feedback compensation, **13**-20
Feedback control design
  μ-synthesis, **15**-16–19
Feedback control hardware, **13**-16
Feedback controller C1
  bode plots of, **15**-18f
Feedback linearization control, **17**-7–8
Feedback sensors, **13**-17–19
Feedforward compensation, **13**-21
  5% model errors effect on, **9**-18f
  10% model errors effect on, **9**-19f
Feedforward control
  action, **9**-15–16
  conversion to command shaping, **9**-23–24
Feedforward controllers, **9**-4
Fictitious constraints, **6**-16
Fieldbuses
  advanced process control, **26**-11
  capabilities, **26**-13f
Filippov solutions, **17**-15
Finite element representations, **24**-25
First joint
  flexible dynamics, **15**-11f
  sensitivity functions for, **15**-16f
First U.S. robot patent, **1**-5
Fixturing errors, **10**-6t
FK, **14**-2
  map, **5**-4, **17**-3–4
Flexible arm
  kinematics of, **24**-20
Flexible exhaust hose, **21**-3
Flexible robot arms, **24**-1–42
  design and operational strategies, **24**-39–41
  open and loop feedforward control
    command filtering, **24**-32–35

Flexible robots trajectory planning, **9**-1–25
  applications, **9**-13–14
Flight simulation, **23**-2
Fluid power actuators, **12**-17–18
Folded back, **3**-2
Food processing, **11**-3
Force(s)
  endeffector, **11**-11–13
  and torques
    acting on link n, **4**-3f
    between interacting bodies, **7**-15–16
  and velocity, **5**-3–4
Force and metrology loops, **13**-5–6
Force and torque, **12**-9
Force computation, **5**-8–9
Force control block diagram, **16**-11f
Force controlled hydraulic manipulator, **21**-18f
Force controller
  with feed-forward compensation, **18**-3f
Force feedback, **19**-18–19
Force sensing, **11**-18, **23**-3
Force sensing resistors (FSR), **11**-18
Force sensors, **11**-17
Force step-input, **16**-11–12
Forward dynamics form, **23**-6
Forward dynamics solver, **23**-20
Forward kinematics (FK), **14**-2
  map, **5**-4, **17**-3–4
Forward-path
  block diagram of, **19**-8f
Forward recursion, **4**-2
Foundation Fieldbus, **26**-11
Foundation Trilogy, **1**-3
Four bar linkage jaws, **11**-10
Four bar linkages gripper arms, **11**-4f
4x4 homogeneous transformation, **4**-1
Fowardkinematics.c, **3**-18, **3**-24–25
Frames of reference
  assigning, **2**-7
Frankenstein, **1**-2
Frankenstein, Victor, **1**-2
Free-body approach, **4**-3
Freedom robot army manipulator, **8**-9f
Frequency domain solutions, **24**-16–19
Frequency response and impulse response, **24**-19
FRFs
  magnitude plots of, **15**-13f
Friction
  in dynamics, **7**-21–22
  and grasping forces, **11**-12–13
Frictional forces, **19**-2f
Friction forces, **7**-16–17
  as result of contact, **7**-22f
Friction modeling, **14**-5–6
Friction modeling and estimation, **14**-19
Friction model validation
  torque applied to third joint, **14**-20f
Friction parameters estimation, **14**-6–7
Friction system
  with feedforward compensation
    block diagram of, **9**-20f
  control effort for, **9**-22f
  response of, **9**-22f
  without feedforward compensation
    control effort in, **9**-21f
    mass response in, **9**-21f
FSR, **11**-18
FTA, **25**-4
Function block diagram (FBD), **26**-15
Furby, **1**-11

G

GAAT, **21**-3
Gauss-Jordan elimination, **3**-26–28
Generalized active force, **6**-4
Generalized conditions, **17**-5
Generalized inertia force, **6**-4
Generalized inertia matrix, **5**-6
General Motors (GM), **1**-2, **1**-5, **1**-7
Generating zero vibration commands, **9**-5–9
Generic system
  block diagram, **9**-4f
Generic trajectory command
  input shaping, **9**-9f
Geodesic equation, **5**-10
Geometric interpretation, **5**-10–11
Geometric model, **23**-17
Geometric vision
  survey, **22**-1–22
Global proximity test, **23**-18
Global warming, **1**-3
GM, **1**-2, **1**-5, **1**-7
Golem, **1**-1, **1**-2
Grafton, Craig, **21**-2
Graphical animation, **21**-12–13
Graphical user interface (GUI), **26**-6
Graphical visualization tools, **21**-1
Grasping forces
  and friction, **11**-12–13
Grasping modes
  endeffector, **11**-11–13
Grasping stability, **11**-11–12
Grasp types
  for human hands, **11**-12f
Greek mythology, **1**-1
Gripper and jaw design geometry, **11**-13
Gripper arms
  four bar linkages, **11**-4f
Gripper design
  case study, **11**-14–15
  products, **11**-13–14
Gripper forces and moments, **11**-12f
Gripper jaw design algorithms, **11**-15–16
Gripper kinematics
  endeffector, **11**-9–11
Grounded, **23**-3
Guaranteed stability of uncertain systems, **17**-14
GUI, **26**-6
Gunite and associated tank hardware, **21**-4f
Gunite and Associated Tanks (GAAT), **21**-3

Index

H

Hair transplantation robot, **25**-12
Hall effect sensor, **12**-8, **12**-8f
Haptic interface to virtual environments, **23**-1–21, **23**-2f
    applications, **23**-3–4
    characterizing human user, **23**-5
    classification, **23**-2–3
    design, **23**-7–9
    related technologies, **23**-1–2
    specification and design of, **23**-5–7
    system network diagram and block diagram, **23**-5f
    system performance metrics and specifications, **23**-4–9
Haptic perception in the blind, **23**-11
Haptic rendering
    block diagram, **23**-8f
    schematic representation, **23**-7f
Haptics
    history, **23**-10–11
Haptic terms
    taxonomy of, **23**-3f
HART
    sensor-level communications protocol, **26**-9–10
HAT controller model
    details, **21**-19f
HAT manipulator model
    details, **21**-19f
HAT operator, **22**-3
HAT simulation model, **21**-18f
Hazard analysis, **25**-4–5
    initial and final risk legend, **25**-5
    likelihood determination, **25**-5
    risk acceptability, **25**-5
    severity determination, **25**-5
    verification and validation, **25**-4
Hazardous environments, **11**-3
Headers
    C-code, **3**-29
Hebrew mythology, **1**-1
HelpMate Robotics, **1**-10
Hexaglide mechanism, **9**-2f
High end robot simulation packages, **21**-7–8
Highway addressable remote transducer (HART)
    sensor-level communications protocol, **26**-9–10, **26**-10f
HMA, **21**-3
HMI, **26**-6–8
Hohn, Richard, **1**-8
Holding torque, **12**-14
Holonomic constraints, **5**-11, **16**-14–16
Homogeneous matrix, **5**-2
Homogeneous transformation, **2**-6, **2**-7
    computes
        C-code, **3**-24–25, **3**-25–26
Homogeneous transformation, 4x4, **4**-1
Homogeneoustransformation.c, **3**-18, **3**-25–26
Homogeneous transformation matrices (HTM), **10**-8, **10**-9, **10**-10
    algorithm for determining, **8**-6–8
Homogeneous vector, **5**-2
Homunculus, **1**-2
Honda, **1**-11

Hooke's law, **24**-2
Hose management arm (HMA), **21**-3
HTM. *See* Homogeneous transformation matrices (HTM)
Human and automatic controller, **23**-4
Human force without compensation, **21**-20f
Human haptics, **23**-9–13
Human-machine interface (HMI), **26**-6–8
    gas delivery subsystem menu example, **26**-7f
Human user
    haptic interface to virtual environments, **23**-5
Hybrid control, **17**-20
Hybrid controller, **26**-5
Hybrid impedance control, **16**-9–14
    type, **16**-9–10
Hybrid impedance controller, **16**-13f
Hybrid position/force control, **16**-6–9, **16**-8f
Hybrid system, **17**-20
Hybrid type of control algorithms, **20**-6
Hydraulic actuators, **12**-17. *See also* HAT controller model
Hydraulic fluid power, **11**-8

I

I, Robot, **1**-3
Idealized structures and loading, **24**-5
IEA, **26**-12
IGRIP, **21**-7, **21**-8
IK. *See* Inverse kinematics (IK)
Image formation, **22**-2–3
Impact equation, **5**-13–14
Impedance
    vs. admittance regulation, **19**-9–10
    and interaction control, **19**-1–23
Impedance design
    for handling an object, **20**-7–9
Impulses, **9**-6
    canceling vibration, **9**-6f
Incremental position sensors, **13**-11–12
Independent proportional plus derivative joint control, **24**-27–29
Inductive (eddy current) sensors, **12**-5
Industrial Ethernet Association (IEA), **26**-12
Industrial Open Ethernet Association (IOANA), **26**-12
Industrial protocol (IP), **26**-12
Industrial robot
    birth of, **1**-4–5
    invention, **1**-2
Inertia activity, **6**-4
Inertial damping controller
    arm degrees of freedom augmentation, **24**-40
    quenching flexible base oscillations, **24**-41f
    three axis arm as micromanipulator for, **24**-41f
Inertial force, **6**-4, **19**-2f
Inertial reference frame, **4**-2
Inertia matrix, **17**-5
Inertia tensor, **4**-9, **5**-6
Infinitesimal motions
    and associated Jacobian matrices, **2**-8–12
    rigid-body, **2**-11–12
    screw like, **2**-11
Infinitesimal twist, **2**-11

Information networks, **26**-12
Inner loop control, **17**-8
Inner loop/outer loop, **17**-8
    architecture, **17**-8f
Input/output, **26**-8–9, **26**-8f
Input shapers, **13**-21
    sensitivity curves of, **9**-10f
Instruction list (IL), **26**-16
Integrated end effector attachment, **11**-4
Integrated Surgical Systems, Inc., **1**-10
Interacting rigid bodies systems dynamics, **7**-1–23
Interaction
    control implementation, **19**-14–15
    as disturbance rejection, **19**-5
    effect on performance and stability, **19**-2–3
    as modeling uncertainty, **19**-5
    port admittance, **19**-12f
    port connection causal analysis, **19**-8–9
Interaction calculator, **23**-17, **23**-19–20
    interconnection flowchart, **23**-18f
Interchangeable endeffectors, **11**-16
International Space Station (ISS), **1**-9
Inuit legend, **1**-1
Invasive robotic surgery, **25**-11
Inverse dynamics, **17**-8
    computational issues, **4**-8
Inverse dynamics form of equations, **24**-26
Inverse kinematics (IK), **3**-1–30, **14**-2
    analytical solution techniques, **3**-4
    dialytical elimination, **3**-13
    difficulty, **3**-1–3
    existence and uniqueness of solutions, **3**-2–3
    map, **17**-3–4
    numerically solves
        n degree of freedom robotic manipulator, **3**-19–22
    reduction to subproblems, **3**-4
    solutions, **3**-2f
        infinite numbers, **3**-3f
    solution using Newton's method, **3**-14–16
    utilizing numerical techniques, **3**-13–16
    zero reference position method, **3**-13
Inversekinematics.c, **3**-18–30
Inversekinematics.h, **3**-18, **3**-30
Inverse matrix
    computes
        C-code, **3**-26–28
IOANA, **26**-12
IP, **26**-12
Isocenter, **25**-9
Isolated link
    force and torque balance, **4**-3–4
Isolate invariants, **23**-12
ISS, **1**-9
Ith arm coordinate system, **20**-3f
It's Been a Good Life, **1**-4

J

Jacobian(s)
    associated with parametrized rotations
        angular velocity, **2**-8–10

constructs approximate
    C-code, **3**-23–24
manipulator, **17**-4
six by six, **3**-14, **3**-23–24
for ZXZ Euler angles, **2**-10–11
Jacobian matrices
    adjoint, **2**-12
    associated
        and infinitesimal motions, **2**-8–12
    body manipulator, **5**-5
Jacobian singularities, **3**-13
Jacquard, Joseph, **1**-2
Japanese Industrial Robot Association (JIRA), **1**-7–8
Japanese manufacturers, **1**-7
Jaws
    design geometry, **11**-13
    four bar linkage, **11**-10
    with grasped object, **11**-15f
JIRA, **1**-7–8
Johnson, Harry, **1**-7
Joint errors
    ranges of, **15**-20f
    variances of, **15**-21t
Joint motions
    online reconstruction of, **14**-9–10
7-joint robot manipulator, **8**-15–18
Joint space, **17**-3
    inverse dynamics, **17**-8–9
    model, **16**-2–3
    trajectory
        for writing task, **14**-18f
Joint torques, **4**-8
Joint variables, **5**-4

K

Kalman filter
    bode plots of, **14**-10f
Kalman filtering technique, **14**-7
Kane, Thomas, **6**-1
Kane's dynamical equations, **6**-3
Kane's equations, **6**-4
    in robotic literature, **6**-22–25
Kane's method, **4**-2, **6**-1–29
    commercial software packages related, **6**-25–29
    description, **6**-3–4
    discrete general steps, **6**-5
    kinematics, **6**-18–22
    preliminaries, **6**-16–18
Kinematic(s), **17**-3–4, **24**-9–11
    chain, **17**-2
        closed, **24**-10
    deformation, **24**-10
    design, **13**-6
    and dynamic models in closed form, **14**-15–17
    interfaces, **23**-3
    Kane's method, **6**-18–22
    modeling, **10**-7–12, **14**-3–4
    simulation, **21**-1
Kronecker product of two vectors, **22**-5
Kron's method of subspaces, **7**-14

L

Ladder diagram, 26-14f
Ladder logic diagrams (LLD), 26-13, 26-14–15, 26-14f
Lagrange-d'Alembert principle, 5-13
Lagrange-Euler (L-E) method, 4-2
Lagrange multipliers, 5-12, 7-19
Lagrange's equations of motion of the first kind, 6-4
Lagrange's formalism
  advantages, 5-11
Lagrange's form of d'Alembert's principle, 6-4
Lagrangian dynamics, 5-1–14
Lagrangian function, 5-6
Language selection, 26-16
Laplace-transformed impedance and admittance functions for mechanical events, 19-6t
Laser interferometers, 13-18
Law of motion, 6-2
LCS, 10-7
Leader-follower type control algorithm, 20-11–12
Lead screw drive
  lead errors associated with, 10-7f
Lego MINDSTORMS robotic toys, 1-11
L-E method, 4-2
Levi-Civita connection, 5-10
Levinson, David, 6-27
Lie algebra, 5-2, 5-4
Life safety systems, 26-18f
Light curtains, 12-12
Limit switches and sensors, 12-12
Linear and rotary bearings, 13-14
Linear axes
  errors for, 10-6t
Linear encoders, 13-17–18
Linear error motions, 10-6t
Linear feedback motion control, 15-1–22
  with nonlinear model-based dynamic compensators, 15-5–10
Linear incremental encoders, 12-1
Linearization
  Kane's method, 6-19
Linearized equations
  Kane's method, 6-13–14
Linear motions jaws, 11-9–10, 11-9f
Linear reconstruction algorithm
  coplanar, 22-13
Linear solenoid concept, 12-13f
Linear variable differential transformer (LVDT), 12-4–5, 12-4f
Link parameters, 3-4
Load capacity, 20-2f
Load cells, 12-9
Load induced deformation, 10-6t
Load sharing problem, 20-7
Local coordinate systems (LCS), 10-7
Logic-based switching control, 17-20
Long reach manipulator RALF, 9-2f
Loop feedforward control
  command filtering, 24-32–35
  learning control, 24-36
  trajectory design inverse dynamics, 24-36–39, 24-38f
  trajectory specifications, 24-32, 24-33f
Loop-shaping, 15-8
Low cost robot simulation packages, 21-8–9
Low-impedance performance
  improving, 19-18–19
Low pass filtering, 24-33
LuGre model, 14-7
Lumped inertia, 24-12
Lumped masses
  dynamics of, 24-13–15
Lumped models, 24-11–13
Lumped springs, 24-12
LVDT, 12-4–5, 12-4f
Lyapunov's second method, 17-14–15

M

Machine accuracy, 10-1
Machine components imperfections, 10-1
Magellan, 1-9
Magnetically Attached General Purpose Inspection Engine (MAGPIE), 1-6
Magnetostrictive materials, 11-9
MAGPIE, 1-6
Manipulators
  background, 17-2–6
  inertia matrix, 5-7
  Jacobian, 17-4
  kinetic energy, 17-5
  potential energy, 17-5
  robust and adaptive motion control of, 17-1–21
  tasks, 20-2f
Manufacturing automation, 26-1–18
  control elements, 26-6–8
  controllers, 26-4–6
  hierarchy of control, 26-2–4, 26-3f
  history, 26-2–4
  industrial case study, 26-17–18
  networking and interfacing, 26-9–13
  process questions for control, 26-1–2
  programming, 26-13–16
  terminology, 26-2
Manufacturing management information flow, 26-3f
Maple, 21-15
Mariner 2, 1-8
Mariner 10, 1-9
Mars, 1-9
Massachusetts Institute of Technology (MIT), 1-5
Mass distribution properties
  of link, 4-8
Massless elastic links
  dynamics of, 24-13–15
Master manipulator, 23-1
Master-slave type of control algorithms, 20-5–6, 20-5f
Material properties, 24-3–4
Mates, 21-10
Mathematica, 21-15
Matlab

code
   3-DOF system full sea state, **21**-24–27
   single DOF example, **21**-23–24
  cost, **21**-11
Matrix exponential, **2**-4
Matrixinverse.c, **3**-18, **3**-26–28
Matrixproduct.c, **3**-18, **3**-28–29
McCarthy, John, **1**-6
Mechanical Hand-1 (MH-1), **1**-5
Mechanical impedance and admittance, **19**-6–7
Mechatronic systems, **13**-8–21
  definition of, **13**-8–9
Mercury, **1**-9
Metrology loops, **13**-5–6
MH-1, **1**-5
Microbot Alpha II, **11**-4
Milenkovic, Veljko, **1**-7
MIL-STD 2000A, **10**-4
Minimally invasive surgical (MIS)
  procedures, **1**-10
  robotic, **25**-9–10
Minimum distance tracking algorithms, **23**-19
Minsky, Marvin, **1**-6
MIS procedures, **1**-10
MIT, **1**-5
MIT Artificial Intelligence Laboratory, **1**-6
Mitiguy, Paul, **6**-28
Mitsubishi PA-10 robot arm, **8**-15–18
  D-H parameters, **8**-15t
  schematic, **8**-16f
Mobile manipulators
  use, **20**-11
MODBUS, **26**-11
Model(s)
  establishing correctness of, **14**-17–19
  parameters estimation, **14**-6–10
  validations, **14**-11
Modeling, **24**-2–27
  errors
    mass response with, **9**-23f
    and slower trajectory, **9**-23f
  material removal processes, **10**-15–19
Modified light duty utility arm (MLDUA), **21**-3
Moment of inertia, **4**-8
Morison, Robert S., **1**-6
Motion controller, **26**-5
Motion control system
  environmental considerations, **13**-8
  serviceability and maintenance, **13**-8
Motion equation
  object supported by multiple manipulators, **20**-3
Motion estimation algorithms
  comparison, **22**-19f
Motion of object and control
  of internal force moment, **20**-5–7
Motion planning, **17**-7
Motion reference tracking accuracy, **15**-1
Motivation based on higher performance, **24**-1
Motor sizing
  simplified plant model for, **13**-20f
Moving-bridge coordinate measuring machine, **9**-3f

MSC Software's Adams, **21**-10
Multibody dynamic packages, **21**-10–11
Multi-bus system architecture, **26**-9f
Multi-component end effectors, **11**-11
Multi-Input Multi-Output, **9**-14
Multi-jaw chuck axes, **11**-11f
Multi-jaw gripper design, **11**-15f
Multi-mode input shaping, **9**-11
Multiple-body epipolar constraint, **22**-8
Multiple-body motion, **22**-8
Multiple images
  3-D point X in m camera frames, **22**-9f
Multiple jaw/chuck style, **11**-10–11
Multiple manipulators
  coordinated motion control, **20**-1–12
  mobile, **20**-10–11
    coordination, **20**-11f
    decentralized motion control, **20**-10–12
Multiple-model-based hybrid control architecture, **17**-20f
Multiple-model control, **17**-20
Multiple-view geometry, **22**-8–13
Multiple-view matrix
  point features, **22**-9
    rank condition, **22**-9–10
    theorem, **22**-10
Multiple-view rank condition
  comparison, **22**-19f
Multiple-view reconstruction
  factorization algorithm, **22**-11–13
Multi-tool endeffector, **11**-17f
Mu-synthesis feedback control design, **15**-16–19
Mythical creatures
  motion picture influence, **1**-2

N

Narrow phase, **23**-19
National Aeronautics and Space Administration (NASA), **1**-6
National Science Foundation (NSF), **1**-6
Natural admittance control, **19**-19–20
Natural pairing, **5**-4
Nature of impacted systems, **24**-1–2
N-E equations, **4**-2–3
N-E method, **4**-2
Networks
  selection of, **26**-13
Neural-network friction model, **14**-6
Newton-Euler (N-E) equations, **4**-2–3
Newton-Euler (N-E) method, **4**-2
Newtonium, **21**-8
Newton's equation of motion, **4**-3f
Newton's law, **7**-2–5
  in constrained space, **7**-5–8
  covariant derivative, **7**-3–5, **7**-4f
Newton's method, **3**-14
  C code
    implementation, **3**-18–30
    six degree of freedom manipulator, **3**-18–30
  convergence, **3**-17
  theorems relating to, **3**-17–18

Newton's second law, **4**-2, **4**-3f
Nodic impedance, **19**-14–15, **19**-15f
Nominal complementary sensitivity functions
    magnitude plots of, **15**-19f
Nominal data
    bode plots, **15**-13f
Nominal plant model, **15**-12–13
Noncontact digital sensors, **12**-10–11
Nonholonomic constraints, **5**-11
    forces, **7**-18–19
Noninvasive robotic surgery, **25**-6–9
Nonlinear friction
    feedforward control of, **9**-19–22
Normal force control component, **16**-7–8
Norway, **1**-7
NSF, **1**-6
Nuclear waste remediation simulation, **21**-3
Numerical problems and optimization, **22**-8
Numerical simulation, **21**-13–21
Nyquist plane, **19**-12f
Nyquist Sampling Theorem, **13**-9

O

Oak Ridge National Laboratory (ORNL), **21**-3
OAT filter, **24**-35
    vs. joint PID and repetitive learning, **24**-38f
Object
    coordinate system, **20**-3f
    dynamics-based control algorithms, **20**-6–7, **20**-6f
    manipulation, **20**-2–5, **20**-3f
ODVA, **26**-12
Odyssey IIB submersible robot, **1**-11
Online gradient estimator
    of BPS, **14**-8
Open and loop feedforward control
    command filtering, **24**-32–35
    learning control, **24**-36
    trajectory design inverse dynamics, **24**-36–39, **24**-38f
    trajectory specifications, **24**-32, **24**-33f
Open DeviceNet Vendor Association (ODVA), **26**-12
OpenGL interface, **21**-12
Open loop and feedforward control, **24**-31–39
Open-loop gains
    for first joint, **15**-16f
Operational space control, **17**-10
Optical sensors, **12**-6–7
    dielectric variation in, **12**-6f
Optical time-of-flight, **12**-7
Optical triangulation, **12**-6–7
    displacement sensor, **12**-7f
Oriented bounding boxes, **23**-18
Orlandea, Nick, **6**-27
ORNL, **21**-3
Orthogonal matrices, **2**-2
Orthographic projection, **22**-4, **22**-13
Orthonormal coordinate frames
    assigning to pair of adjacent links, **8**-1
    schematic, **8**-2
Our Angry Earth, **1**-3
Outer loop, **17**-8

architecture, **17**-8f
control, **17**-8
Overhead bridge crane, **9**-5f
Ozone depletion, **1**-3

P

Painting robot, **9**-14f
Paracelsus, **1**-2
Parallel axis/linear motions jaws, **11**-9–10, **11**-9f
Parallelism, **10**-6t
Partial velocities, **6**-4
Part orienting gripper design, **11**-16f
Passive, **17**-6
Passive damping, **24**-39, **24**-40f
    sectioned constraining layer, **24**-39f
Passive touch, 23x11
Passivity, **19**-10–13
Passivity applied to haptic interface, **23**-15–17
Passivity-based adaptive control, **17**-19
Passivity-based approach, **17**-18
Passivity-based robust control, **17**-18–19
Passivity property, **5**-8, **17**-6
Patient safety
    CyberKnife stereotactic radiosurgery system, **25**-9
Paul, Howard, **1**-10
Payload, **11**-5–6
Payload capacity
    endeffector, **11**-3–7
Payload force analysis, **11**-6–7, **11**-7f
Payload response moving through obstacle field, **9**-5f
PC-based open controller, **26**-6
PD. *See* Proportional and derivative (PD)
pdf, **10**-2, **10**-3, **10**-3f
Penalty contact model, **23**-19–20
Penalty method, **23**-19–20
Performance index, **10**-4, **10**-5
Performance weightings
    magnitude plots for, **15**-17f
Persistency of excitation, **17**-18
Persistent disturbances, **17**-11
Personal computer (PC)
    open controller, **26**-6
Perturbed complementary sensitivity functions
    magnitude plots of, **15**-19f
Physical environment, **23**-1
PID control, **26**-3
Pieper's method, **3**-13
Pieper's solution, **3**-7–11
Piezoelectric, **11**-9
    and strain gage accelerometer designs, **12**-9f
Piezoelectric actuation for damping
    arm degrees of freedom augmentation, **24**-41
Piezoresistor force sensors, **11**-18
Pinhole imaging model, **22**-2f
Piper's solution, **3**-4
Pipettes, **11**-16
Pitch, **5**-3
Pivoting/rotary action jaws, **11**-10
Planar symmetry, **22**-16
Planar two-link robot, **4**-5

Planets
  explored, 1-9
PLC, 26-3, 26-4–5, 26-4f
Pneumatic actuators, 12-17–18
Pneumatic valve connections
  safety, 11-8f
Pointer
  returns to matrix c
    C-code, 3-28–29
Port behavior and transfer functions, 19-7–8
Position control block diagram, 16-11f
Position/orientation
  errors, 20-11
Position-synchronized output (PSO), 13-11
Post-World War II technology, 1-5
Potentiometers, 12-4
Power amplifiers, 13-16–17
Precision
  definitions of, 13-2–3
  machine, 13-14–16
    design fundamentals, 13-2–8
    structure, 13-15
    vibration isolation, 13-15–16
  positioning
    of rotary and linear systems, 13-1–22
Predator UAV (unmanned aerial vehicle), 1-10
Pressure sense, 23-11
Primera Sedan car, 21-2
Prismatic joints, 17-3
Probability density function (pdf), 10-2, 10-3, 10-3f
Procedicus MIST, 21-4, 21-4f
Process capability index, 10-4
Process flow chart, 11-2f
Processing steps interactions, 10-14
Product of Exponentials Formula, 5-5
Pro/ENGINEER simulation
  Kane's method, 6-26
Profibus DP, 26-10
Profibus-FMS, 26-11, 26-12
Profibus-PA, 26-11
ProgramCC
  cost, 21-11
Programmable logic controllers (PLC), 26-3, 26-4–5, 26-4f
Programmable Universal Machine for Assembly (PUMA), 1-8
Pro/MECHANICA
  Kane's method, 6-26
Proportional and derivative (PD)
  controller, 9-1
  position errors, 15-20, 15-20f
Proportional integral and derivative (PID) control, 26-3
Prosthetics, 1-11
Proximity sensors, 11-17, 12-11–12
Pseudo-velocities, 5-12
PSO, 13-11
Psychophysics, 23-11
Pull-back, 5-9
Pull type solenoids, 12-13
Pulse-width-modulation (PWM), 13-16–17
PUMA, 1-8
PUMA 560
  iterative evolution, 3-16
  manipulator, 3-11–13
PUMA 600 robot arm, 8-18-21
  D-H parameters, 8-18t
  schematic, 8-19f
PWM, 13-16–17
Pygmalion, 1-1

Q

Quadrature encoders, 12-2–3
  clockwise motion, 12-2f
  counterclockwise motion, 12-2f
Quantization, 13-11–12
Quaternions, 17-4

R

Radiosurgery, 25-6
Radiotherapy, 25-6
RALF, 24x32f
Random variable, 10-2
Rank condition
  multiple-view matrix, 22-8
RANSAC type of algorithms, 22-3
RCC, 11-5, 11-6f, 20x9f
RCC dynamics
  impedance design, 20-9–10
Readability, 3-18
Real time implementation, 4-8, 9-12–13
Real time input shaping, 9-13f
Reconstructed friction torques, 14-21f
Reconstructed structure
  two views, 22-12f
Reconstruction
  building, 22-21f
  from multiple images, 22-3
  using multiple-view geometry, 22-3
Reconstruction pipeline
  three-D, 22-3
Recursive formulation, 4-2
Recursive IK solutions
  vs. closed-form solutions, 14-18f
Reduced order controller design, 16-15–16
Reduced order model, 16-15
Reduced order position/force control, 16-14–17, 16-16f
  along slanted surface, 16-16–17
Reference configuration, 5-5
Reference motion task, 15-19f
Reference trajectory
  in task space, 14-11f
Reflective symmetry transformation, 22-14f
Regressor, 17-6
Regulating dynamic behavior, 19-5–13
Remote compliance centers (RCC), 11-5, 11-6f, 20-9f
Remote controlled vehicle invention, 1-2
Repeatability, 13-3f
  definition of, 13-2–3
Residual payload motion, 9-4
Resistance temperature transducers (RTD), 26-8

Resolution, **13**-3
  definition of, **13**-2–3
Resolved acceleration control, **17**-10
Resolvers, **12**-5
Revolute joints, **17**-3
Riemannian connection, **7**-3
Riemannian manifold, **7**-4
Riemannian metrics, **5**-14, **7**-6
Riemannian structure, **7**-2
Rigid body
  dynamics modeling, **14**-4–5, **14**-12-14, **14**-22–23
    torques differences, **14**-19f
  inertial properties, **5**-6–7
  kinematics, **2**-1–12
  motion
    velocity, **5**-3–4
Rigidity, **20**-2f
Rigid linkages
  Euler-language equations, **5**-7–8
Rigid-link rigid-joint robot interacting with constrain surface, **18**-3f
Rigid motions, **17**-3
Rigid robot dynamics properties, **17**-5–6
ROBODOC Surgical Assistant, **25**-11, **25**-11f
Robot
  arm end, **11**-5f
  army dynamics
    governing equations, **4**-2
  assembling electronic package onto printed wiring board, **10**-13f
  attachment and payload capacity
    endeffector, **11**-3–7
  control problem
    block diagram of, **17**-7f
  defined, **1**-1
  design packages, **21**-5–6
  dynamic analysis, **4**-1–9
  dynamic model
    experimental validation of, **14**-12f
  dynamic simulation, **21**-9–10
  first use of word, **1**-3
  kinematics, **4**-1
  motion
    animation, **21**-7–9
  motion control modeling, **14**-3–6
    and identification, **14**-1–24
  Newton-Euler dynamics, **4**-1–9
  simulation, **21**-1–27
    high end packages, **21**-7–8
    options, **21**-5–11
  SolidWorks model, **21**-11f
  theoretical foundations, **4**-2-8
Robo-therapy, **1**-11
Robotic(s), **1**-2
  applications and frontiers, **1**-11–12
  example applications, **21**-2–4
  first use of word, **1**-3–4
  history, **1**-1–12
  in industry, **1**-7–8
  inventions leading to, **1**-2
  medical applications, **1**-10–11, **25**-1–25
    advantages of, **25**-1–2
    design issues, **25**-2–3
    hazard analysis, **25**-4–5
    research and development process, **25**-3, **25**-4f
    upcoming products, **25**-12
  military and law enforcement applications, **1**-9–10
  mythology influence, **1**-1–2
  in research laboratories, **1**-5–7
  space exploration, **1**-8–9
Robotic Arm Large and Flexible (RALF), **24**-32f
Robotic arm manipulator with five joints, **8**-8
Robotic catheter system, **25**-12
Robotic hair transplant system, **25**-12f
Robotic limbs, **1**-11
Robotic manipulator
  force/impedance control, **16**-1–18
  sliding mode control, **18**-1–8
Robotic manipulator motion control
  by continuous sliding mode laws, **18**-6–8
  problem sliding mode formulation, **18**-6–7
  sliding mode manifolds, **18**-7t
Robotic simulation
  types of software packages, **21**-5
Robotic toys, **1**-11
RoboWorks, **21**-8
Robust feedback linearization, **17**-11–16
Robustness, **15**-2
  to modeling errors, **9**-10
Robust ZVD shaper, **9**-10, **9**-10f
Rochester, Nat, **1**-6
Rodrigues' formula, **5**-3
Rolled throughput yield, **10**-5
Root lock for three proportional gains, **24**-28f
Rosen, Charles, **1**-5
Rosenthal, Dan, **6**-26
Rotary axes
  errors for, **10**-6t
Rotary bearings, **13**-14
Rotary encoders, **12**-1, **13**-17
Rotary solenoids, **12**-13
Rotating axes/pivoting jaws, **11**-10f
Rotating axes pneumatic gripper, **11**-10f
Rotational component, **5**-6
Rotational dynamics, **7**-8–11
Rotation matrix, **8**-3
  submatrix
    independent elements, **3**-14
Rotations
  rules for composing, **2**-3
  in three dimensions, **2**-1–4
Routine maintenance, **10**-1
RRR robot, **14**-15f, **15**-11f
  DH parameters of, **14**-14f
  direct-drive manipulator
    case study, **15**-10–21
  PD control of, **15**-15f
  rigid-body dynamic model, **14**-16
RTD, **26**-8
Russian Mir space station, **1**-9

S

SAIL, **1**-6
Sampled and held force
   vs. displacement curve for virtual wall, **23**-14f
SCADA, **26**-6
SCARA. *See* Selective Compliance Assembly Robot Arm (SCARA)
Schaechter, David, **6**-27
Scheinman, Victor, **1**-6, **1**-8, **8**-13
Schilling Titan II
   ORNL's RoboWorks model, **21**-9f
Screw, **5**-3
   magnitude of, **5**-3
Screw axis, **2**-6
Screw machine invention, **1**-2
Screw motions, **2**-6
SD/FAST
   Kane's method, **6**-26
Selective Compliance Assembly Robot Arm (SCARA), **1**-8, **8**-11–12
   D-H parameters for, **8**-11f
   error motions, **10**-11t
   kinematic modeling, **10**-10, **10**-10f
   schematic, **8**-11f
Semiautomatic building mapping and reconstruction, **22**-21–22
Semiconductor manufacturing, **11**-3
Semiglobal, **17**-11
Sensing modalities, **22**-1
Sensitive directions, **10**-13
Sensor-level input/output protocol, **26**-9–10
Sensors and actuators, **12**-1–18
Sequential flow chart (SFC), **26**-16, **26**-17f
Serial linkages
   kinematics, **5**-4–5
Serial link manipulator, **17**-3f
Serial manipulator
   with n joints, **14**-3f
Series dynamics, **19**-20–21
Servo controlled joints
   dynamics of, **24**-13–15
Servo control system
   for joint i, **15**-7f
Servo design
   using μ-synthesis, **15**-9f
7-joint robot manipulator, **8**-15–18
SFC, **26**-16, **26**-17f
SGI, **21**-12
Shafts, **24**-5–6
   distributed elements, **24**-15
Shaky the Robot, **1**-5
Shannon, Claude E., **1**-6
Shaped square trajectory
   response to, **9**-15f
Shape memory alloys, **11**-9
Shaping filter, **24**-34
Shear modulus, **24**-3–4
Shelley, Mary Wollstonecraft, **1**-2
Sherman, Michael, **6**-26
Silicon Graphics, Inc. (SGI), **21**-12

Silma, **21**-7
Simbionix LapMentor software, **21**-4
Simbionix virtual patient, **21**-5f
Similarity, **22**-3
SimMechanics, **21**-10
   cost, **21**-11
Simple impedance control, **19**-15–17
Simple kinematic pairs, **24**-10
Simulated mechanical contact, **23**-1
Simulated workcell, **21**-7f
Simulation block diagram, **21**-14f
Simulation capabilities
   build your own, **21**-11–21
Simulation forms of equation, **24**-25–26
Simulation packages
   robot
      high end, **21**-7–8
Simulink, **21**-10, **21**-13
   cost, **21**-11
Sine error, **13**-5f
Single-axis tuning
   simplified plant model for, **13**-20f
Single DOF example
   Matlab code, **21**-23–24
Single jaw gripper design, **11**-14f
Single pole double throw switch (SPDT), **12**-10, **12**-10f
Single-resonance model, **19**-21f, **19**-22f
   equivalent physical system for, **19**-19f
Single structural resonance model, **19**-4f
6-axis robot manipulator with five revolute joints, **8**-13
Six by six Jacobian, **3**-14, **3**-23–24
Six degree of freedom manipulator, **3**-8, **3**-13–16
Six degree of freedom system, **3**-14
Skew-symmetric matrix, **5**-6–7
Slanted surface
   hybrid impedance control along, **16**-13–14
   hybrid position/force control, **16**-8–9
   manipulator moving along, **16**-4f
   task-space formulation for, **16**-3–4
Slave manipulator, **23**-2
Sliding modes, **17**-15–16
   controller design, **18**-7–8
   formulation of robot manipulator, **18**-2–4
Sliding surface, **17**-15–16, **17**-17f
Small baseline motion and continuous motion, **22**-8
Small Gain Theorem, **17**-11
Small motions, **2**-8, **2**-11
Smooth function tracking
   with feedforward compensation, **9**-18f
   without feedforward compensation, **9**-17f
Sojourner Truth, **1**-9
Solenoids, **12**-12–13
Solid state output, **12**-11
SolidWorks, **21**-10
   cost, **21**-10
   robot model, **21**-11f
Sony, **1**-11
Space Station Remote Manipulator System (SSRMS), **1**-9
Spatial distribution of errors, **10**-14–15
Spatial dynamics, **4**-8–9
Spatial information, **23**-11

Spatial velocity, **5**-3–4
SPDT, **12**-10, **12**-10f
Special Euclidean group, **17**-3
Special purpose end effectors/complementary tools, **11**-16
Spectrum analysis technique, **14**-13
Speeds
    online reconstruction of, **14**-9–10
Spencer, Christopher Miner, **1**-2
Sphere
    ANSI definition of circularity, **10**-4f
Spherical wrist center, **3**-9–10
    height, **3**-10
Spring-and-mass environment
    stable and unstable parameter values for, **19**-21f
Spring-mass response
    shaped step commands, **9**-12f
Squareness, **10**-6t
SRI International, **1**-5
SSRMS, **1**-9
Stability, **15**-2
    endeffector, **11**-11–13
Stable factorizations, **17**-11
Standard deviation, **10**-3
Stanford arm, **1**-6, **8**-13–15, **8**-13f
    D-H parameters, **8**-14t
Stanford Artificial Intelligence Lab (SAIL), **1**-6
Stanford cart, **1**-6
Stanford manipulator
    link frame attachments, **3**-7f
    variation, **3**-7f
Stanford Research Institute, **1**-5
Statics, **24**-2–9
Stepper motors, **12**-13–15
Stereotactic radiosurgery system, **25**-6–9
Stiffness control, **16**-5–6
Stiffness of series of links, **24**-12–13
Straightness, **10**-6t
Strain gauge sensor, **12**-8
    applied to structure, **12**-9f
Strains sensors, **12**-8–9
Strength, **24**-4
Stress *vs.* strain, **24**-2–3
Structural compliance, **10**-1
Structured text, **26**-15
    example, **26**-15t
Supervisory control, **17**-20
Supervisory control and data acquisition system (SCADA), **26**-6
Surface grinder
    local coordinate systems, **10**-7f
Surgical simulation, **21**-3–4
Sweden, **1**-8
Swept envelope, **10**-15
Switches
    as digital sensors, **12**-10
Switzerland, **1**-8
Symbolic packages, **21**-15
Symmetric multiple-view matrix, **22**-15
Symmetric multiple-view rank condition, **22**-14–15, **22**-15
Symmetry, **22**-13–17
    reconstruction from, **22**-15
    statistical context, **22**-16
    surfaces and curves, **22**-16
    and vision, **22**-16
Symmetry-based algorithm
    building reconstructed, **22**-22f
Symmetry-based reconstruction
    for rectangular object, **22**-16
Symmetry cells
    detected and extracted, **22**-20f
    feature extraction, **22**-18
    feature matching, **22**-20f
    matching, **22**-20f
    reconstruction, **22**-20f
SystemBuild, **21**-10, **21**-13
System characteristic behavior, **24**-26–27
System modeling, **13**-19–20
System with time delay
    feedforward compensation, **9**-16–18, **9**-16f

T

Tachometers, **12**-1
Tactile feedback/force sensing, **11**-18, **23**-3
Tactile force control, **11**-18–19
Taliban forces, **1**-10
Tangential position control component, **16**-7
Tangent map, **5**-9
Task space, **17**-3
    inverse dynamics, **17**-9–10
    model and environmental forces, **16**-3
Taylor series expansion, **2**-4, **2**-11
Telerobot, **23**-2
Tentacle Arm, **1**-7
Tesla, Nikola, **1**-2
Thermal deformation, **10**-6t
Thermally induced deflections, **10**-1
Thermal management, **13**-7
Theta.dat, **3**-18, **3**-30
Third joint
    flexible dynamics, **15**-12f
Three axis arm as micromanipulator for inertial damping, **24**-41f
Three-dimensional sensitivity curve, **9**-11f
3-DOF system
    full sea state
        Matlab code, **21**-24–27
3-D reconstruction pipeline, **22**-3
Three Laws of Robotics, **1**-4
Three-phase DC brushless motor, **12**-16f
Three term OAT command shaping filter, **24**-34f
Tiger Electronics, **1**-11
Time delay filtering, **24**-34, **24**-35, **24**-35f
Time-delay system without feedforward compensation
    step response of, **9**-16f
Time-domain technique, **14**-13
Tip force without compensation, **21**-20f
Titan 3 servo-hydraulic manipulator, **12**-18f
Tolerances
    defined, **10**-4
    of form, **10**-4

of size and location, **10**-4
on surface finish, **10**-4
Tomorrow Tool, **1**-8
Tool related errors, **10**-6t
Torques and forces
between interacting bodies, **7**-15–16
Torsion, **24**-5–6
Torsional buckling, **24**-9
Trajectory generation, **17**-7
Trajectory planning for flexible robots, **9**-3
Trajectory tracking, **17**-7
Trallfa Nils Underhaug, **1**-7
Trallfa robot, **1**-7
Transfer matrix representation, **24**-16, **24**-18
Transformation matrix, **24**-9–10
Transition Research Corporation, **1**-10
Translating link released from supports, **6**-17f
Translational component, **5**-6
Translational displacement, **4**-9
Transmission transfer function
block diagram of, **19**-8f
Tupilaq, **1**-1–2
Turret lathe invention, **1**-2
Twist coordinates, **5**-2
Twists, **5**-2
Two DOF planar robot
grasping object, **6**-15f
Two DOF planar robot with one revolute joint and one prismatic joint, **6**-8–13, **6**-9f
acceleration, **6**-11
equations of motion, **6**-13
generalized active forces, **6**-13
generalized coordinates and speeds, **6**-9–10
generalized inertia forces, **6**-12
linearized partial velocities, **6**-20t
partial velocities, **6**-11
preliminaries, **6**-9
velocities, **6**-10
Two DOF planar robot with two revolute joints, **6**-4–8
equations of motion, **6**-7
generalized active forces, **6**-7–8
generalized coordinates and speeds, **6**-6
generalized inertia forces, **6**-7
partial velocities, **6**-6–7
preliminaries, **6**-5–6
velocities, **6**-6
Two inverse kinematic solutions, **3**-2f
Two link manipulator, **3**-2f
Two-link robot
with two revolute joints, **4**-5f
Two-link robot example, **4**-4–7
Two-mode shaper
forming through convolution, **9**-12f
Two-part phase stepper motor power sequence, **12**-14f
Two-view geometry, **22**-4–8

U

Ultrasonic sensors, **12**-8
Uncalibrated camera, **22**-8
Uncertain double integrator system, **17**-11f
Unconstrained system
Kane's method, **6**-16
Ungrounded, **23**-3
Unified dynamic approach, **4**-2
Unimate, **1**-5
Unimation, **1**-4
Unimation, Inc., **1**-5
Universal automation, **1**-4
Universal multiple-view matrix
rank conditions, **22**-13
Unmanned aerial vehicle, **1**-10
automatic landing, **22**-17
Unrestrained motions, **6**-21
Unshaped square trajectory
response to, **9**-14f

V

Vacuum, **11**-8
Vacuum pickups, **11**-16
Variability, **10**-1
Vehicle and arm
OpenSim simulation, **21**-13f
Velocity, **4**-9
and forces, **5**-3–4
kinematics, **17**-4
step-input, **16**-10–11
Venera 13, **1**-8
Venus, **1**-8
Vibration reduction
extension beyond, **9**-14–15
Vicarm, **1**-8
Vicarm, Inc., **1**-8
Viking 1, **1**-9
Viking 2, **1**-9
Virtual coupler, **23**-7, **23**-8
Virtual damper, **23**-14
Virtual environments, **23**-9, **23**-17–20
and haptic interface, **23**-1–21
characterizing human user, **23**-5
Virtual fixtures, **23**-3
Virtual trajectory, **19**-14–15, **19**-15f
Virtual wall, **23**-14f, **23**-15
Vision, **12**-12, **22**-1
Voyager missions, **1**-9

W

Water clock invention, **1**-2
Weak perspective projection, **22**-4
Weaver, Warren, **1**-6
Weber's law, **23**-10
Weighting function
magnitude plots for, **15**-17f
Whirlwind, **1**-5
Whittaker, William "Red," **1**-7

Working Model
    Kane's method, **6**-28–29
World frame, **17**-3
World War II, **1**-4
Wrench, **5**-4
Wrist compliance, **11**-5
Writing task, **15**-21f

X

X tip direction, **21**-21f

Y

Yamanashi University, **1**-8
Young's modulus, **24**-2
Y tip direction, **21**-21f

Z

Zero-order-hold reconstruction filter
    magnitude and phase of, **13**-13f
    stairstep version signal, **13**-14f
Zero phase error tracking control (ZPETC), **9**-22–23, **13**-21
    as command generator, **9**-24
Zeroth Law, **1**-4
Zero-vibration impulse sequences
    generating zero-vibration commands, **9**-9
Zero-vibration shaper, **9**-10
ZEUS Robotic Surgical System, **1**-11
Ziegler-Nichols PID tuning, **11**-18
ZPETC, **9**-22–23, **13**-21
    as command generator, **9**-24
Z tip direction, **21**-22f
ZVD shaper, **9**-10, **9**-10f